T0321759

Encyclopedia of the History of Astronomy and Astrophysics

Beginning with its ancient origins world-wide, this comprehensive encyclopedia traces the key advances in astronomy up to the latest developments in astrophysics and space-based research. The initial articles, which are largely organised to cover astronomical developments chronologically, are followed by thematic historical articles on the solar system, types of stars, stellar evolution, active galaxies, cosmology, and so on. These are followed by articles on tools and techniques from the history of spectroscopy to that of adaptive optics. The last part of the encyclopedia is devoted to the history of ground- and space-based telescopes and observatories, covering the full spectral range from gamma-rays through the optical waveband to radio waves. Informative and accessibly written, each article is followed by an extensive bibliography to facilitate further research, whilst consistent coverage from ancient times to the present makes this an ideal resource for scholars, students and amateur astronomers alike.

David Leverington received his first degree in Physics from Oxford University in 1963. Since then he has held a number of senior positions in the space industry, working for both the European Space Agency and British Aerospace before taking early retirement in 1992. Subsequently he has written three books: *A History of Astronomy from 1890 to the Present* (1996), *New Cosmic Horizons; Space Astronomy from the V2 to the Hubble Space Telescope* (Cambridge University Press, 2000), and *Babylon to Voyager and Beyond; A History of Planetary Astronomy* (Cambridge University Press, 2003). He was also technical consultant for ABC-CLIO's *Space Exploration and Humanity; A Historical Encyclopedia* (2010), supported by the History Committee of the American Astronautical Society.

Encyclopedia of the History of Astronomy and Astrophysics

David Leverington

Shaftesbury Road, Cambridge CB2 8EA, United Kingdom

One Liberty Plaza, 20th Floor, New York, NY 10006, USA

477 Williamstown Road, Port Melbourne, VIC 3207, Australia

314–321, 3rd Floor, Plot 3, Splendor Forum, Jasola District Centre, New Delhi – 110025, India

103 Penang Road, #05–06/07, Visioncrest Commercial, Singapore 238467

Cambridge University Press is part of Cambridge University Press & Assessment, a department of the University of Cambridge.

We share the University's mission to contribute to society through the pursuit of education, learning and research at the highest international levels of excellence.

www.cambridge.org
Information on this title: www.cambridge.org/9780521899949

First published 2013

A catalogue record for this publication is available from the British Library

Library of Congress Cataloging-in-Publication data
Leverington, David, 1941–
Encyclopedia of the history of astronomy and astrophysics / David
Leverington
      pages cm
Includes bibliographical references and index.
ISBN 978-0-521-89994-9 (hardback)
1. Astronomy – History – Encyclopedias.   2. Astrophysics – History –
Encyclopedias.   I. Title.
QB15.L388   2013
520.3–dc23   2012036493

ISBN   978-0-521-89994-9   Hardback

Contents

vi   Contents

Preface

This encyclopedia starts with a series of articles on ancient astronomy, followed by articles summarising the main developments in astronomy in a time-based series from the early seventeenth century to the present. These are then followed by the main part of the encyclopedia, which includes historical articles on astronomical objects in Parts 2 to 4, astronomical tools and techniques in Part 5, and finally by those on individual ground- and space-based telescopes and observatories operating in all wavebands from gamma rays to radio waves in Parts 6 to 10. These latter parts also contain overview or summary articles outlining the main developments of their particular type of telescope or observatory over time. Consequently information on the development of our knowledge of individual astronomical objects or the development of individual telescopes or observatories can be found in a number of places in this encyclopedia. But, naturally, the main article will be the one devoted to that individual object or the observatory concerned.

As a result of this arrangement, this book can be read either as a straight reference book by going directly to the detailed subject of interest or as a summary history by reading the period overviews in Part 1, supplemented by the other overview articles at the beginning of Parts 6 to 10.

Evidently a book of this length can do little more than introduce the vast majority of subjects covered. Consequently a bibliography is provided after each article to enable the reader to delve further as required.

Acknowledgments

Some spacecraft articles are based on my articles in *Space Exploration and Humanity: A Historical Encyclopedia* (Santa Barbara, CA: ABC-CLIO 2010). Excerpts reprinted with permission of ABC-CLIO, Inc.

Parts of the period overview articles from the 'Seventeenth century' to the 'Twentieth century prior to the Space Age' in Chapter 2 of Part 1 are based on my chapter in *Encyclopedia of the Solar System*, Second Edition (Academic Press, 2007). Excerpts reprinted with permission.

PART 1
GENERAL ASTRONOMY

1 · Ancient (pre-telescopic) astronomy

ANCIENT ASTRONOMY OVERVIEW

Pre-telescopic astronomy is considered in this survey article in four broad geographical areas based on cultural contacts, regional developments and astronomical traditions. This article is then followed by individual, more detailed articles as indicated in the text.

Africa, Europe, Middle East and India

Not only have the earliest human remains been found in Africa, but so has the earliest physical evidence of astronomical activity. This appears to be a lunar calendar found on an Ishango bone which has been dated at about 23,000 to 18,000 BC (see *Sub-Saharan Africa*). The earliest megalithic structure that appears to be of astronomical significance is that at Nabta Playa, also in Africa, which has been dated to around 5000 to 4000 BC (see *Megalithic astronomy*). The earliest parts of Stonehenge in England, on the other hand, are at least 1,000 years younger, as is the passage tomb at Newgrange in Ireland.

There are thousands of megaliths all over the world, some of which have astronomical significance, built by civilisations of widely different cultures. These civilisations had no contact with one another, as far as we know. But that is not true of ancient Egypt, which was later linked to Babylon and Greece. In these three civilisations we have written records to assist our understanding, which are missing for the early megalithic builders.

The Egyptians had studied the sky since before 3000 BC for both religious and practical reasons (see *Egyptian astronomy*). Their most important gods were the Sun god, Ra, and Nut, the goddess of the sky, who was depicted stretched across the sky as the Milky Way. The heliacal rising of Sirius (Sothis) was also important to the Egyptians, as they used it to predict the annual Nile floods.

The earliest known Egyptian calendar, which was used for both religious and agricultural purposes, consisted of 12 months of 29 or 30 days divided into three 4-month seasons of 'inundation', or flooding of the Nile, 'growth' and 'harvest'. An extra month was added every two or three years in Lower Egypt to ensure that the festival of the birth of Ra occurred in the last month of the year. The same was done in Upper Egypt to ensure that the festival of the heliacal rising of Sothis was in the last month. When Upper and Lower Egypt were unified in about 3000 BC the Upper Egyptian scheme was used.

Astrology was the main driver behind early Babylonian astronomy of the Hammurabi dynasty of about 1895 to 1595 BC, particularly that connected with the visibility of the Moon and planets (see *Babylonian astronomy*). This required the priests to carefully observe the motions of these bodies across the sky, so they could be analysed and predictions made of their future positions. Over the centuries, the Babylonians developed many techniques for these predictions, in the process producing highly accurate estimates of the synodic and sidereal periods of the Sun, Moon and planets. Those of Jupiter, for example, in the second century BC, were accurate to within 0.01%.

The Babylonians saw the sky in two dimensions, whereas the Greeks, who were philosophers, saw it in three (see *Greek astronomy*). They wanted to understand the structure of the universe and explain why the celestial objects moved in the way they did. Thales of Miletus was apparently the first to do this in the sixth century BC, and he is thought by some ancient sources to have discovered the cause of eclipses. A little later, the Pythagoreans hypothesised that the universe consisted of a spherical, non-spinning Earth surrounded by a series of concentric, crystalline spheres that carried the Sun, Moon, planets and stars.

In the fifth century BC, Hicetas of Syracuse suggested that the Earth spun on its axis, but this idea was generally rejected. About two hundred years later Aristarchus of Samos suggested that the Sun, not the Earth, was at the centre of the universe, but this idea was also generally ignored until Copernicus resurrected it about 1,700 years later.

Many of the later Greeks, including Hipparchus and Ptolemy, lived in Egypt, where Eratosthenes measured the diameter of the Earth to a few percent of its correct value in the third century BC. In the next century, Hipparchus estimated the rate of precession of the equinoxes and produced the most accurate star catalogue to date. Three hundred years later, Ptolemy wrote his *Almagest*, which included his star catalogue, and *Planetary Hypotheses*, in which he proposed a geocentric model of the universe based on epicycles and equants. Its largest deficiency was probably that it predicted that the Moon's apparent diameter, as seen from Earth, would vary by a factor of two, which it clearly did not do.

It is unclear how much Babylonian work was known to the Greeks. But the victory of Alexander the Great over the Persian empire in 331 BC certainly improved communications between Greece and Babylon for a while, until the Seleucid empire collapsed in the following century. In the case of India, however, there had been some contact between India and Babylon prior to Darius I's conquest of north-west India in 515 BC, but this conquest naturally produced a greater influx of Babylonian ideas into India (see *Indian astronomy*). So in the fifth century BC, for example, Indian astronomy was a mixture of indigenous astronomy, which involved the sky being divided up into 27 lunar mansions, and methods for predicting the movement of celestial bodies which clearly came from the Babylonians.

In 326 BC, Alexander the Great brought Greek influence to India when he conquered the north-west of the country. Over the next few centuries, Indians translated various Greek and Babylonian works into Sanskrit. Then, in about 425 AD, Indian astronomers devised a new model of the solar system that eliminated Ptolemy's equant. A little later Aryabhata wrote a book, later called the *Aryabhatiya*, in which he recognised that the apparent rotation of the heavens was caused by the Earth rotating on its axis. He produced an accurate estimate of the size of the Earth, and determined the length of the sidereal year to within just 4 minutes.

The prophet Muhammad was born in about 570. Over the next 150 years the Islamic empire expanded to include Spain in the west, and north-west India in the east (see *Islamic astronomy*). The caliph Al Mansūr decided in 762 to move his capital to Baghdad. He then started a major attempt to obtain copies of as many astronomical texts as possible and have them translated into Arabic. This resulted in many Indian and Greek texts being translated. As a result, early Islamic astronomy was a mixture of indigenous work, together with Indian, Persian and Greek astronomy.

In the tenth century, Abd al-Rahman al-Sūfī produced the first significant revision of Ptolemy's star catalogue, followed a little later by Ibn al-Zarqāllu, working in Spain, who produced the *Toledo Planetary Tables*. At about this time a number of Arab astronomers started questioning whether the Earth really was at the centre of the universe, and also expressed a strong dislike of Ptolemy's equant. Then, in the fourteenth century, Ibn al-Shātir succeeded in getting rid of Ptolemy's equant. Interestingly, the movement of the Sun in al-Shātir's geocentric universe was similar to that of the Earth in Copernicus' heliocentric universe.

Unfortunately most ancient Greek astronomical texts disappeared from Europe after the fall of the Roman empire in the fifth century, and it wasn't until the eleventh century that copies began to arrive back in Europe through Islamic Spain (see *European astronomy in the Middle Ages*). The route by which these Arab translations reached Europe was severed in the thirteenth century with the overthrow of the Moors in Spain.

Aristotle's philosophy was taught in European universities in the Middle Ages. That, and the constraints placed on astronomy by the teachings of the Christian Church, had a profound effect on the development of the subject. But some thinkers attacked various of Aristotle's teachings. For example, in the fourteenth century, Thomas Bradwardine attacked the Aristotelian idea that the universe was finite in size, and Jean Buridan dismissed Aristotle's idea that the planets were in motion only because they were each subjected to a continuous force. Instead, Buridan suggested that the planets had been set in motion at their creation and were still moving as they were subject to no resistance.

In 1543, Copernicus published *De Revolutionibus Orbium Caelestium* in which he described his heliocentric theory of the universe. His idea of a spinning Earth in a heliocentric universe was not new, having been proposed by Aristarchus in the third century BC. But the time was now ripe in a Renaissance that was eager for new ideas. Copernicus' theory was based on circular motion and, like Ptolemy's theory, depended on epicycles, although he had deleted the equant. But Copernicus had broken with the Aristotelian concept of a non-spinning Earth at the center of the universe. Then in 1577 Tycho Brahe disproved another of Aristotle's ideas. Aristotle had believed that comets were in the Earth's atmosphere, but Tycho was unable to measure any clear parallax for the comet of 1577. So it could not have been in the atmosphere, and must have been appreciably further away than the Moon.

China and Japan

A rock carving on a cliff at Jiangjumya, which depicts the Milky Way, shows that Chinese astronomy dates back to at least 2000 BC (see *Chinese astronomy*). Oracle bone fragments from An Yang show that by 1400 BC the Chinese had adopted a lunisolar calendar, which consisted of 12 months of alternately 29 and 30 days. Every now and again the Chinese added an intercalary month to keep the lunar and solar years in step. By the late sixth century BC, they realised that the phases of the Moon recur on the same day of the solar year every 19 years. A similar discovery had been made in the West at about the same time. But it was not to be formalised, by Meton of Athens, for another hundred years.

The Chinese had divided the sky into 28 lunar mansions from at least the sixth century BC, whilst the Indians seem to have adopted this concept a little later. The origin of the lunar mansions idea is obscure, however, with many conflicting theories ascribing it to China, India, Mesopotamia, Persia or Egypt.

The Chinese were keen to observe and record any unusual celestial events, and they have left us with the longest unbroken set of astronomical records in the world, dating back to about the sixth century BC. These included extensive records of solar eclipses, comets and new stars. By the first century BC the Chinese understood the real cause of eclipses, and by the third century AD Yang Wei was able to predict the timing of total solar eclipses.

Astronomy was gradually introduced into Japan from China, via Korea, in the sixth century AD, and by the end of the next century the Japanese had adopted the Chinese calendar (see *Japanese astronomy*). Excavation of seventh and eighth century Japanese tombs have revealed ceilings with star charts divided into the 28 lunar mansions. In one case the ceiling showed both the ecliptic and the celestial equator. Both the Japanese and Chinese recorded the new star or supernova of 1054. Thereafter, astronomy gradually stagnated in Japan until the arrival of Jesuit missionaries in the sixteenth century.

The American Continent

The pre-conquest peoples of South America did not develop a written language, and so their history is difficult to interpret (see *South America and the Incas*). The earliest known astronomical alignments in the Americas are at a temple at Buena Vista, near Lima, Peru, which had alignments to both the summer and winter solstices. The temple dates from about 2200 BC. A more complex set of alignments of thirteen towers was found in 2007 at Chankillo, Peru, dating to about 300 BC. Fifteen hundred years later, the Inca built similar structures as part of their horizon-based solar calendar. They, like many other civilisations, also linked the visibility of the Pleiades to their agricultural calendar.

Unlike the Incas, whose known culture seems to have dated from about 1200 AD, the Mayan civilization of Central America dates back a further fifteen hundred years (see *Mayan astronomy*). Also, unlike the Incas, the Maya had developed a written language. This showed that their astronomy had much in common with that of Babylon even earlier still. Both were interested in analysing observations to produce numerically-based predictions of the movements of celestial objects.

For a long time the Maya had two calendars running in parallel: a religious calendar of 260 days and a solar calendar of 365 days, consisting of eighteen months of twenty days, plus five 'nameless days'. As $52 \times 365 = 73 \times 260$, the two calendars repeated after exactly 52 solar years in what was called the 'calendar round'.

Venus had a very important position in Mayan religious observance, as it was seen as a companion to the Sun. Human beings were sacrificed on Venus' first appearance after superior conjunction, and wars were often started based on the Mayan Venus calendar. The Maya recognised that Venus made its heliacal rising almost exactly on the same date in the solar calendar at eight year intervals. In the thirteenth century Dresden codex, the extensive Mayan Venus table made predictions accurate to within one day at the end of 481 years.

It is not clear which civilisation built Teotihuacán, near Mexico City, which is a large pyramid complex probably built between about 200 BC and 100 AD (see *Central Mexico and the Aztecs*). By the time of the Spanish invasion in the sixteenth century, however, the area had been settled by the Aztecs who continued with the practice, also carried out by the pyramids' builders, of human sacrifices to Venus at its heliacal rising. The Aztec calendars were virtually identical to the 260 day and 365 day Mayan calendars. Every 52 years the Aztecs held a ceremony, called the 'Binding of the Years', when their own 260 and 365 day calendars became temporarily in step.

Evidence of any astronomical activity in North America before the arrival of the Europeans is very sparse (see *North America*). Some of the indigenous Indian tribes, like the Hopi people, used horizon-based solar calendars. In addition, some of the petroglyphs (carvings) and pictographs (paintings) at Chaco Canyon, New Mexico also appear to have astronomical significance, although their exact nature is disputed.

The Pacific Basin

The astronomical culture of the Australian Aborigines (see *Australian Aborigines*) goes back thousands of years. Like indigenous cultures all over the world, these Aborigines used the movement of the Sun and the risings and settings of various stars to regulate their agricultural calendar. But, unusually, they were often more interested in the colours and patterns of the stars than in their intensities. They were particularly interested in the Milky Way, which they called the Emu in the Sky, and recognised planets, comets and meteors. One aboriginal tribe noticed the link between the tides and the phases of the Moon, and another realised that a total solar eclipse was caused by the Moon passing in front of the Sun – but, in the latter case, this was interpreted mythologically.

The Polynesians progressively colonised their small Pacific Islands from about 1500 BC to 400 AD, navigating by the stars over vast distances (see *Polynesian and Maori astronomy*). They moved on to New Zealand later. The Polynesian and Maori people, like many other ancient civilisations, observed comets and meteors, and the Hawaiians had names for the celestial paths followed by the Sun at the solstices. Interestingly, the Polynesians called the Sun Ra, like the Egyptians, which seems a remarkable coincidence, as there is no evidence of any contacts between the two cultures.

Bibliography

Hoskin, Michael (ed.), *Cambridge Illustrated History of Astronomy*, Cambridge University Press, 1997.

Leverington, David, *Babylon to Voyager and Beyond: A History of Planetary Astronomy*, Cambridge University Press, 2003.

North, John, *The Fontana History of Astronomy and Cosmology*, Fontana, 1994.

Pannekoek, Anton, *A History of Astronomy*, Allen & Unwin, 1961 (Dover reprint, 1989).

Ruggles, Clive, *Ancient Astronomy*, ABC-CLIO, 2005.

Selin, Helaine (ed.), *Astronomy Across Cultures; The History of Non-Western Astronomy*, Kluwer Academic Publishers, 2000.

Thurston, Hugh, *Early Astronomy*, Springer-Verlag, 1996.

Walker, Christopher (ed.), *Astronomy Before the Telescope*, British Museum Press, 1996.

ANCIENT (PRE-TELESCOPIC) INSTRUMENTS

Armillary sphere

The armillary sphere was a development of the equinoctial armilla, which consisted of a single ring fixed in the plane of the equator, which was used to determine the arrival of the equinoxes. To this was added a ring fixed in the plane of the meridian to make a solstitial armilla to measure solar altitudes. In its final form, the full armillary sphere had numerous rings, including those representing the tropics, polar circles, and the ecliptic. In the observational armillary the number of rings was kept to a minimum, and some of the rings were partial with sights and angular markings, whereas in the demonstration armillary, which was used for teaching, the rings were complete. Usually a ball representing the Earth or, later, the Sun was placed at its centre, the first instrument being called a Ptolemaic armillary and the second a Copernican.

It is thought that Eratosthenes (c. 276−195 BC) used a solstitial armilla for measuring the obliquity of the ecliptic, and that Hipparchus (c. 185−120 BC) probably used an armillary sphere of four rings. Ptolemy (c. 100−170 AD) produced an astrolabon, which was a form of armillary, to determine the location of celestial bodies in ecliptic coordinates. Observational devices were used at the Marāgha observatory in Persia in the thirteenth century, and at Samarkand in the fifteenth century. In Europe, Bernhard Walther (1430−1504) undertook numerous measurements of the latitude and longitude of the planets using an armillary. Tycho Brahe (1546−1601) made a 1.2 m diameter zodiacal armillary for measuring latitude and longitude in ecliptic coordinates. He later made a number of equatorial armillaries, which could be larger as they had fewer rings, for measuring right ascensions and declinations. His largest such device was 2.7 m in diameter.

The Chinese also developed the armillary sphere, or *hun yi* ('celestial sphere instrument'). In 52 BC Geng Shouchang introduced the first permanently fixed equatorial ring, and in 84 AD Fu An and Zia Kui added an ecliptic ring. A water-driven armillary sphere was apparently built by Zhang Heng in 132 AD. Another, much larger armillary sphere was built at Kaifeng by Su Song in 1088, linked to a large, water-driven, public mechanical clock, which allowed the observer to track celestial objects across the sky. Then in 1270 Guo Shoujing produced an equatorially-mounted armillary ring, or *jian yi* ('simplified instrument'), which rotated about an axis pointing to the celestial pole. The detailed arrangement was a forerunner of what is now called the 'English Mounting' used for some astronomical telescopes.

There are various mentions of armillary spheres in historical writings, but some of the descriptions are unclear, so it is not absolutely clear that they are describing an armillary. Ptolemy mentioned armillaries in his *Almagest* of about 150 AD. In the eighth century the Islamic astronomer al-Fazārī wrote a treatise on the armillary sphere, which he called *dhāt al-halaq* ('instrument with rings').

Astrolabe

There are a number of types of astrolabe, of which the plane or planispheric astrolabe was by far the most common amongst astronomers. It was a flat, circular wooden or brass instrument, suspended by a ring. On one side was a moveable sighting bar which was used to measure the altitude of celestial objects. On the other side was a stereographic projection of both the heavens and the altazimuth coordinates for a particular latitude. It was used, amongst other things, to determine the rising and setting times of the Sun and stars, and to determine the time during daylight or at night. It could also be used as an analogue computer to solve mathematical problems.

The astrolabe appears to have been invented by the Greeks. Apollonius of Perga (c. 265−190 BC) and Hipparchus (c. 185−120 BC) undertook significant work on mathematical projections, and Hipparchus may well have made the first instrument, but the evidence is circumstantial.

Theon of Alexandria (c. 335−400) wrote a treatise on the astrolabe that is now lost. It appears to have been the basis of John Philoponus (also called Joannes Grammaticus) of Alexandria's treatise of the sixth century, and that of Severus Sebokht of Syria in the following century. The astrolabe was developed by the Islamic Arabs who used it for astronomical and astrological purposes, as well as to schedule morning prayers. According to Ibn al-Nadīm in the tenth century, the first person to build an astrolabe in the Islamic world was al-Fazārī in the eighth century. Al-Battânî (or Albategnius) (c. 855−929) certainly used one at ar-Raqqah in Syria at the end of the ninth century. The earliest surviving astrolabe is dated 927/8.

Astrolabes came to Europe via Islamic Spain in the tenth century. There Ibn al-Zarqāllu (or Azarquiel) (c. 1029−1087) made a major improvement in their design when he produced a universal astrolabe, called a saphea, that could be used at different latitudes. It was further developed by Ibn al-Sarrāj of Aleppo in Syria when he made an even more universal instrument in the early fourteenth century. This device, which was far more sophisticated than any of the later European

Renaissance instruments, was to be the high point of Islamic astrolabe making.

The astrolabe was not common in Europe until the thirteenth and fourteenth centuries, when it was also used as an educational tool. In 1391 Geoffrey Chaucer (c. 1343–1400) wrote a treatise on the astrolabe, which was the first technical treatise on any subject to be written in the English language. The astrolabe peaked in popularity in Europe in the fifteenth and sixteenth centuries.

In 1370 Mahendra Sūri translated a Persian text on the astrolabe into Sanskrit. Then in 1393 Parameśvara started a long series of eclipse observations, using an astrolabe, in Southern India. This was the first use of an astrolabe, as far as we know, in the south of the subcontinent. Astrolabes continued to be made in India until the end of the nineteenth century.

Cross-staff

The cross-staff was a simple device for measuring the angle between two celestial objects. It consisted of a sighting pole or staff, with one or more cross-pieces that could be slid along it. It was first mentioned by the French astronomer Rabbi Levi ben Gerson in 1328, who used it to measure the altitude and separation of the stars and planets, and the diameter of the Sun and Moon.

Quadrant

There were several types of quadrant, including a sine quadrant, which was used to solve trigonometrical problems; an horary quadrant, for measuring the time of day with the Sun; a geometric quadrant; and a mural quadrant. In the geometric or 'old' quadrant, the celestial object was viewed along one edge of the metal quadrant, and a plumb bob hanging over a calibrated scale measured its altitude. In the mural quadrant, the measuring scale was fixed to a wall, and the celestial object viewed from the scale through the centre of the quadrant. There was also a 'new' or astrolabe/almucantor quadrant, described by Jacob ben Mahir in the thirteenth century, which was a mixture of a quadrant and an astrolabe.

The origin of the various quadrant designs is unclear, but they were clearly used by Arab Islamic astronomers. Probably the best known instruments to pre-date the telescopic age were both built by Tycho Brahe. In about 1569 he built a 4.5 metre radius, wooden quadrant near Augsburg that could be rotated in azimuth for measuring altitudes. Then in 1582 he built a 2 metre radius great mural quadrant which was aligned with the meridian and mounted on a wall at his Uraniborg observatory.

Sextant

There were two basic forms of astronomical sextant prior to the invention of the telescope – the mural sextant and framed or frame-based sextant. They were called 'sextant' as they covered an arc of one-sixth of a circle or sixty degrees. The mural version was by far the earliest.

The first known mural sextant was constructed in Rayy, Iran by Abū Mahmūd al-Khujandī in 994. Called the al-Fakhrī sextant, after his patron, it covered a sixty degree arc on a wall, had a radius of 20 metres and was aligned with the meridian. He used it to measure the obliquity of the ecliptic. Ulugh Beg also constructed a Fakhrī sextant at Samarkand in 1420 with a radius of 40 metres, also aligned with the meridian. It was used by a team of astronomers led by Ulugh Beg to produce a star catalogue. They also determined the obliquity of the ecliptic.

Tycho Brahe seems to have invented the frame-based sextant to measure the separation of astronomical objects, because of problems with the cross-staff. He made a number of frame-based sextants, improving their design over time. Basically two observers were used to measure the separation of objects, viewing from the calibrated arc of the sextant through a pointer at the centre of the arc. One person observed along the fixed radius at one object, and the other observed along the movable radius at the other. The angular separation was then read off from the calibrated arc. Tycho's sextants were made of wood and brass and usually mounted on a type of universal joint.

The design of the sextant was radically changed after the invention of the telescope, but these are outside the scope of this article.

Sundial

The gnomon, a simple vertical post, was the first device to enable people to tell the time of day using the Sun. About 3000 BC the Egyptians built obelisks, which were tall, four-sided, tapering stone monuments which enabled them to tell the time by the position and length of their shadows. The obelisks could also be used to tell the solstices from the lengths of their shadow at mid-day. In the *MUL.APIN* of about 1200 BC the Babylonians gave the length of the shadow cast by a vertical rod one cubit (about 45 cm) high at various times of year. Many centuries later, Eratosthenes of Cyrene (c. 276–195 BC) estimated the diameter of the Earth by measuring the different angles of the Sun to the vertical on midsummer's day in Alexandria and Syene (now called Aswan) using a gnomon.

The shadow clock seems to have come into use in Egypt in about 1500 BC. It consisted of a vertical 'T' piece, with the long top of the 'T' horizontal. The 'T' was attached to the end of a long horizontal beam on which the top of the 'T' cast a shadow of varying lengths. This horizontal beam, which was pointed due west (with the 'T' piece at the east end) in the morning and due east in the afternoon, had 5 hourly markings on it.

This enabled the Egyptians to divide the day into 10 'hours', plus one twilight 'hour' at either end. The shadow clock had the advantage over the simple gnomon in that it was portable, although the time measured was only very approximate.

The earliest description of a sundial in approximately its modern arrangement comes from Berossus, a Babylonian priest, in about 290 BC. It is a half-spherical bowl cut out of a block of stone with a small gnomon in the centre and twelve markings to one side of the gnomon to show the hours. A short while afterwards, a sundial, which had been captured from the Samnites, was set up in Rome.

As time progressed different designs of sundials proliferated. Their gnomons were usually either vertical with a horizontal receiving plate, horizontal with a vertical plate, or pointing to the north celestial pole. In the latter case the Sun rotated uniformly around the gnomon, so the hour lines were equally spaced on a plate perpendicular to the gnomon. People also experimented with various shaped receiving plates for the shadow. In about 25 BC Vitruvius, in Book IX of his *De Architectura*, listed many different types of sundials and their inventors, most of them Greek. Fifteen years later the Solarium Augusti, a giant sundial built by the Emperor Augustus, was dedicated to the Sun. It was built in the Campus Martius to commemorate his victory over Egypt in 30 BC, and used a 22 metre high obelisk imported from Heliopolis as the gnomon. Both the obelisk and some of the inscribed marble pavement that had surrounded it still exist, although the obelisk was re-erected in the Piazza di Montecitorio at the end of the eighteenth century.

Apparently the Islamic Caliph Umar ibn 'Abd al-'Azīz (682−720) used a sundial in about 718 to regulate the times of prayers. In about 820 the Islamic mathematician Al-Khwārizmī (780−850), in his treatise on sundials, produced extensive tables showing the polar coordinates of the intersections of the hour lines with the shadows on horizontal sundials for different latitudes. Shortly afterwards Thâbit ibn Qurra (836−901), in his treatise on sundial theory, gave the mathematical theory for constructing sundials in any plane. Eventually most of the major mosques had their own sundials to enable them to time daily prayers. In 1371−2 Ibn al-Shātir (1304−1375) designed a magnificent sundial for the main minaret of the Umayyad Mosque in Damascus that could be used to measure time relative to any of the five daily prayers. It was accidentally broken in the nineteenth century, but fragments of the original and a copy still exist.

Water clock

The water clock was an important tool in the ancient world for timing astronomical phenomena. Our first direct evidence of such a device comes from the inscription in an Egyptian tomb of about 1520 BC, although it was probably used in both Egypt and Babylon before then.

The early Egyptian water clock, which was an outflow type, was used to measure time at night. It was like a stone bucket with sloping sides. Water was poured into the top, and it dripped out via a very small hole near the bottom. The Egyptians divided the night into twelve hours throughout the year, so the hours were shorter in summer than winter. To allow for this, there were a number of vertical scales inside the water clock that denoted the hours of night throughout the year. During daylight, the Egyptians measured time using shadow clocks, although water clocks may also have been used.

Early water clocks in Babylon were also of the outflow type. They were cylindrical in shape and, instead of having internal scales, the hours were determined by the volume of water coming out. In the earliest times their main use seems to have been to measure the length of the three night watches. A measured amount of water was put into the cylinder at the start of the watch, and the watch ended when it was empty. As the nights varied in length over the course of a year, different amounts of water were required. Initially, the amount used was varied only four times a year, but by about 500 BC it was changed every 5 days. Water clocks of similar design were in use in India at about this time. The Greeks started to use water clocks in about 300 BC, and the Romans a little later.

The Chinese used outflow type water clocks known as *lou lou* ('drip vessel') or *ke lou* ('graduated leak') from at least the seventh century BC. In about 200 BC they changed to an inflow clock in which a bowl, with a hole in it, floated in a water container and was timed to sink.

One of the problems with the outflow water clock is that the speed at which the water leaves the container reduces as it gets emptier. To solve this problem the Chinese used a series of header tanks to maintain a constant flow, and measured the amount of water coming out of the clock to measure the time. The Chinese also realised that problems were caused both by evaporation and by the increase in viscosity of water as it got colder. They eventually solved the latter problem by using mercury instead.

Starting in about 250 BC, the Greeks and Romans began to devise water clocks that drove mechanisms of various sorts. These water-driven mechanical clocks displayed the passage of time by ringing bells or moving pointers or dials. In about 132 AD, Zhang Heng used a water clock in China to drive an armillary sphere. This concept was developed over time, and in about 725 Yi Xing invented the escapement mechanism. Then in 1088 Su Song built a 10 m high astronomical clock tower at Kaifeng, then the capital of China, where a water wheel drove two armillary spheres via an escapement mechanism. This was the world's first public mechanically driven clock.

Bibliography

Hoskin, Michael (ed.), *Cambridge Illustrated History of Astronomy*, Cambridge University Press, 1997.

Needham, J., Wang, L. and De Solla Price, D. J., *Heavenly Clock-work: The Great Astronomical Clocks of Medieval China*, 2nd ed., Cambridge University Press, 2008.

North, John, *The Fontana History of Astronomy and Cosmology*, Fontana, 1994.

Pannekoek, Anton, *A History of Astronomy*, Allen & Unwin, 1961 (Dover reprint, 1989).

Stephenson, B., Bolt, M. and Friedman, A. F., *The Universe Unveiled: Instruments and Images through History*, Cambridge University Press, 2000.

Walker, Christopher, (ed.), *Astronomy Before the Telescope*, British Museum Press, 1996.

AUSTRALIAN ABORIGINES

Australia is a vast country which had over 400 different indigenous or aboriginal cultures at the time of the arrival of the first Europeans about 200 years ago. Some of these Aboriginal tribes or peoples became extinct shortly after their first contact with the Europeans, but others survived, and still do so.

The Australian Aborigines have an astronomical culture going back thousands or tens of thousands of years. Most of this has been handed down by word of mouth, although some tribes have left drawings or other artefacts. The Aborigines recognised the planets, comets, meteors, many of the brighter stars and some of the dimmer ones. But they were more interested in star patterns and the colours of stars than their brightness, sometimes ignoring bright stars near much dimmer ones. The Aranda people of Central Australia recognised white, blue, yellow and red stars. Antares was described as *tataka indora*, or very red, whereas stars in the Hyades were described as *tataka* (red) or *tjilkera* (white). Virtually all the Aboriginal people recognised the Milky Way, which was often referred to as the Emu in the Sky, the Coal Sack (the Emu's head) and the Magellanic Clouds.

The Yolngu people of Arnhem Land in the tropical far north described how *Walu*, the Sun-woman, lit a small fire at dawn and decorated herself with red ochre, some of which spilt onto the clouds to produce the red sunrise. She then lit her torch, crossed the sky, and descended to the western horizon, where some of the red ochre spilt onto the clouds to produce the red sunset. She then put out her torch and travelled underground to reappear in the east at dawn. This is typical of Aborigines' astronomy, in which the Aborigines observed an astronomical phenomenon and described it by an imaginative story or myth.

The Yolngu also noticed the link between the Earth's tides and the phases of the Moon. They explained that when the Moon rose at dusk, tides were high and water filled the Moon. But water then ran out of the Moon so that when the Moon was high in the sky at dusk or dawn, the tides fell, leaving the Moon empty. In addition they held a 'Morning Star Ceremony' to celebrate the appearance of Venus in the morning sky. They believed that a rope connected Venus to the Sun, preventing her from moving too far away. The Warlpiri people realised that a total solar eclipse was caused by the Moon passing in front of the Sun. But they interpreted this as the Moon-man making love to the Sun-woman.

Many Aborigines used astronomical phenomena as a guide to the seasons. For example, those people of the tropical far north used the heliacal rising of Arcturus to signal that it was time to harvest the *rakia* or spike-rush plants, which they used to make fish traps and baskets. To the Boorong people of Victoria in the south, the appearance of Arcturus told them that the wood ant larvae were ready to be harvested. The Boorong also linked the appearance of Lyra in March to the Mallee fowl building their nests, and when Lyra disappeared in October they knew that the eggs were ready to be collected. The Pitjantjatjara of the Western Desert linked the appearance of the Pleiades to the season for culling dingo puppies, which were a valuable part of their diet.

A number of Aboriginal artefacts still exist. For example, on the banks of the Murray river, north of Adelaide, there is a site called Ngaut Ngaut which belongs to the Nganguraku people. The rock carving there shows the Sun and Moon and a series of dots and lines, but their significance has not yet been decoded. An analysis of 97 Aboriginal stone alignments in New South Wales has shown that, in the majority of cases, they are aligned either north-south or east-west, indicating that they have been aligned astronomically. There are also a number of stone arrangements, often roughly circular in shape. For example, the 50 m diameter Wurdi Youang stone arrangement in Victoria, built by the Wathaurung people, is approximately egg-shaped. Its major axis is almost exactly east-west, with other of alignments of its stones apparently indicating the solstices.

Bibliography

Haynes, Raymond, et al., *Explorers of the Southern Sky: A History of Australian Astronomy*, Cambridge University Press, 1996.

Norris, Ray P., *Searching for the Astronomy of Aboriginal Australians*, in *Astronomy and Cosmology in Folk Traditions and Cultural Heritage*, 15th SEAC and Oxford VIII Conference, Klaipeda, Lithuania, 2007, published in *Archaeologica Baltica*, Special Issue 11, 2009.

BABYLONIAN ASTRONOMY

Babylonian astronomy can usefully be divided into four historical periods: the Old Babylonian, Assyrian, New Babylonian and Late Babylonian periods. The Old Babylonian is generally taken to cover the Hammurabi dynasty of about 1895 to 1595 BC, which ended with the invasion of the Hittites. The Assyrians first captured Babylon in about 1230 BC, but their occupation did not last long. The Assyrian period is taken to last from about 820 BC, when the Assyrians became dominant in

Mesopotamia once more, to 612 BC, when their empire finally collapsed. The New Babylonian period follows that, with the Late Babylonian covering the last three centuries BC.

Old Babylonian period

Omens, some of which were connected with the visibility of the Moon and Venus, were very important to the Babylonians during the Hammurabi dynasty. These Babylonians also developed a calendar based on their astronomical observations. Initially their year consisted of 12 months with a 13 month added, or intercalated, whenever the months seemed to be getting out of step with the agricultural year. The months were started when the new Moon was first seen by the priests.

This early intercalation scheme was very hit and miss, however. For example, a thirteenth month would sometimes be added to two consecutive years, and on one occasion an extra month was added after the sixth month, instead of after the twelfth, which was the norm. At some stage, the intercalation scheme seems to have been improved by observing the heliacal rising of various stars and constellations to decide which year needed a thirteenth month. The rules were spelt out in a later astronomical compendium called *MUL.APIN* of about 1200 BC, copies of which were recovered from the ruins of Assurbanipal's library of the seventh century BC.

The Babylonians of the Hammurabi dynasty made a special study of Venus, which was variously called Nindaranna ('mistress of the heavens') or the star of the goddess Ishtar. They often listed Venus with the Sun and Moon, separate from the other four planets, and they appear to have discovered that it went through phases, like the Moon. Venus' periods of visibility near the western and eastern horizons, and its periods when it could not be seen (as it was too close to the Sun), have been recorded in an Assyrian text that was based on an old Babylonian text of about 1600 BC. The Assyrian text indicates that the Babylonians had correctly identified Venus' synodic period of 584 days.

Assyrian period

The Assyrians, who dominated and later occupied Babylon in the Assyrian period, were very much concerned with the interpretation of omens. The Sun (except for its eclipses) and the stars were of little interest to them, as they were predictable. They were much more interested in the Moon and planets. Red Mars was thought to be an evil star, whereas Jupiter was thought to be a lucky object. The Assyrians observed the colour of the Moon, particularly during eclipses, the intensity of its Earth-shine, its apparent corona or halo, and so on. They were able to predict lunar eclipses to some extent, but were initially surprised when some lunar eclipses were missed because, as they later discovered, some took place in daylight. In addition, they produced rules for the rising and setting of the Moon as a function of phase.

The most important service that the Assyrians of this period gave to astronomy was the library assembled on the orders of Assurbanipal at their capital, Nineveh. It contained copies on clay tablets of the old Babylonian texts supplemented by commentaries and new items. Thirteen thousand fragments from this library, which were discovered in the mid-nineteenth century, are now in the British Museum.

New Babylonian period

In 612 BC the Assyrian empire finally collapsed, and Babylon became the centre of a new empire under Nebuchadnezzar, but in 539 BC it became part of the Persian empire under Cyrus the Great. As time progressed, the Babylonians became less and less interested in interpreting omens and became more interested in trying to detect patterns in planetary and lunar movements, to enable astronomical predictions to be made. Water clocks were used to measure time, and in a text of 523 BC the relative timings of sunrise and sunset, and moonrise and moonset, are recorded to an accuracy of about a minute. The Babylonians measured the positions of the planets relative to the stars, and deduced their synodic periods. So in the case of Jupiter, for example, they observed that its synodic period was 1.09 years, resulting in there being almost exactly 65 of Jupiter's synodic periods in 71 years. The Babylonians also recorded both partial and total lunar eclipses, and observed that the cycle of eclipses repeated itself almost exactly every 223 synodic months, a period now called a 'saros'.

Late Babylonian period

Babylon had been part of the Persian empire for about two hundred years, but in 331 BC it became part of the empire of Alexander the Great. Alexander's conquest resulted in the arrival of the Greek influence in the so-called Seleucid period which lasted until 247 BC. During this period communications between Babylon and Greece naturally improved. But over the following sixty years there was much disruption in the area, and communications between Babylon and Greece became spasmodic once more. Then in 181 BC the Parthians took control, and successfully withstood invasion attempts for almost 300 years. This Parthian domination cut off the Babylonians from the Mediterranean civilisations of Greece, Egypt and Rome, whilst also stifling their local culture. But the priests still continued with their astronomical observations and analysis.

Babylonian priests saw the planets as gods moving in a two dimensional sky, not, like the Greeks, as celestial bodies moving in three dimensional space. Our knowledge of their work in the late Babylonian period is contained in about 300 cuneiform tablets. These contain extensive tables with intricate calculations of the movements of the Sun, Moon and planets. By now their positions were measured in ecliptical longitude and latitude coordinates. The longitudes were recorded in one of twelve zodiacal signs, each covering 30° longitude, with their subdivisions being recorded in the sexagesimal notation based on units of sixty.

The main objective of the Babylonian analyses of the positions of the Sun and Moon was to predict when the crescent Moon would first be visible each month. Two systems were used to make these predictions. The first assumed that the Sun moved along the ecliptic at a constant angular velocity for 194° of the ecliptic, and at another constant angular velocity for the remaining 166°. Their analysis produced a sidereal year which was only six minutes too long.

The second system, which appears to have been developed a little later, assumed that the Sun's angular velocity varied in a zig-zag fashion over the year, changing at the rate of 0° 18′ per synodic month. This required the Sun to take 12.368851 synodic months to complete a full 360° sidereal year. But the analysis also produced a slightly different time, now called the anomalistic year, of 12.369136 synodic months for the Sun's velocity to go from maximum to minimum and back again. In this time the Sun would move 360° 0′ 29.8″ along the ecliptic. So the Babylonians appear to have realised that the Sun moves slightly more than 360° in an anomalistic year, although they did not, as far as we know, question why this was so.

In the second century BC, the Babylonians had, by observing the Moon for many years, deduced the length of the average synodic month to high accuracy. Then using a zig-zag function they calculated the length of the average anomalistic month (from maximum to maximum orbital speed of the Moon) to within 2.7 seconds of its correct value.

The Babylonian calculations were shown on extensive lunar tables, the most complete of which ran to 18 columns of numbers. These tables were produced not only to predict the visibility of the crescent Moon, but also to predict lunar eclipses. To do this, astronomers realised that it was vital to get a good estimate of the Moon's latitude above or below the ecliptic. It was recognised that a straightforward zig-zag function of latitude versus time would not be sufficient, so they used a modified version. This enabled them to calculate an index which predicted not only when there would be a full or partial lunar eclipse, but also approximately how long it would last.

The Babylonians' planetary tables gave data on the heliacal rising and setting of the planets, and on the first and second stations (stationary points) and oppositions of the superior planets. The data on Jupiter were particularly extensive.

Babylonian priests wished to predict the position of Jupiter in the sky based on their records of its position going back many years. In their simplest arithmetical system they assumed that Jupiter moved at a constant speed in ecliptic longitudes per

synodic period for part of its movement across the sky, and at a different constant speed for the remainder of the ecliptic. This gave a sidereal period of 11.8611 years and a synodic period of 1.0921 years, which are correct to within 0.01%. Their alternative zig-zag system produced the same synodic and sidereal periods, but both systems produced results for the first and second stations that differed from the observed positions by up to 2° over the course of one sidereal period.

The Babylonians produced similar data for the other planets, giving synodic periods which compare very well with the values known today. However, the predicted positions of the planets at their stations and heliacal risings or settings were, like those for Jupiter, rather inaccurate.

It is not clear how much of the Babylonians' astronomical work was known to the Greeks, but it is thought that a number of the Babylonian constellations were known to Eudoxus of Cnidus in the fourth century BC. Hipparchus in the second century BC was the first influential Greek to use the Babylonian degree notation. He was undoubtedly familiar with much Babylonian work, which assisted his analysis of the precession of the equinoxes. By the time of Ptolemy, 300 years later, the Greeks were using the Babylonian zodiacal reference system, and the degree as the basic angular measure. They also had access to a wealth of Babylonian observational data including lunar eclipse observations going back to at least the eighth century BC. It only became clear gradually in the late nineteenth and twentieth centuries how much the Greeks were indebted to the Babylonians. In fact those links are still being assessed.

Bibliography

Hunger, Hermann, and Pingree, David, *Astral Sciences in Mesopotamia*, Brill, 1999.

Neugebauer, O., *A History of Ancient Mathematical Astronomy*, 3 vols., Springer-Verlag, 1975.

Swerdlow, Noel (ed.), *Ancient Astronomy and Celestial Divination*, MIT Press, 2000.

Swerdlow, Noel, *The Babylonian Theory of the Planets*, Princeton University Press, 1998.

CENTRAL MEXICO AND THE AZTECS

Teotihuaćan, about 40 km (25 miles) northeast of Mexico City, is a large pyramid complex probably built between 200 BC and 100 AD. It was part of a city that occupied an area of about 30 square kilometres, with a population in the region of 150,000 to 200,000 by 300 AD. For some unknown reason, the city went into decline and was abandoned in about the seventh century. By the time of the Spanish invasion by Cortés in 1519, the area had already been settled by the Aztecs who appear to have been in the area for about 200 years. The Aztecs had taken over the magnificent monuments at Teotihuaćan and continued with the practice, also carried out by their builders, of human sacrifices. For example, the Aztecs sacrificed their prisoners to Venus at its heliacal rising.

The two main pyramids at Teotihuacán are the 66 m high Pyramid of the Sun, to one side of the 2.5 km (1.5 miles) long Avenue of the Dead, and the smaller Pyramid of the Moon at the northern end of the avenue. The avenue was oriented 15.5° clockwise off north-south, with crossed roads at a similar angle off east-west. Interestingly, the crossed roads were aligned in the west with the horizon setting position of the Pleiades. In fact, the Pleiades passed directly overhead at the latitude of Teotihuacán, and on the day when the Pleiades rose heliacally the Sun passed through the zenith. But it is not clear if this western alignment of the crossed roads with the Pleiades was fortuitous. An alternative explanation is that they were aligned with the setting Sun on August 13 and April 30, two dates separated by the Aztec 260 day sacred cycle.

The Aztec calendar was based on the 260 day and 360 day Mayan calendars, with the latter consisting of eighteen months of twenty days. Like the Maya, the Aztecs added five extra days to the 360 days to try to keep their calendar in step with the Sun. They also recognised that as $52 \times 365 = 73 \times 260$, the day and month names would be the same in both calendars every 52 years. The Aztecs held a ceremony called the 'Binding of the Years' every 52 years, when this happened, at which they observed the Pleiades to see that they did not stop at the zenith. When they were observed not to stop, the Aztecs were then sure that the world would not end for at least another 52 years.

Bibliography

Aveni, Anthony F., *Astronomy in the Americas*, in Walker, Christopher (ed.), *Astronomy Before the Telescope*, British Museum Press, 1996.
Aveni, Anthony F., *Skywatchers: A Revised and Updated Version of Skywatchers of Ancient Mexico*, University of Texas Press, 2001.

CHINESE ASTRONOMY

The early development of Chinese astronomy was based on the idea that the country's ruler was the 'Emperor of All under Heaven'. His was a divine appointment, and if his rule was good and just, the heavens would behave normally, but if his rule was bad, strange things would appear, like comets and new stars. So Chinese astronomers were keen observers of the heavens, noting down anything unusual. As a result the Chinese have produced the longest unbroken astronomical records in the world, dating from about the sixth century BC.

Oracle bone fragments from AnYang show that by 1400 BC, during the Shang dynasty, the Chinese had adopted a lunisolar calendar with alternating months of 29 and 30 days. Most years contained 12 lunar months, but every now and again an extra long or short month was added to keep the lunar year in step with the solar year. By the late sixth century BC the Chinese realised that the phases of the Moon recur on the same day of the solar year after exactly 19 years. This so-called Metonic cycle of 235 synodic months or 19 tropical years was not clearly recognised in the West until about 100 years later by Meton of Athens.

In the first century BC, a new system of intercalation was used based on *er-shi-si jie-qi* or 'twenty-four fortnightly periods' each of which corresponded to a 15° movement of the Sun along the ecliptic, which takes about 15.2 days. As a lunar month is less than two of these fortnights, every now and again a lunar month would not contain the start of two fortnights. When that happened an intercalary month was added.

The Chinese have been observing and recording total solar eclipses for well over four thousand years. Dozens of eclipses have been recorded from the Chou dynasty and Warring States period (1050–221 BC). In the fourth century BC, Shi Shen realised that the Moon was somehow implicated in solar eclipses, but he thought that they were due to the power of the Moon temporarily suppressing the power of the Sun. By the first century BC, however, the Chinese understood the real cause of eclipses, and by 20 BC they were making eclipse predictions based on a recurrence period of 135 months. Then by the third century AD, Yang Wei was able to predict the timing of the first and last lunar contact during solar eclipses. According to the Book of Han, the Chinese observed their first sunspots in 28 BC.

In 330 AD, Yu Xi appears to have been the first to realise that the solar and stellar years were different. This *sui zha* or annual difference, as it was called, is due to the precession of the equinoxes that had been discovered about five hundred years earlier in the West by Hipparchus.

From at least the sixth century BC the Chinese had divided the sky into 28 *xiu* or lunar mansions of unequal widths, like the segments of an orange going from pole to pole. They also divided the ecliptic into twelve Jupiter stations, as they had observed that Jupiter took about 12 years to come back to its original position in the sky. The earliest known descriptions of the motions of the planets, which were very schematic with little detail, dated from the third century BC. But by about 25 AD much more detail was available. By then the synodic periods of each of the planets had been determined to within about half a day, and the synodic month had been estimated as 29 d 12 h 44 m 27 s, which is just 23 seconds too large.

A rock carving, which is at least 4,000 years old, on a cliff at Jiangjumya depicts the Milky Way. Shi Shen, Gan De and Wu Xian produced, between 370 and 270 BC, a catalogue of 1,464 stars. This predated Hipparchus' star catalogue by almost two centuries. The Chinese divided the ecliptic into 365/14; *du*, with one *du* for each day of the year.

In the fifth century BC the Chinese produced the Book of Silk, which contained details of 29 comets or 'broom stars', as they were called, spread over 300 years. The Chinese were the first to realise that the tails point away from the Sun. The oldest record of a guest star, which may have been a supernova, has been found on an oracle bone dated about 1300 BC. It records a 'great new star . . . in company with Antares'. They recorded another supernova in 185 AD, which the Roman chronicler Herodias said 'shone continuously by day'. Then much later, in 1054 AD, Chinese and Japanese astronomers observed the supernova that was to create the Crab Nebula. Surprisingly, there is no surviving record of that supernova being seen in Europe.

Between 721 and 725 AD Yi Xing and Nangong Yue measured the length of the Sun's shadow at the winter and summer solstices using a gnomon at a number of locations, covering a distance of about 3,000 km (2,000 miles) on an approximately north-south line. They concluded that about 1° of latitude was equivalent to about 155 km (true value 111 km).

The Chinese developed three basic cosmological theories: the Gai Tian, Hun Tian and Xuan Ye schemes. Gai Tian, which first seems to have appeared somewhere between the fourth and second century BC, pictured a gently curved sky above a gently curved Earth. The heavens rotated about an axis through the centre of the Earth. The Hun Tian scheme, which was outlined by Loxia Hong in about 120 BC, consisted of the Earth floating on water, with the heavens supported above by vapour. Finally a later scheme called Xuan Ye, which was described by Qi Meng in the second century AD, saw the universe as an infinite empty space. All the celestial objects floated freely in it and were blown about by a wind.

The Chinese built a number of observatories over the centuries and used a number of different types of instruments; the designs of some of the later ones seem to have been influenced by the Arabs. The main instruments that they used were the gnomon (from at least the seventh century BC), the armillary sphere (the first simple version variously estimated as from the fourth to the second century BC), and the water clock. In 132 AD Zhang Heng devised a water-powered armillary sphere, and in 1270 AD Guo Shoujing devised an equatorial mount for an armillary sphere. This appears to be the first recorded equatorial mount anywhere in the world. Shortly afterwards Guo Shoujing designed a 12 m high brick built gnomon that allowed him to measure the length of the tropical year to within 26 seconds.

Bibliography

Nakayama, Shigeru, *A History of Japanese Astronomy: Chinese Background and Western Impact*, Harvard University Press, 1969.

Needham, Joseph, *Science and Civilisation in China, Volume 3, Mathematics and the Sciences of the Heavens and the Earth*, Cambridge University Press, 1970.

Ronan, Colin A., *The Shorter Science and Civilisation in China; Volume 2*, Cambridge University Press, 1982.

Sivin, Nathan, *Cosmos and Computation in Early Chinese Mathematical Astronomy*, E. J. Brill, 1969.

Xiaochun, Sun, *Crossing the Boundaries Between Heaven and Man: Astronomy in Ancient China*, in Selin, Helaine (ed.), *Astronomy Across Cultures: The History of Non-Western Astronomy*, Kluwer Academic Publishers, 2000.

EGYPTIAN ASTRONOMY

The Egyptian calendar

Ra (or Re), the Sun god, and Nut, the goddess of the sky, were the most important gods to the ancient Egyptians. Nut was depicted as being stretched across the sky as the Milky Way, with her mouth in Gemini and her birth canal at Daneb (α Cygni) where the Milky Way begins to divide into two parts, which were her legs. Ra was depicted as being conceived by entering Nut's mouth at the spring equinox and 272 days (or about 9 months) later, as being born from her birth canal on the morning of the winter solstice. The star Sothis (Sirius) was also highly revered as, in the early years, its heliacal rising presaged the annual Nile floods, which were vital for Egyptian agriculture.

The Egyptians of both Lower and Upper Egypt initially used a 354 day agricultural and religious calendar based on the Moon, with 12 months of 29 or 30 days. There were three 4 month seasons of 'inundation' (or flooding of the Nile), 'growth' and 'harvest'. Every two or three years an extra or intercalary month was added. In the north or Lower Egypt this intercalary month was added as necessary to ensure that the festival of the birth of Ra occurred in the last month of the year. In Upper Egypt it was added to keep the festival of the heliacal rising of Sothis in the last month. When Lower and Upper Egypt were unified in about 3000 BC the Upper Egyptian scheme using the heliacal rising of Sothis was gradually adopted across the whole of the country, with the 'Overseers of the Hours' determining the start of each month.

In about 2900−2800 BC the Egyptians devised a simpler civil calendar that ran in parallel with the agricultural/religious calendar. It was easier to administer, as each year had twelve identical 30 day months, plus 5 extra or epagomenal days to make a 365 day year. Each month was divided into three 10 day weeks.

Because the 365 day year is about one quarter of a day too short when compared with a year based on the stars (Sothis in this case), the year-ends of the two calendars gradually got out of step. So, later, a third parallel calendar was introduced for religious purposes, whilst retaining the old lunistellar calendar for agriculture. In this new religious calendar, which also had months based on the Moon, the intercalary month was added, not based on the heliacal rising of Sothis, but every time the first day of its year came before the first day of the civil year.

The Egyptians also measured hours. In the Ninth Dynasty (c. 2130–2080 BC) they started to use 36 rising stars, called decans, each successively marking an 'hour' for a period of ten days. So the 'hours' were, on average, only 40 minutes long. Eighteen decans marked the interval from sunset to sunrise, of which 3 were assigned to twilight at each end of the night. So the period of total darkness had 12 decans. However, as the length of night varied with the time of year, these decans did not measure equal lengths of time. At Egyptian latitudes the night in midwinter is almost 50% longer than the night in midsummer, but how much the Egyptians were aware of the magnitude of this difference we do not know. It was not until the invention of the water clock, of which the first known example was used in the New Kingdom about 1530 BC, that they had any accurate measure of the effect. But by then the Egyptians had changed from using 18 rising stars at night to using 13 stars crossing the meridian, with one of the 13 stars marking the beginning of night.

It seems strange that the Egyptians, who clearly studied the heavens in some detail, did not leave regular records of planetary movements, nor of solar or lunar eclipses. So they did not try to predict the movement of the Sun, Moon and planets, like the Babylonians, nor try to explain how they moved, like the Greeks. But the Egyptians were aware by about the fourth century BC that the Moon went through 309 synodic months in exactly twenty-five of their 365-day years. This gives a synodic month of 29.5307 days, compared with the correct value of 29.5306 days.

The pyramids

The Fourth Dynasty (c. 2610–2490 BC) pyramids at Giza, including the Great Pyramid, have their entrances in their north faces, and their corridors slope up from the inside so that the lower culmination of the north circumpolar stars could be seen from them. This alignment was so the souls of the deceased pharaoh could reach these stars, which were called *ikhemu-sek* or 'the ones not knowing destruction' as they never set. The brightest such star at that time was Thuban (α Draconis).

The Great Pyramid or Pyramid of Khufu is thought to have been built around 2530 BC. At that time the northern and southern corridors of both the king's and queen's chambers were oriented with stars that were known to be important in Egyptian life. They were Thuban, associated with gestation, and Al Nitak in Orion, associated with the fertility god Osiris, for the north and south corridors of the king's chamber. For the north and south corridors of the queen's chamber, they were Kocab, associated with immortality, and Sothis, associated with Isis, the queen of the gods. So it is clear that not only did the

Egyptians know the stars in the heavens very well, they also apparently took their celestial positions into account in building their pyramids because of their religious significance.

The orientation of the four faces of the Great Pyramid was also remarkably close to that of the cardinal points, with the south-north axis being only 3′ off true north. Likewise the south-north axis of the large Pyramid of Khafre was only 6′ off true north. It is not known how the Egyptians were able to orient their pyramids to such great accuracy. They could have observed the position of the rising and setting Sun and bisected the two positions, or they could have bisected the two extreme east-west positions of a star.

Bibliography

Edwards, I. E. S., *The Pyramids of Egypt with New Material*, revised edition, Penguin Books, 1993.

Neugebauer, O., *A History of Ancient Mathematical Astronomy*, 3 vols., Springer-Verlag, 1975.

Wells, Ronald A., *Astronomy in Egypt*, in Walker, Christopher, *Astronomy Before the Telescope*, British Museum Press, 1996.

EUROPEAN ASTRONOMY IN THE MIDDLE AGES

Many early Christians believed that all truth lay in the Bible, and so the structure of the universe could be deduced only with reference to Biblical texts. However, as Thomas Aquinas (1225–1274) pointed out in his *Summa theologiae*, St Augustine (354–430) had taught that where science and Scripture disagree, scientific ideas could be adopted over Scriptural interpretation, but only if they could be demonstrated to be unambiguously correct, which was rarely possible. So science did have a role to play in interpreting the cosmos in the Middle Ages, but it was heavily constrained by Christian Scripture.

The early Christian Church frowned on astrology, in favour of a more rational approach to the structure of the universe. This was supported in the early seventh century by Bishop Isidore of Seville (c. 560–636) who published his *Etymologiae*, showing how natural phenomena could be explained logically and devoid of superstition.

Most ancient Greek astronomical texts had been lost to Europe after the fall of the Roman empire, and it wasn't until the eleventh and twelfth centuries that copies began to percolate back into Europe through Islamic Spain. The route by which these Arab translations reached Europe was, unfortunately, severed in the thirteenth century with the overthrow of the Moors in Spain. Nevertheless, the influential *Toledo Planetary Tables* had been produced in Spain in the eleventh century by Ibn al-Zarqāllu (Azarquiel), based on his own observations and those made by earlier Islamic scholars. These tables were followed in 1272 by the *Alfonsine Tables* produced under the patronage of Alfonso the Wise, King of Castile.

Meanwhile, elsewhere in Europe, universities had been founded in Bologna (c. 1080), Paris (c. 1160), Oxford (c. 1180), Padua (1222) and Naples (1224). The Church had become more organised and wanted a more-accurate calendar to define the date of Easter. Great thinkers now came to the fore like the Dominican Thomas Aquinas, who argued that reason is able to operate within faith and who lectured on Aristotle at the University of Paris, and the Franciscan Roger Bacon (c. 1220–1292), who argued for and developed experimentation in science. Unfortunately, the thirteenth century not only revived Aristotelian cosmology and mathematical astronomy, but it also signalled a revival of astrology. Astrology and some of Aristotle's teachings were condemned as heretical by Étienne Tempier, Bishop of Paris (d. 1279) in his Condemnation of 1277. Then in about 1310 Dante wrote his *Divine Comedy*, which combined theology and astrology with a modified Aristotelian description of the universe.

In about 1344, in his *De causa Dei contra Pelagium*, Thomas Bradwardine (c. 1290–1349), one of the Oxford Calculators, attacked the Aristotelian idea that the universe was finite in size, arguing that the universe was infinite in extent as God himself was ubiquitous. Likewise, at about the same time, Jean Buridan (1300–1358) in Paris dismissed Aristotle's idea that the planets were in motion only because they were each subject to a continuous force. On the contrary, Buridan argued in his impetus theory that God had given them motion at their creation, and they were still moving as they were subject to no resistance.

Nicole Oresme (c. 1323–1382), Buridan's pupil, wrote a commentary on Aristotle's *On the Heavens* in 1377. In this he adopted both Buridan's impetus theory and Bradwardine's idea of an infinite universe. He also discussed the possible rotation of the Earth on its axis, and showed that it was impossible to determine observationally whether it spins on its axis or not. He finally concluded for theological reasons, however, that it does not spin. Nicholas of Cusa (1401–1464) argued that the universe was infinite and full of stars, and that, as the universe was infinite, the Earth could not be at its centre. He further argued that the Earth moves, but he did not make clear whether he meant that it spins on its axis or that it orbits the Sun, or both. He also thought that there was life on other planets spread throughout the universe. His views were largely ignored at the time, although Giordano Bruno, Kepler and others became aware of them later.

The University of Paris had lost its dominant position on the continent during the Hundred Years War (1337–1453). At about the same time, there was a great flowering of the arts and sciences in Italy and in German-speaking Europe, and new universities were founded in Prague (1348), Vienna (1365) and Heidelberg (1386). It was, however, the invention of the printing press in the middle of the fifteenth century that was to have the most far-reaching consequences for astronomy.

No more would astronomers have to rely on texts that had been copied successively by scribes, who often did not really understand what they were copying, and whose mistakes were generally not corrected at the next transcription, thus adding errors to errors.

In 1460 Cardinal Johannes Bessarion (1403–1472) asked Georg Peurbach (1423–1461) and Johannes Müller of Königsberg (1436–1476), otherwise known as Regiomontanus, to produce an abridged version of Ptolemy's *Almagest*. Peurbach died soon afterwards and Regiomontanus was left to write almost the whole work. However, the resultant *Epitome of the Almagest* of 1462, which was used extensively until the early seventeenth century, was far more than an abridged text, as Regiomontanus added much explanatory detail. Over the next few years Regiomontanus calculated the positions of the Sun, Moon and planets using the *Alfonsine Tables* and compared them with his observations. As a result he found that the *Alfonsine Tables* were seriously in error. Then in 1471–2 he set up a printing press, an observatory and a workshop to build instruments, all paid for by his associate Bernhard Walther (1430–1504). Regiomontanus became the first to print mathematical and scientific books and, after his death, Walther used his observatory to produce the first systematic astronomical observations in Europe, some of which were later used by Copernicus.

By the fifteenth century the European religious calendar was in much need of reform, as the vernal equinox was by then ten days in advance of its correct date, so the computation of Easter Day was compromised. This led Pope Sixtus IV to summon Regiomontanus to Rome in 1475 to seek his advice on what should be done. He advised the pope that new observations were needed before an accurate system could be put in place. But, unfortunately, Regiomontanus died the following year before the new observations could be undertaken, and it was not until 1582 that the calendar of the Roman Catholic Church was finally reformed.

Navigation also called for a wider understanding of astronomy at this time, when explorers were leaving the relative security of the Mediterranean and the coastal waters of western Europe and Asia for almost the first time. In 1488, for example, Dias rounded the Cape of Good Hope. At these previously unexplored southern latitudes, navigators were able to observe new stellar constellations, particularly the Southern Cross. Not only were they travelling further south, however, but in 1492 Columbus crossed the Atlantic to try to find a western route to India. This added a new requirement on astronomy to help to measure longitude accurately, which was not to be solved for well over two hundred years.

Although there had been occasional attempts to improve Ptolemaic cosmology in the Middle Ages, it was still taught at universities in late fifteenth century Europe. There were, however, both philosophical and practical objections to it: the

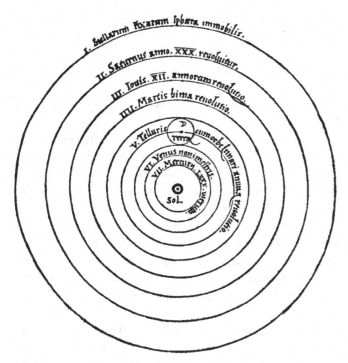

Figure 1.1. Copernicus' heliocentric view of the universe in which the planets orbit the Sun (Sol) and the Moon orbits the Earth (Terra). Although not shown here, Copernicus' system was based not only on circular orbits, but also on epicycles.

concept of the equant was generally disliked, and the large apparent size variations of the Moon, which was a consequence of his model, clearly did not exist.

Copernicus

In 1512 Nicolaus Copernicus (1473−1543) wrote a brief, anonymous text *Commentariolus* in which he pointed out the deficiency, as he saw it, of the Ptolemaic system. Instead he outlined a system still, like Ptolemy, based on circular orbits and epicycles, but without his equant. More importantly, the planets in Copernicus' system orbited the Sun (see Figure 1.1), rather than the Earth, and the Earth spun on its axis, rather than being static. But Copernicus was not the first person to suggest that the Earth spun on its axis, that was Hicetas two thousand years earlier, nor was he the first to propose a Sun-centred universe, that was Aristarchus in the third century BC.

Copernicus explained that his hypothesised motion of the Earth round the Sun could not be detected by looking at the stars, as they were too far away for parallax to be observable. He also explained that the orbital motion of the Earth and planets around the Sun naturally produced retrograde loops for the outer planets, as seen from Earth.

Interestingly, in the light of Galileo's subsequent problems with the Roman Catholic Church, the Church did not raise any serious objections to Copernicus' theory. As a result Copernicus was persuaded to allow his full work, *De Revolutionibus*

Orbium Caelestium (*On the Revolutions of the Heavenly Spheres*), which was based on his *Commentariolus*, to be published. This was completed in 1543, the year he died. It was well received, possibly aided in theological circles by an anonymous Foreword that is thought to have been written by the theologian Andreas Osiander (1498−1552). This explained that the book described a mathematical model of the universe, rather than the universe itself.

Copernicus substantially improved the calculated orbit of the Moon, implying an apparent variation in the Moon's diameter, as seen from Earth, of only ± 13%, which was much closer to the truth, compared with the ± 30% of Ptolemy's model. Unfortunately, however, the planetary latitudes in *De Revolutionibus* were still as inaccurate as those in the *Almagest*, although the longitude predictions were quite good.

Tycho Brahe

Eight years after Copernicus' death, Erasmus Reinhold (1511−1553) published a new set of tables, based on Copernican parameters, to replace the obsolete *Alfonsine Tables*. These tables, called the *Prutenic* (Prussian) *Tables*, were a significant improvement on the *Alfonsine Tables*. But in 1563 Tycho Brahe (1546−1601) found that neither the *Prutenic* nor the *Alfonsine Tables* accurately predicted the date of the conjunction of Jupiter and Saturn of that year. This was a crucial problem to someone like Tycho who was interested in astrology, where such conjunctions were very important. It encouraged him, therefore, to try to measure the positions of the heavenly bodies much more accurately than had previous observers, to enable more accurate predictions to be made.

No one had any serious reason to doubt Aristotle's doctrine of the unchangeability of the universe beyond the Moon until November 1572, when a bright new object was observed. Michael Mästlin (1550−1631), Thomas Digges (c. 1546−1595) and Tycho all independently proved that the star did not move against the other stars, so it could not be a comet. Tycho could also find no parallax, so the new object must be much further away than the Moon, and it twinkled, so it could not be a planet. This led him to conclude that it was a new star. The object gradually decreased in brightness and changed colour, and then in 1574 it finally disappeared.

The appearance of this new star, now called Tycho's star or supernova, persuaded Tycho Brahe to found an astronomical observatory, to observe and measure the heavens to unprecedented accuracy. King Frederick II of Denmark gave Tycho the small island of Hven near Copenhagen and offered to provide money to build the observatory. In 1576 Tycho started to build his observatory of Uraniborg on Hven, which he equipped with a range of state-of-the-art instruments.

In 1577 Tycho disproved another of Aristotle's theories when he observed the comet of that year. Aristotle had taught that comets were in the Earth's upper atmosphere, but Tycho was unable to measure any clear parallax for the comet of 1577. This meant that the comet was appreciably further away than the Moon, proving that comets were astronomical rather than atmospheric phenomena. Along with the new star of a few years earlier, it provided convincing evidence that the universe was not immutable. Tycho's calculations eventually showed that the comet was in an almost circular orbit around the Sun, outside the orbit of Venus.

On cosmology Tycho disagreed with both Ptolemy and Copernicus, as he disliked Ptolemy's equant, and he disagreed with Copernicus that the Earth was spinning on its axis. Tycho was also unable to measure any stellar parallax, indicating that the stars would have to be more than 700 times as far away from Earth as Saturn. He regarded this as ridiculous as, according to Tycho, God would not have wasted so much space.

By 1578, therefore, Tycho had decided to adopt the intermediate cosmological system advocated by Heracleides in the fourth century BC and by Martianus Capella of Carthage (c. 365–440 AD), in which Mercury, Venus and, in Tycho's case, the comet of 1577 orbit the Sun, whilst the Sun, Moon and the other three planets orbit the Earth. Like Aristotle, Tycho thought that all these bodies were carried on real spheres as they orbited their parent body. Five years later, however, Tycho was beginning to have second thoughts, thinking it likely that all the planets orbit the Sun, as the Sun orbits the Earth. Unfortunately, this implied that the sphere that carried Mars around the Sun would have to intercept the sphere that carried the Sun around the Earth. This was clearly impossible if both were crystalline spheres, as proposed by Aristotle, so he was forced to reject yet another of Aristotle's theories. Therefore in his book *De mundi etheri recentioribus phaenomenis* of 1588, in which he described his work on the 1577 comet, he proposed a system (see Figure 1.2) in which all the planets orbit the Sun, as the Sun orbits the Earth once per day. In this system the stars, which he concluded are just outside the orbit of Saturn, also orbit the Earth once per day.

Bibliography

Christianson, John Robert, *On Tycho's Island; Tycho Brahe and His Assistants*, Cambridge University Press, 2000.

Gingerich, Owen, *The Eye of Heaven: Ptolemy, Copernicus, Kepler*, American Institute of Physics, 1993.

Grant, Edward, *Planets, Stars, and Orbs: The Medieval Cosmos (1200–1687)*, Cambridge University Press, 1996.

Koestler, Arthur, *The Sleepwalkers: A History of Man's Changing Vision of the Universe*, Penguin-Arkana, 1989.

Koyré, Alexandre, *The Astronomical Revolution: Copernicus-Kepler-Borelli*, Cornell University Press, 1973 (Dover reprint, 1992).

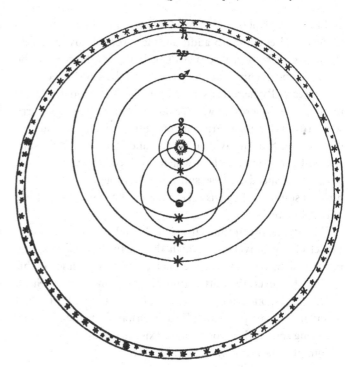

Figure 1.2. Tycho Brahe's geocentric model of the universe in which the Moon, Sun and stars orbit the Earth, whilst all the planets, except the Earth, orbit the Sun.

GREEK ASTRONOMY

The Greeks were philosophers who wanted to understand and explain what they observed in the sky in physical terms and, to do this, they produced conceptual three-dimensional models. Unfortunately, however, they used perishable materials to record their work, so most of the early Greek material is lost. As a result we have to rely on second or third hand accounts, with the inevitable contradictions and confusions that this creates.

Thales of Miletus (c. 625–547 BC) was, according to Aristotle, the founder of Greek natural philosophy. He hypothesised that the Sun and stars were made of water, and that the Earth was a disc that floated on water which evaporated to produce the air. Some disputed Greek sources say that he later recognised that the Earth was spherical, and that he understood the cause of lunar and solar eclipses. Anaxagoras (c. 500–428 BC) was the first to clearly state that the Moon shone with scattered sunlight and so explain the causes of eclipses.

Anaximander of Miletus (c. 610–545 BC), one of Thales' pupils, thought that the Earth was a cylindrical column surrounded by air at the centre of an infinite universe. The Earth was surrounded by two flame-filled tyres, with a hole in each to produce the Sun and Moon, and by a sphere with a flame-filled surface with holes to produce the stars.

It is unclear how much of the work and ideas sometimes attributed to Pythagoras (c. 580–500 BC) came from him or from his followers, the Pythagoreans. The Pythagoreans were probably the first to appreciate that the Earth was spherical, and that the planets each moved in separate orbits inclined to the celestial equator. In addition, they were almost certainly the first Greeks to realise that Phosphorus and Hesperus, the morning and evening stars, were one and the same object.

The early Pythagoreans introduced the doctrine of the 'Harmony of the Spheres' in which a non-spinning, spherical Earth was surrounded by a series of concentric, crystalline spheres. The Moon, Sun, individual planets and stars were each supported by a sphere, each of which revolved around the Earth at different speeds. These spheres were thought to produce a musical sound as they passed each other, known as the 'Music of the Spheres'.

The Pythagorean Philolaus (c. 450−400 BC) devised a new model of the universe in which the Earth was also moving, instead of being static at the centre of the universe. In his scheme, however, the Earth orbited a central fire (not the Sun), called Hestia, once per day in a circular orbit. The Earth kept its inhabited part permanently facing away from the fire, thus requiring the Earth to rotate on its axis once per day. The Sun, Moon, planets and stars also orbited Hestia.

Although Philolaus had required the Earth to spin on its axis once per day, he did not specifically mention this. Another Pythagorean, Hicetas of Syracuse (fl. 5th c. BC), seems to have been the first person to have done so when he proposed that the Earth spins on its axis at the centre of the universe. Heracleides (c. 388−315 BC) developed this idea by suggesting that Mercury and Venus orbited the Sun, as the Sun orbited a spinning Earth, whilst Mars, Jupiter and Saturn orbited the Earth directly. This was later given the misnomer of the 'The Egyptian System'.

Aristarchus of Samos (c. 310−230 BC), who was one of the last Pythagoreans, went one step further and proposed a heliocentric universe in which all the planets orbit the Sun in the (correct) order of Mercury, Venus, Earth, Mars, Jupiter, and Saturn, with the Moon orbiting a spinning Earth. Aristarchus was also the first to produce a realistic estimate of the Earth-Moon distance, although his estimate of the Earth-Sun distance was an order of magnitude too low. Seleucus of Seleucia (fl. 2nd c. BC) appears to have been the only person to have adopted Aristarchus' ideas of a heliocentric universe, until Copernicus resurrected it about 1,700 years later.

In the meantime, a new school of thought had been developed by Plato (c. 427−347 BC). It is difficult to know how to interpret his ideas, as the descriptions given in his books are allegorical and ambiguous. Nevertheless, Plato was a highly respected philosopher whose ideas on a wealth of subjects have had a lasting effect on Western thought. He taught that all the heavenly bodies must be spherical, as that was the perfect shape, and they must move in uniform circular orbits, for the same reason. But his geocentric model of the universe had major deficiencies. He ascribed the same angular velocity to Mercury and Venus as to the Sun, so his model could not explain their

oscillations around the Sun, as seen from Earth. However, he appears to have recognised this problem, as he suggested that Mercury and Venus were subject to "a force opposed to the Sun" or "a contrary tendency to the Sun" (translations vary), but historians of astronomy are still unsure what he meant. And his model could not explain the temporary reversal in directions of Mars, Jupiter and Saturn at their first and second stationary points.

Eudoxus (c. 400−350 BC), one of Plato's pupils, tried to produce a more realistic model of the universe based on Plato's principle of uniform circular motion. To do so he assumed that each planet was on the surface of transparent spheres, with a non-spinning Earth at their common centre. He assumed that there were four spheres per planet. For a given planet, each of the four spheres rotated about axes inclined at different angles to produce the observed motion. The biggest problem with this theory was that it could not explain the intensity variations of the planets, as, in his theory, their distances from the Earth remained constant.

Aristotle (384−322 BC), one of the greatest of Greek philosophers, was also a pupil of Plato. Like his master he devoted only a small amount of time to astronomy. But his adoption of Plato's conviction that all the heavenly bodies were spherical and moved in circular orbits around the Earth was to have a profound effect on Western thought until well into the Middle Ages.

Aristotle's model of the universe was based on that of Eudoxus, which had, in the meantime, been developed by Callippus of Cyzicus (c. 370−300 BC). In his geocentric scheme Aristotle had one sphere for the stars, five for the Moon, and seven or nine spheres for each of the Sun and planets, to produce a total of 57 spheres. His scheme was complicated and inelegant and still possessed the problem of both Eudoxus' and Callippus' models in that it could not explain the intensity variations of the planets over their orbits. Nor could it explain the variations in the apparent size of the Moon, as seen from the Earth, particularly from one annular solar eclipse to another.

The centre of Greek thought now moved to Egypt following Alexander the Great's over-wintering there in 332−331 BC and his defeat of the Persian empire a few months later. Alexander then founded a new city called Alexandria on the coast of Egypt to use as a base for future military operations in the eastern Mediterranean. Ptolemy (not the astronomer of the same name), Alexander's successor, made Alexandria his capital and founded a great library and centre of learning there, which proved to be a great attraction to such brilliant people as Euclid and Archimedes.

Eratosthenes of Cyrene (c. 276−195 BC), a contemporary of Archimedes, was a director of the library of Alexandria. He estimated the diameter of the Earth to within a few percent of its correct value, by measuring the different angles of the Sun to the vertical on midsummer's day at Alexandria and Syene (now called Aswan). He also appears to have determined the obliquity of the ecliptic to within 8′ of its true value at that time, although some historians dispute this, and to have produced a catalogue of 675 stars.

Hipparchus

It had gradually become clear to the Greeks that if 7 extra lunar months were added in total over 19 years to years of 12 lunar months each, then the lunar phase on the first day of the new year was almost exactly the same as it had been 19 years previously. That is 19 tropical years was almost exactly the same as 235 synodic months. This so-called Metonic cycle had first been formally proposed by Meton (c. 460−400 BC) in about 430 BC as the basis for a calendar. Callippus modified this cycle by reducing four Metonic cycles by one day in total, making a synodic month of 29.53085 days and a tropical year of exactly 365.25 days. We now know that both these values, although slightly more accurate than those proposed by Meton, were still fractionally too large. Two hundred years later, Hipparchus (c. 185−120 BC) improved the estimates even further.

Both the Babylonians and the Egyptians had been aware for some time that the sidereal and tropical years were not the same, because they had noticed the long-term drift of the stars compared with the solstices. Hipparchus was clearly aware of much of this Babylonian work, but he appears to have been the first to quantify the drift in his book *On the Length of the Year*. According to Ptolemy the astronomer (we do not have a copy of Hipparchus' original book), Hipparchus concluded that the equinoxes are moving relative to the fixed stars by at least 1° per hundred years, or at least 36″ per year.

In about 330 BC, Callippus had measured the length of the seasons between the solstices and equinoxes as 94 (for spring), 92, 89 and 90 days, indicating an apparent variable velocity of the Sun in its orbit. The Babylonians produced similar figures, as did Hipparchus some time later but, unlike Callippus and the Babylonians, Hipparchus tried to produce an orbital scheme to explain this effect. There were basically two alternatives assuming a geocentric universe based on circular orbits. Either the Earth was at the centre of the Sun's orbit, in which case the velocity of the Sun had to vary in its orbit, or the Earth was off-centre, and then the Sun's velocity could be constant. So Hipparchus, in accepting Aristotle's philosophy of uniform circular motion, concluded that the Earth was not at the centre of the Sun's orbit. His calculation of the amount of offset of $\frac{1}{24}$ of the orbit's radius was far too large, but his direction of the apogee was in error by only 35′.

Hipparchus also tried to estimate the parallax of the Sun and Moon by analysing the appearance of solar and lunar eclipses from different locations in the Greek world. As a result he concluded that the distance of the Moon from Earth varied from 62 to $72\frac{2}{3}$ Earth radii (correct values 55 to 63 Earth radii), but he could not detect any solar parallax, as the Sun was too far away.

Ptolemy

The first person to write about the epicycle theory, as far as we know, was Apollonius of Perga (c. 265−190 BC). In this, a planet orbits a point in space, as that point orbits the Earth. The orbit of the planet around the 'empty' point was called an epicycle, and that of the empty point around the Earth was called a deferent. Both orbits were circular.

Ptolemy (c. 100−170 AD) developed the epicycle theory to try to explain the detailed movements of the Sun, Moon and planets. He carried out his astronomical observations from Alexandria between 127 and 141 AD, when it was under Roman rule. His most famous book, now called the *Almagest*, covered both earlier Greek astronomy as well as his own extensive contributions. It contained a catalogue of over 1,000 stars and was to be the work of reference in astronomy for well over a thousand years. His later book *Planetary Hypotheses* extended his planetary work.

In Ptolemy's geocentric model of the universe (see Figure 1.3), the Moon, Sun and planets each described an epicycle, as the centre of that epicycle described a deferent around a non-spinning Earth. Because the inferior planets of Mercury and Venus each appear almost symmetrically on both sides of the Sun at maximum elongation, Ptolemy assumed that the centres of their epicycles were always on a line joining the Earth and Sun. For the superior planets of Mars, Jupiter and Saturn he assumed that the lines linking these with the centre of their epicycles were always parallel to the Earth-Sun line. The main advantage of this theory was that it could easily explain the temporary reversal in directions observed for Mars, Jupiter and Saturn, as seen from the Earth. It could also explain the apparent change in intensity of these planets as they moved to and from the Earth, which Aristotle's theory, for example, was unable to do.

Unfortunately, this simple system did not provide accurate enough position estimates, so Ptolemy introduced a number of modifications. In the case of the Moon, he made the centre of the Moon's deferent describe a circle whose centre was the Earth. This caused the distance of the Moon from the Earth to vary from 33 to 64 Earth radii. It should have caused the apparent size of the Moon from Earth to vary also by a factor of two, which is clearly not the case, but Ptolemy does not seem to have been concerned about this. For the superior planets he moved the Earth from the centre of their deferents, and defined a point, later called the equant, that he placed at an equal distance to the Earth on the opposite side of the centre. The equant was the point about which the centre of the planet's epicycle appeared to move at a constant angular velocity, which meant that it no longer moved at a constant

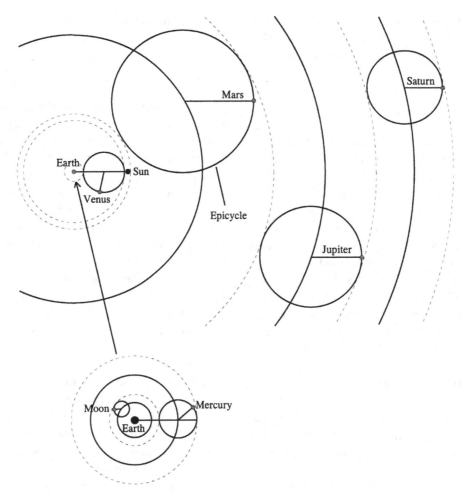

Figure 1.3. Ptolemy's model of the universe based on circular deferents and epicycles. Unlike Heracleides, he assumed that the epicycles of Venus and Mercury were centred not on the Sun but, like the other planets, on empty space.

velocity along the circular deferent. This was controversial, as it violated Aristotle's principle that all celestial objects move at uniform velocity. Other modifications were also required, but by the time Ptolemy had finished, he was able to make quite accurate position estimates for all but the Moon and Mercury.

Assuming that there were no gaps between the furthest part of one epicycle and the nearest part of the next, Ptolemy was able to produce an estimate for the size of the solar system of about 20,000 times the radius of the Earth, or about 120 million kilometres. Although this was a gross underestimate, it gave, for the first time, an idea of how large the solar system really was.

Ptolemy believed that, whilst the apogees of planetary orbits moved along the ecliptic at the same rate as the precession of the equinoxes, the apogee of the Sun's orbit was fixed with respect to the ecliptic. We now know that this is not correct. Ptolemy also concluded that the eccentricity of the Sun's orbit was the same as that previously calculated by Hipparchus, which was too large, and measured the obliquity of the ecliptic as being between 23° 50′ and 23° 52′ 30″, compared with Hipparchus' 23° 51′ 20″. Because of the closeness of his figures

for the obliquity to that of Hipparchus, he decided to take the latter's figure. Again this was incorrect, as, unrecognised by Ptolemy, the obliquity had, in fact, changed over the three hundred years between Hipparchus' and his own observations.

Bibliography

Dicks, D. R., *Early Greek Astronomy to Aristotle*, Cornell University Press, 1985.

Dreyer, J. L. E., *History of the Planetary Systems from Thales to Kepler*, Cambridge University Press, 1906, republished as *A History of Astronomy from Thales to Kepler*, Dover, 1953.

Heath, Sir Thomas, *Aristarchus of Samos, the Ancient Copernicus, A History of Greek Astronomy to Aristarchus Together with Aristarchus's Treatise on the Sizes and Distances of the Sun and Moon, A New Greek Text with Translation and Notes*, Clarendon Press, Oxford, 1913, republished as *Aristarchus of Samos: The Ancient Copernicus*, Dover, 1981.

Heath, Sir Thomas L., *Greek Astronomy*, Dent & Sons, 1932 (Dover reprint, 1991).

Neugebauer, O., *A History of Ancient Mathematical Astronomy*, 3 vols., Springer-Verlag, 1975.

INDIAN ASTRONOMY

The ancient religious Vedic texts show evidence of contact with Mesopotamia prior to Darius I's conquest of north-west India in 515 BC. In these there are some statements that can be clearly traced back to the Babylonian *MUL.APIN*. The Vedic texts used various time spans or *yugas* of two, three, four, five and six years, where a year is twelve months of 30 days each.

Indian astronomy after the north-west's conquest by Darius was described in a text *Vedanga Jyotishya* or *Jyotisavedānga*, written by Lagadha in the fifth century BC. In it he listed the 27 *nakshatras* or lunar mansions, beginning with the Pleiades, through which the Moon moved over the course of a month. He also described the use of zig-zag functions and the water clock, and rules for predicting the motion of the Sun and Moon, all of which were clearly the result of Babylonian influence. But his intercalation cycle was modified to take account of the Vedic preference for a five year repeat cycle of 1860 *tithis*. A *tithi* was exactly one thirtieth of a mean synodic month, so the five year cycle had 62 synodic months, and the twenty five year cycle, which was also mentioned, was of 310 months (rather than the 309 months used by the Egyptians).

A long Greek astrological and astronomical treatise was translated by Yavaneśvara (a leader of the Greek colony in western India) into Sanskrit in about 150 AD. It was later turned into verse in the *Yavanajātaka* (Sanskrit for 'Saying [*jātaka*] of the Greeks [*Yavana*]') by Sphujidhvaja. The original Greek text is thought to have been written in about 100 BC in Alexandria. The modified text in the *Yavanajātaka* provided details of the Babylonian calendar and defined the length of the tropical year, exactly the same as recognised by Hipparchus and Ptolemy.

In about 550, Varāhaminhira provided summaries in his *Pañcasiddhāntikā* of five *siddihāntas* or treatises of previous centuries, based on Greek and Babylonian work. The lunar and solar eclipse calculations, including those of lunar parallax, used adaptations of Greek geometric models. He mentioned a rate of precession of the equinoxes of 54″ per year, which was much closer to the correct value of 50.3″ per year than the 36″ quoted by Ptolemy. He also quoted the difference in longitude between Ujjayinī and Vārānasī in India and Alexandria in Egypt as 44° and 54°. The correct figures are 45° 50′ and 53° 7′. This surprising accuracy could have been achieved only by accurately timing lunar eclipses to establish the difference in local times.

In the *Paitāmahasiddhānta* of about 425, in a text based on the *Brāhmapaksa*, Indian astronomers devised a new theory of planetary movements based on epicycles, but without Ptolemy's equant. Then in 499 the great Indian mathematician Aryabhata (476–550) wrote his mathematical and astronomical masterpiece in 118 verses, later called the *Aryabhatiya*, which was partly based on the *Paitāmahasiddhānta*. The *Aryabhatiya* was translated into Arabic as *Zij al-Arjabhar* in the eighth century, and was later translated into Latin in the thirteenth century.

Aryabhata proposed, in his *Aryabhatiya*, that the Earth was a sphere of circumference 4967 *yojanas* or 39,967 km (24,835 miles), which is within 0.3% of its correct value. He calculated the Earth's sidereal rotation period to within 0.1 sec of its correct value, and that of the sidereal year to within 4 minutes. Aryabhata recognised that the apparent rotation of the heavens was because the Earth rotated on its axis, and that the Moon and planets shine by reflected sunlight. He correctly explained the cause of solar and lunar eclipses and calculated the duration of lunar eclipses to a good accuracy. In his model of the solar system he introduced the concept of pulsing planetary epicycles. There have been suggestions that Aryabhata believed that the solar system was heliocentric and that the orbits of the planets were ellipses, but this has been hotly disputed. Bhaskara I, a follower of Aryabhata, modified Aryabhata's planetary scheme by introducing an equant, which was probably derived from Greek sources, and a scheme in which the epicycles were tilted.

The famous mathematician, Brahmagupta (598–668), who was a contemporary of Bhaskara, was highly critical of Aryabhata's and others' work, although he accepted Aryabhata's hypothesis that the Earth spins on its axis. He was head of the astronomical observatory at Ujjayinī and was the first to apply algebra to the solution of astronomical problems. He wrote a number of texts on mathematics and astronomy, of which the most important was the *Brāhmasphutasiddhānta*, which he completed in 628. This contained many simplifications in order to calculate planetary longitudes and predict eclipses, and used epicycles that pulsed in size, like those of Aryabhata.

In 762 the caliph Abū Jafar, known as Al Mansūr, decided to build a new Islamic capital called Baghdad, and make it a centre of learning. To assist, he invited an Indian delegation to Baghdad which included Kankah, a scholar of Ujjayinī, who was well versed in astronomy. He apparently brought a Sanskrit text to Baghdad called *Mahāsiddhānta*, which was based on the *Brāhmasphutasiddhānta*, to help to explain Indian astronomy. As a result, the caliph ordered the astronomer al-Fazārī to translate this text into Arabic. It was later modified by Al-Khwārizmī and called *Zīj al-Sindhind*, and was further modified and translated into Latin by Adelard of Bath in 1126.

Bibliography

Kak, Subhash C., *Birth and Early Development of Indian Astronomy*, in Selin, Helaine (ed.), *Astronomy Across Cultures: The History of Non-Western Astronomy*, Kluwer Academic Publishers, 2000.

Pingree, David, *Astronomy in India*, in Walker, Christopher, (ed.), *Astronomy Before the Telescope*, British Museum Press, 1996.

Plunket, Emmeline M., *Ancient Indian Astronomy*, Kessinger Publishing, 2005.

Sen, S. N., and Shukla, K. S., *History of Astronomy in India*, 2nd rev. ed., Indian National Science Academy, 2000.

ISLAMIC ASTRONOMY

The pre-Islamic Arabs of the Arabian peninsula developed a folklore based on the rising and settings of the stars. They apparently adopted the lunar mansion concept from India, and observed the passage of the Sun through the twelve signs of the zodiac. Amongst other things, this gave them their time of year and assisted in their agricultural planning.

Following Muhammad's death in 632, the Islamic empire expanded rapidly, initially including Mesopotamia, Egypt and Libya. It finally reached Kabul in the Hindu Kush in 664 and Spain in 711. Fifty years later the caliph Abū Jafar, known as Al Mansūr, decided to build a new capital, Baghdad, and start a major attempt to obtain copies of as many astronomical texts as possible and have them translated into Arabic. The *Mahāsiddhānta*, an important Indian text, was translated almost immediately, whilst Ptolemy's *Syntaxis* or *Almagest* ('the greatest' in Arabic) was translated in the House of Wisdom, which had been established by the caliph al-Ma'mūn in the early ninth century. As a result, early Islamic astronomy was a mixture of pre-Islamic Arabic star lore, and Indian, Persian and, later, Greek astronomy. Eventually many of the Arabic translations reached Spain, where they were translated into Latin in the twelfth century.

Muslims had three basic astronomically based requirements. One was to determine the *qibla* or the direction of Mecca from any location, to enable them to orient their mosques and pray in the correct direction. Another was to determine the correct times of day to pray, and the third was to predict when the crescent Moon would first be visible, so they could start their month. These required knowing the size of the Earth, and the observer's location on its surface, together with being able to transform astronomical predictions made in three dimensions to those in relation to the local horizon. This not only required a three dimensional model of the solar system, like that of Ptolemy, but it also required knowing how to make transformations from that model using spherical geometry and trigonometric functions. This put a great emphasis on not only astronomy but also mathematics, which the Arabs also developed starting with Greek and Indian sources.

Theon of Alexandria (c. 335–400), the last director of the library of Alexandria, had produced a manual to explain Ptolemy's *Handy Tables* in about 380. In this he introduced the erroneous concept of 'trepidation' or oscillation in longitude of the equinoctial points, as he thought that the rate of precession of the equinoxes was variable. Thâbit ibn Qurra (836–901), who worked in Baghdad as a mathematician and astronomer, also worked in the House of Wisdom as a translator of Greek texts. He developed a theory of trepidation, which also explained the reduction in the obliquity of the ecliptic from the 23° 51′ 20″ measured by Hipparchus to about 23° 34′ or 23° 35′ measured in his time.

Al-Battânî, otherwise known as Albategnius (c. 855–929) undertook numerous observations at ar-Raqqah in Syria between 877 and 918. He concluded that the longitude of the Sun's apogee was 82° 17′, with an orbital eccentricity of 0.0346, which was correct. He noted that the apogee's longitude was 16° 47′ larger than deduced by Ptolemy, but he did not suggest that it was gradually increasing. The difference was considerably larger than that to be expected by precession (about 11° 12′, using his figure of 54.5″ per year). So he had found that, not only was the Sun's apogee moving, its movement was faster than attributable to precession, but he did not specifically point this out. Al-Bīrūnī did this in his *Kitab al-Qanun al-Mas'udi* in 1031, but it was left to Ibn al-Zarqāllu, otherwise know as Azarquiel (c. 1029–1087), who lived in Spain, to give different numerical values for both the rate of change of the solar apogee and the precession of the equinoxes.

The first significant revision of Ptolemy's star catalogue was produced by Abd al-Rahman al-Sūfi (903–986) who worked in both Persia and Baghdad. In his *Kitab su-war al-kawakib* (Book on the Constellations of Fixed Stars) he improved Ptolemy's stellar magnitudes, and used Arabic identifications, but he did not add significantly to the list of stars, nor did he correct Ptolemy's position estimates. In the eleventh century Ibn al-Zarqāllu produced the *Toledo Planetary Tables*, which were followed in 1272 by the *Alfonsine Tables*, sponsored by Alfonso the Wise, King of Castile. The latter were in use for the next three hundred years.

It was generally accepted amongst most Arab astronomers that the Earth was at the centre of the universe. But in 1030 Al-Bīrūnī pointed out that observations could not determine whether the Earth or the Sun was at the centre. The problem was, in his view, a philosophical one. Of more concern to Arab astronomers was Ptolemy's equant. In particular, Ibn al-Haytham (965–1039) or Alhazen, in his *Al-Shukūk 'alā Batlamyūs* (Doubts Concerning Ptolemy) of 1028, criticised many of Ptolemy's works, and complained that the equant did not satisfy the requirement for uniform circular motion. His biggest complaint was that Ptolemy's model of the universe could not exist in reality. Nasīr al-Dīn al-Tusī (1201–1274) also found the equant unsatisfactory, and replaced it in his *al-Tadhkira fi'ilm al-hay'a* (Memoir on Astronomy) with two small epicycles in an arrangement now called a Tusi couple which transformed two circular motions into a linear one.

Finally, Ibn al-Shātir (1304–1375) working in Damascus, succeeded in getting rid of Ptolemy's equant by using a variation of al-Tusi's arrangement. Al-Shātir's scheme had the big advantage, compared with the Ptolemaic system, of being able to explain the motion of the Moon without the large apparent size variations implied by the latter. Interestingly, the movement of the Sun in al-Shātir's geocentric scheme was similar to that of the Earth in Copernicus' heliocentric scheme, and yet there is no evidence that Copernicus was aware of a al-Shātir's work.

Observatories

The caliph al-Ma'mūn (ruled 813—833) encouraged astronomers to undertake celestial observations of their own, instead of just using Ptolemy's adjusted for precession. These new observations were designed mainly to determine the position of places on the surface of the Earth and the Earth's diameter, the obliquity of the ecliptic and more accurate celestial positions, particularly those of the Sun and Moon. Observations were initially made in Baghdad, where Yahyā ibn Abī Mansūr recorded them in a *zīj*, or book of tables, called *al-Mumtahan*. Following his death, observations were also made in Damascus under the leadership of Khālid al-Marwarrūdhī.

To undertake these observations, various instruments were used including an armillary sphere, a marble mural quadrant of 5 metre radius, and a 5 metre high iron gnomon. Although a number of astronomical instruments had been invented by this time, Arab astronomers developed them, and built larger and larger versions over the years, assuming that this would improve their accuracy. But this was not always the case, owing to problems in construction.

The Arabs built a number of observatories but many were destroyed, as they were considered to be connected with astrology, which was against the strict Islamic code. As a result, only two observatories had more than a short lifetime, namely those at Marāgha in Persia and at Samarkand.

In 1259 Nasīr al-Dīn al-Tusī persuaded Hūlāgū il Khān, a grandson of Genghis Khān, to build an observatory at Marāgha. It was equipped with an extensive library, a mural quadrant of 4 metre radius, an armillary sphere, parallactic rulers and quadrants. Then in 1420 the great observatory of Ulugh Beg (1394—1449), the grandson of Tamerlane, was built at Samarkand. Its main instrument was a stone sextant of 40 metre radius faced in marble. It was built mostly in a trench which was cut into the hillside and oriented in the plane of the meridian.

The astronomers of the Marāgha school produced the Īlkhānī tables, which were completed in 1272, but much of this work appears to have been copied from other sources. On the other hand, Ulugh Beg's astronomers produced a new catalogue of about 1,000 stars observed between 1420 and 1437. This *Zīj-i Sultani* is the only known Arab star catalogue that had been produced up to then without using Ptolemy's catalogue as a base.

Bibliography

Kennedy, E. S. et al., *Studies in the Islamic Exact Sciences*, American University of Beirut Press, 1983.

King, David A., *Islamic Mathematical Astronomy*, 2nd ed., Ashgate Publishing, 1993.

King, David A., *Islamic Astronomy*, in Walker, Christopher (ed.), *Astronomy Before the Telescope*, British Museum Press, 1996.

Saliba, George, *A History of Arabic Astronomy: Planetary Theories during the Golden Age of Islam*, New York University Press, 1994.

JAPANESE ASTRONOMY

The subject of astronomy was gradually introduced into Japan from China, via Korea, starting in the sixth century AD. In 602 AD the Buddhist priest Kwal-luk arrived in Japan from the Paekche kingdom in Korea with tribute books on astronomy and calendar making. The Japanese set up an astronomical observatory in 625, and by the end of that century Chinese calendars had been adopted in Japan. The Japanese used the changing orientations of the Pleiades (*Subaru*), Hyades and Orion to determine when it was best to plant their rice crops.

The Japanese, like the Chinese, studied the heavens because they believed that unexpected events indicated problems on Earth. So eclipse predictions were important, and the appearance of new stars and comets were causes of concern. As a result, the Japanese have left important records of these events, particularly of the 1054 supernova and of a new star, probably a supernova, in 1181. Unfortunately, in spite of their valuable astronomical observations, the science of astronomy stagnated in Japan until their first contact with European scientists in the sixteenth century.

Archaeoastronomy

Our knowledge of early Japanese astronomy has been assisted in recent years by archaeoastronomy. However, this is still in its infancy in Japan, partly because of limited resources and partly because some ancient tombs are considered to be under the protection of the imperial family. Nevertheless, some tombs have been excavated, mostly dating to the seventh and eighth centuries AD and aligned with the cardinal directions.

Takamatsu Zuka Kofun was excavated in 1972, and Kitora Kofun in 2004 following earlier investigations using a remote controlled camera. Both tombs were found to have a star chart on the ceiling depicting the 28 lunar mansions (*shuku*) that originated in China. In Takamatsu Zuka Kofun the Sun was painted in gold leaf on the east wall, and the Moon in silver on the west wall. The three undamaged walls had paintings of the gods of these three cardinal directions, *Genbu* (black tortoise of the north), *Seiryuu* (azure dragon of the east), and *Byakko* (white tiger of the west). Kitora Kofun was similarly adorned, but with the addition of *Suzaku* (red bird of the south) which had probably been originally on the damaged south wall of the other tomb. The star chart of Kitora Kofun was particularly detailed, showing both the ecliptic and celestial equator.

Bibliography

Nakayama, Shigeru, *A History of Japanese Astronomy: Chinese Background and Western Impact*, Harvard University Press, 1969.

Renshaw, Steven L., and Ihara, Saori, *A Cultural History of Astronomy in Japan*, in Selin, Helaine (ed.), *Astronomy Across Cultures: The History of Non-Western Astronomy*, Kluwer Academic Publishers, 2000.

MAYAN ASTRONOMY

When the Spaniards invaded Central America in the sixteenth century they found a Mayan civilization that had existed for about two thousand years in what are now southern Mexico, Guatemala and Belize. The Maya had developed a very sophisticated series of calendars but, unfortunately, these had been written on bark books, many of which were destroyed by their conquerors. Since then most of the surviving books have also disappeared, leaving just four intact, the most well-known of which is the thirteenth century Dresden codex. In addition to this written record, the Maya have left us numerous astronomically related carved inscriptions dating back to the Classical Mayan period of about 200 to 900 AD.

In many ways Mayan astronomy was very similar to that of Babylon, with its emphasis on numerical predictions of the movements of the Sun, Moon and planets, although the Babylonian work flourished about a thousand years earlier. Both used their calendars for religious and secular purposes and, like the Egyptians, the Maya had a separate calendar for each.

The Mayan 260 day religious or *tzol kin* ('count of days') calendar, which is almost certainly the oldest, is still used in some parts of Central America today. The base of the Mayan counting system was twenty, with subunits of five and one. So the 260 day calendar was usually expressed as twenty day names preceded by a number in the range of from one to thirteen, where thirteen was the number of layers of heaven in Mayan cosmology. First above the Earth was the Moon's layer, then a cloud layer, followed by one for the Sun, and one for the stars, and so on. The thirteenth was where the god creator lived.

The exact reason for the 260 day calendar is unclear. It may be based on Mayan cosmology, as just described. But 260 days is also the approximate period of human gestation, the length of the annual Mayan growing season, and the period that the Sun at noon remains south of the zenith at the latitude of Guatemala and the southern tip of Mexico. Whatever the origins, the Maya used this calendar for religious purposes, with each day having a patron spirit who influenced events in different ways.

The *Haab* or Mayan solar calendar consisted of eighteen months of twenty days plus a period of five 'nameless days' known as *Wayeb*, which was thought to be a dangerous period. This produced a year of 365 days. Unfortunately, this did not take into account the extra quarter day (approximately) of the tropical year, so the calendar months, which were named after particular seasons, no longer occurred in those seasons after a few centuries.

Interestingly, the *Haab* and *tzol kin* calendars repeated after exactly 52 *Haab* years as $52 \times 365 = 73 \times 260$ in what was called the 'calendar round'. So a given day had the same day and 'month' names in both calendars as exactly 52 years earlier. The Maya did not number their years, so to date an historical event, a longer period than 52 years was required. To do this they used another year or *tun* of $18 \times 20 = 360$ days, adding further digits of value 20, 20 and 13 to produce what was called a 'long count' system which repeated after a 'great cycle' period of $13 \times 20 \times 20 \times 18 \times 20 = 1,872,000$ days or about 5,125 solar years. From their inscriptions it is possible to date the first year of their great cycle as somewhere in the range 11 to 14 August 3114 BC. The reason for this choice of date is unknown.

Venus was seen as a companion to the Sun and had a very important position in Mayan religious observance. For example, human beings were sacrificed on Venus' first appearance after superior conjunction, and wars were often started based on the Mayan Venus calendar. Interestingly, an image near a sacred lake at Chich'en Itzá, on the Yucatan peninsular, shows a date in Western reckoning of 15 December 1145 when Venus is known to have transited the Sun. Also at Chich'en Itzá is El Caracol, a cylindrical tower built in the ninth and tenth centuries AD on a large, skewed, two terraced rectangular platform. The tower of El Caracol has been thought to be an astronomical observatory, with suggested Venus alignments, but the evidence is somewhat circumstantial. More convincing Venus alignments occur with the House of the Governor at Uxmal and the Temple of Venus at Copan.

The Maya recognised that Venus' synodic period was about 584 days, and that, as $8 \times 365 = 5 \times 584$, it would make an heliacal rising almost exactly on the same date in the *Haab* calendar at eight year intervals. The extensive Venus table in the Dresden codex covers 65 synodic periods with corrections that make its predictions accurate to within 1 day at the end of 481 years. In this codex 65 synodic periods were seen to be equal to 146 *tzol kin* years as $65 \times 584 = 146 \times 260$. This codex also includes a table for predicting solar and lunar eclipses, and a Mars table which makes use of the fact that Mars' synodic period of about 780 days is 3 × 260, or three *tzol kin* years.

Bibliography

Aveni, Anthony F., *Astronomy in the Americas*, in Walker, Christopher (ed.), *Astronomy Before the Telescope*, British Museum Press, 1996.

Aveni, Anthony F., *Skywatchers: A Revised and Updated Version of Skywatchers of Ancient Mexico*, University of Texas Press, 2001.

Ruggles, Clive, and Urton, Gary (eds.), *Skywatching in the Ancient World: New Perspectives in Cultural Astronomy*, University Press of Colorado, 2007.

MEGALITHIC ASTRONOMY

The earliest megalithic structure that appears to be of astronomical significance is that at Nabta Playa in the Nubian Desert about 800 km (500 miles) south of what is now Cairo. The remains, which stretch in a roughly north–south line for about 2.5 km (1.5 miles), were discovered by a team led by anthropologist Fred Wendorf. Although they were found in 1973, their full significance was realised only in 1992. The main site consists of a stone circle surrounding six large stones near the centre, with two sight lines, one of which points towards the position of the rising Sun at the summer solstice of about 4000 BC. Nearby pieces of charcoal were radiocarbon dated to between 3600 and 3500 BC, giving credibility to this interpretation of the summer solstice alignment. In addition, paleoastronomer J. McKim Malville concluded that the alignment of some of the other nearby megaliths was towards the brightest star of the Plough (Big Dipper) as it rose between 4700 and 4200 BC, with others towards Sirius and the two brightest stars in Orion's belt at about the same period. Astrophysicist Thomas Brophy has recently disputed these findings, suggesting other alignments about 1,000 years earlier.

Such disputes are not unusual in this relatively new science of archaeoastronomy. As a result there is currently very little agreement between interpretations of alignments for all but the most obvious cases. This is partly because of the difficulties in interpreting structures built by people who have left no written records, and partly because of the radically different disciplines of the researchers involved. These researchers often have a limited understanding of the other disciplines (archaeology, anthropology, history and astronomy) required to interpret the structures.

In 1991, an aerial survey photograph showed signs of circular ridges under a wheat field in Goseck, Germany. Later excavation revealed four concentric circles with three gateways, but no megaliths. The ridge structure was dated to about 4800 BC from fragments of pottery found at the site. Two of the gateways were found to have been aligned with the rising and setting Sun at the winter solstice, as seen from the centre of the concentric circles.

Stonehenge in England is probably the best known and best preserved megalithic structure. It was built in three phases of which the first, a large circular earthwork with no megaliths, has been dated at about 3000 BC. The second was a timber structure, and the third, which is the one that we see today, was built progressively from about 2500 to 1600 BC. There are also a number of long barrows in the area, some of which predate Stonehenge, a woodhenge, and extensive ancient earthworks. It appears from the dating of these various structures as though there had been a change from lunar to solar alignments. In addition, the general orientation of Stonehenge appears to have been adjusted slightly over time to follow the sunrise at summer solstice and sunset at winter solstice. Other more precise alignments with astronomical events are disputed.

The distinguished astronomer Sir Norman Lockyer helped to create the modern interest in archaeoastronomy by writing a series of publications around the turn of the nineteenth century on megalithic and other ancient structures. This interest was considerably extended in the 1960s by Alexander Thom, a retired engineer, who surveyed many megalithic sites in Britain, Ireland and Brittany. He concluded that many were precisely arranged and located to facilitate accurate astronomical observations. At about the same time the astronomer Gerald Hawkins caused considerable controversy with his theory that Stonehenge was a large astronomical computer and observatory, a view that the distinguished astronomer Fred Hoyle developed. These theories were largely condemned by archaeologists.

There are numerous megalithic structures in Western Europe, particularly in the countries bordering the Atlantic and North Sea. Most of these consist of megaliths arranged in rings, of which there are more than 900 in the British Isles, or arranged in lines. In addition, there are groups of burial tombs or cairns, some of which have clear lunar or solar alignments, which often pre-date these megalithic structures. There are also passage tombs, like that at Newgrange in Ireland, which was built around 3000 BC. It has an opening above the entrance which let the rising Sun at the winter solstice shine down the 19 m entrance passage to illuminate bones in the central chamber.

There are also a large number of stone circles in northeastern Scotland which include a horizontal or recumbent stone flanked by two vertical stones. They date to about 2000 BC. Viewed from the centre of the circle, the recumbent stone and two flankers usually drew attention to a prominent hilltop whose declination seems to be aligned with the limit of the Moon's monthly movements over its 18.6 year nodal period. Taken as a group these alignments appear to be clear, although taken individually there is quite a variability in their alignments.

Short rows of standing stones in Scotland also seem to have similar alignments, but with similar variabilities.

Bibliography

Burl, Aubrey, *The Stone Circles of Britain, Ireland and Brittany*, revised edition, Yale University Press, 2000.

Kelly, David H. and Milone, Eugene F., *Exploring Ancient Skies: An Encyclopedic Survey of Archaeoastronomy*, Springer-Verlag, 2004.

Ruggles, Clive, *Astronomy in Prehistoric Britain and Ireland*, Yale University Press, 1999.

NORTH AMERICA

The indigenous peoples of North America have left no written records, so it is difficult to produce a consistent picture of their early astronomical interests. Nevertheless, there are a number of archaeological sites that give some indication of their relatively recent activities. In Chaco Canyon, New Mexico, for example, there is a petroglyph [carving] called the 'Sun dagger' that shows the position of the Sun at both solstices and equinoxes. There is also a pictograph [painting] showing a crescent Moon, a star and a handprint. The star has been interpreted as an image of the supernova of 1054 AD, but this interpretation is highly speculative. A more probable explanation is that this was simply a Sun-watching site used by the Zuni people for calendrical purposes. The crescent, star and handprint were used by them, along with a sundisk, to indicate such sites, and there does appear to be a very faint sundisk in the Chaco pictograph.

At the site of the ancient city of Cahokia near St Louis there are a number of mounds, and the remains of a series of wooden circles thought to have been built around 900 to 1100 AD. Various alignments have been proposed for these posts, but the alignments seem to be more fortuitous than deliberate. The Bighorn Medicine Wheel in Wyoming, built by the Plains Indians, is a 25 metre diameter circle of small stones with 28 spokes, also made of stones. Not only is there a central cairn, but five other cairns are arranged around the wheel's circumference. Alignments have been suggested for some of these cairns showing the heliacal rising of Aldebaran, Rigel and Sirius. Another cairn outside of the circle also lines up with the central cairn to indicate the position of the summer solstice sunrise. More recent work has suggested that these alignments were fortuitous, however. The wheel has been dated to about 1200 to 1700 AD.

The Hopi people of Walpi village used a calendar based on the rising and setting positions of the Sun against distant landmarks on the horizon, from different observing positions around the village. The landmarks that marked the rising and setting Sun at the summer and winter solstice were thought by them to be sacred. The Lakota people, on the border of South Dakota and Wyoming, also considered parts of their landscape to be sacred. They regulated their year by observing the position of the Sun against various constellations on or near the ecliptic.

It should not be forgotten, however, that most, if not all, of this astronomical activity was being undertaken when the large cathedrals were being built in Europe, and some millennia after the work of the Megalithic builders, the builders of the Egyptian pyramids, and so on.

Bibliography

Aveni, Anthony F. (ed.), *Archaeoastronomy in the New World: American Primitive Astronomy*, Cambridge University Press, 1982.

Chamberlain, Von Del, *Native American Astronomy: Traditions, Symbols, Ceremonies, Calendars, and Ruins*, in Selin, Helaine (ed.), *Astronomy Across Cultures: The History of Non-Western Astronomy*, Kluwer Academic Publishers, 2000.

Ruggles, Clive, *Ancient Astronomy: An Encyclopedia of Cosmologies and Myth*, ABC-CLIO, 2005.

Williamson, Ray A., *Living the Sky: The Cosmos of the American Indian*, University of Oklahoma Press, new edition, 1987.

POLYNESIAN AND MAORI ASTRONOMY

Polynesia is a roughly triangular region of the Pacific from Hawaii in the north, to Easter Island in the east and New Zealand in the west. The Polynesian people appear to have inhabited their small Pacific Islands progressively from about 1500 BC to 400 AD, having migrated from the west. Some then moved to New Zealand in about 800 AD, with a second major influx about 1350. The 'Classic' Maori culture dates from this second migration.

All this migration in the vastness of the Pacific Ocean meant that the Polynesians were great seafarers. Not only did they migrate, but they regularly undertook two-way voyages. To do this they used the stars, memorising which stars were visible in which locations at which time of year. The Polynesians had no optical instruments to measure angles in the sky, so they used the zenith and horizon stars for navigation.

The Polynesians called the Sun Ra, like the Egyptians, which seems a remarkable coincidence, as there is no evidence of any contacts between the two cultures. The Maori and Polynesian astronomers also had names for the planets, bright stars, Milky Way, Coal Sack, Magellanic Clouds, and the zodiacal light. They also observed comets and meteors. The Hawaiians also had names for the northerly and southerly paths followed by the Sun across the sky at the solstices. The tiny uninhabitable Hawaiian island of Necker has an astonishing number of temple platforms, which may be due to the fact that it is situated exactly on the Tropic of Cancer. Temple enclosures on the Hawaiian island of Maui are aligned with the rising position of the Pleiades. In New Zealand the Maori noted that the Pleiades appeared at the start of their birding season, whilst the appearance of Vega indicated that it was time to harvest their *kumara*, which was a staple crop.

There is very little evidence that the famous large statues or *moai* on Easter Island, most of which are on the coast, have any deliberate astronomical alignments. An exception appears to be a single *moai* standing on a platform or *ahu* some way inland. It faces the Sun as it rises behind a hill at the June solstice, with another nearby hilltop marking the setting Sun at the equinoxes.

In the nineteenth century, a Belgian priest, Honoré Laval, lived for a time on the island of Mangareva in French Polynesia. There he observed that the rising Sun at the solstice rose between two dressed stones set up side by side on a nearby hill, as seen from a flat stone platform in the middle of the village. In the following century, Peter Buck found other similar alignments on the island for both the rising and setting Sun, using both natural and artificial markers.

There were marked similarities and differences between the Maori and Polynesian calendars. Both generally used a 12 month year of 30 days per month. Presumably the 30th day was omitted in alternate months to bring the months in line with the phase of the Moon. In New Zealand the Maori year started in May or June, whereas elsewhere in Polynesia it started in December. The actual starting date in New Zealand depended on the appearance of the first new Moon after heliacal rising of the Pleiades or Rigel. In this it is similar to the method for choosing Easter in Western Christianity, which is based on the first full Moon after the spring equinox.

Bibliography

Liller, William, *Ancient Astronomical Monuments in Polynesia*, and Orchiston, Wayne, *A Polynesian Astronomical Perspective: the Maori of New Zealand*, in Selin, Helaine (ed.), *Astronomy Across Cultures: The History of Non-Western Astronomy*, Kluwer Academic Publishers, 2000.

Orchiston, Wayne, *Australian, Aboriginal, Polynesian and Maori Astronomy*, in Walker, Christopher, *Astronomy Before the Telescope*, British Museum Press, 1996.

Ruggles, Clive, *Ancient Astronomy: An Encyclopedia of Cosmologies and Myth*, ABC-CLIO, 2005.

SOUTH AMERICA AND THE INCAS

The Incas and their predecessors in South America left no written records. As a result, our knowledge of their history and astronomical practices has to rely on archaeological evidence, and on written accounts made after the sixteenth century Spanish conquest.

In 2006 the anthropologist Robert Benfer announced the discovery of a temple at Buena Vista, north of Lima, Peru, which had the earliest known astronomical alignments in either South or North America. The temple, which dated from about 2200 BC, had alignments to both summer and winter solstices. The summer solstice, in particular, was important, as it was the time when the Rio Chillon began its annual floods, and was the time for planting crops. Then in 2007 Ivan Ghezzi and Clive Ruggles deciphered structures at Chankillo, about 400 km (250 miles) north of Lima, that dated from about 300 BC. Thirteen towers, which are spaced about 5 metres apart in a curve along a ridge, had been known for some time. But Ghezzi and Ruggles found that viewing them from two pre-existing viewing sites, one to the west and one to the east, showed that the towers at each end of the row lined up with the rising and setting Sun at the winter and summer solstices.

The Incas, who began their conquests in South America in about 1200 AD, were devout worshipers of the Sun, or *Apu Inti*, which they considered was the source of the ruling Inca's divine power. They, like their predecessors, also recognised that the Sun could be used to produce a horizon-based solar calendar for agricultural purposes. This helped them to decide when to plant and harvest their crops at different altitudes in different parts of their empire. To do this they erected wooden posts or vertical stones, like the thirteen towers of Chankillo, from which the rising or setting Sun could be observed.

The Incas also venerated the Moon, which, like the early Chinese, they thought was being eaten at a lunar eclipse. Venus, as a morning and evening star, was thought to guard the Sun. The Incas linked the visibility of the Pleiades to their agricultural calendar, as the 37 day absence of this star group from their night sky coincided with the end of the harvest and the start of the next planting season.

Mayu, the Incas' name for the Milky Way, was seen as the great celestial river and the equivalent of the terrestrial Urubamba river. They believed that these two great rivers, one celestial and one terrestrial, were joined in a great cosmic sea which surrounded the Earth. The Incas not only recognised constellations of bright stars, but also recognised dark cloud constellations in the shapes of a llama, fox, toad and so on on the edge of the Milky Way.

Bibliography

Benfer, Robert, and Adkins, Larry, *The Americas' Oldest Observatory*, Astronomy Magazine, August 2007, pp. 40–43.

Bauer, Brian S., and Dearborn, David S. P., *Astronomy and Empire in the Ancient Andes: The Cultural Origins of Inca Sky Watching*, University of Texas Press, 1995.

Ziolkowski, Marivsz S., and Sadowski, Robert M. (eds.), *Time and Calendars in the Inca Empire*, British Archaeological Reports, International Series, No. 479, 1989.

SUB-SAHARAN AFRICA

Although the history of sub-Saharan African astronomy has been researched for over a hundred years, there has been an acceleration of our knowledge more recently. Even so, much more needs to be done, as only a small fraction of archaeological sites have been surveyed. For example, there are over 1,500 stone circles in Gambia, Senegal and Togo that remain largely unexplored.

The oldest lunar calendar found so far in sub-Saharan Africa is on an Ishango bone discovered on the shores of Lake Edward on the Zaire/Uganda border. It was originally dated by Marshack to 6500 BC, who concluded that it showed a six month lunar calendar. Later work has yielded a date between 18,000 and 23,000 BC.

One of the most intriguing calendars is that developed by the Borana of Ethiopia and Northern Kenya, which was first described by Legesse in 1973. It was based on the 'conjunction' of seven stars or star groups with the new Moon, the stars or groups being Triangulum, Pleiades, Aldebaran, Bellatrix, Orion's Belt, Saiph and Sirius. 'Conjunction', in this case, meant that the new Moon rose 'side by side' with the relevant star or star group. The new year started when a new Moon was 'in conjunction' with Triangulum, the second month started when the new Moon was in conjunction with the Pleiades, and so on. Orion's belt and Saif were treated together, and the seventh to twelfth months used Triangulum in conjunction with various phases of the Moon.

The Borana recognised that the sidereal month was about 27 days long (27.3 days actually), as they had only 27 day names. But the period from new Moon to new Moon (the synodic month, described above) is 29.5 days. Instead of adding two or three new day names every month, the Borana restarted using the 27 day name sequence on day 28. So the various months started with different day names.

In 1978 Lynch and Robbins discovered nineteen stone pillars, now called Namoratunga II, near Lake Turkana in Kenya, which appear to date from about 300 BC. They suggested that various alignments of these stones could be used to derive the Borana calendar. Later more detailed analysis has made this doubtful.

In about 300 BC the capital of the Nubian kingdom of Kush was moved from Napata to Meroë, which are both on the Nile in Sudan. There they built numerous pyramids. Many of their entrances have been found to face the heliacal rising of Sirius.

The Great Zimbabwe stone ruin, which was built by the Karanga people between 400 and 1350 AD, is the most complete of over 1,000 megalithic sites in this part of Africa. Many early claims for the precision of its building and alignments have been shown to be incorrect, partly because it was found to have been modified. But there is a clear alignment with the rising Sun at the summer solstice, and possible alignments for the equinoxes and the Milky Way. Elsewhere in Zimbabwe some structures have been aligned to the solstices and equinoxes, whilst in Togo and Benin the Batamalimba people have the cross-beams of their houses aligned to the sunrise and sunsets at the equinoxes.

Star lore and myths tell us something about the astronomical observations of many African tribes. For example, Sirius and the Pleiades are important to the people of Ethiopia and Mali in the north of Africa, whilst Canopus, the Magellanic Clouds and the Southern Cross are important to people in the south. In addition, the Milky Way, Venus and the belt of Orion appear to be recognised all over Africa. Most tribes did not recognise that the morning and evening appearances of Venus were of the same object, but the Karanga, Khoikhoi and Xhosa were notable exceptions.

The heliacal rising of the Pleiades was taken to herald the start of the planting season by many tribes in Southern Africa. Likewise the setting of the Pleiades signified the start of the planting season for the Kikuyu and Masai of Kenya and Tanzania. In Mali the Bozo people moved along the Niger river delta when the Pleiades transited overhead, and started their fishing season when the Pleiades left the night sky. The people of Zaire recognised that the Milky Way was oriented north-south during their dry season and east-west during their wet season, and the Khoikhoi used the Pleiades as a herald of their rainy season.

Bibliography

Aveni, Anthony F., *Ancient Astronomers*, Smithsonian, 1993.

Doyle, Laurance R. and Frank, Edward W., *Astronomy of Africa*, in Selin, Helaine (ed.), *Encyclopaedia of the History of Science, Technology and Medicine in Non-Western Cultures*, Kluwer, 1997.

Snedegar, Keith, *Astronomical Practices in Africa South of the Sahara*, in Selin, Helaine (ed.), *Astronomy Across Cultures: The History of Non-Western Astronomy*, Kluwer Academic Publishers, 2000.

2 · Period overviews

SEVENTEENTH CENTURY

Astronomy was changed out of all recognition during the seventeenth century by the invention of the telescope and by the development of physics in the Scientific Revolution. Kepler and Galileo were both key people in these developments at the start of the century.

Kepler

Johannes Kepler (1571–1630) had a deep conviction that the heliocentric solar system exhibited a harmonic structure that God had devised at Creation. This led to the idea, which he published in his *Mysterium Cosmographicum* of 1596, that the separation between the six planets could be explained by successive nesting of the five regular Euclidian solids. But the fit was not perfect. So he examined what would happen if, instead of using the centre of the Earth's orbit as the centre of the planetary orbits, as Copernicus had done, he used the Sun. Although the fit was no better, from this point on Kepler assumed that the Sun was at the centre of planetary orbits.

Kepler looked at the universe in an entirely different way to his predecessors. The Babylonians had examined it arithmetically, and the Greeks and later astronomers had considered it in geometrical terms. Kepler, on the other hand, tried to understand the structure of the solar system by considering physical forces. He conceived of a force emanating from the Sun which pushed the planets around in their orbits of the Sun, such that if the force stopped so would the planetary movements.

Shortly after writing his *Mysterium Cosmographicum*, Kepler joined Tycho Brahe as one of his assistants, to gain access to Tycho's accurate observational data. His first task was to analyze Mars' orbit, which he continued after Tycho's death in 1601. Kepler published his results in 1609 in his *Astronomia Nova*, in which he announced for the first time that the orbit of Mars was an ellipse, with the Sun at one focus. Kepler also considered what type of force was driving the planets in their orbits, and concluded that the basic circular motion was produced by vortices generated by a rotating Sun. Magnetic forces then made the orbits elliptical. So Kepler thought that the Sun rotated on its axis, and that the planets and Sun were magnetic.

Initially, Kepler had shown only that Mars moved in an ellipse, but in his *Epitome* of 1618–1621 he showed that this

was the case for all the planets, as well as the Moon and the satellites of Jupiter. He also stated what we now know as his third law, that the square of the periods of the planets are proportional to the cubes of their mean distances from the Sun. Finally, in his *Rudolphine Tables* of 1627 he listed detailed predictions for planetary positions, and predicted the transits of Mercury and Venus across the Sun's disc.

On 9 October 1604 a new object had been observed in the sky by Ilario Altobelli (1560–1637) that was as bright as Jupiter for a time. Fortuitously, this was the day for the conjunction of Mars and Jupiter in the same part of the sky. This agreement in timing was regarded by astrologers as highly significant, leading them to conclude that the new object was somehow connected with the two planets, and so must be in the solar system. Kepler first saw the new object on 17 October, the day it reached maximum brightness, and Galileo saw it a little later. Both Kepler and Galileo instantly recognised that its sudden appearance, like that of Tycho's star in 1572, disproved Aristotle's doctrine of the unchangeability of the universe beyond the Moon. The new object was visible for a whole year, and through this time no parallax could be measured. As a result, Kepler concluded that it was a new star.

Galileo

Galileo Galilei (1564–1642) made his first telescopes in 1609 and started his first telescopic observations of the Moon in November of that year. He noticed that its terminator was very irregular and concluded that this was because the Moon had mountains and valleys. It was quite unlike the pure spherical body of Aristotle's cosmology.

Galileo also undertook a series of observations of Jupiter in January 1610, and found that it had four moons that changed their positions from night to night (see Figure 1.4). He presented his early Moon and Jupiter observations in his *Sidereus Nuncius* published in March 1610. Then in July 1610 Galileo saw what he took to be two moons on either side of Saturn, but for some reason they did not move. Later in 1610 he observed the phases of Venus, proving once and for all that Ptolemy's structure of the solar system was incorrect. Only then did Galileo settle on the Copernican heliocentric system.

In March 1611, Galileo went to Rome to explain to the Catholic Church authorities what he had observed and what the

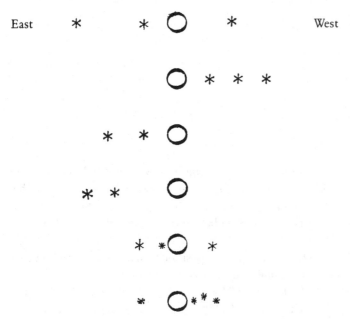

Figure 1.4. Galileo's observations of the moons of Jupiter on consecutive nights from 7 to 13 January 1610 (except 9 January) as shown in his book *Sidereus Nuncius*.

implications were for cosmology. Initially, the Church seemed prepared to tolerate his arguments for Copernicus' heliocentric cosmology, provided Galileo presented this cosmology as a working hypothesis, rather than as a universal truth. But Galileo was not happy with this, and entered into a theological argument with the Church, which he could not win. Eventually, the Church put him on trial, where, contrary to popular accounts, he was treated very well. Nevertheless, he was forced in 1633 to recant his views, and was then placed under house arrest for the remaining nine years of his life.

Sunspots had been seen from time to time in antiquity, but most people took them to be something between the Earth and Sun. Although Thomas Harriot (1560–1621) and Galileo had both seen sunspots telescopically in 1610, it was Johann Fabricius (1587–1616) who first published his observations in June 1611. He concluded that they were spots on the surface of the Sun, and that their movement indicated that the Sun was rotating. This was completely against the teachings of Aristotle that the Sun was a perfect body.

Although Galileo thought that the Moon had an atmosphere, he concluded that there was very little water on the surface, as there were no clouds. His early telescopes were not sufficiently powerful, however, to show much surface detail. But by mid-century it was clear that there were numerous craters on the Moon, and in 1665 Robert Hooke (1635–1703) speculated on their cause in his *Micrographia*. He performed a series of experiments and found that, if round objects were dropped into a viscous mixture of clay and water, features were produced that resembled lunar craters. But he could not think of the source of large objects in space that could impact the Moon. However,

he also found that he could produce crater-like features if he boiled dry alabaster powder in a container. As a result he concluded that lunar craters were the result of lava eruptions on the surface of the Moon.

Huygens

Christiaan Huygens (1629–1695) discovered Saturn's first moon, Titan, which he announced in his *De Saturni* of 1656. Huygens had also mentioned in *De Saturni* that he had solved the problem of Saturn's two 'moons' observed by Galileo. In fact, the behaviour of these 'moons' had been very odd, as they had both completely disappeared in November 1612, reappearing again in mid-1613. Since then their shape had gradually changed. In 1650 Francesco Grimaldi (1618–1663) had discovered Saturn's polar flattening, but still the behaviour of the 'moons', then called 'ansae', was unexplained. Finally Huygens announced, in his *Systema Saturnium* of 1659, that the ansae were actually a thin, flat, solid ring, which was inclined to the ecliptic, and so changed its appearance with time. Then in 1675 Jean Dominique (otherwise known as Giovanni Domenico) Cassini (1625–1712) noticed that Saturn's ring was divided in two by a dark line, now called the Cassini division, going all the way around the planet. Cassini speculated that the two rings were not solid but composed of swarms of small satellites.

Descartes

Kepler had thought that the planets were being pushed round their orbits by a vortex emanating from the Sun, whilst in 1630 he attributed the tides on Earth to the combined attraction of the Sun and Moon. Oddly he did not think of this attractive force from the Sun as having some effect on the orbits of the planets. René Descartes (1596–1650) also developed a vortex theory to explain the motion of the planets. In his theory the vortices were in the ether, which was a frictionless fluid filling the universe. In his *Principia* of 1644, Descartes stated that each planet had two 'tendencies': one tangential to its orbit and one away from the orbit's centre. It was the pressure in the vortex that counterbalanced the latter and kept the planet in its orbit.

Newton

In 1664 Isaac Newton (1642–1727) started to consider the motion of a body in a circle. As a result he was able to prove that the force on a planet moving in a circular orbit is inversely proportional to the square of its distance from the centre. Newton realized that this outward centrifugal force on a planet must be counterbalanced by an equal and opposite centripetal force, but it was not obvious at that time that this force was gravity.

At that time it was known that gravity acted on objects on the Earth's surface, but it was not known how far from Earth that force extended. To get a better understanding of this Newton devised his so-called 'Moon test'. In this he compared the force acting on the Moon, because of its motion in a circle, with the force of the Earth's gravity at the Moon's orbit. He found that although they were similar in magnitude, they were not the same. The difference was not large, but it was sufficient to cause Newton to stop work on gravity. At that time Newton appears to have thought that the centripetal force was a mixture of the gravitational force and the force created by vortices in the ether, so he may not have been too surprised by his result.

Eventually Newton was prompted to return to the subject of gravity by an exchange of letters with Robert Hooke in 1679. In the following year Newton proved that, assuming an inverse square law of attraction, planets and moons will orbit a central body in an ellipse, with the central body at one focus. Then in 1684 he finally rejected the idea of etherial vortices and started to develop his theory of universal gravitation.

It was during this period that the comet of 1680 appeared. At that time most astronomers, including Newton, believed that comets described rectilinear orbits. John Flamsteed (1646–1719), on the other hand, believed that comets described closed orbits. He suggested that the 1680 comet had changed direction before it had reached the Sun, and so had passed in front of it. Newton thought that there had been two comets, one approaching the Sun and one retreating. There was no way of determining who was correct at the time, so Newton dropped the subject temporarily. Eventually, however, he returned to the subject, and by 1686 he had changed his position entirely. He proved that cometary orbits were highly elliptical or parabolic, to a first approximation, so the 1680 comet had been one comet after all. Newton now felt, having solved the problem of cometary orbits, that he could complete his *Principia*, which was published in 1687.

Newton developed his universal theory of gravitation in his *Principia*, which ran to three editions. He used Venus to 'weigh' the Sun, and used planetary moons to weigh their parent planets. So by the third edition he had deduced the masses and densities for the Earth, Jupiter and Saturn relative to the Sun. This was the first time that it had been possible to make such estimates which, in most cases, were very close to the values known today.

Newton realized that if gravity was really universal then, not only would the Sun's gravity affect the orbit of a planet, and the planet's gravity affect the orbit of its moons, but the Sun would affect the orbits of the moons, and one planet would affect the orbits of other planets. In particular, Newton calculated that Jupiter, at its closest approach to Saturn, would have about $\frac{1}{217}$ times the gravitational attraction of the Sun. So he was delighted when Flamsteed told him that Saturn's orbit did not seem to fit exactly the orbit that it should if it was influenced only by the Sun. Gravity really did appear to be universal.

During their attempts to measure the solar parallax in 1672, Richer, Cassini and Picard had found that the Earth appeared to have an equatorial bulge. Newton realised that the Earth's axial rotation would create such a bulge, and fifteen years later he used his new gravitational theory to calculate a theoretical value for the Earth's oblateness of $\frac{1}{230}$ (modern value $\frac{1}{298}$). He then considered the gravitational attraction of the Moon and Sun on the oblate Earth and calculated that the Earth's spin axis should precess at a rate of about 50.0″ per annum (modern value 50.3″). This explained the precession of the equinoxes.

Other major astronomical developments of the seventeenth century, not mentioned above, are listed in Table 1.1.

Bibliography

Caspar, Max, *Kepler*, Abelard-Schuman, London, 1959 (Dover reprint, 1993).

Drake, Stillman, *Galileo At Work: His Scientific Biography*, University of Chicago Press, 1978 (Dover reprint, 1995).

Hall, A. Rupert, *From Galileo to Newton, 1630–1720*, Harper & Row, 1963, republished as *From Galileo to Newton*, Dover, 1981.

Hall, A. Rupert, *Isaac Newton: Adventurer in Thought*, Cambridge University Press, 1996.

Leverington, David, *Babylon to Voyager and Beyond: A History of Planetary Astronomy*, Cambridge University Press, 2003.

Machamer, Peter (ed.), *The Cambridge Companion to Galileo*, Cambridge University Press, 1998.

Pannekoek, Anton, *A History of Astronomy*, Allen & Unwin, 1961 (Dover reprint, 1989).

Small, Robert, and Stahlman, William D., *An Account of the Astronomical Discoveries of Kepler*, University of Wisconsin Press, 1963.

Taton, René, and Wilson, Curtis (eds.), *Planetary Astronomy from the Renaissance to the Rise of Astrophysics; Part A: Tycho Brahe to Newton*, Cambridge University Press, 1989.

EIGHTEENTH CENTURY

Halley's comet

Edmond Halley (1656–1742) published his *Synopsis of the Astronomy of Comets* in 1705 in which he gave the orbital elements of 24 comets, which had been observed between 1337 and 1698, using Newton's gravitational theory. None of the orbits appeared to be hyperbolic, so the comets were all permanent members of the solar system. Halley also concluded that the comets of 1531, 1607 and 1682 were successive appearances of the same comet, as their orbital elements were very similar. But he noted that the time intervals between successive perihelia were not identical. He attributed this to the perturbing effect of Jupiter. Taking this effect into account, he predicted in 1717 that the comet, now called Halley's comet, would return in late 1758 or early 1759.

Table 1.1. *Seventeenth century*

Key discoveries and observations *in addition* to those described in the text

Sun-Earth distance

1672	Richer, Cassini and Picard deduce a Sun-Earth distance of 140 million km from observations of the parallax of Mars. John Flamsteed independently deduces a similar value.

Moon

1611	Harriot observes the libration in latitude.
1632	Galileo observes the libration in longitude.
1647	The first definitive map of the Moon published by Hevelius.
1693	Halley discovers the secular acceleration of the Moon.

Mercury

1631	First observation of a transit of Mercury by Gassendi, Remus and Cysat, independently. It occurred on the date predicted by Kepler.
1639	Phases of Mercury first observed by Zupus.

Venus

1639	First observation of a transit of Venus by Horrocks and Crabtree.
1646	Fontana observes that Venus' terminator is uneven, attributing the cause to high mountains.
1667	J. D. Cassini deduces a rotation period of about 24 hours (now known to be incorrect).

Mars

1659	Huygens observes Syrtis Major and deduces a planetary rotation period of about 24 hours.
1666	J. D. Cassini discovers the bright polar caps.

Jupiter

1630	Torricelli discovers the main belts.
1643	Riccioli observes the shadows of the Galilean satellites on Jupiter's disc.
1663	J. D. Cassini deduces a Jupiter rotation period of 9 h 56 min.
1665	J. D. Cassini observes a prominent spot which may be an early appearance of the Great Red Spot.
1690	J. D. Cassini observes the differential rotation of Jupiter.
1691	J. D. Cassini observes Jupiter's polar flattening, which he estimates to be about 7%.

Saturn

1671	J. D. Cassini discovers Iapetus.
1672	J. D. Cassini discovers Rhea.
1684	J. D. Cassini discovers Tethys and Dione.

Comets

1625	Kepler explains that comets' tails point away from the Sun because the material emitted from their heads is pushed back by solar emissions. (It is unclear whether he meant radiation or particle emissions.)

The wider universe

1600	Blaeu discovers Nova Cygni 1600. It turned out to be a variable star, P Cygni, which flared up again in 1655.
1610	De Peiresc is the first to mention the Orion nebula.
1612	Marius is the first to record telescopic observations of the Andromeda nebula.
1667	Montanari is the first to record the variability of Algol (β Persei).
1670	Anthelme discovers Nova 1670 Vulpeculae.

Alexis Clairaut (1713–1765) attempted to produce a more accurate prediction of the date of Halley's comet's next perihelion. He used a new approximate solution to the three-body problem that allowed him to take account of planetary perturbations. This showed that the comet's return would be delayed by 518 days due to Jupiter and 100 days due to Saturn. As a result, he predicted in November 1758 that Halley's comet would reach perihelion on about 15 April 1759 ± 1 month. It did so on 13 March 1759.

The successful prediction of the return of Halley's comet by Halley and Clairaut was a major triumph for Newtonian dynamics.

Transit of Venus

In 1678 Halley had reviewed possible methods of measuring the solar parallax, and hence the distance of the Earth from the Sun. He concluded that transits of Venus would produce the most accurate results, assuming that they were observed at at least two different points on Earth, preferably as far apart as possible. Unfortunately for Halley, the next transits of Venus would not take place until almost one hundred years later, in 1761 and 1769.

Joseph Delisle (1688–1768) tried to persuade the international community to undertake coordinated observations of the 1761 transit. In the event, France sent observers to Vienna, Siberia, India and an island in the Indian Ocean, whilst other countries sent observers to St Helena, Indonesia, Newfoundland and Norway. Unfortunately, precise timing of Venus' contacts on the Sun proved much more difficult than expected, resulting in solar parallaxes ranging from 8.3″ to 10.6″. Several observers noticed that Venus appeared to be surrounded by a luminous ring when the planet was partially on the Sun. Mikhail Lomonosov (1711–1765) correctly concluded that this showed that Venus was surrounded by an extensive atmosphere.

The lessons learned from the 1761 transit were invaluable in observing the next transit in 1769. This was undertaken from over 70 different sites, producing a best estimate of 8.6″ for the solar parallax. In 1835 Johann Encke (1791–1865) re-analysed the 1769 transit data and produced a new estimate of 8.571″ ± 0.037″, equivalent to a Sun–Earth distance of 153.5 ± 0.7 × 10^6 kilometres, which is within 2.5% of the correct value.

Secular acceleration of the Moon

In 1693 Halley had noticed that the Moon appeared to have accelerated in its orbit compared with its position deduced from ancient eclipse records going back 2,500 years. This secular acceleration of the Moon was confirmed in 1749 by Richard Dunthorne (1711–1775), who calculated a rate of about 10″/century². Four years later, Tobias Mayer (1723–1762) published a figure of 7″/century², which he revised to 9″/century² in 1770. The natural consequence of such an acceleration was that the Moon's orbit was getting smaller and, if it continued, the Moon would eventually collide with the Earth.

Pierre Laplace (1749–1827) suggested that, instead of the Moon accelerating in its orbit, maybe the Earth's spin rate was gradually slowing down. He had previously shown that planetary perturbations were causing the eccentricity of the Earth's orbit to have been reducing for many thousands of years. In 1787 he pointed out that this would result in the mean distance of the Earth from the Sun increasing slightly, causing the effect of the Sun on the Moon's orbit to be marginally reduced, thus allowing the Moon to orbit the Earth slightly faster in a slightly

smaller orbit. He showed that, although this effect is cyclical, its period is several hundred thousand years, so over recorded time the lunar acceleration has been constant. Laplace calculated the magnitude of the effect to be 10.2″/century², in good agreement with observations. So, yet again, Newton's universal gravitational theory had been successful, giving astronomers confidence that they would eventually be able to explain all the detailed, intricate motions of the solar system using this gravitational theory.

Laplace went on to publish an extensive theoretical analysis of the motion of the Moon and planets in his five-volume masterpiece, *Traité de Mécanique Céleste*, which was progressively published from 1799 to 1825. In it he derived the perturbations of eccentricity, inclination, apsides and nodes for the orbits of the Moon and planets over hundreds of thousands of years. In the process he showed that the solar system was deterministic and stable. That is, once the system had been set in motion, it would continue to move in a completely predictable and regular way. Laplace then went on to derive a figure of $\frac{1}{305}$ for the flattening of the Earth and 8.6″ for the solar parallax, from the effect of these parameters on the orbit of the Moon. This was a further remarkable confirmation of Newton's theory of universal gravitation.

Discovery of Uranus

On 13 March 1781 William Herschel (1738–1822), whilst looking for double stars, noticed what he thought was a comet. Four days later, when he next saw the object, it had clearly moved, confirming Herschel's suspicion. He wrote to Nevil Maskelyne (1732–1811), the Astronomer Royal, notifying him of his discovery. As a result, Maskelyne observed the object on a number of occasions, but thought that it could be either a planet or comet.

Over the next few weeks a number of astronomers observed the object and calculated its orbit, which was found to be essentially circular. So it was a planet, now called Uranus, not a comet. It was the first planet to be discovered since ancient times, and its discovery had a profound effect on the astronomical community, indicating that there may yet be more undiscovered planets in the solar system.

Anders Lexell (1740–1784) estimated the radius of Uranus' orbit to be 18.93 AU, which was within 1.4% of its true value, and Pierre-François Méchain (1744–1804) estimated its orbital inclination to the ecliptic as only 46′, which was exactly correct. Then in 1788 Herschel estimated the diameter of Uranus as about 34,000 miles (54,700 km), which is about 8% too high. Herschel also observed that Uranus was not spherical but flattened at the pole, and so concluded that it must be rotating fast.

In 1787 Herschel discovered the first two of Uranus' satellites, now called Titania and Oberon, with orbits at about 90°

to the ecliptic. Eventually he deduced an orbital inclination of about 101° 2′, making these the first satellites known to have retrograde orbits. Using their orbital periods he was able to show that Uranus was intermediate in mass between that of Jupiter and Saturn on the one hand, and the Earth on the other.

Saturn's rings

Jean Dominique Cassini (1625–1712) had suggested in 1705 that the rings of Saturn may consist of swarms of small satellites. Ten years later his son Jacques came to the same conclusion, and yet in the eighteenth century the vast majority of astronomers still considered them as solid.

In 1785 Laplace published his memoir, *Théorie des attractions des sphéroids et de la figure des planètes*, in which he analysed the stability of Saturn's rings mathematically. At that time it was unclear as to whether the Cassini division was a division between two rings, or whether it was a dark marking on one ring. Either way, Laplace showed that the one- or two-ring system could not be stable if the ring or rings were solid, as the ring(s) had to rotate around Saturn to retain their stability. In addition, the particles nearer to the planet would have a higher angular velocity than those further away, causing large, rigid rings to fragment. As a result, Laplace concluded that Saturn had two rings, each consisting of many narrower, concentric rings, separated radially by gaps that were similar in nature to the Cassini division, but much narrower. He further showed that each narrow ring could not be uniform, if they were solid, as the perturbing forces operating on them would cause uniform rings to become unstable. So he proposed that each solid, narrow ring had the density and thickness variations that were just enough for them to retain their stability. Although this seemed highly unlikely, Laplace's theory of Saturn's rings lead the field until James Clerk Maxwell produced his theory in 1857 of particulate rings like those envisaged by Cassini.

Shortly after Laplace had published his theory, Herschel concluded that, as the width of the Cassini division was the same on both the rings' north and south faces, the division was a true gap dividing the one ring into two. But he could not find any convincing observational evidence of the multiple, narrow rings predicted by Laplace.

Origin of the solar system

Immanuel Kant (1724–1804) outlined his theory of the origin of the solar system in his *Universal Natural History* of 1755. In this he suggested that the solar system had condensed out of a nebulous mass of gas, which had developed into a flat rotating disc as it contracted. As it continued to contract, it spun faster and faster, throwing off masses of gas that cooled to form the planets. However, Kant had difficulty in explaining how a nebula with random internal motions could start rotating when it started to contract.

Forty years later Laplace independently produced a similar but more detailed theory in his *Exposition du Système du Monde*. In this theory, the mass of gas was rotating before it started contracting. As it contracted, it spun faster, progressively throwing off rings of material from its outer edge that condensed to form the planets. Laplace suggested that the planetary satellites formed in a similar way from condensing rings of material around each of the protoplanets. Saturn's rings did not condense to form a satellite as they were too close to the planet.

Nebulae

Diffuse objects, or nebulae, had been observed in the sky since the invention of the telescope in the early seventeenth century, and in 1781 Charles Messier (1730–1817) produced a catalogue of nebulae to avoid them being confused with comets. Two years later, William Herschel decided to list all non-stellar objects that could be seen with his telescopes and catalogued over 2,500 of them over the next twenty years.

Immanuel Kant had suggested that nebulae were systems of stars like the Milky Way, but were so far away that they could not be resolved into their individual stars. Initially, Herschel believed Kant's theory, but Herschel's discovery in November 1790 of a planetary nebula (NGC 1514) with a central star caused him to change his mind. As a result he concluded that at least some nebulae consisted of luminous, non-stellar material which would eventually condense to form stars.

Other major astronomical developments of the eighteenth century, not mentioned above, are listed in Table 1.2.

Bibliography

Hoskin, Michael (ed.), *Cambridge Illustrated History of Astronomy*, Cambridge University Press, 1997.

Leverington, David, *Babylon to Voyager and Beyond: A History of Planetary Astronomy*, Cambridge University Press, 2003.

Maor, Eli, *June 8, 2004, Venus in Transit*, Princeton University Press, 2000.

North, John, *The Fontana History of Astronomy and Cosmology*, Fontana, 1994.

Pannekoek, Anton, *A History of Astronomy*, Allen & Unwin, 1961 (Dover reprint, 1989).

Taton, René, and Wilson, Curtis (eds.), *Planetary Astronomy from the Renaissance to the Rise of Astrophysics; Part B: The Eighteenth and Nineteenth Centuries*, Cambridge University Press, 1995.

NINETEENTH CENTURY

The Industrial and Technological Revolutions of the nineteenth century had a profound effect on the development of astronomy, of which advances in the new fields of photog-

Table 1.2. *Eighteenth century*

Key discoveries and observations *in addition* to those described in the text

1718	Halley discovers the proper motion of stars.
1722	Graham discovers the daily variation in the Earth's magnetic field.
1728	Bradley discovers the 'aberration of light' due to the finite velocity of light. He measures the effect as being 20.2″ half-angle with a period of one year.
1733	Vassenius discovers solar prominences during a solar eclipse but thinks they are clouds in the Moon's atmosphere.
1741	Celsius and Hiorter note the correlation between aurorae and magnetic disturbances.
1748	Bradley discovers the nutation of the Earth's axis which causes the apparent position of stars to vary by ± 9″ over a period of about 18 years.
1750	Wright suggests that the Milky Way is a collection of stars rotating about a common centre.
1754	Kant suggests that tidal friction between the Moon and Earth is slowing down the Earth's axial rotation. It is also the reason that the Moon's axial and orbital rotation periods are synchronised.
1762	Michell concludes that there are too many double stars to be explained simply as a line of sight effect, and suggests that some of them may be gravitationally connected.
1769	Wilson discovers that the umbrae of sunspots are depressions below the visible surface of the Sun.
1778	Buffon suggests that Jupiter has not cooled down since its formation and so is still emitting heat.
1782	Goodricke suggests that the variation in Algol's intensity may be due to its partial eclipse by an invisible companion.
1783	Herschel discovers that the Sun is moving in the Milky Way towards the constellation Hercules, by measuring the proper motion of 13 stars.
1783	Michell suggests that light would be unable to escape from very heavy stars. This is the first suggestion of the existence of black holes.
1789	William Herschel discovers Enceladus and Mimas, the sixth and seventh satellites of Saturn to be discovered.
1789	William Herschel measures the polar flattening of Saturn as 1.107, which indicates a rapid rotation. Five years later he measures a rotation period of 10 h 16 min ± 2 min, which is almost exactly correct.
1798	Brandes and Benzenberg show that meteors are travelling with planetary velocities at high altitude in the Earth's atmosphere.

raphy and spectroscopy were probably the most influential. In addition, the gradual increase in wealth of the industrial countries, and the expansion of education, particularly in the universities, meant that progress was no longer dependent on a few brilliant scientists and people with private incomes.

The solar system

The nature of sunspots was still very much a mystery at the start of the nineteenth century. William Herschel, like Alexander Wilson before him, concluded that sunspots were holes in the Sun's atmosphere through which we could see the darker areas below. But many astronomers were puzzled because they thought that the lower layers of the Sun should be bright, and not dark, if they were hotter. Angelo Secchi (1818–1878) proposed a new theory of the nature of sunspots in 1872 to explain why they were dark. He suggested that matter was ejected from the surface of the Sun at the edges of a sunspot. This matter then cooled and fell back into the centre of the spot, so producing its dark central region.

In 1843 Heinrich Schwabe (1789–1875) found that the number of sunspots varied with a period of about 10 years. Nine years later, Sabine, Wolf and Gautier independently concluded that there was a correlation between sunspots and disturbances in the Earth's magnetic field. There were also various unsuccessful attempts to link the sunspot cycle to the Earth's weather. But towards the end of the century, Walter Maunder (1851–1928) pointed out that there had been a lack of sunspots between about 1645 and 1715. He suggested that this longer period, now called the Maunder Minimum, could have had a more profound effect on the Earth's weather than the shorter solar cycle.

Richard Carrington (1826–1875) discovered in 1858 that the latitude of sunspots changed over the solar cycle. Then, in the following year, he found that sunspots near the solar equator moved faster than those at higher latitudes, showing that the Sun was not a solid body. This led Secchi to conclude that the Sun was gaseous. On 1 September 1859 Carrington and Hodgson independently observed two white-light solar flares moving over the surface of a large sunspot. About 36 hours later these white-light solar flares were followed by a major geomagnetic storm.

Astronomy was revolutionized in the nineteenth century by Kirchhoff's and Bunsen's development of spectroscopy starting in 1859 which, for the first time, enabled astronomers to determine the chemical composition of celestial objects. Kirchhoff measured thousands of dark Fraunhofer lines in the solar spectrum, and recognized the lines of sodium and iron. By the end of the century about 40 different elements had been discovered on the Sun. Unfortunately, however, spectroscopic observations of the planets were difficult to interpret because the intervening Earth's atmosphere produced absorption lines of its own.

A number of astronomers detected markings on Mercury's disc in the nineteenth century and concluded that the planet's period was about 24 hours. On the other hand, Daniel Kirkwood (1814–1895) maintained that Mercury should have a synchronous rotation period because of tidal effects on its crust produced by the relatively near Sun. In the 1880s Giovanni Schiaparelli (1835–1910) confirmed this synchronous rotation observationally, and in 1897 Percival Lowell (1855–1916) came to the same conclusion. So, at the end of the century, synchronous rotation was thought to be the most likely.

In the eighteenth century, Venus was thought to have an axial rotation rate of about 24 hours. And a 24-hour period was generally accepted, until in the 1890s Schiaparelli and others concluded that it, like Mercury, had a synchronous rotation period. But at the end of the century there was no general agreement on this or any other period.

Laplace had shown in 1787 that the secular acceleration of the Moon, of about $10''/century^2$, could be completely explained by planetary perturbations. But in 1853 John Couch Adams (1819–1892) included some of Laplace's second order terms, which he had omitted, so reducing the calculated figure from $10''/century^2$ to just $6''/century^2$. Charles Delaunay (1816–1872) suggested that the missing amount was probably due to tidal friction on the Earth, but it was impossible, at that time, to produce a reasonably accurate estimate of its effect. In the early twentieth century, however, Taylor and Jeffreys showed that the cause was tidal friction in certain shallow seas.

In 1879 George Darwin (1845–1912) developed his theory of the origin of the Moon. In this, the proto-Earth had gradually contracted and increased its spin rate as it cooled. Then, when the spin rate had reached about 3 to 5 hours per revolution, it had broken into two unequal parts: the Earth and the Moon. After break-up, tidal forces had caused the Earth's spin rate to slow down and the Moon's orbit to gradually increase in size. A major problem was that the Earth would have had a tendency to break up the Moon shortly after separation, and it was not clear whether the Moon could have passed through the danger zone before this could have happened.

Karl Friedrich Küstner (1856–1936) undertook precise position measurements of a number of stars in 1884 and 1885 from the Berlin Observatory. When he analyzed his results, however, he found that the latitude of the observatory appeared to be varying. As a result, the International Commission for Geodesy (ICG) decided to organize a series of observations around the world to define the effect more precisely. These results indicated that the Earth's spin axis was moving, relative to its surface, with a period of about 12 or 13 months.

Seth Chandler (1846–1913) had also noticed slight variations in the latitude of the Harvard College Observatory. After reviewing his own measurements and those of others, he concluded that the observed effect had two components. One had a period of 14 months, and was due to the non-rigid Earth not spinning around its shortest diameter. The other, which had a period of a year, was due to the seasonal movement of water and air from one hemisphere to the other and back.

It was generally believed by astronomers in the mid-nineteenth century that there must be some form of life on Mars, even if it was only plant life, as the planet clearly had an atmosphere and a surface that exhibited seasonal effects. The polar caps were apparently made of ice or snow, and there were dark areas on the surface that may be seas.

Schiaparelli produced a map of Mars, following its 1877 opposition, that showed a network of linear features that he called *canali* or channels. Schiaparelli and others saw more *canali* in subsequent years (see Figure 1.5), but other, equally competent observers could not see them at all. Lowell then went further in not only observing many *canali*, but interpreting them to be a network of artificial irrigation channels. At the end of the century, the debate as to whether these *canali* really existed was still in full swing.

Both polar caps were observed to melt substantially in their local summers. But calculations showed that the temperature of Mars was too low for this to happen if the caps were made of water ice or snow. In 1898 Ranyard and Stoney suggested that they could be made of frozen carbon dioxide. But there appeared to be a melt band at the edge of the caps in spring, yet carbon dioxide should sublimate directly into gas on Mars.

Jupiter's Great Red Spot (GRS) was first clearly observed in the 1870s. Then in 1880 an unusually bright, white equatorial spot appeared, which rotated around Jupiter over five minutes faster than the GRS. This gave a differential velocity of about 400 km/h. But the rotation rates of both the white spot and the GRS were not constant, indicating that neither could be surface features, as some astronomers had supposed. White and dark spots were continuously appearing and disappearing on Jupiter, suggesting that they were probably clouds. But the GRS was different as, although it changed its appearance and size over time, it was still there at the end of the century. This longevity led astronomers to wonder if it could really be a cloud system.

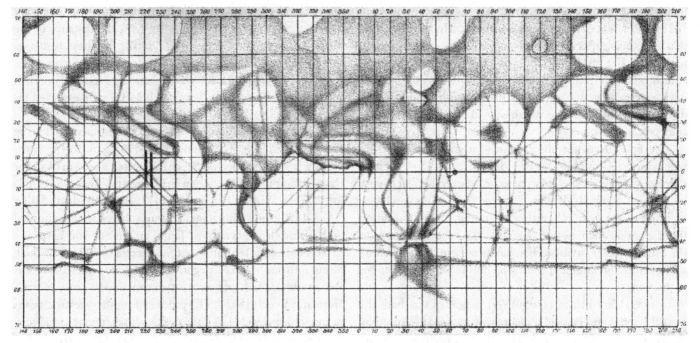

Figure 1.5. Schiaparelli's map of Mars produced after the 1881 opposition showing a large number of canali, some of them double. (From *The Story of the Heavens*, by Robert Ball, 1897, plate XVIII.)

Buffon (1707–1788) had suggested in the late eighteenth century that rapid changes in Jupiter's appearance showed that it had not completely cooled down since its formation. In the nineteenth century, Jupiter's differential rotation and low density, which were both similar in nature to those of the Sun, caused some astronomers to go even further and wonder if Jupiter was self-luminous. Although this was considered unlikely, the idea had not been completely ruled out by the end of the century.

Johann Encke (1791–1865) found in 1837 that Saturn's A ring was divided into two by a clear gap, now called the Encke gap. A little over ten years later, W. C. and G. P. Bond discovered a third ring, now called the C ring, inside the B ring. The new ring was very dark and partly transparent. Then in 1857 James Clerk Maxwell (1831–1879) proved mathematically that Saturn's rings could not be solid or liquid, as many had thought, but must be composed of numerous small particles. Ten years later Kirkwood suggested that orbital resonances between some of Saturn's larger moons and any particles in the Cassini division would have cleared the division of particles.

In 1821 Alexis Bouvard (1767–1843) found it impossible to calculate an orbit for Uranus using both pre-discovery and post-discovery observations. The best he could manage was an orbit based on only post-discovery observations. But, unfortunately, Uranus soon deviated from even this orbit, maybe because it was being disturbed by yet another planet.

In the early 1840s John Couch Adams (1819–1892) and Urbain Le Verrier (1811–1877) independently calculated the orbit of the planet that seemed to be disturbing the orbit of Uranus. Le Verrier published his results in August 1846 and asked Johann Galle (1812–1910) of the Berlin Observatory to undertake a telescopic search for it. Galle and his assistant, d'Arrest, found the planet, now called Neptune, almost immediately on 23 September 1846. However, Neptune's real orbit turned out to be quite different from that predicted by either Le Verrier or Adams. So its discovery had been somewhat fortuitous.

William Lassell (1799–1880) observed an object close to Neptune less than a month after the planet's discovery, which he thought might be a satellite. But it was not until the following July that he was able to confirm his discovery of Neptune's first satellite, now called Triton. Triton was later found to have a retrograde orbit inclined at approximately 30° to the ecliptic.

A number of small bodies, which William Herschel called 'asteroids', were discovered in the nineteenth century orbiting the Sun between Mars and Jupiter. The first, Ceres (now redesignated as a dwarf planet), was found by Giuseppe Piazzi (1746–1826) in January 1801, but it was not until 1845 that the fifth asteroid was found. The discovery rate then increased rapidly, so that nearly 500 asteroids were known by the end of the century. As the number increased, Kirkwood noticed that there were none with various fractional periods of Jupiter's orbital period. This he attributed to resonance interactions with Jupiter.

Jean Louis Pons (1761–1831) discovered a comet in 1818 which, on further investigation, proved to have been seen near previous perihelia. In the following year Johann Encke showed that the comet, now called Encke's comet, had an orbit that took it inside the orbit of Mercury. When the comet returned in 1822, Encke noticed that it was a few hours early, and suggested that it was being affected by some sort of resistive medium close to the Sun. But in 1882 a comet passed even closer to the Sun, and showed no effect of Encke's medium. It was not until the next century that the true cause of the timing anomaly was discovered as being due to variable jet-like emissions from the cometary nucleus.

The first successful observation of a cometary spectrum was made by Giovanni Donati (1826–1873) in 1864. Quite a number of cometary spectra were recorded over the next twenty years. They generally exhibited a broad continuous spectrum, like that of the Sun, when they were first found, indicating that they were then scattering sunlight. As they got closer to the Sun, however, hydrocarbon bands appeared, followed by sodium and iron lines, depending on how close the comets got to the Sun.

In 1839 Adolf Erman (1806–1877) suggested that both the annual Leonid and Perseid meteor showers were produced by the Earth passing through swarms of small particles orbiting the Sun. Twenty-five years later, Hubert Newton (1830–1896) found that the node of the Leonids' orbit was precessing at about 52″/year. John Couch Adams then showed that only a particle in a 33.25-year orbit would have this nodal precession. So the Leonids were orbiting the Sun in a diffuse cloud every 33.25 years, which explained why the most intense showers occurred with this frequency. But there were stragglers all around the orbit, which explained why we saw the Leonids on an annual basis. Then in 1867 Carl Peters (1844–1894) recognized that the source of the Leonid meteor stream was a periodic comet called Tempel-Tuttle. This was just after Schiaparelli had linked the Perseid meteor shower to another periodic comet, Swift-Tuttle.

The wider universe

For many centuries, astronomers had tried unsuccessfully to observe the parallax of stars to determine their distance. The first successful measurements were in 1838, when Friedrich Bessel (1784–1846) determined a parallax of 0.31″ (correct value 0.29″) for the star 61 Cygni, giving a distance of 10.4 light years, and in the following year when Thomas Henderson (1798–1844) announced a parallax of 0.93″ (correct value 0.75″) for α Centauri, equivalent to a distance of 3.5 light years.

In 1844 Bessel found that both Sirius and Procyon oscillated in position, indicating that they had unseen companions. In 1850 Christian Peters (1806–1880) determined the period of revolution of the Sirius binary, designated Sirius A and B, to be about 50 years. Twelve years later, Alvan Clark (1804–1887) and Alvan Graham Clark (1832–1897) accidentally observed Sirius B when they were testing a new telescope. It was about one ten-thousandth the intensity of its companion, thus indicating that stars have a wide variety of intensity, contrary to the general view of the time. By 1890 the distance of the Sirius binary had been measured, allowing the absolute masses to be determined as about 2 and 1 solar masses for Sirius A and B, respectively. So Sirius A was found to be a relatively normal star, whereas Sirius B was very much darker than expected for its mass.

Spectroscopic studies of stars were undertaken following Kirchhoff and Bunsen's seminal work starting in 1859 on interpreting laboratory spectra. Secchi produced a system of stellar classification in 1863 based on their spectra. Then, in the following year, William Huggins (1824–1910) and William Miller (1817–1870) showed that stellar atmospheres were composed of the same elements as those found on the Sun and Earth.

Secchi had by 1867 analysed the spectra of 400 stars, which he put into one of three categories. In the following year, he was able to add a fourth category, which contained 17 carbon stars. Hermann Vogel (1842–1907) produced another stellar classification scheme in 1874, but the definitive scheme was produced by Edward Pickering (1846–1919) in 1890 in the *Draper Memorial Catalogue*, which included the spectra of 10,351 stars.

In 1866 Huggins and Miller were the first to study a nova's spectrum when they observed that of Nova T Coronae Borealis. It had bright hydrogen emission lines. Ten years later, Nova Cygni was found to have a similar spectrum, but it changed completely in the following year to resemble that of a planetary nebula with a single bright green emission line. In 1892 Nova T Aurigae was the first nova to have its spectrum photographed. It had bright hydrogen, helium and sodium lines, together with a continuous spectrum and dark absorption lines. Six months after its initial outburst, Nova T Aurigae's spectrum also included a strong green emission line plus other emission lines of planetary nebulae. This indicated that there was some evolutionary connection between novae and planetary nebulae.

William Herschel had found by 1790 that he could resolve some nebulae into stars, although planetary nebulae seemed to be a special case in which there was a central star surrounded by luminous, non-stellar material. As the nineteenth century progressed, and telescopes became more powerful, more and more nebulae could be resolved into stars. This led some astronomers to believe that, if one had a large enough telescope, all nebulae, with the exception of planetary nebulae, would be resolved into star systems like the Milky Way. But in 1864 Huggins found that the spectra of all eight nebulae that he observed had bright emission lines, indicating that they were all gaseous. The brightest emission line, which was green, was initially attributed to nitrogen. But it later became clear that this attribution was incorrect, and it was thought, instead, to be due to a new element called nebulium.

In 1868 Huggins found, with a much larger sample of nebulae than that of four years earlier, that one-third were gaseous, but two-thirds, including the Andromeda nebula, had continuous spectra, indicating that they were composed of stars. But the appearance of Nova S Andromedae in 1885 in the Andromeda nebula caused confusion, as, if this nebula was a very distant galaxy of stars like the Milky Way, the nova must be incredibly bright.

Observatories

The main observatories at the start of the nineteenth century were in Europe, but around the middle of the century new major observatories were founded on both sides of the Atlantic, in Pulkovo (Russia, 1835), Washington (1842) and Harvard (USA, 1843). Later in the century the two great astrophysical observatories of Potsdam (1874) and Meudon (France, 1875) were established. Various observatories were also founded in the southern hemisphere in South Africa (the Cape Observatory, 1820), Australia (Paramatta, New South Wales, 1822), and South America (Brazil, 1827, Chile, 1852 and Argentina, 1870). The southern or Boyden station of the Harvard College Observatory was also set up at Arequipa, Peru in 1891.

The main observatories in Europe in the nineteenth century were state-financed institutions, although wealthy amateur astronomers like William Herschel and William Parsons (third Earl of Rosse) often set up their own observatories. In the United States, however, the U.S. Naval Observatory was the only nationally funded enterprise, with the others being funded by wealthy citizens, like Lick (in 1875) and Yerkes (in 1893), and/or operating as university departments. Around the end of the century, the United States took the lead from Europe in the quality of their equipment and observing sites.

The questions of climate and altitude were beginning to be seriously discussed by the major observatories towards the end of the nineteenth century. William Lassell left the UK for Malta in 1852 because of its better weather and atmosphere, for example, and James Lick (1796–1876) chose the 1,300 m (4,250 ft) high Mount Hamilton for his observatory in 1875 because of its clear skies. Then in 1891 the Harvard Boyden station was set up in Peru at an altitude of 2,450 m (8,040 ft).

Other major astronomical developments of the nineteenth century, not mentioned above, are listed in Table 1.3.

Bibliography

Clerke, Agnes, *A Popular History of Astronomy during the Nineteenth Century*, 4th ed., Adam and Charles Black, 1902.

Gingerich, Owen (ed.), *Astrophysics and Twentieth-Century Astronomy to 1950; Part A*, Cambridge University Press, 1984.

Hoskin, Michael (ed.), *Cambridge Illustrated History of Astronomy*, Cambridge University Press, 1997.

Leverington, David, *Babylon to Voyager and Beyond: A History of Planetary Astronomy*, Cambridge University Press, 2003.

North, John, *The Fontana History of Astronomy and Cosmology*, Fontana, 1994.

Pannekoek, Anton, *A History of Astronomy*, Allen & Unwin, 1961 (Dover reprint, 1989).

Taton, René, and Wilson, Curtis (eds.), *Planetary Astronomy from the Renaissance to the Rise of Astrophysics; Part B: The Eighteenth and Nineteenth Centuries*, Cambridge University Press, 1995.

TWENTIETH CENTURY PRIOR TO THE SPACE AGE

The solar system

Most physicists had thought in the nineteenth century that heat was transported from the interior of the Sun by convection. But in 1894 Ralph A. Sampson (1866–1939) suggested that the primary mechanism for heat transfer through the solar atmosphere was by radiation. Arthur Eddington (1882–1944) extended this idea of radiative transfer to the whole Sun, and in 1926 calculated a central temperature of about 39 million K. At about the same time, Cecilia Payne (1900–1979) showed that hydrogen and helium were the most abundant elements in stellar atmospheres. Subsequently Eddington assumed similar abundances for the solar interior and reduced his estimate for the Sun's central temperature to about 19 million K.

Eddington's calculations made no assumption of how the Sun's heat was produced, which was still unknown at that time. But in 1938, potential methods of energy production in stars were outlined by Charles Critchfield (1910–1994), who described the proton-proton (p-p) cycle, and Hans Bethe (1906–2005) and Carl von Weizsäcker (1912–2007), who independently discovered the carbon cycle. It was initially thought that the Sun's energy was produced by the carbon cycle, but it was eventually concluded in the 1950s that the Sun used the p-p cycle, with only stars of greater than about 1.5 solar masses using the carbon cycle.

The Sun's corona had been found in the nineteenth century to have a faint continuous spectrum crossed by a number of bright emission lines. One of these was a bright green emission line that had been found by Young and Harkness during the 1869 solar eclipse. It was later attributed to a new element called coronium, as no known element appeared to produce a line at precisely this wavelength.

It was assumed in the early twentieth century that the temperature of the Sun and its corona gradually reduced from the centre of the Sun moving outwards. But in 1934 Walter Grotrian (1890–1954) analysed the coronal spectrum and concluded that the coronal temperature must be about 350,000 K, even though the photosphere's temperature was only about 6,000 K. A few years later Bengt Edlén (1906–1993) showed that the coronal emission lines were produced by highly ionized elements at a temperature of over a million K. The 'coronium'

Table 1.3. *Nineteenth century*

	Key discoveries and observations *in addition* to those described in the text
1802	Olbers discovers Pallas, the second asteroid to be found.
1802	William Herschel produces the first tentative evidence that some stars are true binaries orbiting around their common centre of gravity.
1827	Savary calculates the first orbit of a binary star pair, ξ Ursae Majoris.
1845	Herrick and Bradley observe that Biela's comet has broken into two.
1848	W. C. and G. P. Bond and William Lassell independently discover Hyperion, Saturn's eighth satellite.
1851	Lassell unambiguously discovers Ariel and Umbriel, the third and fourth satellites of Uranus.
1858/9	Le Verrier concludes that the perihelion of Mercury's orbit is precessing at about 38″/century more than can be accounted for by known forces. So he suggests that there is an unknown planet or ring of small planets orbiting the Sun inside the orbit of Mercury. The extra precession was found in 1915 to be a relativistic effect.
1860	Observations during a solar eclipse show that prominences are connected with the Sun, not the Moon.
1864	Donati discovers that Tempel's comet (1864 II) is self-luminous, and does not just shine by scattering sunlight.
1868	Janssen, Lockyer and others find that solar prominences consist of hydrogen and another, unknown gas that Lockyer calls helium. This was confirmed in 1895 when helium was discovered on Earth by William Ramsay.
1872	Draper produces the first useful photographic spectrum of a star, Vega.
1877	Hall discovers Phobos and Deimos, the two satellites of Mars.
1878	Lockyer suggests that, as a star becomes hotter, its constituents break down into atoms, and at higher temperatures, they break down further into what he called 'proto-elements'. These 'proto-elements' are what are now called ionized atoms.
1882/3	The nucleus of the Great September Comet is observed to break up progressively into five pieces that orbit the Sun in line.
1885	Nova S Andromedae is observed. It is the first nova to be discovered in a nebula.
1889	Edward Pickering discovers the first spectroscopic binary, Mizar A.
1892	Barnard discovers Amalthea, Jupiter's fifth satellite. It was the last satellite of any planet to be discovered visually.
1898	W. Pickering discovers Phoebe, Saturn's ninth satellite. It was the first satellite of any planet to be discovered photographically.
1898	Witt and Charlois independently discover Eros, an asteroid that can approach within about 0.15 AU of Earth.

line, in particular, was due to highly ionized iron. How the temperature of the corona could be so high, when the photosphere temperature was so relatively low, was a mystery, which has not been completely resolved even today.

George Ellery Hale (1868–1938) and Walter Adams (1876–1956) found in 1908 that photographs of the Sun taken in the light of the 656.3 nm hydrogen line showed patterns that looked like iron filings in a magnetic field. Hale then examined sunspot spectra at high resolution, and found that they exhibited Zeeman splitting. The magnitude indicated that sunspots were the home of magnetic fields of the order of 3,000 gauss.

At about the same time, Walter Maunder (1851–1928) found that large magnetic storms on Earth started about 30 hours after large sunspots crossed the centre of the solar disc. In 1927 Charles Chree and James Stagg found that smaller storms, which did not seem to be associated with sunspots, tended to recur at the Sun's synodic period of 27 days. Julius Bartels (1899–1964) called the invisible source on the Sun of these smaller storms 'M regions'. Both the so-called 'flare storms' and the 'M storms' were assumed to be caused by particles ejected from the Sun.

Ludwig Biermann (1907–1986) calculated in 1951 that solar radiation pressure was not strong enough to explain the behaviour of cometary tails. He suggested, instead, that tails must be the result of comets being subjected to a continuous stream of charged particles emitted by the Sun.

At the end of the nineteenth century, the key objection to the impact theory for the formation of lunar craters had been that craters were generally circular, when they should have been elliptical as most of the impacts would not be vertical. But after the First World War it was realized that the shape of the lunar craters resembled shell craters. The shell craters were formed by shock waves, so a non-vertical impact could still produce a circular crater. Nevertheless, not all lunar craters have the same general appearance. So, by the start of the space age it was still unclear if they had been produced by volcanic action, meteorite impact, or both.

In 1897 Emil Wiechert (1861–1928) suggested that the Earth has a dense metallic core, mostly of iron, surrounded by a mantle of lighter rock. A little later Richard Oldham (1858–1936) found clear evidence for the existence of the core from earthquake data. Then in 1914 Beno Gutenberg (1889–1960) showed that the interface between the mantle and the core, now called the Wiechert-Gutenberg discontinuity, was at about 0.545r (3,480 km) from the centre of the Earth (where r is its radius).

A number of theories were proposed to try to define and explain the internal structure of the Earth, the other terrestrial planets and the Moon. In the early twentieth century it appeared as though the density of these objects increased with size, with the Earth being the densest and the Moon least dense. In 1934 Harold Jeffreys (1892–1989) produced a theory that assumed that all the terrestrial planets and the Moon have a core of liquid metals, mostly iron, together with a silicate mantle. But his theory could not explain how those planets with the smallest cores could have retained a higher percentage of lighter material in their mantles.

William Ramsey (1922–1965) solved this problem in 1948 when he proposed that the whole of the interior of the terrestrial planets consisted of silicates, with the internal pressure in the largest planets causing the silicates near the centre to become metallic. So the smallest planets are the least dense and have the smallest cores, but proportionately have the largest mantles. But the theory was found to be invalid in 1950 when Eugene Rabe (1913–1974) discovered that Mercury's density was much higher than originally thought. Its density was even higher than that of Venus and Mars, which were much larger planets.

In the mid-twentieth century, most astronomers believed that the planets had been hot when they had first been formed from the solar nebula. But in 1949 Harold Urey (1893–1981) suggested that the solar nebula had been cold. Then, in the case of the Earth-Moon system, the proto-Earth had picked up more of the heavier elements, like iron, from the solar nebula than had the proto-Moon, because the proto-Earth was heavier. But both bodies had been continuously heated by radioactive decay. Internal convection had started on the Earth as iron and other heavy elements had gradually settled into the core, but there had been no such movement on the much smaller Moon. So the Moon was basically homogenous.

At the turn of the nineteenth century it was thought that radio waves travelled in a straight line. So it was a great surprise when Guglielmo Marconi (1874–1937) showed in 1902 that radio waves could be successfully transmitted across the Atlantic. In the following year Oliver Heaviside (1850–1925)

and Arthur Kennelly (1861–1939) independently suggested that the waves were being reflected off an electrically conducting layer in the upper atmosphere.

The structure of what is now called the 120 km high E or Heaviside layer, and of other layers discovered in the ionosphere, was gradually clarified over the next twenty years or so. It was also found that solar flares can cause a major disruption to the ionosphere. Then in 1949 Herbert Friedman (1916–2000) showed that the Sun emitted X-rays, and three years later he showed that the intensity of these X-rays was sufficient to sustain the E layer. In 1956 he found that the Earth's ionosphere was being disrupted by very high energy X-rays emitted by solar flares.

The surface of Mars was still largely unknown in the first half of the twentieth century. Although the polar caps were clear, along with one or two other features, it was thought unlikely that the linear markings called *canali* really existed. But they were still recorded from time to time by respected observers. In addition, some astronomers thought that the bluish-green areas on Mars were some form of vegetation.

In 1909 Georges Fournier (1881–1954) and Eugène Antoniadi (1870–1944) found that the surface of Mars appeared to be completely covered for a while by what appeared to be yellow clouds in the Martian atmosphere. Later, Antoniadi found that these yellow clouds tended to occur around perihelion when the solar heating was greatest. As a result, he concluded that they were clouds of fine dust thrown into the atmosphere by thermally generated winds. Thirty years later Gérard de Vaucouleurs (1918–1995) measured these wind velocities as being typically in the range of 60 to 90 km/h when the clouds first formed.

It was known in the nineteenth century that the densities of Jupiter, Saturn, Uranus and Neptune were similar to that of the Sun, and were much less than those of the terrestrial planets. At that time it was thought that Jupiter, and probably Saturn, had not yet fully cooled down since their formation from the solar nebula. As a result, they were probably emitting more energy than they received from the Sun. Then in 1924 Donald Menzel (1901–1976) showed that most, if not all of Jupiter's and Saturn's heat at formation had disappeared.

In 1924 Jeffreys modelled the internal structures of Jupiter and Saturn, assuming that they each had a rocky core surrounded by a layer of ice and solid carbon dioxide. As a result, the atmospheres on Jupiter and Saturn would have depths of $0.09R_J$ and $0.23R_S$, respectively (where R_J and R_S are the planetary radii). He thought that these atmospheres would consist mainly of hydrogen, nitrogen, oxygen, helium and maybe methane.

Ramsey developed his own theory of planetary structures in the early 1950s, assuming that the giant planets were made of hydrogen. He then added helium and other ingredients until their densities and moments of inertia were those observed. He concluded that Jupiter and Saturn were composed of 76% and 62% hydrogen, by mass, with central pressures of 32 and 6×10^6 bar. At these pressures most of the hydrogen would be metallic.

Vesto Slipher (1875–1969) observed numerous bands in the spectra of Jupiter, Saturn, Uranus and Neptune in the early part of the twentieth century, but had trouble interpreting them. In 1932 Rupert Wildt (1905–1976) concluded that a number of bands in all four planets were due to ammonia and methane. But subsequent work by Reinhard Mecke and others showed that some of the lines had been mis-attributed, so there was no ammonia in the atmospheres of Uranus and Neptune. Henry Norris Russell (1877–1957) pointed out that ammonia would have been frozen out at their lower temperatures. Adel and Slipher also concluded that the methane concentration reduced in going from Neptune to Uranus to Saturn to Jupiter.

In 1955 Bernard Burke (b. 1928) and Kenneth Franklin (1923–2007) made the unexpected discovery that Jupiter was emitting radio waves at 22.2 MHz. Subsequently it was found that Jupiter emitted energy at many radio frequencies. Some of it was thermal energy, with an effective temperature of 145 K, but some was clearly non-thermal. The latter was taken to indicate that Jupiter had an intense magnetic field, with radiation belts similar to those that had, by then, been found around the Earth.

A number of white spots had been observed on Saturn since 1876. The velocities of these spots showed that there was an equatorial current on Saturn, similar to that on Jupiter. But that on Saturn had a velocity of about 1,400 km/h, compared with just 400 km/h for Jupiter. It was unclear why Saturn, which is farther from the Sun, and so receives less heat than Jupiter, should have a much faster equatorial current.

Gerard Kuiper (1905–1973) photographed the spectrum of the ten largest satellites of the solar system in 1943–44, and found evidence for an atmosphere on Titan and possibly Triton. He could find no such evidence for the Galilean satellites of Jupiter, however.

In the nineteenth century, Triton had been found to orbit Neptune in a retrograde sense, and it was unclear at the time whether Neptune's spin was also retrograde or not. But, in 1928 Joseph Moore (1878–1949) and Donald Menzel found, by observing the Doppler shift of its spectral lines, that Neptune's spin was direct or prograde. So Neptune's large satellite was orbiting the planet in the opposite sense to the planet's spin. This was the first time that this had been observed in the solar system for a major satellite.

Kuiper discovered Uranus' fifth satellite, Miranda, in 1948. It was orbiting the planet in an approximately circular orbit inside those of the other four satellites. Then, in the following year, he discovered Neptune's second satellite, Nereid, orbiting Neptune in the opposite sense to Triton. Nereid was in a highly elliptical orbit well outside the orbit of Triton. So Nereid was the 'normal' satellite in orbiting Neptune in the direct or prograde sense, whereas the larger Triton, which was nearer to Neptune in an almost circular orbit, appeared to be the abnormal one.

The discoveries of the planets Uranus and Neptune made astronomers realize that there may well be other planets even farther out from the Sun. As Neptune had only been discovered in 1846, and as it was moving very slowly, its orbit was not very well known in the second half of the nineteenth century. But astronomers had much better information on Uranus' orbit, so they re-examined it to see if there were any unexplained deviations which may indicate where a new planet may be found. Such deviations were soon observed and a number of possible locations for the new planet were proposed by various astronomers, including Percival Lowell (1855–1916). A photographic search was started at Lowell's observatory for the new planet, but it was abandoned when he died in 1916. It was not resumed until 1929, where in February of the following year Clyde Tombaugh (1906–1997) discovered Pluto. Its orbit was found to be highly eccentric, with the largest inclination of any planet.

It was subsequently found, however, that although Pluto's orbit was very similar to that predicted by Lowell, Pluto was not nearly heavy enough to have perturbed Uranus in the way that Lowell had estimated. In 1955 Merle Walker and Robert Hardie deduced a rotation period of 6d 9h 17min from regular fluctuations in Pluto's intensity. Little more was known about the planet when the space age started.

In 1918 Kiyotsugu Hirayama (1874–1943) identified three families of asteroids based on their orbital radius, eccentricity and inclination. He suggested that these three families were each the remnants of a larger asteroid that had fractured. This resurrected, in modified form, the theories of Thomas Wright (1711–1786) and Wilhelm Olbers (1758–1840) in the eighteenth and nineteenth centuries, who believed that there had once been a planet between the orbits of Mars and Jupiter that had broken up.

William Huggins (1824–1910) had concluded in the nineteenth century that there were hydrocarbon compounds in the heads of comets, but he was not able to specify exactly which hydrocarbons were involved. Molecular carbon (C_2) was first identified just after the turn of the nineteenth century, and by the mid-1950s a number of compounds including those containing carbon, hydrogen, oxygen and nitrogen had been found in the heads of comets.

Henri Deslandres and others detected molecular bands in the tail of Daniel's comet in 1907 and of Morehouse's comet in the following year. Some of these bands were later identified by Alfred Fowler (1868–1940) as being due to ionized carbon monoxide (CO^+) and N_2^+. Then in 1948 CO_2^+ was found in the tail of a comet by Pol Swings (1906–1983) and Thornton Page (1913–1996).

In the 1930s, Karl Wurm (1899–1975) observed that many of the molecules found in comets were chemically very active, so they could not have been present there for very long. He suggested, instead, that they had come from the more stable parent molecules cyanogen $(CN)_2$, H_2O, and CH_4 (methane). Pol Swings concluded in 1948 that the parent molecules were water, methane, ammonia (NH_3), molecular nitrogen (N_2), carbon monoxide and carbon dioxide, all of which had been in the form of ice before being heated by the Sun.

In 1950 and 1951 Fred Whipple (1906–2004) proposed his icy-conglomerate model (better known as his 'dirty snowball' theory), in which the nucleus of a comet was composed of ices, such as methane, with meteoric material embedded within it. Unfortunately, some of the parent molecules were highly volatile, so they should have evaporated long ago. But in 1952 Armand Delsemme (b. 1918) and Pol Swings suggested that these highly volatile elements would be better able to resist solar heating if they were trapped within the crystalline structure of water ice.

Ernst Öpik (1893–1985) concluded in the early 1930s, from an analysis of stellar perturbations, that comets could remain bound to the Sun at distances of up to 10^6 AU. Some years later Adrianus van Woerkom (1915–1991) showed that there must be a continuous source of new, near-parabolic comets to explain the numbers of such comets observed. Then in 1950 Jan Oort (1900–1992) showed that the orbits of ten comets, with near-parabolic orbits, had an average aphelion distance of about 100,000 AU. As a result, he suggested that all long-period comets originate in what is now called the Oort cloud, about 50,000 to 150,000 AU from the Sun.

In the early decades of the twentieth century, theories of the origin of the solar system generally focused on the effect of collisions, and close encounters of another star to the Sun. But all of these theories were found to have significant problems, so Laplace's theory of a condensing nebula was reconsidered. Laplace's theory had been previously rejected in the nineteenth century because the original solar nebula did not appear to have had enough angular momentum. However, in the 1930s William McCrea (1904–1999) showed that this would not be a problem if the original nebula had been turbulent.

In 1943 Carl von Weizsäcker (1912–2007) suggested that cells of circulating convection currents, or vortices, formed in the solar nebula after the Sun had condensed. These vortices produced planetesimals that grew to form planets by accretion. Unfortunately, as Subrahmanyan Chandrasekhar (1910–1995) and Gerard Kuiper showed, the vortices would not have been stable enough to allow condensation to have taken place. Kuiper then produced his own theory, as did Viktor Safronov (1917–1999) and others, with the common theme of planetesimals merging to form planets, but none was fully satisfactory.

The wider universe

The Harvard stellar classification of 1890 had been devised by Edward Pickering (1846–1919) to indicate the place of the various stars in the evolutionary process going from the hottest to the coolest. At that time, Pickering thought that the A-type stars were the youngest, followed by B-types, then going down alphabetically to O-type stars, which were the oldest. But in 1897 Antonia Maury (1866–1952) concluded that the evolutionary sequence should start with B-type stars followed by A-types. Maury also introduced an extra dimension to her classification system using the letters a, b and c to indicate the sharpness of their spectral lines. Then four years later Annie Cannon, at Maury's suggestion, placed the O-type stars before the B-types.

Edward Pickering had found in 1896 that the Wolf-Rayet, O-type star ζ Puppis had a most unusual set of spectral lines. Subsequently other O-type stars were found to have similar 'Pickering lines' as they were called. Pickering thought that they were due to some form of hydrogen, but in 1913 Niels Bohr (1885–1962) showed, using his new theory of atomic structure, that they were due to ionised helium. As B-type stars had un-ionised helium, O-type stars must be hotter than B-type stars, as Maury had suspected.

Ejnar Hertzsprung (1873–1967) discovered over the period 1905–1907 that Maury's c-type stars had very small proper motions compared with other stars of the same colour. He concluded, as a result, that the c-type stars were, on average, much further away with much higher luminosities than these non-c-type stars, and consequently they had much larger surface areas. So there were two different populations of high- and low-luminosity stars, which he called 'giants' and 'dwarfs'.

Henry Norris Russell (1877–1957) was not aware of Hertzsprung's work when he also proposed in 1910 that there were two different populations of stars of the same colour – one of large bright stars, the other of small dim ones. Four years later he published his colour-magnitude or 'Russell' diagram (later called the Hertzsprung-Russell diagram) showing both a main sequence of dwarf stars and a giant branch.

In 1920 Michelson, Pease and Anderson succeeded in measuring the diameters of stars for the first time, using a Michelson interferometer at the Mount Wilson observatory. The results clearly showed the immense size of giant stars. Antares, for example, was found to have a diameter of about 600 million kilometres, or 450 times that of the Sun.

In addition to the main sequence of dwarf stars, and a giant branch, Russell's initial colour-magnitude diagram showed one star, 40 Eridani B, all on its own as an A-type star of exceptionally low luminosity. In 1914 Walter Adams confirmed that it was a very low luminosity white star and, in the following year, found that Sirius B was not a red M-type star, as had been supposed, but was also a white A-type, like 40 Eridani B. Sirius B was also of exceptionally low luminosity for a white star. It was known that A-type stars have a very high surface brightness, so if the very low luminosity of both stars was solely due to their size, they must be very small and very dense. In 1925 Adams found that the spectral lines of Sirius B had a relativity shift, confirming that it was very dense, and proving that it was what Eddington called a 'white dwarf' star.

Henrietta Leavitt (1868–1921) began a survey of stars in the Magellanic Clouds in the early years of the twentieth century. By 1912 she had measured the periods and intensity variations of 25 Cepheid variables in the Small Magellanic Cloud (SMC), finding that the brightest Cepheids had the longest periods. In the following year Ejnar Hertzsprung estimated the distances of Cepheids which were relatively close to the Sun to determine their absolute luminosities. He was then able to use Henrietta Leavitt's period-luminosity relationship to determine the distance of the SMC. Harlow Shapley (1885–1972) later used Cepheid variables to determine the distances of globular clusters.

Albert Einstein (1879–1955) solved the equations of his general theory of relativity and found that they predicted that the universe was either expanding or contracting. As he was convinced that the universe was static, in 1917 he introduced a so-called cosmological constant into his equations to produce a static universe. A few years later, Alexander Friedmann (1888–1925) developed Einstein's theory and showed that, under certain conditions, there was a critical density of the universe which determined whether the universe would expand or contract. But Friedmann was a theoretician, and did not seem too interested in whether his mathematical solutions referred to the known universe or not. Because of this, his work had little impact at the time.

In 1912 Vesto Slipher (1875–1969) measured spectra of the Andromeda nebula and discovered, much to his surprise, that it was approaching the Milky Way at the unprecedented velocity of 300 km/s. But over the next five years he measured the line-of-sight velocities of 25 spiral nebulae and found that 21 of them were moving away from the Milky Way at very high speeds.

At that time it was unclear as to whether spiral nebulae were connected with the Milky Way, or whether they were stellar systems like the Milky Way but at very great distances. In April 1920 Harlow Shapley and Heber Curtis (1872–1942) debated the issue at the U.S. National Academy of Sciences. But the debate was inconclusive, and it was not until 1924 that Edwin

Hubble (1889–1953) resolved the problem when he measured the distances of the Andromeda nebula (M31) and Triangulum nebula (M33) using Cepheid variables. This showed that they were both stellar systems like the Milky Way at distances that placed them clearly outside the Milky Way.

Georges Lemaître (1894–1966) became aware of Slipher's galactic redshift measurements, and produced a paper in 1927 connecting the observed redshifts with an expanding universe that he had deduced using general relativity. Two years later Edwin Hubble showed that the recession velocity or redshift of distant galaxies increased linearly with increasing distance at the rate of about 500 km/s/Mpc. In 1931 Lemaître took the final step of winding the clock back and concluded that the universe must, at one time, have all been together in one place in a primeval atom, which had exploded. This was the first appearance of what we now call the Big Bang theory. In the same year, Einstein dropped his cosmological constant, later referring to its original introduction as his biggest blunder.

At the rate of expansion deduced by Hubble the universe could only be 2 billion years old. Yet in the 1940s it became clear, using radioactive dating, that the Earth was at least 3.35 billion years old. Bondi, Gold and Hoyle resolved this inconsistency in 1948 by rejecting the Big Bang theory and replacing it with their Steady State theory, in which the universe had been in existence forever. In their theory matter was created out of nothing at a rate that kept the density of the universe constant as it expanded. Then in 1952 Walter Baade (1893–1960) showed that there were two different types of Cepheid variables, and the distances of the galaxies had been estimated using the wrong type. Correcting this increased the age of the universe to about 4 billion years. Although this still seemed too small, at least it was greater than the age of the Earth estimated at that time. So in the early 1950s there were two basic, plausible, cosmological theories – the Big Bang and the Steady State theories.

Early radio astronomy and sounding rocket research

Karl Jansky (1905–1950) detected radio waves being emitted by a cosmic source when he was investigating radio interference in 1931 and 1932. He speculated that the source was either the centre of the Milky Way, or it was in the direction in space towards which the Sun was moving. Grote Reber (1911–2002) followed up Jansky's observations starting in 1938, and was able by 1944 to produce a map of radio emissions showing the main peak at about the galactic centre in Sagittarius, together with subsidiary peaks in the general direction of Cygnus and Cassiopeia. Meanwhile, in 1942 James Hey (1909–2000) and George Southworth (1890–1972) had independently discovered that the Sun was a radio source.

There were many trained radio engineers and a great deal of surplus radio equipment when the Second World War ended in 1945. As a result, a number of research groups were set up, particularly in the UK, Australia and the Netherlands, to undertake radio observations of the cosmos. In 1946 Hey, Parsons and Phillips detected radio signals coming from a region in Cygnus. Martin Ryle (1918–1984) and F. Graham Smith (b. 1923) confirmed observations of this radio source, now called Cyg A, and found another powerful source, Cas A. John Bolton (1922–1993) and Gordon Stanley (b. 1922) showed that Cyg A was less than 8 arcmin in diameter, and also detected Tau A, Cen A and Vir A. In 1949 Bolton, Stanley and Slee correlated the position of Tau A with the Crab nebula. Two years later, Harold Ewen (b. 1922) and Edward Purcell (1912–1997) discovered the 21 cm emission line of neutral interstellar hydrogen that had been predicted in 1944 by Hendrik van de Hulst (1918–2000).

Over fifty radio sources had been catalogued by 1950, but it was unclear at that time whether they were generally inside or outside the Milky Way. It had been suggested in 1949 by Bolton, Stanley and Slee that Cen A and Vir A were associated with other galaxies, but that was disputed. Then in 1953 Cyg A was identified as an inconspicuous sixteenth-magnitude galaxy, after Graham Smith had been able to locate the radio source to within 1 arcmin. But Roger Jennison (b. 1922) and Mrinal Das Gupta (b. 1923) found, surprisingly, that the radio emission came not from the galaxy but from two diffuse patches on either side of it. Over the next few years it became gradually clear that radio sources were both inside and outside the Milky Way.

The United States and Soviet Union began an arms race at the end of the Second World War using, in particular, developments of the German V2 rocket. The Americans also had a small sounding rocket available, the WAC Corporal, which the Jet Propulsion Laboratory had developed. This was subsequently used as the basis for the more performant Aerobee sounding rocket. Scientific experiments were flown using all three rockets in the 1940s and 1950s, with the Aerobee gradually taking over from the V2.

In 1948 Robert Burnight found evidence, using a V2 experiment, that the Sun emitted X-rays. Herbert Friedman (1916–2000) proved this using another V2 experiment in the following year. Friedman's group continued to study the Sun over the next few years, and found that the majority of soft solar X-rays were emitted by the corona. Then in 1956 they found that solar flares produced very high energy X-rays.

Immediately after the war, the early radio astronomers and sounding rocket researchers were virtually ignored by the astronomical community, who had been trained in optical observations. By the 1950s, however, radio astronomy was beginning to show that the universe was more complex than previously thought, so astronomers began to take note of their results. On the other hand, sounding rockets were still seen as expensive, providing only a few minutes of observing time, being

prone to failure, and producing little in the way of results, considering the amount of effort involved. But the launch of the first satellite in October 1957 was to give astronomical research a valuable new dimension.

Telescopes and observatories

(a) Optical telescopes and observatories

The refractor had been the optical telescope of choice for the majority of nineteenth-century astronomers, but the increasing use of photography and spectroscopy towards the end of the century changed the situation dramatically. No longer was high magnification the main requirement, but fast telescopes to reduce exposure times were the order of the day. This was achieved using reflectors.

The early decades of the twentieth century saw the emergence of the United States as the world leader in the construction of large optical telescopes. In 1904 George Ellery Hale (1868–1938) obtained funding from the Carnegie Foundation to build a 60 inch (1.5 m) diameter reflector on Mount Wilson in California. When it saw first light in 1908, it was the most powerful telescope in the world. This was followed by a 100 inch (2.5 m) reflector, which was completed on Mount Wilson in 1917. In the meantime, two large solar telescopes had also been built at the same location: the 60 ft tower telescope completed in 1908, and the 150 ft completed four years later. As a result, Mount Wilson was the premier observatory in the world from 1908 until 1948, when the 200 inch (5 m) reflector and 48/72 inch (1.2/1.8 m) Schmidt were inaugurated on Palomar Mountain.

(b) Radio telescopes and observatories

The first radio telescope in the world was Karl Jansky's 30 m long antenna that rotated horizontally every twenty minutes and operated at a frequency of 20.5 MHz. Grote Reber's radio telescope, on the other hand, was a 10 m diameter parabolic dish that operated over the frequency range from 160 MHz to 3.3 GHz. It could be moved in declination, but not in right ascension.

After the Second World War there were a large number of different designs of radio telescopes, many of the early ones using military surplus equipment. Dishes were required for observations in the GHz region, but much less accurate antennas could be used at tens of MHz.

The largest problem facing early radio astronomers was that of accurately locating astronomical sources, because radio wavelengths are much larger than optical wavelengths. So interferometers were designed using either long wire antennas or a series of small dishes. In 1945 Pawsey, McCready and Payne-Scott used yet another type of interferometer near Sydney, Australia. It consisted of a 200 MHz wartime radar antenna on a cliff overlooking the sea. This allowed interference fringes to be detected between direct observation of a source and its specular reflection off the sea. A few years later Ryle and Graham Smith built an interferometer at Cambridge, UK consisting of two 7.5 m diameter parabolic antennas 280 m apart, operating mostly at 214 MHz. In 1954 the Mills Cross came into operation near Sydney, using the technique of unfilled apertures. It consisted of two 460 m linear arrays at right angles to each other, operating at 85 MHz.

The 75 m (250 ft) diameter parabolic antenna at Jodrell Bank, UK, was the first large radio telescope dish to be completed. It had originally been designed to operate at metre wavelengths (300 MHz), and so was to be built with a wire mesh surface. But the 21 cm (1.4 GHz) emission line of neutral interstellar hydrogen was discovered during the telescope's design phase, so the dish was built with a more accurate solid surface instead. The telescope was completed in 1957, just in time, as it turned out, for the start of the Space Age.

Other major astronomical developments of the twentieth century prior to the space age, not mentioned above, are listed in Table 1.4.

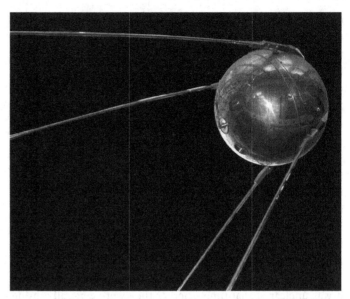

Figure 1.6. The world's first artificial satellite, the 85 kg Sputnik 1, launched by the Soviet Union on 4 October 1957. (Courtesy NASA/Goddard Space Flight Center.)

Bibliography

Ferris, Timothy, *Coming of Age in the Milky Way*, Vintage, 1990.

Gingerich, Owen (ed.), *Astrophysics and Twentieth-Century Astronomy to 1950; Part A*, Cambridge University Press, 1984.

Leverington, David, *A History of Astronomy from 1890 to the Present*, Springer-Verlag, 1995.

Leverington, David, *Babylon to Voyager and Beyond: A History of Planetary Astronomy*, Cambridge University Press, 2003.

Longair, Malcolm, *The Cosmic Century: A History of Astrophysics and Cosmology*, Cambridge University Press, 2006.

Pannekoek, Anton, *A History of Astronomy*, Allen & Unwin, 1961 (Dover reprint, 1989).

Struve, Otto, and Zebergs, Velta, *Astronomy of the 20th Century*, Macmillan, 1962.

Sullivan, W.T., (ed.), *The Early Years of Radio Astronomy: Reflections Fifty Years after Jansky's Discovery*, Cambridge University Press, 1984.

Tassoul, Jean-Louis, and Tassoul, Monique, *A Concise History of Solar and Stellar Physics*, Princeton University Press, 2004.

ASTRONOMY IN THE SPACE AGE

The solar system

The Soviet Union launched the first artificial satellite, Sputnik 1 (see Figure 1.6), on 4 October 1957, but it was the American Explorer 1 and 3 spacecraft, launched in the following year, that made the first significant discovery of the space age. James Van Allen's (1914–2006) Geiger counter on board showed surprising variations in radiation levels, which he attributed to a belt of elementary particles around the Earth. Later American and Soviet spacecraft showed that there were basically two radiation belts, whose positions and intensities varied with solar activity.

Eugene Parker (b. 1927) developed his theory of the solar wind in 1957 and 1958. He pointed out that Biermann's proposed continuous stream of charged particles emitted by the Sun would draw magnetic field lines in the corona out into the solar system. These particles, of the so-called solar wind, were first detected in 1959 outside the influence of the Earth's magnetic field by Luna 2.

Subsequent spacecraft investigated the complex interactions between the solar wind and the Earth's magnetic field in some detail. They showed that there is a region (the magnetosphere) around the Earth, bordered by the magnetopause, which is dominated by the Earth's magnetic field. Further out is the bow shock, which was first detected by the IMP-1 spacecraft in 1963, where the Earth's magnetic field loses its influence. Later spacecraft, including Cluster and TIMED of the twenty-first century, have helped our understanding of how solar particles, which are normally deflected by the Earth's magnetosphere, penetrate the Earth's upper atmosphere.

Horace and Harold Babcock found in the 1950s that there were both bipolar and unipolar regions on the Sun's surface. They suggested that ions and electrons leaving the Sun in bipolar regions would follow the field lines and collide over the Sun, generating radio noise and forming prominences and flares. But those particles emitted from unipolar regions would stream away from the Sun, never to return. The Babcocks suggested that these unipolar regions were Julius Bartels' (1899–1964)

Table 1.4. *Twentieth century prior to the Space Age*

Key discoveries and observations *in addition* to those described in the text

1901	Light emitted from Nova Persei is observed to illuminate a pre-existing dust cloud around the star.
1908	Fowler and Hale independently prove that sunspots are cooler than the surrounding Sun.
1910	Stebbins detects the secondary minimum for the Algol binary.
1912	Hale and colleagues discover that the polarities of the first sunspots of the new solar cycle were reversed in both solar hemispheres.
1912	Wegener suggests that there had originally been one continent, called Pangaea, on Earth, which had broken apart about 200 to 250 million years ago.
1915	Annie Cannon completes the classification of the spectra of 225,300 stars which were published between 1918 and 1924 in nine volumes as the Henry Draper Catalogue.
1917	Shapley concludes, by observing the distribution of globular clusters, that the Sun is not at the centre of the Milky Way.
1920	Saha shows how the degree of ionisation of an atom depends on temperature and pressure.
1922	The Harvard stellar classification system is accepted by the newly founded International Astronomical Union (IAU) as the official system.
1923	Kramers publishes his stellar opacity theory.
1924	Eddington publishes his mass-luminosity diagram for stars. It implied that the gas in both main sequence and giant stars behaves as a perfect gas.
1924	Cecilia Payne estimates the surface temperatures of B- and O-type stars as ranging from 15,000 to 30,000 K.
1926	Fowler develops his theory of degenerate matter, confirming that white dwarfs are largely composed of degenerate matter.
1928	Hubble measures the expansion rate of the Crab nebula, and shows that it is consistent with it being the remnant of the supernova recorded by the Chinese in 1054.
1929	Lyot concludes, from his polarisation measurements, that the surface of the Moon is probably covered in volcanic ash.
1931	Chandrasekhar shows that the maximum theoretical mass of a white dwarf is about 1.4 solar masses, depending on the type of nuclei in the star.
1934	Baade and Zwicky postulate the existence of neutron stars.
1935–7	Dellinger finds a correlation between radio fade-outs and solar flares.
1941	Minkowski concludes that there are two different types of supernova, depending on their luminosity and spectra. Type I have broad emission bands and are brighter than Type II at maximum. Type II have hydrogen emission lines.
1943	Seyfert discovers a number of galaxies, now called Seyfert galaxies, with small, intensely bright, very blue nuclei.
1944	Kuiper detects methane in Titan's atmosphere.
1946	U.S. Army Signal Corps detect radar signals reflected by the Moon.
1947	Kuiper finds clear evidence for carbon dioxide in Mars' atmosphere.
1948	Alpher and Herman suggest that there is a relic radiation left over from the Big Bang present in the universe today at a temperature of about 5 K.
1952	Rense et al image the first solar spectrum showing the Lyman-α emission line in the ultraviolet, using an Aerobee sounding rocket.
1956	Mayer, McCullough and Sloanaker discover that Venus is a radio source, and deduce a surface temperature of about 600 K from the intensity of its radio emissions.
1956	Deutsch concludes that red giant stars are losing a significant amount of mass per year.
1957	Sandage estimates that the age of various star clusters range from about one million years for NGC 2362 to five billion years for M67.
1957	Geoffrey and Margaret Burbidge, Fowler and Hoyle publish their seminal paper on the production of medium-weight and heavy elements in stars.

M regions. They were later found to have a strong, but not perfect, correlation with coronal holes seen in X-rays.

The OSO-7 spacecraft detected clouds of protons leaving the Sun in 1971 in what are now called coronal mass ejections (CMEs). These clouds, which were hundreds of thousands of kilometres in diameter, were leaving the Sun with a velocity of about 1,000 km/s. Large numbers of CMEs have been observed since then. For example, a CME was observed on 6 January 1997 heading towards Earth, where it was detected three and a half days later travelling at 450 km/s by the SOHO and WIND spacecraft. The POLAR spacecraft found that the intensity of the Earth's radiation belts increased by over a hundred times. On the following day, the Telstar-401 spacecraft failed, possibly due to these high particle radiation levels.

Robert Leighton and colleagues discovered five-minute oscillations of the Sun at the Mount Wilson Observatory in 1961. Subsequent observations of these and other vibration modes by the SOHO spacecraft, and the GONG and BiSON ground-based solar networks, have enabled the internal structure of the Sun to be determined. They have shown vertical and horizontal motion in the upper convective layer, and the existence of a turbulent region, the tachocline, at a depth of about 200,000 km.

In the late 1960s Raymond Davis (1914–2006) found that the number of electron neutrinos coming from the Sun was significantly lower than expected. This indicated that nuclear theories explaining how the Sun generates its heat might be wrong. In 1969 Bruno Pontecorvo (1913–1993) suggested, as an alternative, that some of the electron neutrinos emitted by the Sun may be changing to a different type of neutrino en route to Earth and so be missed by Davis' detector. Finally, over thirty years later, the Sudbury Neutrino Observatory in Canada, which was sensitive to all types of neutrinos, showed that Pontecorvo had been correct, as the total number of all types of neutrinos that it detected was as predicted by solar nuclear physics.

Gustav Spörer (1922–1995) and Walter Maunder (1851–1928) had pointed out, towards the end of the nineteenth century, that there had been a lack of sunspots between about 1645 and 1715, a period now called the Maunder Minimum. John Eddy (1931–2009) pointed out in 1976 that this lack of sunspots in the Maunder Minimum correlated with a period of cold winters in Europe. But the correlation between sunspot activity and the amount of solar energy incident on Earth was unclear, because ground-based measurements of the solar constant were complicated by varying atmospheric conditions. It was clearly essential to undertake measurements above the Earth's atmosphere.

Instrument variations bedevilled early sounding rocket and spacecraft measurements of the solar constant. But in 1980 Richard Willson (b. 1937) found, using the Solar Max

spacecraft, that the solar constant *reduced* slightly as large sunspot groups crossed the central solar meridian. It then gradually became clear that the solar constant *reduced* slightly as the Sun approached sunspot *minimum*. In 1988 Peter Foukal and Judith Lean explained these apparently contradictory results by suggesting that the faculae were overcompensating the blocking effect of sunspots over the course of a solar cycle.

The initial spacecraft exploration of the Moon prior to the Apollo 11 landing of 1969 showed that it had a very weak magnetic field, indicting that the core was either solid, or liquid with little internal motion. The Moon's atmosphere seemed to be virtually non-existent, and there appeared to be relatively little dust on its surface. Mass concentrations, or mascons, were found under the lunar maria. At this stage it was still unclear how many of the Moon's features were due to volcanism and how many due to meteorite impacts.

The Apollo 11 astronauts landed on the Sea of Tranquillity in July 1969. The lunar material they brought back was found to be generally basalt, indicating that the maria are of volcanic origin. Breccias (impact rocks) were also found which pre-dated the Tranquillity lavas. Somewhat surprisingly, there was no trace of water found at the time in any of the rocks.

Other Apollo missions showed that the original lunar crust of light anorthosite rock had formed about 4.5 billion years ago. The Moon seemed to be about the same age as the Earth. The numerous lunar craters appear to have been formed by meteorite impacts, although some craters may be volcanic. The lunar soil or regolith was about 2 to 8 m thick over the maria, and about twice that over the uplands. The crust was, on average, about 60 km thick, and below that was the mantle of iron- and magnesium-rich rocks which had been the source of the maria basalts about 4 billion years ago. The core had a temperature of at least 1,200 K. The Moon appears to have had an appreciable magnetic field about 3.8 billion years ago.

William Hartmann and Donald Davis suggested in 1975 that the Moon had been formed as the result of the off-centre impact of a body the size of Mars with the proto-Earth about $4\frac{1}{2}$ billion years ago. The material in the core of the impacting body had been incorporated into the Earth's mantle, but mantle debris from both the impactor and proto-Earth had subsequently aggregated to form the Moon. This explained why the constituents of the Earth and Moon were different.

The subject of continental drift on the Earth has a long history, going back to Francis Bacon (1561–1626), who observed that the coastlines on both sides of the Atlantic seemed as if they could fit into one another. In 1912 Alfred Wegener (1880–1930) developed a theory of continental drift in which all of the Earth's continents had once been together in one continent called Pangaea, which broke apart about 200 to 250 million years ago. This idea was rejected by most geologists at the time, but in the 1950s Stanley Runcorn (1922–1995) and others found paleomagnetic evidence that supported Wegener's theory. At about the same time Marie Tharp (b. 1920) found a rift valley down the centre of the mid-Atlantic ridge, and found that earthquakes appeared to have originated from just below the floor of this valley. During the International Geophysical Year of 1957–8 geologists uncovered a global system of ocean ridges coinciding with major earthquake zones. Then, somewhat later, drill samples obtained using the ship *Glomar Challenger* showed that the Atlantic was broadening by about 2 cm/yr. Eventually, evidence supporting the modern theory of continental drift or plate tectonics was obtained from sites all over the world.

A synchronous rotation period had been generally accepted for Mercury until, in 1962, William Howard and colleagues found that the night side was warmer than it should have been if it was permanently in shadow. Then in 1965 Gordon Pettengill and Rolf Dyce found that Mercury's rotation period was 58.65 days, or exactly two-thirds of its orbital period, by measuring the Doppler shift of radar signals transmitted by the Arecibo radio telescope.

Only two spacecraft so far have visited Mercury, Mariner 10 in the 1970s, and MESSENGER, which had its first fly-by in 2008. Mariner 10 imaged a surface that looked a little like the Moon with a large number of impact craters, but there were apparently no maria. The largest feature was the 1,500 km Caloris basin, whose age, based on cratering density, was estimated at about 3.8 billion years. Surprisingly, for a planet with a slow axial rotation and an apparently solid core, Mercury was found to have an appreciable, probably dipolar magnetic field, together with a very small magnetosphere.

MESSENGER's measurements of Mercury's magnetic field indicated that it was most likely produced by dynamo action in Mercury's outer core, which recent Earth-based measurements had indicated was still probably fluid. MESSENGER found evidence of water in Mercury's exosphere. The spacecraft's images also showed that at least part of the surface was of volcanic origin.

Estimates of Venus' rotation period had varied widely until in 1957 Charles Boyer (1911–1989) observed a distinctive V-shaped cloud pattern that repeated every four days. So the rotation period of the clouds, at least, appeared to be four days, and the rotation was retrograde. It was not until six years later that Richard Goldstein (b. 1927) and Roland Carpenter measured the body rotation rate of Venus using radar as about 240 days retrograde.

In 1956 Cornell Mayer and colleagues concluded, from the intensity of its radio emission, that the surface temperature of Venus was about 600 K (or about 300°C). Shortly afterwards Carl Sagan (1934–1996) deduced a surface atmospheric pressure of 100 bar. Both the temperature and pressure seemed unreasonably high to the majority of astronomers. But, in 1962 Mariner 2, the first spacecraft to visit Venus, measured a surface temperature of about 700 K and a surface atmospheric pressure of 20 bar. Five years later, Mariner 5 found that the surface temperature was at least 700 K, with a surface atmospheric pressure in the range 75 to 100 bar. At the same time, the Venera 4 landing capsule found the atmosphere to be about 96% carbon dioxide. This high concentration of carbon dioxide seems to have produced a runaway greenhouse effect, which was causing the very high temperatures.

Later spacecraft showed that Venus' clouds were composed mainly of sulphuric acid. Spacecraft radar images showed that the surface was relatively flat, with two-thirds within ±500 m of the mean level, although there were exceptions, like the 11,500 m high Maxwell Montes and the 2,000 m deep Diana Chasma. The surface had thousands of volcanoes, none of which appeared to be active, and many impact craters. Analysis of the cratering density showed that most of the surface was only about 400 million years old, with no feature more than 900 million years old. There was evidence of past tectonic activity in the highland regions, but there was no evidence of planet-wide tectonics.

Mars' surface atmospheric pressure was estimated in 1954 by Gérard de Vaucouleurs (1918–1995) to be about 85 millibars. Ten years later, Hyron Spinrad and colleagues concluded that the surface pressure was only about 25 ± 15 millibars, with a partial pressure of carbon dioxide of about 4 millibars. They also concluded that the average amount of precipitable water on Mars was just 14 microns. No oxygen was detected. So Mars appeared to be much drier, and with an even thinner atmosphere, than previously thought.

In 1965 Mariner 4 was the first successful spacecraft to visit Mars. It imaged a surface apparently covered in craters, and measured a surface atmospheric pressure of only 4 to 7 millibars. This crater-covered surface and very low atmospheric pressure were significant blows to those astronomers who thought that there was some form of life on Mars. The next two spacecraft, Mariners 6 and 7, confirmed these findings, and found that the atmosphere was almost 100% carbon dioxide, with only minute amounts of water vapour. However, these spacecraft had, in total, imaged only about 10% of the surface, whereas the next spacecraft, Mariner 9, was to image virtually the whole of the planet. The Mariner 9 images were a revelation, showing a number of large volcanoes and a canyon system (Valles Marineris) that was 4,000 km long and up to 6,000 m deep. Mariner 9 also found what looked like dried-up river beds in the Chryse region (see Figure 1.7).

The two Viking spacecrafts' attempts to detect microbial life on Mars in 1976 produced highly contentious results, with most scientists concluding that no such life existed. But the Viking images showed clear signs of flash floods in the Chryse region. Later evidence from Martian meteorites found on Earth also produced strong evidence that they had been subjected to a water environment sometime in their history. Some scientists interpreted the results from one such meteorite, ALH 84001,

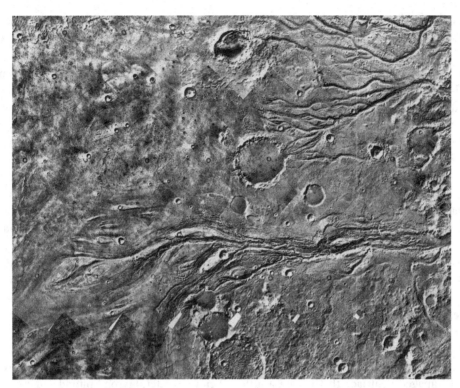

Figure 1.7. A Mariner 9 image of channels in the Chryse region of Mars that appear to have been made by running water. The large crater just above centre is about 75 km in diameter. (Courtesy NASA; Experiment Team Leader, Dr Michael H. Carr, US Geological Survey.)

as showing signs of ancient life, but this interpretation has been hotly contested.

The Mars Global Surveyor, which achieved Mars orbit in 1997, found abundant evidence that there had been liquid water on Mars during the first billion years of its history. The Opportunity rover found stratigraphic features and hematite that had probably been formed in liquid water, and Mars Odyssey found evidence that water ice may still exist in the top metre or so of the Martian surface polewards of about 60^3 N and S. When the Phoenix Mars Lander landed in one of these regions in 2008 it found water ice just below the surface.

Pioneer 10, the first spacecraft to visit the outer planets, flew past Jupiter in 1973. It detected high-energy electrons from Jupiter some distance in front of the planet's bow shock. Pioneer crossed the bow shock a number of times as the latter moved to and from Jupiter, reacting to the varying pressure of the solar wind. The spacecraft found that Jupiter's radiation belts, consisting of protons and high-energy electrons, were about 10,000 times as intense as the Earth's Van Allen belts. For the first time it detected helium in Jupiter's atmosphere, which was found to consist of about 99% hydrogen and helium.

In 1974 Pioneer 11 detected a reduction in high-energy particles near its closest approach to Jupiter. This led Mario Acuña (1940–2009) and Norman Ness to suggest that this reduction may have been caused by an unknown ring or satellite. Although this idea was generally dismissed at the time, a search was

undertaken using the Voyager 1 spacecraft in 1979 and a thin ring was found. Voyager 1 also imaged the four Galilean satellites, of which Io proved to be the most interesting. It was found to have a fresh-looking surface, which included numerous volcanic calderas and rivers of lava. Eight of the volcanoes were found to be active during the Voyager 1 encounter, emitting plumes up to 300 km high, with vent velocities as high as 1.0 km/s.

Voyager 2, which flew past Jupiter four months after Voyager 1, was used to study Jupiter's ring in more detail. It detected three components, now called the Halo, Main, and Gossamer rings. Two previously undetected small satellites, called Metis and Adrastea, were also discovered near the outer edge of the main ring, possibly constraining its outward expansion and/or helping to supply it with material. Voyager 2's images of Europa showed numerous long, dark stripes on a virtually flat surface. This led some astronomers to speculate that these linear markings were cracks in the icy surface, which may be floating on a subsurface water ocean.

In 1995 the Galileo spacecraft was the first to orbit Jupiter and send a probe into the planet's atmosphere. The probe found that the wind velocity increased as it descended through the atmosphere, indicating that Jupiter's weather system is driven by internal heat. The helium-to-hydrogen ratio was almost the same as for the Sun, although the abundances of methane, ammonia and hydrogen sulphide clearly exceeded their solar values, showing that Jupiter had had significant enrichment since its formation.

Galileo found that Ganymede has an intrinsic magnetic field, which implied that it has a molten or semi-molten core, which was a surprise for a relatively small body so far from the Sun. Europa, on the other hand, was found to have a variable magnetic field caused by Jupiter's magnetic field inducing subsurface currents. Astronomers thought that these currents may be in a salty ocean only a few kilometres beneath Europa's icy surface. A similarly induced magnetic field was found on Callisto, where the salty ocean was expected to be about 200 km below the surface. Following this discovery of Callisto's variable magnetic field, Ganymede's magnetic field was re-examined and was found to have an induced component, also implying a sub-surface ocean.

Pioneer 11 was the first spacecraft to visit Saturn when it flew past in 1979. It discovered Saturn's magnetic field and its associated magnetosphere and radiation belts, and found that the ring system prevented belts from forming in the inner magnetosphere. Pioneer discovered the narrow F ring, just outside the A ring, and one new satellite, now called Epimetheus. Both the ring and satellite produced gaps in the electron population of Saturn's magnetosphere.

In the following year, infrared measurements with Voyager 1 showed that Saturn emits 80% more energy than it receives from the Sun, which is a similar percentage to that for Jupiter. But it was not clear why Saturn, which receives much less solar energy than Jupiter, should have a much more powerful equatorial jet. Spokes were discovered in Saturn's B ring, and two new satellites, Prometheus and Pandora, were found on either side of the narrow F ring, apparently shepherding or stabilising it. Peter Goldreich (b. 1939) and Scott Tremaine (b. 1950) had suggested that such shepherding satellites could explain Uranus' narrow rings, but these shepherding satellites of Saturn were the first to be found. Saturn's main rings were found to be composed of hundreds of thin concentric rings, with even the Cassini division containing over a hundred narrow rings.

Although Saturn's largest satellite, Titan, was known to have methane in its atmosphere, Donald Hunten suggested that the main atmospheric constituent was nitrogen. Measurements by John Caldwell and colleagues in 1979, with the VLA radio telescope, implied a surface temperature of 87 K. This led Hunten to deduce a surface atmospheric pressure of about 2,000 millibars. In the event, Voyager 1 vindicated Hunten, finding that Titan's atmosphere was about 90% nitrogen, the surface atmospheric pressure was about 1,500 millibars, and the surface temperature 94 ± 2 K, which is very near methane's triple point of 90.7 K. As a result, it was thought that Titan may have oceans of methane, and cliffs of methane in its polar regions.

The Cassini-Huygens spacecraft that arrived at Saturn in 2004 found methane and nitrogen in Titan's atmosphere, with

methane clouds at an altitude of about 20 km. The surface had dendritic channels and large hydrocarbon lakes. Enceladus was found to have a thin atmosphere and active water ice plumes coming from warm fissures in its polar regions. These ice particles were found to be modifying Saturn's local magnetic field.

In March 1977, Elliot, Dunham and Mink observed the occultation of a relatively bright star by Uranus, using the Kuiper Airborne Observatory to study Uranus' atmosphere. Much to their surprise, they found very brief dips in intensity on both sides of the planet, which they attributed to a series of very narrow rings. Other groups of astronomers produced less complete results, but when all the data were integrated, it appeared as though Uranus had nine narrow rings, one of which, the outermost, was noticeably eccentric. Nine years later, when Voyager 2 flew past Uranus, the spacecraft discovered two more rings. All the rings were very dark, with a reflectivity of a few percent, at most.

Voyager 2 found that Uranus has a magnetic field, with its magnetic axis tilted at 60° to the spin axis, and a magnetic centre displaced from the geometric centre by about 0.3 R_U (radius of Uranus). The helium abundance was found to be the same as that of the Sun, within error, and any internal generated heat was found to be small. The most interesting of Uranus' satellites was found to be the 480 km diameter Miranda. Its complex structure was a big surprise, with its chevron-shaped feature, large ovoids, and cliffs up to 20 km high.

By 1980 the discovery of rings around Uranus and Jupiter meant that Neptune was apparently the only gas giant planet without rings. A number of stellar occultation observations by Neptune in the early 1980s produced ambiguous results. Some of the detected intensity dips were attributed to unknown satellites and some to partial rings or ring arcs. Voyager 2 clarified matters in 1989 when it imaged three narrow rings and two broad diffuse rings, with one of the narrow rings, the Adams ring, being very clumpy. But one of the intensity dips previously observed from Earth was found by Voyager to be due to a 200 km diameter satellite, not to a ring. The satellite was later named Larissa.

Voyager 2 found that Neptune emits 170% more energy than it receives from the Sun, by far the largest percentage for the gas giants, and its equatorial jet was found to have a velocity of an incredible 1,800 km/h. Voyager also found that Neptune has a magnetic field with the magnetic axis tilted at 47° to the spin axis, and a magnetic centre displaced from the geometric centre by about 0.55 R_N (radius of Neptune).

The reflectivity of Neptune's large satellite Triton was found by Voyager to be a very high 85%, resulting in a surface temperature of 38 K, some 20 K lower than expected. As a result, most of the atmosphere had condensed onto the surface, leaving a very thin atmosphere of mostly nitrogen. The surface had a southern polar cap of nitrogen ice, with trace amounts of methane, carbon monoxide, carbon dioxide and water ice, extending three-quarters of the way to the equator. A number of active plumes were discovered ejecting nitrogen or methane gas and trapped surface material about 8 km into the atmosphere.

In 1978 James Christy (b. 1938) discovered Pluto's first satellite, Charon, only about 20,000 km from Pluto. Charon's diameter was later found to be about 50% of that of Pluto. In 2005 two very much smaller satellites, Nix and Hydra, were discovered by Harold Weaver and colleagues using the Hubble Space Telescope.

Dale Cruikshank and colleagues detected methane ice on the surface of Pluto in 1976. Sixteen years later, nitrogen and carbon monoxide were also discovered on the surface of Pluto, and later measurements of atmospheric methane and carbon monoxide indicated that they were minority atmospheric constituents. In the meantime, Voyager 2 had discovered that Triton's atmosphere was mainly nitrogen, leading astronomers to theorise that that nitrogen was probably also Pluto's major atmospheric constituent. The main surface ice on Charon was found to be water ice but, so far, there has been no clear detection of an atmosphere.

Kenneth Edgeworth (1880–1972) and Gerard Kuiper had independently suggested in the mid-twentieth century that there should be many planetesimals, left over after the formation of the planets, still orbiting the Sun at the edge of the solar system. In 1992 David Jewitt (b. 1958) and Jane Luu (b. 1963) discovered the first of these trans-Neptunian or Kuiper belt objects, which was later found to be about 160 km in diameter. Since then, over 1,000 of these trans-Neptunian objects have been found, some of them, like Pluto, locked in a 2:3 resonance orbit with Neptune.

In 1996 David Jewitt and colleagues also discovered an object, 1996 TL_{66}, in a highly eccentric and inclined orbit with a perihelion of 35 AU. It was about 600 km in diameter. Since then many more of these so-called scattered disk objects have been found, with very large and highly eccentric orbits, generally outside the classical Kuiper belt. A classical Kuiper belt object, now called Quaoar, was discovered in 2002 by Chad Trujillo (b. 1973) and Mike Brown (b. 1965). It had a diameter of about 1,150 km. In the following year, Mike Brown and colleagues discovered Sedna, with a 40% larger diameter. Then in 2005 they found a scattered disk object, Eris, with a diameter of about 2,400 km, or roughly the same as Pluto. Evidently, more similar sized or larger objects may exist in the outer solar system. As a result, in 2006 the International Astronomical Union devised a new class of object, called a dwarf planet. The initial list of such objects included both Pluto and Eris.

A flotilla of five spacecraft flew past Halley's comet in 1986 to produce the first detailed in-situ analysis of a comet. They found the bow shock about 1 million kilometres from the nucleus, and the ionopause, where the interplanetary magnetic field and solar wind fall to zero, about 5,000 km from the nucleus. The latter was seen to be an irregular $16 \times 8 \times 8$ km in size, with an albedo of only about 3%. There was one clear crater, and a number of bright jets streaming from the nucleus towards the Sun. Halley's inner coma was found to consist mainly of water, with lesser amounts of carbon monoxide, carbon dioxide, methane, ammonia and polymerised formaldehyde.

Two decades later, an impactor from the Deep Impact spacecraft collided with the nucleus of Comet Tempel 1. The mother spacecraft, other spacecraft, and Earth-based observatories observed the resultant plume, detecting superheated steam and carbon dioxide at temperatures in excess of 1,000 K. The ejecta included amorphous carbon, carbonates, silicates, polyaromatic hydrocarbons and methane.

Theories of the origin of the solar system have been given a massive boost over the last forty years by observations of stellar systems and dust clouds elsewhere in the Milky Way. In 1967, for example, Eric Becklin (b. 1940) and Gerry Neugebauer (b. 1932) discovered an object, now called the BN object, in the Orion nebula, which had a temperature of just 600 K. It was a dust cloud surrounding a hot, young star. Then in 1983, H. H. Aumann and Fred Gillett (1937–2001) found that Vega appeared to be surrounded by a dust cloud at a temperature of only 80 K. Measurements with the IRAS spacecraft indicated that the dust grains were rather large, indicating that the Vega cloud may be in the process of forming protoplanets. Subsequent work on dust clouds around other stars, the discovery of protoplanetary discs or proplyds in the Orion nebula, and the discovery of exoplanets (or extrasolar planets) has further assisted the development of theories of the origin of the solar system.

The starting point in the development of the solar system is now thought to have been a large molecular cloud at a temperature of the order of 10 to 50 K. This gradually collapsed over the order of 100,000 years to form the protosun, leaving a disc of material rotating around it. Over the next 10 million years or so the protosun gradually accumulated material from the surrounding disc, heating up in the process and starting to shine by nuclear reactions. Dust grains that remained in the surrounding disc gradually coalesced to eventually form planetesimals, which then formed planets following numerous collisions. Physical, chemical and isotopic analyses of meteorites and samples returned to Earth by space missions have helped in the development of this basic theory.

In 1991 Aleksander Wolszczan (b. 1946) and Dale Frail found evidence that two planets were orbiting the millisecond pulsar PSR B1257 + 12. Two years later they confirmed this and found evidence for a third planet orbiting the pulsar. This discovery of these first exoplanets was accidental. But the discovery of the first extrasolar planet around a 'normal' star was the result of a deliberate search, in which Michel Mayor

(b. 1942) and Didier Queloz (b. 1966) looked for Doppler shifts in their spectra. In 1995 they found evidence for a planet, which had a mass of at least $1/2 M_J$ but an orbital period of just 4 days, orbiting 51 Pegasi. Although this very short orbital period seemed unlikely, Geoffrey Marcy (b. 1954) and Paul Butler provided independent confirmation shortly after the planet's discovery. Since then hundreds of exoplanets have been found.

High-energy astrophysics

(a) Early X-ray observations of cosmic sources

Early sounding rocket experiments had shown that the Sun emitted X-rays, but the Sun was very close compared with the stars, and most astronomers thought that the stars were too far away for any X-ray emission to be detected with the technology available in the late 1950s. However, in 1958, Giacconi, Clark and Rossi suggested that certain exceptional objects, such as very hot stars, rapidly spinning stars with large magnetic fields, flare stars and supernova remnants, may emit measurable X-rays.

In 1962 Riccardo Giacconi (b. 1931) and his team detected the first source, Sco X-1, of non-solar X-rays using a sounding rocket experiment. This was the first indication of the enormous amount of energy that could be produced in previously unknown astrophysical processes. It was later found that the X-ray luminosity of Sco X-1 was 10^3 times its luminosity in the visible waveband, compared with a ratio of only 10^{-6} for the Sun, and that the intrinsic luminosity of Sco X-1 was 10^3 that of the Sun. Giacconi's group also found a diffuse source of X-rays across that part of the sky that they scanned.

Giacconi's group found two new X-ray sources with their next sounding rocket flight of 1962, one of which was in the general direction of the Crab nebula, which was a well-known supernova remnant. In the following year, Herbert Friedman (1916–2000) and C. Stuart Bowyer were able to show that the X-ray source was clearly associated with the Crab nebula. Then in 1964 Herbert Friedman's group used the occultation of the Crab nebula by the Moon to show that the Crab's X-ray source was not a point source, and so could not be a neutron star, as some astronomers had suggested. Finally, in 1964 Giacconi's group measured the X-ray spectrum of Sco X-1 and showed that it was also not a neutron star.

(b) Active galaxies

The early 1960s were an important period in the development of astrophysics. In 1963 Maarten Schmidt (b. 1929) discovered the first quasi-stellar radio source, or 'quasar', when he found that the spectrum of a blue star, which was a part of the radio source 3C 273, had, for then, a very large red shift of 0.16. This indicated that it was at a distance of about 470 Mpc (1.5 billion light years), assuming a Hubble constant of 100 km/s/Mpc, so it must be intrinsically very intense. Jesse Greenstein (1909–2002) and Thomas Matthews (b. 1927) immediately checked the spectrum of the radio source, 3C 48, and found that it had an even larger redshift of 0.37, indicating that it was at a distance of about 1,100 Mpc. The radio brightness of 3C 48 was also found to vary over the course of a day, indicating that it was only about 1 light day across. Over the next few years, many more of these intrinsically high-intensity objects or quasars were discovered at even greater distances. Some also exhibited short period radio fluctuations, indicating their small physical size.

It was suggested in the mid-1960s that quasars were the more violent relations of Seyfert galaxies, which were known to have very energetic nuclei ejecting large amounts of gas. In 1967 W. S. Fitch and colleagues found a variation in the optical intensity of the Seyfert galaxy NGC 4151, and in the following year William Morgan and colleagues found that its spectrum had changed since 1956. This was explained by Beverley Oke (1928–2004) and Wallace Sargent (b. 1935) as being due to rapidly moving clouds of gas in the galaxy's central region.

BL Lacertae had been identified as a variable star in 1929. Then in 1968 John Schmitt identified it as a variable radio source and detected a faint nebulosity surrounding the star in optical images. BL Lac's optical spectrum was featureless, but in 1974 Beverley Oke and James Gunn (b. 1935) measured the redshift of its associated nebulosity as 0.07, showing that BL Lac was not a star but an object outside the Milky Way. It, and similar BL Lac objects, are now thought to be, like quasars, the nuclei of active galaxies, but oriented so that we are looking virtually straight down their relativistic jet.

(c) Pulsars, neutron stars and black holes

In the mid-1960s, Antony Hewish (b. 1924) and his team built a radio telescope to measure the radio scintillation of quasars to determine their size. In late 1967, shortly after the start of observations, Jocelyn Bell (b. 1943), one of Hewish's research students, found that one source was pulsing with a regular frequency of just 1.34 s. Bell and Hewish found three more of these pulsing radio sources – or pulsars, as they are now called – over the next two months.

Just before the announcement of the pulsar discovery, Franco Pacini (b. 1939) had shown that a rapidly rotating neutron star with a strong magnetic field could transform rotational energy into electromagnetic radiation and accelerate particles to high energies. Just after the pulsar announcement, Thomas Gold (1920–2004), without knowing of Pacini's work, suggested that a pulsar was a rapidly spinning, magnetised neutron star. Gold explained that it was emitting a beam of radiation along its magnetic axis that was not coincident with its rotational axis, hence producing the pulse observed from Earth. Both Gold and Pacini predicted that a pulsar's spin rate would gradually slow down as it lost energy.

Many years earlier, Walter Baade (1893–1960) and Fritz Zwicky (1898–1974) had suggested that a neutron star may be the stellar remnant left behind after a supernova explosion. So astronomers speculated that pulsars may be surrounded by supernova remnants. In 1968 Michael Large and colleagues discovered a radio pulsar, PSR 0833–45, with a period of just 89 ms in the Vela supernova remnant. This period meant that the pulsar must be a neutron star, as a white dwarf would have broken up at such a rapid rotation rate. So Baade and Zwicky's theory was vindicated.

A month later, David Staelin and Edward Reifenstein found a pulsar, NP 0532, in the Crab supernova remnant. The pulsar was found to have a period of just 33 ms. Shortly afterwards, David Richards and John Comella showed that the pulse rate was slowing down, as predicted by Pacini and Gold, and used the rate of slowdown to deduce an age of about 1,000 years. This was consistent with it being the residual star of the supernova explosion observed by Chinese and Japanese astronomers in 1054. Then in 1969 William Cocke and colleagues found optical pulsations in NP 0532, the first to be observed in a pulsar, whilst Gilbert Fritz and colleagues also found X-ray pulsations.

All the early pulsars discovered were single stars. Then in 1971 the Uhuru X-ray spacecraft found that Cen X-3 was emitting X-rays with a regular periodicity of 4.84 s. But every 2.1 days the X-ray source disappeared for about 12 hours, indicating that it was in a binary with another star. This was confirmed when it was found that the pulse arrival times at Earth also varied over a period of 2.1 days. In 1973 Wojciech Krzeminski (b. 1933) found that the pulsar's companion was a 20 $M_\odot$ (solar mass) blue star, as it also exhibited the same period of 2.1 days in visible light.

Saul Rappaport and colleagues used a sounding rocket experiment in 1971 to measure the very fast intensity variations of the X-ray source Cyg X-1. These indicated its very small size. From its position estimate, Paul Murdin (b. 1942) and Louise Webster (1941–1990) suggested that its optical counterpart was the blue supergiant HD 226868, which had a periodicity of 5.6 days. Then in 1975 S. S. Holt and colleagues found, using the British Ariel 5 spacecraft, that Cyg X-1 also had a periodicity of 5.6 days, confirming that it was the companion of the blue supergiant. Analysis of the orbital dynamics indicated that the mass of Cyg X-1 must be about 10 $M_\odot$, which is too large for a neutron star. So Cyg X-1 appeared to be a black hole.

Joseph Taylor (b. 1941) and Russell Hulse (b. 1950) discovered a binary pulsar (PSR 1913 + 16) in 1974 with a binary period of 7.75 hours. Over the years this binary pulsar has provided an excellent check on Einstein's theory of relativity, as both stars appear to be neutron stars orbiting very close to each other. The periastron of the binary pulsar's orbit was found to precess at 4.2°/year (which was four orders of magnitude faster than that of Mercury's perihelion), in excellent agreement with

Einstein's theory. Analysis of the energy loss of the system was also consistent with the existence of gravity waves, although the waves themselves have not been detected.

Uhuru discovered about 300 new sources of X-rays in the early 1970s (see Figure 1.8), including globular clusters, spiral galaxies, Seyfert galaxies and clusters of galaxies. The Dutch ANS spacecraft was launched by NASA in 1974. It detected X-ray emission from two flare stars, and discovered an X-ray burster, 4U 1820–30. Unknown to the astronomical community, X-ray bursters had been previously discovered by the U.S. military Vela spacecraft, but their discovery was only announced later. In the mid-1970s, the Ariel 5 spacecraft found that the X-ray intensity of some Seyfert galaxies varied in about a day. So by the late 1970s the universe was seen no longer as the relatively predictable place of the early twentieth century, with just one or more solar systems, stars and galaxies and the occasional nova or supernova, but one with exotic objects, including pulsars, neutron stars, quasars and black holes.

Astronomers had speculated for some time that very large, or supermassive, black holes exist in the centre of galaxies, and in 1978 Wallace Sargent and colleagues found controversial evidence of a black hole of mass $\sim 5 \times 10^9$ $M_\odot$ at the centre of galaxy M87. In 1994 Holland Ford (b. 1940) and colleagues came to a similar conclusion, using a different measurement technique with the Hubble Space Telescope. But a more convincing case for a supermassive black hole was published in the following year when L. J. Greenhill and colleagues concluded, using very long baseline interferometer observations, that the source at the centre of the galaxy NGC 4258 had a mass density of at least 3.5×10^9 $M_\odot/pc^3$. In parallel, Makoto Miyoshi (b. 1962) and colleagues came to a similar conclusion. This density indicates that there is probably a supermassive black hole at the centre of this galaxy. But it is just possible that the density could be provided by a very dense cluster of stars, although, if so, its ultra-high density would cause the cluster to disintegrate over ~ 1 billion years.

Teams led by Reinhard Genzel and And rea Ghez (b. 1965) have been observing the movement of a number of individual stars near the centre of the Milky Way over the last fifteen years or so. These observations show that most of the unseen mass of $\sim 4 \times 10^6$ $M_\odot$ must be within about 100 AU of Sgr A* at the centre, implying a mass density of $>8 \times 10^{15}$ $M_\odot/pc^3$. Recent millimetre wavelength observations of Sgr A* have shown that it is ~ 0.1 AU in diameter, which is about the same order of magnitude as the Schwarzschild radius for a 4×10^6 $M_\odot$ black hole. These results indicate that there is almost certainly a supermassive black hole at the centre of the Milky Way.

(d) Further X-ray observations

The Einstein observatory spacecraft, launched in 1978, was the first to be able to image X-ray sources, supernova remnants in particular. It also found that not only did violent objects emit

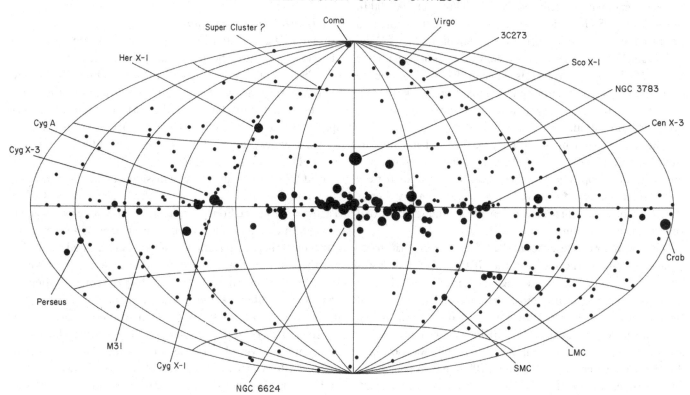

THE FOURTH UHURU CATALOG

Figure 1.8. X-ray sources detected by Uhuru are shown on this map in galactic coordinates centred on the centre of the Milky Way. The Milky Way crosses the plot at 0° latitude. (Courtesy Harvard-Smithsonian Center for Astrophysics.)

X-rays, but, surprisingly, so did cool, optically dim, red dwarf stars. Einstein also undertook two deep sky surveys to try to resolve the X-ray background radiation into discrete objects. Astronomers found that most of the resolved objects in these surveys were active galactic nuclei, including quasars, Seyfert galaxies and BL Lac objects, but a significant amount of the X-ray background radiation still remained unresolved.

The Rosat spacecraft launched in 1990 increased the number of known X-ray sources from 6,000 to over 150,000. It produced the first high-resolution image of the Cygnus Superbubble, and enabled maps to be produced of the X-ray brightness of nearby clusters of galaxies. NASA's Chandra X-ray Observatory and the European Space Agency's (ESA's) XMM-Newton were both launched in 1999, as complementary X-ray observatories. Chandra found a neutron star whose progenitor star appears to have had a mass of at least 40 $M_{\odot}$, which was much heavier than theory predicted. The spacecraft also found a group of massive stars less than a light year away from the Milky Way's central black hole, Sgr A*, implying that they were formed in situ, rather than having migrated there from further out. In addition, Chandra resolved the enigmatic X-ray background radiation into individual X-ray sources. XMM-Newton, for its part, found supporting evidence for the existence of intermediate mass black holes, detected individual hot spots

on rotating neutron stars, and found the first black hole candidate in a globular cluster. It also found the most distant cluster of galaxies (XMMXCS 2215–1738) then known, which the Keck telescope determined had a red shift of 1.45, equivalent to a distance of about 3,000 Mpc (10 billion light years). The cluster contained hundreds of galaxies surrounded by an ultra-high temperature gas.

(e) Gamma-ray observations

The first dedicated gamma-ray observatory spacecraft, Explorer 11, was launched in 1961. In four months it detected just 22 cosmic gamma rays. Six years later, the Orbiting Solar Observatory OSO-3 detected about 600 cosmic gamma rays. Although OSO-3 could not detect individual sources, it showed that the galactic equator was a clear source, with the galactic centre the most intense part. In 1971 a balloon-borne experiment identified the Crab pulsar as a gamma-ray source. Some years later it was found to pulse in gamma rays also.

In 1972 NASA launched the Small Astronomy Satellite SAS-2, which produced the first detailed gamma-ray survey of the sky, confirming that the galactic centre was a strong source of gamma rays. It also provided evidence for a diffuse extragalactic gamma-ray background. In 1975 NASA launched the European COS–B spacecraft, which increased the number of cosmic gamma rays detected from 8,000 to over 200,000 in the 30 MeV to 5 GeV energy range. It enabled astronomers to produce the first complete gamma-ray map of the Milky Way, and locate 25 discrete gamma-ray sources, most of which were unidentified in other wavebands. Then in 1977 a balloon-borne experiment detected gamma-ray emission at 511 keV from the centre of the Milky Way, which was caused by electron-positron annihilation.

The U.S. Department of Defense began launching its advanced Vela series of satellites in 1963. Although not intended for astrophysical studies, these Vela spacecraft were the first to detect short-lived, intense gamma-ray bursts (GRBs). These were first detected in 1967, although their discovery was not announced until 1973.

The gamma-ray telescopes of the 1970s were too inaccurate to be able to locate the source of the GRBs. So a network was set up in 1978 of spacecraft acting together to enable the first reasonably accurate locations to be determined. By the time that the Compton Gamma Ray Observatory (CGRO) was launched in 1991, the position of about 200 GRBs had been determined accurately enough to show that their distribution was isotropic. The CGRO, which detected about 2,700 bursts over its nine-year lifetime, showed that the bursts fell into two categories: short-duration, hard-spectrum bursts and long-duration, soft-spectrum bursts, with a changeover from one to the other at a duration of about two seconds.

The BeppoSAX spacecraft, which was launched in 1996, was designed, amongst other things, to observe GRBs in X-rays, if possible, and give their approximate location rapidly to ground-based optical telescopes. BeppoSAX detected over 50 GRBs over its six-year lifetime, many of which were subsequently observed optically. This showed for the first time that the GRBs originated from well outside the Milky Way.

(f) Supernova 1987A

The nearest supernova to be observed since 1604 was discovered, independently, in the Large Magellanic Cloud on 24 February 1987 by Ian Shelton and Oscar Duhalde. A few hours before their visual observations, a total of 25 neutrinos had been detected from the same source over a period of 12 seconds by neutrino telescopes in Japan, the USA and Russia. Pre-explosion photographs showed that the progenitor star was a blue supergiant, Sanduleak −69°202, which was a surprise, as it was thought that red, not blue, supergiants developed into supernovae. This was the first time that a progenitor star had been found for a supernova. Post-explosion observations by numerous ground- and space-based observatories have now enabled astronomers to improve their understanding of supernova processes.

Low-energy astrophysics

(a) Ultraviolet observations

Ultraviolet light is attenuated by the Earth's atmosphere, so observatories need to be located in space to undertake ultraviolet observations. A few such astronomical observations were made by experiments on sounding rockets and on spacecraft, but it was not until 1968 that the first dedicated ultraviolet spacecraft was successfully launched to undertake an ultraviolet astronomical survey. This was the second Orbiting Astronomical Observatory, OAO-2.

OAO-2 detected ultraviolet emission from more than 5,000 stars and found a huge cloud of hydrogen around Comet Tago-Sato-Kosaka. Four years later, OAO-3, otherwise known as Copernicus, took high-resolution spectra of hundreds of stars, detecting deuterium in the interstellar medium. This was the first time that deuterium had been detected anywhere in the universe except on Earth. Its observed concentration, of 15 parts per million, is an important constraint on whether the universe is open or closed. Copernicus showed that massive, hot stars, such as Wolf-Rayet stars, emit powerful stellar winds much more powerful than the solar wind. The IUE (International Ultraviolet Explorer) spacecraft, which was launched in 1978, showed that the velocity of these stellar winds was in the region of 1,500 to 3,500 km/s. At this rate, these stars would lose 1 solar mass of material in about 25,000 years. IUE was the first to detect absorption lines in a ring nebula (NGC 6888) associated with a hot Wolf-Rayet star (HD 192163). Analysis of the nebula's spectrum showed that it was being heated up to 60,000 K by the shock wave caused by the interaction of the Wolf-Rayet's stellar wind and the interstellar gas.

In the 1970s it gradually became clear that the solar system is located in the 100 parsec diameter Local Bubble, in which the interstellar medium is appreciably less dense than normal. As a result, it should be possible to detect extreme ultraviolet sources within the bubble, but possibly not many outside because of the attenuation of the extreme ultraviolet in the denser regions of the Milky Way. It was something of a surprise, therefore, when in 1992 the EUVE (Extreme Ultraviolet Explorer) spacecraft, shortly after launch, detected the BL Lac object PKS 2155–304, which was the first extreme ultraviolet source found outside the Milky Way. By the end of its third year, EUVE had detected 37 extragalactic objects, including AGN (active galactic nuclei) and Seyfert galaxies. Since then a number of other spacecraft have charted the structure of the Local Bubble and of other inhomogeneities in the Milky Way's interstellar gas. GALEX (Galaxy Evolution Explorer), an ultraviolet spacecraft launched in 2003, has also studied star formation rates and has detected UV emission from the extended tidal tails of interacting galaxies.

(b) Infrared observations

Gerry Neugebauer (b. 1932) and Robert Leighton (1919–1997) carried out the first major infrared sky survey in the 1960s. They used a ground-based telescope to observe through the 2.2 μm atmospheric window, detecting about 20,000 infrared sources. Their work was extended up to 27 μm, using sounding rockets, in the 1970s. Objects and interstellar dust were detected with temperatures of only about 100 K. Then in 1983 the first infrared spacecraft, IRAS (Infrared Astronomical Satellite), was launched. It undertook an all-sky survey and detected protostars in interstellar dust clouds, stars still surrounded by remnants of their original dust clouds, and dust shells emitted by stars, like Betelgeuse, near the end of their lives. IRAS also found ultraluminous infrared galaxies, some of which were emitting over 90% of their energy in the infrared. The main IRAS catalogue, published in 1984, contained about 245,000 sources.

The next infrared spacecraft, ISO (Infrared Space Observatory), launched in 1995, found that water molecules are common in the Milky Way. It also found star-forming regions in both normal and colliding galaxies, where star formation had been triggered by the collision shock wave. The earliest phase of cloud collapse was found in a 2 $M_\odot$ dark cloud that had a temperature of just 13 K. Since its launch in 2003, the Spitzer infrared spacecraft has been the first to detect light emitted by an exoplanet. It has also imaged possible planet-forming gas clouds around stars. Spitzer participated, along with the Herschel Space Observatory (launched 2009) and the ground-based JCMT (James Clerk Maxwell Telescope), in observing all prominent star-forming regions within about 500 parsec (1,600 light years) of the solar system, in the so-called Gould's Belt Survey.

Cosmology

In the early 1950s there were two basic cosmological theories, namely the Big Bang and Steady State theories. But the Big Bang theory was in some trouble, as Walter Baade had deduced an age of the universe of 4 billion years in 1952, based on a Hubble constant H_0 of 250 km/s/Mpc. This age seemed improbable, as it was too near the estimated age of the Earth. However, in 1958 Allan Sandage (b. 1926) reduced the estimated value of H_0 to 75 km/s/Mpc, implying a far more plausible age for the universe of 13 billion years. Unfortunately, shortly afterwards Allan Sandage deduced the age of two globular clusters to be about 25 billion years, which was significantly more than his new estimated age of the universe.

Sandage produced an estimate of the age of the universe in 1961 of 7.4 billion years, using new values of the cosmological parameters. But this made the problem of the age of the universe even worse. He suggested that the only way to solve the problem was either to adopt a Steady State universe, or an Einstein-Lemaître universe with improbable values for its parameters.

In 1974 Richard Gott and colleagues produced a paper arguing that the density of the universe today, ρ, is ≤0.1 times the critical density ρ_{crit} for a flat universe. But, according to the Big Bang theory, any deviation from the critical density would have increased rapidly over time since the expansion of the universe began. So if ρ is ≤0.1 ρ_{crit} today, it must have been within 10^{-15} of ρ_{crit} one second after the Big Bang. It seemed unlikely that there had been such a tiny deviation from the critical density originally. Consequently it was thought that ρ is equal to the critical density now and had always been so. As a result, the universe would always have been flat with the density parameter $\Omega = \rho/\rho_{crit} = 1$. If that was the case, ≤10% of all matter had been detected, assuming Gott's analysis was correct, the remainder being dark matter.

Robert Dicke (1916–1997) had raised two key questions in 1961: Why is the universe so uniform on large scales, the so-called 'horizon problem', and why is the density of the universe apparently so close to the critical density, the 'flatness problem'? Twenty years later, Alan Guth (b. 1947) proposed a major modification to the Big Bang theory which solved both these problems. In his inflation theory, the radius of the universe expanded by an enormous factor almost immediately after the Big Bang. So the universe would have been small enough for it to have reached a uniform homogenous state before the expansion began. In addition, this rapid expansion would have straightened out the geometry of the early universe to produce a flat Euclidean geometry.

By the later 1990s most observationally-based values of the Hubble constant were in the range 55 to 75 km/s/Mpc, equivalent to an age of the universe from about 17 to 13 billion years for a flat universe. At about the same time, a team led by Wendy Freedman (b. 1957) had started a major investigation using the Hubble Space Telescope to try to achieve a definitive value of the Hubble constant.

But whilst Freedman and colleagues were working on the problem, there was a surprising development in our understanding of the structure of the universe. In 1998 both Saul Perlmutter (b. 1959) and colleagues and Brian Schmidt (b. 1967) and colleagues independently concluded, after analysing a number of Type 1a supernovae, that Einstein's cosmological constant was not zero, but positive. As a result, the expansion of the universe was accelerating due to its vacuum or dark energy. For a flat cosmology $\Omega_m + \Omega_\Lambda = 1$, where Ω_m is the density parameter of matter, and Ω_Λ is the density parameter of dark energy. Perlmutter and colleagues found $\Omega_m = 0.28$, and an age of the universe of 15 ± 1 billion years.

Then in 2001 Freedman and colleagues published a value of H_0 of 72 ± 8 km/s/Mpc, which produced an expansion age of the universe of about 13 ± 1 billion years for a flat universe if $\Omega_m = 0.3$, and $\Omega_\Lambda = 0.7$. This was consistent with the best estimate of the age of the oldest globular clusters, which had, by that time, been reduced to about 12.5 billion years.

Cosmic microwave background radiation

Ralph Alpher (1921–2007) and Robert Herman (1914–1997) had suggested in 1948 that there should be radiation left over from the Big Bang present in the universe today at a temperature of about 5 K. This was accidentally discovered in 1965 by Arno Penzias (b. 1933) and Robert Wilson (b. 1936) in the microwave band with a black body temperature of about 3.5 K. The discovery marked the beginning of the end of the Steady State theory, which was virtually abandoned by the end of that decade. In 1976 David Wilkinson (1935–2002) and colleagues found an anisotropy in this cosmic microwave background radiation due to the motion of the Milky Way, but they could find no variation in the true background radiation itself over the celestial sphere.

If the microwave or cosmic background radiation is a true indication of the early universe, it should be anisotropic, as otherwise it would be difficult to see how the galaxies could subsequently have formed. The COBE (Cosmic Background Explorer) spacecraft was launched in 1989 to try to detect this structure. In the following year, John Mather (b. 1946) and colleagues found, using COBE, that the spectrum of the cosmic background radiation was that of a black body at a temperature of 2.735 K. The match with the theoretical black body curve was incredibly accurate. Then two years later, George Smoot (b. 1945) and colleagues found the hoped-for anisotropy, of just one part in 100,000. Data from the spacecraft were also used to produce a map showing the typical variability of this background radiation over the whole sky.

The WMAP (Wilkinson Microwave Anisotropy Probe) spacecraft was launched in 2001 to continue the work started by COBE. It produced a more accurate map of the variation of the cosmic background radiation over the celestial sphere, but also allowed astronomers to deduce the value of various cosmological parameters. After analysing five years of in-orbit results, Gary Hinshaw and colleagues concluded in 2008 that its results are consistent with a flat universe dominated by dark matter and dark energy. In particular, the universe is 13.7 ± 0.1 billion years old and H_0 is 72 ± 3 km/s/Mpc. In addition the universe consists of about 4% normal matter, 22% dark matter, and 74% dark energy.

Ground-based telescopes and observatories

(a) Optical telescopes and observatories

There were two basic problems in the provision of optical telescope facilities in the USA after the Second World War.

First, light pollution was becoming a problem, and second, the best facilities, particularly those on Mount Wilson and Palomar, were privately owned, and therefore not available to the majority of professional American astronomers.

The availability problem was addressed by the establishment of AURA (Association of Universities for Research in Astronomy) in 1957, funded by the National Science Foundation (NSF), with an initial membership of seven universities. AURA then set up the National Optical Observatory on Kitt Peak, at an altitude of 2,100 m. Its first medium-sized telescope, a 0.9 m, was completed in 1960. Since then a number of AURA optical telescopes have been installed on Kitt Peak, the largest of which, the 4.0 m Mayall, was completed in 1973. Also on Kitt Peak is the National Solar Observatory's McMath-Pierce solar telescope managed by AURA, which was dedicated in 1962. In addition there are a number of optical telescopes, not owned by AURA, on Kitt Peak. The first of these was the Steward Observatory's 0.9 m reflector, which was moved from Tucson in 1962 because of light and atmospheric pollution.

AURA also examined the case for building a southern observatory in Chile, following an approach by Federico Rutllant of the University of Chile in 1958. This resulted in the establishment of an AURA observatory, the Cerro Tololo Inter-American Observatory, on Cerro Tololo at an altitude of 2,200 m in the Andes. Its largest telescope to date is the 4.0 m Victor M. Blanco, which saw first light in 1974. As at Kitt Peak, there are now non-AURA telescopes at Cerro Tololo, including the 1.3 m 2MASS telescope that was used to produce a two micron sky survey starting in 1998. Its northern twin is located on Mount Hopkins in Arizona.

The initial impetus to set up an observatory on the 4,200 m high Mauna Kea, Hawaii, was to assist the local economy following a tsunami in 1960. Its first medium-sized telescope was the 2.2 m University of Hawaii instrument, completed in 1970, which was the largest telescope on the mountain for almost a decade. Then in 1979 the 3.8 m United Kingdom Infrared Telescope (UKIRT), the 3.6 m Canada-France-Hawaii (CFH) Telescope, and the 3.0 m NASA Infrared Telescope Facility (IRTF) all saw first light, taking advantage of Mauna Kea's good infrared viewing environment because of its exceptional altitude. Since then a number of very large optical/infrared, submillimetre and radio telescopes have been installed on Mauna Kea. These include the two 9.8 m Keck telescopes, the 8.3 m Japanese Subaru telescope, and the multinational 8.1 m Gemini North telescope.

Major collaborative European institutions were created after the Second World War. These included the European Iron and Steel Community, which was formed in 1951, CERN (European Organization for Nuclear Research, 1954), ELDO (European Launcher Development Organisation, April 1962), and ESRO (European Space Research Organisation, June

1962). In September 1962 European collaboration was extended to astronomical telescopes in the southern hemisphere when the convention setting up the European Southern Observatory (ESO) was initialled by Belgium, West Germany, France, the Netherlands and Sweden. The UK, on the other hand, decided to extend their Commonwealth collaboration by participating in an Anglo-Australian Observatory, and Spain decided to set up a new high altitude observatory on the Canary Islands. Over the years, the membership of the ESO has gradually increased, so it now includes 14 nations, including the UK and Spain. Today, the ESO's most important instrument is the Very Large Telescope at an altitude of 2,640 m on Cerro Paranal in Chile's Atacama desert. It consists of four 8.2 m and four 1.8 m telescopes that can be used individually or linked together to create an interferometer.

The construction and operation of very large optical/infrared telescopes has improved dramatically since the Multiple Mirror Telescope (MMT) saw first light in 1979. It used six 1.8 m telescopes on a common mount inside an observatory building that rotated with the telescope. This building had a very large front opening, allowing natural ventilation to cool the mirror to the temperature of the outside environment. This reduced thermal image distortion. The mount was altazimuth, and the mirrors were controlled by a form of active optics. Inevitably some of the innovations of the MMT worked better than others, but it helped to show the way forward. After the MMT, the UK's 4.2 m William Herschel Telescope (first light 1987) was the first large optical telescope outside the Soviet Union to use an altazimuth mount, the 2.6 m Nordic Optical Telescope (1989) was the first to use active optics, and the 3.6 m ESO's New Technology Telescope (1989) was the first to use an open building that rotated with the telescope.

Other innovations followed. The 9.8 m Keck I, which saw first light in 1992 with all its mirror segments in place, was the first large telescope to use a segmented mirror. At about the same time, the USA declassified adaptive optics, under pressure from American astronomers who saw the Europeans developing their own systems. These had already been tested at the Haute-Provence Observatory in France in 1989, where they had produced the first diffraction limited images of a faint astronomical object. These first adaptive optics systems, particularly those operating at a wavelength below 3 μm, needed a nearby bright star to operate correctly. To solve this problem, the Lick Observatory's 3 m Shane telescope was fitted in 1996 with an adaptive optics system using a laser guide star.

(b) Radio telescopes and observatories

Somewhat surprisingly, radio astronomy in the United States took some time to get into its stride after the Second World War, and concern was expressed in the United States that they would continue to fall further and further behind unless a positive approach was taken. As a result a conference was held

in 1954 jointly sponsored by the National Science Foundation, Caltech and the Department of Terrestrial Magnetism to examine the situation and propose a way forward. This resulted two years later in the foundation of the American National Radio Astronomy Observatory (NRAO) at Green Bank, West Virginia. In 1959 its first major radio telescope, the 26 m (85 ft) diameter Howard Tatel instrument, became operational. Then in the mid-1960s this dish became the fixed element in the NRAO three-element Green Bank Interferometer (GBI).

A 90 m (300 ft) transit telescope was also commissioned at Green Bank in 1962. It was covered with an aluminium mesh and so was able to operate only down to 21 cm, compared with 3.75 cm (8 GHz) for the Howard Tatel telescope. The dish collapsed in 1988 due to metal fatigue, and was replaced by a 100 × 110 m altazimuth dish, the Robert C. Byrd Green Bank Telescope (GBT). It has a computer-controlled active surface that allows operation down to 3 mm (100 GHz).

The Arecibo radio telescope facility was built in Puerto Rico for the U.S. Department of Defense as a radar facility for both military and civilian projects. It had radio astronomy as a secondary goal. When completed in 1963, the Arecibo facility had a 300 m (1,000 ft) diameter fixed dish made of wire mesh. It was designed to operate as a transit telescope at wavelengths longer than 50 cm. The mesh was replaced in the early 1970s by perforated aluminium panels, enabling it to operate down to 6 cm (5 GHz). The civilian radar system was originally designed mainly for ionospheric and planetary research, but the dish has also since been used for radio astronomy projects.

The RATAN-600 (Radio Astronomical Telescope of the Academy of Sciences), built near Zelenchukskaya in the Caucasus, was also a large transit radio telescope. It consisted of a 576 m diameter ring of 7.4 m high panels, which operated between 8 mm (~40 GHz) and 30 cm. It had about one-sixth of the collecting area of the Arecibo radio telescope. Construction began in 1968, with first light six years later.

Australian radio astronomers had designed and built a number of radio interferometers after the Second World War, but they were also keen to have a large moveable dish, like that built at Jodrell Bank in the UK. This resulted in the construction of the 64 m (210 ft) Parkes radio telescope in Australia, which began operation in 1961 with a minimum operating wavelength of 10 cm. It was the first large radio telescope dish in the southern hemisphere. Just over twenty-five years later this Parkes antenna became part of the Australia Telescope, which was a large radio interferometric array. After several upgrades to the dish and electronics, the 64 m telescope can now operate at up to 43 GHz (7 mm).

The West Germans had been restricted in their development of radio astronomy after the Second World War because of its links to radio and radar research. These restrictions were lifted in 1950, but it was not until the early 1960s that large projects could be funded. This resulted in the decision in 1966 to build

Figure 1.9. The VLA near Socorro, New Mexico. (Courtesy Dave Finley, AUI, NRAO, NSF.)

a 100 m steerable dish at Effelsberg, near Bonn, to operate at as a high a frequency as possible. When construction was completed in 1972 it was the largest such dish in the world, and it remained so for about thirty years. Initial operations were at 10.6 GHz (3 cm), but since then the dish surface has been substantially improved to allow it to operate at up to 86 GHz (3.5 mm). The telescope is now linked to European and other radio telescopes further afield in the European Very Long Baseline Interferometer (VLBI) Network managed from Dwingeloo in the Netherlands.

Radio telescopes, even those with vary large dishes, have relatively poor resolution compared with optical telescopes because of the relatively long wavelength of radio waves. Radio interferometers were built to overcome this limitation. One of the most successful has been the 27-dish Very Large Array (VLA) (see Figure 1.9), whose construction was started in 1974 at an altitude of 2,200 m near Socorro, New Mexico. The Y-shaped array was completed in 1981, allowing the simulation of apertures of up to 36 km. It initially operated at wavelengths of between 21 and 1.3 cm with angular resolutions of between 2.1 and 0.13 arcsec.

Even longer baseline radio interferometers have been used. For example, in 1966 radio telescopes at Jodrell Bank and Defford, separated by 127 km in the UK, were linked by a microwave radio link to give a resolution of about 0.025 arcsec at 6 cm. In the following year, Canadian astronomers were the first to use atomic clocks and data tape recording to extend the baseline to about 3,000 km. And by 1970 NASA had extended the baseline from Goldstone, California to Canberra, Australia, a distance of over 10,000 km, and had achieved a resolution of better than 1 milliarcsec (mas) at 13 cm.

In more recent years, the European VLBI Network (EVN) has been created to link a number of radio telescopes in Europe to some in China, South Africa and Puerto Rico. At about the same time the United States built a radio interferometer system, called the Very Long Baseline Array (VLBA), that extended about 10,000 km across the Earth's surface. It consisted of ten identical 25 m diameter dishes operating over a frequency range from 325 MHz to 43 GHz (90 cm to 7 mm). It has achieved resolutions at the high frequency end of about 0.2 mas.

The size of the Earth is not a fundamental limitation on the maximum baseline of a radio interferometer as radio telescopes on Earth can, in principle, be linked to those in space. So in 1997 Japan launched the HALCA (Highly Advanced Laboratory for Communications and Astronomy) radio astronomy spacecraft, which was specifically designed to be part of a radio inter- ferometer. It was linked with ground-based radio telescopes worldwide to achieve a resolution of 0.2 mas at 5 GHz.

A number of radio telescopes have also been built to observe at millimetre and submillimetre wavelengths. The NRAO built an 11 m (36 ft) millimetre-wave telescope on Kitt Peak in 1967, which was heavily modified in the 1980s to cover all atmospheric windows from 68 to 180 GHz (4.4 to 1.7 mm wavelength). A 30 m diameter submillimetre dish was built at Pico Veleta in Spain in the early 1980s, followed by the James Clerk Maxwell submillimetre Telescope (JCMT) and the Caltech submillimetre dish, both located on Mauna Kea. Then in the early 1990s the Heinrich Hertz SubMillimeter Telescope, which can operate up to 1,000 GHz (0.3 mm), was built on Mount Graham, Arizona.

Millimetre and submillimetre interferometers have also been built. Caltech built a millimetre-wave interferometer at Owens Valley, California in the 1980s. This has now been expanded and relocated to the high altitude Cedar Flat loca- tion. In the 1990s the SubMillimeter Array (SMA) was built on Mauna Kea near to the JCMT and Caltech submillimetre

Table 1.5. *The Space Age*

Key discoveries and observations *in addition* to those described in the text

1959	Luna 3 takes the first photographs of the far side of the Moon, showing that it is broadly similar to the visible side, but with fewer maria and more craters.
1961	Raff and Mason discover alternating stripes of high and low magnetic intensity in the floor of the eastern Pacific.
1961	Watson, Murray and Brown suggest that there may be water ice deposits in permanently shaded craters at the Moon's poles.
1963	Cox, Doell and Dalrymple find clear evidence of at least three reversals of the Earth's magnetic field in the last 4 million years.
1965	Ness and Wilcox discover the basic sector structure of the interplanetary magnetic field.
1968	Neugebauer and Westphal find that η Carinae is the brightest star in the sky at 20 μm. The heat was later found to be coming from a surrounding dust cloud.
1969	Reichley and Downs discover discontinuous changes, or glitches, in the slow down rate of the Vela pulsar.
1969	Vaiana discovers X-ray bright points on the Sun.
1969	Gubbay et al. discover apparent superluminal motion in 3C 273.
1971	Shepard and Mitchell of Apollo 14 discover rocks that imply that the Imbrium impact occurred about 3.85 billion years ago.
1971	Scott and Irwin of Apollo 15 discover a sample of original lunar crust 4.5 billion years old.
1972	OSO-7 discovers γ-ray emission from a solar flare.
1972	R. Brown discovers Io's neutral sodium cloud.
1972	The Venera 8 landing capsule finds volcanic rock on Venus.
1973	Pioneer 10 finds that the cloud tops of Jupiter's Great Red Spot are colder than those of the adjacent South Tropical Zone, indicating that it is a high-pressure system rising above the surrounding area.
1975	The Venera 9 and 10 landers transmit the first images from the surface of Venus, showing numerous rocks, their shapes indicating a relatively young surface.
1975	Larson et al. detect water vapour in Jupiter's atmosphere using the Kuiper Airborne Observatory.
1976	Pioneer 11 finds that the magnetic sectors of the interplanetary magnetic field disappear when the spacecraft is above 15° heliographic latitude.
1976	Kieffer et al. find, using the Viking spacecraft, that Mars' residual north polar cap in summer consists of water ice, not carbon dioxide ice that overlies it in winter.
1976	Lewin et al. discover the Rapid Burster, MXB 1730–335, using the SAS-3 spacecraft.
1977	AM Herculis is characterised as the first polar.
1977	Kowal discovers Chiron, an asteroid whose orbit crosses that of Saturn and almost reaches that of Uranus. Later observations showed the intensity of Chiron, which was later characterised as a Centaur, varies significantly over time.
1978	Cruikshank and Silvaggio discover methane in Triton's atmosphere.
1979	Voyager 1 images auroral emissions in Jupiter's polar regions and numerous 'superbolts' of lightning.
1979	Walsh, Carswell and Weymann discover the first gravitationally lensed quasar, 0957+561.
1980	Voyager 1 finds that Saturn's narrow F ring is three intertwined or braided rings.
1981	Voyager 2 finds numerous spiral density wave trains in Saturn's A ring.
1982	Backer et al. discover the first millisecond pulsar, PSR 1937 + 214, with a pulse rate of 1.56 milliseconds.
1983	Cruikshank, Clark and Hamilton Brown discover nitrogen on Triton.
1984	Smith and Terrile image what appears to be a disc-shaped dust cloud around β-Pictoris.
1985	Van der Klis et al. discover quasi-periodic oscillations in the galactic bulge X-ray source GX 5–1.
1986	Voyager 2 discovers two satellites, Cordelia and Ophelia, shepherding Uranus' ε ring.
1989	Voyager 2 discovers Neptune's Great Dark Spot which, like Jupiter's Great Red Spot, is a high pressure atmospheric feature.

(continued)

Table 1.5 (*continued*)

1989	Voyager 2 discovers six new satellites of Neptune with orbits inside that of Triton.
1991	The Galileo spacecraft produces the first close-up images of an asteroid, the $18 \times 11 \times 9$ km Gaspra, showing its highly irregular shape, with dozens of small craters on its surface.
1992	The Yohkoh spacecraft finds evidence of magnetic reconnection at the top of a flaring loop.
1993	The Ulysses spacecraft detects the high speed solar wind coming from the Sun's south polar coronal hole.
1994	Harch discovers the first satellite, Dactyl, of an asteroid, Ida, using the Galileo spacecraft.
1996	Rosat discovers X-rays coming from comet Hyakutake.
1997	The SOHO spacecraft discovers the Sun's magnetic carpet.
1998	Geissler et al. discover Io's aurora using the Galileo spacecraft.
1998	McEwen et al. detect high-temperature silicate volcanism on Io using the Galileo spacecraft.
1999	Butler et al. discover the first extrasolar, multiple planetary system around a Sun-like star, υ Andromedae.
1999	Charbonneau et al. detect the first extrasolar planetary transit of its parent star, HD 209458.
2000	The NEAR Shoemaker spacecraft is the first to land on an asteroid, Eros. It showed a relative scarcity of small craters on the $33 \times 13 \times 13$ km surface.
2000	Charbonneau et al. detect the first atmosphere of an extrasolar planet, the planet orbiting HD 209458.
2004	Schwartz et al. find the first observational evidence of cracks in a neutron star's crust using the Double Star and Cluster spacecraft.
2006	Tiscareno et al. discover propeller-shaped disturbances in Saturn's A ring using the Cassini spacecraft, indicating the existence of intermediate size moonlets.
2007	Retinò et al. detect magnetic reconnection in the turbulent plasma of the Earth's magnetopause using the Cluster spacecraft.
2009	Water is discovered by the Lunar Reconnaissance Orbiter/LCROSS spacecraft in the ejecta when its Centaur upper stage impacts a polar crater on the Moon.

dish. This SMA operates from 180 to 900 GHz, producing resolutions of from 0.5 to 0.1 arcsec.

Other major astronomical developments of the space age, not mentioned above, are listed in Table 1.5.

Bibliography

Leverington, David, *Babylon to Voyager and Beyond: A History of Planetary Astronomy*, Cambridge University Press, 2003.

Leverington, David, *New Cosmic Horizons: Space Astronomy from the V2 to the Hubble Space Telescope*, Cambridge University Press, 2000.

Longair, Malcolm, *The Cosmic Century: A History of Astrophysics and Cosmology*, Cambridge University Press, 2006.

McFadden, Lucy-Ann, Weissman, Paul R. and Johnson, Torrence V. (eds.), *Encyclopedia of the Solar System*, Second Edition, Academic Press, 2007.

Tassoul, Jean-Louis, and Tassoul, Monique, *A Concise History of Solar and Stellar Physics*, Princeton University Press, 2004.

Verschuur, Gerrit L., *The Invisible Universe Revealed: The Story of Radio Astronomy*, Springer, 1987.

Zirker, J.B., *An Acre of Glass: A History and Forecast of the Telescope*, Johns Hopkins University Press, 2005.

3 · International Astronomical Union

In 1904 the International Union for Cooperation in Solar Research (IUCSR) was founded, at George Ellery Hale's initiative, at the International Congress of Science held in St Louis. Six years later, the IUCSR's work was extended to include the promotion of all astrophysical studies.

The First World War interrupted international scientific cooperation, but immediately afterwards the International Research Council was founded to cover all types of scientific research. Germany, as the war's belligerent power, was specifically excluded as a potential member. The International Astronomical Union (IAU) was formed in 1919 as a member of this International Research Council, which was later extended to become the International Council of Scientific Unions. The IAU absorbed the IUCSR, the *Carte du Ciel* project, the Bureau Internationale de l'Heure and other groups. Germany was finally allowed to join the IAU in 1926.

The mission of the IAU, through its various Divisions and Commissions, is to promote astronomy through international cooperation. Every three years it holds a General Assembly, and every year it sponsors various international symposia. It is responsible for defining astronomical nomeclature and for assigning designations to celestial bodies and surface features on them. The IAU is a permanent observer on the United Nations Committee on the Peaceful Uses of Outer Space, and is an authority on light pollution. It also speaks out for radio astronomers in discussions on the international allocation of radio frequency bands. The IAU currently (2010) has 70 national members and over 10,000 individual members. It recently came to the public's attention with its decision to 'downgrade' Pluto from a planet to a dwarf planet. This was a new category of object defined by the IAU.

Bibliography

Adams, Walter S., *The History of the International Astronomical Union*, *Publications of the Astronomical Society of the Pacific*, **61** (1949), 5–12.

Blaauw, Adriaan, *History of the IAU: The Birth and First Half-Century of the International Astronomical Union*, Kluwer Academic Publishers, 1994.

PART 2
THE SOLAR SYSTEM

1 · Overview — The solar system

STRUCTURE AND SIZE OF THE SOLAR SYSTEM

Early cosmologies

Greek philosophers were the first to try to understand the structure of the universe. Inevitably some of the early ideas were wide of the mark, but in the sixth century BC the Pythagoreans appear to have been the first to appreciate that the Earth is spherical, and that the planets all move in separate orbits inclined to the celestial equator. But the Pythagorean spherical Earth did not spin and was surrounded by a series of concentric, crystalline spheres. The Moon, Sun and individual planets were each supported by a sphere, which revolved around the Earth at different speeds.

The Pythagorean, Hicetas of Syracuse, seems to have been the first to have suggested, in the fifth century BC, that the Earth spins on its axis at the centre of the universe. This model was further developed by Heracleides, who proposed that Mercury and Venus orbited the Sun, as the Sun orbited a spinning Earth. Then Aristarchus, who was also a Pythagorean, went one step further in the third century BC and proposed a heliocentric universe in which all the planets, including the Earth, orbit the Sun in the order we know today, with the Moon orbiting a spinning Earth. This was 1,700 years before Copernicus came up with the same idea. Aristarchus was also the first to produce a realistic estimate of the Earth – Moon distance, although his estimate of the Earth – Sun distance was an order of magnitude too low. Shortly afterwards, Eratosthenes of Cyrene, who was one of the last of the Pythagoreans, estimated the polar diameter of the Earth to within a few percent of its correct value by measuring the different angles of the Sun to the vertical on midsummer's day in Alexandria and Aswan.

While the Pythagoreans were developing their ideas, Plato was developing a completely different school of thought. He taught in the early fourth century BC that all heavenly bodies must be spherical, as that was the perfect shape, and they must move in uniform circular orbits around the Earth. In his cosmology, the innermost object was the Moon, followed by the Sun, Venus, Mercury, Mars, Jupiter and Saturn. Aristotle, a pupil of Plato, adopted Plato's philosophy of spherical bodies in circular orbits. Aristotle was to have a profound effect on Western thought in many areas, including logic, rhetoric, psychology and astronomy, until well into the Middle Ages. His

geocentric model of the universe was highly complex, however, requiring a total of 56 spheres to explain the motions of the Sun, Moon and planets around a non-spinning Earth. But not only was his model very complicated and inelegant, many of its predictions were wrong. Nevertheless it was taught for a number of years after his death.

Although the Babylonians and Egyptians had been aware for some time that the sidereal and tropical years were not the same, Hipparchus was, in the second century BC, the first person to quantify the difference. He concluded that the equinoxes were moving relative to the 'fixed stars', as he called them, by at least 1° per hundred years. Hipparchus was also aware that the Sun's velocity along the ecliptic was not constant, which he attributed to the Earth not being at the centre of the Sun's orbit. He also tried to estimate the parallax of the Sun and Moon by analysing solar and lunar eclipses observed from different locations. As a result he concluded that the distance of the Moon from Earth varied from 62 to $72\frac{2}{3}$ Earth radii (correct values 55 to 63 Earth radii), but he could not detect any solar parallax, as the Sun was too far away.

Apollonius of Perga appears to have been the first person to write about epicycles applied to the solar system, in the third century BC. He described a planet orbiting a point in space in an epicycle, as that point orbits the Earth in a deferent, both orbits being circular. Ptolemy developed this epicycle theory in the second century AD of the Sun, Moon and planets orbiting a non-spinning Earth. In his cosmology, the order of the bodies from the Earth was the Moon, Mercury, Venus, the Sun, Mars, Jupiter and Saturn. Because the inferior planets of Mercury and Venus each appear almost symmetrically on both sides of the Sun at maximum elongation, Ptolemy assumed that the centres of their epicycles were always on a line joining the Earth and Sun. For the superior planets he assumed that the Earth was not at the centre of their deferents and defined a point, called the equant, that he placed at an equal distance to the Earth on the opposite side of the centre. He then assumed that the centre of the planet's epicycle appeared to move at a constant velocity as seen from the equant.

Ptolemy's theory could easily explain the temporary reversal in directions observed for superior planets as seen from the Earth, and could also explain the apparent change in intensity of these planets as they moved to and from the Earth. On the other hand, it produced a much larger variation in the size of the Moon in its orbit of the Earth than observed. His theory enabled Ptolemy to produce an estimate for the size of the solar system of about 20,000 times the radius of the Earth, or about 120 million kilometres. Although this was a gross underestimate, it gave, for the first time, an idea of how large the solar system really was.

A number of Indian and Arabic astronomers developed various Greek models of the universe. For example, in the fourteenth century Ibn al-Shātir succeeded in getting rid of Ptolemy's equant. Al-Shātir's scheme had the big advantage, compared with the Ptolemaic system, of being able to explain the motion of the Moon without large apparent size variations. Interestingly, the movement of the Sun in al-Shātir's geocentric scheme was similar to that of the Earth in Copernicus' later heliocentric scheme, yet there is no evidence that Copernicus was aware of al-Shātir's work.

Nicolaus Copernicus first outlined his heliocentric model of the universe in 1512. Although his system was still based on epicycles, he used combinations of uniform circular motion without Ptolemy's equant. But, more importantly, the planets in his system orbited the Sun, rather than the Earth, and the Earth spun on its axis.

Copernicus finally described his theory in full in his *De Revolutionibus Orbium Caelestium* (*On the Revolutions of the Heavenly Spheres*), published in 1543. In an early draft he had acknowledged that his heliocentric model was not new, having first been proposed by Aristarchus almost two thousand years earlier. He also acknowledged that he was not the first person to propose a spinning Earth. But his theory was not very much simpler than that of Ptolemy, and his results were not much more accurate. In fact the key to his importance was not his model of the universe as such, but the fact that he had resurrected some older ideas which he had integrated into a consistent model. He had then published it in a persuasive book at the height of a Renaissance that was eager for new ideas.

Seventeenth century

Johannes Kepler spent a number of years observing the motion of Mars, and in 1609 showed in his *Astronomia Nova* that its orbit was an ellipse, with the Sun at one focus. Kepler also concluded that the basic circular motion of the planets around the Sun was produced by vortices generated by a rotating Sun. Magnetic forces then made the orbits elliptical. So Kepler thought that the Sun rotated on its axis, and that the planets and Sun were magnetic.

Initially, Kepler had only shown that Mars moved in an ellipse, but he later showed that this was the case for all the planets, as well as for the Moon and the satellites of Jupiter. He then listed detailed predictions for planetary positions and was the first to predict the transits of Mercury and Venus across the Sun's disc.

Galileo Galilei observed Jupiter in January 1610, and found that it had four moons that changed their positions from night to night. Then in July Galileo saw what he thought were two moons on either side of Saturn, but for some reason they did not move. Finally, later in 1610, he observed the phases of Venus, proving that Ptolemy's structure of the solar system was incorrect.

Christiaan Huygens discovered Saturn's first moon, Titan, in 1655. Four years later he announced that the two objects observed by Galileo on either side of Saturn was really a thin, flat, solid ring surrounding the planet, inclined to the ecliptic. Then in 1675 Jean Dominique (Giovanni Domenico) Cassini noticed that Saturn's ring was divided in two by a dark line, now called the Cassini division, going all the way around the planet. Cassini suggested that the two rings were not solid but composed of swarms of small satellites. James Clerk Maxwell proved this theoretically in the nineteenth century.

Isaac Newton described his universal theory of gravitation in his *Principia*, which was first published in 1687. He proved theoretically that planets and moons orbit a central body in an ellipse, with the central body at one focus, and that cometary orbits are highly elliptical or parabolic. He used Venus to 'weigh' the Sun, and used planetary moons to weigh their parent planets. So by the third edition he had deduced the masses and densities for the Earth, Jupiter and Saturn relative to the Sun.

Richer, Cassini and Picard had found evidence in 1672 that the Earth had an equatorial bulge. Newton used his new gravitational theory to calculate a theoretical value for this oblateness of 1/230 (modern value 1/298). He then considered the gravitational attraction of the Moon and Sun on the oblate Earth and calculated that the Earth's spin axis should precess at a rate of about 50.0″ per annum (modern value 50.3″), thus explaining the precession of the equinoxes. In 1672 Richer, Cassini and Picard also deduced a Sun–Earth distance of 140 million km from observations of the parallax of Mars. John Flamsteed independently deduced a similar value.

Eighteenth and nineteenth centuries

In 1678 Edmond Halley suggested that observing the transits of Venus from at least two different places on Earth, as far apart as possible, would produce the most accurate measurement of solar parallax, and hence the distance of the Earth from the Sun. The next transit of Venus took place almost one hundred years later in 1761. It was observed from many different places on Earth, but the precise timing of Venus' contacts on the Sun proved rather difficult, resulting in a range of solar parallaxes from 8.3″ to 10.6″. However, the lessons learned were invaluable in observing the next transit in 1769. This was undertaken from over 70 different sites to produce a best estimate of 8.6″. Johann Encke re-analysed this 1769 transit data in 1835 and produced a new estimate of 8.571″ ± 0.037″, equivalent to a Sun − Earth distance of 153.5 ± 0.7 × 10⁶ kilometres, which is within 2.5% of the correct value.

A number of moons had been discovered around Jupiter and Saturn by the mid-eighteenth century, and a number of new comets had been observed, but, apart from these, the known constituents of the solar system had not increased since ancient times. Then, in 1781, William Herschel discovered the first new planet, Uranus, whilst looking for double stars. Its discovery had a profound effect on the astronomical community, indicating that there may yet be more undiscovered planets in the solar system. In 1787 Herschel also discovered the first two of Uranus' satellites, Titania and Oberon. He later found that they both had an orbital inclination of about 101° 2′, making these the first satellites known to have retrograde orbits. Pierre Laplace later concluded that this angle implied that the spin axis of Uranus must be almost in the plane of its orbit around the Sun.

The next new planet, Neptune (see Figure 2.1), was discovered by Johann Galle and Heinrich d'Arrest in September 1846, based on predictions of its location by Urbain Le Verrier.

Actually, its discovery was fortuitous, as its real orbit turned out to be quite different from that predicted by Le Verrier.

Meteorites had been known since ancient times but their source was the subject of much confusion, which lasted well into the eighteenth century. Some people thought that they were atmospheric phenomena, but others thought that they had a cosmic origin. In 1794 Ernst Chladni analysed a number of meteorites and concluded that they were clearly extraterrestrial, appearing as fireballs as they descended through the Earth's atmosphere. Although doubted at the time, his theory was generally accepted in the early nineteenth century.

The first asteroid, Ceres (now redesignated as a dwarf planet), was found by Giuseppe Piazzi in January 1801. But it was not until 1845 that the fifth asteroid was found. The discovery rate then increased rapidly, so that nearly 500 asteroids were known by the end of the nineteenth century orbiting the Sun between Mars and Jupiter. As the number increased, Daniel Kirkwood noticed that there were none with certain fractional periods of Jupiter's orbital period, the so-called Kirkwood gaps, because of resonance interactions with Jupiter.

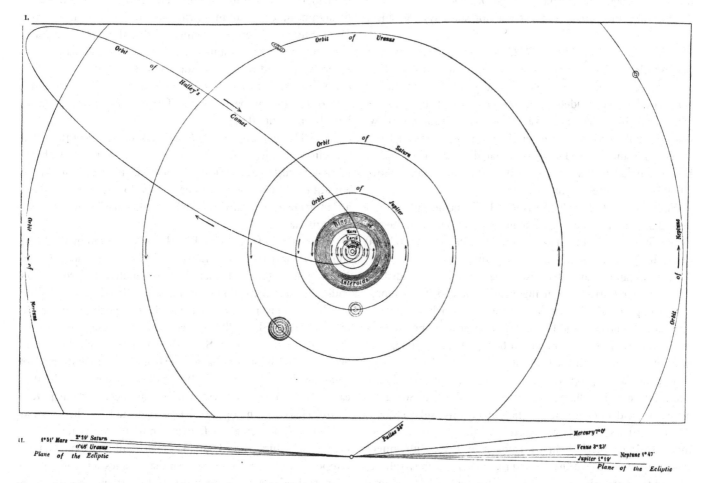

Figure 2.1. The solar system as envisaged in the second half of the nineteenth century after the discovery of Neptune and the first asteroids. (From *The Heavens*, by Amédée Guillemin, 1871, plate 1.)

Adolf Erman suggested in 1839 that both the annual Leonid and Perseid meteor showers were produced by the Earth passing through swarms of small particles orbiting the Sun and spread out along their orbit. Then in 1861 Daniel Kirkwood suggested that periodic meteors were caused by the debris of old comets whose matter had become distributed along their orbits. Shortly afterward Giovanni Schiaparelli showed that the Perseid meteor shower's orbit was the same as that of comet Swift-Tuttle. Then in 1872 there was an impressive display of meteors in place of Biela's comet which had partially disintegrated in the 1840s, confirming the link between meteor showers and decayed comets.

Twentieth century

Pluto, discovered by Clyde Tombaugh in 1930, was the first new planet to be discovered since Neptune. Pluto's orbit was found to be highly eccentric, with the largest inclination of any planet. It was initially thought to have a mass of between 1 and 0.1 M_E (mass of Earth), but this estimate has been gradually reduced over time until today it is only 0.002 M_E. Consequently, in 2006 the International Astronomical Union (IAU) downgraded Pluto to a new category of solar system object called a dwarf planet.

Ernst Öpik concluded in 1932, from an analysis of stellar perturbations, that comets could remain bound to the Sun at distances of up to 10^6 AU. Some years later Adrianus van Woerkom concluded, from the observed distribution of cometary orbits, that there was a continuous source of new, near-parabolic comets. Then in 1950 Jan Oort extended van Woerkom's analysis and suggested that all long period comets had originated in what is now called the Oort cloud, about 50,000 to 150,000 AU from the Sun.

So by the start of the Space Age in 1957 one new planet had been discovered in the solar system. In addition, the influence of the Sun seemed to extend to the Oort cloud, about 1,000 times further away than this remote planet.

In 1958 James Van Allen discovered variations in radiation levels in near-Earth space using experiments on the Explorer 1 and 3 spacecraft. He attributed these to a belt of elementary particles around the Earth. Later spacecraft showed that there were basically two radiation belts, whose positions and intensities varied with solar activity.

Eugene Parker proposed his theory of the solar wind in the late 1950s in which charged particles emitted by the Sun drew magnetic field lines in the corona out into the solar system. These particles were first detected outside the influence of the Earth's magnetic field by Luna 2 in 1959. Subsequent spacecraft showed that there was a region (the magnetosphere) around the Earth that was dominated by the Earth's magnetic field. Further out was the bow shock, which was first detected by the IMP-1 spacecraft in 1963, where the Earth's magnetic field loses its influence. Later spacecraft showed that there were also magnetospheres around other planets.

This work showed that the space between the planets was not just occupied by asteroids and comets but it was permeated by the solar wind. In addition, planetary magnetic fields modified the solar wind in their vicinity.

James Elliot, Robert Millis and colleagues observed a stellar occultation by Uranus in 1977 which fortuitously led to the discovery of nine narrow rings around the planet. Then two years later the Voyager 1 spacecraft found a thin ring around Jupiter. A number of stellar occultations by Neptune were subsequently observed by astronomers hoping to find evidence of rings. They produced ambiguous results, however, with some of the detected intensity dips being attributed to unknown satellites and some to partial rings or ring arcs. Voyager 2 clarified matters in 1989 when it imaged five rings around Neptune. So no longer was Saturn the only planet in the solar system to have a ring system, all of the other three large planets had one.

Daniel Barringer had concluded in 1903 that a 1.2 km diameter crater in Arizona had been produced by the impact of a large metallic meteorite. In spite of drilling to a depth of 420 m no significant pieces of meteorite were found. But in the early 1960s Eugene Shoemaker confirmed that the Barringer crater had been produced by a meteorite when he found samples of severely shocked rock. This was the first definite proof of an extraterrestrial impact on the Earth's surface. The nickel-iron meteorite which impacted about 50,000 years ago is thought to have been about 50 m across.

In 1982 a meteorite, ALH 81005, discovered near the Allen Hills in Antarctica, was found – by analysing its mineral, chemical and isotopic composition – to be of lunar origin. Since then a number of lunar meteorites have been found, most of which are anorthositic regolith breccias from the lunar highlands, with a few maria basalts.

Shoemaker, Hackman and Eggleton suggested in 1962 that some meteorites found on Earth could have been ejected from Mars by an asteroid impact. No such meteorites from Mars could be identified, however, until in 1983 Donald Bogard and Pratt Johnson discovered that the composition of the gas trapped in the EET 79001 meteorite closely resembled that of the Martian atmosphere. Since then a number of other meteorites have been found to have come from Mars, including ALH 84001, which was thought in 1996 to have evidence within it of early life on Mars. That evidence has since been hotly disputed.

In 1943 Kenneth Edgeworth had suggested that there should be many planetesimals, left over from the formation of the planets, orbiting the Sun outside the orbit of Neptune. Gerard Kuiper independently made a similar suggestion eight years later. In 1992 David Jewitt and Jane Luu found the first such trans-Neptunian or Kuiper belt object. Since then over 1,000 of these trans-Neptunian objects have been found, some of them, like Pluto, locked in a 2:3 resonance with Neptune. These so-called Plutinos basically define the inner edge of the Kuiper belt.

David Jewitt and colleagues discovered an object in 1996 with a perihelion distance of 35 AU and aphelion of 132 AU, giving it an orbital period of about 760 years. Its orbit was inclined at about 24° to the ecliptic. Since then, many more of these scattered disk objects have been found, with very large and highly eccentric orbits, generally outside the classical Kuiper belt. It appears as though these scattered disk objects have been scattered out of their original orbits by gravitational interactions with Neptune.

In 2002 a classical Kuiper belt object, now called Quaoar, which has a diameter of about 1,150 km, was discovered by Chad Trujillo and Mike Brown. Then, in the following year, Sedna was discovered with a diameter of about 1,600 km. It was in a highly eccentric orbit with an aphelion of about 940 AU.

Subsequently, a number of other large objects have been discovered by a team led by Mike Brown, including Eris, a scattered disk object with a diameter of about 2,400 km, which is about that of Pluto. Evidently, more similar sized or larger objects may exist in the outer solar system. As a result, as mentioned above, in 2006 the IAU devised a new class of object, called a dwarf planet. The initial list of such objects included both Pluto and Eris.

The major components of the solar system are listed in Table 2.1.

Bibliography

Caspar, Max, *Kepler*, Abelard-Schuman, London, 1959 (Dover reprint, 1993).

Drake, Stillman, *Galileo At Work; His Scientific Biography*, University of Chicago Press, 1978 (Dover reprint, 1995).

Clerke, Agnes, *A Popular History of Astronomy during the Nineteenth Century*, 4th ed., Adam and Charles Black, 1902.

Dreyer, J. L. E., *History of the Planetary Systems from Thales to Kepler*, Cambridge University Press, 1906, republished as *A History of Astronomy from Thales to Kepler*, Dover, 1953.

Gingerich, Owen, *The Eye of Heaven: Ptolemy, Copernicus, Kepler*, American Institute of Physics, 1993.

Hall, A. Rupert, *From Galileo to Newton, 1630–1720*, Harper & Row, 1963, republished as *From Galileo to Newton*, Dover, 1981.

Hall, A. Rupert, *Isaac Newton: Adventurer in Thought*, Cambridge University Press, 1996.

Heath, Sir Thomas, *Aristarchus of Samos, the Ancient Copernicus, A History of Greek Astronomy to Aristarchus Together with Aristarchus's Treatise on the Sizes and Distances of the Sun and Moon, A New Greek Text with Translation and Notes*, Clarendon Press, Oxford, 1913, republished as *Aristarchus of Samos: The Ancient Copernicus*, Dover, 1981.

Heath, Sir Thomas L., *Greek Astronomy*, Dent & Sons, 1932 (Dover reprint, 1991).

Koyré, Alexandre, *The Astronomical Revolution: Copernicus-Kepler-Borelli*, Cornell University Press, 1973 (Dover reprint, 1992).

Leverington, David, *Babylon to Voyager and Beyond: A History of Planetary Astronomy*, Cambridge University Press, 2003.

Machamer, Peter (ed.), *The Cambridge Companion to Galileo*, Cambridge University Press, 1998.

McFadden, Lucy-Ann, Weissman, Paul R. and Johnson, Torrence V. (eds.), *Encyclopedia of the Solar System*, 2nd ed., Academic Press, 2007.

Pannekoek, Anton, *A History of Astronomy*, Allen & Unwin, 1961 (Dover reprint, 1989).

Small, Robert, and Stahlman, William D., *An Account of the Astronomical Discoveries of Kepler*, University of Wisconsin Press, 1963.

Taton, René, and Wilson, Curtis (eds.), *Planetary Astronomy from the Renaissance to the Rise of Astrophysics; Part A: Tycho Brahe to Newton*, Cambridge University Press, 1989, and *Part B: The Eighteenth and Nineteenth Centuries*, Cambridge University Press, 1995.

ORIGIN OF THE SOLAR SYSTEM

Immanuel Kant suggested in 1755 that the solar system had condensed out of a nebulous mass of gas which had developed into a flat rotating disc as it contracted. This disc rotated faster and faster as it continued to contract, and, eventually, it started throwing off masses of gas as it became unstable. These masses cooled to form the planets, and the core of the disc condensed to form the Sun. A little later Georges-Louis Leclerc, Compte de Buffon proposed an alternative in which matter was dragged out of the Sun by a passing comet. This matter later condensed to form the planets.

Pierre Simon, Marquis de Laplace first proposed his theory of the origin of the solar system in his *Exposition du Système du Monde*, first published in 1796. In it he considered a rotating Sun surrounded by an extensive atmosphere larger than the current solar system. This atmosphere gradually contracted as it cooled, and the matter near its outer edge separated to form a ring, which eventually coalesced to form the outermost planet. This process continued, with the solar atmosphere throwing off ring after ring as it contracted and speeded up, thus producing rings that condensed to produce the other planets. Laplace suggested that the planetary satellites formed in an analogous way, from condensing rings of material, around each of the proto-planets, which had been thrown off as they cooled, contracted and speeded up.

Laplace expanded this theory in later editions of his book, encouraged by William Herschel's detailed observations of distant nebulae, which Herschel thought were stars at various stages of formation. Some of the nebulae observed by Herschel were almost structureless masses of very tenuous material,

Table 2.1. *The solar system*

(A) The planets

(a) Orbits

	Discovery	Semimajor axis (AU)	Orbital period (years)	Orbital eccentricity	Orbital inclination to ecliptic
Mercury	Prehistoric	0.39	0.24 (88.0 days)	0.206	7.00°
Venus	Prehistoric	0.72	0.62 (224.7 days)	0.007	3.39°
Earth		1.00	1.00 (365.2 days)	0.017	0.00°
Mars	Prehistoric	1.52	1.88	0.093	1.85°
Jupiter	Prehistoric	5.20	11.86	0.048	1.31°
Saturn	Prehistoric	9.54	29.46	0.054	2.49°
Uranus	1781; William Herschel	19.19	84.02	0.047	0.77°
Neptune	1846; Galle, d'Arrest, Le Verrier	30.07	164.78	0.009	1.77°
Pluto[a]	1930; Tombaugh	39.48	248.4	0.249	17.14°

[a] Since 2006 called a dwarf planet.

(b) Physical parameters

	Equatorial radius (km)	Equatorial radius (Earth = 1)	Mean density (g/cm³)	Sidereal rotation period	Inclination of equator to orbit
Mercury	2,439	0.38	5.43	58.65 d	0.1°
Venus	6,051	0.95	5.24	243.0 d	177.4°
Earth	6,375	1.00	5.52	23.93 d	23.4°
Mars	3,395	0.53	3.94	24.62 d	25.2°
Jupiter	71,500	11.21	1.33	9.93 h	3.1°
Saturn	60,300	9.45	0.70	10.66 h	26.7°
Uranus	25,560	94.01	1.30	17.24 h	97.9°
Neptune	24,760	3.88	1.76	16.11 h	28.8°
Pluto	1,160	0.18	2.0	6.39 d	119.6°

(c) Magnetism and magnetospheres

	Magnetic field at equator (gauss)	Dipole moment (Earth = 1)	Angle between magnetic and spin axis	Dipole offset (radii)	Average sunward distance of magnetopause (planetary radii)
Mercury	3×10^{-3}	5×10^{-4}	14°		1.5
Venus	$< 2 \times 10^{-5}$	$< 4 \times 10^{-4}$	–	–	–
Earth	0.31	1	11.0°	0.07	10.4
Mars	$< 3 \times 10^{-4}$	$< 2 \times 10^{-4}$	–	–	–
Jupiter	4.28	19,000	9.6°	0.10	70
Saturn	0.22	580	0°	0.04	19
Uranus	0.25	50	58.6°	0.30	18
Neptune	0.14	28	47.0°	0.55	24
Pluto	?				

(B) Main planetary satellites

Earth

Name	Year of discovery	Discovered by	Semimajor axis (10^3 km)	Orbital inclination	Mean radius (km)
Moon			384	18.3° − 28.6°	1,738

Mars

	Name	Year of discovery	Discovered by	Semimajor axis (10^3 km)	Orbital inclination	Mean radius (km)
M1	Phobos	1877	Hall	9.38	1.1°	11
M2	Deimos	1877	Hall	23.46	1.8°	6.6

Jupiter

	Name	Year of discovery	Discovered by	Semimajor axis (10^3 km)	Orbital inclination	Mean radius (km)
J5	Amalthea	1892	Barnard	181	0°	84
J1	Io	1610	Galileo	422	0°	1822
J2	Europa	1610	Galileo	671	0°	1561
J3	Ganymede	1610	Galileo	1,070	0°	2631
J4	Callisto	1610	Galileo	1,883	0°	2410

Note: For a more complete list, see Jupiter article.

Saturn

	Name	Year of discovery	Discovered by	Semimajor axis (10^3 km)	Orbital inclination)	Mean radius (km)
S1	Mimas	1789	Herschel	186	2°	199
S2	Enceladus	1789	Herschel	238	0°	252
S3	Tethys	1684	Cassini	295	1°	536
S4	Dione	1684	Cassini	377	0°	563
S5	Rhea	1672	Cassini	527	0°	765
S6	Titan	1655	Huygens	1,222	0°	2,576
S7	Hyperion	1848	W. C. & G. P. Bond and Lassell	1,481	1°	140
S8	Iapetus	1671	Cassini	3,561	8°	735
S9	Phoebe	1898	Pickering	12,948	175°	110

Note: For a more complete list, see Saturn article.

Uranus

	Name	Year of discovery	Discovered by	Semimajor axis (10^3 km)	Orbital inclination	Mean radius (km)
U5	Miranda	1948	Kuiper	130	4°	236
U1	Ariel	1851	Lassell	191	0°	579
U2	Umbriel	1851	Lassell	266	0°	585
U3	Titania	1787	Herschel	436	0°	789
U4	Oberon	1787	Herschel	584	0°	761

Note: For a more complete list, see Uranus article.

Table 2.1 (*continued*)

Neptune

	Name	Year of discovery	Discovered by	Semimajor axis (10^3 km)	Orbital inclination	Mean radius (km)
N1	Triton	1846	Lassell	355	157°	1,353
N2	Nereid	1949	Kuiper	5,513	27°	170

Note: For a more complete list, see Neptune article.

Pluto

	Name	Year of discovery	Discovered by	Semimajor axis (10^3 km)	Orbital inclination	Mean radius (km)
P1	Charon	1978	Christy	19.4	0°	593

(C) Saturn's rings

Ring component	Radial location (km)	Discovered by	Date
(edge of planet)	*(60,300)*		
C	74,500– 92,000	W. C. & G. P. Bond and W.R. Dawes	1850
B	92,000–117,580	Galileo and Huygens	1610 and 1659
Cassini division	117,580–122,200	Cassini	1675
A	122,200–136,780	Galileo and Huygens	1610 and 1659
Encke gap	133,570	Encke	1837
Keeler gap	136,530		

Notes: (a) For a more complete list of Saturn's rings, see Saturn article. (b) For details of the rings around other planets, see the tables attached to their individual articles.

(D) Selected asteroids

Number	Name	Year of discovery	Discovered by	Mean radius (km)	Semimajor axis (AU)	Notes
1	Ceres	1801	Piazzi	470	2.767	
2	Pallas	1802	Olbers	270	2.771	
3	Juno	1804	Harding	134	2.670	
4	Vesta	1807	Olbers	270	2.361	
5	Astrea	1845	Hencke	60	2.573	
6	Hebe	1847	Hencke	93	2.425	
8	Flora	1847	Hind	64	0.156	Head of an asteroid family
433	Eros	1898	Witt and Charlois	12	1.458	First asteroid known to have orbit that comes close to Earth

Note: Ceres is now classified as a dwarf planet.

(E) Selected Trans-Neptunian Objects

Name	Year of discovery	Discovered by	Mean radius (km)	Semi-major axis (AU)	Orbital eccentricity	Orbital period (years)	Comments
1992 QB$_1$	1992	Jewitt and Luu	80	44.0	0.072	292	First Kuiper Belt Object (KBO) to be discovered
1993 RO	1993	Jewitt and Luu	45	39.3	0.200	246	First Resonance KBO to be discovered
1996 TL$_{66}$	1996	Luu and Jewitt	300	83.3	0.583	760	First Scattered Disk Object (SDO) to be discovered

(a) Large Trans-Neptunian Objects

Name	Year of discovery	Discovered by	Mean radius (km)	Semi-major axis (AU)	Orbital eccentricity	Orbital period (years)	Comments
Eris	2005	Brown et al.	1,200	67.7	0.442	557	SDO
Pluto	1930	Tombaugh	1,160	39.5	0.249	248	Resonance KBO
Sedna	2003	Brown et al.	800	486	0.853	10,700	Detached object
Makemake	2005	Brown et al.	750	45.8	0.159	310	Classical KBO
Haumea	2004	Brown et al.	600	43.3	0.195	285	Classical KBO
Quaoar	2002	Trujillo and Brown	570	43.6	0.039	288	Classical KBO

Note: Eris, Pluto, Makemake and Haumea are classified as dwarf planets.

(F) Selected comets

Comet	Year first seen	Perihelion distance (AU)	Orbital period (years)	Orbital inclination
Halley	240 BC	0.587 (in 1986)	76.1	162.2°
Great 1668 comet	1668	0.066	*	144°
Great 1680 comet	1680	0.006	~ 9,000	60°
Lexell	1770	0.68	5.6	1.6°
Biela	1772	0.861	6.6	12.6°
Encke	1786	0.333	3.3	11.9°

* = virtually infinite

(G) Sources of selected meteor showers

Shower	Peak of annual shower (approx.) as of 2010	Parent object	Comet or asteroid period (years)	Comet or asteroid perihelion distance (AU)	Link discovered by, with year
Perseids	12 August	Comet Swift-Tuttle	133	0.960	Schiaparelli, 1866
Leonids	17 November	Comet Tempel-Tuttle	33.3	0.977	Peters, 1867

whereas others showed clear signs of a central condensation, and yet others looked like stars surrounded by a very tenuous envelope. Laplace saw all these as examples of stars, like the Sun, gradually condensing from nebulae.

In these later editions Laplace explained that the four asteroids known at the time were the result of an incomplete process of condensation to form a planet. In addition, Saturn's rings were the remains of the original Saturn nebula that had not condensed to form satellites. In 1833 William Whewell appears to have been the first to refer to Laplace's theory, linked, as it was by then, to Herschel's nebulae observations, as the 'nebula hypothesis'.

Laplace had suggested that the planets had condensed from the Sun's extended atmosphere. But he had also suggested that Herschel's nebulae, some with and some without central condensations, were examples of the early stages of formation of stars like the Sun. So, initially, for the solar system there had just been a revolving primeval nebula, from which the Sun had formed, with the remainder of the nebula attracted to the Sun as the Sun's extended atmosphere. The angular momentum of the primeval nebula would have been very similar to that of the current solar system, as angular momentum must have been conserved. But in 1861 Jacques Babinet showed that, knowing the current angular momentum of the solar system, the primeval nebula could not have had enough angular momentum to cause it to spin off material as it contracted. And even if the rings had been spun off as Laplace had proposed, James Clerk Maxwell showed that they would have been basically stable. So they would probably not have condensed to form planets. It was also difficult to explain how the Sun, with about 99.9% of the solar system's mass, could have retained only about 2% of the solar system's angular momentum.

Other objections were raised to Laplace's nebula hypothesis in the nineteenth century. Nevertheless, it remained popular in various modified forms, in the absence of a convincing alternative, even though it gradually became clear that Herschel's nebulae were not stars in the early stages of formation.

In 1880 Alexander Bickerton suggested, as an alternative to the nebula hypothesis, that the solar system had been formed when another star had a glancing collision with the Sun. W. F. Sedgwick suggested in 1898 that the star didn't need to hit the Sun, as it could have drawn out a large amount of material by tidal attraction if it had passed close to the Sun. James Jeans put forward a similar idea a little later, and worked out the mathematics in 1916. He showed how a long tongue of gas, that had been pulled out of the Sun, would break up into individual gas clouds, which would condense to form the planets. Jeans calculated that such close encounters between two stars would be extremely rare, so that only about one star in 100,000 would today have a planetary system. This was quite a change from the nineteenth-century view where

planets were thought to be present around the majority of stars.

Thomas Chamberlin and Forest Moulton proposed an alternative star encounter theory in 1905, based on the idea that the Sun originally had much larger prominences than now. They suggested that these prominences had been enormously amplified by a passing star, which caused the Sun to eject a great number of small clouds of gas. These cooled to form small solid bodies that they called 'planetesimals', which in turn cooled and coalesced to form the planets. Unfortunately, Jeans showed that the small clouds of gas would have dissipated before they had had time to cool to form the planetesimals.

The rapid explosion of similar encounter theories in the early part of the twentieth century gradually fizzled out, as every new theory was found to have serious drawbacks. For example, Lyman Spitzer analysed Jeans' close encounter theory in 1939, and showed that the filament of gas drawn out of the Sun would not have condensed, but would have formed a permanent gaseous nebula surrounding the Sun.

Condensing nebulae re-examined

In the early twentieth century it became clear that nebulae were often turbulent, with streams of gas moving at velocities of the order of 5 to 10 km/s. Such nebulae would have a net angular momentum, even though they did not appear to be rotating as a whole. William McCrea showed that if a dense nebula with streams of gas had condensed to form the solar system, it could have had more than enough angular momentum to produce the solar system that we see today.

Carl von Weizsäcker suggested in 1943 that cells of circulating convection currents, or vortices, formed in the solar nebula as the Sun had condensed. These vortices then caused small chunks of material to condense into planetesimals in the regions bordered by vortices rotating in opposite senses. The planetesimals then grew to form the planets by accretion. The planetary satellites were formed analogously from small nebulae surrounding each of the planets. Von Weizsäcker showed that the turbulence in his system removed angular momentum from the Sun, solving one of the problems with most other nebula theories. Unfortunately, shortly afterwards Subrahmanyan Chandrasekhar and Gerard Kuiper showed that the vortices would not have been stable enough to allow planetesimal condensations to take place.

Starting in 1948, Gerard Kuiper developed a theory that had its origins in his earlier work with Otto Struve on double stars. Kuiper pointed out that double and multiple star systems are not unusual but are rather common, and suggested that planetary systems would be found around those stars whose initial nebula was not large enough to form a companion star. In Kuiper's theory the eight protoplanets formed in a region of gravitational instability in the solar nebula. They eventually became planets by accretion from the nebula remaining after the Sun had formed.

Kuiper examined the statistics of the condensing nebula and concluded that any condensing nebula must originally have had, after the Sun had condensed, about 100 times more mass, largely in the form of hydrogen and helium, than that currently residing in the planets; otherwise, it would have dispersed rather than condensed. So only about 1% of the nebula condensed to form the planets, the remainder being gradually lost to the solar system.

Harold Urey used the planetesimal theory in the 1950s to explain the depletion of heavy inert gases on Earth compared with the Sun. He reasoned that if the Earth had formed directly from the same nebula as the Sun, as Kuiper proposed, the Earth's gravity would have been sufficient to retain heavy inert gases. But if the solar nebula had produced only relatively small planetesimals, their gravity would have been insufficient to retain these gases. By the time these planetesimals had grown sufficiently to be able to retain the heavy inert gases, those gases would have largely disappeared.

In parallel with Urey's work, Viktor Safronov investigated how planetesimals could aggregate to produce planets. Safronov showed how planets could form with almost circular orbits, provided they were formed by numerous collisions between small bodies. So, although there was not, in the late 1950s, a generally accepted theory of the origin of the solar system, the concept of planets being built by the collision and merging of planetesimals appeared to be the most promising.

All early work on the origin of the solar system had been undertaken with little observational evidence from other stars to assist the astronomers in their theories and speculations. But that changed in the latter part of the twentieth century when observations of other stars began to help our understanding of how the solar system was probably formed. For example, in the 1980s T Tauri stars and various other stars were found to be surrounded by dust clouds, and in a number of cases the structure of the dust clouds and the size of the dust particles could be determined. In the case of Vega, for example, the dust particles were at least 1,000 times larger than the size of interstellar dust grains. In that case it was thought that we could be observing a cloud of dust which may be in the process of forming planets. Improved infrared detectors also enabled a number of protostars to be detected and examined.

In the nineteenth century Henry Sorby had undertaken a detailed analysis of meteorites, concluding that they may have been formed out of the same nebula that had produced the Sun and planets. Further analysis of meteorites in the twentieth century had helped to improve our knowledge of the early solar system. In particular, the analysis of chondrules in chondrite meteorites showed that the chondrules had been formed at a temperature of at least 1,600 K in the inner solar system.

The discovery of planets around other stars in the late twentieth and early twenty-first centuries has also helped astronomers to understand how solar systems can be configured. However, in spite of all this new information on T Tauri stars, meteorites, and other solar systems, there were still many questions outstanding in trying to derive a satisfactory theory of the origin of our solar system.

The currently favoured theory describes our solar system as being formed from a large molecular cloud at a temperature of the order of 10 to 50 K. As time progressed, the initially small variations in gas density throughout the cloud resulted in gas concentrations that became favoured sites of gravitational collapse, with one major centre of collapse. The gravitational attraction of this major centre, or cloud core, in the early solar system eventually overwhelmed the resistance caused by the magnetic fields in the molecular cloud. The cloud material then cascaded onto the cloud core, turning it into a protostar. As a result, the protostar's pressure and temperature increased, eventually causing it to radiate a significant amount of energy.

Initially the random motion of the gas molecules in the cloud would not have completely cancelled each other out, and there would have been a net angular momentum of the cloud. As more and more of the gas was drawn into the protostar, however, it and its surrounding gas cloud would have started to spin faster and faster. As a result, the cloud particles, as well as the local concentrations within the cloud, would eventually form a disk of material orbiting the central, major concentration or protostar.

Details of the above processes are not completely understood but very young stars are known to be surrounded by large dust clouds. In addition, radio telescopes have detected strong winds of material blowing away from protostars in two opposite directions. These bipolar outflows eventually clear the cloud material from over both poles of the protostar. So, when the star is about 1 million years old (for a sunlike star), there is a disk of material flowing in towards the star in the equatorial regions, together with gas- and dust-free regions over each pole. The star remains in this so-called T Tauri stage for about 10 million years, gradually accumulating material and becoming hotter, before its core temperature is high enough for hydrogen burning to commence.

So it appears as though the solar nebula, just before the creation of the planets, would have been a disk of gas and dust whose temperature, density and velocity were largest nearest the Sun. In the inner region of the nebula, it would have been too warm for water to condense as ice, and so the planets that formed there – the terrestrial planets – consist of silicate material and other elements like magnesium and iron, the so-called refractory elements, that condense into solids at high temperatures. In the middle region of the nebula, water ice would have been stable, and the large outer planets of Jupiter, Saturn, Uranus and Neptune were formed, possibly around a nucleus of water ice. Finally, the density of material in the outer part of the nebula was too small for sizeable planets to be produced. Instead, the planetesimals in that region would have remained small, as the forerunners of the comets and asteroids of the Kuiper belt.

Bibliography

Brush, Stephen, *Fruitful Encounters: The Origin of the Solar System and of the Moon from Chamberlin to Apollo*, Cambridge University Press, 1996.

Brush, Stephen, *Nebulous Earth: The Origin of the Solar System and the Core of the Earth from Laplace to Jeffreys*, Cambridge University Press, 1996.

Doel, Ronald E., *Solar System Astronomy in America: Communities, Patronage, and Interdisciplinary Science, 1920–1960*, Cambridge University Press, 1996.

Jeans, Sir James, *The Universe Around Us*, 4th ed., rev., Cambridge University Press, 1960.

Laplace, M. Le Marquis, (Harte, H. H., trans.), *The System of the World*, Longman and Co., 1830.

Laplace, P. S., (Pond, J., trans.), *The System of the World*, Richard Phillips, London, 1809.

Urey, Harold C., *The Planets: Their Origin and Development*, Yale University Press, 1952.

Wood, John A., *Origin of the Solar System in The New Solar System*, 4th ed., Beatty, Petersen & Chaikin (eds.), Sky Publishing and Cambridge University Press, 1999.

Woolfson, Michael M., *The Origin and Evolution of the Solar System*, Taylor and Francis, 2000.

2 · Sun, Earth and Moon

SUN AND SOLAR WIND

Aristotle had taught in the fourth century BC that the Sun was a perfect body. Naked-eye sunspots had been seen from time to time in antiquity, but most people took them to be something between the Earth and Sun. Thomas Harriot and Galileo Galilei had both seen sunspots telescopically in 1610, but it was Johann Fabricius who first published his results in June 1611. He correctly concluded that sunspots were on the surface of the Sun, and that their movement indicated that the Sun was rotating.

Sunspots (see Figure 2.2) were still an enigma by the nineteenth century. Many astronomers thought that they were holes in the photosphere, but were puzzled because they were dark, rather than bright, as they should have been if they revealed the hotter lower layers of the Sun. Then, in 1872, Angelo Secchi suggested that matter was ejected from the surface of the Sun at the edges of a sunspot. This matter then cooled and fell back into the centre of the spot, so producing its dark central region. Just over thirty years later, Alfred Fowler and George Ellery Hale independently showed, using spectroscopy, that sunspots are cooler than the surrounding Sun.

Heinrich Schwabe had found in 1843 that the number of sunspots varied with a period of about 10 years. Rudolf Wolf then showed, by looking back though historical records, that the period varied from 7 to 17 years, with an average of about 11 years. Then in 1852, Sabine, Wolf and Gautier independently concluded that there was a clear correlation between the appearance of large sunspots and disturbances in the Earth's magnetic field.

In 1858 Richard Carrington discovered that the latitude of sunspots changed over a solar cycle starting at about 30° to 40° latitude immediately after sunspot minimum, and gradually reducing as the cycle progressed. Shortly afterwards he found that sunspots near the solar equator moved faster than those at higher latitudes, showing that the Sun did not rotate as a rigid body. This differential rotation of the Sun was interpreted by Secchi as indicating that the Sun was gaseous. Then in 1859 Carrington and Richard Hodgson independently observed two white light solar flares moving over the surface of a large sunspot. This was followed by a major geomagnetic storm on Earth that peaked about 36 hours later.

Gustav Kirchhoff published the first analysis of thousands of Fraunhofer absorption lines in the solar spectrum in 1861, recognising the lines of sodium and iron. By the end of the century about 40 different elements had been discovered in the solar spectrum, showing that there were the same elements on the Sun as on the Earth.

Solar prominences had been observed during a total solar eclipse in 1733, but it was not until 1860 that they were found to be connected with the Sun, rather than the Moon. Then in 1868 Jules Janssen, Norman Lockyer and others found that they consisted of hydrogen and an unknown element that Lockyer called helium. Helium was found on Earth by William Ramsay in 1895.

Estimates of the surface temperature of the Sun in the nineteenth century were severely hampered by the lack of understanding of the laws of radiation. The temperature was estimated in the 1880s as $\sim$10,000 K using the Stefan-Boltzmann law of radiation. Then in 1893 Wilhelm Wien found that the wavelength of peak radiation emitted by a black body was inversely proportional to its absolute temperature. This enabled the effective temperature of the Sun to be estimated as about 6,000 K.

It had been generally thought in the nineteenth century that heat was transported from the interior of the Sun by convection. But Ralph Sampson suggested in 1894 that the primary mechanism for heat transfer through the solar atmosphere was by radiation. Twenty years later, Arthur Eddington extended this idea of radiative transfer to the whole Sun, and in 1926 used the concept of radiative equilibrium to calculate a central temperature of about 39 million K. At about the same time, Cecilia Payne showed that hydrogen and helium were the most abundant elements in stellar atmospheres. In 1932 Eddington assumed similar abundances for the solar interior, and shortly afterwards reduced his estimate for the central temperature of the Sun to 19 million K.

In these calculations Eddington made no assumption on how the Sun's heat was produced, which was still unknown at the time. In 1938, however, Charles Critchfield explained how energy could be produced at very high temperatures by a chain reaction starting with proton-proton collisions and ending with the synthesis of helium nuclei. Hans Bethe then collaborated with Critchfield to develop this idea. But Bethe also examined an alternative mechanism that relied on carbon as

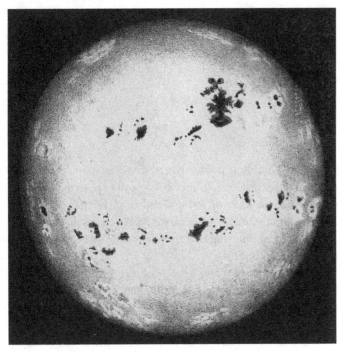

Figure 2.2. Drawing of the Sun made in 1870 by Tacchini at Palermo showing sunspots and faculae. (From *Memoirs of the Italian Spectroscopical Society*, Vol. vi.)

a catalyst to produce helium from hydrogen, in the so-called carbon cycle. Carl von Weizsäcker independently developed this same scheme. Which mechanism was predominant in the Sun depended crucially on temperature, and it was not until the 1950s that it became clear that it was the proton–proton chain.

In the nineteenth century, the Sun's corona had been found to have a faint continuous spectrum crossed by a number of bright emission lines. Of particular interest was a bright green emission line that had been found by Charles Young and William Harkness during the total solar eclipse of 1869. It was originally attributed to iron, but this was later found to be of a slightly different wavelength. No other element was found to produce a line of the required wavelength, so the coronal line was attributed to a new element called coronium.

At that time it was assumed that the temperature of the Sun and its corona gradually reduced with distance from the centre of the Sun. But in 1934 Walter Grotrian analysed the coronal spectrum and concluded that the temperature was an astonishing 350,000 K. A few years later Bengt Edlén, in a seminal paper, showed that the coronal emission lines are produced by highly ionized iron, calcium and nickel at a temperature of over a million K. The 'coronium' line, in particular, was due to highly ionized iron. It was not clear how the temperature of the corona could be so high, when that of the photosphere was only about 6,000 K.

Charles Young had discovered in 1894 that, at very high dispersions, many absorption lines in sunspot spectra appeared to have a sharp bright line in their centres. In 1908 Hale and Walter Adams found that photographs of the Sun taken in the light of the 656.3 nm hydrogen line showed magnetic field-like patterns. Consequently Hale decided to examine sunspot spectra in more detail. As a result he found that the Young effect was caused by Zeeman splitting of spectral lines in a magnetic field, which was of the order of 3,000 gauss. So sunspots exhibited very high magnetic fields.

Hale then started to examine the polarities of sunspots, and found that spots generally occur in pairs, with the polarity of the lead spot, as they crossed the disc, being different in the two hemispheres. This pattern was well established by 1912 when the polarities reversed at solar minimum. They reversed again at the next solar minimum in 1923, showing that the solar cycle was really about 22 years, not 11.

Walter Maunder had found in 1913 that large magnetic storms on Earth start about 30 hours after a large sunspot crossed the centre of the solar disc. Later work showed that the most intense storms were often associated with solar flares. In 1927 Charles Chree and James Stagg found that smaller storms, which did not seem to be associated with sunspots, tended to recur at the Sun's synodic period of 27 days. Julius Bartels called the invisible source on the Sun of these smaller storms, M regions. Both the so-called 'flare storms' and the 'M storms' were assumed to be caused by particles ejected from the Sun.

Post-second world war developments

In 1942 James Hey and George Southworth independently discovered that the Sun was a radio source. Hey detected intermittent radio emissions in the metre waveband, which appeared to be caused by an active sunspot group. Southworth, on the other hand, detected continuous radio emission in the centimetre waveband, which was attributed to the chromosphere.

Four years later, using a V2 experiment, Richard Tousey's Naval Research Laboratory (NRL) group photographed the ultraviolet spectrum of the Sun down to 230 nm for the first time. Then in 1948 Robert Burnight found the first indication that the Sun may be emitting X-rays. Much clearer evidence was found by Herbert Friedman in the following year. Friedman's group continued to study the Sun over the next few years, and found that the majority of soft solar X-rays were due to the corona. Then in 1956 they found that solar flares produced high-energy X-rays, and four years later both radio emissions and X-rays were detected from a solar flare by a ground-based receiving station and the SOLRAD-1 spacecraft. In 1967 the OSO-3 spacecraft measured the spectrum of a solar flare, which indicated that it had heated the local plasma to a temperature of about 30 million K. In 1972 OSO-7 discovered gamma-ray emission from a flare, the main gamma-ray line at 0.5 MeV being due to electron/positron annihilation. This showed that positrons must have been produced in the flare.

Horace and Harold Babcock had studied the magnetic field of the Sun in the 1950s, and found that there were both bipolar and unipolar regions on its surface. In the bipolar regions the magnetic flux leaving the Sun was about equal to that entering it. The Babcocks suggested that ions and electrons leaving the Sun in these regions would follow the field lines and collide over the Sun, generating radio noise and forming prominences and flares. On the other hand, those particles emitted from unipolar regions would stream away from the Sun, never to return. The Babcocks suggested that these unipolar regions were Bartels' M regions that emitted particles during 'M storms'.

In 1957 and 1958 Eugene Parker developed his theory of the solar wind, in which charged particles emitted by the Sun drew magnetic field lines in the corona out into the solar system. Parker envisaged that these charged particles would be emitted by the Sun continuously, and that the field lines would spiral out from the Sun because of the combined effect of radial flow and solar rotation. The particles were first detected outside the influence of the Earth's magnetic field by Luna 2 in 1959, but Luna 2 could not measure their velocity. Three years later the Mariner 2 spacecraft found that their velocity ranged from about 400 to 700 km/s, and occasionally exceeded 1,200 km/s.

Norman Ness found in 1964, using the IMP-1 spacecraft, that the magnetic field lines in interplanetary space made an angle of about 45° to the Sun-Earth line, as suggested by Parker. But, surprisingly, the magnetic field lines changed direction by about 180° every few days, effectively producing magnetic sectors of alternate polarity. Ness and Wilcox then found that the interplanetary magnetic field's polarity correlated with that of the Sun's photosphere at the Sun's equator with a lag of about 4.5 days. This implied an average solar wind velocity of 380 km/s. So the interplanetary magnetic field was an extension of the Sun's via the solar wind. The three-dimensional configuration of the magnetic sectors was unclear, however, but in 1976 Pioneer 11 found that the sectors disappeared when the spacecraft's heliographic latitude exceeded 15°. This indicated that the sector lines were the intersection of the ecliptic with a warped magnetically neutral sheet.

In 1957 Max Waldmeier discovered coronal holes in visible light as gaps in the corona on the solar limb. About ten years later the OSO-4 spacecraft detected coronal holes in the extreme ultraviolet, and in 1973 Skylab imaged many coronal holes in X-rays (see Figure 2.3). In the same year, Krieger, Timothy and Roelof found a good correlation between coronal holes seen in X-rays at the solar equator, and high solar wind velocities measured by the Vela and Pioneer 6 spacecraft. This showed that coronal holes were Bartels' M regions which emitted the fast solar wind. Skylab also showed that coronal holes were unipolar regions where the magnetic field lines ran freely into space, vindicating the Babcocks' theory in which energetic particles were emitted from unipolar regions. Skylab imaged numerous coronal holes, but also imaged coronal loops, linking

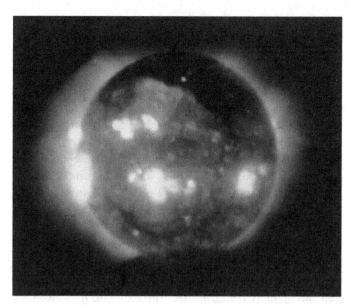

Figure 2.3. A large polar coronal hole is clearly seen as the dark region at the top of this Skylab X-ray image taken in 1973. Skylab images also showed many X-ray bright points, one of which is found in the coronal hole. (Courtesy Harvard-Smithsonian Center for Astrophysics.)

areas of opposite polarity in the photosphere, and hundreds of X-ray coronal bright points, with lifetimes generally in the range of from two hours to two days.

Two Helios spacecraft were placed into orbits around the Sun in the mid-1970s. They confirmed earlier measurements showing a slow and fast solar wind with average velocities of about 350 km/s and 750 km/s. The Helios spacecraft confirmed that the fast wind was emitted by coronal holes. Twenty years later the Ulysses spacecraft, which orbited the Sun out of the ecliptic, found that, around solar minimum, the slow solar wind was constrained to equatorial regions. But the fast solar wind, which originated from the Sun's polar and other coronal holes, filled the majority of the heliosphere. In 2007 the Hinode spacecraft discovered that the low speed solar wind originated from around a cluster of bright coronal loop footprints.

In 1971 the OSO-7 spacecraft detected clouds of protons leaving the Sun in what are now called coronal mass ejections (CMEs). The clouds were hundreds of thousands of kilometres in diameter, with a temperature of about 1 million K, and were leaving the Sun at velocities of about 1,000 km/s. Numerous CMEs have since been observed, particularly by Skylab, Solar Max and the SOHO spacecraft. SOHO, in particular, found that some CMEs had a constant velocity as they moved away from the Sun, whilst others were continuously accelerated, and yet others had a sudden acceleration a few solar radii from the Sun. Since 2006 the STEREO spacecraft has imaged CMEs spectroscopically in an attempt to understand their origin and development in detail.

Five-minute oscillations of the Sun had been discovered by Leighton, Noyes and Simon in 1961. Astronomers in the early

1970s found that this was not a local effect but represented vibration of the Sun as a whole. Later observations of these and other vibration modes by the SOHO spacecraft, and the GONG and BiSON ground-based solar networks, enabled the internal structure of the Sun to be determined. They provided the first images of structures and flows below the surface, and confirmed that the decrease in angular velocity with latitude, measured at the surface, extended through the convective zone. They found that at the base of the convective zone, about one-third of the way into the Sun, was an adjustment region, or tachocline, leading to the more orderly radiative interior. The tachocline appeared to be the source of the Sun's magnetic field.

There had been some surprise in the late 1960s when Raymond Davis found, using the Homestake neutrino detector, that the number of electron neutrinos coming from the Sun was no more than half that expected. This indicated that there was either something wrong with his neutrino detection system, or there was something seriously wrong with the theories explaining how the Sun generated its heat. Davis' results were later confirmed by the Kamiokande-II neutrino detector in Japan. But in the late 1960s Bruno Pontecorvo had suggested that some of the electron neutrinos emitted by the Sun may transform to muon neutrinos en route to the Earth and so be missed by Davis' detector. Later the existence of tau neutrinos was postulated, giving another possible transformation route for electron neutrinos. In 2001 the Sudbury Neutrino Observatory in Canada, which was sensitive to all types of neutrinos, showed that Pontecorvo had been correct, as the total number of all types of neutrinos emitted by the Sun was as predicted by solar physics. The neutrinos were changing type en route to the Earth.

Gustav Spörer and Walter Maunder had pointed out towards the end of the nineteenth century that there had been a lack of sunspots between about 1645 and 1715, a period now called the Maunder Minimum.

Scott Forbush found in the 1940s that the number of cosmic rays observed at the Earth's surface varied with the eleven-year solar cycle, being lowest at solar maximum. Cosmic rays change atmospheric nitrogen to carbon 14, so carbon 14 can be used as a surrogate measure of sunspot activity. Measuring carbon 14 in tree rings clearly showed that the Maunder Minimum was a period of high carbon 14 levels, as expected. Hessel de Vries also found in 1958 that there was another peak in carbon 14 levels from about 1460 to 1540, now called the Spörer Minimum. John Eddy pointed out in 1976 that both the Maunder and Spörer minima correlated with periods of cold winters in Europe. So the lack of sunspots in the Maunder and Spörer minima was seen to correlate with periods of cold winters. But the correlation between sunspot activity and the amount of solar energy incident on Earth, the so-called solar constant, was unclear, because ground-based measurements of the solar

constant were complicated by varying atmospheric conditions. The solution was to use instruments on spacecraft above the atmosphere.

In 1980 Richard Willson showed, using the Solar Max spacecraft, that the solar constant *reduced* slightly as large sunspot groups crossed the central solar meridian. It then became clear, over the next few years, that the solar constant *reduced* slightly as the Sun approached solar minimum. In 1988 Peter Foukal and Judith Lean explained this apparent inconsistency as due to faculae that overcompensated the blocking effect of sunspots over the course of a solar cycle.

Solar Max also provided valuable information on the development of solar flares, showing that electrons and protons are accelerated simultaneously in flares. Hard X-rays brightened simultaneously at both ends of the magnetic loop containing the flare, whilst soft X-rays were emitted by active regions that extended well into the corona. In 1992 the Yohkoh spacecraft showed that hard X-rays were also emitted from the apex of flaring loops, probably as a result of magnetic reconnection. About ten years later the RHESSI spacecraft confirmed that this was the case.

There has been an ongoing investigation over the last fifty years or so into possible mechanisms of heating the solar corona to temperatures of millions of degrees K. In 1997 the SOHO spacecraft found evidence of the upward transfer of magnetic energy from the Sun's surface to the corona though a magnetic carpet, in which the observed magnetic flux was found to be highly mobile. This upward transfer of energy was thought to be a major source of coronal heating. Ten years later the Hinode spacecraft found that the chromosphere is permeated by Alfvén waves, which are sufficiently strong to accelerate the solar wind and possibly to heat the quiet corona. This was confirmed in 2009 by observations made by the Swedish Solar Telescope in the Canary Islands.

Bibliography

Balogh, A., Lanzerotti, L. J., and Suess, S. T., *The Heliosphere through the Solar Activity Cycle*, Springer, 2007.

Bothmer, Volker, and Daglis, Ioannis A., *Space Weather: Physics and Effects*, Springer, 2006.

Brody, Judit, *The Enigma of Sunspots: A Story of Discovery and Scientific Revolution*, Floris Books, 2002.

Golub, Leon, and Pasachoff, Jay, *The Solar Corona*, Cambridge University Press, 1997.

Hufbauer, Karl, *Exploring the Sun: Solar Science since Galileo*, Johns Hopkins University Press, 1991.

Kuiper, Gerard P. (ed.), *The Solar System, Vol. I: The Sun*, University of Chicago Press, 1953.

Lang, Kenneth R., *The Sun from Space*, 2nd ed., Springer Verlag, 2008.

Leverington, David, *New Cosmic Horizons: Space Astronomy from the V2 to the Hubble Space Telescope*, Cambridge University Press, 2000.

Meadows, A. J., *Early Solar Physics*, Pergammon Press, 1970.

Phillips, Kenneth J. H., *Guide to the Sun*, Cambridge University Press, 1992.

See also: Helioseismology.

EARTH'S MAGNETOSPHERE AND IONOSPHERE

Early geomagnetic research

The Chinese discovered, sometime in the first millennium AD, that when a lodestone or a magnetised needle is placed on a float it always pointed north-south. It then gradually became clear over the centuries that a freely pivoted magnetic needle did not point due north, and that the force on the needle was not horizontal, as its north-pointing end always slanted downwards. William Gilbert undertook many experiments in magnetism in the late sixteenth century and in his *De Magnete*, published in 1600, concluded that the Earth behaves like a large magnet.

In 1722 George Graham discovered a diurnal variation in the compass direction, with a swing of about $10'$ at moderate latitudes. He later observed that, at times, the needle moved erratically, indicating a geomagnetic disturbance. Then on 5 April 1741 Graham in London and Anders Celsius and Olof Hiorter in Sweden observed that such a geomagnetic disturbance occurred at the same time as a polar aurora. Later that century Henry Cavendish used triangulation to estimate the height of aurorae as between about 85 and 115 km.

Heinrich Schwabe discovered in 1843 that the number of sunspots seemed to vary with a period of about 10 years. Nine years later, Edward Sabine, Rudolf Wolf and Alfred Gautier independently concluded that there was a correlation between sunspots and disturbances in the Earth's magnetic field. On 1 September 1859 Richard Carrington and Richard Hodgson independently observed two white light solar flares moving over the surface of a large sunspot. This was followed about 18 hours later by the start of a spectacular auroral display, which was seen as far south as the West Indies, peaking about 36 hours after the white light flares had been seen. The white light solar flares had lasted just five minutes, but at exactly the same time Kew Observatory had recorded a similar very brief magnetic disturbance, followed later by a prolonged disturbance coincidence with the aurorae. The simultaneous occurrence of the flare and the magnetic disturbance on Earth was a big surprise, as it implied that the disturbance had been caused by something that had been emitted by the flare at the speed of light.

Kristian Birkeland researched the auroral phenomena at the turn of the nineteenth century. He suggested that the aurora was caused by electrons from the Sun guided by the Earth's magnetic field lines towards the Earth's magnetic poles. Shortly afterwards Carl Størmer undertook a theoretical analysis of the motion of charged particles in the Earth's magnetic field, and showed that such a particle from the Sun can reach the Earth only in two narrow zones centred near the Earth's magnetic poles. Charged particles would be trapped in these zones, travelling from one polar region to the other, and back again. In the 1930s Sydney Chapman and Vincent Ferraro developed a theory in which the Sun emitted a huge cloud of plasma at the time of a solar flare. They concluded that this cloud initially compressed the Earth's magnetic field, and somehow generated a westward-flowing ring current some Earth radii above the equator. The ring current then produced a weak magnetic field contrary in direction to the Earth's, which explained why the Earth's magnetic field at the equator became slightly weaker during magnetic storms.

In 1949 Leiv Harang calculated, assuming aurorae were produced by electrons, that their energies were in the range $15-30$ keV. The particles were confirmed to be electrons by a sounding rocket experiment in 1954.

Unfortunately, Størmer's theory and that of Chapman and Ferraro had a number of problems. Størmer's theory worked well for extremely high energy particles, but not for low energy particles like those in the aurora, and Chapman's and Ferraro's could not explain satisfactorily how the ring current was produced. In 1957, S. Fred Singer developed an improved theory that could explain how low energy particles could be captured by the Earth's magnetic field and produce a ring current.

Meanwhile, in 1951, Ludwig Biermann had suggested that the Sun emitted charged particles continuously, not just during solar storms. Then, in 1957/58, Eugene Parker proposed his solar wind theory, in which charged particles emitted continuously by the Sun drew magnetic field lines in the corona out into the solar system.

Early ionospheric research

As early as 1839, Carl Friedrich Gauss had suggested that the daily variations in the Earth's magnetic field could be due to electric currents in an electrically conducting layer high in the Earth's atmosphere. This idea was extended by Balfour Stewart in 1882 when he proposed that convection caused by solar heating of this conductive layer could explain seasonal variations in the Earth's geomagnetism. Arthur Schuster developed these ideas further and concluded that it was solar ionising radiation which was causing the upper atmosphere to be conductive.

In 1902 Guglielmo Marconi found that he could transmit radio waves across the Atlantic. This was a surprise, as it was thought at the time that radio waves were transmitted in straight lines, except for possibly minor deviations caused by differences in air density. But Oliver Heaviside and Arthur Kennelly independently explained the effect by suggesting that the radio waves had been reflected off an electrically conduct-ing layer, in what is now called the ionosphere, about 100 km above the Earth's surface.

Edward Appleton discovered in 1924 that there was not one layer in the ionosphere, but two. The lower layer, now called the E or Heaviside layer, was at an altitude of about 100 km. It allowed long distance radio transmission at night. The second layer, now called the F_2 or Appleton layer at about 300 km altitude, allowed long distance transmission during the day.

Further work over the next decade or so by numerous researchers succeeded in showing that there are actually three main layers in the ionosphere – the D, E and F layers, centred at about 80, 120 and 300 km altitude – all of whose reflectiv-ity varied over 24 hours. During daytime, the Sun apparently created the D and E layers, and changed the configuration of the F layer into an F_1 and F_2 layer. In the 1930s these changes were generally attributed to ultraviolet solar radiation. But Lars Vegard and E. O. Hulburt independently speculated in 1938 that the source could be solar X-rays.

Howard Dellinger showed in the 1930s that problems with the reception of short-wave radio waves were often, but not always, associated with solar flares seen on the Sun a short time before. Then in 1939 Thomas Johnson and Serge Korff found that the ionisation of the Earth's atmosphere was quickly disrupted at the time of solar flares, and suggested that the cause of the disruption was solar X-rays.

The Space Age

Herbert Friedman found, using a V2 experiment in 1949, that the Sun emitted X-rays. Three years later he was able to show that the X-ray intensity was sufficient to sustain the E layer. At about the same time Richard Tousey and Friedman were able to show independently that the intensity of the Lyman-α ultra-violet solar radiation was sufficient to maintain the D layer. In 1956 Friedman's group found that solar flares were the source of very high energy X-rays that disrupted the ionosphere.

On 1 May 1958, James Van Allen announced that he had discovered a belt of elementary particles, now called the inner Van Allen belt, around the Earth above an altitude of about 1,000 km, from his experiments on the Explorer 1 and 3 space-craft. Later that year he discovered a second radiation belt (see Figure 2.4) at an altitude of about 15,000 km with an experi-ment on Pioneer 3. Shortly afterwards photographic emulsions carried on a Thor-Able rocket proved, for the first time, that most of the particle radiation in the inner Van Allen belt was due to protons, with energies of at least 75 MeV. Then, in 1962, Explorer 12 detected electrons with energies of about 4 MeV in the outer Van Allen belt.

Luna 2 in 1959 was the first spacecraft to detect the solar wind that had been predicted by Parker, outside the influence of the Earth's magnetic field. Subsequent spacecraft showed that there is a region, called the magnetosphere, around the Earth that is dominated by the Earth's magnetic field. In March 1961, Explorer 10 provided the first indication of the tail-like struc-ture of the outer edge of this magnetosphere, called the magne-topause, around local midnight. Later in 1961, Explorer 12 was the first spacecraft to clearly detect the magnetopause near the noon meridian. Then in 1963 John W. Freeman, Van Allen, and Laurence Cahill announced the discovery of the magne-tosheath region outside the magnetopause using Explorer 12. The following year Freeman found that the distance of the mag-netopause from Earth at local noon varied from 8 R_E (Earth radii) to beyond Explorer 12's perigee of 13 R_E, depending on the intensity of the solar wind.

In 1963, IMP-1 provided the first clear-cut evidence of the bow shock at the outer edge of the magnetosheath. The spacecraft crossed this bow shock many times at least about 13 R_E from Earth, depending on the bow shock's position relative to the Sun-Earth line. On the night side of Earth, IMP-1 detected a geomagnetic tail that appeared to extend well beyond the spacecraft's apogee of 32 R_E. IMP-1 also dis-covered a narrow region, only about 1,000 km thick, around local midnight where the geomagnetic field fell almost to zero. This so-called neutral sheet was sandwiched between a region to its north, where the geomagnetic field was pointing towards the Earth, and a region to its south, where the field lines pointed in the opposite direction. In 1966 Samuel J. Bame et al found, using particle data from the Vela spacecraft, that the neutral sheet was embedded in a plasma sheet about 6 R_E thick.

The HEOS 1 and 2 spacecraft were the first, starting in 1969, to explore the polar cusp, which is a V-shaped region between field lines in the Sun-facing part of the magnetosphere and those in the tail. They found that the cusp is a region of disor-dered weak magnetic fields, where magnetosheath plasma can descend to the Earth's ionosphere to produce a characteristic red aurora. Thirty years later the POLAR spacecraft found that the invariant latitude of the centre of the polar cusp varied from 86° to 70° depending on solar wind conditions, the angle of the solar wind to the Earth's axis, and the time of day.

In 1971 the OSO-7 spacecraft detected clouds of protons leaving the Sun in coronal mass ejections. These are similar to the clouds envisaged by Chapman and Ferraro in the 1930s, but their clouds had an equal amount of positive and negative charge, making them electrically neutral.

The ISEE-3 spacecraft began a number of passes through the geomagnetic tail in 1982, the furthest of which was about 240 R_E from Earth. It found that, whilst the near-Earth plasma flow was mostly Earthward, there was a clear outflow in the distant plasma sheet, which appeared to accelerate as it got beyond about 80 R_E from Earth. This acceleration was thought to have been caused by the reconnection of magnetic field

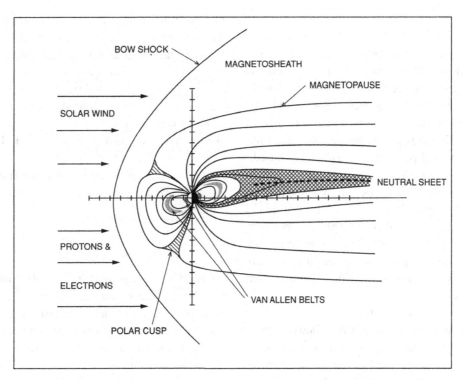

Figure 2.4. The inner and outer Van Allen radiation belts shown within the general configuration of the Earth's magnetosphere. The exact size and configuration of this magnetospheric system, which varied with solar activity, was gradually determined during the first ten years of the space age.

lines near the neutral line. Both ISEE-3 and, later, the Geotail spacecraft showed that the tailward velocity of plasma in the distant tail increased with distance until it matched the velocity of the solar wind.

In 1995 Geotail provided evidence of magnetic reconnection in the geomagnetic tail when it detected intense bursts of energetic particles at about 13 R_E downwind in the magnetosphere. Then four years later the WIND spacecraft became the first to clearly detect magnetic reconnection, when it observed a change in direction of 180° in jets of plasma about 60 R_E downwind in the geomagnetic tail. In 2000 the POLAR spacecraft detected Alfvén waves, propagating along magnetic field lines from the geomagnetic tail towards Earth, which had enough energy to create aurorae during magnetospheric substorms.

Three consecutive substorms were observed simultaneously by the four Cluster and two Double Star spacecraft on 26 September 2005. They showed that magnetic reconnection coincided in space and time with that of the current disruption process, which is characterised by large amplitude and turbulent magnetic field fluctuations. A few tens of seconds later, the IMAGE spacecraft imaged aurorae. So it appeared as though magnetic reconnection in the geomagnetic tail was a cause of aurorae.

On 23 March 2007 the THEMIS constellation of five spacecraft measured the rapid expansion of the plasma sheet, and the simultaneous westward expansion of the visible aurora at about 15° longitude/minute. This was the first time that such observations had been made simultaneously by any observatories. In the following year the THEMIS spacecraft detected magnetic reconnection about 20 R_E from Earth in the geomagnetic tail, followed 90 seconds later by the brightening of the aurora observed by a ground station network, and 90 seconds after that by near-Earth current disruption.

The key to this and other recent work in this area has been simultaneous observations by a number of spacecraft and ground stations.

Bibliography

Hess, Wilmot N., *The Radiation Belt and Magnetosphere*, Blaisdell Publishing Company, 1968.

Hufbauer, Karl, *Exploring the Sun: Solar Science since Galileo*, Johns Hopkins University Press, 1991.

Stern, David P., *A Brief History of Magnetospheric Physics before the Spaceflight Era*, Reviews of Geophysics, **27**, (1989), 103–114.

Stern, David P., *A Brief History of Magnetospheric Physics during the Space Age*, Reviews of Geophysics, **34**, (1996), 1–31.

Van Allen, James A., *Origins of Magnetospheric Physics*, University of Iowa Press, 2004.

EARTH

The philosophers of Ancient Greece proposed various ideas about the configuration of the Earth. The Pythagoreans of about 500 BC were probably the first to appreciate that the Earth is spherical. And around a hundred years later Hicetas of Syracuse seems to have been the first to conclude that the Earth spins on its axis, although this was not generally accepted. For example, the influential philosopher Aristotle favoured a non-spinning Earth, as did Ptolemy.

In the third century BC Eratosthenes of Cyrene estimated the size of the Earth to a remarkable accuracy. He did this by measuring the altitude of the Sun at noon on midsummer's day at two different places on approximately the same meridian. In the following century Hipparchus was the first to draw attention to the fact that lengths of the sidereal and tropical years were not the same.

The cause of this precession of the equinoxes observed by Hipparchus was first explained almost two thousand years later by Isaac Newton. Jean Dominique Cassini and colleagues had found evidence in 1672 that the Earth was not spherical. Newton realised that this was due to the Earth's axial rotation, and he used his theory of gravity in 1687 to calculate the size of its resulting equatorial bulge. He then showed that the gravitational attractions of the Sun and Moon on this bulge was the cause of the precession of the equinoxes. He calculated a value of about $50''$/year, giving a period of about 26,000 years. This was fortuitously correct as he had overestimated the effect of the Moon and underestimated that of the Sun, but the two errors had cancelled each other out.

Polar motion

James Bradley found in the mid-eighteenth century that the Earth's spin axis not only precessed, as explained by Newton, but also nutated by about $\pm 9''$, with a period of 18.6 years. In the early nineteenth century Pierre Laplace estimated the oblateness, or polar flattening, of the Earth from its effect on the orbit of the Moon. But as the nineteenth century progressed it became clear that the shape of the Earth was not an ellipsoid, but an irregular shape caused by the non-symmetrical distribution of mass in the Earth's interior. In 1765 Leonhard Euler showed theoretically that, in such a body, the spin axis would move relative to the surface of the Earth with a period of about 10 months. All attempts to observe this effect failed until it was discovered by accident in 1888.

In that year Karl Küstner concluded, by measuring the position of stars, that the latitude of the Berlin Observatory was varying. Similar results were obtained from other observatories. In 1891 Seth Chandler analysed all these results and concluded that the Earth's pole was moving in a circle on the surface of the Earth with a radius of about $0.3''$. But the period was about 427 days, not the 10 months calculated by Euler. However, as Simon Newcomb pointed out, this was probably due to the fact that Euler had calculated the situation for a rigid body, whereas the Earth is not rigid. Chandler then extended his analysis and concluded that the observed polar movement was due to two superimposed effects, one with a period of about 14 months and one with a period of 12 months. The 14-month period was that predicted by Euler, but modified for a non-rigid Earth, whilst the 12-month period was caused by meteorological effects, such as the movement of water and air masses between the northern and southern hemispheres on an annual basis.

Internal structure

William Gilbert had suggested in 1600 that the Earth behaved like a large magnet. But in 1635 Henry Gellibrand found that the Earth's magnetic field was drifting slowly westward with time. In 1686 Edmond Halley explained this by proposing that the Earth consists of a series of nested spheres. Each sphere was magnetic, but they rotated around the Earth's centre at slightly different rates, producing a net westward drift.

It was known in the early nineteenth century, by taking measurements in deep mines, that the temperature inside the upper part of the Earth increased at the rate of about $1°C$ for every 60 to 80 metres of depth. If this rate of increase continued, rocks would be molten at depths of about 80 km or so, clearly indicating that the Earth's interior is molten. This idea of a molten interior was strengthened by the existence of volcanoes, which eject molten lava at temperatures of between $800°C$ and $1,500°C$. However in 1842 William Hopkins pointed out that, since the melting point of rocks increases with pressure, the solid crust could be much thicker than 80 km. He concluded, from the rate of precession of the Earth's axis, that its crust was at least 1,250 km thick.

It had been thought that the Earth had a small magnetic iron core, but if this was molten it could no longer be magnetic, as iron loses its magnetic properties when heated to a high temperature. But electromagnetic theory was being developed in the nineteenth century, and this showed that the Earth's magnetic field could be produced by electric currents in its interior if it was fluid. So the fact that the Earth had a magnetic field could be explained by either a solid or fluid interior.

In the 1860s Lord Kelvin developed his theory of the cooling of the Earth since its creation. He suggested that as the surface cooled and solidified, it was constantly being broken up by gravitational perturbations, with large solid pieces falling to the centre. These gradually built up into a large solid core which increased in size until the Earth was largely solid, but with pockets of molten material near the surface which erupted from volcanoes. Kelvin estimated that the age of the Earth was in the range of 20 to 100 million years, which was consistent with Helmholtz's 1854 estimated age of the Sun of about 22 million years. But these ages created a problem with geologists of the time, who estimated that the geological processes on Earth had taken a much longer period, and Darwin's theory of natural selection, which also required a longer period.

These discrepancies with the age of the Earth were eliminated in the early twentieth century when a number of physicists used the newly discovered effect of radioactivity to estimate the age of various surface rocks. By 1915 an age of 1,600 million years for the Earth's crust was widely accepted. But the problem with the relatively smaller age of the Sun was not clearly resolved until the late 1930s following research into possible thermonuclear reactions.

Returning to the structure of the Earth, Kelvin's theory of a largely solid Earth was the primary theory at the end of the nineteenth century. In 1897 Emil Wiechert had suggested that the Earth consisted of a dense metallic core, mostly of iron, surrounded by a lighter rocky layer, now called the mantle. Dynamical considerations, based on a mean moment of inertia of the Earth, led him to conclude that the radius of the core was about 4,970 km. In 1906 Richard Oldham found, by studying the seismic records of earthquakes, that the Earth definitely has a core, and in 1914, Beno Gutenberg showed that the interface between the core and mantle, now called the Wiechert-Gutenberg Discontinuity, was at about 3,470 km from the centre of the Earth.

In 1909, Andrija Mohorovičić discovered the boundary between the Earth's crust and its mantle by studying the 1909 Croatian earthquake. It was later found that this boundary, now called the Mohorovičić Discontinuity, varied in depth from about 5 km under the deep oceans to about 70 km under some of the mountain ranges.

Seismologists of the early twentieth century believed, like Kelvin, that the interior of the Earth was basically solid, but in 1926 Harold Jeffreys showed that at least part of the core must be fluid. In 1934 he tried to link the structure of the Earth to that of the other terrestrial planets and the Moon. He suggested that the Earth had a core of liquid metals, mostly iron, surrounded by a solid shell of silicates, mostly olivine, about 2,900 km thick. Jeffreys further proposed that the other terrestrial planets and the Moon had a similar structure, but with the dense metallic core being smaller as the overall densities decrease from Earth to Venus to Mars to Mercury (the density of Mercury was thought to be much less than it was later found to be) and to the Moon.

Inge Lehmann concluded in 1936 that there was a seismic discontinuity inside the Earth's core. His analysis indicated that the core was separated into an inner and outer core at a radius of about 1,400 km. A little later Francis Birch suggested that, although the outer core was fluid, the inner core was solid iron, because of the very high pressures prevailing there.

It also became evident in the early 1930s that there was a density discontinuity in the mantle at a depth of about 400 km. In 1936 John D. Bernal suggested that olivine, the favoured material for the mantle, may change its crystal structure under pressure, which may be the cause of the discontinuity. Unfortunately it was not possible at that time to produce such pressures in the laboratory, and so it was not until many years later that experimental evidence was forthcoming. It was found that α-olivine transforms to β-spinel at a pressure corresponding to a depth of about 400 km. This in turn transforms to γ-spinel at the pressure at about 500 km, and this changes its structure yet again at the pressure experienced at about 660 km. Seismological evidence in the 1960s showed that there were two discontinuities, one at about 400 km and one at 670 km. These were caused by olivine transitions, modified in all probability by the presence of other materials.

Urey's cold accretion theory

Early in the twentieth century it became clear that radioactive heating could play an important part in keeping the interior of the Earth hot. In 1906 Lord Rayleigh showed that the amount of heat produced by the radioactive decay of radium in the Earth's crust was sufficient to account for the entire flow of heat from the Earth. Then in 1949 Harold Urey calculated, using radioactive abundances from meteorites, that the interior of the Earth was heating up, rather than cooling down. This led Urey to propose a radical new theory for the origin of the Earth and Moon, which came to be called the cold accretion theory.

Urey proposed that the Earth and Moon had formed separately as a pair of objects from the condensing solar nebula. He then suggested that the only way that the Moon could have retained its Earth-facing bulge was if it had been formed cold and stayed cold. His cold Moon theory, and by association his cold Earth theory, was disproved by the rock samples brought back to Earth by Apollo astronauts in the late 1960s/1970s.

Dynamics

Edmond Halley discovered in 1693 that the Moon's position in the sky was in advance of where it should be based on ancient eclipse records. This secular acceleration of the Moon could be either because the Moon was accelerating in its orbit, and/or because the Earth's rotation rate was slowing down. In 1787 Laplace showed that this effect could be completely explained by the Moon's orbit gradually being reduced in size, caused by a reduction in the eccentricity of the Earth's orbit. But in 1853 John Couch Adams showed that the effect analysed by Laplace could only explain half of the observed effect. Charles Delaunay suggested in 1866 that tidal friction, which was gradually slowing down the Earth's spin rate, could probably account for the other half of the effect. But in 1909 Thomas Chamberlin concluded that tidal friction would not be enough to do this. However, Geoffrey Taylor analysed the ocean geometry in more detail in 1919, and found that shallow seas were much more important than the oceans in slowing down the Earth. Then in the following year Jeffreys showed that the energy dissipated in the shallow seas of the Earth could explain the observed effect.

Simon Newcomb showed in 1878 that the Moon's position appeared to deviate by a small amount in a cyclical fashion with a period of 273 years, but the cause was unknown. He thought that it could be due to variations in the Earth's rotation rate, but he dropped the idea in 1903. If the effect was due to variations in the Earth's rotation rate, it should be evident in the apparent movement of the Sun and planets, as well as that of the Moon. In 1914, Ernest W. Brown showed that there was such an effect in the apparent motions of the Sun and Mercury. This was later confirmed by Willem de Sitter, who noticed a similar effect for the satellites of Jupiter. But he found that the Earth was showing random, rather than cyclical, irregularities in its axial rotation. In 1937 N. Stoyko found a seasonal variation in the Earth's rotation period, superimposed on all other variations, which was later attributed by W. H. Munk and R. L. Miller to seasonal changes in atmospheric and oceanic circulation.

Continental drift/plate tectonics

Francis Bacon had drawn attention in 1620 to the fact that the shore lines on both sides of the Atlantic Ocean seemed to have a similar shape, as if one could fit into the other. In the early nineteenth century Alexander von Humboldt found that there were striking similarities between the geological strata on the African and South American coasts of the South Atlantic. Shortly afterwards Elie de Beaumont and Henry De La Beche suggested that the crust of the early Earth had broken up as the Earth had cooled, thus producing mountain chains. Then in 1858 Antonio Snider-Pellegrini pointed out that there were identical fossil plants in both North American and European coal deposits, which he took as evidence that the two continents had once been united.

In the second half of the nineteenth century Eduard Suess considered the effect of a continually cooling Earth on its solid surface. He thought that this would cause the surface to wrinkle like the skin of a desiccated apple, and he suggested that this would result in continents and ocean basins. He theorised that as the contraction continued, the continents would be undermined and collapse to form ocean basins, whilst the earlier ocean basins would become dry land. This explained how marine fossils could now be found on dry land and how various identical species, which could not have crossed the oceans, were found on both sides of a current ocean basin. A little earlier James Dana had suggested that, after the initial continents

and ocean basins had formed as the Earth contracted, they had stayed in place on further contraction. His theory explained a number of features of the Earth's surface somewhat better than Suess'. But Dana could not explain how identical species could have crossed the oceans.

Frank Taylor suggested in 1908 that the mountain belts of Asia and Europe were the result of the lateral movement of the continents. He also suggested that the mid-Atlantic ridge, part of which had been discovered in the 1870s, was the line of rifting between Africa and South America, which were gradually moving apart. But his idea was generally ignored.

Alfred Wegener was the first to develop a reasonably detailed theory of continental drift. In 1912, he suggested, based on geological, palaeontological and biological evidence, that there had originally been only one continent, which he called Pangaea. Then in the late Palaeozoic era, about 200−250 million years ago, this continent had broken apart, with South America and Southern North America moving away from Africa and Southern Europe, and with Antarctica and Australia moving away from Africa and India. Later, about 100 million years ago, Northern North America separated from Northern Europe, India from Africa, and Australia from Antarctica. Although Wegener amassed much supporting evidence for his theory it had been rejected by most geologists by the 1930s.

There was a resurgence of interest in the concept of continental drift in the 1950s when Stanley Runcorn and colleagues found, from palaeomagnetic evidence, that, over the last 600 hundred million years, the Earth's north magnetic pole had moved. This pole had moved, as seen from Europe, from near the present geographical coordinates of Hawaii to Japan, and then via Siberia to its present position in Northern Canada. This could be because either the Earth's magnetic pole had moved, or Europe had moved, or both. It was then found that the remanent magnetism of rocks in North America implied a route for the magnetic pole that was displaced mainly in longitude from the one based on rocks from Europe. The two routes could be made to coincide, however, if it was assumed that North America and Europe had been joined together until about 100 million years ago, when they had started to drift apart, as proposed by Wegener. Similar palaeomagnetic data from other continents indicated that they had also, as Wegener had proposed, once been joined together in one super-continent. Nevertheless, because of the difficulty of decoding the magnetic signatures from rocks, and a natural scepticism, most geologists of the time remained unconvinced as to the validity of these results.

Meanwhile a different kind of evidence for continental drift was being collected through a gradual understanding of the structure of the ocean floor.

A part of the mid-Atlantic ridge had been discovered by Charles Wyville Thomson on the *Challenger* expedition of 1872. Deep sea surveying was very time consuming at the time, but it became much easier when sonar was introduced in the 1920s. Since then, a great central ridge has been found that runs the length of the Atlantic Ocean, and a mid-ocean ridge has also been found in the Indian Ocean. The Pacific has been found to have a more complex ridge system, together with a number of very deep trenches around its periphery. A number of trenches have also been found in the Atlantic and Indian Oceans.

In the 1950s Marie Tharp undertook a detailed examination of the mid-Atlantic ridge, finding evidence of a rift valley down its centre. In addition, she found that most of the earthquakes in the mid ocean region had originated from just below the floor of this rift valley. Then, during the International Geophysical Year in 1957–58, work was undertaken on a world-wide scale, which uncovered a global system of oceanic ridges coinciding with major earthquake zones (the earthquake zones are shown in Figure 2.5a).

There had been a number of problems with Wegener's theory of continental drift, probably the most serious of which was the lack of a driving force sufficiently powerful to move the continents. In 1929 Arthur Holmes had suggested that convective currents in the mantle could rise up to the crust and carry the continents along with them. Then as new evidence for continental drift was accumulated in the 1950s, Harry Hess used Holmes' concept as the basis for his own theory of 1960.

Hess suggested, in his so-called seafloor spreading model, that the mid-ocean ridges were caused by mantle material breaking through the crust. This ex-mantle material then cooled, subsided, and moved away from the ridges, finally being carried back to the mantle in the ocean trenches. This explained why earthquakes originated on the mid-ocean ridges, and why volcanoes also exist on those ridges, causing them to rise above the ocean surface in Iceland and the Azores, for example. It also explained why relatively young rocks had been found on the ocean floor, compared with on the continents. Hess' theory predicted that the age of the ocean floor should increase with distance from the mid-ocean ridge, and this was confirmed almost immediately.

Bernard Brunhes had reported, as long ago as 1909, that he had found evidence in ancient lava flows that the Earth's magnetic field had reversed direction in the past. His discovery was basically ignored at the time as it was not thought to be credible. But in 1963 Allan Cox, Richard Doell and Brent Dalrymple found clear, consistent evidence of at least three field reversals in the last four million years, using data from rocks in North America, Europe, Africa and Hawaii.

In 1961 Arthur Raff and Ronald Mason measured a strange pattern of alternating stripes of material of high and low magnetic intensity in the eastern Pacific, off the coast of North America, running approximately north-south. A similar pattern was found the following year in the Indian Ocean. Then in 1966 Frederick Vine and Drummond Matthews, concluded,

having analysed data from a number of oceans, that stripes of reversed magnetisation occurred symmetrically on both sides of ocean ridges. This clearly showed that there was a lateral movement of the extruded mantle material from the ocean ridges in a magnetic field that reversed direction from time to time, giving powerful support to Hess' theory of seafloor spreading.

Further confirmation of Hess' theory was obtained by the drilling ship *Glomar Challenger*, which took samples of the ocean floor in the Atlantic starting in 1968. The ages of these samples, determined from an analysis of their fossils, increased uniformly with distance from the mid-ocean ridge, and agreed with the ages derived from magnetic variations. The spreading rate was found to be about 2 cm/yr for the Atlantic. Later work in the other oceans produced rates of from 1 to 5 cm/yr.

The details of this theory of continental drift or plate tectonics, as it is now called, were progressively worked out over the next few years. In essence, it is believed that the Earth's crust and upper mantle, down to about 60 km, are composed of relatively strong rock material in the so-called lithosphere. This layer is split into a number of plates (see Figure 2.5b) that float on the asthenosphere, which is made of material close to its melting point, and which can flow under stress.

At some of the boundaries between plates, like the mid-Atlantic ridge, new material is being forced up from below, driving the plates apart. At others, such as the Kermadec trench, north of New Zealand, one plate is being pushed or subducted under another plate, whilst at the well-known San Andreas fault in California one plate is sliding past another. Most of the collision zones between two plates occur at the edge of, or underneath, the ocean. But, in a few places, two plates are colliding on land, and then, instead of one plate sliding under another, they push each other higher, as in the case of the Himalayas, where the Indo-Australian and Eurasian plates are colliding.

Extremophiles and hydrothermal vents

In 1966 Thomas Brock found microorganisms in the geysers of Yellowstone National Park that could tolerate temperatures in excess of 60°C. Since then a number of these thermophiles have been discovered in geothermal features all over the world including Iceland, New Zealand and Italy.

John Corliss, Robert Ballard and colleagues discovered hydrothermal vents in 1977, 2,500 m down at the bottom of the Pacific Ocean near the Galapagos Islands. Around these vents they found giant clams and tubeworms. Since then hydrothermal vents have been found in the mid-Atlantic ridge, and many other active areas, emitting water at a temperature of up to about 400°C into the cold ocean. They support a plethora of life, including eyeless shrimp, two metre long tubeworms, armoured snails, and Pompeii worms capable of withstanding

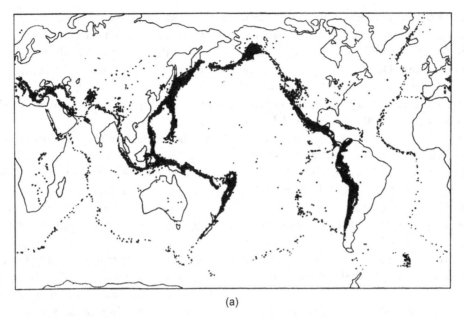

(a)

Figure 2.5a. Earthquake epicentres, shown as black dots, were generally found to occur along clearly defined fault lines on the Earth's surface. Many epicentres were found along oceanic ridges and subduction zones or oceanic trenches (compare Figures 2.5a and 2.5b).

temperatures of up to 80°C. These all exist in the dark in high-temperature/high-pressure environments. Their discovery has had a fundamental effect on our understanding of conditions required to support life. So far over 500 vent species have been discovered, many of which thrive, either directly or indirectly via bacteria, on the minerals and hydrogen sulphide gas emitted by the vents. In 2003 a hyperthermophile, called Strain 121, a single-cell microbe, was found in a hydrothermal vent in the north-east Pacific. It was able to thrive and reproduce at 121°C, and survive for a short period of time at 130°C.

Other extremophiles have been found which flourish in many different types of hostile environment, including cryophiles which prefer temperatures below 15°C, acidophiles which thrive in an acid environment, hypoliths which live inside cold desert rocks, and radioresistant organisms, which

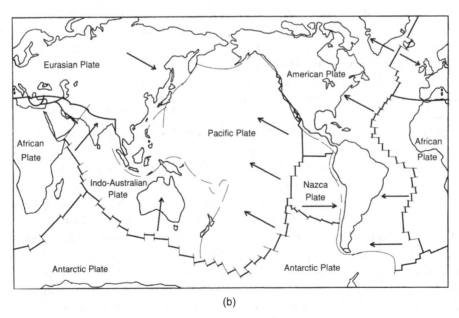

(b)

Figure 2.5b. The major tectonic plates are shown separated by ridges (solid lines) and subduction zones (light dashed lines). The Pacific plate is seen to be almost ringed by subduction zones, which are not present in the Atlantic. The movement of the plates is shown by arrows.

are resistant to high levels of ionizing radiation. Living prokaryotic cells have even been found in core samples taken from 1,600 m below the sea floor off Newfoundland. Then in 2010 a bacteria was found in an arsenic-rich lake in California that thrived on arsenic, apparently substituting arsenate for phosphate in its DNA.

The discovery of all these organisms flourishing in extreme environments has encouraged those hoping to find life elsewhere in the solar system.

Bibliography

Brush, S. G., and Gillmor, C.S., Geophysics, in *Twentieth Century Physics Vol. III*, Brown, Pais and Pippard (eds.), Institute of Physics Publishing and American Institute of Physics Press, 1995.

Doel, Ronald E., *Solar System Astronomy in America: Communities, Patronage, Interdisciplinary Research, 1920–1960*, Cambridge University Press, 1996.

Jeffreys, Harold, *The Earth: Its Origin, History and Physical Constitution*, 4th ed., Cambridge University Press, 1962.

Oreskes, Naomi, *The Rejection of Continental Drift: Theory and Method in American Earth Science*, Oxford University Press, 1999.

Oreskes, Naomi, *Plate Tectonics: An Insider's History of the Modern Theory of the Earth*, Westview Press, 2003.

Spencer Jones, Harold, *Dimensions and Rotation*, in *The Solar System, Vol. II: The Earth as a Planet*, Kuiper, Gerard P. (ed.), University of Chicago Press, 1954.

Urey, Harold C., *The Planets: Their Origin and Development*, Yale University Press, 1952.

Ward, Peter D., and Brownlee, Donald, *Rare Earth: Why Complex Life Is Uncommon in the Universe*, Springer-Verlag, 2000.

See also: Earth's magnetosphere and ionosphere

MOON

In the third century BC Aristarchus of Samos estimated the size and distance of the Moon, in terms of Earth diameters. His result was accurate to within about 15%.

Galileo started to observe the Moon with one of his first telescopes in late 1609. He noted that the terminator was not the uniform curved line that it should be if the Moon was the perfectly smooth sphere of Aristotle's cosmology. Instead, the terminator was very irregular with bright points in the dark area, from which he concluded that the Moon had mountains and valleys. Galileo also thought that the Moon had an atmosphere, but there appeared to be very little water as there were no clouds. So he reasoned that any life that may exist on its surface would be very different from that on Earth.

In 1611 Thomas Harriot was the first astronomer to record the libration in latitude of the Moon, which has a period of one month. This occurs because the Moon's spin axis is not perpendicular to its orbit around the Earth. Some years later Galileo detected a libration in longitude, which he thought had a period of one day. In fact it has a period of one month, and is caused by the eccentricity of the Moon's orbit.

Johannes Hevelius published his *Selenographia* in 1647, which was to be the standard work on lunar cartography for well over a hundred years. Unlike previous works by other astronomers, his full Moon maps showed the effect of librations in both latitude and longitude. Then, a few years later, Giovanni Riccioli published a map that included names of features. These names are those generally in use today.

By the mid-seventeenth century it was clear that there were numerous craters on the Moon, and in 1665 Robert Hooke speculated on their cause. He undertook laboratory-like experiments and noted that, if round objects were dropped into a viscous mixture of clay and water, features were temporarily produced that resembled lunar craters. But he could not think of a source of large objects in space impacting the Moon. However, he also found that he could produce crater-like features if he boiled dry alabaster powder in a container. As a result, he concluded that lunar craters were produced by the collapsed blisters of warm viscous lava in volcanic-like processes.

In the late eighteenth century Johann Schröter recorded what he thought were a number of small surface changes on the Moon, and detected what he thought were signs of lunar twilight, indicating to him that the Moon had an atmosphere. Schröter also believed that the Moon supported vegetation in places, and that it was probably inhabited. William Herschel believed, like Hevelius before him, that 'lunarians' existed. Then in 1816 Franz Gruithuisen reported that he had seen clouds on the Moon, and six years later he announced that he had detected a city. He thought that lunar rilles were dry riverbeds or roads.

Friedrich Bessel tried and failed in the 1830s to observe the refraction of starlight just before or just after a lunar occultation. As a result, he concluded that density of the Moon's atmosphere, if it had one, could be no more than about 0.2% that of the Earth. Three years later Wilhelm Beer and Johann Mädler produced a highly influential book, *Der Mond*, which almost killed off interest in the Moon as a dynamic body. They suggested that virtually all of the reported surface changes on the Moon could be explained by tricks of the light. They concluded that the Moon was geologically dead, and if it had an atmosphere it must be very thin. As a consequence the lunar surface must be dry, with no vegetation. This concept of a geologically dead Moon was generally accepted following Beer and Mädler's work, even though small changes were reportedly seen on the Moon from time to time, as none were generally confirmed.

The impact theory of the formation of lunar craters was resurrected at the start of the nineteenth century after the discovery of the first asteroids and a number of meteorites, as there now seemed to be a source of impacting bodies. The

Bieberstein brothers, in particular, suggested that lunar craters may have been formed by meteorite impact. Franz Gruithuisen had a similar theory in 1824, as did Richard Proctor in 1873. But what puzzled the meteorite protagonists was that lunar craters were almost all circular, whereas most meteorite impacts would have been at an angle to the vertical, so the craters should have been elliptical. As a result most astronomers in the nineteenth century thought that lunar craters were caused by volcanic eruptions.

There were a number of problems with the volcanic theory, however. For example, Grove K. Gilbert pointed out in 1892 that there were morphological differences between volcanoes on Earth and lunar craters. So Gilbert proposed that the craters were of impact origin; the central peaks were the result of rebound following impact, and the lunar rays were composed of material ejected by the impact. He further explained the circularity of the craters by assuming that impacting bodies were small natural Earth satellites, which had only small velocities relative to the Moon, and so would have fallen onto the Moon almost vertically. However, after the First World War it was realised that the shape of lunar craters resembled shell craters, and that the craters were formed by the shock wave of the impact. So a non-vertical impact could still produce a circular crater.

Ralph Baldwin analysed bomb craters from the Second World War in some detail which he published in his book *The Face of the Moon* in 1949. This persuaded most American astronomers that lunar craters were formed by high-speed impacts. But this consensus was not shared in Europe, where Ralph Baldwin was largely unknown. Harold Jeffreys, in particular, who was highly respected in the UK, still favoured the volcanic theory, claiming that the impact supporters had not properly analysed the various forms of volcanism on Earth. By the late 1950s the European opposition to the impact theory had been reduced to some extent, but the question was still open when the first spacecraft were successfully sent to the Moon in 1959.

The question as to whether there is life on the Moon was still unresolved until well into the twentieth century. William Pickering had suggested in 1903 that there may be plants, snow and river beds on the Moon. Twenty years later he suggested that there may be a low form of vegetation that completes its life cycle in the 14 days of sunlight, being fed by carbon dioxide released from rock fissures. But over the decades the idea of there being such relatively simple lifeforms was mostly abandoned. Nevertheless, it was still thought possible in the 1960s, when the Americans were planning their manned lunar landings, that there may be some very elementary forms of life, like bacteria, on the Moon. As a result the early American lunar astronauts were put in quarantine on their return to Earth, just in case they had picked up some forms of lunar life on their spacesuits.

Bernard Lyot had concluded in 1929, from polarisation measurements, that the Moon was probably covered by volcanic ash. This conclusion was consistent with the fall in temperature observed for the lunar surface during the 1927 total lunar eclipse. Then in the 1950s Thomas Gold suggested that the lunar maria may be covered in volcanic ash or dust up to a few metres deep. But other estimates put the thickness at no more than a few centimetres. If Gold was correct, however, this would cause major problems with any spacecraft landings.

Early spacecraft missions

The Soviet Luna 1 spacecraft, which started mankind's exploration of the Moon in January 1959, could detect no magnetic field. So the Moon's magnetic field, if it had one, must be very weak. Some astronomers took this as evidence that the Moon had a solid core, as then there could be no dynamo producing a magnetic field. However other astronomers suggested that the Moon's core could still be molten if it rotated very slowly. Later in 1959 Luna 3 took the first photographs of the far side of the Moon, which showed that it had fewer mare regions and more craters than the near side.

In 1962 Eugene Shoemaker and Robert Hackman analysed the distribution of widespread ejecta produced by the formation of large impact basins, like Imbrium, and craters. In some places ejecta from different impacts had overlaid each other, enabling relative dating of the impacts to be produced.

Ranger 7 was, in 1964, the first spacecraft to successfully impact the Moon. Its images showed craters as small as one metre in diameter and indicated that the surface dust was not very deep. William Hartmann used crater counts to deduce the age of maria around the Ranger 7 impact site as 3.6 billion years. Luna 9, which was the first spacecraft to soft-land a capsule on the Moon (in 1966), showed a rough textured surface apparently of volcanic origin, at its landing site in the Ocean of Storms (Oceanus Procellarum). There seemed to be no dust, thus confirming the Ranger results. Luna 10, which was, in 1966, the first spacecraft to orbit the Moon, detected a weak magnetic field that did not change with altitude. This indicated that it was measuring not the magnetic field of the Moon itself, but rather that of interplanetary space. Radio occultation measurements with Luna 10 showed that the Moon had no detectable atmosphere. In addition, the spacecraft showed that the Moon was slightly pear-shaped, with an elongation facing away from the Earth. Shortly afterwards, the American Lunar Orbiter 1 showed that there were mass concentrations, or 'mascons', under the Moon's maria.

In 1967 Surveyors 5 and 6 found that the chemical composition of the soil at their landing sites in the Sea of Tranquillity (Mare Tranquillitatis) and the Central Bay (of the Sinus Medii) was similar to that of terrestrial basalt. In the following year,

Surveyor 7 landed on the ejecta near the rim of the crater Tycho. It detected anorthositic gabbro with an iron content about half that measured in the maria areas. So the maria, under which there were mass concentrations, appeared to have a basalt surface, whilst other parts of the Moon may have a different composition. It was still unclear, prior to Apollo, how many of the Moon's surface features were due to volcanism and how many were due to meteorite impacts.

Scientific results from Apollo

Neil Armstrong and Buzz Aldrin brought back 22 kg of rocks and debris from the Sea of Tranquillity (Mare Tranquillitatis) following their landing in Apollo 11 on 20 July 1969 (see Figure 2.6). The material was found to be generally igneous basalt, which confirmed previous indications that the Moon's dark maria were of volcanic origin. The Tranquillity lavas appeared to be about 3.65 billion years old, or about 1 billion years after the Moon was thought to have been formed. So the Moon had once been geologically active, with a molten core. Breccias (impact rocks) were also found, many of which pre-dated the Tranquillity lavas. There appeared to be no trace of any water or organic material in any of the rocks brought back by Apollo 11. So the Moon seemed, from these limited samples at least, to have had no opportunity to support even the most basic forms of life. Basalt samples returned from the next mission, Apollo 12, which landed in the Ocean of Storms (Oceanus Procellarum), were about 500 million years younger than the

Apollo 11 basalts, showing that the maria did not all form at the same time.

The next successful mission, Apollo 14, landed in a region of hills and craters on the edge of the Mare Imbrium. Here the astronauts collected breccias that indicated that the impact that had created the Imbrium basin had occurred about 3.85 billion years ago. Rocks of KREEPy lavas, rich in potassium (K), rare Earth elements (REE), and phosphorous (P), were also found that appeared to have been produced by partial melting of the primitive crust.

Apollo 15 landed near the Hadley Rille at the foot of the lunar Apennines in July 1971. The astronauts discovered what came to be known as the Genesis Rock, which was, at about 4.5 billion years old, an example of the original lunar crust, just 100 million years younger than the Moon itself. Other rock samples confirmed the date of 3.85 billion years, deduced from Apollo 14 samples, for the Imbrium impact that had created the lunar Apennines. The astronauts also brought back samples of lunar regolith containing green, red, brown and yellow pyroclastic glass beads that had, apparently, been brought to the lunar surface from hundreds of kilometres down by fire fountains. On the surface of the beads, which were about 3.3 billion years old, were volatile elements, like zinc, lead, sulphur and chlorine, which were derived from volcanic vapours during the eruptions.

So the history of the Moon was gradually being pieced together as the Apollo 16 mission was being planned. The Moon had apparently formed about 4.6 billion years ago and, within

Figure 2.6. Buzz Aldrin photographed by Neil Armstrong on Apollo 11 near the passive seismic experiment that they had set up on the Moon. The lunar module is seen in the background. The surface of the Mare Tranquillitatis, where they had landed, was seen to be remarkably flat. (Courtesy NASA.)

about 100 million years, a crust had formed of light anorthosite rock as the magma ocean had cooled, with the heavier iron- and magnesium-rich materials sinking. The ALSEP (Apollo Lunar Surface Experiments Package) experiment packages, left on the Moon by the various Apollo missions, had shown that the loose lunar soil, or regolith, was about 2 to 8 metres thick over the maria, and some 15 metres or more over the uplands. This regolith covered the crust, which was now, on average, some 60 km thick. Below that was the mantle of iron- and magnesium-rich rocks which had been the source of the maria (volcanic) basalts when the mantle had still been molten. The numerous craters seemed to have been formed by mete- orite impacts, although the existence of some volcanic craters could not be ruled out. Some rock samples of about 3.8 billion years old were found to have a remanent magnetic field of over 100 times that of the Moon today, indicating that the Moon must have had, at that time, an appreciable magnetic field, possibly caused by dynamo action in a large liquid core. Tem- perature and heat conductivity measurements carried out by ALSEP packages showed that there was still an appreciable heat flow at the surface, which implied that the Moon has, even today, a core temperature of at least 1,200 K.

Apollo 16 landed in the Descartes highlands, as photo- graphic evidence had indicated that the Cayley plains in these highlands were of volcanic origin. In the event, most of the rocks that the Apollo 16 astronauts brought back in 1972 were breccias, with a few anorthosite remnants of the ancient crust. There were no examples of volcanic rock.

The area selected for the Apollo 17 landing, the last Apollo mission, in December 1972 was the Taurus-Littrow valley on the edge of the Sea of Serenity (Mare Serenitatis). Harrison Schmitt, the only geologist to land on the Moon, discovered a layer of orange soil under the surface dust near the edge of a 100 m diameter impact crater. This orange soil proved to be made of very small beads of glass (like those from Apollo 15), 3.7 billion years old, with a high titanium content (of at least 9%). The material had originated deep inside the Moon, but it had been brought to the surface by a fire fountain, produced by the impact that had made the crater 19 million years ago. In addition, the black colour of the valley proved to have been due to volcanic glass generated in a similar way, the difference in colour being due to a different mineral content.

The Apollo ALSEP experiments had shown that the very tenuous lunar atmosphere consisted mostly of helium, which appeared to come from the solar wind, and argon. Then in 1987 Drew Potter and Tom Morgan found minute amounts of sodium and potassium in the lunar atmosphere. Eleven years later Michael Mendillo, Jeffrey Baumgardner, and colleagues found that the Moon had a very tenuous sodium tail, similar to that of a comet, pointing away from the Sun. It appeared to be the result of the bombardment of the lunar surface by micrometeorites.

In 2008 Alberto Saal and colleagues re-examined the vol- canic beads brought back by Apollo 15 and found evidence of water. In the same year magnetic measurements made by Ian Garrick-Bethell and colleagues of a 4.2 billion year old rock returned by Apollo 17 indicated that the Moon had once had a global magnetic field. This implied that it had once had a liquid core and dynamo.

Clementine and Lunar Prospector

Although Apollo provided substantial new evidence on the structure and composition of the Moon, it only produced rock samples from a few small areas of the lunar surface. Equally, although the Lunar Orbiters, and the astronauts in the Apollo command modules, produced unprecedentedly high resolu- tion images of the lunar surface, they were unable to provide a detailed topographic or relief map of the Moon. All this changed in 1994, however, when for $2\frac{1}{2}$ months the Clementine space- craft observed the Moon in eleven different wavebands between 415 nm and 2.78 μm. It also carried a laser altimeter to measure topography.

William Hartmann and Gerard Kuiper had suggested in 1962 that there is probably a large crater or basin on the far side of the Moon that had produced the mountains seen on the Moon's south-western limb. Evidence for something of the sort had been found in the 1970s, and by the Galileo spacecraft in 1990, but in 1994 Clementine showed the full extent of this so-called South Pole-Aitken basin for the first time. It turned out to have a diameter of about 2,500 km and be about 12 km deep.

Clementine enabled a map to be drawn of gravitational anomalies, including a better delineation of the mascons first identified by Lunar Orbiter 1. Putting the gravity anomaly and topographic data together enabled a map to be produced giving an estimate of the thickness of the Moon's crust all over its surface. This indicated that the crustal thickness ranged from about 120 km for parts of the far side, to about 10 km on the near side. The centre of mass of the whole Moon appeared to be offset by about 2 km from the geometric centre in the direction of the Earth.

Analysis of the relative reflectance of the lunar surface in the various wavebands enabled a mineralogical map of the Moon to be produced. This confirmed that anorthosite is the dominant constituent of the lunar highlands, and that the near-side maria have been flooded with iron rich lavas.

It had been pointed out in 1961 by Kenneth Watson, Bruce Murray and Harrison Brown of Caltech that, as the Sun never deviates by more than 1.6° from the Moon's equatorial plane, the floor of some craters very close to the Moon's poles could be in permanent darkness. As a result, they suggested that there may be water ice deposits in such regions, left there by cometary impacts. Clementine's highly inclined orbit was ideal to undertake a search for such deposits, but its results were ambiguous.

In early 1998 the Lunar Prospector spacecraft was put into orbit around the Moon. It showed a reduction in epithermal neutrons, but no decrease in fast neutrons, at both poles. Although not proving the existence of ice, the results were consistent with deposits of water ice, covered with a thin layer of regolith, in the permanently shaded craters at the poles. Multi-waveband analysis indicated that KREEP-rich material had been excavated from the Moon's crust/mantle boundary by the Imbrium impact. The material had been distributed around the Moon, where it has been partially buried by subsequent impacts and by volcanism. The impact that created the South Pole-Aitken basin had also exposed KREEP-rich material. Lunar Prospector found a relatively strong magnetic field at the antipodes of the Mare Imbrium and Mare Serenitatis. Gravity mapping identified a number of new mascons and indicated that the Moon has a 650 km diameter iron rich core.

At the end of its mission in 1999, the Lunar Prospector spacecraft was targeted to impact the Moon in the permanently shadowed area of a polar crater, hoping to liberate water vapour. But no such water vapour was detected by Earth-based telescopes. Radar signals from the Arecibo radio telescope also failed to detect any water ice at the poles in 2006. But three years later water vapour and water ice were detected by the Lunar Reconnaissance Orbiter/LCROSS (Lunar Craft Observation and Sensing Satellite) spacecraft in a dust plume following the impact of its Centaur upper stage with a polar crater.

Bibliography

Beattie, Donald A., *Taking Science to the Moon: Lunar Experiments and the Apollo Program*, Johns Hopkins University Press, 2001.

Chaikin, Andrew, *A Man on the Moon: The Voyages of the Apollo Astronauts*, Penguin Books, 1995.

Doel, Ronald E., *Solar System Astronomy in America: Communities, Patronage and Interdisciplinary Research, 1920–1960*, Cambridge University Press, 1996.

Sheehan, William, and Dobbins, Thomas A., *Epic Moon: A History of Lunar Exploration in the Age of the Telescope*, Willmann-Bell, 2001.

Taton, René, and Wilson, Curtis, (eds.), *Planetary Astronomy from the Renaissance to the Rise of Astrophysics: Part A: Tycho Brahe to Newton*, Cambridge University Press, 1989.

Wilhelms, Don E., *To a Rocky Moon: A Geologist's History of Lunar Exploration*, University of Arizona Press, 1993.

ORIGIN OF THE MOON

Pierre Laplace suggested in his *Exposition du Système du Monde* of 1796 that the protoplanets of the early solar system had thrown off rings of material as they had each cooled, contracted and speeded up. These rings of material had then condensed to form the planetary moons, including that orbiting the Earth.

Eighty years later George Darwin proposed that the Moon had once been part of the early molten Earth. This proto-Earth had gradually contracted as it cooled, and when its rotation rate had increased to about one revolution every 3 to 5 hours, it had broken into two unequal parts: the larger was the Earth and other the Moon. Darwin realised that the rotation of the molten proto-Earth would not have been sufficient on its own to have caused rupture, however. So he suggested that the break-up had been facilitated by resonance coupling between tidal forces acting on the proto-Earth by the Sun, and the natural oscillation frequency of the molten proto-Earth.

After break-up, both the Earth's rotation on its axis, and the Moon's rotation around the Earth, had slowed down because of tidal forces, as both bodies were still molten. This caused the Moon's orbit to gradually become larger. The gravitational attraction of the Earth caused the Moon to have a small Earth-facing bulge, which remained facing the Earth, even after the Moon had solidified.

In 1885 James Nolan pointed out that the Earth would have fractured the Moon by tidal forces when it was still very close, as it would be within the Roche limit. But Darwin pointed out that, if this was the case, the pieces could have coalesced to form the Moon once the pieces were outside the Roche limit.

Darwin's theory was shown, in the early part of the twentieth century, to be untenable. For example, Forest R. Moulton pointed out, in 1909, that the viscosity of the proto-Earth would have been high enough to stop it from breaking up, given the known angular momentum of the Earth-Moon system. If, on the other hand, in some unknown way, the spin rate had been high enough to achieve separation, so much angular momentum would have been transferred from the Earth to the Moon, that the Moon would have escaped completely from the Earth's gravitational pull, and would have become another planet of the Sun. Other objections to Darwin's theory surfaced over the years, so that by the middle of the twentieth century it had generally been abandoned. At that time most astronomers assumed that the Earth and Moon had been formed, in some way, as a pair out of the original solar nebula.

It was thought in the early twentieth century that the Earth and Moon were still cooling down after formation, however they had been formed. Then in 1949 Harold Urey calculated, using radioactive abundances from meteorites, that the interior of the Earth was heating up, rather than cooling down. As a result Urey proposed a radical new theory for the origin of the Earth and Moon, now called the cold accretion theory.

Urey believed that the Earth and Moon had formed separately as a pair of objects from the condensing solar nebula. He then argued that the only way that the Moon could have retained its Earth-facing bulge and supported its high mountains was if it had been formed cold and stayed cold. Urey theorised that the Moon's mare lavas were the result of melting due to impacts, and not due to the upwelling of molten lava from beneath the surface. Furthermore, as the Moon had always been cold there could be no volcanoes on the Moon. His cold Moon theory was disproved by rock samples brought back to Earth by the Apollo astronauts in the late 1960s/early 1970s.

Thomas See suggested in 1909 that the Moon had been captured by the Earth, but his theory was ignored. Then in 1955 Horst Gerstenkorn proposed a similar capture theory. Although capture was feasible, the Moon would have had to approach the Earth down a narrowly defined corridor. If the Moon had approached too fast, it would not have been captured by the Earth, whereas if it had been too slow, it would have collided with the Earth and have been fractured. So, although it was not impossible, the probability of a successful capture would be very low.

Analysis of lunar rocks brought back by the Apollo astronauts showed that the Earth and the Moon have similar relative amounts of oxygen isotopes, indicating that they were both formed in the same part of the solar system. In addition, the relative amounts of these oxygen isotopes for the Earth and Moon were found to be different from those of Mars and the meteorites. This gave powerful evidence for the common origin of the Earth and Moon.

In 1975, William Hartmann and Donald Davis suggested that the Moon had been formed as a result of the off-centre impact of a large planetesimal with the Earth about $4\frac{1}{2}$ billion years ago. A similar theory had been proposed thirty years earlier by Reginald Daly, but his suggestion had been ignored at the time, and another similar theory was proposed in 1976 by Alastair Cameron and William Ward. Hartmann and Davis suggested that the material in the core of the impacting body had been incorporated into the Earth's mantle, but that a cloud of mantle debris from both the impactor and the Earth had been ejected by the collision, and this debris had subsequently aggregated to form the Moon. This theory solved the angular momentum problem of spontaneous fracture, and was consistent with contemporary theories of the origin of the solar system. It also explained why the Moon has proportionally less volatile elements than the Earth, as they were lost by the mantle debris that formed the Moon at the high temperatures generated by the collision.

Various modifications and expansions have been made to this basic collision theory since it was first proposed. But, although none of these versions are completely satisfactory, the collision theory is still the most favoured theory of the origin of the Moon.

Bibliography

Brush, Stephen G., *Fruitful Encounters: The Origin of the Solar System and of the Moon from Chamberlin to Apollo*, Cambridge University Press, 1996.

Canup, Robin M., and Righter, Kevin (eds.), *Origin of the Earth and Moon*, University of Arizona Press, 2000.

Hartmann, W.K., Phillips, R.J., and Taylor, G.J. (eds.), *Origin of the Moon*, Lunar and Planetary Institute, 1986.

Mackenzie, Dana, *The Big Splat, or How Our Moon Came to Be*, John Wiley and Sons, 2003.

3 · Inner solar system

VULCAN

In 1840 Dominique Arago suggested to Urbain Le Verrier that he tackle the problem of Mercury's orbit as it did not seem to be consistent with Newton's gravitational theory. Le Verrier started work on this problem, but shortly afterwards he decided to tackle the problem of Uranus' orbit, which led to the discovery of Neptune. This discovery proved to Le Verrier that Newton's gravitational law was valid, so he decided to return to the problem of Mercury's orbit.

Le Verrier examined the detailed observational data on 14 of Mercury's solar transits from 1697 to 1848, finding that the disagreement between its theoretical and observed orbit was getting larger with time. He calculated that the perihelion of Mercury's orbit was precessing by about $38''$/century more than expected.

One possible explanation was that Mercury's orbit was being perturbed by an unknown planet nearer to the Sun than Mercury. But Le Verrier's analysis indicated that this planet should be easily visible. As a result, he suggested in September 1859 that there may be a ring of small planets between the Sun and Mercury, of such a size that they either had not been observed or had been mistaken for sunspots. So he suggested that astronomers should undertake a careful study of sunspots to see if any of them may be one of these intramercurial planets.

Le Verrier was not the first astronomer to think that there may be a planet or planets inside the orbit of Mercury. For example, in 1826 Karl Harding had suggested that regular observing of the Sun may reveal an intramercurial planet, but 30 years of observing had yielded nothing.

Shortly after Le Verrier inferred the possible existence of intramercurial planets, he received a letter in December 1859 announcing the discovery of such a planet from Edmond Lescarbault, an amateur astronomer, who claimed to have seen a small perfectly circular object crossing the face of the Sun on 26 March 1859. Le Verrier calculated the orbit of Lescarbault's new planet, called Vulcan, assuming that the orbit was circular. He deduced a mass of about 6% that of Mercury, an orbital radius of 0.147 AU, and a period of 19 days 17 hours. Transits should occur around 3 April and 6 October each year. So something like twenty Vulcan-sized planets would be required

to explain Mercury's motion. It seemed doubtful that these could have escaped detection.

Rudolf Wolf had already produced a list of fast-moving spots on the Sun, which could have been observations of intramercurial planets. From this list he suggested that three could be prior observations of Vulcan. J. C. R. Radau used two of these observations, along with those of Lescarbault, to deduce an orbit for Vulcan. His orbit, which was completely different from Le Verrier's, implied that transits would occur on four dates between 29 March and 7 April 1860. But observations by numerous astronomers on these days, and on those predicted by Le Verrier, yielded nothing. Likewise observations during the total solar eclipse of July 1860 yielded nothing either, so more astronomers began to doubt the existence of Vulcan. Instead they proposed other possible causes of the anomalous precession of Mercury's orbit, which Simon Newcomb had calculated in 1880 to be $41'' \pm 2''$/century. But none of these were really satisfactory in explaining the effect.

Doubts still prevailed into the twentieth century as to whether there were any intramercurial planets or not. Then in 1915 Albert Einstein showed, using his new general theory of relativity, that he could explain the anomalous precession of Mercury's perihelion without recourse to such planets. His calculated value of $43''$/century was, within error, the same as that calculated by Le Verrier and Newcomb. This showed that Newton's theory now needed supplementing in certain cases by Einstein's general theory of relativity.

Bibliography

Baum, Richard, and Sheehan, William, *In Search of Planet Vulcan: The Ghost in Newton's Clockwork Universe*, Plenum Press, 1997.

MERCURY

Mercury is so small and close to the Sun that it was impossible for early telescopic astronomers to detect any visible features, although Giovanni Zupus did apparently observe its phases in 1639. Its orbit, which seemed to be at variance with Newton's law of gravity (see *Vulcan* article), gained a great deal of attention, particularly in the nineteenth century. But this problem was not solved until the early twentieth century when

Mercury's anomalous precession was explained by Einstein's general theory of relativity.

The discovery of Encke's comet in the early nineteenth century enabled Mercury's mass to be determined, for the first time, by observing deviations in the comet's orbit when it was close to the planet. As a result, in 1835 Johann Encke was able to calculate an approximate mass for Mercury.

John Flamsteed had detected a luminous ring around Mercury during its solar transit of 5 May 1707, which he attributed to a thick atmosphere surrounding the planet. In 1792 Johann Schröter also concluded, from his observations of the gradual reduction of light across Mercury's partially illuminated disc, that the planet had an atmosphere. He was more confident of this when he, and a number of other observers, saw a luminous ring around the planet during the transit of 1799. Observation of the ring was only occasionally reported during the nineteenth century, however, so that by the end of that century most, but not all, astronomers had concluded that the ring was an optical illusion.

In 1871 Hermann Vogel observed Mercury's spectrum and concluded that it had water vapour in its atmosphere. In the same year Angelo Secchi suggested that Mercury had clouds in a dense atmosphere. But nineteenth-century photometric observations, particularly those of Johann Zöllner and Gustav Müller, tended to suggest that Mercury had, at most, a thin, mainly transparent atmosphere.

Observations by Johann Schröter in 1800, Leeson Prince in 1867, and William Denning in 1882 seemed to indicate that Mercury's rotation period was about 24 hours. But observations by Giovanni Schiaparelli between 1882 and 1889, and by Percival Lowell towards the end of the nineteenth century indicated that Mercury's rotation was synchronous at 88 days. Interestingly, Daniel Kirkwood had predicted this synchronous rotation in the 1860s, because of the tidal effects on its crust by the relatively close Sun. Schiaparelli concluded that Mercury had a cloudy atmosphere, whereas Lowell thought that if Mercury had an atmosphere, it must be very thin.

So by the end of the nineteenth century the idea of a synchronous rotation period was gaining acceptance, but there was still no agreement on whether Mercury had an atmosphere or not, with or without clouds.

In 1929 Eugène Antoniadi confirmed a synchronous rotation period for Mercury after 5 years of intensive observations, during which he also saw evidence of occasional dust clouds. Later, Bernard Lyot and Audouin Dollfus independently confirmed the synchronous rotation period. So a synchronous rotation period was generally accepted by astronomers until William E. Howard and colleagues found in 1962 that Mercury's night side seemed to be warmer than it should be if it was permanently in shadow. Three years later, Gordon Pettengill and Rolf Dyce measured the Doppler shift of radar signals, and concluded that Mercury's rotation period was not 88 days but 59 days,

later updated to 58.65 days, or exactly two-thirds of its orbital rotation period. So Mercury rotated exactly $1\frac{1}{2}$ times on its axis per Mercurian year. This meant that there were only two positions on Mercury's equator that could be directly facing the Sun at perihelion. These sub-solar points, which were on opposite sides of the planet, alternated as sub-solar points at successive perihelia.

Gustav Müller's photometric measurements of Mercury indicated in 1893 that its surface was somewhat darker and rougher than the Moon. Then, in 1924, Bernard Lyot deduced, by measuring variations in polarised light scattered from its surface, that Mercury was covered in volcanic ash. At about the same time, Seth B. Nicholson and Edison Pettit measured the temperature of Mercury. They found a value of about 690 K (about 420°C) at full phase and about 600 K at half phase.

The American Mariner 10, the first spacecraft to visit Mercury, flew past the planet on three consecutive spacecraft orbits during 1974–1975. Unfortunately, the orbital dynamics were such that these intercepts took place at intervals of exactly two Mercurian years. This meant that the spacecraft imaged the same illuminated half of the planet at each fly-by, because the planet rotated exactly three times between successive intercepts.

Mariner 10 showed that Mercury had a surface that looked, at first glance, very much like that of the Moon with a large number of impact craters, although extensive lunar-like seas were noticeably absent. Mercury's primary impact craters appeared to be somewhat shallower than those on the Moon, because of the planet's greater surface gravity. This greater gravitational force had also caused the distance of the ejecta blankets from the crater rims to be smaller than on the Moon. There was no evidence of wind erosion of Mercury's craters, indicating that the planet has not had an appreciable atmosphere for most of its lifetime.

Mariner 10 images showed that Mercury had a 1,350 km diameter basin, called the Caloris basin, whose age, based on cratering density, appeared to be about 3.8 billion years. This basin was ringed by the Caloris mountains, which were about 2,000 to 3,000 m high and up to 100 km across. The floor of the basin seemed to be composed of lava which had been modified to produce a pattern of both concentric and radial ridges and grooves. Robert Strom, Newell Trask and J. E. Guest suggested that the ridges had been produced by compression when the floor of the basin had subsided.

An area called the 'weird' or 'hilly and lineated' terrain was found at the antipodes of the Caloris basin, consisting of hills and mountains up to 1,800 m high arranged in a strange pattern. Peter Schultz and Donald Gault suggested that seismic waves produced by the impact that had produced the Caloris basin had formed these strangely shaped hills and valleys. They further explained how these seismic waves would have been focused by the large planetary core.

Lobate scarps, or thrust faults, up to 500 km long and 2 km high were found to be common on Mercury, providing evidence for the contraction of the planet as a whole by about one to two kilometres radius. This could have been due either to cooling, or to the planet taking on a spherical shape as its spin rate reduced, or both.

Mercury's atmosphere or exosphere was found by Mariner 10 to consist mainly of atomic hydrogen, helium and oxygen at a total pressure of only about 10^{-12} bar. This lack of a meaningful atmosphere, together with Mercury's slow axial rotation, meant that the surface facing the Sun was very hot, but that on the other side just before dawn was very cold. In fact the temperatures measured by Mariner of 430°C (max.) and −183°C (min.) showed the largest temperature range of any planet in the solar system.

In 1985 Andrew Potter and Thomas Morgan, using Earth-based observations, detected sodium in Mercury's exosphere. Later observations detected potassium and calcium. Then in 2001 Potter and colleagues discovered that Mercury had an anti-sunward sodium tail some 40,000 km long. In 2007 Jeffrey Baumgardner and colleagues detected the tail stretching up to 2.5 million kilometres from the planet.

Radar studies of Mercury in 1991 by Duane Muhleman, Martin Slade and Bryan Butler indicated that there may be water ice at its north pole. Later observations showed similar results for the south pole. This completely unexpected observation, of potential water ice on a planet so close to the Sun, could be explained by the fact that the axis of Mercury is virtually perpendicular to the plane of its orbit around the Sun. As a result, the Sun, as seen from Mercury's poles, is permanently on the horizon, and so the floors of craters near the poles are in permanent shadow, and always very cold. David Paige estimated in 1992 that the temperature could be as low as 60 K, well below the temperature of 112 K required to retain water ice on Mercury for billions of years.

Nothing had been known about Mercury's magnetic field and associated magnetosphere until the Mariner 10 fly-bys. It had been expected that, because of Mercury's small size, its core would no longer be fluid. And even if a small part of its core was fluid, it was thought that its slow axial rotation would be too slow to sustain dynamo activity. But the Mariner 10 fly-bys showed that Mercury had an appreciable, probably dipolar magnetic field, together with a very small magnetosphere. Mariner also discovered intense, very short bursts of energetic particles in Mercury's magnetotail, although the planet's magnetic field was too small to maintain a belt of trapped charged particles.

The MESSENGER spacecraft that arrived at Mercury in 2008 confirmed that Mercury's magnetic field was dipolar, with its magnetic axis closely aligned to its spin axis. But its magnetic equator was found to be offset relative to its geographic equator by about 500 km. Mercury's magnetic field was thought to be produced by dynamo action in its outer core, which recent Earth-based measurements had indicated was still probably fluid. MESSENGER found water group ions in Mercury's exosphere, and found that the Caloris basin was about 200 km larger than previously thought. The spacecraft's images also cleared up a problem following the Mariner 10 mission as to whether Mercury's planes were volcanic or not. MESSENGER found clear evidence of volcanic structures along the margins of Caloris and elsewhere (see Figure 2.7), indicating that at least some of the planes were of volcanic origin.

Bibliography

Antoniadi, E. M. (Moore, P., trans.), *The Planet Mercury*, Keith Reid, 1974.

Beatty, J. Kelly, Petersen, Carolyn Collins, and Chaikin, Andrew (eds.), *The New Solar System*, 4th ed., Cambridge University Press, 1999.

Butrica, Andrew J., *To See the Unseen: A History of Planetary Radar Astronomy*, NASA SP-4218, 1996.

Domingue, D. L., and Russell, C. T. (eds.), *The MESSENGER Mission to Mercury*, Springer, 2007.

Dunne, J.A., and Burgess, E., *The Voyage of Mariner 10: Mission to Venus and Mercury*, NASA SP-424, 1978.

McFadden, Lucy-Ann, Weissman, Paul R., and Johnson, Torrence V. (eds.), *Encyclopedia of the Solar System*, 2nd ed., Academic Press, 2007.

Sandner, Werner (Helm, A., trans.) *The Planet Mercury*, Faber and Faber, 1963.

Strom, Robert G., and Sprague, Ann L., *Exploring Mercury: The Iron Planet*, Springer Praxis, 2003.

VENUS

In 1610 Galileo Galilei discovered that Venus exhibited a full set of phases, proving that Ptolemy's structure of the solar system was incorrect. As a result, Galileo settled on the Copernican heliocentric system.

Johannes Kepler predicted in his *Rudolphine Tables* of 1627 that there would be a transit of Venus across the Sun four years later, but it was apparently not observed. Subsequently Jeremiah Horrocks produced his own orbital analysis and predicted that the next transit of Venus would take place on 24 November 1639 (Old Style). Horrocks was keen to observe this transit, as he wanted to check his analysis and obtain an accurate measurement of Venus' angular diameter. In the event, both Horrocks and his friend William Crabtree were able to observe part of the transit, enabling Horrocks to measure Venus' diameter as $72'' \pm 4''$ (modern value about $60''$).

In 1646 Francesco Fontana observed that Venus' terminator sometimes appeared to have an irregular outline, which he erroneously took to be due to mountains. Twenty years later

Figure 2.7. A relatively young volcanic vent imaged on Mercury by the MESSENGER spacecraft. (Courtesy NASA/Johns Hopkins University Applied Physics Laboratory/Carnegie Institution of Washington.)

Gian Domenico Cassini observed the movement of a few spots on Venus, from which he deduced a planetary rotation period of about 24 hours.

James Gregory had suggested in 1663 that observations of a transit of Mercury could be used to determine the solar parallax, and hence the distance of the Sun from the Earth. Such a determination required observations from at least two different places on Earth, separated by as large a distance as possible. In 1677 Edmond Halley observed such a transit from St Helena. But when he returned he found that Jean Gallet seemed to have been the only other person to have recorded the transit. Unfortunately, as Halley acknowledged, there were too many problems in comparing their observations, resulting in a highly inaccurate solar parallax. Halley suggested, instead, that transits of Venus would produce more accurate results. Unfortunately for Halley, the next such transit was not due for over eighty years, on 26 May 1761 (Julian date, or 6 June Gregorian date).

Joseph Delisle took up Halley's suggestion and tried to motivate the astronomical community to undertake coordinated observations of the 1761 transit. Various nations eventually decided to send observers to record the transit, but there were problems in making precise timings of the planetary contacts. This resulted in a range of solar parallaxes from 8.3″ to 10.6″. Nevertheless, the lessons learnt were put to good effect for the next transit due on 3 June 1769.

In all over 150 astronomers set out to view the 1769 transit from over 70 sites, stretching from Siberia and Hudson Bay, to Peking, California, and Tahiti. These observations produced a general consensus that the most likely value was about 8.6″

(modern value 8.79″). In 1835 Johann Encke re-analysed the data, and produced a best estimate of 8.571″ ± 0.037″, which is equivalent to an astronomical unit of $153.5 \pm 0.7 \times 10^6$ kilometres.

Several observers saw a luminous ring around Venus during the transit of 1761, which was interpreted by Mikhail Lomonosov as showing that Venus had an extensive atmosphere. About twenty years later, William Herschel noticed markings on the planet which changed randomly over time, leading him to conclude that Venus had a thick atmosphere. Its existence was essentially confirmed in 1842 when Guthrie observed a thin luminous ring surrounding Venus when it was almost in front of the Sun. This ring was again seen by Prince in 1861, and later by many other observers.

Venus and Mercury were observed so close together by James Nasmyth in 1878 that both could be seen at the same time in his telescope. He noticed that, in spite of the fact that the intensity of the sunlight on Mercury is far greater than on Venus, Venus appeared very much brighter. According to Johann Zöllner, their albedos were about 0.70 for Venus and 0.13 for Mercury, giving another strong indication that Venus is almost covered in clouds.

It was appreciated that if Venus was almost completely covered in cloud, the movement of any dark or light patches may not provide a true estimate of the planet's spin rate, but rather that of its clouds. Any inaccuracy due to this was expected to be small, however. In 1788 Johann Schröter, like Cassini before him, deduced a spin rate of about 24 hours by observing the movement of a 'filmy streak'. Then in the following year Schröter observed that the planet's southern horn appeared to be too short, with a bright point just inside the dark region, which he attributed to a very high mountain. Schröter's idea of a mountainous south polar region lasted well into the nineteenth century, even though it had been dismissed out of hand by William Herschel in 1793.

A number of astronomers in the nineteenth century deduced an estimated spin rate for Venus of about 24 hours, although one observer, Hussey, produced a figure of 24 *days*. Then in 1890 Schiaparelli concluded that Venus' spin was synchronous like that of Mercury, giving a spin rate for Venus of 224.7 days. Shortly afterwards a number of astronomers came to the same conclusion, but others stood by the 24 hour figure. So at the end of the nineteenth century Venus' spin rate was still an open issue, with the two main alternatives being around 24 hours or the synchronous rate of 224.7 days. The idea of synchronous rotation had its theoretical attractions, as it could easily be explained by the long-term effect of tidal friction on the planetary surface.

Spectroscopic observations of Venus yielded conflicting results in the nineteenth century. A number of astronomers, including Huggins, Vogel and Janssen, thought that they had detected oxygen and water vapour lines in its atmosphere, but W. W. Campbell, who used the powerful Lick telescopes, could find no such lines.

There was considerable confusion in the early part of the twentieth century about Venus' rotation period. All sorts of periods were proposed between about 24 hours and synchronous. In 1924 W. Coblentz and Carl Lampland measured the temperature of Venus' light and dark hemispheres as about 320 and 260 K (about 50° and −10°C), respectively. This temperature of the dark hemisphere indicated that it cannot be in permanent darkness, so the spin rate cannot be synchronous. Then in 1957 Charles Boyer found a distinctive V-shaped pattern of Venus' clouds that had a four-day period. When this was announced, Henri Camichel found the same period on photographs taken in 1953. The rotation period of Venus' clouds, at least, appeared to be 4 days retrograde. This finding was not universally accepted at the time, however, and it was not until 1974 that it was proved to be correct by the Mariner 10 spacecraft.

In May 1962, Roland Carpenter detected Venus' rotation using radar. It appeared to be retrograde, but he was unsure of its actual period. In the following year Richard Goldstein and Carpenter deduced a period of 240 ± 50 days retrograde, and in 1965 a more accurate period of 243 days retrograde was measured.

Vesto Slipher concluded in 1908 that there was no oxygen or water vapour in the upper part, at least, of Venus' atmosphere. Then in 1932 Walter Adams and Theodore Dunham, whilst supporting this conclusion, found that carbon dioxide was clearly present. A few years later, Rupert Wildt calculated that the greenhouse heating caused by this carbon dioxide could produce a surface temperature as high as 408 K (135°C).

In 1954 Donald Menzel and Fred Whipple resurrected the popular nineteenth century view that there were oceans of water on Venus, with clouds of water vapour. Menzel and Whipple based their idea that the clouds consisted of water vapour on polarisation measurements by Bernard Lyot in 1929, which water vapour fitted well. They envisaged Venus as being in a state similar to that of the Earth millions of years ago, but Venus would be hotter as it was closer to the Sun. But the key question was, how hot was Venus? If the surface was hotter than the boiling point of water, it must be completely dry, but if it was cooler Menzel and Whipple might be correct.

Mayer, McCullough, and Sloanaker analysed the 3.15 cm radio emissions from Venus and concluded in 1956 that its surface temperature was about 600 K (or approximately 300°C). It appeared to demolish Menzel and Whipple's idea about surface conditions, but many astronomers doubted these very high temperature values, arguing that some of the 3.15 cm radio emissions could be caused by non-thermal processes. A little later Gerard de Vaucouleurs deduced a surface atmospheric pressure of the order of five bar. Then Carl Sagan produced an estimate of 100 bar. So as the space age got under way it was still not known whether Venus was hot and humid, with oceans of water and clouds of water vapour, as envisaged by Menzel and Whipple, or whether it was so hot that the surface was one great desert, with no water to be found anywhere. And was the surface pressure really 100 bar, or was it a much more believable 5 bar? In 1960 both concepts seemed equally valid.

In the 1960s radar observations of Venus from Earth produced relatively crude maps of the surface. In 1967 Ray Jurgens found one high-reflectivity feature called Maxwell. Five years later radio interferometric measurements by Goldstein showed clearly defined craters on Venus for the first time. This was something of a surprise as it had been thought that Venus' thick atmosphere would have broken up incoming objects, or at least slowed them down sufficiently, to avoid them leaving any large craters.

The American Mariner 2 spacecraft was the first spacecraft to visit Venus, flying past the planet in December 1962. It measured a surface temperature of about 450°C (720 K) and a surface atmospheric pressure of 20 bar. So the surface must be a desert, as liquid water could not possibly exist under these conditions. The spacecraft's infrared radiometer showed that the cloud tops were between 60 and 80 km above the surface, with cloud-top temperatures of from −30°C to −50°C. Mariner 2 found that Venus had no measurable magnetic field and no discernable radiation belts.

Venera 4, the first successful Soviet planetary spacecraft, consisted of a main spacecraft with an ejectable probe that was designed to land on the surface of Venus. It arrived at Venus in October 1967. Data transmission from the landing probe stopped, at what Soviet scientists believed to be the surface, when it measured an atmospheric pressure of 20 bar and a temperature of 270°C. The atmosphere was found to be composed of about 96% carbon dioxide, with the remainder being nitrogen, water vapour and oxygen. (The detection of water vapour was later found to be erroneous.)

The American Mariner 5, which flew past Venus just one day after Venera 4's capsule had landed, found that the surface temperature was at least 430°C, with a surface-level atmospheric pressure in the range of 75 to 100 bar. For some time there was concern about the apparent disagreement between the surface temperature and surface atmospheric pressure deduced from the Venera 4 landing probe (270°C and 20 bar) and from Mariner 5 (at least 430°C and 75 to 100 bar). It was later found that this was because the Venera 4 probe had stopped transmitting before it reached the surface. The greenhouse effect, caused by the large concentration of carbon dioxide, was responsible for the very high surface temperatures.

In December 1970 a capsule released from the Soviet Venera 7 spacecraft was the first capsule to land successfully on the surface of Venus and make in situ measurements. Eighteen months later a capsule from Venera 8 also landed on Venus. They both measured a ground-level temperature of about 470°C even though the Venera 7 capsule had landed on the night side of the planet, whilst the other had landed on the daylight side. This was because of the very high density of the atmosphere, whose surface pressure was measured by the Venera 8 lander to be 91 ± 1.5 bar. The atmosphere was found to be largely transparent below about 32 km altitude but, because of the thick overlaying clouds, only about 2 to 3% of the Sun's light reached the surface.

The Venera 8 lander measured a wind speed as it descended of from 360 km/h at 48 km altitude to only 4 km/h below 10 km. Gamma-ray spectroscopy indicated that the surface at the landing site was probably composed of volcanic rock. This suggested that Venus had once been hot enough to produce a differentiated structure.

The Venera 9 and 10 spacecraft which started orbiting Venus in October 1975 both detached capsules to land on the surface. The Venera 9 lander touched down about 2,500 metres above the mean Venusian surface level on the slope of what we now know to be the shield volcano Rhea Mons. The Venera 10 capsule landed at the foot of another shield volcano now called Theia Mons. On the surface the γ-ray spectrometers measured the chemical composition of the rocks which appeared to be similar to slowly cooled basalt, confirming the volcanic origin deduced from the Venera 8 data. But the most excitement was generated by the receipt of a single monochromatic photograph of the surface from each of the landers. They both showed a surface with numerous rocks with little or no sand. The rocks at the Venera 9 site had sharp edges, with no evident erosion, whereas those at the Venera 10 site appeared to be somewhat

older as they showed clear evidence of erosion. Nevertheless, γ-ray analysis showed that the material at both sites was younger than at the Venera 8 site. Overall it appeared as though the surface of Venus was relatively young.

Two American Pioneer-Venus spacecraft arrived at Venus in 1978. One was an orbiter, the Pioneer-Venus Orbiter (PVO), and the other, the Pioneer-Venus Multiprobe (PVM), consisted of four atmospheric probes, none of which were designed to survive their surface impact. The PVO included a radar altimeter designed to provide the first detailed map of the surface of Venus.

The PVM atmospheric probes confirmed the theoretically based predictions made in 1973 by Godfrey Sill, Andrew Young and Louise Young that the Venusian clouds were composed mainly of sulphuric acid. So finally the true nature of the beautiful evening star as seen from Earth had been revealed. Not only was its surface hot enough to melt lead, under an atmospheric pressure of ninety times that on Earth, but its clouds were made of sulphuric acid.

Over the next two years the PVO radar mapped 93% of the surface. It showed a relatively flat surface, although there were exceptions like the 11,500 m high Maxwell Montes (the same Maxwell as previously detected by Earth-based radar) and the 2,000 m deep Diana Chasma. Two large elevated plateaux were found, namely Ishtar Terra in the far north and Aphrodite Terra just south of the equator. Ishtar Terra was found to have in its western half a 2,500 km plain, called Lakshmi Planum, that rises about 4,000 m above the rolling lowland plains. Two shield volcanoes, called Rhea Mons and Theia Mons, which are much larger than any on Earth, were also found in the Beta Regio highland region.

Infrared images from the PVO showed a circumpolar collar of very cold air, about 2,500 km from the pole, 1,000 km across and 10 km deep. The collar, which was at an altitude of about 70 km, was just above the clouds and was about 30°C colder than the atmosphere on either side. Surprisingly, at an altitude of 85 km the PVO found that the Venusian atmosphere was warmer at the poles than at the equator.

The PVO had not imaged the polar regions. So the Soviet Union rectified this for the north polar region with the launch of two radar-carrying spacecraft, Veneras 15 and 16, which arrived at Venus in October 1983. Their surface images showed many ridges and narrow valleys which appear to have been formed by the horizontal motion of the crust. Several ten to forty metre diameter craters were found in the elevated Lakshmi Planum, but few craters were seen in the lowland regions suggesting that they may be young lava plains. Crater counts in some lowland regions gave an age as young as about 300 million years. Two unusual circular features, called Anahit Corona and Pomona Corona, were observed, which appear to have been formed by a mixture of both volcanic and tectonic processes. These two large corona structures, and a number of smaller ones, are unique to Venus as far as we know. Veneras 15 and 16 confirmed that Beta Regio is split by a rift valley.

The Soviet Union launched two spacecraft, Vegas 1 and 2, designed to intercept Venus in 1985 on their way to intercept Halley's comet in the following year. Each spacecraft released a landing module and balloon system as it flew past Venus, the balloon systems being designed to drift in the Venusian atmosphere.

Balloon 1 was released from the Vega 1 lander at a latitude of 7° N. It drifted around the planet at an altitude of 54 km where the horizontal wind velocity was found to be about 240 km/h, with downward gusts of up to 12 km/h. Balloon 2 was released from the Vega 2 lander at a latitude of 7° S. The horizontal velocity was about the same as for Balloon 1, but for the first 20 hours the downdrafts for Balloon 2 were very light. After 33 hours, however, it came into a very turbulent area after it had crossed over a 5 km high mountain peak. This turbulence continued for another 2,000 km.

The American Magellan spacecraft was launched to Venus in May 1989. On board was an imaging radar and a radar altimeter with unprecedented resolutions. Crater counts showed that most of the surface appeared to be about 400 million years old, with no features older than about 900 million years. Magellan found thousands of volcanoes, although none seemed to be active at the time of observation.

Most of the 900 impact craters imaged by Magellan were fresh-looking, suggesting that atmospheric weathering on Venus is a slow process. The largest crater imaged was a highly modified, double-ringed crater about 275 km in diameter. The lack of any larger craters, and the relative paucity of other large double-ringed craters, was due to the relative youthfulness of the planet's surface. Interestingly no craters were found of less than about 3 km in diameter because the small meteorites, which would normally have produced these smaller craters, would have burnt up in the very thick atmosphere. Craters of between 3 and 20 km in diameter were found to have uneven floors with signs of rubble on them, indicating that the meteorites that caused them were breaking up just before impact. Craters in the 20 to 70 km range were found to be generally pristine in appearance, with a central peak and smooth floor of solidified lava. Some craters were found to have an asymmetry in their ejecta blankets. These were most likely caused by the wake of the impacting meteorite in the thick atmosphere preventing the ejecta being deposited in a back-scattering direction.

The solidified lava flows imaged by Magellan showed characteristics not seen on Earth. In some places the lava had obviously been extremely fluid, running down slopes as shallow as 1°, whilst in other places it had been very viscous. On one hand, the fluid lava behaved in many ways like water on Earth cutting channels up to about 7,000 km long in previous lava deposits. This produced features that look more like rivers of water than lava flows. On the other hand the very viscous lava produced a number of pancake-like volcanic domes about 20 to 60 km in diameter and about 100 to 750 m high (see, for example, those shown in Figure 2.8). Some of these pre-dated faults which have now split them, and some cover faults, obviously post-dating them.

Venus shows clear signs of past tectonic activity in the highland regions. The deformational (tectonic) features show the results of both compressional and extensional forces. Rifting of the crust has occurred to produce relatively shallow chasmas and abundant faulting in the Aphrodite Terra and Beta Regio highlands, whereas compressional forces have produced the mountain belts of Ishtar Terra. It is unclear whether there is any limited resurfacing proceeding today due to local tectonic activity or not. There is no global network of faults like on Earth, and no evidence of subduction, showing that there has been no obvious *plate* tectonic activity over the last few hundred million years.

The European Space Agency's Venus Express spacecraft was placed into orbit around Venus in April 2006, primarily to study Venus' atmosphere. Venus Express imaged both the horizontal and vertical structure of the double-eyed vortices over both poles. It found that the difference of wind velocity with height, which was about 150 km/h between about 46 and 66 km at low southern latitudes, disappeared above 70° S, with all atmospheric layers co-rotating with the south polar vortex. The spacecraft discovered hydroxyl in Venus' atmosphere, this being the first time that hydroxyl had been observed in the atmosphere of any planet other than the Earth. It also detected the loss of various ions, including oxygen and helium, from the planet's upper atmosphere, caused by its interaction with emissions from the Sun.

Venus Express provided high resolution images of the atmospheric 'ultraviolet absorbers' which absorbed large amounts of the incident solar ultraviolet. It generated images of Venus' sulphuric acid clouds in three dimensions, allowing their dynamics to be studied, and produced temperature maps of the surface and of the atmosphere at different altitudes.

Bibliography

Butrica, Andrew J., *To See the Unseen: A History of Planetary Radar Astronomy*, NASA SP-4218, 1996.

Chauvin, Michael, *Hōkūloa; The British 1874 Transit of Venus Expedition to Hawai'i*, Bishop Museum Press, 2004.

Cooper, Henry S. F. Jr., *The Evening Star; Venus Observed*, Johns Hopkins University Press, 1994.

Grinspoon, D.H., *Venus Revealed*, Perseus, 1998.

Maor, Eli, *June 8 2004: Venus in Transit*, Princeton University Press, 2000.

Marov, Mikhail Y., and Grinspoon, David H., *The Planet Venus*, Yale University Press, 1998.

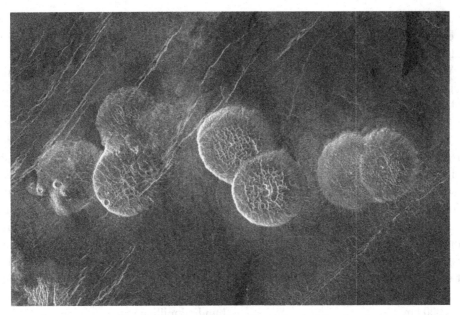

Figure 2.8. These pancake-like domes, imaged on Venus by the Magellan spacecraft, are the result of very viscous lava oozing on to the surface. They have steep sides and flat tops and are about 750 m high and 25 km in diameter. (Courtesy NASA/JPL/Caltech.)

Maunder, Michael, and Moore, Patrick, *Transit: When Planets Cross the Sun*, Springer-Verlag, 2000.

Moore, Patrick, *The Planet Venus*, 2nd ed., Faber and Faber, 1959.

Roth, Ladislav E., and Wall, Stephen D., (ed.), *The Face of Venus: The Magellan Radar-Mapping Mission*, NASA SP-520, 1995.

Surkov, Yuri, *Exploration of Terrestrial Planets from Spacecraft: Instrumentation, Investigation, Interpretation*, 2nd ed., John Wiley, 1997.

MARS

The movement of Mars across the sky had been studied in some detail by a number of civilisations of the first millennium BC. But it was not until Johannes Kepler studied its movement in the first decade of the seventeenth century, using pre-telescopic observations made by Tycho Brahe and himself, that its true orbit was determined. This resulted in Kepler announcing in his *Astronomia Nova* of 1609 that Mars orbited the Sun in an ellipse.

Christiaan Huygens observed what is now called Syrtis Major on Mars in 1659. He used it to deduce a rotation period of Mars of about 24 hours. A few years later Jean Dominique Cassini deduced a more accurate period of 24 h 40 min, which is within three minutes of the correct value. Cassini also discovered the bright Martian polar caps.

The first systematic investigation of Mars' polar caps was undertaken by Giacomo Maraldi, who found that the south polar cap had completely disappeared in late 1719, although it later returned. When the south polar cap was very small, he also noticed that it had a slight oscillation at Mars' rotation period, indicating that it was not exactly centred on the south geographical pole. In 1783 William Herschel also noticed this slight oscillation, and concluded that the centre of the south polar cap was about 9° from the south pole. Herschel suggested that the reason why the polar caps changed size with season was because they consisted of ice and snow that melted in their local summer.

Cassini thought that he had detected the presence of a very large, dense atmosphere on Mars during his observations of 1672, when he noted that a bright star had disappeared 6′ before its occultation by the planet. Herschel repeated the observation about one hundred years later with two other stars and found no such effect. However Herschel did notice that the dark surface markings on Mars appeared to be transient. So he concluded that the surface of Mars was often obscured by clouds in a moderate atmosphere, something like that on Earth. Herschel, like Immanuel Kant before him, was convinced that Mars was inhabited by intelligent beings, as both the Earth and Mars appeared to have many similarities. For example, both appeared to have ice and snow at their poles, and clouds in their atmospheres.

Wilhelm Beer and Johann Mädler, who observed Mars during its favourable opposition of 1830, noticed that the dark features were in fixed locations, but that they were frequently covered in mists, thus changing their contrast and outlines. Syrtis Major was clearly seen, as was a small round dark patch, now called Sinus Meridiani, that had previously been recorded by William Herschel and Johann Schröter. Using this marking, Beer and Mädler were able in 1837 to deduce Mars' spin rate to within about a second of its true value.

In 1830 Beer and Mädler observed that the south polar cap was at its smallest extent of about 6° about a month after the summer solstice in the southern hemisphere, adding to William Herschel's suspicion that it was made of ice and snow. They also noticed in 1837 that the north polar cap was about 12° at its smallest extent. But whilst the north polar cap was larger than the south polar cap during their respective summers, it appeared to be the other way round during their winters. This was attributed to the fact that southern summers, which occurred near perihelion, were warmer than northern summers, whilst southern winters were colder than those in the north.

At the end of the eighteenth century most astronomers had thought that the reddish colour of Mars was due to its atmosphere. But in 1830, John Herschel suggested that it was the true colour of the surface material. Camille Flammarion, on the other hand, concluded that it was the colour of its vegetation.

John Herschel noted that the dark patches on Mars sometimes had a green tint, suggesting to him that they were seas. Then in 1863 Angelo Secchi interpreted the white polar regions as being due to either snow on the ground or clouds over the poles. Either way, this implied to him that there was water on Mars, leading him to conclude, like John Herschel, that the dark areas were seas that were bordering reddish-coloured continents. In 1860, however, Emmanuel Liais suggested that the dark areas were areas of vegetation, whilst the reddish areas were deserts.

Giovanni Schiaparelli produced a map of Mars following its opposition of September 1877 that showed a network of linear features that he called *canali*. This was translated incorrectly into English as canals, which implied that they were built by intelligent beings. Schiaparelli and others saw more *canali* in subsequent years, some of them being double. But other, equally competent, observers could not see the *canali* at all.

Schiaparelli believed that the large, dark areas of Mars were shallow seas, so he was not surprised when he noticed in 1879 that some of their outlines were somewhat different from two years earlier. He also noticed what turned out to be a significant new feature, a very small white patch that he named Nix Olympica (the Snows of Olympus).

In 1892 William Pickering caused a great deal of controversy when he reported seeing a number of lakes on Mars at the junction of some of the canals. Two years later Percival Lowell confirmed the existence of these 'lakes', and reported that their seasonal darkening, and that of the canals, indicated that water was flowing along the canals from the melting polar caps towards the equator. Lowell suggested that Pickering's lakes were oases in the red desert of Mars. Moreover he concluded that the canals, which he observed to be prolific, had been built by intelligent beings to transport precious water from the melting polar caps to irrigate the dark areas, which were areas of vegetation. Such a startling conclusion inevitably caused

more controversy, which continued well into the twentieth century.

Spectroscopic observations of Mars in the late nineteenth century yielded conflicting results. Some astronomers detected oxygen and water vapour lines, whereas W. W. Campbell at the Lick Observatory could find none. There was also a problem with the polar caps. Calculations showed that the average temperature of Mars, given its distance from the Sun, should be about $-34°C$. If that was the case, the polar caps should not melt nearly as much as observed during their spring and summer seasons if they had been made of water ice or snow. So in 1898, A. C. Ranyard and G. Johnstone Stoney suggested that the caps could be made of frozen carbon dioxide. But there appeared to be a melt band at the edge of the caps in spring, and yet carbon dioxide should sublimate directly into gas on Mars.

Two satellites of Mars, now called Phobos and Deimos, were discovered by Asaph Hall in 1877. Their orbits were extremely close to the planet, and the satellites were both very small. As a result they were thought to be captured asteroids.

In the early twentieth century it seemed clear that the yellow clouds seen on Mars were of dust. Then in 1909 Georges Fournier and Eugène Antoniadi found that they appeared to cover almost the whole of the planet for a while. Antoniadi later found that the clouds tended to occur around perihelion, when the solar heating was greatest, and so appeared to be produced by thermally generated winds. Thirty years later, Gérard de Vaucouleurs measured the wind velocities as being typically in the range 60 to 90 km/h when the clouds first formed.

There was a great deal of uncertainty about the surface of Mars in the first half of the twentieth century. It was thought unlikely that the linear *canali* really existed, but they were still recorded from time to time by respected observers. Antoniadi concluded that the large dark bluish-green areas on Mars changed in both colour and shape with the seasons, and agreed with Lowell that they were probably areas of vegetation. Gavril Tikhov compared the spectra of some of these regions with those of localities on Earth, and concluded that the Martian areas had vegetation similar to some subarctic regions on Earth. A number of astronomers speculated that these dark areas of Mars could be covered in a very basic form of life, like moss or lichens.

Coblentz, Lampland and Menzel measured the temperature of Mars during its 1926 opposition and concluded that its average noon surface temperature ranged from about $-30°C$ at 55° N latitude, to $+35°C$ at the subsolar point at 15° S, to 0°C at the south pole. The south pole was warmer than mid-northerly latitudes because it was in permanent sunlight at that time of year. As a result they concluded that the south polar cap was probably composed of water ice and snow, as it was far too warm for solid carbon dioxide.

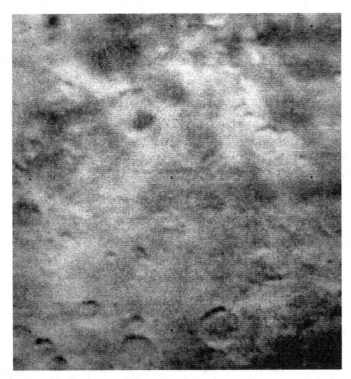

Figure 2.9. This image, taken by Mariner 4 in 1965, was the first to show craters on Mars. (Courtesy NASA.)

In 1947 Gerard Kuiper found clear evidence for a small amount of carbon dioxide in Mars' atmosphere, and in 1963 Audouin Dollfus, and Hyron Spinrad and colleagues, working in parallel, found a trace amount of water vapour. Estimates of the surface atmospheric pressure, made in the first half of the twentieth century, generally ranged from about 25 to 120 millibars. In 1963 a figure of 25 ± 15 millibars was estimated by Kaplan, Münch and Spinrad from Mars' infrared spectrum, with a partial pressure for carbon dioxide of about 4 millibars. No oxygen was detected. In 1964 a NASA-sponsored summer school, using all the most up-to-date evidence, settled on a range of 10 to 80 millibars as the likely surface atmospheric pressure.

The American Mariner 4, which was the first successful spacecraft to fly past Mars, provided a big surprise to most astronomers when it discovered craters on Mars in July 1965 (see Figure 2.9). Instead of a planet with patches of vegetation, like moss or lichens, the spacecraft showed a cratered, desolate world resembling the Moon, with the dark areas visible from Earth simply being low-albedo areas.

Craters had apparently been seen on Mars by Edward Barnard in 1892, and by John Mellish in 1915, although not everyone believed that Barnard and Mellish had seen craters. However, their existence had been predicted by D. L. Cyr in 1944, Clyde Tombaugh, Ernst Öpik, Ralph Baldwin and, most recently, Fred Whipple. Nevertheless, the discovery of craters on Mars was a major surprise to the astronomical community as a whole.

The atmospheric pressure measured by Mariner 4 was also a surprise as it turned out to be only about 4 to 7 millibars at ground level, rather than the 10 to 80 millibars anticipated. This low pressure implied that $\sim$95% of the atmosphere must be carbon dioxide, given Kaplan, Münch and Spinrad's carbon dioxide partial pressure of 4 millibars.

The Mariner 4 images showed no evidence that water had ever existed on Mars, and water could not exist there now with such a low atmospheric pressure and at the low surface temperatures previously measured by Earth-based instruments (Mariner 4 made no temperature measurements). The atmosphere was also so thin that it would not shield the surface from solar ultraviolet radiation which would have killed off any micro-organisms long ago, even if there had been water to sustain them. So conditions were highly unfavourable for life on Mars now, and it appeared probable that there had never been any form of life in the past.

Mariner 4 also found no measurable magnetic field on Mars, indicating that it had no molten magnetic core. The apparent lack of volcanoes also indicated that the layer of rock near the surface had not been molten for a long time. Finally Mariner found no Van Allen radiation belts, which was consistent with there being no measurable magnetic field, so charged particles emitted by the Sun would impact the surface virtually unimpeded, again being a threat to life.

Although the Mariner 4 data indicated that Mars was unlikely to have supported life in the past, some astronomers pointed out that this spacecraft had imaged only about 1% of the surface at a resolution of only three kilometres. The other 99% of the planet may look different, and evidence of the previous existence of water may be seen when the resolution was significantly improved. In addition, the atmosphere may have been significantly denser in the past, which would have allowed significant amounts of water to have existed on the surface. So although Mariner 4's observations made the existence of life unlikely now, they did not rule out the possibility of life in the past.

The next two spacecraft to successfully visit Mars, Mariners 6 and 7, flew past the planet in 1969. Mariner 6 imaged more cratered terrain, which was no longer a surprise. It found that the atmosphere was almost 100% carbon dioxide, which was a disappointment to those people hoping to find conditions suitable for life. It measured a surface temperature that varied typically from about 15°C during the day to −75°C at night, suggesting that the surface was covered in a highly insulating material such as dust.

Mariner 7 imaged more cratered terrain, although the Hellas plain had no discernible craters. So this area was either very sandy, allowing the craters to be filled in shortly after they had been created, or was much younger than the other areas observed. The surface of the south polar ice cap appeared to be covered in frozen carbon dioxide in the form of ice or snow.

Mariner 7 detected only minute amounts of water vapour, with clouds of dust and carbon dioxide crystals, in Mars' almost 100% carbon dioxide atmosphere.

The American Mariner 9 was the first spacecraft to orbit Mars. It arrived in November 1971 in the middle of a planet-wide dust storm. Whilst waiting for the storm to blow itself out, the spacecraft took the first detailed images of the two small Martian satellites, Phobos and Deimos, showing them to be irregularly shaped and covered in craters. The largest crater on the $27 \times 21 \times 19$ km irregularly shaped Phobos was found to be a massive 9 km in diameter, although the craters on Deimos were significantly smaller.

As the dust storm gradually abated, Mariner 9 revealed a completely different Mars from that of its Mariner predecessors. Mars was seen to have massive volcanoes and an enormous canyon system. The largest volcano, Olympus Mons in the Tharsis region, was 25,000 m (80,000 ft) high and 500 km (300 miles) in diameter, which was far larger than any volcano on Earth. Schiaparelli had detected a small whitish patch in this area in 1879, and had christened it Nix Olympica. Mariner 9 imaged three other large volcanoes in the Tharsis region, where a number of ground-based observers had also seen small whitish patches. All these patches were now seen to be due to clouds near the tops of the volcanoes.

Mariner 9 imaged the enormous canyon system (see Figure 2.10), now called Valles Marineris, that was found to be up to 4,000 km (2,500 miles) long, 200 km (120 miles) wide, and 6,000 m (20,000 ft) deep. The spacecraft also discovered sinuous channels in Chryse Planitia. The Hellas plain turned out to be a 2,300 km diameter impact crater or basin whose floor, at the lowest point, was some 5,000 m below the mean Martian surface level. The white deposits often seen in this Hellas basin were found to be carbon dioxide frost. Mariner 9 also showed that the colour changes in the dark areas of Mars were caused by seasonal winds, with velocities of up to 180 km/h (110 mph), alternately covering and uncovering the darker substrate with lighter dust.

The discovery of large volcanoes indicated that there has been no general plate tectonics on Mars. With no plate movement, hot spots in the underlying mantle would have continued to eject lava through the same place in the crust year after year to build up these enormous volcanoes. In addition, geologists thought that Valles Marineris could possibly be where two plates had started to move apart, but were stopped as Mars cooled quickly, shortly after the crust was formed.

Mariner 9 found that the water vapour content over the south polar region varied from about 15 precipitable microns during its summer to zero in winter, whereas the seasonal variation of water vapour over the north polar region was somewhat larger. So there appeared to be water ice at both poles that partially evaporated during their local summers. If there

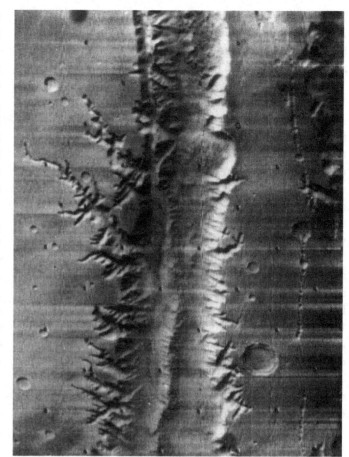

Figure 2.10. A small part of the extensive Valles Marineris on Mars as imaged by Mariner 9 in January 1972. The image covers an area of about 380×480 km. (Courtesy NASA; Principal Investigator Harold Masursky.)

was some water vapour in the atmosphere now, maybe there had been liquid water on the surface earlier in the planet's history.

The American Viking 1 and 2 spacecraft, which were launched in 1975, were designed to search for evidence of life in low-lying areas of Mars, where there may have been surface water in the past. Both spacecraft consisted of an orbiter and a landing module. The landers each included a sophisticated suite of automatic instruments to detect elementary lifeforms. Initially it was thought that these experiments had found signs of such life on Mars. But as time went on, doubts about the interpretation of their results began to increase, so that it is now generally accepted that they showed no such signs.

The Viking 1 lander, which landed at 22.5° N on the western side of the Chryse region, confirmed earlier results that Mars' atmosphere was mostly carbon dioxide. During its descent, it also detected small amounts of nitrogen, argon, carbon monoxide, oxygen and nitric oxide in the upper atmosphere. Shortly after landing, the first-ever image from the surface of Mars was transmitted to Earth. It showed a red, rock-strewn, sandy surface and pink sky. The soil was found to be iron-rich clay, and the atmosphere was found to consist of 95.3% carbon dioxide, 2.7% nitrogen, 1.6% argon and 0.13% oxygen. Interestingly, both the soil and the very fine dust were found to have an abundance of magnetic particles.

It was early summer in the northern hemisphere when the Viking 1 lander touched down, and on the first day the atmospheric temperature varied from −86°C at dawn to −33°C in the early afternoon. The atmospheric pressure was 7.6 millibars, and the wind velocity gusted up to 52 km/h. During the northern winter the atmospheric pressure increased to 9.0 millibars, as the southern polar cap released some of its carbon dioxide. Evidently the southern polar cap was releasing more carbon dioxide in the southern summer than the northern cap was collecting in the northern winter.

The images sent back to Earth by the Viking 2 lander showed a similar red, rock-strewn surface to that seen by Viking 1. The atmospheric temperature at the 48° N Viking 2 site was about five to ten degrees Celsius lower than those at the Viking 1 site. Frost was first observed on the surface near the Viking 2 lander in September 1977 when the temperature was −97°C, after an overnight low of −113°C.

The Viking images showed clear signs of flash floods in the Chryse region and many landslides, some of which seem to have occurred in saturated soil. Dendritic or branching drainage features were also seen resembling terrestrial river systems. Temperature and water vapour sensors showed that the north polar cap was made of water ice in the northern summer at a temperature of about −65°C. It was then covered by carbon dioxide ice in the northern winter. On the other hand, the carbon dioxide ice of the southern cap did not completely melt in the southern summer. There was also evidence of extensive permafrost regions, with the permafrost around the polar caps possibly extending down to a depth of several kilometres in places.

The concentration of water vapour in the Martian atmosphere was found to be highest in low-lying regions and, as with Mariner 9, more water vapour was found during the summer than during the winter, when it was presumably frozen out of the atmosphere. In regions of rough terrain there were marked daily fluctuations in the amount of water vapour, possibly due to changing wind patterns. The greatest concentration of atmospheric water vapour occurred near the edge of the northern polar cap in the northern summer.

The images of both Phobos and Deimos from the Viking orbiters were much sharper than from Mariner 9 and, at one stage, Viking Orbiter 2 was manoeuvred to within only 30 km of the surface of Deimos, producing images with a resolution of 3 m. Phobos was found to have a network of linear grooves, which appear to be fractures caused by the collision that created the 9 km crater called Stickney. On the other hand, Deimos was seen to have a smoother surface than Phobos, and some of its craters appeared to be partially filled by material.

A 1.9 kg meteorite known as ALH 84001 had been picked up in 1984 in the Allen Hills in Antarctica. It was later found to have come from Mars. In August 1996, a team of scientists lead by David McKay and Everett Gibson Jr. announced that they had found evidence in the meteorite for early life on Mars. The team acknowledged that each of their four pieces of evidence could be disputed, but claimed that, taken as a whole, it added up to strong evidence for early life on Mars. They showed an image of elongated, worm-like shapes, which appeared remarkably similar to Earth-based microfossils. The conclusion that this meteorite showed evidence for life on Mars was quickly dismissed by a large number of researchers. These objections have since obtained majority support, although the original researchers are currently still standing by their initial conclusion.

NASA returned to the surface of Mars in 1997 with the Pathfinder spacecraft and its Sojourner rover, which was the first rover vehicle to be placed on Mars. The spacecraft's landing site had been chosen in the Ares Vallis, which seemed to be the site of an ancient flood plain. Sojourner imaged a number of rounded pebbles, some of which were included in the surface of some of the rocks. This configuration was reminiscent of conglomerates on Earth, which are the result of a long-time exposure to running water. Furthermore, a group of large rocks nearby was observed to be generally leaning in the same direction as each other, indicating that they may have been subjected to a strong flow of water in the past.

Most of the rocks analysed had a high silica content, like andesites on Earth. This indicated that they consisted of lava which had been differentiated when molten, with the heavier elements sinking to the bottom. If so, Mars had had a much more active subsurface volcanic system in the past than previously envisaged.

In September 1997 NASA's next Mars spacecraft, Mars Global Surveyor (MGS), arrived at the planet, where it used aerobraking to reach its final Sun-synchronous orbit. MGS confirmed that Mars does not have a dipole magnetic field, but it detected a strong remanent field over the older parts of the Martian surface. This implied that Mars had had a molten core in the distant past.

MGS found that Mars' northern plains were remarkably flat, which led some astronomers to conclude that this area was the dried-up bed of an ancient ocean. This idea was reinforced when it was found that much of the border defining the 'shoreline' was at the same altitude, within ± 280 m, around this hypothesised former ocean. The 2,300 km diameter Hellas impact basin was found to have a total relief of more than 9 kilometres.

The young age of some of the lava flows on Mars was confirmed by MGS. Images of lava flows in the Elysium Planitia region, for example, showed very few impact craters, indicating that they were very young, probably only a few tens of millions of years old. And crater counts on the youngest lava flows on Olympus Mons indicated that these flows were less than 100 million years old.

The spacecraft detected large amounts of grey hematite, which normally formed in the presence of water, in the Meridiani Planum. It also imaged 20 new impact craters in 2006, mostly in the range of from 10 to 30 m in diameter, that had not been present seven years earlier, and found bright new deposits, which also had not been present earlier, of some material in two gullies. In both cases the bright material was thought to have been carried by a fluid that behaved like liquid water, as it had avoided low obstacles in its flow downstream. Its flow could not have been very fast, however, as none of the obstacles that it did impact appeared to have been moved.

Martian meteorites found on Earth have been found to fit into two basic groups. One group consists of meteorites made from lava that solidified on Mars about 1.3 billion years ago, and which were ejected from the planet about 11 million years ago. The second group consists of those that solidified on Mars only 150 to 600 million years ago. Carbonate deposits or clay materials have been found inside these meteorites, indicating that they had been subjected to a water environment before they were ejected from the planet. If liquid water or concentrated water vapour was present on Mars just a few hundred million years ago, it may well be present there today, possibly just under the Martian surface.

Mars Odyssey, which was placed into orbit around Mars in October 2001, was primarily aimed at surface geology. It found a wide diversity of igneous material on the surface of Mars from low-silica basalts to highly evolved lavas. It also found extensive surface areas high in hydrogen abundance polewards of about 60° N and 60° S. This was believed to show that the surface there was well over 50% water ice by volume.

The European Space Agency's Mars Express spacecraft was put into orbit around Mars in December 2003. Mars Express found that the alteration products (phyllosilicates) in the early history of Mars were indicative of abundant water, whereas the mineralogy of the later Mars suggested that the planet was, by then, colder and dryer, with only occasional water on the surface. It found that the interior of the south polar cap was almost completely water ice. Buried water ice at the south polar cap extended down by as much as 3.5 km in places and that, together with buried water ice in the ridged area of Dorsa Argentea nearby, would be enough to produce a 11 m deep global ocean. In addition, Mars Express found a cap of water ice (see Figure 2.11) in a 35 km diameter crater of the Vastitas Borealis region at about 70° N. This water ice is what remained after the overlaying layer of carbon dioxide had sublimated in the local summer. The spacecraft also discovered methane in the Martian atmosphere, sparking a debate as to whether it was caused by some form of life or by volcanic activity. On the other hand, some astronomers put it down to the degradation of olivine.

Crater counts indicated that some lava flows on the flanks of Olympus Mons were only 2.4 million years old, with its caldera floor ranging from 100 to 200 million years old. Lobate deposits near the base of Olympus Mons and of some other large volcanoes suggested that there had been glacial activity as recently as 4 million years ago. Mars Express also detected a number of buried basins elsewhere on the surface.

NASA landed two rovers, called Spirit and Opportunity, onto the surface of Mars in January 2004. Spirit landed in Gusev crater, which was thought to have been flooded by liquid water early in its history, whilst Opportunity landed in the Meridiani Planum, where Mars Global Surveyor had detected a large amount of grey hematite, which is normally formed in the presence of water.

It took some time for Spirit to find any evidence of past water near Gusev crater. Most of the rocks nearby were found to contain olivine, indicating that they were of volcanic origin. However, when Spirit reached the Columbia hills, a short distance away, it found hematite and sulphates, indicating the presence of previous water. Opportunity, on the other hand, soon found a great deal of evidence of previous water. It found large concentrations of sulphates and abundant jarosite, which normally contains water in its crystal lattice. Opportunity also discovered 1 to 5 mm diameter spherules of hematite that had probably been formed by the precipitation of iron in a water-rich environment. Finally its images of layered rocks were reminiscent of those layered rocks that had been formed in shallow water on Earth.

Mars Reconnaissance Orbiter, which was placed into Mars orbit in 2006, detected subsurface water ice under aprons of rocky debris extending from mountains and cliffs in the region between 35° and 60° latitude in both northern and southern hemispheres. Subsurface water ice had not been detected by previous spacecraft at such low latitudes. Mars Reconnaissance Orbiter also detected repetitive layering in Mars' polar caps, and rhythmical patterns in sedimentary rocks which seemed to be linked to Mars' climate cycles. It also found carbonates in the Nili Fossae region, which were closely associated with both phyllosilicate and olivine rocks, and which probably formed about 3.5 billion years ago from the alteration of olivine by water. The presence of carbonates and accompanying clays suggests that the water was either neutral or alkaline at the time of carbonate formation and afterwards, otherwise the carbonates would not have been formed and survived.

Images from Phoenix Mars Lander, which landed in Vastitas Borealis at 68° N in 2008, showed that it had landed on a flat surface cut into polygon shapes, and strewn with pebbles. The spacecraft was the first to directly detect water ice when it heated a soil sample from only 5 cm below the surface. It also detected falling snow which vaporised before it reached the surface. Phoenix found that the soil was alkaline, and detected

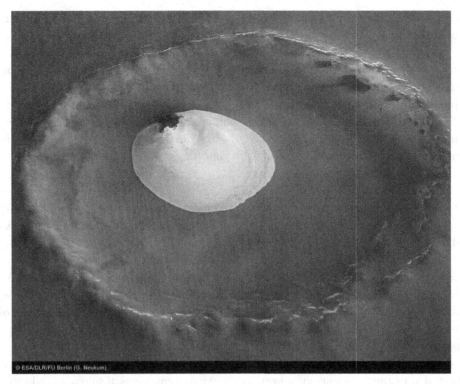

Figure 2.11. A residual cap of water ice imaged at the centre of a crater at about 70° N by the Mars Express spacecraft in 2005. (Courtesy ESA/DLR/FU Berlin (G. Neukum.))

calcium carbonate and phyllosilicates, which are formed in water.

In summary, there is now abundant evidence for subsurface water ice on Mars, water vapour in its tenuous atmosphere, and water ice at its poles. More and more evidence is also being accumulated that there was a substantial amount of liquid water on the surface of Mars in the past. But as yet there is no clear evidence of elementary forms of life on Mars today or in the past.

Bibliography

De Vaucouleurs, Gérard (Moore, P., trans.), *The Planet Mars*, 2nd ed., Faber and Faber, 1951.

Ezell, Edward C., and Ezell, Linda N., *On Mars: Exploration of the Red Planet 1958–1978*, NASA SP−4212, 1984.

Godwin, Robert (ed.), *Mars: The NASA Mission Reports*, Apogee Books, 2000.

Morton, Oliver, *Mapping Mars: Science, Imagination and the Birth of a World*, Fourth Estate, 2002.

Raeburn, Paul, and Golombek, Matt, *Mars: Uncovering the Secrets of the Red Planet*, National Geographic Society, 1998.

Sheehan, William, *The Planet Mars: A History of Observation and Discovery*, University of Arizona Press, 1996.

Surkov, Yuri, *Exploration of Terrestrial Planets from Spacecraft: Instrumentation, Investigation, Interpretation*, 2nd ed., John Wiley, 1997.

4 · Giant planets

ATMOSPHERIC CONSTITUENTS OF THE OUTER PLANETS

Absorption lines of unknown origin had been detected in the nineteenth century in the spectra of Jupiter, Saturn and Uranus. But it required new photographic emulsions, sensitive to the red end of the spectrum, to enable real progress to be made in determining the atmospheric constituents of these outer planets. These photographic emulsions were not available until the early twentieth century.

Vesto Slipher's photographs of the spectra of Uranus and Neptune over the period 1902–04 showed enhanced solar absorption lines at 486.1 nm and 587.6 nm. Slipher thought that these were due to hydrogen and helium, respectively. There were also three absorption bands, of unknown origin, at 510, 543 and 577 nm, all of which were stronger in Neptune's spectrum than in that of Uranus. In addition, Slipher observed an absorption band visually at 618 nm, which also seemed to be stronger in Neptune's spectrum.

In 1905 and 1906, Slipher concentrated on photographing the spectra of Jupiter and Saturn. Comparing these with the earlier spectra of Uranus and Neptune showed strong similarities as well as differences. In particular, he found a strong absorption band in the red at 619 nm in all four spectra, and noted that this band, and that at 543 nm, became progressively weaker in going from Neptune to Uranus to Saturn to Jupiter. He also noted that the band at 577 nm was stronger for Uranus than for Saturn or Jupiter, whilst one at 646 nm was much stronger for Jupiter than Saturn. In general, however, he observed that the spectrum of Jupiter was similar to that of Saturn, in both the existence and intensities of their absorption bands. In addition, that of Uranus was more like that of Neptune, although there were some bands in Neptune's spectrum that were not in the spectrum of any of the other three planets.

Slipher experimented with various sensitising dyes for his photographic plates in 1906 and 1907, and was able to extend their sensitivity up to about 750 nm in the near infrared. In doing so he was able to record strong absorption bands near 719 and 726 nm in the spectra of Jupiter and Saturn, and show that the spectra of Uranus and Neptune were weak above about 690 and 675 nm respectively. Finally, in 1931 he reported numerous absorption bands in the spectrum of Jupiter between 702 and 863 nm, together with a wide and very intense absorption band above about 882 nm.

The breakthrough in understanding the spectra of the four outer planets came in 1931, when Rupert Wildt noticed that two of the absorption bands that he had detected in the spectra of Jupiter and Uranus, centred at about 785 and 899 nm, appeared to correlate with absorption bands at about 792 and 880 nm in the spectrum of ammonia in the laboratory. Encouraged by this, Wildt reviewed Slipher's results, and in 1932 found that six of Slipher's lines in the absorption band at about 646 nm also appeared to coincide with lines in the laboratory spectrum of ammonia.

At that time the only known methane band below about 1 μm was that at 886 nm, which Dennison and Ingram thought was the third harmonic of the methane band at 3.3 μm. If this was so, Wildt calculated that there would be other harmonics at 726, 620 and 544 nm in the methane spectrum. These seemed to correspond to the planetary absorption bands at 726, 619 and 543 nm. So Wildt attributed these bands, and the Jupiter band above 882 nm, to methane.

As a result, in 1932 Wildt appeared to have found evidence for ammonia and methane in the spectra of the four outer planets. The evidence for methane on Uranus and Neptune was circumstantial, however, as no absorption band had yet been reliably detected at 726 nm, and Uranus' band centred on 899 nm detected by Wildt could be due to either ammonia or methane. Clearly what was now required was experimental confirmation that the three bands at 726, 619 and 543 nm exist in the spectrum of methane, and that the 726 nm absorption band, and hopefully that at 886 nm, exist in the spectra of Uranus and Neptune.

In 1933 Reinhard Mecke responded to Wildt's request for experimental evidence by detecting absorption bands in the laboratory spectrum of methane at 890, 860, 840, 784, 725 and 620 nm, the latter two of which appeared to be two of Wildt's hypothesised lines. In parallel, Theodore Dunham measured the spectra of Jupiter and Saturn, and compared these with laboratory spectra. As a result, Dunham proved convincingly that sixty-nine of Jupiter's absorption lines around 645 and 792 nm were due to ammonia and, although Saturn's lines were undoubtedly weaker, again he confirmed the presence of ammonia. In addition, the existence of methane on both

111

planets was confirmed by the correlation of eighteen lines in its absorption spectrum at around 864 nm.

Mecke's work on the methane spectrum called into question Wildt's attribution to ammonia of the bands at 785 and 899 nm that he had recorded in the spectra of Jupiter and Uranus. Ammonia was known to have bands at 792 and 880 nm, and now Mecke had found methane bands at 784 and 890 nm, so maybe the two bands that Wildt had observed on Jupiter and Uranus were due to methane and not ammonia.

The answer was provided by Adel and Slipher in 1934, who showed theoretically that methane should have absorption bands at 886, 725, 619, 543, 486 and 441 nm. Adel and Slipher were also able to confirm the existence of all but the 441 nm band experimentally. So the enhancement of the solar line at 486 nm in Uranus' and Neptune's spectra, as recorded by Slipher in 1902–04, was due not to hydrogen, as he had supposed, but to methane. By 1934, Slipher had also recorded a line at 441 nm in Neptune's spectrum, but not in that of Uranus. It was now seen to be clearly due to methane, but the line's absence for Uranus indicated that there was more methane in Neptune's atmosphere than in that of Uranus.

Furthermore, Adel and Slipher found theoretically and experimentally that methane has absorption bands at 874, 861, 788, 782, 720, 702, 668 and 662 nm. In addition, they showed theoretically, but could not observe experimentally, methane bands at 576 and 509 nm. So, rather than ammonia being the predominant gas spectroscopically on the four outer planets, the predominant gas was methane. It was also clear, from their work, that the methane concentrations reduced in going from Neptune to Uranus to Saturn to Jupiter, which explained the relative intensities on the four planets of the bands observed at 510, 543, 577 and 619 nm.

Interestingly, the band at 646 nm had been observed by Slipher to be stronger for Jupiter than Saturn, but it had not been observable at all for Uranus and Neptune. This band, and the one at 792 nm, which had also not been observable for Uranus and Neptune, were now the only two bands clearly due to ammonia in the spectra of the four planets. The 646 nm band indicated that there was probably more ammonia on Jupiter than Saturn, with no ammonia at all on either Uranus or Neptune. Henry Norris Russell pointed out that this was quite plausible as ammonia would be completely frozen out of the atmospheres of Neptune and Uranus, because of their very low temperatures.

Gerhard Herzberg pointed out in 1938 that, although the homonuclear diatomic molecules of hydrogen and nitrogen had no dipole moment, they would have a quadrupole moment, which varied during the vibration of the molecule. As a result, they would have a quadrupole rotation-vibration spectrum which, although faint, should be observable in the radiation from the outer planets. Herzberg deduced, from the work of James and Coolidge, that the most likely molecular hydro-

gen lines to be detectable were those at 850, 828 and 815 nm. Herzberg also considered the case of similar quadrupole lines in molecular nitrogen, but concluded that they would be much more difficult to detect, mainly because of the presence of nitrogen in the Earth's atmosphere in substantial quantities.

None of the predicted molecular hydrogen and nitrogen lines were observed in the spectra of the outer planets until 1949, when Gerard Kuiper reported detecting a band some 3 to 4 nm wide, centred on 827 nm, in the spectra of Uranus and Neptune, together with a series of lines between 747 and 757 nm. He was unable to attribute any of these lines to known elements, but in 1952 Herzberg suggested that the 827 nm line was due to molecular hydrogen. Herzberg backed up this suggestion by detecting a 4 nm wide line, centred on 826 nm, that had been produced by hydrogen in the laboratory. Then in 1957 Kiess, Corliss, and Kiess detected all three molecular hydrogen lines at 850, 827 and 815 nm in Jupiter's spectrum. In addition, they detected the next line in the series at 805 nm.

Spacecraft have carried out spectroscopic observations of the outer planets over the last fifty years both from Earth's orbit and from planetary intercepts and orbiters. Even so only one spacecraft, the Galileo Jupiter probe, has so far made in situ measurements of one of the outer planets, Jupiter, and even here it made measurements at just one entry point as it descended through the atmosphere. In addition, a number of assumptions have had to be made to determine some of the elemental abundances. Nevertheless modern observations appear to confirm the pre-spacecraft analysis that showed that the mixing ratio of methane reduced going from Neptune to Uranus to Saturn, although that for Saturn and Jupiter appeared to be about the same. In addition the mixing ratio of ammonia reduced going from Jupiter to Saturn, with no ammonia detectable in Uranus or Neptune.

There has been considerable interest over the years in measuring the helium mass fractions for the outer planets, to see how close they are to that of the Sun, to give some idea how they have evolved since the formation of the solar system. Early Pioneer measurements of Jupiter showed that its helium mass fraction was the same as that of the Sun, within error, although the errors involved were quite large. Over the years it has become clear that the helium mass fraction for Jupiter's troposphere is the same as that of the visible part of the Sun. But that similarity is somewhat fortuitous, as the mass fraction of both bodies is less than that estimated for the proto-Sun. And clearly both bodies have undergone radically different processes since the solar system was formed.

Bibliography

Bagenal, F., Dowling, T., and McKinnon, W. (eds.), *Jupiter: The Planet, Satellites and Magnetosphere*, Cambridge University Press, 2004.

Beatty, J. K., Petersen, C. C., and Chaikin, A. (eds.), *The New Solar System*, 4th ed., Cambridge University Press, 1999.

McFadden, L. A., Weissman, P. R., and Johnson, T. V. (eds.), *Encyclopedia of the Solar System*, 2nd ed., Academic Press, 2007.

INTERNAL STRUCTURES OF THE OUTER PLANETS

Isaac Newton had calculated in the early eighteenth century that the densities of the Earth, Jupiter and Saturn were 400, 94.5 and 67 relative to a solar value of 100, in arbitrary units. So Jupiter and Saturn were much less dense than the Earth. He also calculated Jupiter's polar flattening, which was appreciable, caused by its low density and rapid rotation. Newton's result was in agreement with Cassini's observations of 1691. It was not until the end of the eighteenth century, however, that William Herschel measured Saturn's fast rotation and observed its appreciable polar flattening.

The Compte de Buffon suggested in 1778 that the continuous, rapid changes in Jupiter's appearance indicated that it had not cooled down completely since its formation. Some astronomers concluded, in the first half of the nineteenth century, that shadows of the Galilean satellites on Jupiter were not completely black, and suggested that this was because Jupiter was partially self-luminous. Then in 1859 Richard Carrington found that the Sun's angular rotation was faster at its equator than in its middle latitudes, just like Jupiter. Some astronomers took this as confirmation that there were similarities between the Sun and Jupiter, supporting the notion that Jupiter was partially self-luminous. Other astronomers disagreed.

In 1798 Henry Cavendish had measured the density of the Earth as about 5.5 g/cm^3, which implied that the densities of Jupiter and Saturn were similar to that of water, given Newton's relative values. By the end of the nineteenth century the densities of Jupiter, Saturn, Uranus and Neptune were thought to be about 1.38, 0.75, 1.28 and 1.15 g/cm^3, respectively. So in the early years of the twentieth century it was generally thought that these large outer planets were largely gaseous. Jupiter, in particular, was also thought to be quite hot because it was the largest planet and had probably not had enough time to cool down completely since its formation.

This idea that the four outer planets, and Jupiter in particular, were hot gaseous planets was challenged by Harold Jeffreys in 1923. He suggested that they were cold solid bodies, made of very low density materials, whose original heat had virtually completely disappeared. He calculated that, if the density of the solid cores of Jupiter and Saturn were the same as some of their large satellites, the depths of the planetary atmospheres need only be about 20% of their radii to explain their measured densities.

In the same year, Donald Menzel reanalysed the radiometric measurements made of Jupiter and Saturn by Coblentz and Lampland, and concluded that their cloud-top temperatures were about 160 K. Although this temperature is very cold, it is still higher than the 120 K and 90 K for Jupiter and Saturn, respectively, that would be maintained solely by incident solar radiation. So Jupiter and Saturn did appear to have some source of internal heat, although their cloud-top temperatures were still quite low.

Jeffreys extended his work on the internal structure of Jupiter and Saturn in 1924 by analysing their moments of inertia. This indicated that both planets were much more condensed towards the centre than the Earth. This analysis, and Menzel's estimates of Jupiter and Saturn's cloud-top temperatures, led Jeffreys to conclude that Saturn, at least, must have an extensive atmosphere consisting of large quantities of hydrogen and/or helium. Jeffreys also acknowledged that the cloud-top temperatures of Jupiter and Saturn were too high to be just due to solar heating, and so there must be a significant amount of heat being emitted from inside these two planets. He continued to maintain, however, that the original heat at formation must have largely disappeared by now, and suggested that this extra internally generated heat may be due to the decay of radioactive elements near the surface.

To develop his theory of planetary structures further, Jeffreys assumed that the atmospheres of Jupiter and Saturn surrounded a layer of ice and solid carbon dioxide of density 1 g/cm^3, which in turn surrounded a rocky core of density 3 g/cm^3. Using this model, he found that the moments of inertia and densities of Jupiter and Saturn implied that the depth of their atmospheres was about 0.09 R_J (Jupiter radii) and 0.23 R_S (Saturn radii), respectively. He also concluded that these atmospheres probably consisted mainly of hydrogen, nitrogen, oxygen, helium and maybe methane, with clouds possibly consisting of solid carbon dioxide.

In the following year, 1925, Cecilia Payne showed that hydrogen and helium are the most abundant elements in stellar atmospheres. So the idea that Jupiter, Saturn, Uranus and Neptune may also have large amounts of hydrogen in their structures became highly plausible, as the planets were assumed to have formed from the original solar nebula.

It was difficult to take proper account, in developing the theory of planetary structures, of the effects of ultra-high pressure on the materials of which they are composed. But in 1938 Rupert Wildt tried to see if the pressure inside the large outer planets may be sufficient to produce degeneracy. The models of Jupiter and Saturn used by Wildt consisted of a dense core, similar to that of the terrestrial planets, surrounded by a thick layer of ice, which was in turn surrounded by a layer of highly compressed condensed gases, mainly solid hydrogen. Like Jeffreys, Wildt then calculated dimensions of the layers from the overall densities and moments of inertia of the two planets. This gave radii of 0.43 R_J and 0.26 R_S for the cores, and 0.82 R_J and 0.66 R_S for the outside of the ice layers. The pressures at the centre of the core were about 60 Mbar for Jupiter and 15 Mbar for Saturn, which Wildt concluded were probably just below the pressure required to produce degeneracy. He did speculate, however, that maybe some of the hydrogen that was under very high pressure could be in the metallic hydrogen form proposed theoretically by Eugene Wigner and Hillard Huntingdon in 1935. Simultaneously, Daulat Kothari came up with the same suggestion.

Wildt's 1938 model for the internal structure of Jupiter and the other outer planets implied that there was about three times as much hydrogen, in percentage terms, in Saturn as in each of the other three outer planets. This seemed unlikely, assuming that they all formed from the original solar nebula. The model also made no allowance for compressibility, which for these large planets must be considerable.

In 1950 William Ramsey investigated the behaviour of hypothetical, non-rotating planets composed completely of hydrogen, which he calculated became metallic at a pressure of 0.8 Mbar. In so doing, the density of solid hydrogen doubled at the transition from the molecular to the metallic phase. Ramsey calculated that, on this basis, such a planet could only have a core of metallic hydrogen if the planet's mass exceeded 88 M_E (mass of the Earth). At Jupiter's mass of 317 M_E the hydrogen planet would have a radius of 79,400 km compared with Jupiter's actual mean radius of about 69,900 km.

Ramsey modified his model of hypothetical hydrogen planets, in the following year, to include helium and to take account of rotation. As a result, he found that Jupiter and Saturn would be composed of 76% and 62% hydrogen, respectively, by mass (compared, for example, with 74% for the Sun), assuming a uniform mixing of the hydrogen and helium. The central pressures were calculated to be 32 and 6 Mbar for Jupiter and Saturn, respectively, with central densities of 3.66 and 1.91 g/cm^3, the metallic cores of the two planets containing 92% and 67% of their total mass.

Unfortunately, Ramsey's new models of Jupiter and Saturn produced moments of inertia that were somewhat too high. So, in the following year, he and Miles analysed the effect of adding heavier elements, which were partly concentrated in the core and partly distributed uniformly throughout the structure. This yielded better moments of inertia, but had little effect on the proportions of hydrogen, which varied from 76% to 84% by mass for Jupiter, and from 62% to 69% for Saturn, depending on the assumptions made.

Ramsey also considered the likely structure of Uranus and Neptune, and pointed out that they are about an order of magnitude denser than for hydrogen planets of the same mass. As a result, he concluded that they must have lost most of their original hydrogen and helium, and now be composed mainly of water, methane and ammonia, together with terrestrial materials.

Over the last fifty years or so models of the interior of the four large outer planets have been developed, based on the above considerations. Although these models are only theoretical, it is clear that Jupiter and Saturn are, like the Sun, largely composed of hydrogen and helium. But, as concluded earlier by Ramsey, the radii of Uranus and Neptune are too small, given their masses, for them to have such a near-solar composition.

Modern planetary models have to take account the planet's masses, densities, and moments of inertia, mentioned above, as well as new information gleaned over the last fifty years. This includes measurements made by the Pioneer and Voyager spacecraft of the planets' internally generated heat, of their atmospheres, and of their gravity fields, including harmonics. In addition, recent laboratory experiments have provided more data on the behaviour of hydrogen under ultra-high pressure.

As a result it is thought that Jupiter consists of a layer of molecular hydrogen and helium which extends down to about 0.8 R_J, where the pressure reaches about 2 Mbar at a temperature of about 7,000 K. At this pressure the hydrogen becomes metallic. A layer of metallic hydrogen and helium then extends down to about 0.1 R_J where the pressure reaches about 40 Mbar, at a temperature of about 20,000 K. Below that there is probably a small layer of water, ammonia and methane ices, followed by a small rocky core probably of iron and silicates. Alternatively, the ices and rocky elements may all be mixed together in the core.

The structure of Saturn is similar to that of Jupiter, except that the thickness of the metallic and molecular hydrogen layers are thought to be similar to each other in Saturn. It appears as though helium is not soluble in hydrogen at all mass fractions and temperatures. As a result, it is thought that droplets of helium-rich material in Saturn are constantly forming in the molecular to metallic transition region. Because these droplets are denser than their surrounding material they fall towards the centre, until they reach a region where conditions are sufficient to allow mixing again.

The internal pressures in Uranus and Neptune are not high enough for hydrogen to be metallic. So their cross-sections show a small rocky core, a large layer of possibly water, methane and ammonia ices, topped by a smaller molecular hydrogen and helium layer. It is thought that radioactive decay is insufficient to produce the observed energy released by Jupiter, Saturn and Neptune. Instead this energy is released as heavier elements in these planets gravitate towards their centres. Uranus is clearly different in this respect as it has no internally generated heat.

Bibliography

Bagenal, F., Dowling, T., and McKinnon, W. (eds.), *Jupiter: The Planet, Satellites and Magnetosphere*, Cambridge University Press, 2004.

Beatty, J. K., Petersen, C. C., and Chaikin, A. (eds.), *The New Solar System*, 4th ed., Cambridge University Press, 1999.

McFadden, L. A., Weissman, P. R., and Johnson, T. V. (eds.), *Encyclopedia of the Solar System*, 2nd ed., Academic Press, 2007.

JUPITER

General

Galileo discovered the four large moons of Jupiter, now called Io, Europa, Ganymede and Callisto, when he observed the planet for the first time with his telescope in January 1610. Further observations enabled him to conclude, early in the following year, that the periods of these four satellites were approximately in the ratio of 1:2:4:8. By 1612 he had managed to determine their orbital periods within a few minutes of their modern values.

Galileo's pupil Evangelista Torricelli discovered the main belts of Jupiter in 1630, and about thirty years later Giovanni Domenico Cassini proved that Jupiter was rotating. He deduced a rotation period of 9 h 56 min, based on the movement of irregularities in Jupiter's equatorial bands. This remarkably rapid rate causes Jupiter to show a noticeable polar flattening, which Cassini measured in 1691. In the previous year he had also observed that the rotation rate at higher latitudes was about 5 minutes longer than for equatorial regions.

In 1664 Robert Hooke observed a small spot in what is now believed to be the North Equatorial Belt. Then in the following year Cassini recorded a particularly impressive spot, which he called 'the Eye of Jupiter', in what is now called Jupiter's South Tropical Zone. Cassini's spot remained visible until 1713 when it was last recorded by Giacomo Maraldi. It is now thought that it may have been an early appearance of the Great Red Spot (GRS), which was not seen again until 1878. The GRS, which dominated the views of the planet from 1879 to 1882, was measured to be about 40,000 km in length and about 13,000 km in width. Since then it has varied both in size and in the shade and density of colour. The relatively long life of the GRS suggested to astronomers in the 1880s that it may be connected with a permanent feature on the surface of Jupiter. But subsequent measurements indicated that its rotation period about the centre of Jupiter was varying.

In 1914 Paul Guthnick found that the brightness of all four of the Galilean satellites varied in a regular way, correlating with their position in orbit, thus showing that the axial rotation of all four is synchronous.

Jupiter's fifth satellite, Amalthea, was discovered by Edward Emerson Barnard in 1892 only about 110,000 km above the planet's surface. It had a period of a little less than 12 hours, with an estimated size of about 100−200 km diameter. It was the last satellite of any planet to be discovered visually. Four more satellites were discovered between 1904 and 1914, and by the time that the first spacecraft arrived at Jupiter in 1973, three additional satellites had been found, making a grand total of 12. Of these twelve, the four Galilean satellites and Amalthea orbit relatively close to Jupiter, there are three small satellites orbiting at about 11 million km from the planet, followed by four small satellites orbiting at about twice that distance. The latter group all orbit Jupiter in a retrograde sense. In fact this, and the closeness of Jupiter to the asteroids, led Forest Moulton to suggest that the outside group were captured asteroids.

The first two spacecraft to visit Jupiter, Pioneer 10 and 11, flew past the planet in December 1973 and December 1974, respectively. Pioneer 10 detected helium in Jupiter's atmosphere, and found that the day and night sides of the planet were at the same temperature as each other. Jupiter's atmosphere was found to have a temperature of 165 K at a pressure level of 1 bar. The temperature then went through a minimum of 108 K at 0.1 bar about 150 km higher in the atmosphere, before increasing to 150 K at 0.03 bar at cloud top height. This increase was attributed to the absorption of sunlight by a thin haze of dust particles. Pioneer 10 also found that the cloud tops of Jupiter's Great Red Spot (GRS) were colder than those of the adjacent South Tropical Zone, so the GRS was thought to be an enormous high pressure system rising about 8 km above its surrounding area. In 1975 the Kuiper Airborne Observatory detected water vapour in Jupiter's atmosphere.

Before the spacecraft missions to Jupiter, astronomers had detected a pattern of alternating easterly and westerly winds in zones parallel to Jupiter's equator. The only exception to this so-called 'zonal flow' was the anticyclonic flow around the GRS. But the Voyager 1 and 2 images in 1979 showed that, within the zonal system, there was a planet-wide pattern of small-scale vortices (see Figure 2.12) that were forever changing on timescales of only hours. The most spectacular pattern was that surrounding the GRS, where the clouds were ripped apart in regions of very high shear. A comparison between the Voyager 1 and 2 images enabled the relative velocities of Jupiter's various belts and zones to be determined. The maximum wind velocity was found to be about 540 km/h at low latitudes.

Prior to the Voyager 1 encounter it was thought that Io would look like a reddish version of our Moon, but covered with sulphur-coated impact craters. However, in the issue of 'Science' published just three days before Voyager 1's closest approach, Stanton Peale, Patrick Cassen and Ray Reynolds came up with an alternative theory. They pointed out that since Io was subjected to resonant gravitational forces by the other Galilean satellites, its orbit would be eccentric, although over time it would average out as circular. As Io is so close to Jupiter, the planet's gravity would produce powerful tidal forces in Io's crust, causing the satellite to heat up. As a result, they suggested that there would be widespread volcanism on Io.

In the event, Voyager 1 showed that Io did indeed have a fresh-looking landscape (see Figure 2.13) showing numerous volcanic calderas and rivers of lava. Three days after closest approach, Linda Morabito even found an umbrella-shaped plume on the limb of Io, reaching about 270 km above the surface. It was the plume of an erupting volcano. Eventually eight volcanoes were found to have been active during Voyager 1's closest approach, producing plumes ranging from 70 to 300 km in height with vent velocities as high as 1.0 km/s.

Figure 2.12. This Voyager 2 image of Jupiter shows structure in both the Great Red Spot and in Jupiter's complex zonal system. (Courtesy NASA/US Geological Survey.)

Voyager 1 showed that Ganymede, the largest of Jupiter's satellites, had a complex intersecting pattern of parallel grooves and ridges, together with numerous rayed craters, on a basically two-toned surface. The largest area of dark terrain, called Galileo Regio, was an approximately circular feature about 3,000 km in diameter. Many of the rayed craters were seen to be very light in colour, probably because the ice there was fresher and more powdery than on the surface as a whole. Some of the grooves showed lateral offsets, indicating fault lines where the surface had moved laterally.

Prior to the Voyager 1 encounter it had been thought that Callisto, Ganymede and Europa would have ice-covered surfaces. There would be little surface relief, and only small impact craters, as the ice would have flowed and covered all but the most recent craters over time. In the event, Voyager 1 found that the surface of Callisto was almost saturated with what must be very old craters. These craters were quite shallow, however, with the limb of Callisto showing that there was virtually no surface relief. The largest feature was found to be the large bright impact basin, now called Valhalla, which was surrounded by a series of concentric rings or ridges spaced 20 to 100 km apart, extending to a radius of about 2,000 km. These were produced

by shock waves caused by the impact that created the basin. In common with the remainder of Callisto, Valhalla showed virtually no surface relief, presumably because water and ice had flowed back after the impact to fill the depression.

Voyager 1 images showed that the surface of Europa was very bland, with surface markings of low contrast. Numerous dark stripes, tens of kilometres wide and up to thousands of kilometres long, were seen criss-crossing the surface. It was thought that these may be faults or fractures caused by tectonic activity. Voyager 2 showed that some of the stripes, the so-called 'triple bands', had a light centre and dark edges. It imaged a few palimpsests, with diameters of the order of 100 km, which appear to have been caused by large impacts. Near the terminator a few 20 metre diameter craters were also observed. Crater analysis indicated that, although the surface of Europa was younger than that of Ganymede or Callisto, it was still probably a few hundred million years old.

Voyager 2 showed that the surface of Europa had virtually no surface relief above about 100 m high, which is consistent with the pre-Voyager concept of a 100 km thick ice layer covering the surface. Tidal heating and/or radioactive heating could be sufficient to melt the lower levels of this ice crust,

Figure 2.13. The fresh-looking surface of Io is shown in this Voyager 1 image. The circular, doughnut-shaped structure at the centre is the volcano Prometheus, which was seen to be erupting in other images. (Courtesy NASA.)

however, and some astronomers wondered whether the surface of Europa may be composed of pack ice floating on water, with the observed linear patterns showing where the ice had fractured.

Voyager 2 images of Io showed what appeared to be clouds or white surface deposits along scarps or faults, particularly in Io's south polar region, and spectrophotometric analysis indicated that these white particles were composed of sulphur dioxide and sulphur. Sulphur dioxide gas was also discovered near the volcano Loki.

Vent velocities of particles emitted by Io's volcanoes were found to be about a factor of ten higher than those on Earth, indicating that Io's volcanoes are significantly different in nature. They were also more consistent in their output over time than those on Earth, and this, together with their umbrella-shaped plumes, suggested to planetary geologists that they were more like geysers. The driving force for geysers is usually the transition from water to steam, but that on Io must be different because of the lack of water. The most likely material driving the plume eruptions of sulphur appeared to be sulphur dioxide or, in the case of very high plume velocities, possibly sulphur itself.

Voyager measured a maximum noontime temperature on Io's equator of about 120 K, whereas the temperature of a dark feature just south of the volcano Loki appeared to be about 300 K. This 200 km diameter feature was thought to be a lake of lava from the volcano. It was probably made of liquid sulphur and in the lake were lighter features that looked like icebergs

probably of solid sulphur. Near Loki and the volcano Pele there were also very small hot spots with a temperature of about 500 K.

The Galileo spacecraft, which was launched in October 1989, consisted of two parts: a main spacecraft to orbit Jupiter, and an attached probe which would be released, as the main spacecraft approached the planet, to descend through Jupiter's atmosphere. The Galileo orbiter and entry probe arrived at Jupiter in December 1995.

The entry probe started transmitting atmospheric data when it reached the 0.35 bar level in Jupiter's atmosphere, at an atmospheric temperature of 120 K. It continued transmitting until the atmospheric pressure had increased to 23 bar at 425 K, some 160 km further into the planet.

The density of Jupiter's upper atmosphere, a few hundred kilometres above the 1 bar level, was found to be much greater than expected. Surprisingly, the probe also found that the wind velocity increased from about 350 to 720 km/h as the probe descended through the atmosphere. This increase was probably due to Jupiter's internal heat source.

One key part of the probe mission was to measure the helium mass fraction for Jupiter. The Pioneer and Voyager spacecraft had only been able to measure this remotely for the top of Jupiter's atmosphere, whereas the Galileo probe would be able to measure it in situ for more than just the outside layer. The probe measured a value of 0.24 which was, as expected, similar to that of the proto-Sun.

Before the Galileo mission it had been thought that there were three cloud layers on Jupiter, namely an outer layer of ammonia ice, which is seen from Earth, followed by one of ammonium hydrosulphide crystals, and finally one of water ice or possibly water droplets, at about the 8 bar level. It was expected that the atmosphere below this last cloud layer would be relatively clear.

The Galileo probe detected the top cloud layer at about 0.5 bar, at a temperature of 145 K, some 15 km above the 1 bar reference. The next cloud layer was between 0.8 and 1.3 bar, followed by a vertically thin layer at 1.6 bar and 200 K, 30 km below the reference level. The temperature of this latter layer implied that it was the expected layer of ammonium hydrosulphide. All these cloud layers were thinner and more transparent than expected, with none of them being more opaque than a light fog or mist. Surprisingly, the expected cloud layer of water ice or droplets did not appear to exist.

There then followed an extensive debate as to whether the probe site was exceptionally dry compared with the rest of Jupiter, or whether there was really relatively little water in Jupiter's atmosphere as a whole. However, the Galileo orbiter showed that the probe had entered Jupiter's atmosphere in an infrared hot spot. So the most likely explanation seemed to be that this was an exceptionally dry part of Jupiter, as there appeared to be ample water vapour elsewhere on the planet.

Ganymede was the first of Jupiter's Galilean satellites to be observed close up by the main Galileo spacecraft after it had been put into orbit around Jupiter. The spacecraft's first pass was at a distance of just 850 km above the surface, some 70 times closer than during the Voyager fly-bys. Voyager had imaged the large, dark, almost circular Galileo Regio feature that was separated from the smaller, dark Marius Regio by a light, irregular young feature known as Uruk Sulcus. Galileo now produced high resolution images of the surface of Uruk Sulcus, and showed that its ridges and troughs found by Voyager, which were up to some hundreds of kilometres long and tens of kilometres wide, actually consisted of a much finer system of ridges parallel to the main ridges. This appeared to be due to tectonic activity caused by the crustal water ice expanding. The variation in height between the ridges and troughs was about a few hundred metres, with the tops of the ridges appearing light and the troughs dark.

The generally dark surface of Galileo Regio was found to be littered with craters of all sizes, with crater counts indicating a basic surface age of about 4.2 billion years. But its surface had clearly been substantially modified since its formation to produce a hummocky terrain of bright hills and dark planes, caused by a mixture of tectonics and cryovolcanism. Sun-facing slopes were found to be generally dark, apparently because the Sun had caused the ice to sublimate, leaving the darker substrate visible.

Spectroscopic analysis showed that the bright regions on Ganymede were rich in water ice with patches of carbon dioxide ice, whilst the dark surface material was found to contain abundant clays and tholins, which are rich in carbon, hydrogen, oxygen and nitrogen. Daytime temperatures were found to vary from 90 to 160 K depending on the time of day and latitude.

Ganymede was the first planetary satellite found to possess an intrinsic magnetic field. It had a strength at the equator of about 0.02 gauss. The existence of such an intrinsic magnetic field implied that Ganymede has a molten or semi-molten core of iron and/or iron sulphide, at a temperature of at least 1,300 K. This was surprising on two counts: first that the satellite was differentiated with a clearly defined core, and second that that core was still at least partly molten in such a relatively small body so far from the Sun. Galileo's magnetic field measurements also indicated that there was an induced component, implying that Ganymede has a salty ocean tens of kilometres thick, about 170 km below the surface.

Galileo also detected a magnetosphere along with a thin ionosphere, indicating that Ganymede probably also has a tenuous atmosphere. This was apparently confirmed by Earth-based infrared observations that detected molecular oxygen, and by Hubble Space Telescope observations in the ultraviolet that also detected signs of ozone, although it was not absolutely clear if this oxygen was on the surface or in the atmosphere. On the other hand, Galileo found clear evidence of a tenuous atmosphere of atomic hydrogen, with ionised hydrogen streaming away at high speed. Apparently charged particles in Jupiter's magnetosphere were impacting the water ice on Ganymede's surface, driving off atoms of hydrogen and oxygen.

Callisto was the second of Jupiter's satellites to be imaged by the Galileo orbiter. Voyager had shown that this, the outermost of Jupiter's Galilean satellites, had a dark surface almost completely saturated with craters. It was generally expected that, with the higher resolution of Galileo, a large number of smaller craters would be seen. But to everyone's surprise, Galileo showed that there was an almost complete lack of craters less than about one kilometre in diameter. Callisto has virtually no atmosphere, so the surface must have been impacted by innumerable very small objects, but their craters seemed to have disappeared.

Callisto seemed to be covered with a dark material that created a very smooth surface, quite unlike that seen anywhere else in the solar system. This dark material, which could be several metres thick, may well be covering the small craters. The central area of the large Valhalla basin was also covered in the dark material, so it could not have been ejecta from the Valhalla impact. Galileo's spectrometer showed that this 'soil' consisted of clays and tholins, but so does the soil on Ganymede and it is completely different in appearance. Even now the reason for this difference in appearance is not clear.

Early, pre-encounter spectra showed that there was carbon dioxide frost on Callisto's surface but, strangely, the poles had less frost. Instead it tended to be concentrated on Callisto's trailing hemisphere, which is subjected to particle bombardment as Jupiter's magnetosphere overtakes Callisto in its orbit. Sulphur dioxide frost was also found on Callisto's surface, but its distribution was more patchy than that of carbon dioxide. Unlike the latter, the sulphur dioxide frost showed no tendency to concentrate on Callisto's trailing hemisphere, suggesting that it does not originate from the sulphur ions in Jupiter's magnetosphere which come from Io's volcanoes. On the contrary, there was a tendency for the sulphur dioxide frost to be deposited on Callisto's leading hemisphere, leading to the suggestion that it came from the impact of numerous micrometeorites.

As in the case of Ganymede, Galileo detected hydrogen atoms escaping from Callisto's surface. This implied that there must also be oxygen somewhere on Callisto from the break-up of water molecules of its icy surface. As Callisto is further out from Jupiter than Ganymede, however, this break up is probably caused by the bombardment of the surface with solar ultraviolet light, rather than by charged particles in Jupiter's magnetosphere. In fact Galileo soon found evidence of the oxygen probably as both a frost and as a gas in Callisto's very thin atmosphere, which appeared to consist of hydrogen, oxygen and carbon dioxide.

Initially Galileo could find no evidence of an overall magnetic field on Callisto. Later analysis showed, however, that it had a variable magnetic field which was caused by Jupiter's magnetic field inducing subsurface electric currents. Calculations showed that these currents could be in a 10 to 20 km deep salty ocean, about 200 km beneath the surface. The heat to keep this ocean liquid, or at least slushy, must be coming from inside the satellite. Callisto appeared to have a cold, undifferentiated interior, however, and the gravitational stresses on its interior were relatively small at its distance from Jupiter. So it was concluded that the heat must be coming from radioactive decay.

The Galileo orbiter's next target was Europa, which Voyager had shown to be a remarkably smooth, ice-covered satellite with numerous long dark stripes, some of them, the so-called triple bands, with lines down the middle. The Galileo images showed that the edges of the stripes or bands on Europa were not clear-cut, as they had appeared to be in the Voyager images, but were rather indistinct. Close-up views under side lighting showed that some of the stripes consisted of prominent ridges with a fracture running down the middle. This appeared to be the result of repeated extrusions of ice or water along fractures in the surface ice sheet. The triple bands appeared to have a similar structure, although there were generally more dark and light bands in their cross sections. Some triple bands crossed other triple bands, showing that they were clearly of different ages.

Galileo found that the area of mottled terrain on Europa about 1,000 km due north of the 26 km diameter crater Pwyll was chaotic in structure (see Figure 2.14). This area of chaotic terrain, now called Conamara, was found to consist of numerous polygonal rafts of ice, each about 5 to 10 km across, whose top surfaces were about 20 to 200 m above the surrounding surface. Some of the rafts were tilted, as if they had been subjected to lateral compressional pressure at some stage. On the western fringe of this area, the rafts had been clearly powdered by ice particles ejected by the Pwyll impact. In addition, a number of small craters appeared to have been produced by blocks of ice ejected by the same impact. It appeared as though these rafts are icebergs that at one stage had been floating on sub-surface ice or slush. Some of the rafts were seen to carry segments of ridges, and reconstructions have shown that about half of the original surface is missing because, presumably, it melted during the event that caused this localised surface rearrangement.

Detailed observations of the relatively young crater Pwyll indicated that the collision that formed it was sufficient to puncture the icy crust of Europa, releasing darker material from beneath the surface. The crater's floor was found to lie at the same level as the surrounding terrain, however, suggesting that it filled immediately after formation with slushy material. The impact that created the 30 km crater Manann'an also appeared to have punctured the surface, which was probably just a few kilometres thick at that time.

Dynamics and Doppler measurements made particularly during Galileo's closest approaches indicated that Europa, like Ganymede, has a differentiated structure, with the radius of the core in Europa's case being about half the total radius. Gravitational tidal action caused by Jupiter, and by Europa's orbital resonances with Io and Ganymede, would have kept Europa's interior warm and have caused flexing of its surface. This would have helped to create the surface fractures seen by Voyager and the Galileo orbiter.

As Jupiter's magnetic field is inclined to its spin axis, its direction at Europa changes sinusoidally with a period of 11 hours. As a result, if Europa's magnetic field, which had been previously detected by the Galileo orbiter, is induced, it should change with this frequency. In fact Galileo found that the field reversed direction every 5.5 hours. So it was clearly induced, and further analysis indicated that it was probably generated in a 100 km deep salty ocean that may be only a few kilometres below Europa's surface.

It was difficult to determine the exact temperature of the surface around Io's volcanoes with the Voyager spacecraft, as

Figure 2.14. This heavily fractured Conamara region of Europa is believed to show icebergs that once floated on a subsurface ocean. The icebergs clearly moved laterally after break-up, but they now appear to be frozen in place. (Courtesy NASA/JPL/Caltech.)

the spatial resolution of the radiometers was relatively poor. Nevertheless, very small hot spots had been found on or near the volcanoes Loki and Pele that had temperatures of about 500 K. It was thought that the yellow, red and black colours of Io's surface were due to sulphur at various temperatures, although the sulphur probably included significant impurities which affected its behaviour.

Voyager had found mountains up to 10 km high on Io, which implied that its upper crust must be quite strong, and probably could not, therefore, be made of sulphur. In addition, the walls of some of the volcanic caldera appeared too steep to be made of pure sulphur, as this would have slumped after formation. So it was possible that the crust could be made of silicate rocks, and there may be some silicate/sulphur or silicate volcanism. The key to finding out was to measure the surface temperature across Io, as silicate rocks become molten at much higher temperatures than sulphur.

In 1986 a team of astronomers studying Io from the top of Mauna Kea in Hawaii observed a significant increase in its infrared brightness, indicating that there had been a brief eruption of lava with a temperature of at least 900 K. Since then other Earth-based observations have detected similar events at even higher temperatures, indicating that at least some of Io's volcanic lava is silicate, as the temperature of 900 K is far too high for sulphur.

In 1979 Voyager had detected Pillan Patera as a simple caldera just over 10° away from Pele. Seventeen years later Galileo's initial observations showed that there had been no change in Pillan's appearance, but over the next few orbits Galileo detected that Pillan's albedo was variable, even though its temperature appeared to be approximately constant. Then on 5 July 1997 the Hubble Space Telescope discovered a 200 km high plume being emitted by Pillan Patera. The Galileo spacecraft also imaged the plume. In Galileo's image Pillan's plume was on the limb of Io, and there was another plume over the volcano Prometheus near the terminator. The magnitude of the change produced on Io's surface by the Pillan eruption became clear on Galileo's next orbit, which showed that the volcano had produced a 400 km diameter, dark, pyroclastic blanket that covered part of the reddish halo around Pele. Unlike most other plume deposits on Io, however, which were white, yellow or red, Pillan's deposit was grey, resembling that of Babbar Patera nearby. At a temperature of about 1,600 K it was most likely due to silicate volcanism.

Rings

Pioneer 11 detected a significant dust environment close to Jupiter during its fly-by in 1974, together with a reduction of high energy particles near its closest approach. Mario Acuña and Norman Ness suggested that the latter could have been caused by either an undiscovered ring or satellite. So a single

11 minute exposure was made in 1979 as Voyager 1 passed through Jupiter's equatorial plane, and much to most astronomers' surprise a faint, thin ring was seen.

A few months later, Voyager 2 imaged Jupiter's ring as the spacecraft both approached and retreated from the planet. The ring appeared more than twenty times brighter in forward-scattered light than in back-scattered light, implying that many of the ring's dust particles are only about one to two microns in diameter. Such small particles can exist there for only a short period of time, implying that there must be some resupply mechanism. Mark Showalter et al. found, after further analysis, that they could distinguish three components in Jupiter's ring, namely a main ring, a thick faint 'halo' ring inside the main ring, and a very faint 'gossamer' ring outside it. Two previously undetected small satellites (Metis and Adrastea) were discovered near to the outer edge of the main ring. They were possibly constraining its outward expansion and/or helping to supply it with material, following impacts by either interplanetary particles and/or volcanic dust from Io.

The Galileo spacecraft undertook a number of ring observations starting in 1996, and found that the gossamer ring had two basic components, one inward of Amalthea, and one connected with Thebe. Dust from both of these satellites appeared to have supplied the gossamer ring. The material in the halo ring seemed to have come from the main ring possibly as a result of electrostatic effects.

Magnetism and magnetosphere

In 1955 Bernard Burke and Kenneth Franklin accidentally discovered bursts of 22.2 MHz radiation from Jupiter. Alex Shajn then looked back at records of cosmic noise that he had made at 18.3 MHz in 1950−51. He found clear evidence of signals from Jupiter which had been previously attributed to local interference. Analysis of these records showed that the source had a period similar to Jupiter's rotation period, suggesting it was localised on the planet. However, in 1959 Frank Drake and Hein Hvatum detected radio emissions from Jupiter at 400 MHz which implied that the source was non-thermal. They proposed that these radio emissions were caused by synchrotron radiation generated by relativistic electrons trapped in an intense magnetic field around the planet. This radiation belt was analogous to those around the Earth but was much more intense. This hypothesis was verified in the following year when Venkataraman Radhakrishnan and James Roberts found that Jupiter's radio signals were polarised, and came from an area about three times the size of the planet.

Keith Bigg found in 1964 that Io appeared to modulate Jupiter's decametric radiation. James Roberts and Max Komesaroff also found that the plane of polarisation of the radio signals oscillated through an angle of about ±10°, indicating that Jupiter's magnetic axis was inclined at about 10° to its axis of rotation.

The Pioneer 10 spacecraft first detected high energy electrons from Jupiter at about 300 R_J (Jupiter radii) from the planet as the spacecraft approached Jupiter during its 1973 fly-by. Pioneer 10 then passed through Jupiter's bow shock at 108 R_J from the planet. Apparently some high energy electrons had managed to escape from Jupiter's magnetosphere and somehow cross its bow shock. Pioneer 10 also found that the bow shock's distance from Jupiter varied in response to variations in the intensity of the solar wind.

Pioneer 10 detected a dipole magnetic field with a strength of about 4 gauss at Jupiter's cloud tops. The magnetic axis of this dipole field, which extended to about 20 R_J from Jupiter, was inclined at about 10° to the planet's spin axis. Outside of this dipole field there was a stronger field which was more confined to Jupiter's equatorial plane. In the following year Pioneer 11, which flew past Jupiter closer than Pioneer 10, found that the magnetic field closest to the planet was not a simple dipole field but had multiple harmonics.

Jupiter's radiation belts were found by the Pioneer 10 spacecraft to be about 10,000 times as intense as the Earth's Van Allen belts. The spacecraft also found that Jupiter's magnetic field and associated radiation belts co-rotated with Jupiter, and overtook the four Galilean satellites in their orbits of the planet. These satellites, which were within the belts, swept up protons and electrons from the belts as the belts overtook them.

Pioneer 10 detected an ionosphere around Io extending up to 100 km above its surface. At about the same time, using a ground-based telescope, Robert Brown observed neutral clouds of sodium in the vicinity of Io. Then in 1979 Broadfoot and Bridge independently identified an Io plasma torus from Voyager 1 and 2 data. This dense luminous torus, which was found to include sulphur and oxygen ions, orbited Jupiter at Io's distance from the planet. It was thought that it had been produced by sulphur dioxide emitted by Io's volcanoes. The Galileo spacecraft showed in 1995 that the plasma flow in Io's torus was slowed by Io's ionosphere, redirected around Io, and then reaccelerated in Io's wake. In addition, as Galileo passed Io, the strength of Jupiter's magnetic field was observed to decrease by about 30%, and then increase again to about its original level. Io appeared to be in a bubble in Jupiter's magnetosphere, which some planetary scientists attributed to a magnetic field on Io. However, later Galileo fly-bys showed that the effect can be explained by currents flowing in the plasma torus.

Voyager 1 showed that Io appeared to be linked to Jupiter by north- and south-directed flux tubes following Jupiter's magnetic field lines down to Jupiter. The flux tubes were found to carry a current of about 3 million amperes. In 1995 the Galileo spacecraft made the first in situ measurements of electrons flowing up and down these Io/Jupiter flux tubes. It found that the energy dumped into each of Jupiter's polar regions was enough to produce Jupiter's aurorae.

Note: The impact of the comet Shoemaker-Levy 9 with Jupiter is included in the *Comets* article.

Further information on Jupiter's satellites and rings is given in Table 2.2.

Bibliography

Bagenal, F., Dowling, T., and McKinnon, W. (eds.), *Jupiter: The Planet, Satellites and Magnetosphere*, Cambridge University Press, 2004.

Beatty, J. K., Petersen, C. C., and Chaikin, A. (eds.), *The New Solar System*, 4th ed., Cambridge University Press, 1999.

Godwin, Robert, and Whitfield, Steve, *Deep Space, The NASA Mission Reports*, Apogee Books, 2005.

Leverington, David, *Babylon to Voyager and Beyond: A History of Planetary Astronomy*, Cambridge University Press, 2003.

McFadden, L. A., Weissman, P. R., and Johnson, T. V. (eds.), *Encyclopedia of the Solar System*, 2nd ed., Academic Press, 2007.

Meltzer, Michael, *Mission to Jupiter: A History of the Galileo Project*, NASA SP−2007−4231, 2007.

Miner, E. D., Wessen, R. R., and Cuzzi, J. N, *Planetary Ring Systems*, Springer-Praxis, 2007.

Morrison, D. and Samz, J., *Voyage to Jupiter*, NASA SP-439, 1980.

Peek, B. M., *The Planet Jupiter*, Faber and Faber, 1958.

Shirley, James H., and Fairbridge, Rhodes W. (eds.), *Encyclopedia of Planetary Science*, Chapman and Hall, 1997.

See also: Atmospheric constituents of the outer planets; Comets; Internal structures of the outer planets.

SATURN

General

Galileo Galilei had observed Saturn's rings in 1610, but he thought that he had seen two satellites of Saturn. In 1659, however, Christiaan Huygens correctly interpreted the observations as being those of a ring around the planet (see next section).

On 25 March 1655, Huygens discovered the first satellite of Saturn, now called Titan. A few years later he determined Titan's orbital period within a few seconds of its correct value. Jean Dominique Cassini discovered Saturn's satellite Iapetus in 1671, followed by Rhea in 1672, and Tethys and Dione two years later. Cassini noted the intensity variation of Iapetus with time, and concluded that it had two hemispheres of greatly different reflectivity, and that its spin rate was synchronous. In 1707, however, Cassini dropped his theory of synchronous rotation. About a hundred years later, William Herschel resurrected the idea after observing its intensity variations over ten orbits of Saturn.

Table 2.2. *Jupiter's main satellites and rings*

(a) Jupiter's satellites

	Name	Year of discovery	Discovered by	Semimajor axis (10^3 km)	Orbital inclination	Mean radius (km)
J16	Metis	1979	Voyager team	128	0°	22
J15	Adrastea	1979	Voyager team	129	0°	8
J5	Amalthea	1892	Barnard	181	0°	84
J14	Thebe	1979	Voyager team	222	1°	49
J1	Io	1610	Galileo	422	0°	1822
J2	Europa	1610	Galileo	671	0°	1561
J3	Ganymede	1610	Galileo	1,070	0°	2631
J4	Callisto	1610	Galileo	1,883	0°	2410
J13	Leda	1974	Kowal	11,165	27°	10
J6	Himalia	1904	Perrine	11,461	27°	85
J10	Lysithea	1938	Nicholson	11,717	28°	18
J7	Elara	1905	Perrine	11,741	27°	43
J12	Ananke	1951	Nicholson	21,276	149°	14
J11	Carme	1938	Nicholson	23,404	165°	23
J8	Pasiphae	1908	Melotte	23,624	151°	30
J9	Sinope	1914	Nicholson	23,939	158°	19

(b) Jupiter's rings

Ring component	Radial location (km)	Discovered by	Date
(edge of planet)	*(71,500)*		
Halo	90,000−123,000	Voyager 2/Showalter	1979
Main	123,000−129,000	Voyager 1	1979
Amalthea Ring	140,000?−181,000	Galileo spacecraft	1997
Thebe Ring	140,000?−221,900	Galileo spacecraft	1997
Thebe Extension	221,900−280,000	Galileo spacecraft	1997

William Herschel discovered Saturn's sixth satellite, Enceladus, in 1789 in line with the rings and Tethys, Dione, Rhea and Titan. The rings were virtually edge on at that time, and so were less bright than normal, making it easier to see dim satellites close to the planet. In the following month, Herschel detected an even dimmer satellite, now called Mimas.

Saturn's eighth satellite, now called Hyperion, was discovered simultaneously by William C. and George P. Bond, and William Lassell in 1848. Fifty years later, William Pickering discovered Phoebe, the first satellite of any planet to be discovered photographically. Its orbit, which was some distance from Saturn, was found to be retrograde This, and its large orbital eccentricity, indicated that it was a captured object, rather than having been originally part of the Saturn system.

It had been difficult in the eighteenth century to measure the rotation period of Saturn because of the lack of clear markings on its visible surface. But in 1789 Herschel measured Saturn's appreciable polar flattening, which indicated that the planet must have a fast spin rate. A few years later he was able to measure this as 10 h 16 min ± 2 min by observing the motion of subtle changes on Saturn. This was only about two minutes longer than its true equatorial period.

No significant progress was made in trying to understand the rotational behaviour of Saturn until Asaph Hall noticed a well-defined, white equatorial spot on the planet in December 1876. Subsequent observations showed that its rotation period around Saturn was 10 h 14 min 23.8 s ± 2.5 s. Hall was cautious enough to point out that this was the rotation period of the spot around Saturn, and not necessarily that of the planet as a whole. It was thought most likely, in view of Saturn's very low density, that Hall was observing a spot in Saturn's atmosphere, rather than one on its surface. A number of other spots were observed on Saturn in the 1890s with about the same rotation period around the planet.

At that time there had been no clear evidence of differential rotation on Saturn, like that already observed on Jupiter and the

Sun. But in 1903 Edward Barnard observed a prominent white spot at a latitude of about 36° N that appeared to rotate about the centre of Saturn much more slowly than expected. Kasimir Graff's estimated rotation period of 10 h 39 min was initially treated with some scepticism, however, as it was so much longer than normal. But subsequently other astronomers deduced a similar value. As Stanley Williams observed at the time, this implied that there was an equatorial current on Saturn, similar to that on Jupiter. On Jupiter the velocity of the current was estimated to be about 400 km/h, whereas the one on Saturn appeared to have a velocity of about 1,400 km/h.

José Comas Solá noticed in 1907 that Titan, Saturn's largest satellite, showed a pronounced limb darkening effect, and suggested that this meant that it had an atmosphere. No clear, independent evidence was obtained for an atmosphere, however, until the winter of 1943–44 when Gerard Kuiper photographed the spectra of the ten largest satellites of the solar system. He found definite evidence of the 619 and 726 nm methane bands for Titan, and some evidence of the 619 nm band for Triton, Neptune's largest satellite, but he could find no such bands for Jupiter's four Galilean satellites, or for Saturn's satellites Tethys, Dione or Rhea. Kuiper also calculated an index showing, theoretically, the likely stability of planetary or satellite atmospheres against dispersion, over time. This index was based on the escape velocity for such a body, and on its surface temperature. The results gave him, in order of reducing likelihood of retaining an atmosphere, a list of the Earth, Venus, Triton, Mars, Titan, Ganymede, Callisto, Io, and Europa, with the likely boundary between atmosphere and no atmosphere between Titan and Ganymede.

In 1966 Audouin Dollfus photographed a tenth satellite of Saturn called Janus. Eleven years later Stephen Larson and John Fountain announced that they had found another satellite at about the same distance from Saturn as Janus, when re-examining photographic plates taken in 1966. The exact orbits of both Janus and the eleventh satellite, now called Epimetheus, were uncertain, however. But in September 1979 Pioneer 11 accidentally flew within about 2,500 km of Epimetheus, which was about 14,000 km outside the A ring. It was not until March 1980, however, that it became evident that Janus and Epimetheus were in essentially the same 16.67 hour orbit.

The images of Saturn received during the Pioneer 11 encounter in 1979 were disappointing, showing only the same subtle banded structure seen from Earth. The images of Titan were even more frustrating, as they were completely featureless. Iapetus, Rhea and Titan, Saturn's three largest satellites, were found to be of low density, indicating that they were mostly composed of ice.

Jean Lecacheux and P. Laques found another satellite of Saturn in 1980 using a ground-based telescope. This satellite, now called Helene, was subsequently found to be near the 60° Lagrangian point ahead of Dione and co-orbiting with it as

a Dione trojan. At about the same time, two other satellites, now called Telesto and Calypso, had been imaged by Bradford Smith, Dan Pascu and colleagues. They were found to be at the Lagrangian points 60° in front of and 60° behind Tethys, and co-orbital with it, as Tethys trojans.

In the early 1980s, combined data from both Voyagers 1 and 2 enabled the rotation rate of Saturn to be determined for the first time based on variations in Saturn's radio emissions, which were thought to be linked to the rate of rotation of Saturn's core. This produced a period of 10 h 39.4 min, which implied that the 10 h 14 min equatorial spots previously seen had been moving around Saturn at about 1,400 km/h relative to the core.

The Voyager 1 images of Saturn in 1980 showed that the wind velocity at cloud-top height reached a maximum of almost 1,800 km/s practically on the equator. This peak velocity was about three times that on Jupiter, and the equatorial jet stream on Saturn, of which this high-speed movement was part, was found to cover the region from about 30°N to 40°S (where the velocity falls to zero), compared with Jupiter's jet that only covers the region from about 12°N to 16°S. Voyager 1 showed that Saturn emitted about 1.8 times as much energy as it received from the Sun, which is, within error, the same ratio as that deduced for Jupiter. It was not clear why Saturn, which received much less solar energy than Jupiter, should have much the more powerful equatorial jet.

The Voyager 1 images of Titan, by far the largest of Saturn's satellites, were featureless, with only the faintest shadings being visible after extensive computer processing. However, images of the limb showed an extensive haze layer above the impenetrable clouds.

In 1944 Gerard Kuiper had detected methane lines in Titan's spectrum, but before Voyager 1's fly-by no other constituents of its atmosphere were known, although the methane lines were known to suffer from pressure-broadening. Some astronomers thought that Titan's atmosphere probably consisted of methane, with only traces of other gases, at a surface-level pressure of about 20 millibars. This would explain the observed pressure broadening, but Donald Hunten suggested that the same result could be obtained if the main atmospheric constituent was nitrogen, which was undetectable from Earth, with methane being present at a much lower percentage. Hunten's model was somewhat flexible, however, depending on how deep the atmosphere was. Measurements by John Caldwell and colleagues in 1979, with the partially completed VLA radio telescope, implied that the surface temperature of Titan was about 87 K, and this led Hunten to conclude that the surface pressure would be about 2,000 millibars, or about twice that of Earth.

Voyager 1's radio occultation experiment showed that Titan's surface-level atmospheric pressure is actually about 1,500 millibars at a temperature of 94 K, thus vindicating

Hunten's atmospheric model. The ultraviolet spectrometer also showed that Titan's atmosphere was approximately 90% nitrogen, in line with Hunten's prediction. The minimum atmospheric temperature detected was about 68 K at a pressure of 380 millibars about 50 km above the surface. Interestingly, nitrogen could condense into droplets under those conditions and form clouds.

The orange colour of Titan seen from Earth was found to be due to a layer of photochemical smog, which is produced when ultraviolet sunlight breaks up atmospheric methane and nitrogen molecules into their constituent atoms. These then recombined to form various hydrocarbon and nitrogen compounds, including ethane, ethylene, propane and hydrogen cyanide, which were found by Voyager 1 in small quantities in Titan's atmosphere. The surface temperature of 94 ± 2 K was very close to methane's triple point of 90.7 K, at which methane can exist in solid, liquid and gaseous states. So it was thought that the surface of Titan may have oceans of methane, with cliffs of methane in its polar region, with methane on Titan acting like water on Earth in its various forms. Images taken by the Hubble Space Telescope in 1994 showed brightness variations across the surface, indicating that any ocean, if it existed, did not completely cover the surface.

Mimas was found by Voyager 1 to have a 135 km diameter crater on its 390 km diameter surface. Not only was the crater very large compared with the size of the satellite, but it was an incredible 10 km (33,000 ft) deep. Dione was shown to have a dark and a light-coloured hemisphere, with broad white streaks criss-crossing the dark trailing side, that appeared to be relatively smooth. The light side, on the other hand, was seen to have a number of sinuous valleys on its heavily cratered surface. Rhea was found to be similar in appearance to its smaller neighbour Dione, having one heavily cratered hemisphere and a darker hemisphere that was less heavily cratered. Also like Dione it had broad white streaks on its surface, although the contrast difference between the streaks and the adjacent surface was not as high. Hyperion was not well imaged by Voyager 1, but the images were good enough to show that it was very irregular in shape. A satellite of its size (410×220 km) should be spherical, so it was thought to be part of a body that has been broken up by an impact.

The Voyager 2 images of Iapetus in 1981 showed that there were numerous craters on its light-coloured hemisphere, with some of these craters near to the dark hemisphere having dark-coloured floors. Before Voyager 2 it was thought that, as the dark hemisphere is the leading side of Iapetus in its orbit around Saturn, the dark material could have been picked up as the satellite orbited the planet. The discovery of dark-floored craters in the light-covered hemisphere made this unlikely, however.

Unfortunately the best Voyager 2 images of Enceladus were lost, owing to a spacecraft problem, but those that were returned showed remarkable detail on this the most reflective

satellite in the solar system. Large areas were found to be craterless, suggesting a surface age of less than 100 million years, but even the cratered areas showed relatively few large craters, suggesting that that surface was not as old as that of Saturn's other satellites. In addition, there were numerous valleys and ridges up to 1 km high cutting across both the cratered and uncratered regions. All this indicated that Enceladus had been geologically active until the relatively recent past (in astronomical terms) and it may even be geologically active today. This was remarkable in a satellite of only 500 km in diameter, as its initial heat at formation should have been dissipated billions of years ago. Interestingly, neither Mimas nor Tethys, the satellites with orbits on either side of Enceladus, showed anything like this level of activity. Enceladus is intermediate in size between these two satellites, and all three have similar densities, so the cause does not appear to be due to flexure caused by its close proximity to Saturn. A more likely reason was thought to be the 2:1 resonance of its orbit to that of the larger satellite Dione, which would cause tidal flexing.

The Voyager 2 images of Tethys showed an enormous 400 km diameter crater called Odysseus which, relative to the size of the satellite, was even larger than the 135 km diameter crater on Mimas. In addition, Voyager 2 showed that the north-south gorge imaged by Voyager 1, which extended about two-thirds of the way around Tethys, has an average width of about 100 km and depth of about 4 km. It was unique in its size relative to that of Tethys for bodies in the solar system, as far as we know. It has been suggested that this gorge, now called Ithaca Chasma, could have been produced when the watery interior of Tethys froze, causing the satellite to expand. But it was not clear why it appeared to be the only large gorge on Tethys, and why similar gorges did not exist on the other icy satellites of Saturn.

Finally, one of Voyager's strangest discoveries involves the co-orbiting satellites Janus and Epimetheus. Their orbital periods were found to differ by only thirty seconds and, once every four years, they would approach each other so closely that they would swop orbits.

The Cassini–Huygens spacecraft, which arrived at Saturn in 2004, consisted of a Cassini Saturn orbiter and a Huygens probe. The probe was released from its Cassini carrier and descended through Titan's atmosphere in January 2005. It found methane and nitrogen in Titan's stratosphere, with the methane concentration increasing as the probe descended. Clouds of methane were detected at an altitude of about 20 km. The probe discovered dendritic channels during its descent, and imaged large pebbles, probably of dirty water ice, at its landing site, which had a temperature of 93.65 ± 0.25 K.

The Cassini orbiter radar imaged what appeared to be large lakes on Titan. In 2007 one of these lakes, Ontario Lacus (see Figure 2.15), was found to contain liquid hydrocarbons, of which ethane was positively identified. The ethane was thought

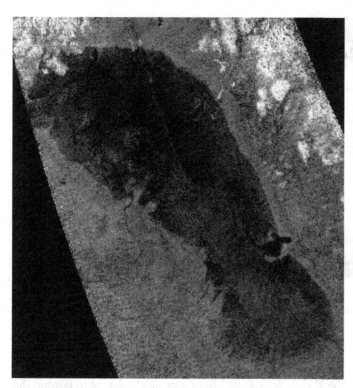

Figure 2.15. Ontario Lacus, the 15,000 square kilometre lake of hydrocarbons imaged by the Cassini radar on Titan in 2010. Other Cassini observations appear to show that its shoreline has changed over time. (Courtesy NASA/JPL/Caltech.)

to be in a liquid solution with methane, other hydrocarbons, and nitrogen.

Early in its mission, Cassini detected a modification in Saturn's local magnetic field by Enceladus, indicating that Enceladus had a thin atmosphere mostly of water vapour. This atmosphere was then positively detected and found to be asymmetric, being most prevalent in the south polar region. Shortly afterwards Cassini discovered plumes of water ice ejected into Saturn's E ring from surface cracks or 'tiger stripes' near Enceladus' south pole. Although the average temperature of the south polar region was about 85 K, that in the tiger stripes was up to about 180 K. It was thought that there were probably pockets of liquid water under the surface fuelling these geyser-like eruptions. In March 2008 Cassini flew through a plume and detected water vapour, carbon dioxide and hydrocarbons.

The Cassini spacecraft found that Saturn's kilometric radio emission varied with a period of 10 h 45 m 45 $\pm$ 36 s. This was some six minutes longer than the period of 10 h 39 m 24 $\pm$ 7 s also deduced from Saturn's radio emissions some years earlier using Voyager. Later analysis of the Cassini data showed that the radio emissions varied at a different rate in the northern and southern hemispheres. The reason for these different rates is currently (in 2012) unclear, leaving the rotation rate of the interior of the planet uncertain.

Rings

Galileo Galilei was the first to observe Saturn's rings when in July 1610 he noticed two objects one on either side of the planet. He thought that they were moons of Saturn, but by November 1612 the two objects had disappeared. They reappeared in the following year, and he later noticed that they appeared to be non-circular in shape.

Over the next forty years or so many astronomers observed Saturn's strange attendants, noticing their varying shapes. It was not until 1659, however, that Christiaan Huygens correctly interpreted them as being a thin flat ring surrounding Saturn, inclined at an angle of about 30° to the ecliptic. The ring configuration, as seen from Earth, varied cyclically as Saturn orbited the Sun, with the rings apparently disappearing when they were seen edge-on.

In 1675 Jean Dominique Cassini observed that Saturn's ring was divided in two by a dark line (now called the Cassini division) going all round the planet. He found that this was a permanent feature dividing a brighter inner ring (now called the B ring) from the outer ring (A ring). However he was unsure as to whether this was a dark marking on one ring, or whether it was a clear division between two rings. In 1705 Cassini suggested that Saturn's ring was not solid, as Huygens had thought, but consisted of swarms of small satellites. Later that century Laplace showed theoretically that Saturn's one or two rings could not be solid, but must consist of many, narrower, solid, concentric rings orbiting the planet.

William Herschel observed both the north and south faces of Saturn's rings and found that the Cassini division was at the same distance from the planet on both. As a result he concluded in 1791 that the Cassini division was a true gap dividing one ring into two, but he could find no evidence of Laplace's narrow rings. He confirmed that the rings were rotating around Saturn, when he measured the rotation period of a bright spot that he thought was on the outer edge of the A ring. The period that he measured, however, seems anomalous as it is the rotation period of the outer edge of the B ring.

In 1850 William C. and George P. Bond and, independently, William Dawes discovered the dark C ring (see Figure 2.16), inside the B ring. Two years later William Jacob and William Lassell independently observed the edge of the planet through the C ring, showing that it was partially transparent. Lassell also saw the globe of Saturn through the Cassini division, thus confirming that it was a true gap between the two bright rings.

James Clerk Maxwell proved mathematically in 1857 that Saturn's rings could not consist of a number of thin solid rings, as proposed by Laplace. Nor could they be fluid, as some astronomers had suggested, as fluid rings would have been unstable. Instead, he concluded that the rings consisted of an indefinite number of particles orbiting the planet, as previously suggested by Cassini. The first indication that Maxwell was

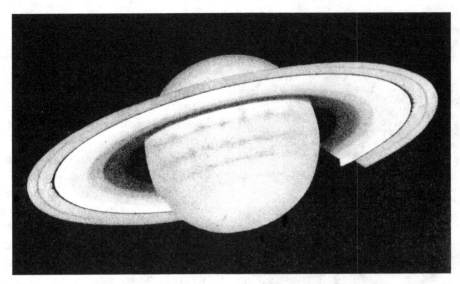

Figure 2.16. Etienne Trouvelot's 1874 drawing of Saturn clearly shows the dark C ring extending from the inner edge of the B ring to about half-way to the planet. His drawing also showed that the edge of Saturn could be seen through the ring. (From *The Sun: Its Planets and Their Satellites*, by Edmund Ledger, 1882, Plate IX.)

correct was produced in 1895 by James Keeler, who measured the Doppler shift of the inside and outside of the rings, showing that their velocities were inconsistent with the rings being solid.

A few years earlier, Edouard Roche had analysed the stability of a fluid satellite of the same density of Saturn orbiting the planet. He showed that it would have been disrupted by tidal forces created by Saturn if it were closer than 2.44 R_S (Saturn radii) to the centre of the planet. Roche hypothesised, therefore, that a fluid satellite had approached too close to Saturn, where it had been broken up into numerous small particles that now form the rings. This Roche distance was consistent with the distances of the rings of Saturn which were known at that time, as they were all within 2.27 R_S of the centre of the planet.

Daniel Kirkwood noted in 1867 that any particles in the Cassini division would have periods of about one-half that of Mimas, one-third that of Enceladus, one-quarter that of Tethys, and one-sixth that of Dione. As a result he concluded that, because of resonances with these satellites, the Cassini division would be cleared of particles.

A number of astronomers reported seeing divisions in the A and B rings in the first half of the nineteenth century, but the reported positions were not consistent. For example, in 1837 Johann Encke observed a gap in the A ring about one third of the way from the inner to the outer edge. But in 1843 Dawes and Lassell both saw the so-called Encke gap outside the middle of the A ring. Later that century, various excellent observers with first-class telescopes could not see the gap at all. Likewise, various astronomers reported seeing a ring inside the C ring, and one outside the A ring. But it was not until 1966 that Walter Feibelman was able to clearly extend the observed ring system when he photographed a very faint ring, now called

the E ring. It extended from outside the A ring to beyond the orbit of Dione, about 380,000 km (6.3 R_S) from the centre of Saturn.

Unexpectedly strong radar echoes were received from Saturn's rings in the early 1970s. These echoes indicated that particles in the rings may be up to a metre in diameter, with the majority falling into the 5 to 30 centimetre range. Some of the particles were thought to be metallic because of their high reflectivity. Radio and infrared measurements indicated that the rings also contained water ice particles, with a mean radius of about one centimetre.

In 1979 the narrow F ring was discovered by the Pioneer 11 spacecraft just 3,000 km outside the A ring. Pioneer also found that the Cassini division was not empty, but appeared to be filled with low density material. In the following year Richard Terrile discovered radial 'spokes' in Saturn's B ring (see Figure 2.17) using the Voyager 1 spacecraft. Radial shadings on the A ring, and occasionally on the B ring, had been observed from time to time in the previous hundred years or so, but it had never been clear whether these shadings were real or not. Andy Collins and Richard Terrile found two new satellites, now called Prometheus and Pandora, one just inside and one just outside the narrow F ring, apparently shepherding or stabilising it. In fact, shepherding satellites had first been proposed by Peter Goldreich and Scott Tremaine in 1979 to explain the narrow rings of Uranus, but these two shepherding satellites of Saturn were the first to be discovered for any planet. Then Terrile found another small satellite, now called Atlas, orbiting Saturn just 800 km outside the outer edge of the A ring, and apparently restricting the ring's outward expansion.

Figure 2.17. This Voyager 2 image shows the radial spokes in Saturn's B ring. Some of the spokes were observed to change their appearance over timescales as short as only 20 or 30 minutes. (Courtesy NASA/JPL/Caltech.)

Voyager 1 found that Saturn's ring system consisted of about 500 to 1,000 narrow rings, with even the Cassini division between the A and B rings containing over 100 narrow rings. The spokes on the B ring were found to rotate around Saturn at the same angular rate as the planet rotated on its axis, implying that they were associated with its magnetic field. Further analysis indicated that they were caused by charged particles elevated from the ring by either magnetic or electrostatic forces.

Voyager 1's radio occultation experiment showed that the particles in the dark C ring were rather large, typically being about 1 metre in diameter. There was a narrow gap between its outer edge and the inner edge of the B ring. A very faint D ring was found inside the C ring. The narrow F ring, discovered by Pioneer 11 just outside the A ring, was found to be three intertwined or braided rings, apparently contravening the laws of dynamics. Like the spokes in the B ring, however, it was suggested that this braided ring could also be responding to Saturn's magnetic field. A very faint, relatively narrow ring, ring G, was found between the F ring and the orbit of Mimas, at the position predicted by the absence of charged particles detected by Pioneer 11. Finally, the very large and faint E ring, first detected by Feibelman in 1966, was recorded stretching 300,000 km from about the orbit of Mimas, past the orbits of Enceladus, Tethys and Dione to almost that of Rhea.

Before Voyager 1 arrived at Saturn in 1980, it was thought that the ring system, which appeared to consist of three broad, featureless rings, plus a very faint E ring, was generally understood. The gaps in the rings were thought to be caused by resonances with Saturn's satellites, Mimas in particular. However, the fine structure of the main A, B and C rings found by Voyager 1, the non-empty Cassini division, the spokes on the B ring, the braided F ring, and the existence of other new rings demanded a fundamental rethink of the theory of Saturn's rings.

By the time that Voyager 2 arrived at Saturn, nine months after Voyager 1, there had been tentative explanations proposed for many of these new phenomena, with the new rôle of shepherding satellites receiving particular attention. Voyager 2's images showed more fine structure in the main rings than its predecessor, and the Voyager 2 photopolarimeter showed even finer structure, with some rings in the main rings being only a few hundred metres wide. Such fine structure implied that the rings must be very thin perpendicular to the plane of their orbits. The A ring, for example, which is about 15,000 km from its inner to outer edge, was estimated to have a thickness of only about 100 m at most. Voyager 2 found no new shepherding satellites.

The theory that the Cassini division was caused by a resonance with Mimas and other satellites had to be modified when Voyager 1 found that the Cassini division was not empty. But Carolyn Porco showed, using Voyager 2 data, that the outer edge of the B ring, which is the inner edge of the Cassini division, is not circular but slightly elliptical, with the major axis precessing once every 22.6 hours. This is the orbital period of Mimas, so the outer edge of the B ring is clearly controlled by Mimas. Similarly the six-lobed petal shape of the outer edge of the A ring showed that it resulted from a 7:6 resonance with Janus.

Goldreich and Tremaine had shown in the 1970s that spiral density waves, like those thought to occur in galaxies, would also exist in circumplanetary discs. In the latter case the density waves would be produced by a satellite orbiting near the disc, producing a pattern of condensations and rarefactions in the disc. Voyager 2 found about fifty such spiral density wave trains in Saturn's A ring, partially explaining its complex structure.

In 1986 Mark Showalter and colleagues analysed the Encke gap in the A ring, and deduced the possible mass, size and orbit of the satellite responsible for producing the gap. This enabled a search to be made for the satellite using the Voyager images. In 1990 it was successful, with the discovery by Showalter of a 20 km diameter satellite, now called Pan, orbiting within the Encke gap.

The spokes seen on Saturn's B ring by Voyagers 1 and 2 were subsequently imaged by the Hubble Space Telescope in Earth orbit. It imaged them until 1998, but they then disappeared. The Cassini spacecraft searched for the spokes when it arrived in orbit around Saturn in 2004, but none were found until the following year. Their existence seemed to depend on the solar illumination angle of the rings.

In 2005 Cassini discovered a 7 km moonlet, now called Daphnis, near the centre of the Keeler gap in the A ring. Its existence had been inferred from its gravitational effect on the outer edge of the gap. In the following year, four much smaller moonlets, of the order of about 100 metres in size, were also discovered in the A ring, characterised by the propeller-shaped disturbances that they made in the ring. They were too small to clear channels in the ring.

On 17 September 2006 Cassini spent 12 hours in the shadow of Saturn and was able to make a number of images of the ring system in silhouette. These revealed a tenuous ring between the F and G rings, associated with Janus and Epimetheus, and another associated with Pallene, a satellite that had been discovered by Cassini in 2004. Jets were also seen to be emitted by Enceladus replenishing the densest part of the E ring.

The Cassini spacecraft discovered a number of new satellites of Saturn, two of which, Methone and Anthe, were associated with ring arcs that had also been detected by the Cassini spacecraft. Anthe was found to move backwards and forwards along its arc, which was about 20° longitude in length. Both arcs seemed to be constrained by resonances with Mimas.

Magnetism and magnetosphere

No radio emission had been detected from Saturn prior to the Pioneer 11 intercept in 1979, so it was unknown at that time if it possessed a magnetosphere and radiation belts. In the event, Pioneer 11 crossed the bow shock at 24 R_S from the centre of Saturn. The planet's magnetic field was found to be about 0.22 gauss at the equator, with a dipole moment of about 600 times that of Earth. The field was dipolar, with a magnetic axis indistinguishable from its spin axis. Titan, Saturn's largest satellite, was found to be generally within the planet's magnetosphere, but varying solar wind pressure drove the magnetopause back and forth across Titan, sometimes leaving it exposed to the solar wind. As expected, Saturn's main rings apparently prevented radiation belts from forming in the inner magnetosphere, although an extensive radiation belt was discovered between about 2.3 and 3.5 R_S from the centre of the planet, outside the A ring. Then in 2004 the Cassini spacecraft discovered a completely unexpected, relatively weak radiation belt between the inner edge of the D ring and the top of Saturn's atmosphere.

In 1980 Voyager 1 found that, as the plasma particles in Saturn's magnetosphere flowed past Titan, their density increased, whilst their velocity slowed down. No magnetic field was detected on Titan. Voyager also found that Titan and the other large satellites inside the magnetosphere, namely Rhea, Dione, Tethys, Enceladus and Mimas, absorbed electrons and protons from the magnetosphere. But, surprisingly, the satellites allowed the inward migration of electrons of specific energies.

Further information on Saturn's satellites and rings is given in Table 2.3.

Bibliography

Alexander, Arthur F. O'D., *The Planet Saturn: A History of Observation, Theory and Discovery*, Faber and Faber, 1962.

Beatty, J. K., Petersen, C. C., and Chaikin, A. (eds.), *The New Solar System*, 4th ed., Cambridge University Press, 1999.

Dougherty, M. K., Esposito, L. W., and Krimigis, S. M. (eds.), *Saturn from Cassini-Huygens*, Springer, 2009.

Godwin, Robert, and Whitfield, Steve, *Deep Space: The NASA Mission Reports*, Apogee Books, 2005.

Harland, David M., *Cassini at Saturn: Huygens Results*, Praxis Publishing, 2007.

Leverington, David, *Babylon to Voyager and Beyond: A History of Planetary Astronomy*, Cambridge University Press, 2003.

McFadden, L. A., Weissman, P. R., and Johnson, T. V. (eds.), *Encyclopedia of the Solar System*, 2nd ed., Academic Press, 2007.

Miner, E. D., Wessen, R. R., and Cuzzi, J. N., *Planetary Ring Systems*, Springer-Praxis, 2007.

Morrison, D., *Voyages to Saturn*, NASA SP−451, 1982.

Shirley, James H., and Fairbridge, Rhodes W. (eds.), *Encyclopedia of Planetary Science*, Chapman and Hall, 1997.

See also: Atmospheric constituents of the outer planets; Eighteenth century: Saturn's rings; Internal structures of the outer planets.

URANUS

General

Uranus was discovered in March 1781 by William Herschel as a fuzzy object that he mistook for a comet. Nevil Maskelyne, the Astronomer Royal, subsequently observed the object, and concluded that it could be either a comet or a planet. He was doubtful that it could be a comet, however, as it was unlike any comet that he had seen previously.

The first orbits were calculated separately by Pierre Méchain and Anders Lexell on the assumption that the object was a comet. But it soon became clear that its orbit was essentially circular, and so the object was a planet. It is not clear who was the first to realise this, but both Lexell and Jean Baptiste de Saron seem to have come to this conclusion independently in May 1781.

It is difficult to exaggerate the effect that the discovery of the first planet since ancient times had on the astronomical community. At a stroke Herschel had doubled the diameter of the known solar system, and encouraged those astronomers like Lexell who thought that there may be yet more planets. In 1781 Lexell estimated that the radius of the new planet's orbit was

Table 2.3. *Saturn's main satellites and rings*

(a) Saturn's satellites

	Name	Year of discovery	Discovered by	Semimajor axis (10^3 km)	Orbital inclination	Mean radius (km)
S18	Pan	1990	Showalter	134	0°	13
S15	Atlas	1980	Terrile	138	0°	17
S16	Prometheus	1980	Collins and Terrile	139	0°	50
S17	Pandora	1980	Collins and Terrile	142	0°	44
S11	Epimetheus	1980	Larson and Fountain	151	0°	55
S10	Janus	1966	Dollfus	151	0°	90
S1	Mimas	1789	Herschel	186	2°	199
S2	Enceladus	1789	Herschel	238	0°	252
S3	Tethys	1684	Cassini	295	1°	536
S13	Telesto	1980	Smith et al.	295	1°	12
S14	Calypso	1980	Pascu et al.	295	1°	10
S4	Dione	1684	Cassini	377	0°	563
S12	Helene	1980	Lecacheux and Laques	377	0°	16
S5	Rhea	1672	Cassini	527	0°	765
S6	Titan	1655	Huygens	1,222	0°	2,576
S7	Hyperion	1848	W. C. & G. P. Bond, and Lassell	1,481	1°	140
S8	Iapetus	1671	Cassini	3,561	8°	735
S9	Phoebe	1898	Pickering	12,948	175°	110

(b) Saturn's rings

Ring component	Radial location (km)	Discovered by	Date
(edge of planet)	*(60,300)*		
D	67,000–74,500	Voyager 1	1980
C	74,500–92,000	W. C. & G. P. Bond and W. R. Dawes	1850
B	92,000–117,580	Galileo and Huygens	1610 and 1659
Cassini division	117,580–122,200	Cassini	1675
A	122,200–136,780	Galileo and Huygens	1610 and 1659
Encke gap	133,570	Encke	1837
Keeler gap	136,530		
F	140,200	Pioneer 11	1979
G	166,000–174,000	Voyager 1	1980
E	181,000–483,000	Feibelman	1966

about 18.93 AU. In the following year Joseph de Lalande, after analysing observations over a period of 13 months, concluded that it was in an approximately circular orbit of the same radius, with a period of 82.37 years.

But Lexell and others found that, as time went on, the planet deviated from a circular orbit. So in 1783 Barnaba Oriani, J. F. Hennert and Pierre-François Méchain independently calculated elliptical orbits that were remarkably similar to each other. Méchain's orbit, for example, had a semi-major axis of 19.08 AU (correct value 19.19 AU) and period of 83.3 years

(correct value 84.02 years). Then in 1788 Herschel estimated the diameter of the new planet to be about 55,000 km, which is now known to be within about 10% of its true value.

Lalande suggested calling the new planet *Herschel*, whilst Johann Bode proposed *Uranus*, as Uranus was the father of Saturn, who in turn was the father of Jupiter. Herschel suggested calling it *Georgium Sidus* (George's Star), after King George III, who was king of England at the time. In the event the new planet was called *Uranus* in Germany, and first *Herschel* and then *Uranus* in France, but it was not until the

1820s that its name was generally changed from the *Georgian Star* or *Georgian Planet* to *Uranus* in Britain.

Uranus had been observed at least 24 times by various astronomers prior to its discovery by Herschel. But its character as a planet had never been realised. The earliest recorded observation had been made by John Flamsteed, the first Astronomer Royal, in 1690, whilst Pierre Le Monnier had observed it nine times between 1764 and 1769, four of these on consecutive nights in January 1769. Unfortunately for Le Monnier, in January 1769 the planet was near the stationary point in its orbit, as seen from Earth. As a result its movement had been slight and so he had missed it.

Herschel detected the first two satellites of Uranus in January 1787: the inner one, now called Titania, and Oberon. Herschel soon realised that the orbits of both satellites, although almost circular, were inclined at about 90° to the ecliptic. But he was unable to deduce when the satellites were on the far side of Uranus and when they were on the near side, as seen from Earth, as the orientation of their orbits was such that neither satellite was eclipsed by the planet at that time. The first eclipses, which occurred in early 1798, allowed Herschel to conclude that the angle of the satellites' orbits to the ecliptic was greater than 90°. This meant that these were the first satellites of the solar system known to have retrograde orbits. He later calculated the inclination of their orbits to that of Uranus to be about 101° 2′, which was just a few degrees out.

Herschel used the two satellites to deduce the mass of Uranus as 17.74 relative to that of the Earth (correct value 14.53). Although this estimate was rather inaccurate, he had shown that Uranus was intermediate in mass between Jupiter and Saturn, on the one hand, and the Earth on the other. Herschel also observed that Uranus was flattened at the poles, like Jupiter and Saturn, and concluded that it must also, like them, have a fast rotation.

William Lassell discovered the next two satellites of Uranus, Ariel and Umbriel, in October 1851, both within the orbit of Titania. He then found that he had seen both satellites previously on 6 November 1847, and, using those observations, was able to calculate their orbital periods to within one minute of their correct values. Further work showed that Lassell had been the first to observe Ariel on 14 September 1847, and that Otto W. Struve had been the first to observe Umbriel on 8 October of the same year.

A number of astronomers observed Uranus in the late nineteenth and early twentieth centuries to try to determine its rotation period. Various values were proposed, including one of about 10 h 45 min retrograde, by Percival Lowell and Vesto Slipher in 1912 by measuring Doppler shifts. Joseph Moore and Donald Menzel repeated Lowell and Slipher's measurements about fifteen years later, and deduced a similar period of 10 h 49 min ± 10 min. As a result a period of about 10 h 50 min was accepted for many years. Then in the 1970s longer periods

of 15 to 17 hours were deduced from intensity fluctuations, and in 1984 R.G. French derived a period of about 15 h 35 min from Uranus' observed oblateness.

Gerard Kuiper discovered Uranus' fifth known satellite, now called Miranda, in 1948. He found that it was orbiting Uranus in an approximately circular orbit in the same plane as the other four satellites.

The Voyager 2 spacecraft detected a strong burst of polarised radio signals from Uranus as it flew past the planet in January 1986, confirming the presence of a magnetic field (see later section). The variations in these radio signals implied a rotation rate of Uranus' interior of 17 h 14 min, which was somewhat longer than expected.

Individual clouds were observed on Uranus by Voyager 2, enabling its wind velocities to be determined. As the Sun was virtually overhead at the south pole during the Voyager 2 fly-by, clouds could only be seen in the southern hemisphere, and even here they were rather sparse. The results showed a 360 km/s easterly equatorial jet, with a strong westerly flow for most of the southern hemisphere. This was contrary to the case for both Jupiter and Saturn, where the equatorial jets were both westerlies. In fact the strongest winds on Uranus were found at medium (southerly) latitudes, again unlike Jupiter and Saturn, where the strongest winds were found to be at or near the equator.

Peter Goldreich and Scott Tremaine had suggested in 1979 that the narrow rings of Uranus, which had been discovered two years earlier (see later section), may be constrained by shepherding satellites. So it had been hoped, before the Voyager intercept, that Voyager would detect a pair of shepherding satellites for each of Uranus' nine known rings. Six new satellites were discovered as Voyager 2 approached the planet, but none were shepherds. However, four days before Voyager's closest approach to Uranus, two small shepherding satellites, Cordelia and Ophelia, were found on either side of the 1 ring. Only Puck, of the satellites discovered by Voyager 2, was large enough to have its diameter measured directly. All the other satellite diameters were estimated from their brightness, assuming that their reflectivities were the same as Puck.

Little was known about Uranus' five largest satellites, which had been discovered using Earth-based telescopes, until the Voyager 2 intercepts. In fact, it was not until the early 1980s that their diameters had been measured with any certainty, ranging from about 1,600 km (1,000 miles) for Oberon and Titania, to about 500 km (300 miles) for Miranda. This meant that they were smaller than Jupiter's Galilean satellites and about the same size as the large satellites of Saturn, excluding Titan. Because of their relatively small size and their great distance from the Sun, it was anticipated that these five Uranian satellites would show little in the way of geological activity, being covered instead with craters dating back to their early formation.

The satellites of Saturn are generally less dense than those of Jupiter, because they contain more water ice, and it was generally anticipated that those of Uranus would contain even more water ice, and be even less dense, as they are further from the Sun. They were darker than Saturn's satellites, however, and the ice spectral bands were not as clear, so it was assumed that the ice was more contaminated on their surfaces. In the event Voyager 2 showed that the densities of the Uranian satellites were generally higher than those of the similar-sized satellites of Saturn, not lower, as they consisted of less ice, not more.

Oberon, the outermost of the Uranian satellites known at the time, was seen by Voyager 2 to have an ancient, heavily cratered surface, with some craters being more than 100 km in diameter. Some of the craters were seen to have smooth, dark floors, which was thought to be due to a mixture of melted ice and carbonaceous rock coming from beneath the surface. On the highest resolution image, an 11,000 m (33,000 ft) high mountain, which was probably the central peak of an impact crater, was seen on the edge of the disc.

Titania, with a diameter of about 1,580 km (980 miles), was found to be the largest of the Uranian satellites. Two very large craters were imaged near its terminator, one of which had three concentric rings, and the other two. A large rift valley system was found that was at least 1,500 km long, with some of the valleys being up to 50 km wide and 5,000 m deep. It was thought that this rift valley system may have been produced when ice in the interior froze, in the same way that the large rift valley was thought to have been produced on Tethys, one of Saturn's satellites. Clearly the ancient surface of Titania had been modified since formation by geological processes, although it was still relatively old.

Umbriel, with an albedo of 0.20, is the darkest of the large Uranian satellites. It was somewhat surprising, therefore, that Ariel, whose size and density are similar, and whose orbit is just inside that of Umbriel, has an albedo of 0.36 and is the brightest of the satellites. Umbriel and Ariel are of almost identical size and density, so it is not clear why one is so much darker than the other. It may be partially explained by the age of their surfaces, as Umbriel was seen to have an ancient, heavily cratered surface like Oberon, which was also quite dark. Umbriel was seen to have just one outstanding feature, a bright, doughnut-shaped ring near the edge of the disc, although there also appeared to be linear features and prominent cliffs on the visible surface.

Ariel was found to be similar to Titania in so far as both had large rift valleys, although those on Ariel covered more of the visible surface. Some of the valleys on Ariel, many of which were flat-bottomed, were over 20,000 m (60,000 ft) deep, which is astonishing for a satellite of only about 1,160 km (720 miles) diameter. Parts of its surface had fewer craters than Titania, indicating that they have been resurfaced more recently by volcanic activity. In fact, there was a relative dearth of large craters on Ariel, and detailed crater counts seemed to indicate that there had been several periods of partial resurfacing extending up to quite recent times. The reasons for this are unclear as, unlike the case of Io, there are no orbital resonances between Ariel and any of the other large satellites, although there may have been some resonances in the past.

The images of Miranda had the highest resolution of all the Uranian satellites. With a diameter of only about 470 km (290 miles), it is the smallest and least dense of the five large satellites and is the closest of these to Uranus. Miranda was found to have a chevron-shaped feature, two large ovoids and enormous cliffs. The light-coloured chevron shape was seen to be surrounded by a large almost rectangular-shaped feature, now called the Inverness Corona, that had several parallel linear features that intersected at right angles. The two large ovoids, now called Elsinore Corona and Arden Corona, had concentric sets of ridges and grooves that made them look like large dirt racetracks. The Inverness and Arden coronae appeared to be bordered by trenches often associated with parallel scarps. In the case of Inverness corona, these scarps continued to the terminator where the enormous cliff face called Verona Rupes was seen, ranging in height from 10,000 to 20,000 m (30,000 to 60,000 ft), which is incredible on such a small satellite. So Miranda was seen to be far removed from the simple cratered world that had been expected.

A number of new satellites have been found since the Voyager 2 encounter, using the Hubble Space Telescope and various ground-based instruments, to bring the total number known (in 2010) to 27.

Rings

The rings of Uranus were accidentally discovered in 1977 by James Elliot, Robert Millis and colleagues who had been observing a stellar occultation by Uranus. The rings were designated, with increasing distance from the planet, 6, 5, 4, α, β (or 3), η, γ (2), δ, and ε (1). All of them, with the exception of the ε ring, appeared to be virtually circular, whilst the ε ring was clearly eccentric.

Numerous stellar occultations by the rings were observed over the next few years. They showed that the nine 'classical' rings were very narrow, with all but the ε ring being 12 km, or less, in width. The η, γ, δ and ε rings were coplanar with Uranus' satellites, whereas the five inner rings had small but measurable inclinations. These inclinations, together with the marked eccentricity of the ε ring, caused these six rings to precess slowly around the planet.

It had been expected that pairs of shepherding satellites would be found for each of Uranus' rings during the Voyager 2 intercept. In the event, two small shepherding satellites (Cordelia and Ophelia) were found on either side of the ε ring (see Figure 2.18), but no further shepherding satellites were found.

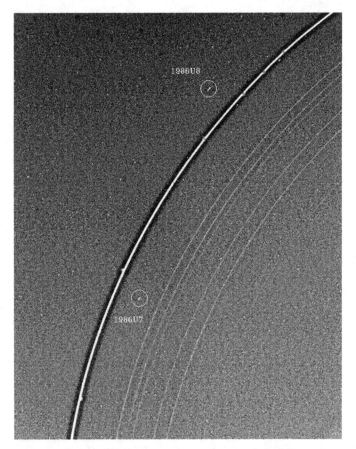

Figure 2.18. Voyager 2's image of Uranus' satellites Cordelia and Ophelia shepherding the bright ε ring. All nine rings detected from Earth in 1977 were visible in the original image. (Courtesy NASA/JPL/Caltech.)

As Voyager 2 approached Uranus, the nine known rings appeared narrow and dark. Then on the day before closest approach a very faint tenth ring, called the λ ring, was found between the ε and δ rings. An image taken after Voyager had passed through the ring plane showed each of the ten narrow rings, with the newly-discovered λ ring being the brightest.

During the Voyager 2 intercept, a faint ring (1986 U2R) about 2,500 km wide was found between the innermost ring and the planet. This U2R ring was not observed again until 2003, when a ring was detected using the Keck telescope. The Keck ring, now called the ζ ring, stretched from 37,850 km to 41,350 km, compared with the 37,000 to 39,500 km of the U2R ring. The ζ ring was observed again during the ring plane crossing of 2007, when it was the brightest of the rings.

A density wave pattern, of the sort found in Saturn's A ring, was found in the highest resolution profiles of Uranus' δ ring taken by Voyager 2. Unfortunately, the satellite causing this density wave pattern has not been found, presumably because it is too small to be imaged. Later analysis showed that the sharp outer edge of the δ ring is in resonance with Cordelia, and that the sharp inner edge of the γ ring is in resonance with Ophelia.

In 2004 a series of 80 four-minute exposures taken by the Hubble Space Telescope, when the rings were almost edge-on, revealed two previously unknown rings, called the ν and μ rings. They were both quite broad, and outside the other rings. The outermost ring appeared to be supplied with dust by a 12 km diameter satellite discovered only the previous year.

Magnetism and magnetosphere

So far Voyager 2 is the only spacecraft to have visited Uranus. Shortly before its closest approach in 1986, Voyager 2 detected a strong burst of polarised radio signals indicating, for the first time, that Uranus had a magnetic field. Its magnetic axis was found to be tilted at about 60° to the planet's rotational axis, and its magnetic centre displaced from its geometric centre by about 0.3 R_U (Uranus radii). The magnetic field at Uranus' equator was found to be about 0.25 gauss, on average, with a dipole moment about 50 times that of Earth.

Voyager 2 crossed Uranus' bow shock about 23 R_U from the planet. All of Uranus's satellites were found to lie within the magnetosphere and, as at Jupiter and Saturn, the large inner satellites, namely Miranda, Ariel and Umbriel, and the particles in Uranus' rings were found to absorb energetic particles from the radiation belts. The 98° orientation of Uranus' equator to the plane of its orbit around the Sun, the 60° inclination of its magnetic axis to its spin axis, and the relatively fast axial rotation rate of just over 17 hours were found to cause a complex diurnal variation of the magnetosphere.

The planet's magnetotail was found to have two lobes of opposite polarities separated by a plasma sheet, as at the Earth. The plasma sheet was in the magnetic equatorial plane near Uranus, but it bent to become parallel to the solar wind flow about 10 R_U from the planet. The whole tail structure was observed to rotate in space approximately about the Uranus–Sun line as the planet spun on its axis.

Further information on Uranus' satellites and rings is given in Table 2.4.

Bibliography

Alexander, Arthur F. O'D., *The Planet Uranus: A History of Observation, Theory and Discovery*, Faber and Faber, 1965.

Beatty, J. K., Petersen, C. C., and Chaikin, A. (eds.), *The New Solar System*, 4th ed., Cambridge University Press, 1999.

Bergstralh, J. T., Miner, E. D., and Matthews, M. S. (eds.), *Uranus*, University of Arizona Press, 1991.

Godwin, Robert, and Whitfield, Steve, *Deep Space: The NASA Mission Reports*, Apogee Books, 2005.

Leverington, David, *Babylon to Voyager and Beyond: A History of Planetary Astronomy*, Cambridge University Press, 2003.

Littmann, Mark, *Planets Beyond: Discovering the Outer Solar System*, John Wiley, 1990.

McFadden, L. A., Weissman, P. R., and Johnson, T. V. (eds.), *Encyclopedia of the Solar System*, 2nd ed., Academic Press, 2007.

Table 2.4. *Uranus' main satellites and rings*

(a) Uranus' satellites

	Name	Year of discovery	Discovered by	Semimajor axis (10^3 km)	Orbital inclination	Mean radius (km)
U6	Cordelia	1986	Voyager team	50	0°	13
U7	Ophelia	1986	Voyager team	54	0°	15
U8	Bianca	1986	Voyager team	59	0°	21
U9	Cressida	1986	Voyager team	62	0°	31
U10	Desdemona	1986	Voyager team	63	0°	27
U11	Juliet	1986	Voyager team	64	0°	42
U12	Portia	1986	Voyager team	66	0°	54
U13	Rosalind	1986	Voyager team	70	0°	27
U14	Belinda	1986	Voyager team	75	0°	33
U15	Puck	1985	Voyager team	86	0°	81
U5	Miranda	1948	Kuiper	130	4°	236
U1	Ariel	1851	Lassell	191	0°	579
U2	Umbriel	1851	Lassell	266	0°	585
U3	Titania	1787	Herschel	436	0°	789
U4	Oberon	1787	Herschel	584	0°	761

(b) Uranus' rings

Ring component	Radial location (km)	Discovered by	Date
(edge of planet)	*(25,560)*		
1986 U2R/ζ	~37,000–41,000	Voyager 2/Keck	1986/2003
6	41,837	Millis et al.	1977
5	42,234	Millis et al.	1977
4	42,571	Millis et al.	1977
α	44,718	Elliot et al.	1977
β	45,661	Elliot, Millis et al.	1977
η	47,176	Elliot et al.	1977
γ	47,627	Elliot, Millis et al.	1977
δ	48,300	Elliot et al.	1977
λ	50,024	Voyager 2	1986
ε	51,149	Elliot, Millis et al.	1977
ν	66,100–69,900	Showalter and Lissauer	2004
μ	86,000–103,000	Showalter and Lissauer	2004

Miner, Ellis D., *Uranus: The Planet, Rings and Satellites*, 2nd ed., John Wiley, 1998.

Miner, E. D., Wessen, R. R., and Cuzzi, J. N, *Planetary Ring Systems*, Springer-Praxis, 2007.

Shirley, James H., and Fairbridge, Rhodes W. (eds.), *Encyclopedia of Planetary Science*, Chapman and Hall, 1997.

See also: Atmospheric constituents of the outer planets; Eighteenth century: Discovery of Uranus; Internal structures of the outer planets.

NEPTUNE

General

Alexis Bouvard tried to calculate an orbit for Uranus in 1820 using both pre-discovery and post-discovery observations. But he could not find a single orbit to fit them. The best that he could produce was an orbit based on only the post-discovery observations, but this implied that some of the pre-discovery observations were in error by up to 65″, which seemed very unlikely. Unfortunately, it did not take long for Uranus to deviate increasingly from even this orbit, so that by 1845 the longitude discrepancy had reached about 2'. One possible explanation was that Uranus was being disturbed by another planet, and in 1836 Friedrich Nicolai suggested, that if the Titius-Bode series was correct, the unknown planet would be about 38 AU from the Sun.

John Couch Adams, an English mathematician, set out in 1843 to try to calculate the orbit of the planet that seemed to be disturbing the orbit of Uranus. By September 1845, he had calculated its orbital elements and its expected position in the sky, assuming that it was in an elliptical orbit with a mean solar distance of 38.4 AU. Over the next year he progressively modified this orbit. Unfortunately, his predictions of its expected location varied wildly, making it impossible to use them for a telescopic search of the suspected planet.

In parallel, and unknown to Adams, Urbain Le Verrier, a French mathematician undertook the same task, starting work in June 1845. He published his final results on 31 August 1846 and three weeks later asked Johann Galle of the Berlin Observatory if he would undertake a telescopic search for it. Galle and his assistant Heinrich d'Arrest found the planet within an hour of starting their search on 23 September. It turned out to be less than 1° from the position predicted by Le Verrier. However, it was not until the following night that they could be sure that they had discovered the planet when they found that it had moved. Johann Encke, the director of the Berlin Observatory then announced its discovery, crediting Galle and himself, but ignoring the young d'Arrest.

Galle in his letter to Le Verrier of 25 September, notifying him of the discovery, suggested calling the new planet *Janus*. Le Verrier, in his reply of 1 October, said that the Bureau des Longitudes had already named the planet *Neptune*. There was then a brief attempt by Le Verrier to have it called after himself, but that failed, so the new planet was called *Neptune*.

The discovery of Neptune was followed by a heated argument between the English and French astronomical establishments on the priority of the orbital predictions. But much of the evidence on the English side was never published, and an 'official line' was agreed. However, that evidence has recently come to light, being found in Chile in 1999. Nevertheless, it had been clear in the 1840s that when Neptune's real orbit was calculated, it turned out to be quite different from the orbits finally predicted by either Le Verrier or Adams. At an average distance of about 30 AU it was appreciably closer to the Sun than either had assumed. So its discovery had been somewhat fortuitous.

Later it transpired that Michel de Lalande, Joseph de Lalande's nephew, had seen Neptune on 8 and 10 May 1795. In his original manuscript he had rejected the first observation and queried the second, as the position of the 'star' had moved slightly between the two nights. But it did not seem to have occurred to him that he might have been observing

a planet. Amazingly, in 1980 Charles T. Kowal and Stillman Drake found, on looking through Galileo's notebooks, that he had also apparently observed Neptune on both 28 December 1612 and 28 January 1613, when it was near to Jupiter.

Less than a month after Neptune's discovery, William Lassell observed an object close to Neptune, which he thought may be a satellite. But bad weather, and the nearness of Neptune to the Sun's glare, meant that it was not until the following July that he was able to confirm that it was a satellite, now called Triton. It was later found to have a retrograde orbit. Triton was seen to be very bright, considering its distance from the Sun. As a result, it was thought that it may be the largest satellite in the solar system. It was also quite close to Neptune in its retrograde orbit, so it was thought probable that Neptune's axial spin would also be retrograde.

The discovery of Triton allowed the mass of Neptune to be determined as about 17.1 times the mass of the Earth. In 1895 Edward Emerson Barnard measured Neptune's diameter as 52,900 km, implying a density of about 1.32 g/cm^3, similar to that of Jupiter and Uranus.

A number of astronomers tried to measure the rotation period of Neptune in the late nineteenth and early twentieth centuries, but they produced wildly different results. Then in 1928 Joseph Moore and Donald Menzel deduced an unambiguous rotation period of 15.8 ± 1 hours, prograde, by measuring its Doppler shift. So the idea that Neptune's axial spin may be retrograde, because of the retrograde orbit of Triton, was found to be incorrect. As a result it was clear that Triton was orbiting its planet in the opposite sense to the planet's spin, being the first major satellite in the solar system to be observed to do so. Then in 1949 Gerard Kuiper discovered Nereid, Neptune's second satellite, and found that it orbited Neptune prograde, or in the opposite sense to Triton. Triton's orbit was almost circular, with a radius of 355,000 km, but Nereid's was highly elliptical, with an apogee of 9.7 and a perigee of 1.3 million km.

The rotation period of 15.8 hours deduced by Moore and Menzel was generally accepted, until in 1977 Sethanne Hayes and Michael Belton deduced a period of 22 hours, based on Doppler shifts. Then in 1981 Robert Hamilton Brown, Dale Cruikshank and Alan Tokunaga found a period of 17.95 hours photometrically.

Bradford Smith, Harold Reitsema and S. M. Larson were able to record clear cloud features in Neptune's atmosphere for the first time in 1979. The images in the 890 nm methane absorption band showed a broad, dark equatorial band, where methane deep in Neptune's atmosphere absorbed sunlight, and bright features in both the northern and southern hemispheres due to high-altitude clouds. Four years later, images taken by Richard Terrile and Bradford Smith showed four clear atmospheric features, which allowed the planet's rotation period to be determined. The resulting period of 17.83 hours, prograde, was virtually the same as that measured photometrically two years earlier by Hamilton Brown and colleagues.

Heidi Hemmel confirmed the changing nature of Neptune's atmosphere, which had been first reported a hundred years before by Maxwell Hall, when she found in 1986 and 1987 that clouds, that had clearly been seen in the northern hemisphere just shortly before, had disappeared. In 1986 and 1987 she also noticed one very bright cloud at latitude 38° S that had a rotation period about the centre of Neptune of 17.83 hours. A year later, however, the only bright cloud was at 30° S, with a rotation period of 17.67 hours.

Dale Cruikshank and Peter Silvaggio detected the 2.3 μm methane absorption band on Triton in 1978, indicating that it had a tenuous methane atmosphere. If that was the case, Triton would be only the third satellite in the solar system known at the time to have an atmosphere, the others being Titan and Io. The 1.7 μm methane band was relatively weak, however, indicating that there was little or no methane frost or ice on the illuminated surface, which was assumed to be largely rocky.

Triton's orbit is inclined at 23° to Neptune's equator, and Neptune's equator is inclined at 29° to the plane of its orbit around the Sun. So the Sun is at the zenith on Triton at 52° latitude on midsummer's day. As Neptune's year is 165 years long, there is plenty of time for one hemisphere of Triton to heat up and the other to cool down, the poles each being without sunlight for 82 years. So Cruikshank and Silvaggio suggested that there may be methane ice or frost on Triton's unilluminated surface, even though they had not detected any elsewhere. Then in 1983 Apt, Carleton and Mackay analysed the visible spectrum of Triton, and concluded that there was clear evidence for methane frost or ice on its surface, although water ice appeared to be largely absent. In the same year Cruikshank, Clark and Hamilton Brown also detected nitrogen on the surface, implying that it would also be present in Triton's tenuous atmosphere, as nitrogen is highly volatile. So by the time that the first spacecraft, Voyager 2, arrived at Neptune in 1989, Triton was thought to possess an atmosphere of methane and nitrogen, with methane ice or frost and nitrogen on its surface. The exact amounts and condition of each constituent would depend critically on Triton's surface temperature, however, which was thought to be in the range 50 to 65 K, depending on albedo and latitude. Liquid nitrogen freezes at about 63 K at zero pressure, so it was thought that most of the nitrogen on Triton's surface would probably be in the solid form, unless Triton's temperature was very near the top end of its expected range.

In 1986 Cruikshank and colleagues found that the methane feature in Triton's spectrum was weakening compared with earlier years. Similarly, whilst observations in 1977 and 1981 had shown that the intensity of Triton varied by 6% as it rotated, those in 1987 showed that the intensity variation was less than 2%. So it was thought that the methane atmosphere was becoming more hazy as the Sun was gradually heating up the southern hemisphere, and there were fears that the atmosphere may be too hazy during the Voyager 2 intercept of 1989 for the spacecraft to image the surface.

The Voyager 2 spacecraft imaged a dark spot, called the Great Dark Spot or GDS, on Neptune, centred at about 22° S latitude and rotating around the planet once every 18.3 hours. Although it was physically smaller than Jupiter's Great Red Spot (GRS), it was about the same size as the GRS relative to the size of their respective planets. Also like the GRS, the GDS rotated counter-clockwise about its centre south of the equator, making it a high-pressure feature. Numerous other features were observed on Neptune during the Voyager encounter, including a second, smaller dark spot called D2 at about 53° S, which like the GDS also appeared to be a high-pressure system. When D2 was discovered it had a rotation period around Neptune of 16.0 hours. Its period then slowed to 16.3 hours as it moved north, before moving south with a period of 15.8 hours. It seemed to be less constrained in latitude than similar features on Jupiter and Saturn.

Voyager detected radio signals from Neptune that were varying with a period of 16.11 hours, which was assumed to be the spin rate of Neptune's interior. Relative to this, Neptune's clouds showed that there was an easterly equatorial jet, like that on Uranus, only those on Neptune were five times as fast at an incredible 1,800 km/h. These Neptune winds were, along with Saturn's equatorial jet, the fastest atmospheric winds in the solar system, which was surprising considering that Neptune receives such a small amount of radiation from the Sun. So the winds were thought to be driven by Neptune's significant internal heat source. Later images by the Hubble Space Telescope showed that both the GDS and D2 had disappeared by 1995.

Before the Voyager 2 intercept, Neptune was known to have two satellites, Triton and Nereid. Triton, which is the closer to Neptune, orbits the planet in a retrograde sense, suggesting that it may have been captured by Neptune. On the other hand its orbit is almost circular, whereas that of Nereid, which is prograde, is much further away from Neptune and is highly elliptical. The satellites' orbits are inclined at 23° and 27°, respectively, to Neptune's equatorial plane, which are unusually large inclinations, so maybe both satellites have been captured, Triton because of its retrograde orbit and Nereid because of its large orbital eccentricity. It was hoped that the discovery of new satellites by Voyager 2 would help to clarify this.

Voyager 2 discovered six new satellites of Neptune, all of which were in circular, prograde orbits in Neptune's equatorial plane, inside of Triton's orbit. One of these satellites, Larissa, had previously been detected by Reitsema and colleagues during a stellar occultation in 1981. But at that time it was not clear whether they had detected a satellite or ring of Neptune (see later section). Proteus and Larissa were the only new satellites to be imaged at high resolution, showing highly-cratered surfaces with reflectivities of only 6%. Proteus, with a diameter of 420 km, was found to have a 160 km diameter crater on its relatively small surface. Contrary to expectations, the presence of these six small satellites in circular, prograde orbits did not help to resolve the question as to whether Triton is an original or captured satellite of Neptune.

The Voyager 2 spacecraft found that Triton had a highly reflective surface with an average reflectivity of about 85%. As a result its measured diameter of about 2,700 km was near the low end of the expected range. Its density was about 2.05 g/cm^3, greater than that of any of the satellites of Saturn or Uranus, and somewhat higher than generally expected. This indicated that Triton has less ice and more rock in its interior than anticipated. Interestingly, Triton's density and size are almost identical to those of Pluto, adding to the idea that Triton was captured by Neptune.

Because of Triton's highly reflective surface, its surface temperature was lower than anticipated at 38 K. This was far too cold for nitrogen to exist in liquid form on the surface, except in possible 'hot' spots. Because of the lower than expected temperature, the surface-level atmospheric pressure of 15 microbars was considerably less than expected, as most of the atmosphere was 'frozen out' on the surface. Fortunately, although there was a haze layer about ten kilometres above the surface, the surface could be clearly seen. The very thin atmosphere was found to consist almost completely of nitrogen, with only trace amounts of methane, carbon monoxide, carbon dioxide and water vapour.

Voyager arrived at Triton during early summer in its southern hemisphere. It revealed a very bright, pinkish-white, southern polar cap of nitrogen ice, with trace amounts of methane, carbon monoxide, carbon dioxide and water ice extending three-quarters of the way from pole to equator. The ice cap appeared to be slowly melting at its edges to reveal a darker, redder surface underneath.

The retreating and advancing polar caps have ensured that the surface of Triton is still subject to change. Dark streaks up to 150 km long were seen all over the south polar cap and, as the ice retreats every summer, those streaks near the edge of the cap must have been produced relatively recently. Their appearance suggested that nitrogen or methane had been ejected at one end of the streak in some sort of explosion, and carried by the wind in Triton's tenuous atmosphere.

Robert Hamilton Brown proposed a mechanism for these eruptions, likening them to geysers on Earth. He suggested that sunlight penetrated the almost-transparent nitrogen ice of the polar cap, where about two metres down it was absorbed by frozen methane, which had been darkened by exposure to ultraviolet light. The heat was trapped by the nitrogen ice, as it was a poor conductor, and as the heat built up the subsurface

Figure 2.19. This image of Triton, which is about 500 km across, shows a marked lack of impact craters implying a relatively young surface. It also shows two depression, possibly old impact basins, that have since been flooded by cryovolcanic fluids. (Courtesy NASA/JPL/Caltech.)

nitrogen ice was turned into gas. Eventually the gas pressure was too much to be resisted by the surface ice, and the nitrogen gas exploded through the surface, carrying with it the Sun-darkened methane, to produce a geyser-like eruption.

A month after the Voyager encounter, Lawrence Soderblom and Tammy Becker were stereoscopically examining images of the dark streaks on the south polar cap, when they were amazed to discover an image of an eruption in progress. It produced a plume 8 km high, which extended horizontally for about 150 km. They then found an image of a second eruption in progress. These two eruptions, and three other suspected eruptions, were all near to the sub-solar point, indicating that they were caused by solar heating, giving some support to Robert Hamilton Brown's theory of geyser-like eruptions.

There was only a relatively small number of craters seen on Triton (see Figure 2.19, for example), and the largest was only 27 km in diameter, indicating that none of Triton's original surface had survived. So there must have been considerable geological activity in its early lifetime, possibly caused by gravitational stresses as a result of its hypothesised capture by Neptune. In addition, resurfacing must have continued almost up to the present in geological terms, and some resurfacing may still be occurring today.

In the equatorial regions there were large areas of relatively young, dimpled terrain, quite unlike anything seen elsewhere in the solar system. This so-called cantaloupe terrain was criss-crossed in places by shallow linear ridges up to 30 km wide and 1,000 km long, probably caused when water ice in Triton's mantle froze and expanded. Within this cantaloupe terrain and to its north there were areas of frozen lakes and calderas, showing evidence of liquid flows, possibly as a result of volcanic activity. Some of these frozen liquid flows were relatively old with many craters, but some looked fresh and were almost crater-free. These lakes and calderas cannot be made of nitrogen and/or methane, as the surface temperature is too near their melting points for these ices to be able to support the crater walls, which, in some places, are over 1,000 m high. Instead it appears as if the subsurface here is made of a mixture of water ice, nitrogen and/or methane, with the water ice providing the strength, and the nitrogen and/or methane reducing the freezing point of water to enable it to flow at low temperatures.

Since the Voyager 2 encounter a further 5 small satellites have been discovered, all of which have orbits outside that of Nereid. Three of these are retrograde and two are prograde orbits. Two of the satellites, Psamathe and Neso, have the largest orbits of any planetary satellites discovered to date, with semimajor axes just under 50 million km.

Rings

Harold Reitsema and colleagues attempted to find a ring or rings around Neptune in 1981, following the discovery of rings around Uranus and Jupiter in the late 1970s. They observed a stellar occultation by Neptune that year, but it showed an intensity reduction on only one side of the planet, so they could not have discovered a ring. Reitsema concluded instead that they had fortuitously discovered a satellite, at least 180 km in diameter and about 50,000 to 70,000 km from the centre of the planet.

On 22 July 1984 the first unambiguous discovery of a partial ring was found during a stellar occultation. Although the occultation occurred on only one side of Neptune, like that in 1981, this time it was observed by two teams of astronomers at two different observatories about 95 km apart. If this was due to a satellite, it would have been so large as to be easily visible from Earth. So they concluded that the occultation was due to a partial ring or ring arc about 67,000 km from the centre of Neptune. This discovery raised the question as to whether Reitsema and colleagues had also seen a ring arc in 1981. A further occultation in 1985 also showed that there was probably another ring arc 63,000 km from the centre of Neptune.

Voyager 2's intercept of Neptune in 1989 finally resolved the question of rings around Neptune, when it found five continuous rings. The Adams, Arago and LeVerrier rings were narrow and distinct, whereas the Lassell and Galle rings were relatively broad and indistinct. It transpired that the occultations of 1984 and 1985 had discovered parts of the Adams ring, which was remarkably clumpy, with density variations of about a factor of ten along its length. The object discovered in 1981, however, was a satellite, now called Larissa, of about 200 km diameter.

Voyager 2 discovered a satellite, Galatea, orbiting Neptune about 1,000 km inside the clumpy Adams ring, and a satellite, Despina, orbiting about 700 km inside the Le Verrier ring, possibly helping to shepherd the rings. Galatea, which had a 42/43 resonance with Adams, was probably controlling its ring arcs.

Magnetism and magnetosphere

In 1989 Voyager 2 detected Neptune's magnetopause about 24 R_N (Neptune radii) from Neptune, and found that all of Neptune's satellites known at that time, except Nereid, were within the planet's largely empty magnetosphere. Neptune's magnetic axis was found to make an angle of about 47° to its spin axis, and its magnetic centre was displaced from its geometric centre by about 0.55 R_N. Its dipole moment was about 28 times that of Earth.

The 30° orientation of Neptune's equator to the plane of its orbit around the Sun meant that the angle between Neptune's magnetic axis and the direction of the solar wind can vary enormously. At some times its magnetic axis can be almost pole-on to the solar wind. These radical variations in orientation of magnetic axis to solar wind meant that the configuration of Neptune's magnetosphere and its radiation belt can vary wildly over the course of one Neptune day of 16 hours.

Further information on Neptune's satellites and rings is given in Table 2.5.

Bibliography

Beatty, J. K., Petersen, C. C., and Chaikin, A. (eds.), *The New Solar System*, 4th ed., Cambridge University Press, 1999.

Godwin, Robert, and Whitfield, Steve, *Deep Space: The NASA Mission Reports*, Apogee Books, 2005.

Littmann, Mark, *Planets Beyond: Discovering the Outer Solar System*, John Wiley, 1990.

McFadden, L. A., Weissman, P. R., and Johnson, T. V. (eds.), *Encyclopedia of the Solar System*, 2nd ed., Academic Press, 2007.

Miner, Ellis D., and Wessen, Randii, *Neptune: The Planet, Rings and Satellites*, Springer-Praxis, 2002.

Miner, E. D., Wessen, R. R., and Cuzzi, J. N., *Planetary Ring Systems*, Springer-Praxis, 2007.

Moore, Patrick, *The Planet Neptune: An Historical Survey Before Voyager*, 2nd ed., Wiley-Praxis, 1996.

Shirley, James H., and Fairbridge, Rhodes W. (eds.), *Encyclopedia of Planetary Science*, Chapman and Hall, 1997.

Standage, Tom, *The Neptune File: A Story of Astronomical Rivalry and the Pioneers of Planet Hunting*, Walker Publishing, 2000.

See also: Atmospheric constituents of the outer planets; Internal structures of the outer planets.

Table 2.5. *Neptune's main satellites and rings*

(a) Neptune's satellites

	Name	Year of discovery	Discovered by	Semimajor axis (10^3 km)	Orbital inclination	Mean radius (km)
N3	Naiad	1989	Voyager team	48	4°	33
N4	Thalassa	1989	Voyager team	50	0°	41
N5	Despina	1989	Voyager team	53	0°	75
N6	Galatea	1989	Voyager team	62	0°	88
N7	Larissa	1981	Reitsema et al.	74	0°	97
N8	Proteus	1989	Voyager team	118	0°	210
N1	Triton	1846	Lassell	355	157°	1,353
N2	Nereid	1949	Kuiper	5,513	27°	170
N10	Psamathe	2003	Jewitt et al.	46,695	137°	14
N13	Neso	2002	Holman et al.	48,387	133°	30

(b) Neptune's rings

Ring component	Radial location (km)	Discovered by	Date
(edge of planet)	*(24,760)*		
Galle	41,000−43,000	Voyager 2	1989
Le Verrier	53,200	Voyager 2	1989
Lassell	53,200−57,200	Voyager 2	1989
Arago	57,200	Voyager 2	1989
Adams	62,930	Manfroid and Hubbard	1984

5 · Smaller objects

ASTEROIDS

Johannes Kepler suggested in 1596 that there may be a planet between Mars and Jupiter to fill what appeared to be an orbital gap in the solar system. In the eighteenth century Thomas Wright and Johann Lambert speculated that there may once have been a planet between Mars and Jupiter, but that it had been broken up by a collision with a comet, according to Wright, or had left the solar system, according to Lambert.

Johann Titius pointed out in 1766 that the relative distances, in tenths of an astronomical unit, of the planets from the Sun was almost consistent with the series, 4 for Mercury, and $4 + 3 \times 2^n$, where $n = 0, 1, 2, 4$ and 5 for Venus, Earth, Mars, Jupiter and Saturn. But, as Titius pointed out, there was no planet for $n = 3$, at a distance of 2.8 AU from the Sun, between the orbits of Mars and Jupiter. A few years later Johann Bode mentioned this mathematical series, convinced that there was an undiscovered planet at this distance from the Sun. Then in 1781 Uranus was discovered at a distance of about 18.9 AU, which was very close to the expected value of $4 + 3 \times 26 = 196$, or 19.6 AU, for the next planet after Saturn.

Franz von Zach was so impressed by the success of this Titius-Bode series, as it is now called, in anticipating the distance of Uranus, that in 1800 he organised a cooperative search for the missing planet at 2.8 AU. This group of astronomers came to be known as the Celestial Police. But in January 1801, before this coordinated search could get under way, Giuseppe Piazzi accidentally discovered what he thought may be a planet or a comet, although it had no tail. Later that year Carl Friedrich Gauss calculated that this new object, now called Ceres, was clearly a planet in an approximately circular orbit. Its mean distance from the Sun of 2.767 AU was virtually exactly the same as predicted by the Titius-Bode series.

No sooner had the orbit of Ceres been determined than Heinrich Wilhelm Olbers found another new object in March 1802 at a mean distance of about 2.771 AU from the Sun. But the orbital inclination of the new object, called Pallas, was 35°, which was more reminiscent of comets than planets. However, like Ceres it had no nebulosity or tail.

In May 1802 William Herschel estimated the diameters of Ceres and Pallas to be only about 250 km, so obviously Ceres and Pallas were nothing like the size of the planets already known. (Although these estimates turned out to be 2 or 3 times too small, the point was still valid.) But they were not comets either, as they had no tails and did not have the extremely eccentric orbits of comets. So they seemed to be very small planets, and Herschel suggested calling these objects *asteroids*, to distinguish them from the planets themselves.

Ceres and Pallas were about the same distance from the Sun as each other. As a result, Olbers suggested that they were really fragments of a full-sized planet that had exploded at some time in the past, either as a result of internal forces, or due to the collision of a comet. Such an explosion could explain the relatively high inclinations and relatively high eccentricities of the orbits of Ceres and Pallas. Then in September 1804, Karl Harding discovered the third asteroid, Juno, which had an orbit consistent with Olbers' theory. Unfortunately, the semi-major axis of the orbit of the fourth asteroid, Vesta, which was discovered by Olbers himself in 1807, was appreciably smaller than that of the other three asteroids. So he abandoned his theory.

No more asteroids were discovered until 1845 when Karl Hencke discovered the fifth asteroid, Astrea, followed in July 1847 by his discovery of Hebe. The discovery rate then increased dramatically with a total of 13 asteroids known at the end of 1850, increasing to 62 at the end of 1860, 219 at the end of 1880, and nearly 500 at the end of 1900. Max Wolf was the first to use photography to discover an asteroid, namely Brucia (323) on 22 December 1891. The use of this new medium then led to a veritable avalanche of discoveries.

In 1866, and later in 1876, Daniel Kirkwood pointed out that there were no asteroids with periods of $\frac{1}{3}$, $\frac{2}{5}$ and $\frac{2}{7}$ of Jupiter's period. Later on Kirkwood discovered gaps at other simple fractions of Jupiter's period (see Figure 2.20), which he attributed to an orbital resonance between the asteroids and Jupiter, which caused asteroids with these periods to be forced out of their orbits by Jupiter. In their new orbits they may have collided with other asteroids causing them to break up, or they may have coalesced with them. Some years later Charles de Freycinet suggested that Jupiter, rather than clearing asteroids from their 'resonance' orbits, may have prevented them from forming there in the first place.

The asteroid Aethra (132), which was discovered in 1873 by James Watson, had perihelion of 1.60 AU. It was the only asteroid known at that time to have at least part of its orbit inside that of Mars. Up to then all asteroids had orbits that were wholly

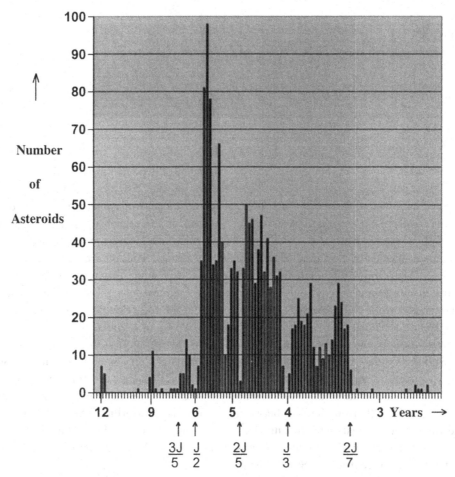

Figure 2.20. This histogram, based on one published in the early twentieth century, shows the number of asteroids with various periods. It clearly shows that there are some periods where there are few or no asteroids. The clearest of these so-called Kirkwood gaps, with periods a simple fraction of Jupiter's period, J, are identified by arrows.

between those of Mars and Jupiter. Then in August 1898 Eros (433) was independently photographed by Gustav Witt and by Auguste Charlois. Its orbit had a perihelion of 1.13 AU, so it could approach very close to the Earth from time to time. In 1931 its closest approach was just 0.174 AU, which allowed the most accurate estimate of the astronomical unit to be produced until the use of radar thirty years later.

In 1906 two asteroids were discovered which travelled in similar orbits to Jupiter, with one (Achilles, 588) 60° in front of Jupiter and one (Patroclus, 617) about 60°- behind. These were the first two Jupiter Trojan asteroids to be discovered in stable orbits on either side of Jupiter, in accordance with the orbital principles first described by Joseph de Lagrange in 1772. About five thousand Jupiter Trojans were known at the time of writing (in 2012).

Orbits

Asteroid families were first identified by Kiyotsugu Hirayama in 1918 based on their orbital radius, eccentricity and inclination. Initially, Hirayama identified three families, namely

Themis (22 members), Eos (21 members) and Koronis (13 members), followed a few years later by the families of Maria and Flora. Hirayama believed that the existence of these families was no accident, but due, in each case, to the fracture of a larger asteroid, thus resurrecting, in modified form, the theories of Thomas Wright and Wilhelm Olbers. More recent work has shown that about 50% of asteroids are associated with one or other of about 35 families which can be distinguished by both their orbital parameters and their colour.

In March 1932, Eugene Delporte discovered a one kilometre sized asteroid, Amor (1221), that came even closer to the Earth than Eros. Its perihelion distance of 1.086 AU brings Amor to within about 16 million km of the Earth every 8 years. Almost four thousand of these so-called Amor asteroids have now been discovered in orbits having a perihelion between 1.017 AU (the aphelion distance of the Earth) and 1.3 AU. So their orbits have perihelia that almost reach the Earth's orbit, but do not cross it, unless their orbits are perturbed.

An asteroid was found in April 1932 that had an orbit that crossed that of the Earth and Venus. This asteroid, provisionally designated 1932 HA and informally called Apollo, was discovered by Karl Reinmuth. It passed within just 0.65 AU of the Sun, but was lost until 1973 when Richard McCrosky and Cheng-Yuan Shao rediscovered it using the 61 inch (1.5 m) reflector at Harvard Observatory's Agassiz Station. This asteroid, now called Apollo (1862), was found to be a binary asteroid by the Arecibo radar in 2005.

Four years after the original discovery of Apollo, Eugene Delporte found another asteroid, designated 1936 CA or Adonis (2101), whose orbit crossed that of the Earth. Its perihelion was even closer to the Sun than Apollo at 0.44 AU. It did not take long for the third member of this Earth-crossing family, now called the Apollo asteroids, to be found. Like Apollo itself it was discovered by Reinmuth. Now called Hermes (69230), it was found in 1937 two days before it was due to pass within just 780,000 km of the Earth, moving across the sky at 5° per hour. This was uncomfortably close as, although Hermes was only a few hundred metres in diameter, the impact of such a body with the Earth would cause a major catastrophe. There are now over four thousand Apollo asteroids known, which, by definition, have a perihelion within the 1.017 AU aphelion of the Earth, and so cross the Earth's orbit.

Icarus (1566), one of the most well-known Apollo asteroids, was discovered by Walter Baade in 1949. It was found to have a perihelion of about 0.18 AU, which is well within the orbit of Mercury, and an aphelion that lies outside of the orbit of Mars. J. J. Gilvarry pointed out a few years later that it has a relativistic advance of its perihelion of 11″/century (compared with 43″/century for Mercury), which gave further proof of Einstein's general theory of relativity.

Yrjö Väisälä had begun an asteroid survey programme in 1935, which continued until 1957, whilst Frank Edmondson had started a similar survey programme in 1949. Gerard Kuiper also undertook such a programme in the early 1950s, but it was not until twenty years later that asteroid survey programmes were carried out specifically to find Near Earth Asteroids. One of these was the Palomar Planet-Crossing Asteroid Survey carried out by Eugene Shoemaker and Eleanor Helin starting in 1973, which used the 18 inch (46 cm) Palomar Schmidt.

One of Helin's first significant discoveries with this survey was asteroid 1976 AA. It was found to have an orbit that had a perihelion of 0.791 AU, an aphelion of 1.141 AU, and a period of just 346.8 days. Now called Aten (2062), this was the first Near Earth Asteroid found to have a period of less than one year. At the time of writing over seven hundred of these so-called Aten asteroids have been discovered.

Amors, Apollos, and Atens are all classed as Near Earth Asteroids.

Spacecraft intercepts

The Galileo spacecraft flew past Gaspra (951) in October 1991 on its way to Jupiter, producing the first close-up images of an asteroid. Prior to the Galileo intercept Claudine Madras

Figure 2.21. The Galileo spacecraft proved conclusively that asteroids can have satellites when it found in 1993 that the asteroid Ida had a small satellite, later called Dactyl. The maximum separation between the asteroid and its satellite was larger than is apparent here. (Courtesy NASA/JPL/Caltech.)

had found Gaspra to be an egg-shaped asteroid with a 7 hour rotation period. Then infrared observations made in 1990 by Jeffrey Goldader implied that metallic iron exists on the asteroid's surface. This seemed to indicate that Gaspra had once been part of the lower mantle of a differentiated larger body that had been shattered by a collision. Noriyuki Namiki and Richard Binzel hypothesised that Gaspra would be found to be irregular in shape, consistent with this idea, and less than one billion years old.

In the event, the Galileo spacecraft, which flew to within about 1,600 km of Gaspra, confirmed that it was highly irregular in shape, with dozens of small craters on its surface. Crater counts confirmed an age of the order of a few hundred million years. At 18 × 11 × 9 km in size, it is slightly larger than expected and similar in size to Mars' satellite Deimos. A number of grooves were seen, about 300 m across and 10 to 20 km deep, which cross the surface for up to a few kilometres, resembling the fractures seen on Mars' larger satellite, Phobos.

Two years later, Galileo imaged the 54 km long Ida (243) which was found to have a small spherical companion (see Figure 2.21), about one kilometre in diameter, now called Dactyl. The discovery of Ida's satellite allowed Ida's density to be determined as about 2.6 g/cm³. This indicated that Ida, which is thought to be a stony asteroid, probably has many voids inside it.

The discovery of Dactyl ended a long argument about whether asteroids were large enough to retain satellites. In 1802 William Herschel had suggested that asteroids were not large enough to do so, and calculations in the twentieth century seemed to confirm this. But in the 1970s and 1980s a number of secondary intensity dips were observed when some asteroids occulted stars, and these dips were attributed by some astronomers to unseen asteroid companions. Now the Galileo spacecraft had shown that even a relatively modest sized asteroid could retain a companion. Then in 2005 Franck Marchis and colleagues, using the European Southern Observatory's 8.2 m Yepun telescope, discovered that the 280 km Sylvia (87) had two satellites. They were about 18 and 7 km in size, orbiting about 1,360 and 710 km from their primary.

Gaspra and Ida are silicaceous, stony or S-type asteroids, which are reddish in colour and common in the inner regions of the asteroid belt. Their albedos typically range from 0.15 to 0.25. But the next asteroid, Mathilde (253), to be visited by a spacecraft was a very dark carbonaceous or C-type, which is more common in the outer regions of the asteroid belt.

In June 1997 the NEAR (Near Earth Asteroid Rendezvous) Shoemaker spacecraft flew within about 1,200 km of Mathilde, whilst the spacecraft was en route to intercept the asteroid Eros (433). According to ground-based observations, Mathilde was a medium sized, virtually black asteroid, being about 60 km in diameter with an albedo of just 0.04.

NEAR Shoemaker showed that Mathilde was 59×47 km in size with four 20 to 30 km diameter craters on the surface that were visible during the encounter. The albedo and colour of Mathilde's surface was the same at the bottom of the large craters as elsewhere on the surface, indicating that it was a uniformly dark, undifferentiated asteroid. Crater counts indicated that it was at least 2 billion years old. Spacecraft orbital deviations around the closest approach showed that Mathilde's density was about 1.3 g/cm^3, suggesting that it is extremely porous. Porous bodies are more likely to survive an impact than more rigid ones, as the porosity allows more 'give' than a rigid surface, soaking up more energy. This explained why Mathilde had managed to survive the impacts that had created the large craters.

NEAR Shoemaker imaged the S-type asteroid Eros during a fly-by in January 1999. It was found to be a $33 \times 13 \times 13$ km object with one large impact crater. There were fewer smaller craters compared with Ida, for example, indicating that Eros had a younger surface. Spacecraft trajectory changes showed that the asteroid's density was about 2.7 g/cm^3, which was significantly lower than the 3.4 g/cm^3 density of chondrite meteorites that have approximately the same composition. Then in February 2000 the spacecraft was put into orbit around Eros, and later it landed on the surface.

Images from orbit showed a surprising variety of geological features. These ranged from odd-looking, squarish craters, to flat 'ponds' of very fine dust in low-lying areas, and boulders of various sizes, some of which had small craters on their surface. There was a relative scarcity of very small craters, possibly because they had been filled in by loose surface material when Eros had suffered large impacts that had caused it to vibrate. NEAR Shoemaker landed on one of the flat 'ponds' and, in its last image, it confirmed the lack of small craters extended down to those of centimetre size. It also showed parts of two small channels, which indicated some sort of slumping of surface material, possibly due to underground cracks.

The LINEAR asteroid survey programme (see below) discovered Itokawa (25143), an S-type asteroid in 1998. Shortly afterwards the Japanese Space Agency (JAXA) decided to send a spacecraft, now called Hayabusa, to land briefly on Itokawa and return a small sample to Earth. In the event the landing did not go according to plan. Nevertheless, Hayabusa's sample return capsule did land successfully in Australia in 2010. Preliminary results indicate that it contained extremely small dust particles of extraterrestrial origin.

Itokawa was found to be just $535 \times 295 \times 210$ m in size with a density of 1.90 g/cm^3, which indicated that it was basically a rubble pile with a microporosity of 40%, twice that of Eros. Its surface had no craters, but it was generally very rough with many large boulders. The remaining 20% of its surface was smooth, being covered with pebbles and dust but no boulders.

The European Space Agency's Rosetta spacecraft flew within 800 km of the asteroid Steins (2867) in 2008, en route to comet 67P/Churyumov-Gerasimenko. Steins was thought to be a rare E-type asteroid poor in iron, with a surface rich in enstatite, which is a silicate mineral produced only by very high temperatures. This was confirmed by Rosetta, indicating that Steins had been once part of a larger differentiated body. The spacecraft also found that the asteroid was an irregular 6×4 km in size, with a 1.5 km crater and a seven-crater chain on its surface.

Near Earth Asteroids and Earth impacts

In 1980 Luis and Walter Alvarez and colleagues discovered an increased concentration of iridium in a 65 million year old geological layer at various places on Earth. They suggested that this had been caused by an asteroid impact, as chondrite meteorites and asteroids have a higher iridium content than the Earth's crust. They further proposed that this impact, of what they estimated to be a 10 km diameter asteroid, had been responsible for the Cretaceous-Tertiary extinction that had occurred at about the same time. The site of the hypothesised impact was later identified as the 180 km diameter Chicxulub Crater discovered on the coast of the Yucatan peninsula.

This impact, whether or not it caused the Cretaceous-Tertiary extinction (which most geologists think it did), increased the drive to find all those asteroids that may impact the Earth in the future. As a result, in 1990 the Spacewatch programme at the University of Arizona began the automatic detection of Amor, Apollo and Aten Near Earth Asteroids (NEAs). Then in 1994 the breakup of the comet Shoemaker-Levy 9, and the collision of its resulting pieces with Jupiter, provided a timely demonstration of the effect of cometary impacts. This has resulted in a number of new searches in the USA and elsewhere for Near Earth Objects (NEOs) of asteroids and comets likely to impact the Earth and cause potential damage.

So far the most successful searches for NEOs have been the LINEAR survey of the MIT Lincoln Laboratory, jointly funded by the U.S. Air Force and NASA, and the Catalina Sky Survey, which is a joint USA/Australian programme. Only 53 NEAs and 44 Near Earth Comets (NECs) were known in 1980, but by 2012 this number had been increased to about 9,000 NEAs and 92 NECs. From this population of NEAs, about 1,325 Potentially Hazardous Asteroids (PHAs) have been identified which are at least 150 m in diameter and which will come within 0.05 AU (7.5 million km) of Earth. So far the closest PHA to be observed was the 200 m diameter 2002 JE$_9$ which passed within about 150,000 km of Earth in 1971.

Further information on the more important asteroids historically is given in Table 2.6.

Bibliography

Bottke, W. F., et al. (eds.), *Asteroids III*, University of Arizona Press, 2002.

Cunningham, Clifford J., *Introduction to Asteroids: The Next Frontier*, Willmann-Bell, 1988.

Gehrels, T., and Matthews, M. S. (eds.), *Asteroids*, University of Arizona Press, 1979.

Gehrels, T., Binzel, R. P., and Matthews, M. S. (eds.), *Asteroids II*, University of Arizona Press, 1989.

Peebles, Curtis, *Asteroids: A History*, Smithsonian Institution Press, 2000.

See also: Trans-Neptunian Objects and Centaurs; Galileo, Hayabusa, NEAR Shoemaker, and Stardust in Part 9.

COMETS

Comets were considered by many ancient civilisations as harbingers of doom, such as defeat in war or the death of a king. However, many philosophers tried to find more rational explanations. For example, the Pythagoreans (sixth and fifth centuries BC) and Hippocrates of Chios (c. 440 BC) believed that there was only one comet and it was a planet. Apollonius of Myndus (fourth century BC) believed that there were a number of comets in highly eccentric orbits that periodically took them near the Earth. But in about 330 BC Aristotle maintained that comets were atmospheric phenomena caused by the Sun or planets warming the Earth. Aristotle's standing as a philosopher meant that his idea was the one generally accepted in Europe for the next two thousand years.

Michael Mästlin and Tycho Brahe independently observed the comet of 1577. Neither could detect any parallax, proving that it was not an atmospheric phenomenon, as proposed by Aristotle. Mästlin published his results in 1578, concluding that the comet was in an almost circular heliocentric orbit just outside that of Venus. Tycho came to a similar conclusion in his a detailed paper ten years later, although he suggested that the orbit could be either circular or oval.

Johannes Kepler and most of his contemporaries in the first half of the seventeenth century thought that comets followed rectilinear orbits. But Johannes Hevelius, who studied a number of comets in the mid-seventeenth century, concluded in 1665 that the orbits of comets were conic sections with the Sun at one focus. A little later he modified this to parabolic or hyperbolic orbits but with the Sun no longer at the focus. Georg Dörffel, who was aware of Hevelius' work, calculated the orbit of the comet of 1680, concluding that it was a parabola with the Sun at its focus.

John Flamsteed, the first Astronomer Royal, thought that the 1680 comet had turned back in its orbit before it reached the Sun. He communicated his ideas to Isaac Newton who initially thought that the 1680 comet was two comets, one approaching and one retreating from the Sun. But in 1686 he changed his mind and concluded that the 1680 comet was one comet in a highly elliptical or parabolic orbit with the Sun at the focus. Newton assumed in his *Principia* of 1687 that the orbits of comets were ellipses of such large eccentricities that they can be assumed, to a first approximation, to be parabolas. He then showed how it was possible to work out the parameters of such a parabolic orbit for any comet using just three positional observations.

In 1705 Edmond Halley published the orbital elements of 24 comets which had been observed between 1337 and 1698. These comets all seemed to be permanent members of the solar system as none of their orbits were hyperbolic. Halley also concluded that the comets of 1531, 1607 and 1682 were successive appearances of the same comet as their orbital elements were very similar. But the time intervals between successive perihelia were not identical, a fact he attributed to the perturbing effect of Jupiter. Taking this into account, in 1717 he predicted that the comet would return in late 1758 or early the following year.

Shortly before the expected return of this comet, now called Halley's comet, Alexis Clairaut attempted to produce a more accurate prediction of its perihelion date, using a new approximate solution to the three-body problem that allowed him to take planetary perturbations into account. This showed that the return would be delayed by 518 days due to Jupiter and 100 days due to Saturn. As a result, he predicted that Halley's comet would reach perihelion on about 15 April 1759 $\pm$ 1 month. In the event it did so on 13 March of that year.

In 1770 Charles Messier discovered a comet, now called Lexell's comet, which passed within just 0.015 AU of the Earth. Anders Lexell calculated that it had an elliptical orbit with a period of just 5.6 years, but, oddly, it had not been seen on any of its previous orbits. Lexell then pointed out that it had passed very close to Jupiter in 1767, and that Jupiter must have

Table 2.6. *Important asteroids historically*

(Early discoveries)

Number	Name	Year of discovery	Discovered by	Mean radius (km)	Semimajor axis (AU)	Orbital inclination	Orbital eccentricity
1	Ceres	1801	Piazzi	470	2.77	10.6°	0.08
2	Pallas	1802	Olbers	270	2.77	34.8°	0.23
3	Juno	1804	Harding	134	2.67	13.0°	0.26
4	Vesta	1807	Olbers	270	2.36	7.1°	0.09
5	Astrea	1845	Hencke	60	2.57	5.4°	0.19
6	Hebe	1847	Hencke	93	2.43	14.8°	0.20

Heads of early asteroid families

Number	Name	Year of discovery	Discovered by	Mean radius (km)	Semimajor axis (AU)	Orbital inclination	Orbital eccentricity
8	Flora	1847	Hind	64	2.20	5.9°	0.16
24	Themis	1853	Gasparis	99	3.13	0.8°	0.13
158	Koronis	1876	Knorre	18	2.87	1.0°	0.06
170	Maria	1877	Perrotin	22	2.55	14.4°	0.07
221	Eos	1882	Palisa	52	3.01	10.9°	0.10

Other important asteroids historically

Number	Name	Year of discovery	Discovered by	Mean radius (km)	Semimajor axis (AU)	Notes
132	Aethra	1873	Watson	21	2.61	First asteroid known to have part of its orbit inside that of Mars
323	Brucia	1891	Wolf	18	2.38	First asteroid to be discovered photographically
433	Eros	1898	Witt and Charlois	12	1.46	First asteroid known to come close to the Earth's orbit. Used in the early twentieth century to estimate the size of the astronomical unit. NEAR Shoemaker spacecraft landed on Eros in 2001.
588	Achilles	1906	Wolf	68	5.19	With Patroclus (below) the first Jupiter Trojan to be discovered
617	Patroclus	1906	Kopff	56	5.23	With Achilles the first Jupiter Trojan to be discovered
1221	Amor	1932	Delporte	0.8	1.92	Asteroid having an orbit that, at perihelion, comes closer to the Earth's orbit than Eros but still does not cross it
1862	Apollo	1932	Reinmuth	0.9	1.47	First asteroid found to have an orbit that, at perihelion, comes well within the Earth's orbit
1566	Icarus	1949	Baade	0.7	1.08	First asteroid found to have an orbit that, at perihelion, comes well within the orbit of Mercury
2062	Aten	1976	Helin	0.5	0.97	First asteroid found to have a period of less than one year
951	Gaspra	1916	Neujmin	6	2.21	Galileo spacecraft flew past in 1991
243	Ida	1884	Palisa	30	2.86	Galileo spacecraft flew past in 1993. First asteroid proved to have a moon
253	Mathilde	1885	Palisa	26	2.65	NEAR Shoemaker spacecraft flew past in 1997
25143	Itokawa	1998	LINEAR	0.2	1.32	Hayabusa spacecraft landed in 2005 and returned a sample to Earth in 2010.
2867	Steins	1969	Chernykh	0.2	2.36	Rosetta spacecraft flew past in 2008

changed the comet's orbit during that very close approach, which was why the comet had not been seen before.

Johann Encke showed in 1819 that comets seen in 1786, 1795, 1805 and 1819 were all appearances of the same comet, now called Encke's comet, in a 3.3 year orbit. This took it to inside the orbit of Mercury at perihelion.

Karl Rümker observed Encke's comet in 1822 on its next return very close to its predicted position. But Encke noticed that it reached perihelion a few hours early, and in 1823 he suggested that this was because it was being affected by a resisting medium. Encke found that the comet's orbital period was getting shorter at the rate of about $2\frac{1}{2}$ hours per orbit, and correctly predicted the times for its return to perihelion from 1825 to 1858. Because of this success, contemporary astronomers tended to accept Encke's hypothesis of a resisting medium. Friedrich Wilhelm Bessel did not agree, however, putting the effect down to the effect of emissions from its nucleus on its orbit, like those that he had seen for Halley's comet in 1835.

The Great September Comet (comet 1882 II) was observed both before and after its very close approach to the Sun in 1882. Its orbit showed no measurable change during this close perihelion passage, however, showing that the resisting medium did not exist. This was confirmed in 1933 when Michael Kamienski found that the return of Wolf's comet was late, rather than early like Encke's comet.

The problem of Encke's and Wolf's comets was finally solved in 1950 by Fred Whipple when he returned to Bessel's solution of a jet-like emission from the cometary nucleus. But Whipple assumed that the nucleus was rotating, so allowing the jet to either slow down or speed up the comet in its orbit.

Early spectra

Giovanni Donati made the first successful observation of a cometary spectrum in 1864. At that time it was thought that comets generally shone by reflected sunlight, but when Donati observed the spectrum of Tempel's comet (1864 II) when it was near the Sun, he found three faint luminous bands, indicating that it was self-luminous. Four years later, William Huggins found that these bands were similar to those produced by hydrocarbon compounds in the laboratory. Meteorites, when heated in experiments, also gave off hydrocarbons, and so it appeared that comets and meteorites were made of similar elements.

Quite a number of cometary spectra were recorded over the next twenty years. As the comets approached the Sun they generally exhibited a broad continuous spectrum like that of the Sun, indicating that they were scattering sunlight. As they got closer to the Sun, however, the hydrocarbon bands appeared. Then in 1882 Wells' comet (1882 I) approached very close to the Sun. Near perihelion its bandlike structure disappeared to be replaced by a bright, double sodium line. In the Great

September Comet (1882 II) of 1882, this double sodium line was also accompanied by several iron lines when the comet was very near the Sun. As the comet receded, these lines faded and the hydrocarbon bands reappeared.

Origin

In the early nineteenth century both William Herschel and Pierre Laplace theorised that non-periodic comets come from interstellar space. But in 1783 Herschel had shown that the Sun was moving in space towards the star λ Herculis. So in 1860 Richard Carrington and Henrik Mohn independently suggested that if non-periodic comets originated in interstellar space, and so were not participating in the general movement of the solar system, the Sun should sweep up more of them in its forward direction. In addition those comets should, on average, have a higher velocity relative to the Sun than those behind it. No such effect could be found, however, and so it appeared as though non-periodic comets must be an integral part of the solar system, even if their aphelia were some distance away from the Sun.

Links with meteor showers

Denison Olmsted suggested in 1834 that the annual November Leonid meteor shower may be caused by particles from a comet-like body being heated by friction in the Earth's atmosphere. From their apparent velocity he concluded that the comet-like source orbited the Sun twice a year, meeting the Earth on 12 November every year near the source's aphelion. Two years later Lambert Quetelet showed that the August Perseid meteors also returned annually. Olmsted's explanation for the November Leonids may have been plausible, but it seemed highly improbable that the August Perseids could also be in such a closely prescribed orbit. In 1839 Adolf Erman proposed, instead, that the meteoric material did not orbit the Sun in a comet-like cloud, but rather in a ring, thus eliminating any requirement to have orbital periods that are a simple fraction of that of the Earth. The showers simply occurred when the Earth passed through the ring on an annual basis.

Giovanni Schiaparelli showed in 1866 that the Perseids' orbit was the same as that of the comet Swift-Tuttle (1862 III) as calculated four years earlier by Theodor von Oppolzer. This was the first identification of a meteor stream with a specific comet. So there was a ring of particles around the Sun, in the same 120 year orbit as the comet, and the Perseid meteors were seen every year when the Earth passed through that ring.

In about 60 AD Seneca had ridiculed Ephorus of Cyme who, in about 370 BC, had observed that a comet had split into two pieces. But in 1845 Edward Herrick and Francis Bradley also observed that Biela's comet had broken in two. Later analysis indicated that Biela's comet may have been fractured during a close encounter with Jupiter in 1841. At their next appearance in 1852 the two Biela comets were about 2 million km apart, but they were never seen again on subsequent orbits. In 1867 Edmund Weiss calculated that the Earth passed very close to the orbit of Biela's comet on 28 November each year. As a result he predicted a Bielid meteor shower on about 28 November 1872 or 1879. He was proved correct, as on 27 November 1872 the Earth was treated to an impressive display of meteors, and on 27 November 1885, approximately two orbits later, there was an even more spectacular display.

Kreutz sungrazing comets

The Great March Comet of 1843 (1843 I) passed within about 0.006 AU of the Sun in February of that year. J. S. Hubbard calculated an orbital inclination of 144° and a period of about 530 years. A similar bright sungrazing comet, called the Great Southern Comet (1880 I), was observed in 1880. The calculations of Gould, Hind and Copeland independently showed that the orbital elements of these two comets were virtually identical. Clearly they couldn't be the same comet, as their orbital periods were far too long. So in 1880 Daniel Kirkwood suggested they were two fragments of the same ancient comet that had fractured, like Biela's comet, with the two remnants remaining in the same long-period orbit some 37 years apart.

Two years later another sungrazing comet, the Great September Comet (1882 II), appeared. Two weeks after perihelion its nucleus was seen to have broken in two, and just over three months later it had five separate nuclei all orbiting the Sun in line looking like 'pearls on a string'.

Towards the end of the nineteenth century Heinrich Kreutz analysed the orbits of many sungrazing comets, finding that 1843 I, 1880 I and 1882 II had similar orbital elements. As a result, he concluded that they were all parts of the same original comet that had been broken up by the Sun near perihelion many years previously. This family was now travelling in similar orbits of large eccentricity with periods of the order of 600 years.

In more recent times a number of analyses have been carried out on possible Kreutz sungrazers, together with possible breakup scenarios. For example in 1967 Brian Marsden analysed the eight generally accepted Kreutz sungrazers known at that time (1668, 1843 I, 1880 I, 1882 II, 1887 I, 1945 VII, 1963 V, and 1965 VIII). They had orbital inclinations generally in the range 142° to 145°, with almost identical perigee coordinates. Marsden concluded that these eight, together with some other possible sungrazers, could be put into two subgroups based on their orbital elements, and suggested that one or both of these subgroups could be remnants of the comet of 1106.

The SOLWIND, Solar Max, and SOHO spacecraft have, in more recent years, found numerous additional small sungrazing comets (see, for example, Figure 2.22) which have tended to smear the differences between the two sungrazing subgroups.

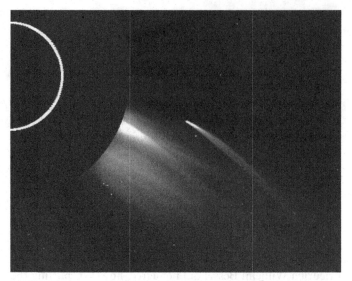

Figure 2.22. A sungrazing comet imaged by the SOHO spacecraft in 2011. The Sun is indicated by the white circle. (Courtesy SOHO [ESA and NASA].)

Zdenek Sekanina and Paul Chodas analysed the orbits of the large known Kreutz sungrazers and concluded in 2004 that their progenitor comet split in two as it was approaching the Sun in the fourth century AD. The two superfragments then passed through perihelion within a week of each other in 356 AD, and split into more fragments in 1100 and 1106 just after a further perihelion crossing. A search of ancient records confirmed the existence of the comet of 1106, but, so far, observations of the comets of 356 or 1100 have not been found.

Chemical composition

William Huggins had shown in the nineteenth century that there were hydrocarbon compounds in the heads of comets. The exact compounds did not become clear until molecular spectra became better understood in the first half of the twentieth century. Molecular carbon, C_2, was identified in the head of a comet just after the turn of the century, and by the mid 1950s C_3, CH, CN, OH, NH and NH_2 had been found in the heads of comets.

Molecular bands were observed in the tail of Daniel's comet (1907 IV) by Deslandres, Bernard, and Evershed in 1907. Similar bands were observed in the tail of Morehouse's comet (1908 III) by Deslandres and Bernard the following year, along with another shorter wavelength band. The bands observed in both comets were later identified by Alfred Fowler as being due to singly ionised carbon monoxide, CO^+. He also identified the shorter wavelength band in Morehouse's comet as being due to N_2^+. Then in 1948 CO_2^+ was found in the tail of a comet.

Karl Wurm observed in the mid 1930s that many of the molecules found in comets, such as C_2, CH, and NH, were chemically very active, and so they could not have been present for very long. He suggested, instead, that these so-called daughter molecules had come from more stable parent molecules, such as cyanogen $(CN)_2$, H_2O, and methane CH_4. In 1948 Pol Swings, in his study of Encke's comet, concluded that the parent molecules were water, ammonia (NH_3), methane, molecular nitrogen (N_2), carbon monoxide and carbon dioxide, all of which had been in the form of ice before being heated by the Sun.

Tail structure

The ancient Chinese discovered that the main tail of a comet streamed away from the Sun no matter in which direction the comet was travelling, a fact discovered in the West by Apianus (otherwise called Peter Apian) in 1532. They also seem to have discovered that occasionally some comets have a short, so-called 'antitail', which appeared to be directed towards the Sun.

Kepler suggested in 1625 that comets' tails were formed from material emitted from the head, which was then pushed away from the Sun by the Sun's rays. Kepler also pointed out that the loss of material into the tail would eventually lead to the comet's death.

In 1873 James Clerk Maxwell showed that the Sun would produce a radiation pressure on any surface on which it shone. This pressure could be the cause of comets' tails streaming away from the Sun. But George Fitzgerald showed theoretically that such pressure acting on a hydrogen molecule would only be just enough to overcome the Sun's gravitational attraction at a distance of 1 AU from the Sun. In 1900, however, the Swedish physicist Svante Arrhenius suggested that comets' tails could be caused by the effect of radiation pressure on cometary dust, rather than on cometary molecules. Shortly afterwards, Karl Schwarzschild followed up this suggestion, and showed that the repulsive force on particles of the order of 0.1 μm could be as much as twenty times as strong as the Sun's gravitational attraction. But it was known that in many comets the repulsive force was much greater than this. So, once again, radiation pressure did not seem to be enough to produce comets' tails.

John Schaeberle suggested in 1893 that particles emitted by the solar corona caused cometary tails. When these coronal particles impinged on the material coming from the cometary nucleus, they slowed down and became denser, whilst accelerating the cometary material. The coronal particles, together with the cometary material, formed the tail.

In spite of these promising ideas, there was no fully satisfactory theory in the first half of the twentieth century for the production of comets' tails. Most astronomers favoured the radiation pressure idea, even though Fitzgerald had shown it insufficient for molecules and Schwarzschild's analysis had shown it insufficient for dust. In 1951 Ludwig Biermann confirmed that radiation pressure could not exert enough force to create the observed tail velocities. On the other hand, he showed

that solar ions and electrons could do so, if their velocities were between 500 and 1,000 km/s, and their density was between 100 and 1,000 /cm^3 at the distance of the Earth. This interplanetary solar wind was discovered by the Luna 2 spacecraft in 1959.

Structure of the nucleus

The linkage between comets and meteor showers that had been found in the nineteenth century had led some astronomers to believe that the nucleus of a comet was simply a gravitationally bound swarm of small particles and absorbed gases orbiting the Sun. As the twentieth century progressed, however, this "flying sandbank" model fell out of favour, due mainly to the relatively long life of sungrazing comets. Then in 1950 and 1951 Fred Whipple proposed his 'icy conglomerate' or 'dirty snowball' theory, in which the nucleus was a single body, composed of various ices, with meteoric material embedded within it. As the nucleus approached the Sun, the surface ices turned to vapour, releasing their entrapped meteoric material which, together with the vapour, streamed away from the Sun.

Whipple envisaged the meteoric material, embedded in the cometary ices, as consisting of sodium, iron and other elements in the form of free atoms or particles from a few microns up to at least a few centimetres in size. Consequently these free atoms had been responsible for the metallic emission lines first detected in the nineteenth century when comets passed very close to the Sun.

Whipple, like Swings, assumed that the nucleus consisted of ices, such as those of methane, which were necessary to explain the existence of some daughter molecules. Unfortunately, some of these parent molecules were also highly volatile, and it was difficult to understand how they could have survived in the comets for more than a few perihelion passes. Then in 1952 Armand Delsemme and Swings suggested that the highly volatile methane and other molecules could be embedded within the crystalline structure of water ice as so-called clathrate hydrates.

Origin continued

It was difficult to determine the orbits of long-period comets because they were only observed for a fraction of their orbit when they were close to the Sun. However, a survey of about 400 cometary orbits observed up to 1910 showed that only eight appeared to be hyperbolic. Elis Strömgren and Gaston Fayet then showed that none of these comets had hyperbolic orbits before they had passed close to Jupiter or Saturn on their approach to the Sun. So the long-period comets appeared to have originated in the solar system.

Ernst Öpik concluded in 1932, from an analysis of the effect of stellar perturbations on comets, that they could remain bound to the Sun at distances of up to 10^6 AU. He found that stellar perturbations would tend to increase the perihelion distance of long-period comets with time, causing them to eventually form a cloud or shell surrounding the Sun at large distances.

In 1948 Adrianus van Woerkom examined possible capture scenarios for comets by the solar system, and concluded that the velocity of the Sun relative to that of any existing interstellar comets would be far too large to allow capture. This showed that long-period comets could not have originated in interstellar space. He also showed that after about one million years, all long-period comets in the solar system would either have become short-period comets in the inner solar system, or have been ejected into interstellar space. In addition, van Woerkom concluded, from the observed distribution of cometary orbits, that that there was a continuous source of new near-parabolic comets. Van Woerkom then went on to point out that a cloud of comets, moving with the Sun through interstellar space, could be such a source.

Jan Oort took van Woerkom's analysis further by analysing the orbits of a number of long-period comets. He concluded that all long-period comets had originated in a cloud of about 2×10^{11} comets, now called the Oort cloud, about 50,000 to 150,000 AU from the Sun where they would be subject to perturbations by passing stars. As the nearest stars surround the Sun in three dimensions, the comets perturbed by these stars would enter the solar system with a wide variety of orbital inclinations, as observed.

Oort suggested that the cloud of comets may have originated, together with asteroids and meteorites, from the explosion of a planet that had been between the orbits of Mars and Jupiter. Those fragments that had almost circular orbits became members of the observable solar system, losing their gaseous constituents because of their continuous exposure to solar radiation, and became asteroids and meteorites. Those fragments with strongly elliptical orbits, on the other hand, had their orbits perturbed by Jupiter and the other major planets. As a result, a number of these fragments were given hyperbolic orbits, and were thus lost to the solar system, but a significant percentage were given orbits with aphelia of about 50,000 to 150,000 AU. Stellar perturbations then distorted these orbits near aphelia, making them more circular, thus producing the Oort cloud.

In 1951 Gerard Kuiper proposed an alternative theory for the production of the Oort cloud. In his theory the comets were produced by the condensation of the original solar nebula outside the orbit of Neptune, between about 35 and 50 AU from the Sun. Kuiper then suggested the outer planets had caused the orbits of these comets to become highly elliptical, causing them to be injected into the Oort cloud, where their orbits had been made more circular by neighbouring stars.

The source of short-period comets was unclear. They did not appear to have come from the Oort cloud because their orbital inclinations were, unlike those of long-period comets, relatively constrained. Kuiper thought that all the comets in the Kuiper belt outside the orbit of Neptune would have been removed by now. But Fred Whipple thought otherwise, although he could find no evidence of their existence. Then in 1980 Julio Fernández found, using a computer simulation, that a comet belt or disc between 35 and 50 AU from the Sun could maintain the current population of short-period comets.

The impact of Comet Shoemaker–Levy 9

Comet Shoemaker–Levy 9 had been discovered by Carolyn Shoemaker in March 1993. Further work showed that the comet had been in orbit around Jupiter for some decades, and had been broken into a number of pieces by its very close encounter with the planet in July 1992. The break-up stresses were calculated to be very low, so the nucleus must have been very fragile. The fragments, which were still in orbit around Jupiter, were observed to break up even more in the months after discovery as they approached the planet once again. They crashed into Jupiter one by one over a few days in July 1994, leaving pronounced visible disturbances of the planet's atmosphere that lasted for well over a week.

Spacecraft intercepts

The International Cometary Explorer (ICE) was the first spacecraft to make in situ measurements of a comet when it flew through the tail of Comet Giacobini–Zinner about 7,800 km from its nucleus in 1985. It detected many water-group ions of H_2O^+ and H_3O^+ in the comet's tail, providing support for Whipple's theory of the comet's nucleus. The velocity of the solar wind decreased from 500 km/s in interplanetary space to 60 km/s at the spacecraft's closest approach. The plasma tail was found to be about 10,000 km across and the current sheet about 1,000 km thick.

Our knowledge of comets was greatly enhanced by the six spacecraft (Giotto, Vega 1 and 2, Suisei, Sakigake and ICE) that flew past Halley's comet after its perihelion in 1986. They all flew past the comet on the sunward side, at closest approach distances ranging from 28 million km for ICE to just 596 km for Giotto.

The Giotto and Vega spacecraft found that Whipple's dirty snowball theory was broadly correct for Halley's comet. The inner coma had a very high water content, and the nucleus had a density of about 0.2 g/cm^3. However, the inactive part of the nucleus had an albedo of only 0.03, so it was really a very dirty porous snowball, being blacker than coal. This black surface was a good absorber of solar radiation, producing a

Figure 2.23. An image of the nucleus of Halley's comet taken by the Giotto spacecraft in 1986. On the original image a number of craters can be seen on the left, Sun-facing side of the nucleus, which is dominated by the emission of two jets of material. (Courtesy ESA/MPAE.)

temperature of 330 K. At this temperature the water ice in the nucleus should sublimate below its dark insulating surface.

Halley's nucleus was found to be $16 \times 8 \times 8$ km in size, covered with bumps and hollows, some of which looked like craters (see Figure 2.23). Bright jets could clearly be seen streaming towards the Sun from about 10% of the total surface. At the time of the spacecraft encounters, the water vapour production rates were measured to be ~20 tons/s, and the dust production rates to be about 5 tons/s. This would allow about 1,000 more orbits of the Sun before Halley's comet would run out of material.

Observations by the OAO-2 spacecraft in 1970 had shown that Comet Tago-Sato-Kosaka was surrounded by a spherical cloud of neutral hydrogen over a million kilometres in diameter. Similar clouds were also found around Bennett's and Encke's comets in 1970, and Kohoutek's comet in 1973, which, in the latter case, was found to be over 10 million km in diameter. Halley's comet was also found to be surrounded by a spherical corona of neutral hydrogen atoms, extending some 10 million km from the nucleus at the time of the spacecraft intercepts.

Halley's bow shock was found to be about 400,000 km from the nucleus at its closest point. The solar wind velocity was observed to decrease from 400 km/s in interplanetary space, to 60 km/s about 150,000 km from the nucleus. Giotto was the only spacecraft to penetrate the ionopause, which was found to be some 4,700 km from the nucleus. At the ionopause the interplanetary magnetic field dropped abruptly to zero, the solar wind disappeared, and a stream of neutral molecules and cold ions was found flowing away from the nucleus. It is these neutral molecules that drag dust particles away with them to form the coma or head of the comet.

Halley's inner coma was found by the Vegas and Giotto spacecraft to consist mainly of water, with lesser amounts of carbon monoxide, carbon dioxide, methane, ammonia and polymerised formaldehyde $(H_2CO)_n$. It was thought that the polymerised formaldehyde may be the reason why the nucleus was so dark. These spacecraft found that the dust particles consisted of carbon, hydrogen, oxygen and nitrogen, the so-called CHON elements, and simple compounds of these elements, or of mineral-forming elements such as silicon, calcium, iron and sodium. The relative abundances of key elements in the material emitted by Halley's comet were close to those in the Sun, rather than in the Earth or in meteorites, indicating that comets consisted of very primitive material, which was depleted only in the volatile elements of hydrogen and nitrogen.

After 6 orbits of the Sun Giotto was retargeted to fly by Comet Grigg-Skjellerup in 1992. Giotto detected pick-up ions from Grigg-Skjellerup when it was still about 600,000 km from the nucleus. These were caused by contamination of the solar wind by cometary gas. Then, about 25,000 km from the nucleus, spacecraft instruments detected a remarkable series of waves in the magnetism of the solar wind with a wavelength of about 1,000 km. These waves were much stronger and more clearly defined than those for Halley. Just before the bow shock, water, carbon monoxide and other ions were detected, and then about 17,000 km from the nucleus Giotto passed through the bow shock itself.

Comet Borrelly was observed from a distance of just 2,200 km by the Deep Space 1 (DS 1) technology spacecraft in September 2001. Discovered in 1904, Comet Borrelly, with a period of 6.9 years, is one of the most active comets that regularly visit the inner solar system. As DS 1 approached Borrelly, it observed that its nucleus was about 8 km long by 3 km wide. The spacecraft also detected three jets of vaporised ice and dust extending towards the Sun, together with smaller, non Sun-pointing jets from other parts of its surface. The main jets appeared to come from relatively bright, smooth patches on the comet's surface. Elsewhere the surface, that had an average albedo of about 0.03, was seen to be mottled with many small, very dark patches. These were thought to be made of an organic rich residue left behind after some of the comet's ices had sublimated into space.

The DS 1 intercept produced one completely unexpected result as, although the solar wind was found to flow symmetrically around the comet's coma, as expected, the nucleus appeared to be off-centre, with the outward streaming cloud of ionised gases offset from the nucleus by about 2,000 km. The cause of this asymmetry has still to be explained.

The Stardust spacecraft was the first to capture cometary material and return it to Earth for analysis. The spacecraft flew to within 240 km of Comet Wild 2 in 2004 at a relative velocity of 6.1 km/s, and its re-entry capsule returned samples to Earth two years later. Comet Wild 2 had been chosen as it had not been exposed to the Sun during many passes through the inner solar system, having only been diverted there by its close encounter with Jupiter in 1974. As a result it was expected that its composition would have been little altered since its original formation.

Wild 2's nucleus was found to be roughly spherical with a diameter of 4 km and an albedo of 0.03. Analysis of its dust samples showed that the comet consisted of a mixture of rocky material such as olivine, which had been formed at very high temperatures, and ices that had been formed at very low temperatures. The existence of such high-temperature material in a comet that had apparently been formed outside the orbit of Neptune was a big surprise. Glycine was also found in the comet's samples. This was the first time that an amino acid had been detected in a comet.

Deep Impact was the first spacecraft to include an impactor to collide with a cometary nucleus, that of Tempel 1, and to observe the resulting crater and plume with a fly-by spacecraft. The main Deep Impact spacecraft, other spacecraft, and Earth-based observatories observed the impact that took place on 4 July 2005. The impactor was designed to penetrate the comet's crust and to expose the relatively fresh material underneath.

Comet Tempel 1 was found to have an irregular 7.6 × 4.9 km nucleus with an albedo that varied from about 0.02 to 0.06. For the first time, a small amount of water ice was detected on a comet's surface. The ejecta, which contained $\sim 10^7$ kg of material, consisted of $1-100$ micron particles. Individual species included water, water ice, amorphous carbon, carbon dioxide, methane, clays, carbonates, polyaromatic hydrocarbons, sulphides and crystalline silicates like olivine. As in the case of Wild 2, some of these species required high temperatures for their formation, and others required low temperatures.

The Deep Impact fly-by spacecraft could not image the impact crater as it was obscured by the ejected plume. However in 2007 NASA decided to redirect its Stardust spacecraft to image the crater, which it did successfully in 2011. At 150 m diameter it was slightly larger than expected. In the meantime the Deep Impact spacecraft had been redirected to fly-by Comet Hartley 2 in 2010 at a distance of 700 km. It showed an active nucleus with numerous jets driven by the heating of subsurface carbon dioxide, rather than of water ice.

Further information on the more important comets historically is given in Table 2.7.

Bibliography

Beatty, J. K., Petersen, C. C., and Chaikin, A. (eds.), *The New Solar System*, 4th ed., Cambridge University Press, 1999.

Brandt, John C., and Chapman, Robert D., *Introduction to Comets*, 2nd ed., Cambridge University Press, 2004.

Clerke, Agnes M., *A Popular History of Astronomy during the Nineteenth Century*, 4th ed. rev. and corr., Adam and Charles Black, 1902.

Heidarzadeh, Tofigh, *A History of Physical Theories of Comets from Aristotle to Whipple*, Springer-Verlag, 2008.

Jewitt, D., Morbidelli, A., and Rauer, H., *Trans-Neptunian Objects and Comets*, Springer, 2010.

Taton, René, and Wilson, Curtis (eds.), *Planetary Astronomy from the Renaissance to the Rise of Astrophysics*, Parts 2A and B, Cambridge University Press, 1989 and 1995.

McFadden, L. A., Weissman, P. R., and Johnson, T. V. (eds.), *Encyclopedia of the Solar System*, 2nd ed., Academic Press, 2007.

Yeomans, Donald K., *Comets: A Chronological History of Observation, Science, Myth and Folklore*, Wiley and Sons, 1991.

METEORITES

Meteorites were known to many ancient civilisations. In 654 BC Livy described a fall of meteorites near Rome. Then in 467 BC a comet was observed, followed by the fall of a large meteorite at Aegospotami in Thrace. Naturally observers linked the meteorite to the appearance of the comet. Aristotle believed that comets were formed during windy conditions on Earth, and that the meteorite had been lifted off the surface by strong winds before it had fallen back to Earth. A number of these "thunderstones" were also venerated by the Greeks and Romans as thunderbolts hurled by angry gods.

A number of meteorite falls were observed in middle age Europe, often accompanied by dark clouds and thunder. This confirmed to most people that they were either acts of God or Earth-based rather than cosmic phenomena. For example, the meteorite that fell in 1492 at Ensisheim, Alsace was hung in the local church as it was believed to be the result of a miracle. Other people believed that meteorites were probably ordinary rocks struck by lightning.

The largest meteorite found in the eighteenth century, with a mass of about 700 kg, was discovered in 1749 near Krasnoyarsk, Siberia. In 1772 the naturalist Pyotr Pallas examined this Krasnoyarsk meteorite, and found that it was partly covered with a black crust. There were also many translucent olivine crystals set in its iron matrix, in an arrangement that he had never seen previously. It was clearly different from anything found elsewhere in the vicinity.

Ernst Chladni wrote a booklet in 1794 discussing the Krasnoyarsk and other meteorites, observing that they had obviously been subject to intense heating and that there were compositional differences from terrestrial rocks. He concluded that

Table 2.7. *Selected comets*

Comet	Year first seen	Perihelion distance (AU)	Orbital period (years)	Orbital inclination	Comments
Halley	240 BC	0.587 (in 1986)	76.1	162.2°	Halley correctly predicted return in 1758 or 1759. Six spacecraft flew past in 1986.
Great 1106 comet	1106				Sungrazer; possible progenitor of family of Kreutz sungrazers
Great 1577 comet	1577	0.18	*	105°	Michael Mästlin and Tycho Brahe prove that the comet is not in Earth's atmosphere
Great 1668 comet	1668	0.066	*	144°	Sungrazing comet
Great 1680 comet	1680	0.006	~ 9,000	60°	Newton concluded that orbit is very nearly parabolic
Swift-Tuttle (1862 III)	1737	0.960	133	113.5°	Linked to Perseid meteor shower
Lexell	1770	0.68	5.6	1.6°	Lexell calculated that it passed very close to Jupiter in 1767
Biela	1772	0.861	6.6	12.6°	Observed to have broken in two in 1845
Encke	1786	0.333	3.3	11.9°	Period gradually reducing
Grigg-Skjellerup	1808	1.117 (in 1999)	5.3 (in 1999)	22.4°	Giotto spacecraft flew past in 1992
Great March Comet (1843 I)	1843	0.006	530	144.4°	Sungrazing comet
Donati (1858 VI)	1858	0.578	~ 2,000	117.0°	First successful photograph of a comet
Tempel (1864 II)	1864	0.909	~4,000	178.1°	First observation of cometary spectrum
Tempel-Tuttle (1866 I)	1866	0.977	33.3	162.5°	Linked to Leonid meteor shower
Tempel 1	1867	1.51 (in 2000)	5.5 (in 2000)	10.5°	Projectile from Deep Impact spacecraft impacted nucleus in 2005. Crater imaged by Stardust spacecraft in 2011.
Great Southern Comet (1880 I)	1880	0.006	*	144.6°	Sungrazing comet
Tebbutt (1881 III)	1881	0.734	~ 3,000	63.4°	First comet to have spectrum recorded photographically
Wells (1882 I)	1882	0.061	*	73.8°	Sodium line appeared double near perihelion
Great September Comet (1882 II)	1882	0.008	800	142.0°	Sungrazing comet; nucleus observed to fracture
Wolf	1884	1.57 (in 1884)	6.8 (in 1884)	27.5°	Period gradually increasing
Giacobini-Zinner	1900	1.034	6.6	31.9°	ICE spacecraft flew past in 1985
Borrelly	1904	1.35	6.8	30.3 °	Deep Space 1 spacecraft flew past in 2001
Daniel (1907 IV)	1907	0.512	*	9.0°	Molecular bands observed in tail spectra
Great January Comet	1910	0.129	*	138.8°	Often confused with Halley's comet's appearance at about the same time
Tago-Sato-Kosaka	1969		*		OAO-2 spacecraft detected huge neutral hydrogen cloud
Wild 2	1978	1.592	6.4	3.2°	Stardust spacecraft flew past in 2004 and returned tail samples to Earth
Shoemaker-Levy 9	1993	Orbited Jupiter			Nucleus fractured. Pieces impacted Jupiter in 1994

* = virtually infinite

meteorites, at least those composed mostly of iron, were of cosmic origin, and that they were the cause of fireballs. However, his conclusions were treated with scepticism and ridicule. Chladni also suggested that meteorites were the result of an impact or explosion of some object or objects in space.

It was also thought that meteorites may have been emitted by volcanoes on Earth. This idea was reinforced by a fall of stones near Siena, Italy in 1794 just eighteen hours after the eruption of Vesuvius. But an analysis of the velocity of rocks ejected by Earth-based volcanoes indicated that this was an unlikely source. Shortly afterwards, Heinrich Olbers suggested that meteorites may have been ejected from volcanoes on the Moon. This was easier to accept as the escape velocity on the Moon is much less than on the Earth.

In 1795 a large stony meteorite fell at Wold Cottage, England out of a clear blue sky with no thunder or lightning, showing that lightning could not be the source of heating. Edward Howard and Jacques de Bournon then carried out a scientific analysis and found grains of nickel-iron, similar in composition to those of iron meteorites, but unlike those in terrestrial rocks, indicating that both stony and iron meteorites were of extraterrestrial origin. Then in 1803 about 3,000 meteorites fell at l'Aigle in France. The French Minister of the Interior asked a distinguished physicist, Jean-Baptiste Biot, to investigate the fall and his report convinced many sceptics of their cosmic origin. This was followed by a meteorite fall at Weston, Connecticut in 1807, which was also subject to detailed scrutiny, convincing many of the remaining sceptics of meteorites' cosmic origin.

In the nineteenth century interest in meteorites escalated and a number of collections were formed. Knives made of meteoritic iron were given to Captain John Ross by Inuits in 1818, but it was not until 1894 that Robert Peary was shown the area where the original meteorite fell at Melville Bay, Greenland. The largest mass, weighing 31 tons, is now preserved in the American Museum of Natural History.

Henry Sorby undertook a detailed analysis of meteorites in the 1860s. He carried out a microscopic examination of chondrites (a type of stony meteorite), in particular, and their spherical inclusions called chondrules, which had first been noticed by Edward Howard in 1802. Sorby observed that the chondrules were composed of very small crystals and glass. As a result he concluded that they had solidified from molten material prior to their inclusion in chondrites. Because of the high temperatures involved he proposed two possible sources for the chondrites: either they were pieces of the Sun ejected in solar prominences, or they had been formed out of the original solar nebula.

Work in the twentieth century showed that chondrites have a very similar chemical composition to that of the Sun, although they contain no free hydrogen and helium. Isotope concentrations indicated that they had been formed within the first few million years of the solar system's birth about 4.56 billion years

ago. The spherical chondrules were clearly once molten, having been flash heated to above 1,800 K, before they rapidly cooled, solidified and dispersed in the solar nebula. This high temperature seems to have limited their formation to the inner part of the solar nebula.

The 1.2 km diameter Barringer Meteorite Crater in Arizona first came to the attention of scientists in the nineteenth century. Although the surrounding area was covered with about 30 tons of large oxidized pieces of meteorite, it was initially thought that the crater was of volcanic origin as no evidence of the main meteoritic mass could be found. There were no magnetic anomalies in the crater, for example. But in 1903 Daniel Barringer concluded that the crater had been produced by the impact of a large iron-metallic meteorite, the majority of which would be buried underneath the surface. In spite of drilling to a depth of 420 m no significant pieces of meteorite were found. (It was not realised at the time that most of the meteorite would have been vaporized during its descent and on impact.)

In the early 1960s Eugene Shoemaker confirmed the Barringer crater's meteoritic origin when he found samples of severely shocked rock there, like that produced by nuclear explosions. This was the first definite proof of an extraterrestrial impact on the Earth's surface. The nickel-iron meteorite which impacted about 50,000 years ago is thought to have been about 50 m across and to have weighed about 300,000 tons. Since then many asteroid and meteorite craters have been confirmed on Earth.

It had been suggested from time to time over the years that meteorites may have come from the Moon. But there was no way of showing if this was true or not until samples of lunar surface material had been returned by the Apollo astronauts and the unmanned Luna landers at the end of the 1960s/early 1970s. No meteorites known at the time fitted with these samples. But in 1982 a meteorite – ALH 81005, discovered near the Allen Hills in Antarctica – was found, by analysing its mineral, chemical and isotopic composition, to be of lunar origin. Since then a number of lunar meteorites have been found, most of which are anorthositic regolith breccias from the lunar highlands, with a few maria basalts.

Cosmic ray exposure ages have shown that some meteorites were ejected from the Moon only a few hundred years ago, showing that significant impacts are still occurring on the Moon. Cosmic ray exposure ages and chemical and mineral compositions have shown that some meteorites, although being found in different parts of the Earth, were, in fact, ejected from the Moon by one impact.

Shoemaker, Hackman and Eggleton suggested in 1962 that some meteorites found on Earth could have been ejected from Mars by an asteroid impact. Crystallization ages for the SNC (Shergottite, Nakhlite, Chassignite) group of meteorites were then found to range from 4.5 to 0.3 billion years or less. This suggested that they had been ejected from a planet which has undergone a series of melting episodes, rather than from an asteroid which would have cooled down long ago. Mars was the prime possibility, having a relatively small escape velocity whilst being relatively close to Earth. But at the time no meteorites were known to have come from the Moon, so the idea that some meteorites could have come from Mars was treated with a great deal of scepticism.

Then the first lunar meteorite was discovered in 1982. In the following year Donald Bogard and Pratt Johnson discovered that the composition of the gas trapped in the Antarctic Shergottite meteorite EET 79001 closely resembled that of the Martian atmosphere measured by the Viking landers in 1976. Since then a number of other meteorites have been found to come from Mars.

In 1984 a 1.9 kg meteorite known as ALH 84001 had been picked up in the Allen Hills in Antarctica. It was initially thought to be a Diogenite, but in 1993 David Mittlefehldt found that this was not so. Further analysis showed that ALH 84001 appeared to have come from Mars, having been thrown into space by a large impact millions of years ago. But this meteorite was considerably older than the other known Martian meteorites. It appeared to have crystallised slowly from molten rock beneath the surface of Mars about 4.5 billion years ago. Isotopic abundances showed that ALH 84001 had spent about 16 million years in space before landing in Antarctica. There it had been covered in snow and ice for something like 13,000 years, until subsurface ice flows had brought it to the surface.

David McKay and Everett Gibson Jr. announced in August 1996 that they had found strong circumstantial evidence in ALH 84001 for early life on Mars. Images taken with a scanning electron microscope showed clusters of elongated, worm-like shapes, no more than 0.1 microns long. These elongated shapes appeared remarkably similar to those of microfossils which had been formed on Earth about 3.5 billion years ago. This and other possible evidence for life from ALH 84001 has been strongly disputed, however.

It has become clear recently that meteorites from Mars, excluding ALH 84001, fit into two basic groups. One group consists of the Nakhlites and Chassignites whose magma crystallised about 1.3 billion years ago. The second group consists of the Shergottites whose ages are difficult to determine, but which appear to range from about 600 to 150 million years. This indicates that igneous activity on Mars probably extends almost to the present day. In addition, carbonate deposits or clay materials have been found inside these SNC meteorites, indicating that they had been subjected to a water environment before they were ejected from the planet.

Bibliography

Ball, Robert S., *The Story of the Heavens*, new and rev. ed., Cassell and Company, 1897.

Humboldt, Alexander von (Otté, E. C., and Paul, B. H., trans.), *Cosmos: A Sketch of a Physical Description of the Universe*, Vol. **IV**, Henry G. Bohn, 1852.

McSween, Harry Y., Jr., *Meteorites and Their Parent Planets*, 2nd ed., Cambridge University Press, 1999.

Middlehurst, Barbara M., and Kuiper, Gerard P. (eds.), *The Solar System, Vol. IV, The Moon, Meteorites and Comets*, University of Chicago Press, 1963.

See also: Extraterrestrial life.

METEORS

Meteors, or shooting stars, were generally thought in the Middle Ages to be purely atmospheric phenomena. But in 1714 Edmond Halley suggested that meteors may be caused when the Earth, in its orbit around the Sun, meets matter formed in the ether. Similar ideas for the cosmological origin of meteors were proposed by John Pringle in 1759, David Rittenhouse in 1786, and Ernst Chladni in 1794, but by the end of the century most scientists still believed that they were purely atmospheric phenomena.

Johann Benzenberg and Heinrich Brandes simultaneously observed 22 meteors in the autumn of 1798 from two locations a few kilometres apart. As a result they found that the heights varied from about 10 to 210 km, clearly indicating that they were in the upper reaches of the atmosphere. This was the first time that the height of meteors had been measured.

The earliest account of a meteor shower that can be linked to a modern meteor shower was of the intense Leonid storm of March 687 BC that was recorded by the Chinese. Many similar records have been found of later meteor showers observed in the Far and Middle East.

Alexander von Humboldt and others observed a spectacular meteor shower on the morning of 12 November 1799, which Humboldt noted appeared to come from one point in the sky, now called the radiant. Thirty-four years later, on the night of 12–13 November 1833, another spectacular display was seen. A number of observers, including Denison Olmsted and Alexander Twining, noted that the radiant was in the constellation of Leo. The radiant was interpreted by both Olmsted and Twining as the point where the Earth intercepted the meteors in their orbit, thus proving their cosmic origin.

Olmsted analysed the various observations of the 1833 November meteors – or Leonids, as they are now called – and noted the similarity between the 1833 shower and that of 1799, as well as with the less intense shower of November 1832. Olmsted suggested that the meteors may be caused by material from a comet-like body being heated by friction in the Earth's atmosphere. From their apparent velocity he concluded that the comet-like source orbited the Sun twice a year, meeting the Earth on about 12 November every year near the source's aphelion.

The display of the November meteors in 1834 and subsequent years was less intense than in 1833. In fact, in 1837 Heinrich Olbers noted not only the extreme variability in the intensity of the display from year to year, but suggested that the most intense showers occurred at intervals of 34 years. As a result he predicted that the next intense display would occur in 1867.

It had been known for some time that there was often a meteor shower in August. John Locke observed that the meteors in the shower of 8 August 1834 seemed to originate from a point in the constellation of Perseus. Then in 1836 Lambert Quetelet showed, by analysing historical records, that the August meteors or Perseids returned annually, and he correctly predicted another shower on or about 9 August 1838.

The discovery of the annual nature of these Perseid meteors proved fatal to Olmsted's theory for the Leonids, however, as it was highly unlikely that the comet-like source of the Perseids was also in such a closely prescribed orbit. But in 1839 Adolf Erman suggested that the meteoric material did not orbit the Sun in a comet-like cloud; the material was strung out in a ring instead. The showers simply occurred when the Earth passed through the ring every year. Then in 1861 Daniel Kirkwood suggested that periodic meteors were caused by the debris of old comets whose matter had become distributed around their orbits.

In 1864 Hubert Newton and Josiah Gibbs collected and analysed data on the Leonid and Perseid meteors, and showed that they returned in periods of one sidereal year rather than one tropical year. As a result, the point of intersection of these meteor streams with the Earth's orbit was changing at the rate of about 1° or 1 day every 70 years. Newton analysed historical records of meteor showers going back to 902 AD and concluded that major Leonid showers occurred every 33.25 years. This led him to predict that the next major shower would occur on 13–14 November 1866. In the event this was correct for Europe, although in the USA the shower of the following year was better.

Newton found that the node of the Leonids' orbit in space was advancing at the rate of 52.4″/year, or 29′ in 33.25 years, for example. John Couch Adams then found that a particle in a 33.25 year orbit would have the observed nodal precession of 29′. So the reason why the Leonid meteor showers peaked every 33.25 years was because that was the period of the meteor particles in their orbit around the Sun. Although there was one main cloud of particles, which the Earth intercepted on average every 33.25 years, there were other particles strung out in the same orbit which the Earth intercepted every year.

Giovanni Schiaparelli showed in 1866 that the Perseids' orbit (see left-hand side of Figure 2.24) was the same as that of the comet Swift-Tuttle (1862 III). This was the first identification of a meteor stream with a specific comet. So there was a ring of particles around the Sun, in the same 133 year orbit as the comet, and the Perseid meteors were seen every year when the Earth passed through that ring.

In 1867 Carl Peters recognised the striking similarity between the orbit calculated by Urbain Le Verrier for the Leonid meteor stream (see right-hand side of Figure 2.24), and that produced by Theodor von Oppolzer for Comet Tempel-Tuttle (1866 I). In the same year Le Verrier concluded that planetary perturbations acting on a cloud of meteor particles would eventually spread the particles uniformly around their orbit. In the case of the Leonids, this had clearly not yet happened, as there was a peak about every 33 years, so the cloud must be still relatively young. His calculations showed that in AD 126 the cloud of particles had gone relatively close to Uranus. This led him to suggest that the precursor to the Leonids had originally been following a much more elliptical orbit before 126, when its orbit had been radically changed by this close planetary encounter.

In 1845 Edward Herrick and Francis Bradley observed that Biela's comet had broken into two pieces. At their next appearance in 1852 the two comets were about 2 million km apart, but they were never seen again on subsequent orbits. In 1867 Edmund Weiss calculated that the Earth passed very close to the orbit of Biela's comet on 28 November each year. As a result he predicted a Bielid meteor shower on about 28 November 1872 or 1879. He was proved correct, as on 27 November 1872 there was an impressive display of meteors, and on 27 November 1885, about two orbits later, there was an even more spectacular display. This shower is now called the Andromedids.

The orbits of the source objects of the Perseids, Leonids and Andromedids are summarised in Table 2.8.

The Apollo asteroid Phaethon (3200) was discovered by Simon Green and John Davies in 1983 using the IRAS spacecraft. Phaethon was found to have a highly eccentric orbit that took it from the main asteroid belt, outside the orbit of Mars, to well within the orbit of Mercury. Shortly after its discovery Fred Whipple pointed out that this orbit was identical to that of the Geminid meteor shower, so Phaethon was the parent body of the Geminids. This was the first time that an asteroid had been found to be the originator of a meteor shower. So astronomers began to question whether Phaethon was really a comet. But no coma or tail was visible, so if it was a comet, it must have been an extinct one.

The first photographic meteor spectrum was obtained serendipitously at the Harvard Observatory Station in Peru in 1897, but by 1930 only a total of nine meteor spectra had been photographed world-wide. Peter Millman then started a programme of objective prism spectra photography in the early 1930s. He found that the spectra fell into two types, Y and Z. The Y type had H and K lines of ionized calcium as the most prominent lines, and the Z types had lines of iron and chromium. The Y type meteors occurred at an altitude above

Table 2.8. *Sources of selected meteor showers*

Shower	Peak of annual shower (approx.) as at 2010	Parent object	Comet or asteroid period (years)	Comet or asteroid perihelion distance (AU)	Link discovered by, with year
Perseids	12 August	Comet Swift-Tuttle	133	0.960	Schiaparelli, 1866
Andromedids	14 November	Comet Biela (now lost)	6.6	0.861	Weiss, 1867
Leonids	17 November	Comet Tempel-Tuttle	33.3	0.977	Peters, 1867
Geminids	14 December	Asteroid Phaethon (3200)	1.43	0.140	Whipple, 1983

80 km, whereas the Z types were below that height. Millman found that the Leonids exhibited Y type spectra, which he attributed to stony bodies.

Since the 1930s two other major spectral types have been added, X and W. The first type has bright lines of sodium and magnesium, and the second type is for meteors of unusual composition. The classifications of X, Z and Y have been found to correlate mainly with the velocity of the meteors, with X types having a velocity of about 15 to 20 km/s, Z types a velocity of ~30 km/s, and the Y types of ~60 km/s.

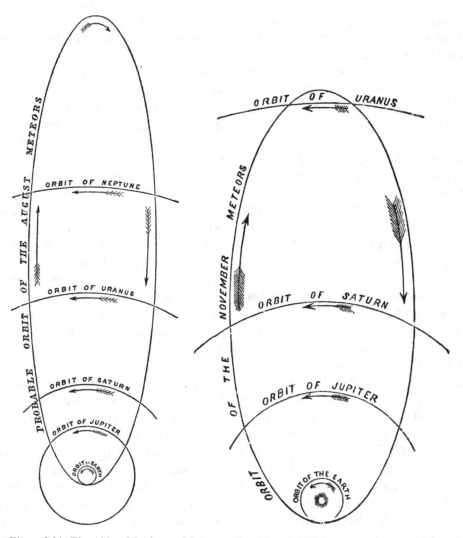

Figure 2.24. The orbits of the August Meteors or Perseids and the November Meteors or Leonids as deduced in the 1860s. (From *Popular Astronomy*, by Simon Newcomb, 1898, pp. 407 and 408.)

Bibliography

Clerke, Agnes M., *A Popular History of Astronomy during the Nineteenth Century*, 4th ed., rev. and corr., Adam and Charles Black, 1902.

Humboldt, Alexander von (Otté, E. C., and Paul, B. H., trans.), *Cosmos: A Sketch of a Physical Description of the Universe*, Vol. IV, Henry G. Bohn, 1852.

Jenniskens, Peter, *Meteor Showers and their Parent Comets*, Cambridge University Press, 2009.

Millman, Peter M., *An Analysis of Meteor Spectra*, Annals of Harvard College Observatory, **82**, No. 6 (1932), 113–146, and **82**, No. 7 (1935), 149–177.

Newcomb, Simon, *Popular Astronomy*, 2nd ed. rev., Macmillan and Co., 1898.

Yeomans, Donald K., *Comets: A Chronological History of Observation, Science, Myth, and Folklore*, John Wiley, 1991.

PLUTO

Urbain Le Verrier speculated on the possible existence of planets further from the Sun than Neptune within a week of Neptune's discovery in 1846. Some thirty years later, David P. Todd analysed Uranus' orbit and concluded that it was being affected by a trans-Neptunian planet, but his search for it was unsuccessful. Over the remainder of the nineteenth century a number of other astronomers searched for a trans-Neptunian planet, but in vain.

Percival Lowell and William Pickering started a parallel, competitive, photographic search for a trans-Neptunian planet in the early years of the twentieth century. Both tried to predict its orbit following their analysis of the apparent orbital deviations of the outer planets. Lowell and his staff at Lowell's Flagstaff Observatory never found Planet X, as he called it, before the search was put on hold following his death in 1916. William Pickering was also unsuccessful.

A new search was started at the Lowell Observatory in April 1929 by Clyde Tombaugh, who had been recruited for the task. By January 1930 he had photographed the whole of the zodiac and in the following month he found Planet X, now called Pluto, using a blink comparator. The discovery was made public on 13 March, the 75th anniversary of Lowell's birth and the 149th anniversary of William Herschel's discovery of Uranus. The Lowell Observatory released only one celestial position of the new planet so that they would have the best chance of defining its orbit. But the first orbit, published on 7 April, was by Armin Leuschner, Ernest Bower and Fred Whipple of the University of California, who had deduced a distance of 41 AU and an orbital inclination of 17°, both very close to the truth, although their orbit was highly eccentric. Percival Lowell's predicted orbit of 1915, using Uranus and Neptune residuals, was found to be quite close to that discovered. But this was fortuitous, as his estimated mass of Pluto of 6.6 M_E (mass of Earth) was far too high.

Pluto did not show a disc in even the largest telescopes in 1930, and it did not appear to have a satellite, making it impossible to produce an accurate estimate of Pluto's mass. The best estimates at that time were between 1 and 0.1 M_E. Over the years Pluto's estimated mass has been gradually reduced, until today it is only 0.002 M_E, or 20% that of the Moon.

Merle Walker and Robert Hardie detected Pluto's axial rotation in 1955 when they discovered regular variations in its intensity with a period of 6.387 days. Over subsequent years Pluto's maximum intensity during its spin period decreased, even though it was approaching perihelion and so getting more light from the Sun. In addition, the variation in the light curve over each spin cycle increased, so the dark part became even darker. Hardie attributed these effects to the sublimation of surface frost as Pluto's surface got warmer, so uncovering a darker surface.

In 1976 Dale Cruikshank, Carl Pilcher, and David Morrison detected methane ice on Pluto's surface. Then on 22 June 1978 James Christy noticed that Pluto's photographic images were not circular. On the following day, he examined a series of five plates, taken over the course of a week, which showed the bulge moving around Pluto with a period of about six days. He had found that Pluto has a satellite, now called Charon.

Charon's orbital period was found to be identical to Pluto's spin rate, so Pluto has one face permanently facing its satellite. The centre-to-centre distance of Pluto to Charon turned out to be only 20,000 km (or 1/20 of the distance of the Earth to the Moon). Charon's orbit showed that the angle between Pluto's equator and Pluto's orbit around the Sun is similar to that of Uranus, with both planets almost spinning on their side.

Mutual eclipses between Pluto and Charon lasting from 1985 to 1991 allowed Pluto's diameter to be accurately determined, for the first time, as 2,300 km. So Charon, with a diameter of about 1,200 km, is the largest satellite, relative to its primary, of any satellite in the solar system. Charon was found to be appreciably darker than Pluto, with an albedo of about 0.35 in the visible waveband, compared with about 0.55 for Pluto. The total eclipses of Charon by Pluto in 1987 allowed the spectrum of Charon to be determined for the first time. This showed that the previously observed methane feature of their joint spectrum was due to Pluto, and that Charon had water ice on its surface.

Pluto's high reflectivity presented something of a problem, as the methane ice on its surface should have been darkened by the solar ultraviolet. If, however, the methane sublimated around perihelion, it could condense as pure white methane frost every time Pluto approached aphelion. In that case, Pluto should have an atmosphere for a few decades at least around perihelion, which was due in 1989.

Pluto's deep but tenuous atmosphere, with a density of only about 10 microbars, was first detected during a stellar occultation on 9 June 1988. It is thought that this atmosphere, which is probably produced by the surface ice sublimating, may exist for only a few decades around perihelion. So it was somewhat of a surprise when in 2002 Pluto's atmospheric pressure appeared to have increased rather than decreased following its perihelion passage. Then in 2006 it was found that Pluto's temperature was 10 K less than expected, so there seems to be a time lag in sublimation of the atmosphere from the ground around perihelion.

In 1992, Tobias Owen, Leslie Young and Dale Cruikshank made the surprising discovery that the ice on Pluto's surface, as on Triton, was mainly nitrogen, with small amounts of carbon monoxide and methane, rather than mainly methane as previously thought. They also concluded that the atmosphere must be about 98% nitrogen, with small amounts of methane and carbon monoxide. Ethane was discovered on Pluto's surface in 2006.

Harold Weaver, S. Alan Stern and colleagues detected two small satellites of Pluto in 2005 of about 30 to 80 km diameter. Now called Nix and Hydra, they are orbiting the planet in the same plane as Charon about 49,000 and 65,000 km from the gravitation centre of the Pluto system.

Raymond Littleton had suggested in 1936 that Pluto had once been a satellite of Neptune, but had been ejected by a close encounter with Neptune's largest satellite Triton. This encounter had also caused Triton to orbit Neptune in a retrograde direction. However, the discovery of Charon made this scenario unlikely, as any disturbance powerful enough to eject Pluto from Neptune would almost certainly have separated Pluto and Charon. In addition, in 1965 Charles Cohen and Clyde Hubbard discovered that Pluto's orbit was in a 2:3 resonance with Neptune's. This resonance, and their radically different orbital inclinations, meant that, if the resonance had always been present, Pluto could never have come closer than 2.7 billion kilometres to Neptune, and so could never have been part of the Neptune system.

The general view today is that Pluto, Charon and Triton are planetesimals left over from the formation of the solar system, and that Triton was captured by Neptune. William McKinnon and Steve Mueller developed a theory to explain the large size of Charon relative to Pluto, and the angular momentum of the Pluto–Charon system. In their theory, which was based on Hartmann's and Davis' theory of the origin of the Moon, Charon was formed from the debris after a large planetesimal had collided with Pluto, which was in orbit around the Sun. In 2005 Robin Canup showed, using a computer simulation, that this was quite feasible. In the most likely scenario Pluto received a glancing impact from a body of 0.3 to 1.0 of Pluto's mass at a relative velocity of about 1 km/s.

A number of bodies of approximately the same size as Pluto were discovered in the Kuiper Belt in the early years of the twenty-first century, raising the question of how to define a planet. As a result, in 2006 the International Astronomical Union devised a new class of object called a dwarf planet that now includes Pluto.

In 2006 the New Horizons spacecraft was the first spacecraft to be launched to Pluto. It is due to fly by the planet on 14 July 2015.

Bibliography

Cruikshank, Dale P., *Triton, Pluto and Charon*, in Beatty, J. K., Petersen, C. C., and Chaikin, A. (eds.), *The New Solar System*, 4th ed., Cambridge University Press, 1999.

Hoyt, William G., *Planet X and Pluto*, University of Arizona Press, 1982.

Levy, David H., *Clyde Tombaugh: Discoverer of Planet Pluto*, University of Arizona Press, 1991.

Stern, S. Alan, *Pluto*, in McFadden, L. A., Weissman, P. R., and Johnson, T. V. (eds.), *Encyclopedia of the Solar System*, 2nd ed., Academic Press, 2007.

Stern, Alan, and Mitton, Jacqueline, *Pluto and Charon: Ice Worlds on the Ragged Edge of the Solar System*, Wiley and Sons, 1999.

Tombaugh, Clyde W., and Moore, Patrick, Out of Darkness, The Planet Pluto, Mentor, 1981.

See also: Trans-Neptunian Objects and Centaurs.

TRANS-NEPTUNIAN OBJECTS AND CENTAURS

Kenneth Edgeworth suggested in 1943 and again in 1949 that there were probably many planetesimals left over from the formation of the planets orbiting the Sun outside the orbit of Neptune. Gerard Kuiper independently made a similar suggestion in 1951, although he thought that these icy planetesimals would have been scattered out of their orbits by Pluto some time ago to either approach the Sun as comets or be sent into deeper space. Fred Whipple thought that a belt of icy comets could still exist just beyond the orbit of Neptune, but could find no evidence of its existence. Then in 1980 Julio Fernandez concluded, following a computer simulation, that a comet belt between 35 and 50 AU from the Sun could maintain the population of short-period comets.

Meanwhile in October 1977 Charles Kowal discovered an object, initially called 1977 UB, using the Palomar Schmidt telescope. Pre-discovery images produced an orbit with a perihelion of 8.5 AU, aphelion of 18.9 AU, inclination of 6.9°, and period of 51 years. Its orbit went so close to Saturn and Uranus, however, as to make it unstable. Kowal suggested calling the object Chiron, who was a Centaur and the mythological son of Saturn and the grandson of Uranus.

When it was discovered, Chiron appeared to be an asteroid with a diameter of about 180 km, although its orbit was more

Table 2.9. *Selected Trans-Neptunian objects and Centaurs*

Name	Year of discovery	Discovered by	Mean radius (km)	Semimajor axis (AU)	Period (years)	Type of object
Classical Kuiper Belt Objects or Cubewanos						
1992 QB$_1$	1992	Jewitt and Luu	80	44.0	292	
1993 FW	1993	Jewitt and Luu	90	43.8	290	
Resonance Kuiper Belt Objects or Plutinos						
1993 RO	1993	Jewitt and Luu	45	39.3	246	
1993 RP	1993	Jewitt and Luu	35	39.3	246	
1993 SB	1993	Williams et al.	65	39.3	246	
1993 SC	1993	Williams et al.	180	39.6	249	
Scattered Disk Objects (SDOs)						
1996 TL$_{66}$	1996	Luu and Jewitt	300	83.3	760	
1999 CV$_{118}$	1999	Jewitt et al.	75	52.5	380	
Large Trans-Neptunian Objects (in order of size)						
Eris	2005	Brown et al.	1,200	67.7	557	SDO
Pluto	1930	Tombaugh	1,160	39.5	248	Resonance KBO
Sedna	2003	Brown et al.	800	486	10,700	Detached object
Makemake	2005	Brown et al.	750	45.8	310	Classical KBO
Haumea	2004	Brown et al.	600	43.3	285	Classical KBO
Quaoar	2002	Trujillo and Brown	570	43.6	288	Classical KBO

Note: Eris, Pluto, Makemake and Haumea are classified as dwarf planets.

			Centaurs			
Chiron	1977	Kowal	90	13.7	51	
Pholus	1992	Rabinowitz	80	20.4	92	

like that of a comet. Then in 1989 Karen Meech and Michael Belton found that Chiron had a coma of ice and dust. In the following year Bobby Bus, Ted Bowell and Mike A'Hearn also detected cyanogen gas surrounding Chiron to a distance of 50,000 km. Cyanogen was known to be a constituent of the ionised gas tails of comets, but this was the first time that it had been detected at such a large distance from the Sun. So Chiron seemed to be a comet, although its nucleus was much larger that that of any known comet. Then it was found that Chiron's activity decreased markedly as it approached perihelion in early 1996, so it is not a comet as we normally understand them, as a comet's activity increases approaching perihelion. On the other hand, with its large coma, it is not a normal asteroid either.

The asteroid 1992 AD, or Pholus (5145), was found in January 1992 by David Rabinowitz, who used the Spacewatch telescope on Kitt Peak, Arizona. Within a few weeks Pholus, which was estimated to have a diameter of about 160 km, was found to be in an even more distant orbit than Chiron, with a perihelion of 8.7 AU and aphelion of 32 AU. Although it was discovered only a few months after perihelion, Pholus showed no evidence of a coma. It was found to be very red for an asteroid (or comet), being over three times as bright in the near infrared waveband as in visible light. Pholus' orbit was, like Chiron, also chaotic, crossing, as it did, the orbits of Saturn, Uranus and Neptune. Such chaotic orbits that are between the orbits of Jupiter and Neptune are the key characteristic of Centaurs. They are thought to have originated from outside Neptune's orbit, but to have been scattered by Neptune to their current location as Neptune's orbit expanded. Chiron was the first Centaur to be discovered and Pholus the second.

Meanwhile, in 1987 David Jewitt and Jane Luu had started to search for the hypothesised objects in the so-called Kuiper belt beyond the orbit of Neptune, using the University of Hawaii's 2.2 m telescope on Mauna Kea. For five years they found nothing, then in August 1992 they discovered the first Trans-Neptunian Object (TNO), called 1992 QB$_1$, with an average distance from the Sun of 44 AU. It is about 160 km in diameter and has a period of about 292 years. In March 1993 Jewitt and Luu found the next TNO, called 1993 FW, in a similar orbit to 1992 QB$_1$. These two TNOs are now also called classical Kuiper Belt Objects (KBOs) or cubewanos (after QB$_1$, or "CuBeWan").

Whilst the first two Trans-Neptunian Objects had orbits that came nowhere near the orbit of Neptune, the next four TNOs to be discovered (1993 RO, RP, SB and SC) had orbits that came much closer and, as Brian Marsden pointed out, had orbital periods in 2:3 resonance with Neptune. This resonance ensures that their orbits, with semi-major axes of about 39.5 AU, are stable. These TNOs are now called resonance Kuiper Belt Objects or Plutinos, as Pluto was the first such object found to have a 2:3 resonance with Neptune.

In 1996 Luu and Jewitt discovered an object, called 1996 TL66, with a perihelion of 35 AU, aphelion of 132 AU, and period of about 760 years. Since then many more of these scattered disk objects (SDOs) have been found, with aphelia of about 100 AU or more, outside the classical Kuiper belt. It is thought that these SDOs have probably been scattered from their original orbits by gravitational interactions with Neptune as its orbit expanded. This scattered disk is thought to be the origin of short-period comets. There are also a few objects, including Sedna (see below), whose perihelia are so large that they cannot have been moved there by Neptune. They are called extended scattered disk or detached objects.

A classical Kuiper belt object, now called Quaoar, was discovered in 2002 by Chad Trujillo and Mike Brown using the Oschin Schmidt telescope on Mount Palomar. Quaoar has an estimated diameter of about 1,150 km. Then in the following year Sedna was discovered with a diameter of about 1,600 km, in a highly eccentric orbit with an aphelion of about 900 AU. Since then a number of other large objects have been discovered by a team led by Mike Brown, including Eris, an SDO with a diameter of about 2,400 km, which is about that of Pluto; Makemake, a classical KBO with a diameter of about 1,500 km; and Haumea, a classical KBO, with a diameter of about 1,200 km. Evidently more similar sized or larger objects may exist in the outer solar system. As a result, in 2006 the International Astronomical Union (IAU) devised a new class of object, called a dwarf planet.

The IAU defined a dwarf planet as an object that (a) orbits the Sun, (b) has sufficient mass to be essentially spherical, (c) has not cleared its neighbourhood of planetesimals and (d) is not a planetary satellite. The initial list of such objects included Pluto, Eris and the asteroid Ceres. Since then Makemake and Haumea have been added.

Further information on selected TNOs and Centaurs is given in Table 2.9.

Bibliography

Davies, John, *Beyond Pluto: Exploring the Outer Limits of the Solar System*, Cambridge University Press, 2001.

Delsanti, Audrey, and Jewitt, David, *The Solar System Beyond the Planets*, in Blondel, Philippe and Mason, John (eds.), *Solar System Update*, Springer, 2006.

Jewitt, D., Morbidelli, A., and Rauer, H., *Trans-Neptunian Objects and Comets*, Springer, 2007.

Morbidelli, Alessandro, and Levison, Harold F., *Kuiper Belt Dynamics*, in McFadden, L. A., Weissman, P. R., and Johnson, T. V. (eds.), *Encyclopedia of the Solar System*, 2nd ed., Academic Press, 2007.

See also: Asteroids; Pluto

6 · Exoplanets

In 1983 the IRAS spacecraft detected an infrared excess emitted by Vega which was attributed to a dust disc extending up to about 80 AU from the star. IRAS later showed evidence of dust discs around other stars, including β Pictoris and Fomalhaut, and in 1984 Bradford Smith and Richard Terrile imaged a 800 AU diameter dust disc around β Pictoris. These dust discs were thought to be the remains of the nebulae from which the stars had been formed, which could condense to form planets.

The first clear evidence of an extrasolar planet or exoplanet was found by chance in 1990/91 by Alex Wolszczan and Dale Frail when they observed timing variations in the millisecond pulsar, PSR B1257 + 12, using the Arecibo radio telescope. There appeared to be two planets orbiting the pulsar with masses of at least 3.4 and 2.8 M_E (mass of the Earth) and orbital periods of 67 and 98 days. A little later they found that there was also a third, much less massive, planet, with a period of 25 days.

Michel Mayor and Didier Queloz started a programme to find exoplanets around 'normal' stars in April 1994 by measuring stellar Doppler shifts from the Haute-Provence Observatory. In January 1995 they found that the Doppler shift of the yellow, G-type star, 51 Pegasi, was varying regularly with a period of just 4 days. The idea that this could be due to a planet was difficult to accept, however, as it appeared to have a mass of at least $^1/_2 M_J$ (mass of Jupiter), and be so close to the star that its temperature would be about 1,300 K. Nevertheless, this seemed to be the most likely explanation, so on 6 October 1995 Mayor and Queloz announced their discovery. Eleven days later Geoffrey Marcy and Paul Butler announced corroborative measurements from the Lick Observatory.

In December 1995 Marcy and Butler found that another solar type star, 47 Ursae Majoris, appeared to have a planet. It had a minimum mass of 2.4 M_J at a distance of 2.1 AU from its parent star, with a period of 3 years. This seemed to be a much more reasonable set of parameters than those of the planet around 51 Pegasi, if the configuration of the solar system was taken as the norm.

Doppler shift measurements only allow the minimum mass of a planet to be determined. But in 1999 David Charbonneau and colleagues made the first observation of the complete transit of a star, HD 209458, by an exoplanet. This enabled the radius of the planet to be accurately determined as 1.27 ± 0.02 R_J, as well as its mass, which turned out to be 0.63 M_J. In the following year Charbonneau and colleagues observed the spectrum of the star using the Hubble Space Telescope during four transits of the planet. This showed that the planet had an atmosphere.

The first image of an exoplanet was made by Gael Chauvin and colleagues in 2004 using the 8.2 m Yepun telescope at the European Southern Observatory. It was detected in the infrared, orbiting the brown dwarf 2M1207 at a distance of 40 AU. The planet was thought to have a mass of about 4 M_J, although there is some evidence that it may be about twice this.

Paul Kalas and colleagues used the Hubble Space Telescope in 2004 to produce the first image of the dust disc around Fomalhaut in visible light. It was in the form of a belt between 133 and 158 AU radius, with a sharp inner edge, that had a centre offset 15 AU from the star's position. The offset and sharp inner edge were thought to indicate the presence of planets. An image of one of these possible planets was found in 2004, and a follow-up image taken two years later showed that the object was clearly a planet. This 3 M_J planet was the first exoplanet to be imaged in visible light.

The number of confirmed exoplanets currently (in mid-2012) exceeds 700, most of which have been detected by radial velocity or transit observations. The majority of these planets have masses in excess of that of Jupiter, but, as the detection techniques have become more sensitive, less and less massive planets have been detected around normal stars, such that in 2006 a 5.5 M_E planet was discovered orbiting a red dwarf. Shortly afterwards an even smaller planet, with a minimum mass of about 1.7 M_E, was discovered orbiting a red dwarf.

The orbits of the planets of our solar system have eccentricities less than 0.1, with the exception of the small planet Mercury which is nearest the Sun. Surprisingly, however, the orbital eccentricities of the exoplanets found so far are, in general, much larger than this, with about 60% having eccentricities greater than 0.1 and about 15% having eccentricities greater than 0.5. Theories of the origin of solar systems in general need to take these statistics into account.

Bibliography

Casoli, Fabienne, and Encrenaz, Thérèse, *The New Worlds: Extrasolar Planets*, Springer-Praxis, 2007.

Croswell, Ken, *Planet Quest: The Epic Discovery of Alien Solar Systems*, Oxford University Press, 1997.

Mason, John (ed.), *Exoplanets: Detection, Formation, Properties, Habitability*, Springer, 2008.

Perryman, Michael, *The Exoplanet Handbook*, Cambridge University Press, 2011.

Seager, Sara (ed.), *Exoplanets*, University of Arizona Press, 2011.

PART 3
STARS

1 · Stars considered individually

STAR ATMOSPHERES

Various elements had been detected in the spectra of the Sun and stars since the 1860s, but it was still not clear fifty years later how the abundance of elements could be determined from their spectra.

Niels Bohr outlined his theory of atomic structure in 1913 explaining how spectral lines were produced. In particular, he showed that the Pickering lines seen in O-type stars were due to ionised helium. But the B-type stars exhibited stronger lines of un-ionised helium, so the B-type stars must be cooler than O-types. Stellar spectra that Lockyer had thought were due to proto-elements were seen to be due to various ionised atoms.

Megh Nad Saha was able to show in 1920 how the degree of ionisation of an atom depended on temperature and pressure. In the following year he showed theoretically that atmospheres of the same composition, when heated to higher and higher temperatures, would exhibit spectra shown in the Harvard stellar sequence M, K, G, F, A, B and O. So the differences in stellar spectra were generally due to temperature rather than compositional differences.

Saha showed that the degree of ionisation increased with decreasing pressure, so the high intensity of the ionized lines of helium in very luminous O-type stars was due to very low density. Similarly, he concluded that the observed correlation between line intensity and luminosity for late-type stars of the same colour, and thus temperature, was probably due to differences in atmospheric density.

In 1923 Arthur Milne and Ralph Fowler showed that the percentage of atoms of a given element responsible for the production of a spectral line could be estimated from the line intensity, if the temperature and pressure of the star's atmosphere were known. Cecilia Payne analysed stellar spectra using Milne and Fowler's theory and showed in 1925 that hydrogen and helium were by far the most abundant elements in stellar atmospheres. But she then rejected this finding as spurious following comments from the highly respected Henry Norris Russell.

Russell, Walter Adams and Charlotte Moore investigated the abundance of elements in the Sun and red giants, and in 1928 concluded that either the atmospheres of the red giants consisted of large amounts of hydrogen, or they were not in thermal equilibrium. Eddington pointed out that the latter was

not a valid explanation. Albrecht Unsöld showed in 1928 that there were large amounts of hydrogen in the Sun. Donald Menzel also analysed the flash spectrum of the Sun, showing that the gas in the chromosphere has an average atomic weight of about 2. This implied that it must have a large amount of hydrogen. So in 1929 Russell finally accepted that there was an enormous amount of hydrogen in the Sun's and stellar atmospheres, and produced an impressive survey paper summarising the evidence.

It was thought in the 1920s that probably all stars had the same elements in approximately the same proportions in their atmospheres. But doubts about this were expressed in the early 1930s by Ralph Curtiss and Russell for the cool K, M, R and N stars, as the K and M stars showed strong titanium oxide bands, whilst the R and N stars showed strong carbon and cyanogen bands.

In 1932 Bengt Edlén found that the majority of the emission bands in Wolf-Rayet stars were due to highly ionised carbon, nitrogen and oxygen. Carlyle Beals then divided Wolf-Rayet stars into two groups: those with intense nitrogen emission lines and those with very strong carbon and oxygen emission lines, but very weak nitrogen emission lines. He suggested that this showed real differences in chemical abundance.

Marcel Minnaert and colleagues constructed the first curves of growth for spectral lines, relating line strength to element abundance, in 1930. In the following year William McCrea produced model atmospheres for hot stars with neutral atomic hydrogen opacity, but this failed for cooler stars. Instead, Rupert Wildt suggested that H^- could be the main source of their opacity. As a result, in 1944 Bengt Strömgren used this H^- opacity in modelling the atmosphere of the Sun and of stars types A5 to G0. For hotter stars he showed how the neutral hydrogen opacity became dominant as the H^- ion lost its extra electron at higher temperatures.

It had gradually become clear in the 1920s that the H-R diagram for stars in open clusters and for stars relatively close to the Sun was different from that of stars in globular clusters. In the 1940s Walter Baade found a similar difference for stars in the spiral arms of galaxies and stars in the nucleus regions of spiral galaxies (which were like stars in globular clusters). He called the former Population I and the latter Population II stars. Baade concluded that population II stars are much older than population I stars, as globular clusters were known to be almost completely devoid of gas. In the following decade a number of independent astronomers found that the population I stars have more metals in their atmospheres than population II, indicating that population II stars are the first generation of stars to be formed in a galaxy. Otherwise the majority of stars had essentially the same composition of elements in their atmospheres as the Sun.

Bibliography

Haramundanis, Katherine (ed.), *Cecilia Payne-Gaposchkin: An Autobiography and Other Recollections*, Cambridge University Press, 1984.

Hearnshaw, J. B., *The Analysis of Starlight; One Hundred and Fifty Years of Astronomical Spectroscopy*, Cambridge University Press, 1986.

Herrmann, Dieter B. (Krisciunas, K., trans. and rev.), *The History of Astronomy from Herschel to Hertzsprung*, Cambridge University Press, 1984.

Struve, Otto, and Zebergs, Velta, *Astronomy of the 20th Century*, Macmillan, 1962.

STAR DISTANCES

The ancient Greeks thought that the stars were fixed on a sphere surrounding the Sun and planets because the stars were not observed to move relative to each other. Even Copernicus and Tycho Brahe thought that the stars were all at the same distance from us. But ideas began to change in the seventeenth century as many of the laws of physics were understood, including those of motion and gravity, and the universe was seen as a dynamic place, with comets, in particular, having virtually parabolic orbits.

Parallax

It was expected in the early seventeenth century that the newly invented telescope should be able to reveal stellar parallaxes caused by the Earth orbiting the Sun, as proposed by Copernicus. John Flamsteed and others thought that they had detected this parallax, but this did not prove to be correct. Then in 1725 James Bradley and Samuel Molyneux started a series of measurements of the star γ Draconis hoping to detect its parallax, but they were unsuccessful. Instead, Bradley discovered the aberration of light which was caused by the Earth's orbital velocity being a detectable percentage of that of the velocity of light. This was the first experimental proof of the Earth's motion about the Sun.

In 1792 Giuseppe Piazzi found that the star 61 Cygni had a large proper motion of about 5.2″/yr, indicating that it was probably close to the Earth. So Friedrich Bessel chose to observe this star in the 1830s when he was trying to detect the first stellar parallax. In December 1838 he was successful, measuring a parallax of 0.31″, which implied that it was at a distance of about 10.4 light years. Bessel's announcement was followed by one from Thomas Henderson in January 1839 of a parallax of 0.93″ (equivalent to a distance of 3.5 light years) for α Centauri, based on measurements made a few years earlier, and one from F. G. Wilhelm Struve in the following year of 0.26″ (12.5 light years) for Vega. Although all these results were not accurate, they were of the correct order of magnitude. So, for the first time, the distance of some of the nearest stars were known, giving some idea of the scale of the nearby universe. The distances were known to about 150 stars by the end of the nineteenth century.

Henry Norris Russell published the first Hertzsprung-Russell diagram in 1914 showing the relationship between a star's Harvard spectral type and its absolute magnitude. This showed both the main sequence and some giant stars. In the same year Walter Adams and Arnold Kohlschütter found that the intensity of certain lines in stellar spectra correlated closely with the stars' absolute luminosity. So using both the Harvard spectral type and the intensity of the spectral lines enabled the absolute intensity of stars to be determined, and from their apparent intensity their true distances could be estimated. This method was misleadingly called the method of 'spectroscopic parallax'. It could be used to determine the distance of stars that were too far away for their geometric parallax to be determined.

Cepheid variables

Henrietta Leavitt had found in 1908 that the brightest of 16 Cepheid variable stars in the Small Magellanic Cloud (SMC) had the longest period. By 1912 she had managed to increase the number of observed Cepheids in the SMC to 25 with the same result. Because all stars in the SMC were at virtually the same distance from us, the differences in apparent luminosity were the same as differences in absolute luminosity, but the distance of the SMC was not known, so the absolute luminosities of these stars were not known.

Then in 1913 Ejnar Hertzsprung estimated the distances of 13 relatively close Cepheids, based on their proper motions, which gave him their absolute luminosities. He also measured their periods, enabling him to normalise Henrietta Leavitt's period-luminosity relationship. So using Henrietta Leavitt's data on *apparent* luminosities and periods for the SMC Cepheids, he was able to determine the distance of the SMC.

Harlow Shapley recalibrated the absolute magnitude scale of Cepheids close to the Sun in 1918 and, as a consequence, increased the estimated distance of the SMC by almost a factor of three. He was also able to use Cepheids to estimate the distances of globular clusters, and in 1924 Edwin Hubble identified Cepheid variables in the Andromeda and M33 nebulae. This enabled him to estimate their distances for the first time.

Both the spectroscopic parallax and Cepheid variable methods of estimating stellar distances assumed that interstellar space was transparent. But in 1930 Robert Trumpler found the first conclusive evidence that this was not so. There were clouds of absorbing dust in the Milky Way which gave the impression that stars were further away than they really were. Joel Stebbins found that those globular clusters in the plane of the Milky Way were redder than those out of the plane due to dust. So the absorption made those globular clusters in the plane of the Milky Way appeared four times further away than they really were, whereas those just 5° above or below the plane only required a correction factor of two. For much larger angles the dust effect was generally small.

Walter Baade found in 1952 that there were two different types of Cepheids, whereas the previous distance estimates had been made assuming that Cepheids were all the same. Taking this difference into account required a modification to the distance estimates for the Magellanic Clouds and the Andromeda nebula.

Parallax continued

The accuracy of stellar parallax measurements for relatively nearby stars has been gradually improved over the years, so by the 1980s stellar parallaxes could be measured to an accuracy of about 50 milliarcseconds (mas) using the best ground-based optical telescopes. The use of spacecraft, like the European Space Agency (ESA) Hipparcos, and the NASA/ESA Hubble Space Telescope have subsequently improved this to about 1.0 mas or better. As a result Hipparcos has enabled astronomers to determine the parallax of about 22,000 stars to an accuracy of about 10% or better out to a distance of about 300 light years.

Bibliography

Hirshfeld, Alan W., *Parallax: The Race to Measure the Cosmos*, W. H. Freeman and Company, 2001.

Hoffleit, D., *The Quest for Stellar Parallax*, Popular Astronomy, **57**, (1949), 259–273.

Pannekoek, Anton, *A History of Astronomy*, Allen & Unwin, 1961 (Dover reprint, 1989).

STAR MAGNETIC FIELDS

George Ellery Hale discovered magnetic fields in sunspots in 1908 when he examined sunspot spectra after they had been passed through a polarizer. He found that the lines exhibited the Zeeman effect, having two components of opposite polarity.

Five years later he also discovered the general magnetic field of the Sun. This discovery of the Sun's general magnetic field encouraged other astronomers to try to detect magnetic fields in other stars.

The attempts to detect magnetic fields in stars were unsuccessful for many years until Horace Babcock began a programme to measure the Zeeman effect in peculiar A-type (Ap) stars. In 1947 he discovered a magnetic field of about 1,500 gauss in the Cr-Eu (chromium-europium) star 78 Virginis. He followed this by detecting a variable magnetic field in a similar star, HD 125248, in which chromium and europium lines were known to vary periodically. Babcock found that the magnetic field in HD 125248 was at its maximum positive intensity when the europium lines were prominent, and that its reversed polarity field was at its maximum when the chromium lines were strongest. He suggested that an oblique rotating magnetic dipole could account for the spectroscopic and magnetic variations observed in this and in Ap-type stars in general.

Babcock published a catalogue of magnetic stars in 1958, most of which were Ap-types with variable magnetic fields. Babcock included in his catalogue seven metallic-line (Am) type stars that he listed as having detectable, but weak, magnetic fields. This apparent detection caused problems in the theory of Am-type stars for a number of years until its was found to be erroneous. No Am-type stars have since been found to have measurable magnetic fields.

Since the publication of Babcock's catalogue, magnetic fields have been found on late-type stars by Mark Giampapa and colleagues in 1981, and on a T Tauri star by Gibor Basri and colleagues about ten years later. The strongest magnetic field measured for a main sequence star has been the 34,000 gauss for HD 215441.

Bibliography

Hearnshaw, J. B., *The Analysis of Starlight; One Hundred and Fifty Years of Astronomical Spectroscopy*, Cambridge University Press, 1986.

Schrijver, C. J., and Zwaan, C., *Solar and Stellar Magnetic Activity*, Cambridge University Press, 2000.

See also: Pulsars; Soft gamma repeaters, magnetars and anomalous X-ray pulsars.

STAR MASSES

Isaac Newton concluded in the third edition of his *Principia* in 1726 that the mass of the Sun was about 170,000 times that of the Earth, which is about half its true value. So the approximate mass of one star was known in the early eighteenth century. But it was necessary to observe the orbit of binary stars, of known parallax, before the mass of another star could be found. This was not possible until the nineteenth century.

Alvan and Alvan Graham Clark discovered Sirius B, the companion to Sirius, in 1862. The period of the binary was found to be about 50 years, and from the relative motion of the two stars it was concluded that the mass of Sirius B was about half that of its companion. The parallax of Sirius measured in the 1890s gave masses for the two stars as about 1 and 2 times $M_{\odot}$ (mass of the Sun).

Henry Norris Russell found in 1912, by examining the light curves of eclipsing binary stars, that the average masses of stars on the main sequence were smaller the further down the main sequence they were. Then in 1924 Arthur Eddington showed theoretically, using the perfect gas laws, that the absolute magnitude of both main sequence and giant stars was determined almost completely by their mass.

The range of stellar masses known in the late 1920s was about 0.2 to 100 $M_{\odot}$. Eddington calculated at that time that the radiation pressure inside a star would exceed the gravitational force in stars with a mass of above about 100 $M_{\odot}$, blowing them apart. This gave the maximum possible mass for a star.

The most massive star known for a long time was HD 47129, or Plaskett's star, found by John Plaskett in 1922. It consisted of two O7-type supergiants in a binary. Plaskett thought that each star had a mass of about 90 $M_{\odot}$, but in 1961 Jorge Sahade estimated their masses as about 60 and 40 $M_{\odot}$.

Currently it is uncertain what is the maximum mass possible for a star, as many candidates, that had been observed to have masses well in excess of about 100 $M_{\odot}$, have fallen by the wayside when they were found to be binary or multiple stars. On the other hand, the minimum mass seems easier to determine as about 0.08 $M_{\odot}$, when hydrogen burning is no longer possible. Below that mass, the objects are deuterium-burning brown dwarfs.

Bibliography

Eddington, Arthur S., *The Internal Constitution of the Stars*, Cambridge University Press, 1926.

Struve, Otto, and Zebergs, Velta, *Astronomy of the 20th Century*, Macmillan, 1962.

STAR ROTATIONS

William Abney speculated in 1877 on the effect of stellar rotation on spectral lines, pointing out that lines would be broadened, possibly causing some of the fainter lines to disappear. Hermann Vogel disagreed with much of Abney's paper, suggesting that it was doubtful that even fast-rotating stars would give measurable line broadening as most of the star's light comes from the central regions of the stellar disc where the Doppler shift would be small. Vogel pointed out that line broadening should be observed in all of a star's spectral lines, yet in a number of stars he had only observed a broadening of the hydrogen lines, so this broadening could not be due to stellar rotation. In addition if some of this line broadening was due to rotation it implied that some stars, such as Altair and Vega, had equatorial velocities of about 100 times the 2 km/s velocity of the Sun, which seemed unreasonable. However, in 1898 Vogel had to rescind his comments when he observed that all of Altair's spectral lines were broad, indicating that it had an exceptionally fast rotational velocity.

In 1909 Frank Schlesinger observed the spectrum of the eclipsing binary δ Librae, which consisted of a bright and faint star. He noted that the radial velocity curve derived from the spectrum of the bright star deviated slightly from the expected curve. Schlesinger attributed this to the rotation of the bright primary. Shortly afterwards he also found a similar effect for the eclipsing binary λ Tauri.

Walter Adams and Alfred Joy observed the spectrum of the eight-hour contact binary W Ursae Majoris in 1919, and concluded that the difference in velocity in the line-of-sight between the two limbs of the star was about 240 km/s. Shortly afterwards, Richard Rossiter and Dean McLaughlin detected the rotation of the primary stars in the binaries β Lyrae (a semidetached binary) and β Persei (the eclipsing binary Algol), respectively. Then in 1929 Grigorij Shajn and Otto Struve analysed the effect of stellar rotation on spectral line profiles in spectroscopic binaries. As a result they concluded that the line broadening observed in short-period spectroscopic binaries was mainly the result of stellar rotation.

Rotational line broadening for single stars was analysed in the early 1930s by Christian Elvey, Otto Struve, Christine Westgate and others. This showed that emission line O- and B-type stars exhibited the fastest equatorial line-of-sight velocities of about 250−500 km/s. For main sequence stars, rotation rates were found to increase from early type stars to about B5, and then reduce with later types so that rotation hardly existed in G-, K- and M-type stars. In the early 1950s, Arne Slettebak found that peculiar A-type (Ap) and metallic-line (Am) stars were much slower rotators than normal A-types, and that supergiants hardly rotated at all.

Horace Babcock measured a strong magnetic field in the Ap-type star 78 Virginis in 1947. He suggested a little later that an oblique rotating magnetic dipole could account for the spectroscopic and magnetic variations observed in Ap-type stars in general. Then in the early 1960s Helmut Abt concluded that all Am-type stars were members of spectroscopic binaries with orbital periods usually of a few days and slow equatorial rotation rates. On the other hand, normal A-type stars in spectroscopic binaries had much larger orbital periods and faster rotation rates. As a result Abt concluded that tidal interactions had reduced the spin rates of Am stars.

In 1992 Joanne Attridge and William Herbst found that distribution of rotation periods of T Tauri-type stars had two peaks. This allowed the stars to be divided into a slow rotator group, with a mean period of about 8.5 days, and a rapid rotator group, with a mean period of 2.2 days. It was later found that classical T Tauri stars, which were surrounded by an accretion disc, were slow rotators, whereas the weak-line T Tauri stars, which had no accretion disc, tended to be fast rotators.

Bibliography

Hearnshaw, J. B., *The Analysis of Starlight: One Hundred and Fifty Years of Astronomical Spectroscopy*, Cambridge University Press, 1986.

Struve, Otto, and Zebergs, Velta, *Astronomy of the 20th Century*, Macmillan, 1962.

Tassoul, Jean-Louis, *Stellar Rotation*, Cambridge University Press, 2007.

Tassoul, Jean-Louis, and Tassoul, Monique, *A Concise History of Solar and Stellar Physics*, Princeton University Press, 2004.

STAR TEMPERATURES

Estimates of the surface temperature of the Sun varied wildly for most of the nineteenth century because the basic laws of radiation were unknown. These laws were progressively outlined by Joseph Stefan in 1879 and by Wilhelm Wien in 1893, enabling the effective temperature of the Sun to be estimated as about 6,000 K. The laws of radiation were given a theoretical basis a little later by Max Planck, who derived curves of radiation versus wavelength, showing their temperature dependency.

Julius Scheiner and Johannes Wilsing started a programme of visual spectrophotometry in 1905 to determine the colour temperature of stars, assuming that they were black bodies. Hans Rosenberg started a similar programme in 1914 using photographic spectrophotometry. The results from the two programmes were inconsistent, with the largest problem being with the early-type stars. Then in 1918 William Wright observed the Balmer jump in the spectrum of Vega, which showed that its emission curve did not follow the Planck radiation curve for a black body. Ch'ing Sung Yü extended this analysis to 91 B- and A-type stars in 1926. He also found that the Planck curves were poor approximations of their energy distribution, partly explaining the inconsistencies found between the estimates of Scheiner and Wilsing, on the one hand, and Hans Rosenberg, on the other.

Edward Pickering had discovered a Wolf-Rayet star, ζ Puppis, in 1896 which had a series of spectral lines that alternated with the Balmer hydrogen lines. Their cause was unknown, but in 1913 Niels Bohr was able to show that they were due to ionised helium. Then in 1920 Megh Nad Saha derived his ionisation formula giving the rate of ionisation of any atom as a function of temperature and pressure. In the following year he applied this to stellar spectra using a technique of marginal appearances of prominent spectral lines. This showed that the differences in spectra between the various Harvard classes of stars (O, B, A, F, etc.) was generally not due to differences in the composition of their atmospheres, as previously assumed, but to temperature which affected their ionisation states. On this basis he estimated that the temperature of O*a* to O*c*-type stars were from 24,000 to 20,000 K, and of M*b* to M*c*-type stars were from 4,500 to 4,000 K.

Ralph Fowler and E. A. Milne developed Saha's theory and used maximum strengths, instead of Saha's marginal appearances, to obtain, in 1923, a temperature sequence from B2-type stars at 16,100 K to K5-types at 3,900 to 4400 K (depending on which lines were used). In the following year they extended the scale to hotter stars to give B2 at 16,500 K, B1 at 19,000 K, B0-O9 at 26,500 K, and O5 at >35,000 K. In the same year Cecilia Payne deduced a similar temperature scale from B5 at 15,000 K to O8 at 30,000 K by measuring the intensities of ionized silicon lines.

There was still a problem with the hotter stars as most of their radiation was in the ultraviolet, which is absorbed by the Earth's atmosphere. O-type stars were often found to be surrounded by either a planetary nebula or to be embedded within a nebula. Herman Zanstra and Ira Bowen found, independently, in the late 1920s that they could estimate the temperature of these O-type stars by the effect that they had on their surrounding nebulae. This led to estimates ranging from 30,000 K for the trapezium stars of the Orion nebula to more than 100,000 K for a few stars.

At the other extreme, the temperature of many red M stars and deep-red N (carbon) stars could not be deduced in the 1920s from their colours, as much of their energy was in the infrared. Their temperatures, which were found to range from about 1,000 to 3,000 K, were deduced by measuring the intensities of their titanium oxide bands or by radiometer measurements. Then in 1938 Gerard Kuiper estimated the effective temperatures of M6 red giants, for example, as 2,750 K, by reanalysing published photometric data, whilst thirty years later Hollis Johnson estimated their temperatures as 2,800 K using visual and infrared photometry. Later estimates, following the availability of more accurate stellar diameters, have tended to be a few hundred of degrees higher for these cool stars. For example in 1980 Stephen Ridgway and colleagues produced an estimate of 3,250 for M6 stars using infrared photometry, whilst in 1998 Guy Perrin and colleagues produced a figure of $3,243 \pm 79$ K based on bolometric measurements.

Bibliography

Hearnshaw, J.B., *The Analysis of Starlight: One Hundred and Fifty Years of Astronomical Spectroscopy*, Cambridge University Press, 1986.

Hearnshaw, J.B., *The Measurement of Starlight: Two Centuries of Astronomical Photometry*, Cambridge University Press, 1996.

2 · Stars considered as a group

HERTZSPRUNG-RUSSELL DIAGRAM

Ejnar Hertzsprung discovered over the period 1905–1907 that Antonia Maury's *c*-type stars (stars with very sharp spectral lines) were very high intensity, distant stars whose luminosity did not seem to change with spectral type. On the other hand, the luminosity of the late-type (i.e., yellow to red) normal stars was very much lower than for the *c*-types of the same colour, indicating that the late-type stars were in two different populations of radically different luminosities. In addition, he found that the luminosity of the low-luminosity stars seemed to reduce with increasing redness from blue to red.

Hertzsprung realised that if the late-type stars of the same colour, and therefore of the same temperature, have different luminosities, they must have different surface areas. So the two different populations of high- and low-luminosity late-type stars corresponded to large and small stars, or *giants* and *dwarfs*. In 1906 Hertzsprung used the newly discovered radiation laws to estimate the diameter of Arcturus, and found it to be about the same as that of the orbit of Mars, confirming that giant stars really did exist. Five years later, Hertzsprung published his colour-magnitude diagrams for stars in the Pleiades and Hyades star clusters using colour indices, deduced by comparing photographic and visual magnitudes of stars, as measurements of colour.

Hertzsprung published his initial results in an obscure magazine, so Henry Norris Russell was in ignorance of Hertzsprung's work when he proposed in 1910 that there were two different populations of G-, K-, and M-type stars, one population being of large bright stars, the other of small dim ones. Four years later Russell published his first 'Russell' colour-magnitude diagram (see Figure 3.1), as it became known. In it he used the Harvard spectral type, in the order defined by Annie Cannon, as the index of colour.

The first diagram produced by Russell showed the main sequence or 'dwarf' branch of stars running from bluish-white B-type stars with an absolute magnitude of about −2, to red M stars with an absolute magnitude of about +10. In addition, there was a giant branch of late-type stars with absolute magnitudes of ~0, the most luminous of these being the *c*-type stars listed by Antonia Maury.

Russell's diagram was much easier to interpret than that produced by Hertzsprung, but Russell was quite happy to acknowl-

edge Hertzsprung's prior analysis. In 1933 Bengt Strömgren referred to the 'Hertzsprung-Russell diagram' when discussing Russell's results, and that (or the H-R diagram) is what it has been called ever since. Hertzsprung, when asked for his view of the late recognition of his work, is reputed to have said, 'Why not call it the colour-magnitude diagram? Then we would all know what it is about.'

Michelson, Pease and Anderson succeeded in measuring stellar diameters directly for the first time in 1920. They used a Michelson interferometer on the newly-completed 100 inch (2.5 m) telescope on Mount Wilson. Betelgeuse was found to have a diameter of 0.047″, Antares 0.040″ and Arcturus 0.024″, equivalent to diameters of about 400, 600 and 30 million kilometres, respectively. This was the first direct evidence for, and was a crucial confirmation of, the existence of giant stars that had been deduced earlier by Hertzsprung and Russell.

The Hertzsprung-Russell diagram is still in use today for plotting the spectral type and luminosity of individual stars, and to indicate their evolutionary paths. It has been extended to include stars of higher and lower luminosities than in the original diagram, but its fundamental format has not changed.

Bibliography

DeVorkin, David, *Stellar Evolution and the Origin of the Hertzsprung-Russell Diagram*, in Gingerich, Owen (ed.), *Astrophysics and Twentieth-Century Astronomy to 1950: Part A*, Cambridge University Press, 1984.

Hearnshaw, J. B., *The Analysis of Starlight; One Hundred and Fifty Years of Astronomical Spectroscopy*, Cambridge University Press, 1986.

Herrmann, Dieter B. (Krisciunas, K., trans.), *The History of Astronomy from Herschel to Hertzsprung*, Cambridge University Press, 1984.

Leverington, David, *A History of Astronomy from 1890 to the Present*, Springer-Verlag, 1996.

Nielsen, Axel V., *Contributions to the History of the Hertzsprung-Russell Diagram*, Centaurus, 9 (June 1964), pp. 219–253.

INTERNAL STRUCTURE OF STARS

It was generally assumed for most of the nineteenth century that the primary heat transport mechanism in the Sun and

Harvard spectral type

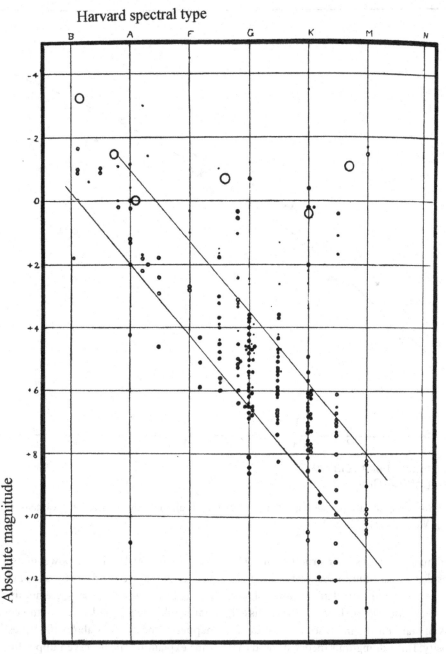

Figure 3.1. The first H-R diagram produced by Russell. The main sequence is indicated by the two inclined parallel lines. (From *Nature*, Vol. 93 (1914), p. 252.)

stars was convection. But in 1894 Ralph Sampson suggested that radiative transfer was more important in the solar atmosphere. Karl Schwarzschild developed a model of the solar atmosphere, assuming a predominance of radiative transfer, and explained the limb-darkening effect in 1906. Ten years later Arthur Eddington extended Schwarzschild's theory to the interior of stars, to try to determine how their maximum internal temperature depended on mass. Eddington's initial work was on low-density giant stars, which he assumed were sufficiently diffuse to allow the use of the perfect gas laws, and for these he calculated a central temperature of 7 million K.

In 1923 Eddington calculated the luminosity of giant stars using Kramers' opacity theory of the same year. He showed, using the perfect gas laws, that the luminosity of a star was determined almost completely by its mass (see curve in Figure 3.2). He was surprised to find, however, when comparing his theoretical predictions with real stars (see points in Figure 3.2), that main sequence stars fitted on exactly the same mass–luminosity curve as the giant stars. This indicated that gas in main sequence stars also behaved as a perfect gas, even though it was highly compressed, because it was highly ionised.

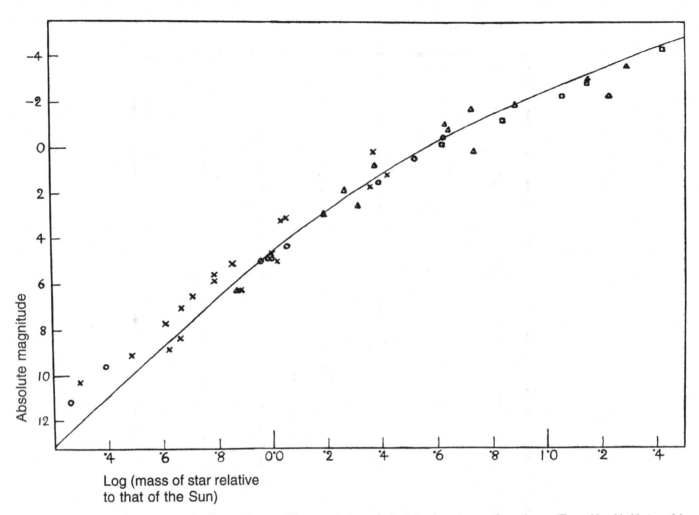

Figure 3.2. Eddington's mass-luminosity diagram for stars. The curve is theoretical, whilst the points are for real stars. (From *Monthly Notices of the Royal Astronomical Society*, Vol. 84 (1924), Plate 8.)

Eddington outlined his analysis of stellar structures in 1924, together with the resulting mass-luminosity correlation for giant and main sequence stars. This correlation was consistent with the effect found observationally by Henry Norris Russell, that the average mass of stars on the main sequence reduced with reducing luminosity. But Eddington's theory showed that the intensities of all but one of the stars that he had considered were all about ten times too high. He put this discrepancy down at the time to a fault in Kramers' opacity theory, and he decided to work backwards and calculate the value of the opacity required to make the calculated and observed luminosities of the star Capella the same. He then applied this correction to his calculated luminosities for all the other stars, and found a perfect fit between theory and observation for all but one star, Sirius B, which was later found to be a white dwarf.

Eddington pointed out that, whilst stars differ enormously in brightness and density, they differ much less in mass. In the mid-1920s, when Eddington was developing his theories, the range of masses known was about 1/5 to 100 times that of the Sun. He reasoned that, as the radiation pressure inside a star

was proportional to the 4th power of its temperature, there must be a temperature inside a star at which the radiation pressure exceeded the gravitational force holding the star together. The star would then explode. Eddington concluded that this would happen if the star was about 100 times $M_\odot$ (mass of the Sun), thus explaining the observed upper limit on stellar masses.

After considerable hesitation, Russell finally accepted in 1929 that hydrogen was by far the most common element in the Sun's and stellar atmospheres. But Eddington's theory of stellar structures had been based on the assumption that the mix of elements in stars was similar to that of the Earth. When Eddington included the high proportion of hydrogen in his model, he found that the calculated and observed luminosities agreed, showing that Kramers' opacity theory was correct. So hydrogen was the most abundant element within stars, as well as in their atmospheres.

There was considerable debate amongst astrophysicists in the 1920s and 1930s about Eddington's theoretical approach as it had required too many assumptions and approximations. For example, he had assumed that energy was transported to the star's surface by radiation, but a number of astrophysicists thought that convection played a key role. Then in 1935 Thomas Cowling constructed the first theoretical model in which energy was released in a convective core, which was surrounded by a radiative envelope.

Ernst Öpik analysed two different models of a main sequence star in 1938. In one of these, the star had a hydrogen-burning core in convective equilibrium surrounded by an inert hydrogen envelope in radiative equilibrium. Provided there was no exchange of material between the core and the envelope, the core would continue to burn hydrogen to helium until its supply of hydrogen was completely exhausted. Hydrogen burning would then continue in a shell surrounding the inert, isothermal core, whilst the core would begin to collapse, releasing gravitational potential energy. The nucleus would become very dense and very hot, and the rapid release of energy would lead to an expansion and cooling of the envelope turning the star into a red giant.

Sources of energy in stars were discovered in 1938 by Charles Critchfield who described the proton-proton (p-p) cycle, and Hans Bethe and Carl von Weizsäcker who independently discovered the carbon cycle. The p-p cycle occurred in lighter stars and the carbon cycle in heavier ones.

It gradually became clear that the p-p cycle was weakly dependent on temperature, indicating that there was no convection in the core of light stars. But a completely radiative model produced too high a temperature. Then in 1953 Donald Osterbrock solved the problem assuming a radiative core and convective envelope for low mass stars. Today it is thought that main sequence stars with masses below about 1.5 $M_\odot$ have a radiative core and convective outer envelope, whereas heavier stars have a convective core and radiative envelope.

Bibliography

DeVorkin, David, *Stellar Evolution and the Origin of the Hertzsprung-Russell Diagram*, in Gingerich, Owen (ed.), *Astrophysics and Twentieth-Century Astronomy to 1950; Part A*, Cambridge University Press, 1984.

Eddington, Arthur S., *The Internal Constitution of the Stars*, Cambridge University Press, 1926.

Struve, Otto, and Zebergs, Velta, *Astronomy of the 20th Century*, Macmillan, 1962.

Tassoul, Jean-Louis, and Tassoul, Monique, *A Concise History of Solar and Stellar Physics*, Princeton University Press, 2004.

See also: Sources of stellar energy.

SOURCES OF STELLAR ENERGY

Julius Mayer and John Waterston independently proposed in the middle of the nineteenth century that the Sun was generating heat by the infall of meteoric material, but the amount of material required seemed much too great. A little later Hermann von Helmholtz proposed that the Sun's energy was produced by its gravitational contraction. It would have to contract by only about 75 m per year, but this would only have kept the Sun producing energy for a few tens of millions of years. Unfortunately geological evidence at that time indicated that the Earth had existed for appreciably longer than that. This contradiction was still unresolved at the end of the nineteenth century.

Albert Einstein gave the first clue of an alternative source of energy in 1905 when he published his special theory of relativity. In this he proposed the equivalence of mass and energy according to the relationship $E = mc^2$. It was immediately realised that, if the whole mass of the Sun or star could be transformed into energy, it would be sufficient to explain lifetimes of many billions of years. But the method of transforming the complete mass into energy was unknown.

In 1926 Arthur Eddington outlined an alternative to the theory of the complete transformation of matter in stars when he suggested that four hydrogen nuclei (i.e., protons) and two electrons could combine to form a helium nucleus. He calculated that the energy released in such a process, in which the star loses about 1% of its mass, would be enough to keep a star shining for a few tens of billions of years. But it soon became clear that the simple amalgamation of four protons and two electrons was not physically possible.

Factors effecting nuclear processes gradually became clearer over the next ten years or so. Then in 1938 sources of energy production in stars were outlined by Charles Critchfield who described the proton-proton cycle, and Hans Bethe and Carl von Weizsäcker who independently discovered the carbon cycle. The p-p cycle occurred in lighter stars and the carbon cycle in heavier ones. Initially it was thought that the Sun's energy was produced by the carbon cycle, but it was eventually concluded that the Sun used the p-p cycle, with only stars of greater than about 1.5 solar masses using the carbon cycle.

The two chain reactions were:

Proton-proton cycle

$$^1H + {}^1H \rightarrow {}^2H + e^+ + v$$
$$^2H + {}^1H \rightarrow {}^3He + \gamma$$
$$^3He + {}^3He \rightarrow {}^4He + {}^1H + {}^1H$$

where v is a neutrino and γ a gamma ray.

Carbon-nitrogen-oxygen (CNO) or carbon cycle

$$^{12}C + {}^1H \rightarrow {}^{13}N + \gamma$$
$$^{13}N \rightarrow {}^{13}C + e^+ + v$$
$$^{13}C + {}^1H \rightarrow {}^{14}N + \gamma$$
$$^{14}N + {}^1H \rightarrow {}^{15}O + \gamma$$

$$^{15}O \rightarrow {}^{15}N + e^+ + \upsilon$$
$$^{15}N + {}^1H \rightarrow {}^4He + {}^{12}C$$

In both cycles four hydrogen nuclei have been transformed into helium, with energy being released mainly in the form of γ-rays. In the second case, carbon acted as a catalyst. These processes could explain the production of helium from hydrogen, but it was unclear how the heavier elements were formed. Carl von Weizsäcker suggested that they could have been made in the Big Bang.

In 1948 Ralph Alpher and George Gamow suggested that the early universe, just after the Big Bang, consisted of a sea of neutrons, some of which decayed to produce protons and electrons. Successive capture of neutrons then produced the heavier elements. Unfortunately, it was soon realised that the chain broke down at helium, as the nucleus produced after the capture of a neutron by a helium nucleus was unstable.

Other astrophysicists wondered if heavy elements were produced in stars. Although there had been an enormous amount of circumstantial evidence of this, the first direct proof was furnished in 1952 by Paul Merrill's discovery of the relatively heavy element technetium in red giants. Technetium has a maximum half-life of a million years, so it could not have been present in red giants when they were formed billions of years earlier. It must have been formed more recently in these stars.

Ernst Öpik and Edwin Salpeter independently suggested that the heavy elements were formed in the centre of hot, very heavy stars, by the fusion of three helium nuclei to form carbon, when the internal temperature reached a few hundred million K. The helium burning would take place as follows in the so-called triple-α process, so-called because an α particle is a helium nucleus:

$$^4He + {}^4He \rightarrow {}^8Be + \gamma$$
$$^8Be + {}^4He \rightarrow {}^{12}C + \gamma$$

Unfortunately, the cross section for the beryllium-to-carbon process seemed too small to create a significant amount of carbon. But in 1954 Fred Hoyle suggested that this cross section would be increased if there was a resonance associated with the formation of carbon 12 in an excited state. This was found experimentally almost immediately at the energy predicted by Hoyle. He then found that helium burning would take place at a temperature of about 100 million K, which was the expected temperature at the centre of red giants at the tip of the giant branch.

Öpik and Salpeter had also suggested elements heavier than carbon could be created by the successive addition of alpha particles, as follows:

$$^{12}C + {}^4He \rightarrow {}^{16}O + \gamma$$
$$^{16}O + {}^4He \rightarrow {}^{20}Ne + \gamma$$

and so on.

In 1954 Hoyle suggested that once helium had been exhausted in the core of a star, very heavy stars would continue to contract, the central temperature would continue to increase, and heavier and heavier elements would be produced. Carbon burning would produce magnesium, and so on until the core consisted of iron, the element with the greatest nuclear binding energy, in the heaviest stars.

The classic B^{2}FH paper, on the production of elements in stars, was published in 1957 by Geoffrey and Margaret Burbidge, Willy Fowler and Fred Hoyle. In it they explained that elements could be produced in stars by a number of processes which, in addition to fusion, included:

- The s process, in which neutrons are added *slowly* to nuclei, which have time to decay before the next neutron is added.
- The r process, in which neutrons are added *rapidly* to nuclei, so that the nuclei do not have time to decay before another neutron is added.
- The p process, in which *protons* are added to nuclei in Type II supernovae.

The role of supernovae in producing elements by explosive nucleosynthesis was also discussed by Alastair Cameron in the same year. Then in 1970 computer simulations of these explosive processes were produced by David Arnett and Donald Clayton, explaining many element abundances.

Modern theories of element production in stars are based on those proposed in the B^{2}FH paper. The s, r and p processes are crucial as they are the only way in which elements heavier than iron can be produced. The r process can occur only in supernovae where enormous numbers of neutrons are available.

Bibliography

Burbidge, E. M., Burbidge, G. R., Fowler, W. A., and Hoyle, F., *Synthesis of the Elements in Stars*, Reviews of Modern Physics, **29**, No. 4 (October 1957), pp. 547–650.

Gamow, George, *The Birth and Death of the Sun*, Macmillan, 1940.

Longair, Malcolm S., *The Cosmic Century: A History of Astrophysics and Cosmology*, Cambridge University Press, 2006.

Tassoul, Jean-Louis, and Tassoul, Monique, *A Concise History of Solar and Stellar Physics*, Princeton University Press, 2004.

SPECTRAL CLASSIFICATION OF STARS

Joseph Fraunhofer noticed in the early nineteenth century that Sirius, Betelgeuse and a few other stars had a different pattern of dark lines crossing their spectra from those of the Sun. In 1859 Gustav Kirchhoff and Robert Bunsen found that bright spectral lines emitted by a hot gas were unique to each element. Kirchhoff then found that when light with a continuous spectrum was passed through a cool gaseous layer (like the outer layers of the Sun), this layer absorbed

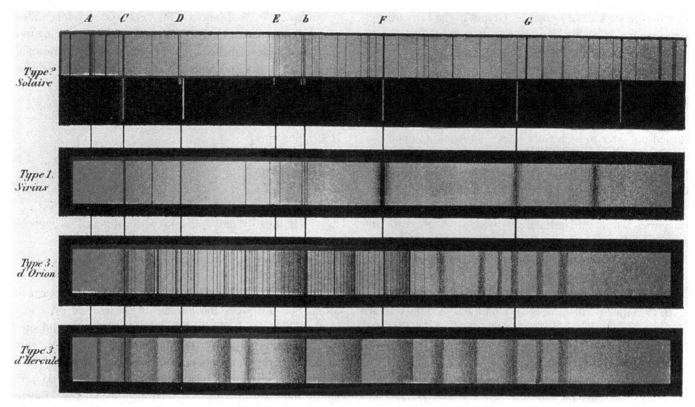

Figure 3.3. The spectra of typical stars in three Secchi categories. From top to bottom: Type 2, the Sun; Type 1, Sirius; Type 3, Betelgeuse; and Type 3, α Herculis. (From *Les Étoiles*, by Angelo Secchi, Vol. 1, 1879, Plate VII.)

its characteristic lines from the continuous spectrum thus creating the dark absorption lines like the Fraunhofer lines of the Sun.

The first Fraunhofer lines to be identified in the solar spectrum were those of sodium and iron. Then in 1864 William Huggins and William Miller identified hydrogen, sodium, iron, magnesium and calcium in stellar spectra. This showed that stellar atmospheres were composed of the same elements as those found on the Sun and Earth.

Angelo Secchi produced a system of stellar classification in 1863, which he developed over the next five years (see Figure 3.3). In 1868 he had four classes of stars, with some stars unclassified. His first three categories were characterised by different relative intensities of hydrogen and metallic lines, with the fourth category consisting of red stars with prominent carbon lines. In 1877 Secchi introduced a fifth class of emission line stars. Hermann Vogel produced his own independent classification system in 1874, which he revised in 1895 following the discovery of helium. Meanwhile in 1890 Edward Pickering and Williamina Fleming had introduced a more comprehensive system, the precursor of what is known today as the Harvard system, in the *Draper Memorial Catalogue*, by modifying and extending Secchi's classification. This catalogue contained the spectral classification of 10,351 stars down to 8th magnitude, nearly all north of −25 degrees declination. They were categorised according to the following spectral types:

A	White stars with strong hydrogen absorption lines
B	Bluish-white stars with hydrogen and helium absorption lines
C	Stars with double lines
D	White or blue stars with emission (i.e., bright) lines
E	Yellow stars with certain hydrogen absorption lines
F & G	As E with weaker hydrogen and stronger metallic lines
H	As F with weak lines at blue wavelengths
I	As H but with additional lines
K	Orange stars with emission lines
L	Stars with peculiar lines
M	Orange-red stars with virtually no hydrogen lines and a complex spectrum (This was Secchi type III)
N	Red stars with prominent carbon lines (Secchi type IV)
O	Wolf-Rayet stars (faint stars with broad emission lines)
P	Planetary nebulae
Q	Miscellaneous

Antonia Maury published her analysis of the spectra of 681 bright northern stars in 1897 in which she introduced a new parameter, whose value was shown by the letters *a*, *b* and *c*, based on the sharpness of the spectral lines. B-type stars associated with the Orion and Pleiades nebulae had relatively simple spectra and were thought by Antonia Maury to be in an early stage of development. So she concluded that B-type stars should precede A-types in the Harvard stellar sequence, implying that she saw that as an evolutionary sequence.

Annie Cannon completed the classification of 1,122 southern stars in 1901. Following a suggestion by Antonia Maury, she placed the O-type stars before the B-types because of the greater intensities of their "additional hydrogen lines" (which turned out later to be helium lines). So Annie Cannon presented the stars in the sequence O, B, A, F, G, K and M, in that order, plus P and Q, as she saw the Harvard sequence from O to M, or possibly from M to O, as an evolutionary sequence. She also found that she was able to sub-divide each of the categories B to K into decimal subdivisions. A few years later Pickering discovered a new category of stars, which had similar absorption bands to those of the red N-type carbon stars, but with far more blue light. Pickering designated these as R-type stars.

In 1913 the International Solar Union accepted the Harvard classification system as a provisional standard, and in 1922 the newly formed International Astronomical Union (IAU) accepted the Harvard classification as the official system.

In the meantime, Annie Cannon started classifying the spectra of 225,300 stars for the new *Henry Draper (HD) Catalogue*. It was published between 1918 and 1924, introducing yet another new category of stars, S-types, with a very complicated absorption and emission line spectrum between 4500 and 4700Å. By this time the Harvard sequence, which was later recognised as basically a temperature sequence from hot to cool, consisted of O, B, A, F, G, K and M. In addition, P (for nebulae), plus S, R, N and Pec (for peculiar), were also used.

Table 3.1 provides modern definitions for the Harvard types. Typical temperature ranges for each type are also listed.

The MK or Yerkes classification system

When the Harvard stellar classification system was first introduced in 1890 it was attacked as being too complicated, but, by the 1930s, developments in spectroscopic and stellar research had overtaken it, and astronomers were beginning to find it too simple.

Walter Adams and Arnold Kohlschütter had showed in 1914 that the absolute luminosities of F-, G- and K-type stars could be determined from the intensity of some of their spectral lines. Then in 1929 Gwyn Williams showed that the strength of the hydrogen absorption lines in B0 to A5 stars was closely related to stellar luminosity. The most luminous stars were found to have relatively weak but sharp hydrogen lines in their spectra, whilst main sequence stars had wide and intense hydrogen lines. So it gradually became clear that, although the Harvard spectral classification system was extremely useful, it would be even more useful if it included another parameter based on the intensity and structure of individual spectral lines.

As a result, W. W. Morgan of the Yerkes Observatory added another dimension to the Harvard system by introducing five luminosity classes, labelled I to V in 1938, to cover the range of luminosities from supergiants to main sequence stars as follows:

Ia Most luminous supergiants
Ib Less luminous supergiants
 II Bright giants
III Normal giants
IV Subgiants
 V Main sequence or dwarf stars

Class I was, initially, only split into two for supergiants between F4 and F8.

In 1943 Morgan, Philip Keenan and Edith Kellman published their *Atlas of Stellar Spectra* describing what became known as the MKK system. It gave examples of the spectra of various Harvard types and luminosity classes, and presented the classification criteria used for each spectral type and subtype. The criteria used for the luminosity classification were line intensity ratios, mainly of neutral to ionised atoms.

The list of standard stars and the classification criteria were slightly modified in a paper by Harold Johnson and Morgan of 1953. Since then the system has been known as the MK or Yerkes classification system.

Bibliography

Hearnshaw, J. B., *The Analysis of Starlight; One Hundred and Fifty Years of Astronomical Spectroscopy*, Cambridge University Press, 1986.

Hearnshaw, J. B., *The Analysis of Starlight: Some Comments on the Development of Astronomical Spectroscopy 1815—1965*, Vistas in Astronomy, **30** (1988), 319—375.

Jaschek, Carlos, and Jaschek, Mercedes, *The Classification of Stars*, Cambridge University Press, 1987.

Kaler, James B., *Stars and their Spectra; An Introduction to the Spectral Sequence*, Cambridge University Press, 1997.

See also: Spectroscopy.

STAR CATALOGUES

The first star catalogue of 1,464 stars appears to have been produced in China between about 370 and 270 BC by Shi Shen, Gan De and Wu Xian. Timocharis of Alexandria and Aristillus produced the first star catalogue in the West in about 280 BC. Then some 150 years later Hipparchus produced his catalogue of about 850 stars. Hipparchus found that the longitude of the stars had changed over time by comparing his stellar positions with those of Timocharis. This enabled him to quantify the rate of precession of the equinoxes.

Table 3.1. *Stellar spectral classification*

(A) Harvard stellar types (updated)

W	Wolf-Rayet stars, which are extremely hot and luminous stars with broad emission lines on a continuous background. They have mostly helium in their atmospheres instead of hydrogen. There are three subclasses: WC has carbon emission lines, WN, nitrogen emission lines, and WO oxygen emission lines. It is thought that Wolf-Rayet stars are the helium-burning cores of stars whose initial mass was in excess of 30 solar masses, and which have lost their hydrogen-rich outer layers due to mass outflow in a very strong stellar wind.
O	Extremely hot and luminous blue stars with most of their energy output in the ultraviolet. The lines of ionized helium decline towards later types (i.e., going from O4 to O9), whereas neutral helium lines and hydrogen lines strengthen towards later types. At O6 He II λ 4686 is in emission in supergiants (i.e., Yerkes type I) and absorption in dwarfs (type V). The relative strengths of selected lines are used to determine Harvard categories O2 to O9.5, and Yerkes categories I to V. Various ionized metallic lines are visible including C IV, C III, Si IV, Si III, O V and N IV.
B	Blue stars not quite as hot as O-types. Their spectra have neutral helium but virtually no ionized helium lines. The neutral helium lines are strongest in B2-type stars, whilst hydrogen lines strengthen towards later types. Ionized metal lines include those of Si II, Mg II, and C II. B-type stars tend to cluster in OB associations associated with giant molecular clouds.
A	Bluish-white or white stars. Their spectra have strong hydrogen lines (strongest in early A-types), and lines of ionised metals, but no neutral helium lines.
F	White or yellowish-white stars. Spectra show both neutral and ionized metal lines with weaker hydrogen lines. The neutral metal lines strengthen towards later F-types whilst the hydrogen lines weaken.
G	Yellowish-white or yellow stars. The Sun is in this category. Neutral metal lines stronger and hydrogen lines weaker than in F-types. CH and CN molecular bands easily visible in later G-types.
K	Orangish stars. Prominent lines of neutral metals and very weak or non-existent hydrogen lines. CH and CN molecular bands visible, and titanium oxide bands visible from K5 onwards.
M	Red stars. Spectra show molecular and neutral metal lines. Titanium oxide bands strengthen with later types, with vanadium oxide bands appearing in later M-type spectra. Eighty percent of nearby main sequence stars are M-types. The vast majority of these M-types are dwarfs, although there are also a few M-type giants and supergiants.
L	Dwarfs. Very dark red in colour and brightest in the infrared. Some of these L-types are massive enough to support hydrogen fusion, but some are not. Metal hydride bands and alkali metal lines are prominent in their spectra. The lower-mass L dwarfs with lithium in their spectra are called brown dwarfs.
T	Cool brown dwarfs. Emission peaks in the infrared. Spectra shows prominent methane absorption bands.
S	S-types are intermediate between M-type and C-type stars. These are nearly all giants or supergiants with lines of zirconium oxide in their spectra. In late S-type stars the titanium oxide lines, which are present in early S-types, are missing. Borderline cases between M and S-type stars are called **MS** stars, and those between S and C-types are called **SC** stars.
C	Carbon stars which were originally classified as **R** and N stars. They are mostly red giants near the ends of their lives, which have an excess of carbon in their atmosphere. **C-R, C-N, C-J, C-H, and C-Hr** are subsets of C stars.
D	White dwarfs. They are divided into **DO, DB, DA, DQ, DZ, DC, DX, DAB, DAO, DAZ, DBZ**-types depending on their spectra. Pulsating white dwarfs are known as **DAV** (variable white dwarfs with a hydrogen envelope), **DBV** (variable white dwarfs with a helium envelope) or **DOV**-types (variable pre-white dwarfs).

Table 3.1 *(cont.)*

(B) Harvard types (updated) – effective temperatures

Modified Harvard type	Approximate temperature range (K)
W	~35,000–85,000
O	25,000–50,000
B	10,000–25,000
A	7,400–10,000
F	6,000–7,400
G	5,000–6,000
K	3,500–5,000
M	2,200–3,500
L	1,400–2,200
T	600–1,400

Hipparchus' catalogue appears to have been used by Ptolemy as the basis of his star catalogue in his *Almagest* of about 150 AD. It contained both the position and brightness or magnitude of 1,022 stars, with the magnitude being specified on a 1 to 6 scale. This system, which is the basis of the one in use today, is thought to have originated with Hipparchus. The Ptolemaic catalogue was in use in both the Western and the Arab worlds for over a thousand years. Then in Samarkand Ulugh Beg's astronomers produced a new catalogue of about 1,000 stars which they had observed between 1420 and 1437. This was the first Arab star catalogue that had been produced using their own observations, rather than simply updating those of Ptolemy. But it was not available in the West until 1665, by which time it had been superseded by European catalogues.

Tycho Brahe's star catalogue was published in 1602. Even though it also relied on naked eye observations it was much more accurate than Ptolemy's and was in regular use for over a century. Edmond Halley published a supplement in 1679 covering the southern stars visible from St Helena. Then in 1725 John Flamsteed's catalogue of about 3,000 stars was published. It was the first comprehensive catalogue to be published using telescopic observations, so producing unprecedented stellar position accuracies.

James Bradley undertook an extensive series of stellar observations between 1750 and his death in 1762 which were progressively published between 1798 and 1805. F. W. Bessel then reduced these measurements and published his catalogue in 1818 based on Bradley's measurements. Friedrich Argelander published his *Uranometria Nova* in 1843 containing the magnitudes of over 3,000 stars down to a magnitude of 6. In 1879 Benjamin Gould produced a similar atlas and catalogue, the *Uranometria Argentina*, based on his observations from Argentina of more southerly stars. He added a

seventh magnitude and specified magnitudes to one place of decimals.

So far brightness estimates had been made purely visually. But in 1836 John Herschel made a visual photometer, his *astrometer*, to assist him in measuring the brightness of the brightest stars observable from the Cape of Good Hope. Two years later he used this instrument to produce a brightness catalogue listing 191 of the brightest stars in both hemispheres. Later in the century Gustav Müller and Paul Kempf used a Zöllner photometer to measure the intensity of stars down to magnitude 7.5 visible from Potsdam. The resulting luminosity catalogue of over 14,000 stars was published progressively as the *Potsdamer Durchmusterung* between 1894 to 1907.

Meanwhile more specialised catalogues were being produced. Argelander's catalogue of proper motions of 390 stars was published in 1830, followed in 1888 by Arthur Auwers and his Auwers-Bradley catalogue of 3,200 proper motions. Then in 1892 Hermann Vogel and Julius Scheiner published a catalogue of 51 radial velocities. Meanwhile Angelo Secchi had produced a catalogue of 316 stellar spectra in 1867. Then in 1890 Edward Pickering and Williamina Fleming produced the *Draper Memorial Catalogue* containing the spectral classification of 10,351 stars down to −25° declination. This introduced the Harvard spectral classification system still in use today. By 1924 the *Henry Draper Catalogue* produced by Annie Jump Cannon and Pickering included the spectral classification of 225,300 stars. Finally, in 1911 Henry Norris Russell had published a catalogue of 52 stellar parallaxes, followed in 1924 by Frank Schlesinger who published his *General Catalogue of Stellar Parallaxes* containing 1,850 trigonometric parallaxes.

The above catalogues give the properties of stars, but there were also catalogues on various types of stars. For example, Johann Bode listed 80 double stars in 1781, followed by William Herschel who produced a catalogue of 848 double stars which was published progressively from 1782 to 1821. In 1786 Edward Pigott had listed twelve known variable stars, plus 39 possibles, whilst in 1850 Argelander listed 24 known variables.

Early catalogues concentrated on recording the position, or proper motion, or parallaxes, or magnitude of stars. But in recent years a few comprehensive catalogues have been produced that include measurements of most or all of these parameters. Which catalogue is used depends, amongst other things, on the sky coverage, intensity coverage, and positional accuracy (both relative and absolute). For example, the Hipparcos catalogue, which was produced using the European Space Agency's Hipparcos spacecraft, listed the positions, proper motions, parallaxes and photometry of about 120,000 stars to an unprecedented accuracy. Position accuracies were to 1 to 3 milliarcseconds (mas), for example. The Tycho-2 catalogue, which also used the Hipparcos spacecraft, extended this database to include about 2.5 million lower-intensity stars to a lower accuracy. Two other similar comprehensive catalogues have also recently been produced, namely the U.S. Naval Observatory (USNO) CCD Astrograph Catalog, 2nd release (or UCAC2) and the USNO B1.0 Catalog. The UCAC2 covered about 86% of the sky and included 48 million stars in the middle magnitude range, with positional accuracies of 20–70 mas, whereas the USNO B1.0 Catalog covered about 1.0 billion stars with positional accuracies of about 200 mas.

Table 3.2 lists a selection of catalogues of historical interest not mentioned above. In addition, there have been numerous catalogues limited to stars selected by location (e.g., polar stars), intensity (e.g., bright star catalogues), waveband (e.g., infrared stars), star types (e.g., white dwarfs), etc., which are beyond the scope of this article.

Bibliography

Debarbat, S., et al. (eds.), *Mapping the Sky: Past Heritage and Future Directions*, Springer-Verlag, 1988.

Eichhorn, Heinrich, *Astronomy of Star Positions: A Critical Investigation of Star Catalogues, the Method of Their Construction, and their Purpose*, Ungar, 1974.

Hearnshaw, J. B., *The Analysis of Starlight: One Hundred and Fifty Years of Astronomical Spectroscopy*, Cambridge University Press, 1986.

Hearnshaw, J. B., *The Measurement of Starlight: Two Centuries of Astronomical Photometry*, Cambridge University Press, 1996.

Hoffleit, D., *The Quest for Stellar Parallax, Popular Astronomy*, **57** (1949), pp. 259–273.

Hoffleit, D., *A History of Variable Star Astronomy to 1900 and Slightly Beyond*, Journal of the American Association of Variable Star Observers, **15** (1986), pp. 77–106.

Hogg, H. S., *Variable Stars*, in Gingerich, O. (ed.), *Astrophysics and Twentieth-Century Astronomy to 1950; Part A*, Cambridge University Press, 1984.

Jaschek, C., *Data in Astronomy*, Cambridge University Press, 1989.

Sterken, C., and Jaschek, C. (eds.), *Light Curves of Variable Stars: A Pictorial Atlas*, Cambridge University Press, 1996.

See also: Spectral classification of stars.

STELLAR EVOLUTION

Astronomers began to develop theories of stellar evolution following the first attempts at classifying stellar spectra in the 1860s. In 1865 Johann Zöllner proposed that a star started its life as a gaseous planetary nebula that contracted and cooled to become a liquid sphere with a solid crust. The sphere then gradually cooled, changing colour from blue to yellow to red. (This was the origin of the terms 'early type' and 'late type' stars used to describe blue and red stars, respectively.) In the following decade Zöllner's theory was developed by Hermann Vogel, who added an early heating phase as the stars evolved from the initial nebula. But he argued that stars moved through this phase so quickly that none could currently be seen in it.

Zöllner's theory was basically conceptual, but Arthur Ritter, in a series of papers published in the late 1870s and early 1880s, developed a more scientific theory based on the behaviour of a perfect gas sphere in convective equilibrium. He proposed that stars began their lives as large cool red stars, which gradually contracted and heated up, behaving like a perfect gas, to become small blue stars. These then gradually cooled and changed colour from blue to yellow to red. So, unlike in Zöllner's theory, there were two types of red star, large young red stars and small old ones. Ritter also proposed that the maximum temperature a star could reach depended on its mass, with the heaviest stars reaching the hottest temperature, as they would contract more than less-massive stars under their own gravity.

Norman Lockyer developed his 'meteoritic hypothesis' for the birth of stars in the late 1880s. In this, meteorites collided with each other to form a nebula which condensed to form a star. Initially, as the star condensed it got hotter and changed colour from red to yellow to blue, but eventually the heat generated by collisions of the meteoritic material was not enough to counterbalance that lost through radiation, and the star began to cool and change colour from blue to yellow to red. Unlike Ritter, who thought that the most massive stars attained the highest temperature during the contraction phase, Lockyer believed that all stars followed the same evolutionary path. Lockyer also suggested that as the stars got hotter their compounds broke down into atoms, which subsequently broke down into 'proto-elements' as they got hotter still. It was not until the twentieth century that Lockyer's proto-elements were found to be ionized atoms.

In 1890 Edward Pickering published the *Draper Memorial Catalogue* containing the spectra of over 10,000 stars. In this catalogue, the alphabetical classification of the early-type stars, for example, was based on the intensity of the Balmer lines of hydrogen, with the intensity decreasing from A-type to B-type stars and so on down the spectral sequence. Then in 1897 Antonia Maury produced her more detailed analysis of the spectra of 681 bright northern stars in which she introduced a new parameter, with values shown by the letters *a*, *b* and *c* based on the sharpness of the spectral lines. She also put the B-type stars in front of the A-types, as they were prevalent in the nebular regions of Orion and the Pleiades and so appeared to be in an earlier stage of development. Antonia Maury put only 18 stars into the *c* category that had the sharpest lines. She commented that she thought that the *c* and non-*c*-type stars were undergoing 'parallel courses of development'.

In 1901 Annie Cannon produced her classification of 1,122 southern stars according to the Harvard alphabetical system. Following a suggestion from Antonia Maury, she put the O-type stars, which had emission lines, ahead of the B-types

Table 3.2. *Star catalogues*

   (A) Selected historically important star catalogues *in addition* to those mentioned in the text

Compiler	Name of catalogue	Type of catalogue	Date issued	Number of stars
Argelander	Bonner Durchmusterung	Stellar positions and luminosities	1859–62	324,188
Schönfeld	Bonner Südliche Durchmusterung	Stellar positions and luminosities	1886	133,659
Gill and Kapteyn	Cape Photographic Durchmusterung	Stellar positions and luminosities	1896–1900	454,875
Thome et al.	Cordoba Durchmusterung	Stellar positions and luminosities	1892–1932	614,000
Pickering	Harvard Photometry Catalogue	Stellar photometry	1884	4,260
Pickering and Bailey	Revised Harvard Photometry Catalogue	Stellar photometry	1908	45,792
Boss	General catalogue	Stellar positions and proper motions	1936	33,342
Haramundanis	Smithsonian Astrophysical Observatory star catalog	Stellar positions and proper motions	1966	258,997
Jenkins	General Catalog of Trigonometric Stellar Parallaxes, 3rd Ed.	Parallaxes	1952	5,822
Van Altena, Lee and Hoffleit	General Catalog of Trigonometric Stellar Parallaxes, 4th Ed.	Parallaxes	1995	8,112
Campbell and Moore	Radial Velocities of Stars Brighter than Visual Magnitude 5.51	Radial velocities	1928	2,771
Moore	A General Catalog of the Radial Velocities of Stars, Nebulae and Clusters	Radial velocities	1931	6,739
Wilson	General Catalog of Stellar Radial Velocities	Radial velocities	1953	15,107
Cannon	Henry Draper (HD) Extension Catalogue	Spectra	1925–36	46,850
Cannon	HD Charts	Spectra	1949	86,932

(B) Historically important double star catalogues *in addition* to those mentioned in the text

Compiler	Name of catalogue	Date issued	Number of stars
Burnham	A General Catalogue of Double Stars within 121° of the North Pole	1906	13,665
Innes	Southern Double Star Catalogue	1927	7,041
Aitken	New General Catalogue of Double Stars within 120° of the North Pole	1932	17,180
Jeffers and van den Bos	Index Catalogue of Visual Double Stars	1963	64,247
Worley and Douglass	Washington Visual Double Star Catalog	1984	73,610
Mason et al.	Washington Double Star Catalog	2001	84,486

(C) Historically important catalogues of variable stars *in addition* to those mentioned in the text

Compiler	Name of catalogue	Date issued	Number of stars
Chandler	Third Catalogue of Variable Stars	1896	393
Pickering and Cannon	Provisional Catalogue of Variable Stars	1903	1,227
Müller and Hartwig	History and Literature of the Light-Changes of Stars Recognised as Variable up to the End of 1915	1918	1,687
Kukarkin and Parenago	General Catalogue of Variable Stars (GCVS), First Edition	1948	10,820
Kholopov et al.	GCVS, Fourth Edition (in three volumes)	1985–88	28,484
Kazarovets et al.	GCVS4 (above three volumes with two extra volumes; includes extragalactic variable stars)	2006	40,215

because she thought, correctly as it turned out, that they were in an earlier stage of development.

A few years later, Ejnar Hertzsprung discovered that Antonia Maury's *c*-type stars were very high-intensity, distant stars, whilst the non-*c*-type stars, particularly of the late types (G, K and M), were much less intense and much nearer. He realised that if stars of the same colour, and therefore of the same temperature, have different intensities, they must have different surface areas. So the high-intensity stars were much larger than the low-intensity stars of the same colour. He called them *giants* and *dwarfs*. Hertzsprung then suggested that stars can evolve along two alternative paths, both of which start with hot, bright, bluish-white stars which gradually become dimmer and redder, one path ending with the red giants and the other with red dwarfs.

In 1910 Henry Norris Russell also proposed, unaware of Hertzsprung's work, that there were two different populations of late-type stars, one of large bright stars, the other of small dim ones. Four years later, Russell published his colour-magnitude diagram (later called the Hertzsprung-Russell or H-R diagram) showing a main sequence, or dwarf branch, running down from highly luminous bluish-white B-type stars to relatively dim red M-types. Above this and to the right there was a giant branch mainly of late-type stars.

Russell initially thought that stars started off life as red giants that contracted and became hotter, according to the perfect gas laws, to become bluish-white B- and A-type stars. The stars then cooled and contracted and followed the main sequence down to eventually end their lives as red dwarfs. But in 1912 he found that the average masses of stars decreased going down the main sequence from blue to red. It seemed unlikely that stars would lose a significant amount of mass as they cooled and progressed down the main sequence. So Russell concluded that, as Ritter had proposed, only the most massive stars on contracting had reached the hottest temperatures. In terms of his colour-magnitude diagram, this meant that only the most massive stars joined the main sequence at the top, with less-massive stars joining it lower down, before all stars proceeded to follow the main sequence down as they contracted and cooled.

Arthur Eddington showed in 1924 that stars on the main sequence still followed the perfect gas laws, even though they

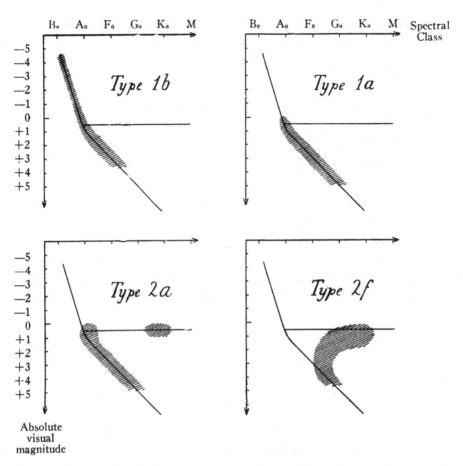

Figure 3.4. Trumpler's H-R diagrams for four stellar clusters. The hatched areas show where the stars are located with reference to an idealised main sequence and horizontal giant branch. (From *Publications of the Astronomical Society of the Pacific*, Vol. 37 (1925), p. 315.)

were highly compressed. So they could not cool down as they contracted and went down the main sequence, as assumed by Russell, unless they lost an appreciable amount of gas on the way, which seemed unlikely. As a result in 1927 Russell finally abandoned his idea that the stars went down the main sequence as they evolved.

Robert Trumpler began a detailed investigation of the distribution of stellar types in open clusters in the early 1920s. He produced separate H-R diagrams for each cluster and noted that, although the results for each cluster were consistent with the shape of the H-R diagram, with a main sequence and a giant branch, the population density of the stars at various places on the main sequence and/or giant branch was significantly different for these open clusters compared with that for stars much closer to the Sun. Some clusters, for example, had virtually no stars cooler than type F5, whereas others had virtually no stars hotter than F0 (see Figure 3.4). All of the open clusters had main sequence stars, but some had no giant stars at all.

In analysing his results, Trumpler assumed in 1925 that all stars in a cluster were born at approximately the same time as each other. At that time it was generally assumed that stars evolved from red giants, at the top right of the H-R diagram,

to meet the main sequence, which they then followed downwards. So Trumpler concluded that the reason for the different population densities of stars in the H-R diagrams for different clusters was not due solely to the different ages of the clusters. Some clusters must also have more massive stars than others.

Cecilia Payne showed in 1925 that hydrogen was the most abundant element in stellar atmospheres. Russell did not agree at the time, but three years later Albrecht Unsöld also showed that hydrogen was the most abundant element on the Sun. This helped to convince Russell that Payne was correct. Accepting the importance of hydrogen in stars, Bengt Strömgren showed in 1932 that a star which had reached the main sequence would move towards the giant branch, in the top right of the diagram, as it gradually used up its hydrogen fuel. His analysis was based on the mean molecular weight of a star, and was not dependent on the nuclear processes inside the star, which were unknown at the time.

In 1936 Gerard Kuiper combined Trumpler's observational results with Strömgren's theoretical analysis, assuming that Trumpler was correct in assuming that all stars in a cluster were about the same age. Kuiper plotted the mean colour-magnitude distributions for each cluster on the same H-R diagram, and noted that the main sequences for each cluster were virtually identical, except for the very early-type and very late-type stars. In addition, the common main sequence was approximately consistent with Strömgren's constant hydrogen curve.

Ernst Öpik analysed two different models of a main sequence star in 1938. In one of these, the star had a hydrogen-burning core in convective equilibrium surrounded by an inert hydrogen envelope in radiative equilibrium. Provided there was no exchange of material between the core and the envelope, the core would continue to burn hydrogen to helium until its supply of hydrogen was completely exhausted. Hydrogen burning would then continue in a shell surrounding the inert, isothermal core, whilst the core would begin to collapse, releasing gravitational potential energy. The nucleus would become very dense and very hot, and the rapid release of energy would lead to an expansion and cooling of the envelope, turning the star into a red giant. The giant phase could last for only a very short time, compared with the main sequence phase, because the red giants had to emit energy at a much greater rate than main sequence stars to account for their very high luminosities.

In 1942 Mario Schönberg and Subrahmanyan Chandrasekhar considered the stability of stars with inert isothermal cores. They found that they could produce no model in which the star was stable if its inert core contained more than about 10% of the mass of the star. This so-called Schönberg-Chandrasekhar limit implied that the lifetime of a star on the main sequence was limited to the time it took to burn 10% of its hydrogen into helium. The star then developed into a red giant, as outlined by Öpik. The subsequent paths followed on the H-R diagram were predicted by Allan Sandage and Martin Schwarzschild in 1952 for globular cluster stars of masses between 1 and 4 $M_\odot$ (mass of the Sun).

Martin Schwarzschild and Richard Härm showed in 1962 that helium began to burn abruptly in the contracting degenerate core of a 1.3 $M_\odot$ red giant to produce carbon. The ignition of this reaction was explosive producing a helium flash, with a temperature of over 300 million K, as the rate of energy production in the core was very much greater than the rate at which energy could be transported outwards. Very little of the helium flash was visible from outside, as the non-degenerate helium part of the core effectively blanketed it. The helium flash stopped the star moving to the upper right of the H-R diagram and caused it to move along the horizontal branch towards the main sequence.

It was clear in the 1950s that, as the heavy, brightest stars stayed on the main sequence for much less time than the dimmer stars, the age of a mature star cluster could be determined by observing which stars in the cluster had left the main sequence, assuming that all the stars in the cluster were formed at approximately the same time as each other. As a result, in 1957 Allan Sandage deduced the ages of various open clusters

from the shape of their H-R diagrams. They ranged from less than 1 *million* years for cluster NGC 2362 to 5 *billion* years for M67.

Turning now to the birth and early stages of a star, in 1947 Bart Bok discovered a number of small dark patches in the Milky Way, which he thought could be the birthplace of stars. He suggested that the material in these Bok globules, as they became known, would gradually contract under their mutual gravitation to heat up and become a star. In 1955 Louis Henyey, R. LeLevier and R. D. Levee calculated the evolutionary tracks for this pre-main sequence phase for stars with masses of between 0.65 and 2.29 $M_\odot$. They found that they reached the main sequence between ~150 million years and 3 million years, respectively.

Very young star clusters were analysed by Pavel Parenago in 1953 who concluded that, although their hottest and most mature stars had reached the main sequence, their less-massive cool stars, which were situated above the main sequence, had not. Merle Walker confirmed this a few years later, and concluded that these less-massive stars were still in their Helmholtz contraction phase. He estimated the ages of these young clusters by observing where the stars had joined the main sequence. This resulted in ages of about 1.8 million years and 3 million years for the open clusters NGC 6611 and NGC 2264, respectively.

It was generally assumed in the 1950s that the pre-main sequence stars had to be in radiative equilibrium. But in 1961 Chushiro Hayashi showed that convection dominated the structure of a proto-star, so that, as the star developed it moved rapidly downwards on the H-R diagram. As the star contracted and heated up, a radiative zone developed near the centre, gradually including more and more mass, causing the star's path on the H-R diagram to turn sharply to the left until it reached the main sequence.

In summary, by the early 1960s the general sequence of events in the lifetime of a star was known. It started life as a nebula which contracted and heated up, and followed the so-called Hayashi Tracks in the H-R diagram to reach the main sequence. After this relatively rapid Helmholtz contraction phase, the star spent something like a thousand times as long on the main sequence, converting hydrogen to helium in its core. When the core of hydrogen had been converted to helium, the star started burning hydrogen in a shell surrounding the core. The star quickly moved to the top right of the H-R diagram where it became a red giant. In this giant phase, helium was converted into heavier elements like carbon and oxygen, but this happened much more rapidly than the production of helium on the main sequence, causing the star to rapidly run out of fuel. It became cooler, so that the pressure of radiation trying to leave the star could no longer counteract the star's gravity, causing the star to contract and move from the top right of the H-R diagram to eventually become a white dwarf in the bottom left. Stars that, at that stage, were heavier than about 1.4 $M_\odot$ would contract further to form neutron stars or black holes, rather than white dwarfs.

The stellar evolution processes outlined in this last paragraph are still basically as we understand them today, although more recent research has led to many changes and additions of detail which are beyond the scope of this article.

The processes involved in the formation of white dwarfs, neutron stars and black holes, and in various unstable or catastrophic phases like those of RR-Lyrae stars, T Tauri stars and supernovae, are considered in individual articles.

Bibliography

DeVorkin, David, *Stellar Evolution and the Origin of the Hertzsprung-Russell Diagram*, in Gingerich, Owen (ed.), *Astrophysics and Twentieth-Century Astronomy to 1950; Part A*, Cambridge University Press, 1984.

Harpaz, Amos, *Stellar Evolution*, A. K. Peters, 1994.

Hearnshaw, J. B., *The Analysis of Starlight; One Hundred and Fifty Years of Astronomical Spectroscopy*, Cambridge University Press, 1986.

Longair, Malcolm, *The Cosmic Century; A History of Astrophysics and Cosmology*, Cambridge University Press, 2006.

Ryan, Sean G., and Norton, Andrew J., *Stellar Evolution and Nucleosynthesis*, Cambridge University Press, 2010.

Struve, Otto, and Zebergs, Velta, *Astronomy of the 20th Century*, Macmillan, 1962.

Tassoul, Jean-Louis, and Tassoul, Monique, *A Concise History of Solar and Stellar Physics*, Princeton University Press, 2004.

See also: Spectral classification of stars.

3 · Types of stars

BINARY STARS

General

The first telescopic discovery of a double star appears to have been made by Giovanni Battista Riccioli in about 1650 when he found that Mizar (ζ Ursae Majoris) had two components. Subsequently many more double stars were discovered. It was unclear for many years as to whether these were simply optical doubles, caused by the accidental alignment of stars at two completely different distances, or whether they were gravitationally connected as a binary pair. In the early eighteenth century it was generally thought that double stars were simply optical doubles. But in 1761 Johann Lambert concluded that there were too many double stars to be solely the result of chance alignments, and six years later John Michell went one step further and proposed that some double stars were physical pairs.

William Herschel tried to measure stellar parallaxes in 1779 using a method first proposed by Galileo. This involved observing double stars, as if one of the two stars was much further away from Earth than the other, it would show much less parallax, and so the distance between the two stars would vary cyclically over a year. It was also much easier to measure the relative rather than the absolute positions of stars.

Herschel had originally thought that the vast majority of double stars were optical doubles, rather than binary stars. However, in 1802 he noticed that some double stars were changing their separations with periods that were not annual, indicating that they were binary stars orbiting their common centre of gravity. Two years later he published the relative movements of 50 binary pairs, but it was not until 1827 that there was sufficient data for the orbit of such a pair to be accurately calculated by Felix Savary. He found that the binary pair ξ Ursae Majoris had a period of about 58 years.

Initially it was thought that binary stars were unusual, but by 1900 over 1,000 were known, and by 1950 over 30,000. In the third edition of the Yale Parallax Catalogue, published in 1952, there were 215 stars within 10 parsecs of the Sun. Of these, 117 were single stars, but the other 98 formed 38 double, 6 triple and one quadruple system. So it was clear that about half of the stars in the neighbourhood of the Sun were constituents of multiple systems, and there was no reason to suppose that this should not be generally true for more distant stars.

Astrometric binaries

Friedrich Wilhelm Bessel found in 1844 that both Sirius (α Canis Majoris) and Procyon (α Canis Minoris) had an oscillatory motion superimposed on their linear proper motions. He concluded that this was caused by the visible star being perturbed by an unseen companion in both cases. These were the first examples of so-called astrometric binaries discovered from their non-linear proper motions. In 1850 Christian Peters deduced the period of the Sirius binary to be about 50 years from Sirius' observed motion, and about ten years later Arthur Auwers deduced the period of the Procyon binary to be about 40 years using the same method.

Sirius B, the companion to Sirius, was first observed by Alvan and Alvan Graham Clark in 1862 using a 18.5 inch (47 cm) diameter lens that they were in the process of testing. (This was, at the time, the largest objective lens in the world). Over the years, Otto W. Struve, S. W. Burnham and others also tried to observe Procyon's companion, but they were unsuccessful. In the event, it was first observed by John M. Schaeberle in 1896, using the 36 inch (90 cm) Lick refractor.

Spectroscopic binaries

Edward Pickering discovered in 1889 that the spectral lines of Mizar A were split in two, with the separation between the two components varying in a cyclical manner. He attributed the two sets of lines to two different stars, and the variation in the line separations to changes in Doppler shifts, as the two stars orbited their common centre of gravity in a binary pair. This was the first so-called spectroscopic binary to be found. In the same year Hermann Vogel found that Algol (β Persei) was a spectroscopic binary, but in its case only one set of spectral lines were seen which oscillated slightly in frequency. The lines of its hypothesised companion could not be seen.

Eclipsing binaries

(a) Algol

Algol was known in pre-telescopic times to vary in brightness. Geminiano Montanari studied these variations in 1667, but it was not until 1783 that their cause was correctly explained by John Goodricke. He noted that the brightness variations were

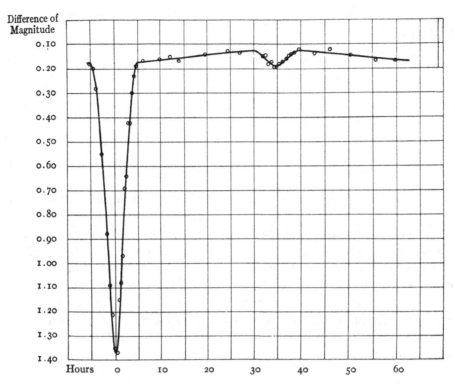

Figure 3.5. Algol's light curve as measured by Stebbins using his prototype selenium photometer. It shows both the primary and secondary minima. (From *The Astrophysical Journal*, Vol. 32 (1910), p. 199.)

regular with a period of 2 d 21 h, and suggested that the visible star may be partially eclipsed by a darker companion in orbit about their common centre of gravity.

As just mentioned, in 1889 Hermann Vogel detected varying Doppler shifts in Algol's spectrum. He also found that these Doppler shifts were in phase with Algol's brightness variations, proving that Algol was a binary. But its companion was too faint to be seen at that time either directly or by means of its spectrum. Vogel estimated, assuming a circular orbit, that Algol A (the main star) had a diameter of about 1.7 million km, its companion a diameter of about 1.3 million km, with a centre-to-centre distance of 5.2 million km. He estimated the relative masses of the two stars, assuming that their densities were equal, as $0.44\,M_\odot$ (mass of the Sun) and $0.22\,M_\odot$, respectively. Although these masses were too low (see below) his analysis showed how close the stars were to each other.

The brightness reduction seen in Algol was due to the eclipse of the brighter star by the dimmer one, but the reduction caused by the eclipse of the dimmer one by the brighter was undetected until 1910. In that year Joel Stebbins detected a 6% dip in the light curve of Algol (see Figure 3.5) using an early prototype of his selenium photometer. Later work showed that the main Algol system consisted of a relatively massive star (Algol A, B-type, $3.6\,M_\odot$) which was still on the main sequence, and a less massive star (Algol B, K-type, $0.8\,M_\odot$) which was already a subgiant. This was contrary to the theories of stellar evolution

in which the more massive star should be more evolved. This so-called Algol paradox was explained by John Crawford in 1955 as being the result of mass transfer between the two stars, which are very close together. The initially more massive star (Algol B) had filled its Roche lobe as it evolved, and had lost mass to its companion (Algol A).

Algol is now considered to be a semi-detached binary as just one of the stars fills its Roche lobe.

(b) β Lyrae

In 1784 Goodricke discovered that β Lyrae (or Sheliak) was variable with two similar but unequal minima. In this case it was thought that there were two stars of different but relatively similar luminosities orbiting their common centre of mass. Over the years the intensity of this binary was found to be continuously variable out of eclipse.

The spectrum of β Lyrae was found to vary by Eugen von Gothard in 1882, and in 1891 Edward Pickering showed that this variation was periodic, having the same periodicity of 12 d 22 h seen in its light output. The variations in its radial velocity were measured by Aristarkh Belopol'skii in 1892 and Norman Lockyer in 1893, and its orbital elements were computed by George Myers in 1898. These showed that it was an extreme case of an eclipsing binary star system, as the two stars, of masses 21 and $9.5\,M_\odot$ were so close together and distorted that they were almost touching. Myers also suggested that the system was surrounded by a gaseous envelope. But there were still many unexplained features in β Lyrae's spectrum.

In 1858 Argelander had found that the binary period of β Lyrae was increasing at about 20 seconds per year. Analysis of its spectrum in the early twentieth century indicated that it was ejecting large amounts of gas. The primary (i.e., brighter) β Lyrae star was thought to be a giant B8 star that filled its Roche lobe. As a result some of its gas was thought to flow from the primary to the secondary through the L_1 Lagrangian point. In 1941 Otto Struve suggested that some of the gas was also lost by both stars to a gradually expanding ring of material around the binary system. This explained the increase in period of the binary.

It gradually became clear by the early 1960s that the dimmer star was heavier than its very bright companion in the β Lyrae binary, thus violating the mass–luminosity relationship. In 1963 Su-Shu Huang interpreted this as showing that there is a relatively opaque accretion disc surrounding just the secondary star, which explained its apparent low luminosity. Thirty years later Petr Harmanec and Gerhard Scholz analysed a great number of published data on β Lyrae and concluded that the stars have masses of about 2.9 $M_\odot$ for the primary or donor star and 13 $M_\odot$ for the gainer. The system was first clearly imaged by Ming Zhao and colleagues in 2008. These images showed the mass donor and a thick disc surrounding the mass gainer. The donor is brighter than the gainer and generally appears elongated, as expected, as it fills its Roche lobe.

(c) Contact binaries

In 1903 Gustav Müller and Paul Kempf discovered that W Ursae Majoris varied in intensity by about 0.7 magnitude, with a period from minimum to minimum of 4 hours. They thought that this might be because W Ursae Majoris was a binary, but the very short time period indicated that the two stars of the binary would have to be almost touching. Müller and Kempf were naturally cautious about making such a suggestion, as it seemed unlikely, although not impossible, that such a configuration could be stable for any length of time. They further theorised that, if the system was such a very close binary, the stars must be very similar in size. Later work has shown that Müller and Kempf were basically correct as the two stars, both of which have F-type spectra, have diameters of about 1.1 and 0.8 $D_\odot$ (diameter of the Sun).

A number of similar so-called W Ursae Majoris variables, or contact binaries, were discovered over the next few decades. Both primary and secondary minima were found to be of almost the same depth, implying very similar effective temperatures or luminosities for both components. By the 1940s it became clear that these contact binaries exhibited a mass–luminosity anomaly, as, although the luminosities of the pair of stars in each binary were virtually identical, there were significant differences in their masses. Otto Struve suggested in 1948 that this may be because there was a common envelope surrounding both stars. This idea was further developed, independently, by Yoji Osaki and Leon Lucy in the 1960s.

W Ursae Majoris variables, or contact binaries, are now known to consist of two stars, each of which fills or overfills its Roche lobe. They are usually late F- to K-type stars, with periods of less than a day. Their light curves vary continuously due to the severe distortion of both components.

(d) ε Aurigae

Johann Fritsch noted the variability of ε Aurigae in 1821. Then in 1847 it was found by Friedrich Wilhelm Argelander and Eduard Heis to have decreased from its normal third magnitude to fourth magnitude, but it recovered its original intensity by the end of the following year. Subsequently it was observed to have short-term as well as long-term variations. ε Aurigae was seen to be of fourth magnitude in 1875, but it again recovered to its original intensity. At the time it was thought to be an irregular variable, but in 1901 it again decreased in intensity from its normal third to fourth magnitude. Two years later Hans Ludendorff suggested that it was an eclipsing binary with a period of about 54 years (which is twice its true period). In the same year Vogel concluded that it was a spectroscopic binary of very long period, but the displacement of its lines was too small to determine its period.

During eclipses ε Aurigae decreased in intensity by about 0.7 magnitude for seven months, remained approximately constant for ten months and then increased back to its normal level over the next seven months. If ε Aurigae was an eclipsing binary, this long eclipse period implied that the stars were exceptionally large. This and the short-term intensity variations caused the binary nature of the system to be doubted for some time, but this was generally accepted by the time of its next eclipse season from 1928 to 1930.

One of the main problems in trying to explain the configuration of the ε Aurigae system is the fact, observed by Ludendorff during the 1901 eclipse season, that the F-type spectrum of the visible star was still visible at the maximum eclipse. In addition, the spectrum of the other star has never been unambiguously detected. Various suggestions have been made over the years to explain this and more detailed effects observable both in and out of eclipse. For example in 1937 Gerard Kuiper, Otto Struve and Bengt Strömgren suggested that the eclipsing star was semitransparent, with the F-type's light being scattered by the eclipsing star's extremely thin atmosphere. In the 1950s Sergei Gaposchkin suggested that the eclipse could be annular. Then in 1965 Su-Shu Huang suggested that the eclipsing body was not a star but an edge-on thick disc. Six years later Robert Wilson suggested that the mid-eclipse brightening could best be explained by a tilted, thin disc with a central opening surrounding a black hole. In 1991 Sean Carroll and colleagues rejected the idea of a black hole at the centre of the disc, because of the lack of any high energy emission. They suggested, instead, that the disc may be a protoplanetary disc with one or possibly two stars at its centre.

So far none of these or other proposed configurations explain all of the observed effects. A new eclipse season began in August 2009 and finished in late 2011, but at the time of writing (2012) analysis of the results is still in its preliminary stages.

Bibliography

Aitken, Robert Grant, *The Binary Stars*, University of California Press, 1918 (Dover reprint 1964).

Burnham, Robert Jr., *Burnham's Celestial Handbook, An Observer's Guide to the Universe Beyond the Solar System; Vols 1, 2 and 3*, Dover Publications, 1977–1978.

Clerke, Agnes, *A Popular History of Astronomy during the Nineteenth Century*, Adam and Charles Black, 1902.

Hearnshaw, J. B., *The Analysis of Starlight: One Hundred and Fifty Years of Astronomical Spectroscopy*, Cambridge University Press, 1986.

Hilditch, R. W., *An Introduction to Close Binary Stars*, Cambridge University Press, 2001.

Hogg, H. S., *Variable Stars*, in Gingerich, O. (ed.), *Astrophysics and Twentieth-Century Astronomy to 1950: Part A*, Cambridge University Press, 1984.

Percy, John R., *Understanding Variable Stars*, Cambridge University Press, 2007.

Struve, Otto, and Zebergs, Velta, *Astronomy of the 20th Century*, Macmillan, 1962.

BLACK HOLES

John Michell had suggested in 1783 that, if stars were heavy enough, light would be prevented by gravity from leaving the surface. Pierre Laplace made a similar suggestion 13 years later. These ideas of 'dark stars' were mostly ignored in the nineteenth century, as light was then thought to be a massless wave not influenced by gravity.

In 1915 Albert Einstein published his general theory of relativity, and in the following year Karl Schwarzschild solved Einstein's field equations for a point mass. He showed that a black hole could theoretically exist by calculating the Schwarzschild radius or event horizon around an extremely dense object. In 1939 J. Robert Oppenheimer, George Volkoff and Hartland Snyder developed the theory of neutron stars and black holes, showing that stars greater than a certain mass would continue to collapse indefinitely to form black holes. In the 1960s Roy Kerr developed the theory of rotating black holes, and in the following decade Stephen Hawking suggested that black holes can emit particles and radiation, and so gradually erode over time.

A number of black hole candidates have been discovered in recent decades. The most convincing cases include those of Cyg X-1, A0620-00 and V404 Cygni.

Cyg X-1

The X-ray source Cyg X-1 was first detected by C. Stuart Bowyer and colleagues using an Aerobee sounding rocket experiment in 1964. When observed in the following year its X-ray intensity was found to have decreased by about 75%. In December 1970 the Uhuru spacecraft found that Cyg X-1 showed variations in its X-ray intensity over many different timescales down to about 100 milliseconds. Rappaport, Doxsey and Zaumen observed even faster intensity variations using a sounding rocket experiment in the following year. In addition, their experiment was able to pin-point the position of Cyg X-1 much more accurately than the Uhuru spacecraft.

In 1971 Paul Murdin and Louise Webster suggested that the blue supergiant HD 226868 was the optical counterpart of Cyg X-1. At about the same time Luc Braes and George Miley and (independently) Robert Hjellming and Campbell Wade found a radio source near HD 226868 within the X-ray error box. The radio source was also found to vary on a timescale of weeks in a similar way to the X-ray source, giving extra credibility to the suggestion that the radio and X-ray sources were basically the same.

Shortly afterwards Webster and Murdin and (independently) Thomas Bolton showed that HD 226868 and the X-ray source were in a 5.6 day binary. The most likely mass of the blue supergiant was thought to be about 20 $M_\odot$ (solar mass), giving a mass of about 10 $M_\odot$ for its unseen binary companion. Using a different method, based on measuring the periodicities of the X-ray emission, gave a mass of the unseen companion of 8.7 ± 0.8 $M_\odot$. So the mass of the unseen companion appeared to be well in excess of the maximum mass of 3 to 5 $M_\odot$ for a neutron star, implying that it is a black hole. It is thought that the X-rays are being emitted by matter that has been emitted by the blue supergiant and is falling towards the black hole via an accretion disc.

A0620-00

In August 1975 Martin Elvis and colleagues detected a strong source of X-rays using the Ariel 5 spacecraft. This source, designated A0620-00, reached a maximum X-ray intensity of about four times that of Sco X-1, the brightest X-ray source in the sky, before starting a gradual decline. The position of this newly observed X-ray transient was found by Forrest Boley and colleagues to coincide with the recurrent nova V616 Mon, which was then a 12th magnitude blue star, having increased from its normal 18th magnitude.

Spectra taken in 1976 and 1978 after outburst showed that the blue star had reverted to a K-type (reddish-orange) star with emission lines indicative of an accretion disc. Then in 1983 Jeff McClintock and colleagues found that the orange star's intensity was varying regularly with a period of 7.8 hours. They later measured the Doppler shift of its spectral lines, and in 1986 concluded that the mass of its compact, X-ray emitting companion must be at least 3.4 $M_\odot$ and maybe more than double that, implying that it was probably a black hole. Then in 1990 Carole Haswell and Allen Shafter measured the Doppler shift of Hα emission from the accretion disc around the dark companion. They concluded that the dark companion had a mass of 10.6 times that of the orange dwarf, implying a minimum mass for the dark companion of 3.8 $M_\odot$. This gave added credibility to the idea that it was a black hole.

V404 Cygni

GS2023+338, a soft X-ray transient, was first detected by the Ginga spacecraft in 1989. A few days later its optical counterpart was identified as a star of visual magnitude 12.8. Subsequent investigations showed that the star had brightened from about 20th magnitude, and Brian Marsden showed that it was a reappearance of the recurrent nova V404 Cygni first recorded in 1938. Then in 1991 Casares, Charles and Naylor discovered periodic variations in the spectrum of V404 Cygni, indicating that it was a binary consisting of a late G or early K-type star orbiting a dark companion every 6.5 days. They calculated that the dark companion had a mass of at least 6.3 $M_\odot$, and concluded that it was a black hole with a probable mass in the range 8–15 $M_\odot$. In 2008 Charles Bradley and colleagues deduced, using the XMM-Newton spacecraft, that the binary consisted of a $\sim$0.7 $M_\odot$, K-type star orbiting an approximately 12 $M_\odot$ mass black hole.

Supermassive black holes

Black holes are not restricted to stellar objects. In fact astronomers had speculated for some time that supermassive black holes may exist at the centre of active galaxies powering their radio jets. But supermassive black holes may also exist at the centre of ordinary galaxies.

(a) The Milky Way

An extremely compact radio source, Sgr A* was discovered in 1974 by Bruce Balick and Robert Brown at the centre of the Milky Way. Eleven years later K. Y. Lo and colleagues observed this source using a very long baseline interferometer (VLBI), which enabled them to set an upper limit on its diameter of just 20 AU. Meanwhile in 1980 J. H. Lacy and colleagues had analysed the movement of gas clouds in the vicinity of Sgr A* using infrared fine-structure line emission of NeII. They concluded that the source probably consisted of a point-like mass of $\sim$3 $\times$ 10^6 $M_\odot$ surrounded by $\sim$3 $\times$ 10^6 $M_\odot$ of stars within 1 parsec (pc) of the centre.

Over the last fifteen years or so, teams led by Reinhard Genzel and Andrea Ghez have been observing the movement of a number of individual stars near the galactic centre at 2 μm in the infrared. One of these stars has been observed over its complete 15.8 year elliptical orbit, and two of the stars have been observed within 100 AU of the centre. The results show that most of the unseen mass of $\sim$4 $\times$ 10^6 $M_\odot$ must be within a radius of 100 AU of Sgr A* at the centre. This implies a mass density of $>$8 $\times$ 10^{15} $M_\odot$/pc^3.

Recent millimetre wavelength radio observations of Sgr A* have shown that it is $\sim$0.1 AU in diameter, which is about the same order of magnitude as the Schwarzschild radius for a 4 $\times$ 10^6 $M_\odot$ black hole. In addition, Sgr A* has been found to be virtually motionless in the centre of the galaxy, indicating that it must contain the vast majority of the central mass. All these results indicate, with a high degree of certainty, that there is a supermassive black hole at the centre of the Milky Way.

(b) Other Galaxies

In 1978 Wallace Sargent and colleagues found evidence of a possible supermassive black hole of mass $\sim$5 $\times$ 10^9 $M_\odot$ at the centre of the galaxy M87, based on redshifts and velocity dispersions as a function of radius. Their work was updated in the 1990s using the Hubble Space Telescope (HST), which implied a similar mass but a greater mass density at the centre of the galaxy of $\sim$1 $\times$ 10^5 $M_\odot$/pc^3. Although high, this density is not above that of stars in dense clusters and does not require a supermassive black hole.

John Kormendy measured the velocity of stars 1 arcsec from the nucleus of the Andromeda galaxy (M31) in 1987. He concluded that their high velocity, together with the abrupt fall-off in velocity with distance from the nucleus, showed that the centre of the galaxy contained a mass of between 10^7 and 10^9 $M_\odot$ (current estimate $\sim$3 $\times$ 10^7 $M_\odot$). Such a large mass concentration in such a relatively small volume appeared to indicate the presence of a supermassive black hole. Then in 1992 a radio source was discovered at the centre of M31, giving support to the black hole theory.

In 1987 the existence of a black hole of 3 $\times$ 10^6 $M_\odot$ was also suspected to be at the centre of the galaxy M32, because of the abrupt increase in the orbital velocities of stars near the centre. Four years later, Tod Lauer, Sandra Faber and colleagues found further evidence for a black hole at the centre, when high-resolution images taken with the HST showed a rapid increase in brightness at the centre of the galaxy.

The HST imaged the nucleus of the elliptical active galaxy NGC 4261 in 1992, and found a dark disc of dust and gas surrounding the centre. The outside of the dark disc was about 100 pc in diameter, and it extended to within less than a light day of the nucleus, which was seen as a bright point. Radio telescopes had previously shown that NGC 4261 has a pair of jets, spanning a distance of 30,000 pc, coming from the nucleus. These radio jets were aligned along the spin axis of the gas disc, indicating the presence of a very hot accretion disc within the bright core of the galaxy, the accretion disc surrounding a black hole. The visible structure seen around the centre of this galaxy, and the radio jets seen coming from the nucleus region, was thought to provide strong circumstantial evidence that there is a $\sim 5 \times 10^8$ $M_\odot$ black hole at the centre of this active galaxy.

In 1995 L. J. Greenhill and colleagues published the first VLBI observations of the water maser in the nucleus of the galaxy NGC 4258. As a result they concluded that there was a central source in the galaxy with a mass density of at least 3.5×10^9 $M_\odot/pc^3$. In parallel Makoto Miyoshi and colleagues also observed rotating gas near the centre of NGC 4258, which indicated, in their view, the presence of about 4×10^7 $M_\odot$ in a region less than 0.13 pc in radius. Both sets of data were consistent with each other and indicated that there may well be a supermassive black hole at the centre. But it is just possible that this mass density could be provided by a very dense cluster of stars, although it would disintegrate over ~ 1 billion years.

These and a number of other investigations into possible supermassive black holes at the centre of galaxies have yielded encouraging results, but none have definitively proved, as yet, that such supermassive black holes actually exist there. The nearest to such a proof is that of a supermassive black hole at the centre of the Milky Way.

Bibliography

Barger, Amy J., *Supermassive Black Holes in the Distant Universe*, Kluwer Academic Publishers, 2004.

Begelman, Mitchell, and Rees, Martin, *Gravity's Fatal Attraction: Black Holes in the Universe*, 2nd ed., Cambridge University Press, 2010.

Camenzind, Max, *Compact Objects in Astrophysics: White Dwarfs, Neutron Stars and Black Holes*, Springer, 2007.

Hawking, Stephen, W., *A Brief History of Time: From the Big Bang to Black Holes*, Bantam Press, 1988.

Melia, Fulvio, *The Edge of Infinity: Supermassive Black Holes in the Universe*, Cambridge University Press, 2003.

Miller, Arthur I., *Empire of the Stars: Friendship, Obsession and Betrayal in the Quest for Black Holes*, Little Brown, 2005.

Penrose, Roger, *Chandrasekhar, Black Holes and Singularities*, in Srinivasan, G. (ed.), *From White Dwarfs to Black Holes: The Legacy of S. Chandrasekhar*, University of Chicago Press, 1996.

Reid, Mark J., *Is There a Supermassive Black Hole at the Center of the Milky Way?*, International Journal of Modern Physics D, 18, 6 (2009), pp. 889–910.

BROWN DWARFS

Shiv Kumar analysed the behaviour of low-mass stars and concluded in 1962 that stars with a mass of about 0.08 $M_\odot$ (solar mass) could not produce energy by hydrogen burning, and so would never reach the main sequence. He called objects with masses below this limit 'black dwarfs', but this was changed to 'brown dwarfs' following a suggestion by Jill Tarter in 1975.

Further theoretical work in the 1980s clarified the effect of deuterium burning, which was found could occur in brown dwarfs with masses down to 0.013 $M_\odot$. This is now generally taken to be the minimum mass of a brown dwarf. Brown dwarfs are not now considered to be stars, but objects in mass between stars and planets.

There were a number of searches for brown dwarfs undertaken in the 1980s and early 1990s, and a number of possible candidates found, before the first clear-cut case of a brown dwarf was identified in 1995. Obviously, dim, cool objects could be stars in the process of formation, so some determinant other than luminosity, colour or surface temperature was required to identify an object as a brown dwarf. The presence of lithium in its atmosphere was one possibility, provided the object is old enough, as stars destroy lithium when hydrogen burning begins. The presence of methane would also indicate a brown dwarf, provided the object was old and massive enough.

In 1995 Tadashi Nakajima and colleagues announced the discovery of a brown dwarf orbiting the M-type dwarf Gliese 229 only about 6 parsec from Earth. Called Gliese 229B, it was clearly a brown dwarf as methane had been detected in its spectrum. The presence of methane indicated that its surface temperature must be less than about 1,400 K, and its extremely low luminosity and the age of its stellar companion implied that it had a mass of about 0.05 $M_\odot$. Recent observations indicate an effective temperature of about 950 K. Gliese 229B is now considered to be the prototype methane or T-type brown dwarf.

Another brown dwarf candidate, PPL 15 in the Pleiades cluster, was observed by Basri, Marcy and Graham in 1995, but their categorization of it as a brown dwarf was disputed. This was because, although lithium had been detected, the lithium appeared to be depleted, and the object's mass was very close to the theoretical maximum for a brown dwarf. Then in 1999 Basri and Martin found that PPL 15 was a binary of two brown dwarfs, each with masses of about 0.06 or 0.07 $M_\odot$.

Finally a third brown dwarf candidate, Teide 1, was discovered in the Pleiades cluster in 1995. Its identity as a brown dwarf was confirmed the following year when Rafael Rebolo and colleagues announced the discovery of lithium in its spectrum. At the same time they announced that another Pleiades star, Calar 3, was also a brown dwarf as it also had lithium in its spectrum. Both Teide 1 and Calar 3 were estimated to have masses of about 0.055 $M_\odot$.

Since then about 800 brown dwarfs have been found, 600 of which are L-types that have metal hydride bands and alkali metal lines in their spectra, and 200 of which are cooler T-types that have methane in their spectra. In 2007/8 a few brown dwarfs were found, including ULAS 0034 and ULAS 1335, with surface temperatures of about 500–600 K and masses around the bottom limit of about 0.013 $M_\odot$ for T-type brown dwarfs.

The subject of brown dwarfs is currently (2010) in a state of flux, as discoveries of brown dwarfs go back only 15 years and much needs to be understood, particularly around the star/brown dwarf and brown dwarf/planet boundaries. It is noticeable that brown dwarfs can exist on their own, indicating that their formation may be very similar to that of stars rather than planets. But this view may change in the future, particularly if large numbers of lone planets are found as search techniques improve.

Bibliography

Basri, Gibor, *A Decade of Brown Dwarfs*, Sky and Telescope, May 2005, pp. 34–40.

Burningham, Ben, et al., *Exploring the Substellar Temperature Regime Down to 550K*, Monthly Notices of the Royal Astronomical Society, **391** (2008), 320–333.

Kumar, Shiv S., *The Structure of Stars of Very Low Mass*, Astrophysical Journal, **137** (1963), 1121–1125.

Reid, Neil, and Hawley, Suzanna, *New Light on Dark Stars: Red Dwarfs, Low-Mass Stars, Brown Stars*, 2nd ed., Springer-Praxis, 2005.

CATACLYSMIC VARIABLES

Classical novae

(a) Early novae

A number of 'new stars' were mentioned in the dynastic chronicles of China, Korea and Japan, but in Europe such records were virtually non-existent until the observations of the 'novae' of 1572 and 1604. These were later found to have been much more powerful supernovae, and so are not considered here (see *Supernova* article).

The first true classical nova to be clearly recorded in the telescope era was Nova CK Vulpeculae 1670. It was discovered by Father Anthelme on 20 June 1670 at third magnitude, and discovered independently by Johannes Hevelius just over a month later. CK Vulpeculae was no longer visible in October, but it reappeared for a few months in both subsequent years, being last observed in May 1672. The nova had been observed by many astronomers over these two years, so its celestial position was quite accurately known. This enabled the remnant to be located in 1981.

No other nova was recorded for over a hundred years until Joseph D'Agelet discovered Nova WY Sagittae 1783 in July of that year at about sixth magnitude, and John Russell Hind discovered Nova V841 Ophiuchi 1848 at about fifth magnitude in April.

Nova T Coronae Borealis 1866 was discovered by John Birmingham on 12 May 1866 when it reached magnitude 2, but it decreased to magnitude 8 within ten days. It was the first nova to have its spectrum examined. William Huggins and William Miller found on 16 May that the spectrum appeared to come from two sources: one, like the Sun, producing an absorption spectrum on a continuum, and the other producing hydrogen and other emission lines. T CrB exhibited another outburst in 1946 to about magnitude 3, and so was described as a recurrent nova. In both 1866 and 1946 there was a faint secondary maximum about 100 days after the initial outburst.

Nova Cygni (Q Cygni) 1876 was discovered by J. F. Julius Schmidt on 24 November 1876 when it reached magnitude 3, before decreasing over the next two weeks to magnitude 6. It showed, in its nova state, a number of bright emission lines due to hydrogen, and probably helium, superimposed on a continuous spectrum with strong absorption. The star was lost to view in March 1877, but when it was observed in September its spectrum had completely changed to resemble that of a planetary nebula with a single bright green emission line.

The Rev. Thomas Anderson discovered Nova T Aurigae on 24 January 1892. Subsequent searches of the astronomical archives showed that the star had been fainter than magnitude 13 as late as 8 December 1891, and that it had been first recorded photographically in its nova state on 10 December at magnitude 5.4, reaching a peak of magnitude 4.2 seven days later. This showed, for the first time, how quickly a nova increased in intensity. Nova T Aurigae stayed close to maximum for a little over two months before suddenly decreasing in intensity in March 1892.

Nova T Aurigae was the first nova to have its spectrum photographed, the best early spectrum being produced on 22 February 1892 by William Huggins and his wife Margaret. It showed bright hydrogen, helium and sodium lines, together with dark absorption lines on a faint continuous spectrum. Doppler shifts indicated that the emitting gases were receding from Earth at 400 km/s, whilst the absorbing gases were approaching Earth at 600 km/s. Nova T Aurigae was last recorded as magnitude 16 on 26 April, but on 17 August it was found to be at magnitude 10, and four days later its spectrum was seen to consist of the strong green and other emission lines of a planetary nebula. Three of the five dimmer novae discovered in the remainder of the nineteenth century showed the same spectral development.

(b) Early theories

Tycho Brahe had thought that novae were new stars which had been created by condensation of 'cosmic vapours' of the Milky Way. In 1713 Isaac Newton suggested that novae were old stars that brightened when impacted by comets. Laplace thought that they were caused by a surface explosion on a single star. E. F. Wilhelm Klinkerfues and Johannes Wilsing suggested that novae were the result of tidal disruptions of one star by the close approach of another. Other theories revolved around the collision of many celestial objects, including meteors, comets, asteroids, planets and other stars.

William Huggins suggested that the emission lines in the spectrum of Nova T CrB 1866 had been produced by the star erupting large quantities of hot gas consisting of a mixture of hydrogen and some other element. Later he explained the spectrum of Nova T Aurigae as being that of two bodies, one with absorption lines approaching Earth and one with hydrogen emission lines moving away from Earth following a close encounter. But the development of its spectrum over the next two or three months, when both types of spectra faded at about the same rate, made that explanation seem unlikely. Then when the spectrum changed to that of a planetary nebula, this was taken to indicate that only one body was involved. Hugo von Seeliger suggested at about this time that a nova was produced when a star entered a cloud of dense gas.

The problem at the end of the nineteenth century was that there was simply not enough information to distinguish between these and other theories.

(c) Novae 1900–1920

The Rev. Thomas Anderson discovered the intense nova, Nova GK Persei, on 22 February 1901 at magnitude 2.7. It was later found that it had brightened from magnitude 12.8 on 20 February. The nova continued to increase in brightness to magnitude 0.1 on 23 February before starting to lose intensity, so that a year later it was at magnitude 7. Nova GK Persei was the first nova to have its spectrum recorded after outburst but before maximum. Just prior to maximum its spectrum was that of a B-type star, with hydrogen and helium absorption lines, but, by the following day, when it was at maximum intensity, it had changed to an A-type spectrum, indicating that a reduction in its surface temperature had taken place. By the following day the familiar hydrogen emission lines had appeared, showing, for the first time, that the emission line spectrum of a nova did not appear when the star started its rapid increase in intensity, but only when the decrease in intensity had begun.

The bright hydrogen lines, although very broad, were, on average, stationary with respect to the Earth, but the material causing the dark lines was approaching the Earth with a velocity of 1,200 km/s. About 3 weeks after discovery, these dark lines were found to be split into many fine dark lines, indicating that the various layers of gas were approaching us with different velocities. A few days later Nova GK Persei started oscillating in intensity with a period of about 4 days, and an amplitude of about 1.5 magnitude, with corresponding changes in its spectrum. When the nova brightened, the continuous spectrum became brighter, the absorption lines became darker, and the emission lines faded, and when the nova faded all these changes reversed. By July the nova had stopped oscillating and it was showing the usual emission line spectrum of a nebula.

On 19 August 1901, six months after the outburst of Nova GK Persei, Camille Flammarion and Eugène Antoniadi found that the star was partially surrounded by a faint nebulous arc of about 6′ in diameter. A month later George Ritchey found that it was a complete circle centred on the nova. On 7 November Charles Perrine found that the circle had become larger and, from the estimated distance of the nova, calculated the expansion rate as about that of the velocity of light. As such high expansion rates are impossible for matter, Jacobus Kapteyn suggested that they were seeing the light from the original outburst being scattered by a pre-existing dust cloud. Confirmation was achieved when the spectrum of the brightest part of the ring was found to be like that of the nova at maximum in February 1901. Later photographs showed that the diameter of the ring was expanding at the rate of 11′ per year.

Fifteen years later Edward Barnard detected a very small faint ring around Nova GK Persei, which by August 1919 was about 10″ or so in diameter, with an outward expansion rate of about 0.4″ per year. The velocity of the gas was known from its Doppler shift, so the expansion of the ring enabled the distance of the nova to be estimated as ∼500 parsec, implying an absolute magnitude of about −8 at maximum brightness.

Nova GK Persei 1901 was later found to be in a class of cataclysmic variables now called 'Intermediate Polars', see later.

Nova V603 Aquilae, discovered on 7 June 1918 at magnitude 6, reached magnitude −1.4 two days later. It was the first nova to have its spectrum recorded some time before maximum in a pre-nova state. Although the earliest spectrum was dated 1888, the best pre-nova spectrum had been taken at Harvard in 1899 which showed a continuous spectrum with hydrogen absorption lines. At that time the star was a bluish-white class A object of magnitude 11. As the star brightened in June 1918, the lines became very narrow, like the c lines of supergiants, and in its nova state the Doppler shifts of the absorption lines indicated gas velocities of up to 3,400 km/s. The gaseous expansion ring, of the sort detected for Nova Persei, was detected for Nova V603 Aquilae by Barnard four months after outburst. Ten years later the gaseous ring was found to be 20.5″ in diameter. This enabled its distance of about 360 parsec, and absolute magnitude of −8 (at maximum), to be determined.

Because Nova V603 Aquilae was so bright, the effect of the changes of its spectral emission lines on its colour were clearly seen. At maximum intensity the nova was white, but as it lost its intensity it changed successively to yellow, pink and cerise.

The change from yellow to pink was due to the red emission line of hydrogen becoming more and more prominent in its spectrum, and the change to cerise was due to the addition of a number of blue emission lines. Finally the nova changed to green when its spectrum changed to the emission line spectrum of a nebula, but at this stage it was below naked-eye visibility.

Based on these and other novae, it became progressively clear in the first two decades of the twentieth century that a nova was a star that threw off two or three successive shells of gas over a few days, starting when the nova was at maximum intensity. These shells produced absorption lines in the light coming from the central star, whilst emitting light of their own in bright emission lines. The centres of these bright lines were generally undisplaced, but the lines were broad because the gas that was emitting them was in a spherical expanding shell, moving in all directions relative to the nova-Earth line. The dark absorption lines were displaced, however, because only that part of the shell that was directly between the central star and the Earth was absorbing the light en route to the Earth.

(d) Novae 1920–1940

The next major step in understanding novae was taken by analysing the behaviour of Nova RR Pictoris, which was discovered by R. Watson in South Africa on 25 May 1925. It took the unprecedented period of 15 days to reach maximum, and, during this time, its intensity increased from 2.3 magnitude by only about 1 magnitude. It then decreased in intensity, but unexpectedly brightened again to have a secondary peak of magnitude 1.9 on 9 August. In spite of its slow rise to maximum, it was later found to have been fainter than twelfth magnitude as late as 18 February 1925.

Nova RR Pictoris' spectrum was found to have remained essentially unchanged during its initial 15 day rise to maximum, indicating that its temperature was constant, and that its increase in luminosity must be due to an increase in size. This was quantitatively confirmed by the displacement of the absorption lines that showed an expansion rate of about 110 km/s. Harold Spencer Jones at the Cape Observatory, South Africa estimated that, when the star was at its maximum size and intensity, it had a diameter similar to that of a supergiant of about 400 times that of the Sun.

The spectrum of Nova RR Pictoris was extensively studied at the Cape both before and after its rise to maximum. It was found that, although the spectrum generally developed in the normal way after maximum, there were many differences of detail from a normal nova development. Various shells of gas were detected at variable velocities of up to about 1,300 km/s. From a careful study of the fading of the continuous spectrum, Spencer Jones calculated that the central star was shrinking at the rate of about 100 km/s. So Nova RR Pictoris was caused by a star that expanded and threw off expanding shells of gas, whilst the star shrank back in size.

On 13 December 1934 J. P. Manning Prentice discovered the next major nova, Nova DQ Herculis, at magnitude 3.4. It reached magnitude 1.3 on 22 December, before fading to magnitude 4 by early February. Starting in early April its brightness suddenly dropped to magnitude 13 in just one month. After a short interval, the nova started increasing in intensity again and, 5 weeks later, it reached magnitude 7, where it remained for over a year before fading once more. On discovery the spectrum was that of a B-type star, but this changed to an A-type as maximum approached, indicating a reduction in its surface temperature. When it faded out of naked-eye visibility in April it was cerise in colour, but when it reappeared in July it was a beautiful emerald green colour caused by the nebula emission lines. It is now recognised, like Nova GK Persei 1901, to be an intermediate polar (see later).

So by 1935 the spectra of novae were beginning to be understood. During the rise to maximum, which could take from a few days to a few weeks, the star swelled in size. Its spectrum was largely unchanged in many cases during this rise, although, in the case of both Nova GK Persei 1901 and Nova DQ Herculis 1934, the star changed from B-type to A-type at or near maximum, indicating a reduction in its surface temperature. At and immediately after maximum the star expelled shells of gas whilst the star reduced in size. The shells of expanding gas temporarily reduced the intensity of the central star, but as the shells expanded and became more diffuse, the central star became visible again. The shells produced a spectrum of their own with broad emission lines and, at the same time, produced absorption lines in the spectrum of the central star. Eventually, after a few months, the bright green spectrum of a tenuous nebula appeared.

It was still not clear what caused a nova to increase in intensity in the first place. It was thought that a nova was either caused by some sort of collision, or by some sort of instability just beneath the surface of a star. There were far too many novae to explain them by stellar collisions or close stellar encounters, however, as they would occur too infrequently. But some sort of collision of a star with a planet or other much smaller object like an asteroid was still thought to be possible. However the favoured theory in the mid-1930s was of some sort of instability inside the star that caused it to eject its outer layers. Unfortunately knowledge of the internal structure of stars was too rudimentary at the time to enable astronomers to propose any convincing cause of this instability.

(e) Novae since 1940

The big breakthrough in the theory of novae was made in 1954 when Merle Walker found that Nova DQ Herculis 1934 was an eclipsing binary with a period of 4 h 39 m. Eight years later Wojciech Krzeminski found that the nova WZ Sagittae was also an eclipsing binary with a period of 1 h 22 m. Then in the following year Walker found that Nova T Aurigae 1891 was an eclipsing binary with a similar period. These results suggested that all novae may be binaries. Robert Kraft examined this possibility in the early 1960s and concluded that most, if not all, novae occurred in binary star systems consisting of a star close to, or at, its white dwarf state and a red main sequence star. Mass was leaving the main sequence star, as it had filled its Roche lobe, and was creating an accretion disc around the white dwarf. Material from this accretion disc was then falling onto the surface of the white dwarf, so creating the nova explosion.

Kraft was unsure as to the exact mechanism that created the explosion. But in the 1970s Sumner Starrfield and colleagues showed that material at the bottom of this accretion disc would be gradually heated and compressed until it became hot enough for hydrogen burning to start. A thermonuclear runaway would then begin which would eject the hydrogen-rich envelope of the white dwarf to produce the observed nova outburst.

Dwarf novae or U Geminorum stars

U Geminorum (UG) was discovered by John Russell Hind at ninth magnitude on 15 December 1855, but within four weeks its intensity had reduced by over three magnitudes. Norman Pogson observed it again on 26 March 1856 at magnitude 9.5, before it again reduced in intensity. Then in 1869 Eduard Schönfeld found that its intensity had increased by three magnitudes in 24 hours. Subsequently U Geminorum was found to spend most of its time at minimum intensity, but there were short peaks in intensity at irregular intervals varying from about 60 to 250 days, as if it were undergoing large-scale eruptions.

No other star had been observed to behave in such a manner until Louisa Wells observed SS Cygni in 1896. It had a photographic magnitude that varied from 11.2 to 7.2, and later work showed that it spent most of the time at minimum intensity with a period between maxima that varied from about 30 to 70 days. A third member of the class, SS Aurigae, was discovered in 1907, and by the early 1920s a handful of similar stars, now called dwarf novae, were known. These increased in intensity by 2 to 6 magnitudes for a few days in semi-periodic intervals, normally of a month or two, before returning to normal. In 1957 Walker found that the intensity of some dwarf novae varied irregularly or flickered on a timescale of minutes.

Walter Adams and Alfred Joy showed in 1922 that the spectra of dwarf novae were similar to novae. Then in 1956 Joy found that SS Cygni was a spectroscopic binary with a period of 6 h 38 min. The two components were eventually found to be a white dwarf and a red main sequence star. Later, U Geminorum was also found to be a similar binary with a period of about 4 h 10 min. In 1973 Saul Rappaport and colleagues discovered soft X-ray emission from SS Cygni, using a sounding rocket experiment. This and subsequent observations of X-ray emission from dwarf novae helped to explain the reason for their outbursts.

All dwarf novae are, like normal novae, close binaries consisting of a white dwarf, surrounded by an accretion disc, and a red main sequence star. The light curves of novae and dwarf novae are completely different, however. It became clear in the mid-1970s that dwarf novae were not just fainter versions of novae with more frequent eruptions, but the mechanism of their outbursts was completely different. In the case of dwarf novae the outburst was produced following the release of potential energy as matter in the disc impacted the surface of the white dwarf. This energy was released mainly in the optical, ultraviolet and X-ray wavebands.

There are currently two alternative theories explaining the variability of accretion responsible for the dwarf nova outbursts. In one, the so-called mass transfer model proposed by Geoffrey Bath in 1973, it is due to a variable mass transfer rate from the red star onto the accretion disc, and in the other, the disc instability model proposed by Yoji Osaki in the following year, it is caused by basic instabilities in the accretion disc itself.

Today over 600 dwarf novae are known, divided into the three main categories of SS Cygni (UGSS), Z Camelopardalis (UGZ), and SU Ursae Majoris (UGSU), depending on the shapes of their intensity curves. Z Camelopardalis stars have a standstill or halt in their light curve, for several weeks to years, in their reduction from maximum. SU Ursae Majoris stars, on the other hand, have brighter and longer supermaxima interspersed with the more normal maxima. They also have small regular so-called 'superhumps' on their supermaxima.

Polars (AM Her systems)

Max Wolf discovered the variable star, AM Herculis, in 1923. For over fifty years it appeared to be a normal irregular variable star with an amplitude of about two magnitudes. But in 1976 Richard Berg and J. Graeme Duthie re-examined the star and found that its intensity varied continuously on timescales of minutes, leading them to suggest that it was a cataclysmic variable as it exhibited some of the detailed features of dwarf novae. They also suggested that it was the optical counterpart of the X-ray source 3U 1809+50 that had been discovered a few years earlier by the Uhuru spacecraft. But it was just outside the error box for the X-ray source's position.

In the meantime, Hearn, Richardson and Clark had detected a soft X-ray source with the SAS-3 spacecraft in 1975–6 at a position near to that of 3U 1809+50. They found that the X-ray flux of their source varied irregularly in a very similar way to the optical variability of AM Herculis, which lay inside the error box of their X-ray source. So the optical and their X-ray source appeared to be the same. It was later found, on reanalysing the Uhuru data, that the Uhuru source 3U 1809+50 and the SAS-3 source were one and the same.

In 1976 Anne Cowley and colleagues found, based on spectroscopic and photometric measurements, that AH Herculis was a binary star with a period of 3 h 6 min. The intensity of the X-ray source varied with exactly the same period, confirming that AM Herculis and 3U 1809+50 were the same object. In the same year Tapia, Stockman and Angel found that the light of AM Her was linearly and circularly polarised in the yellow and red spectral bands, indicating the presence of a compact star, most likely a white dwarf, with a magnetic field of $\sim 2 \times 10^8$ gauss. They also noted that its polarisation varied with a period of 3 h 6 min. Unpolarised light observed in the ultraviolet suggested that a hot source, the companion of the compact source, was unaffected by the magnetic field.

William Priedhorsky measured the optical spectrum of AM Her in 1976 and found radial velocities varying from 200 km/s in recession to 400 km/s in approach, and back again, over the 3 h 6 min period of the binary. From an analysis of the phase of the oscillatory motion he concluded that these velocities were probably those of matter being transferred between the stars.

Further work has shown that there is no accretion disc in the AM Her system because of the strong magnetic field of the white dwarf. So material from the secondary star is accelerated as it spirals down the field lines of the magnetic white dwarf to impact its surface in its polar regions at high velocity. Spiralling electrons produce cyclotron emission, which is seen as polarised light, varying at the rotation period of the white dwarf, whilst the high temperatures on the surface of the white dwarf produced by the gas impact causes it to emit soft X-rays.

AM Her was the first so-called 'polar' to be identified, in which matter is lost directly from a red star to the magnetic poles of its white dwarf companion, and in which the spin rate of the white dwarf is synchronised with the orbital period of the binary by the very strong magnetic field of the white dwarf.

The discovery of the first polar resulted in a search for other cataclysmic variables with large magnetic fields shown by their polarised optical emissions. Two such sources were quickly found, namely AN Ursae Majoris and VV Puppis. The VV Puppis and other polar systems were also found in the 1990s by the Rosat and Extreme Ultraviolet Explorer (EUVE) spacecraft to be sources of extreme ultraviolet radiation. The spectrum and the variation of the extreme UV radiation during the eclipse of VV Puppis, for example, enabled Paula Szkody and colleagues to estimate that the energy was being emitted from a 200 km diameter, 300,000 K temperature region on the surface of the white dwarf.

By 2008 the number of known polars had reached 100.

Asynchronous polars

Nova Cygni 1975 (now called V1500 Cygni) appeared on 29 August 1975 and peaked at magnitude 1.7 two days later, but, within a week or so, it was back to below naked-eye visibility. It was then found that in its pre-nova state V1500 Cygni had been fainter than magnitude 21. Later analysis showed it to be about 1,200 parsec away, with a peak absolute magnitude of about −10. In September 1975 V1500 Cygni was found to vary in brightness with a period of about 3 h 23 min but, within a year, this period had reduced by about 4 minutes. The period continued to vary by a minute or so until it stabilised in August 1977.

In 1987 Stockman, Schmidt and Lamb discovered that the visible light from V1500 Cygni was circularly polarised, and that the polarisation was varying with a period of about 3.5 minutes less than that of the brightness fluctuations. This indicated that V1500 Cygni was a polar, or magnetic binary, similar to AM Herculis. But, unlike AM Herculis, the white dwarf's spin period was not the same as the orbital period of the binary, the difference being 3.5 minutes.

Intermediate polars (DQ Herculis systems)

As mentioned earlier, Merle Walker found in 1954 that Nova DQ Herculis 1934 was an eclipsing binary star with a period of 4 h 39 min. Its light curve was also found to be modulated with a period of just 71 seconds. Walker concluded that the 71 second variations were due to either a stellar surface effect or to the oscillation of one of the stars in the binary. A few years later Robert Kraft concluded from spectroscopic observations that the system consisted of a white dwarf, surrounded by an accretion disc, and a red main sequence star. The latter overflowed its Lagrangian lobe and supplied material to the disc.

In 1974 Nather, Smak and McGraw found that DQ Herculis exhibited linear polarisation which varied at the 71 second period. In the same year D. Q. Lamb and others suggested that this very short period pulsation was due to the rapid rotation of the accreting white dwarf, which was highly magnetic. It was expected that DQ Herculis would produce hard X-rays, but observations with the Einstein Observatory spacecraft in the late 1970s failed to show any such emission. Later work did show that DQ Herculis was an X-ray emitter, but the accretion disc obscured the X-ray source from view. DQ Herculis was called an 'intermediate polar' as, having a very strong magnetic field and an accretion disc, it was intermediate in type between a polar and a dwarf nova.

The next intermediate polar to be recognised was H2252-035, which had been detected as an X-ray source by the HEAO-1 spacecraft in the late 1970s. In 1980 R. E. Griffiths and colleagues identified this source with a 13 magnitude emission line variable star, later named AO Piscium. White, Patterson and colleagues detected X-ray pulsations with a period of 805 s. Then, later that year, Joseph Patterson and Christopher Price detected a 3.59 h period in its optical emissions, which were modulated at a frequency of 859 s. Patterson and Price concluded that H2252-035 was a binary with an orbital period of 3.59 h. They noted that the orbital period was the beat period between the optical and X-ray pulsation periods. As a result they concluded that the X-ray pulsations were produced by an accreting compact star, probably a white dwarf, whilst the optical pulsations were reprocessed radiation originating in the X-ray-heated atmosphere of the secondary star.

As mentioned earlier, Nova GK Persei 1901 was also an intermediate polar. A number of relatively small optical outbursts were observed from GK Persei in the 1960s and 1970s, which temporarily increased its optical intensity from its normal magnitude 13 by one or two magnitudes. In 1978 its optical intensity increased again, and King, Ricketts and Warwick found, using the Ariel 5 spacecraft, that GK Persei also flared in X-rays. Another optical outburst was detected in July 1983, and Watson, King and Osborne found that GK Persei was pulsing in X-rays with a period of 351 seconds. They concluded that GK Persei was an intermediate polar in which the magnetic white dwarf had a spin period of 351 seconds. In 1986 Crampton, Cowley and Fisher found that the period of the binary was 47.9 hours, which was much longer that that of any other intermediate polars known at the time. Subsequent work has shown that GK Persei exhibits modified X-ray pulsations in periods of quiescence, and other oscillations in different wavebands. The detailed structure of this binary is still open to doubt, but the complex variations observed is clearly connected with its relatively large orbital size for a compact binary.

In summary, intermediate polars consist of a normal star that fills its Lagrangian lobe and loses mass to the magnetic poles of its white dwarf companion via an accretion disc around the white dwarf. But, unlike polars, the magnetic field of the white dwarf is not strong enough to synchronise its spin rate to the orbital period of the binary, although it is strong enough to disrupt the inner part of the accretion disc. The exact structure of the accretion disc and the effect of the white dwarf's magnetic field on it, and on the flow of material from the secondary star, depends on the intensity of the magnetic field, the mass of the secondary star, and of the orbital period of the binary, amongst other things. This partly explains the variations in phenomena observed between different intermediate polars.

Bibliography

Burnham, Robert Jr., *Burnham's Celestial Handbook: An Observer's Guide to the Universe Beyond the Solar System, Vols. 1, 2 and 3,* Dover Publications, 1977–1978.

Charles, Philip A., and Seward, Frederick D., *Exploring the X-ray Universe,* Cambridge University Press, 1995.

Clerke, Agnes, *A Popular History of Astronomy during the Nineteenth Century,* Adam and Charles Black, 1902.

Hack, Margherita, and La Dous, Constanze, *Cataclysmic Variables and Related Objects,* NASA SP–507, 1993.

Hellier, Coel, *Cataclysmic Variable Stars: How and Why They Vary,* Springer-Praxis, 2001.

Hogg, H. S., Variable Stars, in Gingerich, O. (ed.), *Astrophysics and Twentieth-Century Astronomy to 1950: Part A,* Cambridge University Press, 1984.

Percy, John R., *Understanding Variable Stars,* Cambridge University Press, 2007.

Warner, Brian, *Cataclysmic Variable Stars,* Cambridge University Press, 2003.

ERUPTIVE VARIABLES

T Tauri-type stars

The variability of T Tauri was first noted by John Russell Hind in 1852. At the same time he also noticed a variable nebula, later called NGC 1555 or Hind's nebula, which almost touched the star. This nebula was later found to be illuminated by the star which caused the nebula's variability. Then in 1890 Sherburne Wesley Burnham found that T Tauri was situated within another, much smaller, nebula.

Twenty-five years later Walter Adams and Francis Pease found that T Tauri's spectrum consisted of a number of well-defined bright lines, including those of hydrogen, helium and the H and K lines of calcium, superimposed on a dark line spectrum. Shortly afterwards another three stars were found that had similar emission features, but it was not until 1945 that Alfred Joy pointed out similarities between these and a few other stars in what he called T Tauri stars. He noted that these stars, which were each associated with nebulosity, were characterised by an irregular light curve, with an amplitude of about 3 magnitudes, low luminosity, and a spectral type in the range F5–G5, with hydrogen, calcium and other emission lines.

A number of astronomers suggested in the late 1940s that the emission line spectrum of a T Tauri star was caused by the passage of a normal, main sequence star through its associated nebula. On the other hand Viktor Ambartsumian suggested that T Tauri stars were low-mass main sequence stars in the process of formation.

George Herbig found in 1952 that T Tauri stars were too bright for their spectroscopic luminosities, and so tended to lie above the main sequence. He also found that they generally had significantly broader absorption lines than main sequence stars of the same spectral types. A few years later Herbig and Merle Walker independently came to the conclusion that T Tauri stars were, as Ambartsumian had suggested, stars in the process of formation. They thought that they were following the evolutionary tracks computed by Henyey, LeLevier and Levée in 1955 en route to the main sequence. New evolutionary tracks, en route to the main sequence, were determined by Chushiro Hayashi in 1961.

An abnormally strong lithium absorption line was detected in the spectrum of T Tauri by Roscoe Sanford in 1947. Ten years later Kurt Hunger confirmed this, and detected a strong lithium line in another T Tauri star, RY Tauri. Unusually high lithium concentrations were then found in a number of other T Tauri stars over the next few years. The cause was not clear, but it was eventually realised that lithium amounts in the T Tauri stars were normal. It was that in the main sequence stars that was low. This was because lithium was destroyed in main sequence stars by temperatures in excess of about a million K.

Over the years evidence had been found for strong mass outflows, sometimes collimated as bi-polar outflows, and accretion discs around some T Tauri stars. In 1992 Joanne Attridge and William Herbst found that the frequency distribution of rotation periods for T Tauri-type stars in the Orion nebula cluster was bimodal. The slow rotator group had a mean period of about 8.5 days, and the rapid rotator group a mean period of 2.2 days. It was later found that classical T Tauri stars (CTTS), which were surrounded by an accretion disc, were slow rotators, whereas the weak-line T Tauri stars (WTTS), which had little or no accretion disc, had rotation periods between 1.5 and 16 days. It is thought that the CTTS evolved into WTTS, which then evolved into main sequence stars.

T Tauri stars, which are still in the process of contraction, are now known with F-, G-, K- and M-type spectra. The FGK stars have luminosities in the range 4 to 50 $L_\odot$ (solar luminosity), radii 2 to 6 $R_\odot$, masses 1 to 3 $M_\odot$, and ages 1 to 10 million years. The M-types have lower masses and luminosities. Mass inflow rates via the accretion discs can be as high as 10^{-8} to 10^{-7} $M_\odot$/year, with mass outflow rates somewhat smaller.

Main sequence variables

(a) Wolf-Rayet stars

Charles Wolf and Georges Rayet discovered three faint stars in Cygnus in 1867, now called Wolf-Rayet stars, that had very broad emission lines on a continuous background. By the mid-1880s a total of 13 Wolf-Rayet stars were known, but only one was included in Harvard's *Draper Memorial Catalogue* of 1890. It was the sole representative of the O star category. Antonia Maury thought that Wolf-Rayet stars were stars in a very early stage of development because of the intensities of their hydrogen lines. As a result, Annie Cannon placed the O-type stars before the B-types in her catalogue of stellar spectra in 1901.

James Keeler had suggested in 1890 that the stars at the centre of planetary nebulae were related to Wolf-Rayet stars. Then in 1914 William H. Wright confirmed that these central stars were, in many cases, Wolf-Rayet stars.

In 1896 Edward Pickering discovered that the Wolf-Rayet star ζ Puppis had a series of spectral lines that alternated with the Balmer hydrogen lines. The cause of these so-called Pickering lines was the subject of much speculation. Pickering

thought that they were due to hydrogen in some unusual physical condition, but in 1913 Niels Bohr showed that they were due to ionised helium. In 1921 Megh Nad Saha deduced, using his ionisation theory, that the effective temperature of Wolf-Rayet stars was about 22,000 K.

Some Wolf-Rayet stars had been found to have emission lines with a blue-shifted absorption component, like that of P Cygni. Carlyle Beals interpreted this in 1929 as being due to rapid mass loss in an extended transparent envelope surrounding the star. In 1932 Bengt Edlén found that the majority of the emission lines in Wolf-Rayet stars were due to highly ionised carbon, nitrogen and oxygen. Beals then identified two groups of Wolf-Rayet stars; those with intense nitrogen emission lines, and those with very strong carbon and oxygen emission lines, but very weak or no nitrogen emission lines. These are now called WN and WC stars. Beals concluded, from the intensities of their various ionised lines, that their effective temperatures were in the range 60,000–100,000 K, which was much hotter than previous thought.

Terminal velocities of the winds emitted by Wolf-Rayet stars were found using high resolution ultraviolet spectra produced by the IUE spacecraft in the late 1970s and 1980s. They were found to vary from about 3,500 km/s for WN3 (WN subtype 3) and WC5 stars down to about 1,500 km/s for WN8 and WC9 stars. Radio or infrared data was used to determine the density of the outermost regions of the gaseous outflow, and this, together with the velocities deduced from IUE spectra, gave average mass outflow rates of about 4×10^{-5} $M_\odot$/year. At this rate a star would lose 1 $M_\odot$ of material in just 25,000 years, so they could not continue to eject mass at this rate for very long.

In 1979 Martin Huber and colleagues analysed the high-resolution IUE spectra of the Wolf-Rayet star HD 192163, which is one of the Wolf-Rayet stars surrounded by a ring nebula. They detected two sets of absorption lines, one undisplaced in wavelength and one blueshifted. The latter lines, which were the weaker, were found to be from the nebula. This was the first detection of absorption lines in a nebula associated with an early-type star.

Morton Roberts had produced a catalogue in 1962 of all known Wolf-Rayet stars within about 2 kiloparsec of the Sun, excluding those at the centre of planetary nebulae. This was updated and corrected by Lindsey Smith a few years later. The distribution of these galactic Wolf-Rayet stars was analysed by Anne Underhill in 1967 who concluded that, contrary to the general view of the time, they were not relatively old massive stars, but most were probably massive Population I stars in the process of evolving onto the main sequence. In the same year Bohdan Paczński suggested that Wolf-Rayet stars were formed following mass exchange in a binary system. However, as Underhill observed, not all Wolf-Rayet stars were in binaries.

Peter Conti suggested in 1976 that massive O-type stars could evolve to become Wolf-Rayet stars as a result of mass loss via strong stellar winds. This is now the preferred evolutionary theory, at least for non-binary Wolf-Rayet stars. The stellar winds are emitted by a star of at least 30 $M_\odot$ to leave behind a helium-burning core, which is the Wolf-Rayet star. The different types of WC, WN and WO simply show the extent to which the outer layers of the original star have been stripped away.

(b) Flare stars

Ejnar Hertzsprung noticed in 1924 that a faint star, now called DH Carinae, suddenly increased in brightness by 1.8 magnitudes in less than an hour. It subsequently faded by 1.1 magnitudes in 80 minutes. Willem J. Luyten then found two stars, V1396 Cyg and AT Microscopii, of spectral type dM3e (dwarf type M3 with emission lines) and dM4e, respectively, which had very bright hydrogen emission lines. They then faded rapidly.

In 1938 Arno Wachmann found that the bright lines of hydrogen and the continuum spectrum of V371 Orionis faded over a period of an hour. He estimated that its intensity had decreased by at least 1.5 magnitudes over that period. Its spectral type was dM3e. In the following year, Adriaan van Maanen noticed that WX Ursae Majoris was 1.5 magnitudes brighter than normal, and he later found that YZ Canis Minoris was also brighter than normal. The spectral types were dM5.5e and dM4.5 e, respectively.

The first real advance in the study of these stars, now called flare stars, occurred whilst Edwin Carpenter was taking a number of short exposures of Luyten 726–8 in 1948. He found that the star had increased in brightness by a factor of 12 (or almost three magnitudes) over a three-minute period, and had then faded more gradually to its previous brightness. Luyten concluded that a violent flare had occurred on the star, similar in type but of much greater magnitude than those seen on the Sun. Luyten 726–8 was found to be a binary star, and it was the fainter star, now called UV Ceti, that had exhibited the flare. Alfred Joy and Milton Humason had also taken a number of spectrograms which showed that, at its peak, the bright emission lines of hydrogen and the continuum spectrum were extremely bright, and helium emission lines appeared. Luyten 726–8 was subsequently found to flare quite frequently. In 1952, for example, Vasilije Oskanjan observed it to increase by 5.8 magnitudes in less than 20 seconds. Then in 1963 a flare was detected at radio wavelengths by Bernard Lovell and colleagues at Jodrell Bank.

The ANS spacecraft was used to make a survey of flare stars in 1974 and 1975 to try to detect flares in X-rays. The search was successful when John Heise and colleagues discovered X-ray emissions from the flare stars UV Ceti and YZ Canis Minoris. A few years later the Einstein spacecraft found, much to most

astronomers' surprise, that flare stars are also strong X-ray emitters in their quiescent state. This implied that their coronal temperatures must be of the order of 10 to 20 million K.

Jay Bookbinder and colleagues used the Hubble Space Telescope in 1992 to observe flare stars during their flaring activity. They found that in the case of the flare star AD Leonis, material was falling towards the star during the flare, not after the flare like in the case of the Sun. The spectrum of AD Leonis indicated that the material concerned had a temperature of about 100,000 K, indicating that it was in the star's transition region.

It is now known that flare stars are red dwarfs with masses in the range 0.06 to 0.6 $M_\odot$, and with effective temperatures from about 2,700 to 4,000 K. They are generally dMe stars like UV Ceti. The Balmer hydrogen lines, which are in emission, increase their intensities strongly during flares. They start to increase before the continuum, but peak later and remain enhanced after the continuum has returned to normal. Many flare stars are found to be in binaries. Many are members of young stellar associations, like that in Orion, although some older flare stars are also known.

Supergiants

(a) Luminous blue variables

The term Luminous Blue Variable (LBV) was first used by Peter Conti in 1984 to describe hot ($\sim$20,000−30,000 K), massive stars, other than Wolf-Rayets. As defined they are often not blue and sometimes not variable. But LBVs are intrinsically very bright, with luminosities $\sim$$10^6$ $L_\odot$, and so they can be detected at considerable distances. They appear to spend only about 50,000 years in the LBV phase, probably en route to becoming Wolf-Rayet stars, so they are very rare.

Although Conti used a minimum mass limit in defining LBVs of 10 $M_\odot$, this appears to be significantly on the low side for stars which exhibit typical LBV characteristics. Intensity variations can be of $\sim$0.1−0.2 magnitudes over periods of days, or of $\sim$1−2 magnitudes over periods of years or decades. In addition, there are rare eruptions, like that of η Carinae, in which the star's brightness increases by 3 magnitudes or more over periods of centuries. LBVs, like Wolf-Rayet stars, also exhibit significant mass loss.

The history of observations and theories connected to two important LBVs, η Carinae and P Cygni, first observed in the seventeenth century, are outlined below.

(i) **η Carinae.** Edmond Halley observed η Carinae from St Helena in 1677 when he was undertaking a survey of the southern stars. At that time it was of magnitude 4, but in 1751 Nicholas Lacaille recorded its magnitude as 2. By the 1810s it was back to magnitude 4, whilst in 1827 Burchell recorded it as magnitude 1. It achieved its maximum brightness in April 1843 when it was the second brightest star in the sky with a magnitude of −0.8. These large swings in brightness over the

Figure 3.6. A modern image of the Homunculus nebula surrounding η Carinae produced using the Hubble Space Telescope. (Courtesy N. Smith, J. A. Morse [U. Colorado] et al., NASA.)

years baffled astronomers as η Carinae did not seem to behave like a nova. After 1843, η Carinae's brightness faded dramatically, except for a brief eruption in 1887, so that by the early part of the twentieth century it was only of 8th magnitude. Its intensity has gradually increased since the 1930s to its current magnitude of 5.

Albert Le Sueur was the first to describe the spectrum of η Carinae in 1870, noting that it was crossed by a number of bright lines, including what were to be later identified as the Balmer Hα and Hβ lines. In the 1890s its spectrum was observed to change from an absorption line spectrum early in the decade to a mainly emission line spectrum by 1895. Annie Cannon commented that the latter reminded her of a nova spectrum. Improved spectral resolution of the early twentieth century showed many more emission lines, many of which were unidentified. But in 1928 Paul Merrill identified many of these as being due to forbidden transitions, which indicated nebulosity. Then in 1953 A. D. Thackeray detected P Cygni profiles (broad emission lines with absorption lines Doppler-shifted to shorter wavelengths), indicative of mass loss. The violet absorption components were shifted by −450 km/s.

Lacaille had noticed in 1751 that η Carinae was located in a nebula. Then in 1914 Robert Innes discovered that η Carinae was surrounded by a much smaller shell of gas and dust, now called the Homunculus nebula (see Figure 3.6), only a few arcseconds in diameter. Gerry Neugebauer and Jim Westphal made the first observations of η Carinae at infrared wavelengths in 1968, and were surprised to find that it was the brightest star in the sky at about 20 μm. In the following year Bernard Pagel suggested that this infrared radiation came from a thick dust

shell surrounding the central star which had heated the dust to a temperature of about 250 K. Kris Davidson recorded the ultraviolet spectrum of the gas cloud in 1981, and found that it was typical of one ejected by a star near to the end of its lifetime.

In 1972 Robert Gehrz and Ed Ney measured the dust and gas cloud of the Homunculus nebula surrounding η Carinae from records made by many observers since its discovery. As a result they were able to confirm previous suggestions that the cloud had been ejected by the star during its brightening in the early 1840s, and that the cloud was still gradually expanding. It was thought that gradual dispersal of this dust cloud may partially explain η Carinae's apparent increase in luminosity since the 1930s.

η Carinae is the most spectacular Luminous Blue Variable known. If it is a single star, it appears to have a mass of about 100−150 $M_\odot$, with a luminosity of a few million $L_\odot$. This is close to or above the Eddington limit of about 120 $M_\odot$, at which a star is unstable against its own radiation pressure. However, A. Damineli concluded in 1996 that η Carinae varies in the visible and near infrared wavebands with a period of 5.52 years, indicating that it is probably a binary. Similar changes have also been observed in the X-ray and radio bands, supporting this hypothesis. But unexpected recent (2008–9) changes have raised new problems with the expected configuration of η Carinae. Nevertheless, the star/stars appear to be losing a substantial amount of gas at a rate of ∼10^{-3} $M_\odot$/yr, which is an even higher rate than that of Wolf-Rayet stars.

In 1989 Roberta Humphreys estimated that η Carinae had lost about 2 to 3 $M_\odot$ during its outburst in the 1840s. More recent estimates have produced even higher estimates of 10 to 15 $M_\odot$, showing that η Carinae is not a normal Luminous Blue Variable.

(ii) P Cygni. Nova Cygni 1600, now called P Cygni, was discovered by William Blaeu in 1600 at magnitude 3. It stayed at this magnitude for a few years before gradually fading to below naked eye visibility (magnitude 6) in 1626. It reached naked eye visibility again in 1654, and peaked at magnitude 3.5 the following year, before fading slowly. Since then it has had frequent relatively small, long-term, light variations of about 1 magnitude.

P Cygni's first known spectrum, in which he found one bright line, was recorded by Edward Maunder in 1888. In the following year James Keeler found a number of bright lines which had dark borders on their short wavelength side. These turned out to be dark absorption bands which Belopol'skii observed in 1899 to be generally displaced to the violet by about 1.5 Å.

Two key papers were published by Carlyle Beals in 1929 and 1934 providing the basis for our present understanding of P Cygni. He concluded from its spectrum that it was ejecting gaseous material, like Wolf-Rayet stars, either in discrete shells or continuously. But he suggested that P Cygni would be more extended, more diffuse and cooler than Wolf–Rayet stars. He estimated the effective temperature of P Cygni, from the intensities of its emission lines, as ~30,000 K (cf. modern estimates 19,000 ± 700 K). In 1935 Otto Struve concluded that there was a shell around P Cygni and that it was expanding outwards at an accelerating rate.

There had been some confusion early in the twentieth century as to whether the spectral lines of P Cygni were changing in appearance and/or position. In 1968 Mart de Groot found evidence that the star was surrounded by three separate shells expanding at rates of about 95, 160 and 210 km/s, with the outermost shell, at a distance of about 3 stellar radii from the surface, pulsating with a period of 114 days. This could possibly account for the earlier confusion about the stability of P Cygni's spectrum. In 2001 de Groot, Sterken and van Genderen also detected periods of 17.3 days and 4.2 years in the intensity of P Cygni. The amplitude of variations was small, however, being only ~0.1 magnitude in both cases.

In 1973 Heinrich Wendker and colleagues discovered a nebula surrounding P Cygni using radio telescopes. Since then a total of three radio-emitting nebulae have been detected around P Cygni. The inner nebula has a radius of about 300 times the star's radius, whereas the outer two are a few hundred times further out.

P Cygni, a luminous blue variable, is now also a prototype for stars with expanding atmospheres. It is variously described as a supergiant or hypergiant that normally loses mass at the rate of ~5×10^{-6} $M_\odot$/yr, but during its two seventeenth-century outbursts it appears to have lost about 3×10^{-2} $M_\odot$. The expansion rate of the shell or shells increases from about 25 km/s just above the photosphere to about 200 km/s or so further out.

(b) R Coronae Borealis type stars

R Coronae Borealis (R CrB) was observed by Edward Pigott in 1795 to suddenly reduce in brightness and recover again. Over the following century it was seen to exhibit similar sudden large reductions in brightness of up to nine magnitudes, but there seemed to be no periodicity. It spent most of its time at maximum. In 1906 Hans Ludendorff found that the Balmer lines of hydrogen were missing from its spectrum. Then in 1923 Alfred Joy and Milton Humason observed the spectrum of R CrB at minimum and found that, as the star's brightness decreased, the previously strong titanium absorption lines faded to become prominent sharp emission lines at minimum.

In 1935 Eppe Loreta suggested that the brightness reductions for R CrB were caused by the emission of dark clouds of material from the star which obscured its surface. In the same year Louis Berman found R CrB was a hydrogen-deficient,

carbon-rich star, indicating that the obscuring clouds were probably of amorphous carbon. Then in 1969 Wayne Stein and colleagues discovered an infrared excess for R CrB which they suggested could be due to a circumstellar dust cloud, probably emitted from the star at the last minimum. Seventeen years later Fred Gillett and colleagues resolved the cool dust cloud around R CrB using the IRAS spacecraft. It was about 8 parsec in diameter with a mass of ~0.3 $M_\odot$ and temperature of just 25–30 K.

Another star, RY Sagittarii, had been found to be variable by Jacobus Kapteyn in 1896. Over subsequent years it was found to be normally about magnitude 6.5 but, every now and again, its magnitude would fall to about 14 within a few weeks, before recovering over a longer period of time. These reductions showed no definite periodicity. Its spectrum showed a hydrogen deficiency and an overabundance of carbon, like R CrB. In 2004 Patrick de Laverny and Djamel Mékarnia discovered that it was also surrounded by a number of dust clouds.

At present about 40 of these R CrB variables are known in the Milky Way, most of which are F- and G-type supergiants.

Be Stars/γ Cassiopeiae variables

Angelo Secchi discovered emission lines in the spectrum of γ Cassiopeiae in the 1860s. These included what came to be recognised as the hydrogen β line and the D_3 line of neutral helium. Various astronomers found that the visibility of these and other emission lines was randomly variable over the remainder of the nineteenth century, whilst the star's intensity hardly appeared to change. In 1894 Norman Lockyer reported that the hydrogen emission lines were double and were superimposed on broader hydrogen absorption bands.

A number of other stars like γ Cassiopeiae with hydrogen emission lines had been discovered since Secchi's first observations, so that by 1895 W. W. Campbell was able to list 32 such 'bright Hβ' type stars, as he called them. He concluded that the intensities of their hydrogen emission lines decreased towards the violet, whilst the intensities of their hydrogen absorption lines decreased towards the red. These characteristics were used by the International Astronomical Union (IAU) in 1922 to define a new category of star designated Be (B-type with emission). The IAU also noted that the hydrogen emission lines were often double and superimposed nearly symmetrically on the equivalent hydrogen absorption lines. By 1949 over one thousand Be stars were known.

Otto Struve had concluded in 1931 that the rapid rotation of Be stars had caused them to become unstable, ejecting matter from their equator to form a ring of material. It was this ring of material, revolving around the star, that produced the observed emission lines. Then in 1933 the spectral lines of γ Cassiopeiae started to change spectacularly, associated with changes in its visual intensity. These changes, which lasted for the best part of ten years, were thought to be the result of stellar eruptions that peaked in 1934, 1937 and 1939, producing shells of gas that gradually dispersed.

In 1972 Riccardo Giacconi and colleagues found an X-ray source near the Be star X Persei using the Uhuru spacecraft. At first it was thought that X Persei was not the source, as γ Cassiopeiae did not appear to be an X-ray source. But increasing evidence over the next two or three years showed that X Persei probably was such a source.

Maraschi, Treves and van den Heuvel concluded in 1975 that a number of X-ray transient sources may be Be stars in binaries with neutron stars, the X-rays being produced when the gas emitted by the Be star impacted the neutron star at high speed. Almost immediately J. Garrett Jernigan discovered that γ Cassiopeiae was an X-ray source using the SAS-3 spacecraft. It later transpired that the Copernicus spacecraft had previously detected X-rays from γ Cassiopeiae, whereas the Uhuru spacecraft had not, indicating that the source was variable in X-rays.

Evidence gradually accumulated over the next few years that Maraschi and colleagues had been correct as many Be stars appeared to have neutron stars as their binary companions. By 2000 there were 38 optically identifiable Be X-ray binaries, with 23 possibles. The binary periods ranged from 16 to 400 days, and the neutron star spin rates varied from 0.07 to 1,413 seconds. γ Cassiopeiae, in particular, was found to be a member of a spectroscopic binary with a period of 204 days.

Bibliography

Burnham, Robert Jr., *Burnham's Celestial Handbook: An Observer's Guide to the Universe Beyond the Solar System, Vols. 1, 2 and 3*, Dover Publications, 1977–1978.

Conti, Peter S., and Underhill, Anne B., *O Stars and Wolf-Rayet Stars*, NASA SP-497, 1988.

Gershberg, Roald E. (Knyazeva, Svetlana, trans.), *The Solar-Type Activity in Main-Sequence Stars*, Springer, 2005.

Hearnshaw, J.B., *The Analysis of Starlight; One Hundred and Fifty Years of Astronomical Spectroscopy*, Cambridge University Press, 1986.

Percy, John R., *Understanding Variable Stars*, Cambridge University Press, 2007.

Tassoul, Jean-Louis, and Tassoul, Monique, *A Concise History of Solar and Stellar Physics*, Princeton University Press, 2004.

GAMMA-RAY BURSTS

A series of Vela military spacecraft detected about twenty short, intense gamma-ray bursts (GRBs) over a five-year period start-ing in 1967, with each burst lasting from 0.1 to 30 seconds. Unfortunately the sources of these bursts, which were all different, could not be pinpointed accurately enough for their optical, radio or X-ray counterparts to be determined.

Individual gamma-ray telescopes of the 1970s were far too inaccurate to locate sources of GRBs. So an international network, called the Inter-Planetary Network (IPN), was created in 1978. It used a number of spacecraft acting together to accurately measure the time of arrival of GRB signals and so locate their source position. The first tentative identification of a GRB source was made following the detection of an intense burst on 5 March 1979 by all nine spacecraft of the IPN. But this source in the Large Magellanic Cloud had a much softer gamma-ray spectrum than that of other GRBs, suggesting that it may not be typical. It was later found to be a different type of object, now called a soft gamma-ray repeater.

One of the key missions of the Compton Gamma Ray Observatory (CGRO) was to detect GRBs. By the time that it was launched in 1991 about 500 GRBs had been detected by various spacecraft. Their positions were known reasonably accurately for about 200 of them to be plotted on an all-sky map. This showed a generally uniform distribution across the sky, which implied that their sources could be very close to us, or be in a halo around the Milky Way, or be at extragalactic distances. Astronomers were split on which of these hypotheses was correct; some thought that they were relatively near to Earth, whilst others thought that they were at extragalactic distances.

The CGRO detected about one GRB per day (2,700 in total) over its nine-year lifetime, confirming that their locations were isotropic. The spacecraft also showed that the bursts fell into two categories, one of short-duration, hard-spectrum bursts, and the other of long-duration, soft-spectrum bursts, with the changeover from short to long duration being at about 2 seconds. None of these GRBs could be detected in other wavebands to assist identification, however.

The first major breakthrough in our understanding of GRBs was made when the BeppoSAX spacecraft detected a gamma-ray burst, GRB 970228, on 28 February 1997. BeppoSAX also detected a simultaneous X-ray burst that enabled the position of the GRB to be measured to within about 1 arcmin. The fading afterglow of the source was then detected optically by the William Herschel Telescope on La Palma, enabling its position to be located to within 1 arcsec. This was the first time that a GRB source had been detected in the optical waveband. Subsequent images from the Keck Observatory and the Hubble Space Telescope appeared to show that the source was in a faint distant galaxy.

Another gamma-ray burst, GRB 970508, was discovered by BeppoSAX in May 1997. An X-ray counterpart was detected within hours and optical observers found an object that was still increasing in intensity. Its optical intensity, which peaked two days later, was bright enough to enable its redshift of 0.835 to be determined. This was the first accurate determination of the distance to a GRB source, proving that it was at a cosmological distance. Ralph Wijers and Titus Galama calculated that it had emitted $\sim 3 \times 10^{52}$ ergs of energy, assuming that it was emitted isotropically, which was an order of magnitude more than that emitted by a core-collapse (Type Ic) supernova.

The next burst to have its redshift determined was GRB 971214 of December 1997, which was found to have an optical counterpart with a redshift of 3.42. This GRB appeared to emit $\sim 3 \times 10^{53}$ ergs of energy, assuming that it was emitted isotropically, which was two orders of magnitude more than emitted by Type Ic supernovae. This seemed unlikely, so astronomers questioned if the energy was really being emitted isotropically, suggesting that it may be emitted in two oppositely directed beams instead. Achromatic breaks in the afterglow light curves were expected if the bursts were directed. These have since been observed in many cases, and have permitted the jet opening angles to be estimated for a number of GRBs ranging from 1° to 10°. So the energy released in GRBs was comparable to that emitted by bright core-collapse supernovae.

Initially it had been thought that GRBs were probably the result of the merger of two compact objects in a binary – for example , two neutron stars or a neutron star and black hole. But the GRBs just mentioned were not in the mature galaxies expected, but were associated with young stars in active star-forming galaxies. Then on 25 April 1998 an unusually weak burst, GRB 980425, was detected in a galaxy only about 40 megaparsec from Earth. Two days later an exceptionally powerful supernova, SN 1998bw, was seen at the same location, providing the first clear link between GRBs and supernovae, which was something of a surprise. Then in 2003 another gamma-ray burst, GRB 030329, of more normal energy this time, was found to be associated with a supernova, SN 2003dh, at a redshift of 0.17. This and later identifications tended to confirm that long-duration, soft-spectrum GRBs (LSBs) occurred simultaneously with core-collapse supernovae. But in 2006 two LSBs were detected by the Swift spacecraft that had sufficiently low redshifts that their associated supernovae should easily have been observed. But they were not. The reason for this is still unclear.

In 2005 the HETE-2 spacecraft's detection of GRB 050709 allowed the Chandra X-ray Observatory and the Hubble Space Telescope to identify its X-ray afterglow and, for the first time, the optical afterglow of a short-duration, hard-spectrum GRB (SHB). This showed the cosmological origin of this short duration GRB, which was associated with a star-forming galaxy with a redshift 0.16. SHBs appear to emit less energy than LSBs and come from a lower-redshift population. They are thought to be caused by the merging of binary neutron stars.

Bibliography

Bloom, Joshua S., *What Are Gamma-Ray Bursts?*, Princeton University Press, 2011.

Lamb, D. Q., *The Distance Scale to Gamma-Ray Bursts*, Publications of the Astronomical Society of the Pacific, **107** (1995), pp. 1152–1166.

Paczyński, Bohdan, *How Far Away Are Gamma-Ray Bursters?*, Publications of the Astronomical Society of the Pacific, **107** (1995), pp. 1167–1175.

Schilling, Govert, *Flash!; The Hunt for the Biggest Explosions in the Universe*, Cambridge University Press, 2002.

Vedrenne, Gilbert, and Atteia, Jean-Luc, *Gamma-Ray Bursts: The Brightest Explosions in the Universe*, Springer-Praxis, 2009.

See also: Soft gamma repeaters, magnetars and anomalous X-ray pulsars.

NEUTRON STARS

Theory

In the early 1930s Subrahmanyan Chandrasekhar showed that the maximum mass of a white dwarf was about 1.4 $M_\odot$ (mass of the Sun), whilst heavier stars would continue to collapse.

James Chadwick announced his discovery of the neutron in 1932. Two years later Walter Baade and Fritz Zwicky suggested that a neutron star may be the remnant left over from a supernova explosion. Then in 1939 J. Robert Oppenheimer and George Volkoff showed that stable static neutron stars have a maximum mass of about 0.7 $M_\odot$. This seemed odd as it was less than the maximum mass for a white dwarf, which would not be as compact. Then in 1959 Alastair Cameron showed that the inclusion of nuclear forces increased the maximum mass of neutron stars to a much more reasonable 2 $M_\odot$.

It then gradually became clear that stars could lose a significant amount of mass in the red giant phase after they left the main sequence. This allowed stars with initial masses of up to ~ 8 $M_\odot$ to become white dwarfs. Those with initial masses between ~ 8 to 25 $M_\odot$ would become neutron stars, and stars with masses above this would become black holes. The maximum theoretical mass of a neutron star is now thought to be between about 3 and 5 $M_\odot$, depending on assumptions, and any compact object heavier than that is thought to be a black hole.

Observations

In 1962 Riccardo Giacconi and colleagues discovered the first cosmic X-ray source, now called Sco X-1, of non-solar origin. A little later the same year they also found an X-ray source in the region of the Crab nebula, which was known to be a supernova remnant. Herbert Friedman, C. Stuart Bowyer and colleagues confirmed the existence of both of these sources the following year. Friedman speculated that a neutron star was the source of the Crab X-rays, and in 1964 he and his colleagues used an occultation of the Crab nebula by the Moon to try to measure the size of the X-ray source. It was found to be far too large to be a neutron star.

The first neutron star was discovered unexpectedly a few years later in the guise of what is now known as a pulsar.

Antony Hewish and Jocelyn Bell discovered the first pulsar when they found a regular radio pulse of 1.34 s coming from a source in November 1967. They found another pulsar in December and two more in January 1968. During that year, other radio observatories found even more pulsars. Thomas Gold suggested that these were rapidly rotating, magnetised neutron stars, with typical diameters of 10 to 20 km, which were emitting beams of synchrotron radiation. Some years earlier it had been suggested that neutron stars were produced by supernova explosions, so astronomers speculated that these pulsars may be the stellar remnants of such explosions.

David Staelin and Edward Reifenstein discovered two pulsating radio sources in the region of the Crab nebula in 1968, one of which was eventually found to be associated with the nebula itself. This Crab pulsar, also known as NP 0532, was subsequently found to have a period of 33 milliseconds. In the following year optical flashes and X-ray pulses were found with the same frequency. This very short period showed that the pulsar was a neutron star, as a white dwarf would have broken up at such a fast rate of rotation.

Meanwhile, in 1967 Iosif Shklovskii had analysed the X-ray and optical observations of Sco X-1 and concluded that a stream of gas was flowing from the secondary star in a close binary system towards the primary, which was a neutron star. The X-rays were emitted when the stream of gas fell onto the latter. The donor star of this *low-mass X-ray binary* (LMXB) was not detected until 2001 when its mass was found to be about 0.4 $M_\odot$, compared with the neutron star's assumed mass of 1.4 $M_\odot$.

Since then neutron stars have been found alone or in various types of binary system as summarised below.

The *soft X-ray transient* Cen X-4 had first been detected in 1969. It was eventually found to be in a LMXB consisting of a neutron star, surrounded by an accretion disc, and a K-type donor star, with a binary period of about 15 hours. Soft X-ray transients, in general, were found to be in binary systems in which a small, usually K–M type star overflowed its Roche lobe, and was the donor of material to an accretion disc which surrounded a black hole or neutron star. The disc generated X-rays when it periodically collapsed onto the compact object.

The *hard X-ray transient* A0538-66 had first been detected in 1977 by the Ariel 5 spacecraft. Its X-ray emission was not only highly variable but exhibited a regular 16 day period indicating that it was in a binary system. The spectrum of the optical source measured in 1981 showed that it was a B-type star surrounded by a shell of gas. In the following year Gerry Skinner and colleagues detected X-ray pulsations with a period of 69 milliseconds showing the other star in the binary system was a rapidly rotating neutron star or pulsar. It appears that the pulsar was in a highly eccentric orbit around the B-type star, passing through this B star's shell once per orbit and creating a burst of X-ray energy.

Hard X-ray transient systems are binaries consisting of a massive Be-type star, which is surrounded by a disc of material, and a neutron star. The donor star in these hard X-ray transient systems is heavier than the compact object, whereas in soft X-ray transients it is lighter. The compact object in hard X-ray transients is a neutron star, whereas in soft X-ray transients it is usually a black hole.

X-ray bursters differ from X-ray transients in having an approximately constant X-ray emission on which very brief bursts of X-rays are superimposed. Jonathan Grindlay and John Heise reported the first observations of such an X-ray burster, 4U 1820–30, in late 1975 which emitted sudden, very-brief bursts of X-rays. Stella, White and Priedhorsky found in the mid 1980s that the X-ray output of 4U 1820–30 varied slightly with a regular period of 685 s. This very short orbital period indicated that the LMXB system, which was thought to consist of a white dwarf losing mass to a 1.5 $M_\odot$ neutron star, was very small.

Walter Lewin and colleagues discovered the *rapid burster*, MXB 1730–335, in early 1976. Burst durations varied from a few seconds to a few minutes, with variable time intervals between bursts. At times it produced ~1,000 bursts/day. The rapid burster was found to be active for a few weeks about every six months or so.

Normal X-ray bursters exhibited Type I bursts with intervals from hours to days, whilst the rapid burster generally exhibited Type II bursts with intervals from seconds to minutes. In Type I bursts a companion star loses hydrogen onto the surface of a neutron star where it burns steadily to form helium. Eventually when the density and temperature are high enough, the helium burns explosively to produce an X-ray burst. Type II bursts are due to an instability in the accretion disc around the neutron star. From time to time material from the disc is suddenly released, falling rapidly onto the neutron star where it generates a Type II burst.

Isolated (i.e., non-binary) neutron stars have been detected as soft gamma repeaters (SGRs), the first of which, SGR 0526-66, was found to emit an intense burst of gamma rays on 5 March 1979. The source was a young, magnetised neutron star with a spin period of about eight seconds in a supernova remnant. In 1992 Robert Duncan and Christopher Thompson suggested that these SGRs were 'magnetars', that is neutron stars powered by ultrastrong magnetic fields. Duncan and Thompson showed that rapid internal convection and dynamo action in a newly formed neutron star could produce a magnetic field of $\sim 10^{15}$ gauss. It was initially thought that the SGR outbursts were due to reconnection instabilities in their magnetospheres, but more recently the role of crustal instabilities has been thought to be highly significant.

The first *anomalous X-ray pulsar* (AXP) was discovered in 1980 as a strong point-like X-ray source, 1E 2259+586, in what appeared to be a supernova remnant. Shortly afterwards the point-like source was found to be an X-ray pulsar with a period of about seven seconds. Since then a total of nine AXPs have been discovered six of which have been found to emit X-ray bursts. AXPs appear to be magnetars which are less active than those in SGRs.

The X-ray and optical afterglow of the short-duration, hard-spectrum *gamma ray burst* (SHBs) GRB 050709 was found in 2005. These showed that it was associated with a star-forming galaxy with a redshift 0.16. SHBs are thought to be caused by the merging of two neutron stars in a binary to form a black hole.

More detail about the different types of objects shown in italics above, each of which is a neutron star on its own or in association with another type of object, is given in individual articles listed in 'See also' below.

Bibliography

Becker, Werner (ed.), *Neutron Stars and Pulsars*, Springer, 2009.

Camenzind, Max, *Compact Objects in Astrophysics: White Dwarfs, Neutron Stars and Black Holes*, Springer, 2007.

Haensel, P., Potekhin, A. Y., and Yakovlev, D. G., *Neutron Stars 1: Equations of State and Structure*, Springer Science, 2007.

Lyne, Andrew G., and Graham-Smith, Francis, *Pulsar Astronomy*, 3rd ed., Cambridge University Press, 2006.

Miller, Arthur I., *Empire of the Stars: Friendship, Obsession and Betrayal in the Quest for Black Holes*, Little, Brown, 2005.

See also: Gamma-ray bursts; Pulsars; Soft gamma repeaters, magnetars and anomalous X-ray pulsars; Supernova remnants; X-ray bursters; X-ray transients.

PLANETARY NEBULAE

Charles Messier recorded the first planetary nebula in 1764. They were later given this name of 'planetary nebula' by William Herschel, who thought that their appearance often resembled that of planets in the telescope. Herschel's observations of planetary nebulae, in which the nebula often surrounded a central star (see, for example, Figure 3.7), led him to conclude in 1791 that they were in the process of condensing to form stars.

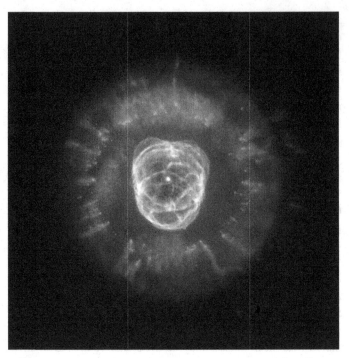

Figure 3.7. A Hubble Space Telescope image of the Eskimo nebula showing material emitted by the central star. This planetary nebula was first observed by William Herschel in 1787. (Courtesy NASA, ESA, Andrew Fruchter and the ERO team [STScI].)

In 1864 William Huggins observed the first spectrum of a planetary nebula, NGC 6543, or the Cat's Eye nebula. Much to his surprise it consisted of a bright green emission line, and apparently nothing else. The line was initially thought to be due to nitrogen, but later measurements showed this was not so. Instead Huggins attributed it to an unknown element that he called 'nebulum' (later generally called 'nebulium'). As spectroscopic observations improved later in the nineteenth century other emission lines, including the H_β line, were found in the spectra of planetary nebulae.

In the early 1920s, following the work of William H. Wright and others, it was recognised that the central stars of planetary nebulae are very hot and so radiate a great amount of energy in the ultraviolet. Henry Norris Russell pointed out in 1921 that the degree of excitation of planetary nebulae decreased with increasing distance from the centre, and suggested that this excitation was due to the central star. At about the same time Edwin Hubble found an approximate correlation between the intensity of the central star and the size of the planetary nebula, giving support to Russell's suggestion.

Herman Zanstra used quantum theory in the mid-1920s to try to explain the spectra of planetary and other nebulae as being caused by the absorption of ultraviolet light emitted by nearby stars. Following a suggestion of Walter Baade, he later included the effect of recombination of hydrogen ions with free electrons, following photoionisation. As a result Zanstra concluded in 1927 that temperatures of the central stars of planetary nebulae must be ~30,000 K to produce the observed ionisation of the nebulae.

All this time there was a nagging doubt created by the lack of identification of many of the emission lines, particularly those of nebulium. But in 1928 Ira Bowen showed that nebulium did not exist. Instead, its lines were produced by forbidden transitions between low-lying metastable states of nitrogen and oxygen, for example, and their ground states. The transition probabilities were too low in normal laboratory conditions, but high enough in low-density gaseous nebulae. In 1941 Donald Menzel and Lawrence Aller showed that the mechanism of production of the forbidden lines created strong cooling in the planetary nebula, limiting its electron temperature to 20,000 K, no matter how hot the central star was.

There was considerable confusion in the 1920s about the movements of planetary nebulae following the discovery that many of their spectroscopic lines were broad or split. It was thought by some astronomers that they were rotating, and estimates were made of the masses of the central stars as a result. But in 1929 Charles Perrine correctly interpreted the lines as showing that the nebulae were expanding.

There were considerable difficulties in determining the distances to a significant number of planetary nebulae. But in 1956 Iosif Shklovskii proposed a new method of distance determination which produced absolute magnitudes of the central stars generally from +5 to +10, with mean densities similar to those of white dwarfs. As a result, he concluded that many of these stars were 'over-heated' white dwarfs which, over time, would cool to become normal white dwarfs. He proposed, therefore, that planetary nebulae are in the short, ~5×10^4 year, transition phase of the evolution of stars, possibly red giants, en route to becoming white dwarfs. He further suggested that the red giants would lose their outer atmosphere gradually, and that planetary nebulae were the most powerful supplier of gas to interstellar space.

Ten years later George Abell and Peter Goldreich argued, quantitatively, that red giants lose their mass by a sudden but not catastrophic ejection, rather than in the gradual way envisaged by Shklovskii. They also showed statistically that most if not all stars of ~1.2 solar mass now leaving the main sequence will produce planetary nebulae.

More recent work on planetary nebulae has shown that the temperature of their central stars range from ~25,000 K to 250,000 K, although these temperatures are rather uncertain, with luminosities of about 10 to 10^4 times solar. Their radii range from those typical of hot white dwarfs to those a little larger than the Sun. They seem to have a carbon-rich core, surrounded by a helium envelope, sometimes overlaid by a hydrogen skin. All this indicates that they are in the process of becoming white dwarfs. Their spectra show some similarities with those of Wolf-Rayet and O-type stars, although their physical structures are very different from these 'classical' stars.

It is now thought that planetary nebulae are produced in two main stages. First, a pulsating red giant loses its outer layers in the form of a strong stellar wind. At this stage the star's carbon core is surrounded by a hydrogen- or helium-burning shell. Second, this shell breaks free, leaving a hot, inert core. The shell then catches up with the earlier emitted gas to produce the observed planetary nebula.

Bibliography

Gurzadyan, Grigor A., *The Physics and Dynamics of Planetary Nebulae*, Springer, 1997.

Kaler, James B., *Stars and Their Spectra: An Introduction to the Spectral Sequence*, Cambridge University Press, 1997.

Kwok, Sun, *The Origin and Evolution of Planetary Nebulae*, Cambridge University Press, 2000.

Osterbrock, Donald E., *Herman Zanstra, Donald H. Menzel, and the Zanstra Method of Nebular Astrophysics*, Journal of the History of Astronomy, **32** (2001), 93–108.

Stanghellini, L., Welsh, J. R., and Douglas, N. G. (eds.), Planetary Nebulae Beyond the Milky Way; Proceedings of the ESO Workshop held at Garching, Germany, 19–21 May, 2004, Springer, 2006.

PULSARS

Jocelyn Bell and Antony Hewish discovered the first pulsar, PSR 1919+21, in November 1967 when they found a regular radio pulse lasting 16 ms, with a pulsation period of 1.34 s. This period was regular to about one part in 10^7. They quickly established that the source was not terrestrial but they delayed publication until February 1968, unsure of what they had found. They discovered another pulsing source or pulsar, as these sources were soon called, in December, and two more in January 1968. During 1968 other radio observatories found even more pulsars.

Franco Pacini had shown, just before the announcement of the pulsar discovery, that a rapidly rotating neutron star with a strong dipole magnetic field could efficiently transform rotational energy into electromagnetic radiation and accelerate particles to high energies. As a result it could provide energy to any surrounding nebula like the Crab nebula.

After the discovery of pulsars, Thomas Gold, without knowing of Pacini's work, suggested that pulsars were isolated, rapidly rotating, magnetised neutron stars. He reasoned that if a large star collapsed at the end of its life to form a neutron star, then it must rotate very fast to conserve angular momentum. He also pointed out that not only would the star's rotational rate increase when it collapsed, but its magnetic field would also increase substantially to about 10^{12} gauss. This would produce highly directional beams of radio emission emitted along the magnetic axis. If the magnetic axis was not coincident with the spin axis, then these directional radio beams would sweep across the sky. If the Earth was in the path of such a beam, we would see a pulse of radio energy at the rotational rate of the star as the beam swept past the Earth. Gold, like Pacini, pointed out that the spin rate would gradually slow down as the star lost energy through particle emission from its magnetic poles.

This theory was initially treated with considerable scepticism, but in October 1968 a radio pulsar, PSR 0833-45, was discovered by Michael Large, Alan Vaughan and Bernard Mills in the Vela supernova remnant. It had a period of only 89 ms, the shortest then known. This very short period showed that the pulsar must be a neutron star, as a white dwarf would have been broken up at such a fast rotational rate. This was a crucial discovery linking as it did a pulsar, or pulsating neutron star, with a supernova remnant. It vindicated Walter Baade and Fritz Zwicky's theory of 1934 in which they had suggested that a neutron star may be produced by a supernova explosion, and Pacini's suggestion made in 1967 that a highly magnetic, rapidly rotating neutron star could be the source of energy in supernova remnants.

David Staelin and Edward Reifenstein discovered two pulsating radio sources in the region of the Crab nebula in late 1968, one of which, the Crab pulsar NP 0532, was subsequently found to have a period of 33 ms. David Richards and John Comella then found that the pulse rate of this Crab pulsar had lengthened, as proposed by Gold and Pacini, by about 4×10^{-8} s/day. This deceleration rate was consistent with the idea that the Crab nebula was the remains of the supernova explosion observed in the year 1054.

In 1969 the periods of the Vela and Crab pulsars suddenly reduced, and since then similar "glitches" have been observed for other pulsars. Ruderman, Baym and colleagues pointed out that the surface crust of a neutron star would be oblate because of the star's rapid spin rate, but this oblateness would tend to reduce as the star spins down. They suggested that the glitches could be due to the neutron star readjusting itself discontinuously to the star's smooth spin down by a periodic fracture of the crust.

William Cocke, Michael Disney and Donald Taylor were the first to observe optical pulses emitted by a pulsar when, in 1969, they discovered optical flashes coming from the Crab pulsar with exactly the same frequency as the radio pulses. In the same year the Crab pulsar was observed to be pulsing in X-rays, and a few years later it was found to be pulsing in gamma rays.

A new type of source was discovered by the Uhuru spacecraft in 1971, when it found that Cen X-3 was emitting X-ray pulses with a period of 4.84 s. In addition, every 2.1 days the source disappeared for about 12 hours, indicating that this X-ray pulsar was part of an eclipsing binary system. This

was proved by accurately measuring the arrival times of the pulses that were also seen to vary with a period of 2.1 days. Wojciech Krzeminski then detected the pulsar's companion as a blue star whose mass was later estimated to be about 20 $M_\odot$ (mass of the Sun). It was concluded that some of the matter emitted by the blue star was being channelled onto the poles of its compact companion where it was producing the X-ray radiation.

Her X-1 was also found by Uhuru to exhibit a similar behaviour to Cen X-3, but with a pulse period of 1.24 s and a binary period of 1.7 days. Its visible companion was found to be the blue variable HZ Herculis. But in this case the spectrum of the optical source varied, and the X-ray pulsation behaviour varied with a period of 35 days. It is thought that the X-ray source is accreting material from HZ Herculis, and the orbit of the binary is precessing with a period of 35 days.

Joe Taylor and Russell Hulse discovered a pulsar, PSR 1913+16, that had a pulsation rate of 59 ms, using the Arecibo radio telescope in 1974. This rate varied with a period of 7 h 45 min, indicating that the pulsar was in a binary orbit. Later analysis showed that the pulsar is a 1.4 $M_\odot$ neutron star in orbit around the common centre of mass of an invisible star of about the same mass. This binary pair provided an excellent check of Einstein's general theory of relativity, as the two stars were reasonably heavy and orbiting very close to each other. This resulted in a precession of 4.2°/year, which was in good agreement with theory.

In 1982 Don Backer and colleagues found a pulsar, PSR 1937+214, with a pulsation rate of just 1.56 ms. This, the first millisecond pulsar to be discovered, appeared to have a characteristic age two orders of magnitude greater than that of typical pulsars. Since this discovery of PSR 1937+214 about 200 millisecond pulsars have been found, most of which are in globular clusters. Some are in low-mass X-ray binaries, but some seem to be isolated stars. The origins of millisecond pulsars are still unclear.

Current theories predict that neutron stars would break up at a spin rate of ~1,500 revolutions/second (a pulsation rate of ~0.7 ms), and that they would lose energy by gravitational radiation which would probably limit their maximum spin rate to ~1,000 revs/s. The fastest pulsar currently known has a spin rate of ~1,100 revs/s (pulsation rate ~0.9 ms).

In 2003 Marta Burgay and colleagues discovered a millisecond pulsar, PSR J0737−3039, in a binary orbit with a neutron star. A little later this neutron star was also found to be a pulsar, making this the first double pulsar system to be discovered. The second pulsar had a period of 2.8 s, the binary period was just 2.4 hr, and the relativistic advance in the periastron was found to be about 17°/yr, which was by far the highest value ever measured. It is expected that these two stars will coalesce in about 85 million years.

Bibliography

Becker, Werner (ed.,), *Neutron Stars and Pulsars*, Springer, 2009.

Camenzind, Max, *Compact Objects in Astrophysics: White Dwarfs, Neutron Stars and Black Holes*, Springer, 2007.

Lorimer, Duncan, and Kramer, Michael, *Handbook of Pulsar Astronomy*, Cambridge University Press, 2005.

Lyne, Andrew G., and Graham-Smith, Francis, *Pulsar Astronomy*, 3rd ed., Cambridge University Press, 2006.

See also: Neutron stars; Soft gamma repeaters, magnetars and anomalous X-ray pulsars.

PULSATING VARIABLES

Cepheid variables

Edward Pigott discovered in 1784 that η Aquilae was varying in brightness with a period of 7 days, and that it increased in brightness more rapidly than it decreased. A few weeks later, John Goodricke found that δ Cephei was exhibiting similar fluctuations with a period of about 5 days.

Over the next century many more of these so-called Cepheid variables were discovered. Some astronomers thought that they were eclipsing binaries with highly eccentric orbits, but the shape of their light curves was difficult to explain. Then in 1879 Arthur Ritter suggested that they could be stars undergoing radial oscillations, that is stars oscillating in size. But this idea did not seem plausible to astronomers of the time.

In 1894 Aristarkh Belopol'skii found that the radial velocity of δ Cephei, measured spectroscopically, varied with the same period as its intensity, indicating that it may be a spectroscopic binary, but its minimum brightness occurred a day earlier than expected from its velocity curve. Three years later he found that η Aquilae's maximum radial velocity almost coincided with its minimum intensity, so its intensity variations could not be caused by eclipses. Nevertheless, it was possible that Cepheids could be non-eclipsing spectroscopic binaries. This was the favoured theory of Cepheids in the early twentieth century, in spite of numerous problems of detail.

Karl Schwarzschild found in 1899 that the photographic intensity range was 50% greater than the visual range for Cepheids, indicating that they were bluer, and therefore hotter, at maximum intensity than at minimum. Eight years later Sebastian Albrecht confirmed spectroscopically that the variations in luminosity of Cepheids appeared to be caused by temperature. These temperature changes were attributed to some sort of tidal effect on the Cepheid caused by a dark binary companion. But when the tidal behaviour of the star was analysed, the results did not correlate with observations.

An alternative theory was suggested by Robert Emden in 1906 and Forest Moulton in 1909. They suggested that Cepheids underwent non-radial oscillations, that is they oscillated in shape from prolate to oblate. Unfortunately this theory produced a radial velocity period that was twice its luminosity period, whereas they were observed to be the same. So, at this stage, all the theories about the cause of the Cepheids' variability had failed in some way or other.

The Magellanic Clouds were surveyed in the early years of the twentieth century by Henrietta Leavitt. In 1908 she published a list of 1,777 variables in the two clouds and by 1912 she was able to measure the periods of 25 Cepheid variables in the Small Magellanic Cloud (SMC), finding that the brightest stars had the longest periods (see Figure 3.8). Because all stars in the SMC were at virtually the same distance from us, the differences in apparent luminosity were essentially the same as differences in absolute luminosity. So the periods of the Cepheid variables were clearly correlated with their absolute luminosities.

In 1913 Ejnar Hertzsprung estimated the distances of 13 relatively close Cepheids, based on their proper motions. This gave their absolute luminosities. He also measured their periods, which enable him to normalise Henrietta Leavitt's period-luminosity relationship. As a result he was able to estimate the distance of the SMC using Henrietta Leavitt's data on apparent luminosities and periods for the SMC Cepheids. In the same year he and Henry Norris Russell independently concluded that Cepheid variable stars were of very high luminosity, which, given their spectral type, implied that they were giant stars of low density.

Meanwhile by 1897 Solon Bailey had found 310 short-period variables in 16 globular clusters, most of which had periods of less than one day. But, unlike the case of Cepheid variables, these short-period or cluster variables, as they were called, were of about the same magnitude, no matter what their period, within any given cluster. Two years later Williamina Fleming also discovered that the star RR Lyrae in the Milky Way had a period of less than one day. In 1912 C. C. Kiess concluded that RR Lyrae, and fifteen similar stars not in clusters, showed the same type of variability as Bailey's cluster variables. He also pointed out that they were similar in many ways to Cepheid variables.

In 1914 Shapley reviewed the theories that tried to explain the behaviour of Cepheid variables and concluded that both the Cepheids and the cluster variables (since called RR Lyrae variables), were radially pulsating stars. But it was not until Arthur Eddington published his theory of pulsating stars over the period 1918–1919 that this theory started to become accepted. Eddington showed theoretically that $P\sqrt{\rho}$ was virtually constant for all Cepheids, where P is the period and ρ the central density. This was found observationally to be correct.

Eddington pointed out that the driving mechanism, which at that time was unknown, must add heat to the star during compression, enhancing the subsequent expansion. Then in

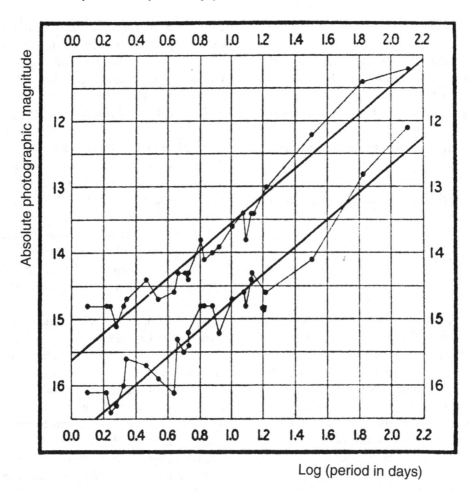

Log (period in days)

Figure 3.8. Henrietta Leavitt's period-luminosity diagram for 25 Cepheid variables in the Small Magellanic Cloud. The two lines indicate maximum and minimum photographic magnitudes. (From *Harvard College Circular*, No. 173 (1912).)

1941 he suggested that the thin outer layer of the star, where hydrogen changes from neutral to ionised, may be the location of the driving mechanism. Subsequent work by Sergei Zhevakin, John Cox, Charles Whitney, and others showed that the region of the second ionisation of helium (where He^+ is ionised to become He^{++}) was the source of the drive mechanism for pulsations of Cepheids and RR Lyrae variables.

In 1952 Walter Baade found that what he called the Population I, or so-called classical Cepheids, were about 1.5 magnitudes brighter at a given period than the Population II Cepheids (or W Virginis variables) in globular clusters. This required a modification to the distance estimates for the Magellanic Clouds and the Andromeda nebula.

It was later found that, although the W Virginis and classical Cepheids were both pulsating stars, the W Virginis variables were older and less massive than the classical Cepheids. W Virginis variables also tended to have a lower metal abundance than the Sun, and sometimes exhibited emission lines in their spectra caused by shock waves passing through their low-density atmospheres.

RR Lyrae variables

As mentioned above, near the end of the nineteenth century Solon Bailey discovered numerous cluster variables, having periods of less than one day, in globular clusters. In 1902 he divided these cluster variables into three subclasses depending on the amplitude and shape of their light curves. At about the same time Williamina Fleming discovered that the star RR Lyrae in the Milky Way also had a period of less than one day. As more and more of these short-period variables were found outside globular clusters, it seemed inappropriate to still call them cluster variables. So in 1948 the International Astronomical Union agreed to call them RR Lyrae variables, no matter where they were found.

In 1914 Harlow Shapley concluded that both the Cepheids and cluster (RR Lyrae) variables were radially pulsating stars. But there were observational differences between the two different types, in addition to their different periods. In addition, Ejnar Hertzsprung pointed out that Cepheids were concentrated in the plane of the Milky Way, whilst RR Lyraes were found at all galactic latitudes. So eventually RR Lyraes and Cepheids came to be treated as separate types of stars.

By 1916 Richard Prager and Harlow Shapley had independently discovered that the amplitude and shape of RR Lyrae's light curve varied cyclically with a period of about 40 days. This so-called Blazhko effect, which has been detected in a number of RR Lyrae variables, has still not been satisfactorily explained.

Over the years it has gradually become clear that, although both RR Lyraes and classical Cepheids are radial pulsators, there are many differences between the two types of stars. The RR Lyraes, which are very much older than classical Cepheids, have exhausted the hydrogen in their core and are burning helium. They occupy a horizontal branch on the H-R diagram as, although they have a wide range of temperatures, they have only a narrow range of luminosities. Because of this they are particularly useful as distance indicators, even though they are not as bright as classical Cepheids. In the Milky Way they are found near the galactic centre and in the galactic halo, as well as in globular clusters. They are physically smaller than classical Cepheids having radii of about 4–6 $R_\odot$ (solar radii), compared with 14–200 $R_\odot$ for Cepheids. Their masses are usually about 0.6 $M_\odot$, compared with Cepheids whose masses are more than that of the Sun. The metal abundances of RR Lyraes, which is very low, varies substantially from star to star, depending on age.

β Cephei/β Canis Majoris variables

The variability of β Cephei was first observed spectroscopically by Edwin Frost in 1902. He found that that it had a period, which he published in 1906, of only 0.19 days. At the time it was assumed to be, like δ Cephei, a spectroscopic binary, but its variability was much faster than any other spectroscopic binary known at the time. A few years later Paul Guthnick found that its intensity varied by just 0.05 magnitude. β Cephei was initially thought to be a Cepheid variable, but later results showed that, unlike the case of δ Cephei, maximum luminosity occurred when the star was most highly compressed.

Over the years a number of these β Cephei variables have been found. They are characterised by large variations in radial velocity but small changes in luminosity, generally ranging from 0.01 to 0.2 magnitudes. Their periods are usually within the range 0.1 to 0.3 days. β Cepheids, which generally have spectral types from B0 to B3, are very hot stars on or just above the main sequence, and are a long way from the Cepheid instability strip. Their masses usually range from 10 to 15 $M_\odot$.

A number of these β Cephei variables, including β Canis Majoris, have been found to be multiperiodic. Paul Ledoux suggested in 1951 that that this was due to nonradial oscillations. But in 1962 Subrahmanyan Chandrasekhar and Norman

Lebovitz thought that it would be unlikely that purely nonradial oscillations would be excited in preference to radial modes, as the former would be relatively highly damped. They suggested, instead, that the multiperiodic behaviour may be due to a coupling between radial and nonradial modes caused by the stars' rotation. For a spherical nonrotating star, the nonradial periods would be degenerate and have the same values, whereas rotation would remove degeneracy, producing nonradial modes with slightly different periods. Yoji Osaki showed in the early 1970s that nonradial oscillations could explain most of the observed properties of β Cephei variables. Myron Smith then showed that, although the main pulsation is purely radial, both radial and nonradial modes must be involved in those stars that exhibit multiperiodic behaviour.

The cause of the β Cephei variables' pulsations was uncertain for most of the twentieth century. But in 1982 Norman Simon pointed out that they could be explained if the opacities of heavy elements in the stars were higher than they were then calculated to be. These calculated opacities were re-examined by Iglesias, Rogers and Wilson and were shown to be incorrect five years later. As a result it became clear that the ionisation of iron at a temperature of the order of 150,000 K deep within the star was sufficient to drive the observed pulsations.

δ Scuti variables

δ Scuti stars were first recognised in the early twentieth century as variable stars with short periods and generally small magnitude variations. Those with the largest amplitudes were classified as RR Lyrae stars, whilst those with the smallest amplitudes tended to be put in a miscellaneous category. Harlan Smith analysed RR Lyrae stars and concluded in 1955 that those with periods of less than 0.2 day were probably not true RR Lyrae stars. Since then, most of the small amplitude, short period variables were called δ Scuti stars, whilst those with amplitudes greater than 0.3 magnitude were called dwarf Cepheids or RR Lyrae stars. However some variables with amplitudes less than 0.3 magnitude were also called dwarf Cepheids.

In 1979 Michel Breger showed that the majority of dwarf Cepheids were not fundamentally different from δ Scuti stars, so he abandoned the term dwarf Cepheid and included those stars within the δ Scuti group. A sub-group was then devised, called SX Phoenicis stars, which included those δ Scuti stars which have low metal abundances and are older than the parent group. This is the arrangement prevailing today.

This complex story shows how difficult it is to categorise variable stars and, in particular, where to draw the borderline, particularly when stellar structures and the physical processes are so poorly understood in many cases.

The Cepheid instability strip on the H-R diagram, above the main sequence, contains various classes of pulsating stars from classical Cepheids and W Virginis variables to RR Lyrae variables. The δ Scuti stars, which are usually in the spectral range A5 to F2, were found to occupy that part of the instability strip where it meets the main sequence. The δ Scuti stars have periods usually in the range from 0.02 to 0.3 day, which is the range of natural radial pulsation frequencies for stars of their spectral type. Their amplitude variations are generally from 0.01 to 0.3 magnitudes. Many δ Scuti stars are multiperiodic, pulsating in a number of radial and non-radial modes.

Mira and long-period variables

Mira (o Ceti) was the first star, apart from Algol, found to vary in brightness. David Fabricius saw the star disappear in 1596, and Phocylides Holwarda observed its variability over the period 1638−39. In 1667 Ismaël Boulliau showed that it was varying with a period of about 334 days. It was the first long-period variable to be discovered.

Mira behaved quite differently from Algol as its intensity variation was neither constant in magnitude nor in period. It was clearly not an eclipsing binary, but was a star that appeared to be genuinely varying in luminosity. Boulliau suggested that its variability may be due to its rotation alternately showing a spotted and unspotted hemisphere. In 1852 Rudolf Wolf, noting the similarity between the irregular behaviour of long-period variables, including Mira, and the recently discovered irregular sunspot cycle, also suggested that the stellar variations may be due to star spots. Then in 1879 Arthur Ritter made the radical suggestion that these irregular variables could be stars oscillating in size. But his idea was ignored for some time, until it was resurrected in the early twentieth century.

By 1890 almost 100 long-period variables had been discovered, having periods of from 90 to just over 600 days. The amplitude of their light variations was very large, averaging about 6 magnitudes. They were all cool red stars. Bright hydrogen lines near maximum intensity, which were found to be characteristic of these variables, had been first discovered in 1869 by Angelo Secchi for the long-period variable R Geminorum. Henry Norris Russell showed in 1918 that these long-period variables were red giants of low density.

Ritter's theory of radial stellar oscillations was developed in the first half of the twentieth century by Arthur Eddington and others. So by about 1950 it was generally accepted that the observed luminosity variations were caused by periodic radial expansions and contractions. Eddington suggested in 1941 that the thin outer layer of the star, where hydrogen changes from neutral to ionised, may be the location of the driving mechanism. Subsequent work seems to have confirmed that this is the dominant driving source for most long-period variables.

Bibliography

Burnham, Robert Jr., *Burnham's Celestial Handbook, An Observer's Guide to the Universe Beyond the Solar System; Vols. 1, 2 and 3*, Dover Publications, 1977–1978.

Hearnshaw, J.B., *The Measurement of Starlight; Two Centuries of Astronomical Photometry*, Cambridge University Press, 1996.

Hogg, H.S., *Variable Stars*, in Gingerich, O. (ed.), *Astrophysics and Twentieth-Century Astronomy to 1950: Part A*, Cambridge University Press, 1984.

Percy, John R., *Understanding Variable Stars*, Cambridge University Press, 2007.

Smith, Horace A., *RR Lyrae Stars*, Cambridge University Press, 2003.

Tassoul, Jean-Louis, and Tassoul, Monique, *A Concise History of Solar and Stellar Physics*, Princeton University Press, 2004.

SOFT GAMMA REPEATERS, MAGNETARS AND ANOMALOUS X-RAY PULSARS

Soft gamma repeaters

An intense burst of gamma rays was detected on 5 March 1979 by nine spacecraft. Accurate timing of the burst from the various spacecraft enabled the source position to be located in the supernova remnant (SNR) N49 in the Large Magellanic Cloud, although it was thought, at first, that it may have been a foreground object because of its exceptionally high gamma-ray intensity. The initial gamma-ray burst from what was to be called SGR (soft gamma repeater) 0526-66 had a rise time of <0.2 ms and an initial peak of just ~200 ms. It was then followed by a series of pulses about 8.0 s apart of softer gamma rays detectable for about three minutes. Additional but weaker bursts were also detected 0.6, 29 and 50 days after the first event.

The source SGR 0526-66 was behaving very differently from that of gamma-ray bursters known at the time as its rise time was exceptionally fast, its intensity was exceptionally bright, its high intensity period was abnormally brief, its pulse structure was unusual, its gamma-ray spectrum was softer than normal and it had further gamma-ray bursts some time after the initial event. The clear 8.0 s gamma-ray pulsations and the source's apparent association with a SNR indicated that the source was a young, magnetised neutron star with a spin period of 8.0 s.

In 1982 E. P. Mazets and colleagues published a review of sources of short gamma-ray bursts, a number of which were in soft gamma rays. The paper was mainly concerned with details of the 5 March 1979 source, but it also included details of two other sources of brief, soft gamma-ray pulses. The first had emitted a single 200 ms gamma-ray burst on 7 January 1979, whilst the second, listed as B 1900+14, had emitted three pulses in four days in late March of the same year.

The 7 January 1979 source subsequently emitted pulses on thirteen days between March and December 1983. Four years later J.-L. Atteia and colleagues concluded that this source, which they called SGR 1806–20, was intermediate in type between X-ray bursters and gamma-ray bursters. They located it on the galactic equator about $10°$ from the galactic centre. They agreed with Mazets and colleagues that it was similar in type to SGR 0526–66 and B 1900+14, now called SGR 1900+14.

SGR 1806–20 had been quite active since it was first detected in January 1979. In the 1980s it produced more than 100 outbursts, then in 1994 Toshio Murakami and colleagues found that it emitted X-rays, even in gamma-ray quiescence, and that its X-ray intensity increased dramatically at the same time as its gamma-ray bursts. Four years later Chryssa Kouveliotou and colleagues found that its X-ray intensity was pulsating with a period of 7.5 s, and that that rate was gradually reducing at 2.6×10^{-3} s/yr.

Shrinivas Kulkarni and Dale Frail had suggested in 1993 that SGR 1806–20 was associated with the radio nebula and apparent supernova remnant G10.0-0.3. But this association was called into question in 2002 when the position of the SGR source was measured more accurately using the Chandra spacecraft. It was then clear that the SGR source was not coincident with the non-thermal core of G10.0-0.3. It is currently thought to be associated with the molecular cloud of which G10.0-0.3 is part.

An exceptionally bright gamma-ray flare was observed from SGR 1806–20 on 27 December 2004 that lasted for 380 s. A precursor was detected 142 s before the main peak. The main flare created a massive disturbance in the Earth's daytime lower ionosphere, increasing the electron density by about two or three orders of magnitude. The SGR's X-ray behaviour, which was observed several months later, showed a different spin-down rate, a different pulse profile and a softer spectrum than before the major flare.

SGR 1900+14 was, like SGR 1806–20, also found to be an X-ray source. It emitted a giant gamma-ray flare on 27 August 1998. After the peak, its intensity gradually reduced so that flare activity stopped after about 380 s. Its gamma-ray emissions had a period of 5.16 s, which was the period of its previously detected X-ray emissions. This was thought to be the spin rate of the SGR. The outburst was so powerful that it strongly ionised the Earth's upper atmosphere, causing the ionosphere on the night side to reduce in height to its daytime level, thus affecting radio communications.

So the above three SGRs have very similar spin rates, and each of them has emitted a giant gamma-ray flare over the last few decades. It was initially thought that each SGR was associated with a supernova remnant, but today (2010) that link is by no means certain. Since the discovery of the above three SGRs, only a few more have been found.

Magnetars

The spin rate of SGRs was too slow to power their X-ray and gamma-ray emissions. So in 1992 Robert Duncan and Christopher Thompson suggested that these were powered by ultrastrong magnetic fields. They showed that a newly formed neutron star could have a magnetic field of $\sim 10^{15}$ gauss caused by rapid internal convection and dynamo action. They also calculated that the magnetic field of SGR 0526–66 would need to be about 6×10^{14} gauss to have caused its spin rate to reduce to its observed period of 8 s in the estimated age of $\sim 10,000$ years of its associated SNR N49. Then in 1998 Kouveliotou and colleagues estimated a magnetic field strength of 8×10^{14} gauss for SGR 1806–20, based on its observed spin-down rate. In the following year they also estimated a magnetic field strength of 2 to 8×10^{14} gauss for SGR 1900+14 based on its observed spin-down rate of 3.5×10^{-3} s/yr. The similarities of these magnetic field values to that estimated by Duncan and Thompson gave a clear boost to their theory of what they called 'magnetars'.

It was initially thought that the SGR outbursts were due to reconnection instabilities in their magnetospheres, but more recently the role of crustal instabilities has been thought to be highly significant. One possibility is that the stresses created by the internal magnetic field could cause the stellar crust to crack, allowing significant magnetic energy to be released. This could initiate magnetic reconnections in the star's magnetosphere. The theory of magnetars is still very much 'work in progress'.

Anomalous X-ray pulsars

In 1980 Philip Gregory and Greg Fahlman discovered a strong point-like X-ray source, now called 1E 2259+586, at the centre of curvature of a semicircular diffuse X-ray source which they suggested was a supernova remnant. Shortly afterwards the point-like source was found to be an X-ray pulsar with a pulse period of 6.9786 s. Although this period was found to gradually increase over time, the release of rotational energy was far too slow to power the pulsar's X-ray luminosity. No companion was found so it could not be gravity powered like an X-ray binary. So Thompson and Duncan suggested that the source could be a magnetar powered by its ultrastrong magnetic field. By 1995 two other similar sources had been found of what Duncan called 'anomalous X-ray pulsars' (AXPs).

Van Paradijs, Taam and van den Heuvel suggested in 1995 that AXPs may be isolated (i.e., not binary) neutron stars surrounded by discs of material. In this case the magnetic fields need not be as high as in magnetars, and the emissions would be powered by material from the discs falling onto the star. They suggested that the material was left over from the initial process that had formed the neutron star.

Faint visible and infrared counterparts were then found to some AXPs, but they seemed to be too faint to be consistent with the disc model. Then in 2001 Gavriil, Kaspi and Woods detected two X-ray bursts from AXP 1E 1048.1-5937, and in the following year Victoria Kaspi and colleagues detected over 80 X-ray bursts in just four hours from AXP 1E 2259+586. Today nine AXPs are known of which six have been found to emit bursts. The faintness of their visible and infrared counterparts and the existence of these bursts was taken to indicate that AXPs are magnetars, rather than isolated neutron stars surrounded by an accretion disc.

Bibliography

Becker, Werner (ed.), *Neutron Stars and Pulsars*, Springer, 2009.

Duncan, Robert C., *Anomalous X-Ray Pulsars: Mystery Solved?*, Sky and Telescope, January 2005, pp. 34–42.

Woods, P. M., and Thompson, C., *Soft Gamma Repeaters and Anomalous X-ray Pulsars: Magnetar Candidates*, in Lewin, Walter, and van der Klis, Michiel,(eds.), *Compact Stellar X-ray Sources*, Cambridge University Press, 2006.

Zhang, Bing, *Magnetars and Pulsars: A Missing Link*, in Cheng, K. S., Leung, K. C., and Li, T. P. (eds.), *Stellar Astrophysics - A Tribute to Helmut A. Abt*, Kluwer Academic Publishers, 2010.

Zirker, J. B., *The Magnetic Universe:; The Elusive Traces of an Invisible Force*, Johns Hopkins University Press, 2009.

SUPERNOVAE

Observations

A number of possible supernovae were recorded by various civilisations before Tycho's supernova of 1572. For example, the Chinese recorded what appears to be a supernova in AD 185 and AD 393. Both stars were visible for at least eight months. Another new star that appears to have been a supernova was reported by Japanese, Chinese, Arab, and European observers in 1006. Analysis of the records suggest that it was probably the brightest supernova on record at about magnitude −9. It was reported to have been so bright that it cast shadows. Then in 1054 the Chinese, Japanese, and Arabs recorded a new star in the constellation of Taurus, which the Chinese recorded was visible in daylight. It is thought to have been of magnitude −6. Finally in 1181 a supernova was observed by the Japanese and Chinese that appeared to have been visible for about six months in the night sky.

The next unambiguous discovery of a supernova was made on 6 November 1572 by Francesco Maurolyco and Wolfgang Schüler, who independently observed this bright new object. Other people observed it in subsequent days, including Tycho Brahe a few days later. Tycho could find no diurnal parallax, so the new object must have been much further away than the Moon, and it twinkled so it could not be a planet. He and other observers found that the star did not move against the other stars and so it could not be a comet. This led him to conclude that it was a new star. He was reluctant to say that it was a 'brand new' star, however, because of the immutability of the stellar regions preached by Aristotle. So he suggested that the star had probably always been there, but that God had only just made it visible to humans. After November 1572 it gradually decreased in brightness and changed colour, and then in 1574 it finally disappeared.

Ilario Altobelli observed another bright new object in the sky on 9 October 1604 that was as bright as Jupiter for a time. Unlike Tycho's star of 1572, however, this new object was seen before it reached maximum intensity, because astronomers and astrologers had been observing the conjunction of Mars and Jupiter that had been predicted to take place in October 1604 in the same part of the sky. Surprisingly, the new object was first seen on the day of closest approach of the two planets to each other. Astrologers saw this as being especially significant. If the new object was connected with the two planets, which seemed likely, then it appeared to indicate that it was in the solar system, and was not a new star. Kepler first saw the object eight days after its discovery when it reached maximum brightness. It then gradually reduced in intensity until October of the following year when it was too dim to be seen.

Galileo first observed this new object on 28 October. Both Galileo and Kepler recognised that its sudden appearance, like that of Tycho's star in 1572, disproved Aristotle's doctrine of the unchangeability of the universe. Neither Altobelli nor Galileo could detect any annual parallax. So either it was a distant star, or Copernicus was wrong and the Earth did not go around the Sun. Interestingly, Galileo did not come out clearly in favour of either explanation. Kepler, on the other hand, took the object's lack of parallax to show that it was a new star, like Tycho's new star of 1572.

In the twentieth century it was finally shown that both the 1572 and 1604 new stars were not 'novae' but 'supernovae' (see below). They were the last two supernovae to be observed in the Milky Way. In the meantime, however, supernovae have been observed in the Large Magellanic Cloud and many galaxies. The first of these was 'Nova' S Andromedae, discovered independently by Ernst Hartwig, Ludovic Gully and others in August 1885 in the Andromeda nebula. It was of 9th magnitude on 20 August and it brightened to 7th magnitude on 31 August, from which it declined to 16th magnitude by February of the following year, and thence into obscurity. Some observers thought that it had a continuous spectrum, but others reported seeing bright emission lines like those of previous novae. At the time there was considerable discussion as to the true nature of S Andromedae and, in particular, whether it was really in the Andromeda nebula or whether it was only along our line of sight. It was also unclear at that time whether the

Andromeda and other nebulae were internal or external to the Milky Way.

Ten years later Williamina Fleming discovered the next 'nova' in a nebula when she found Z Centauri in NGC 5253. Annie Cannon found that its spectrum was unlike that of any other nova except S Andromedae, increasing the probability that both 'novae' really were in nebulae. Then in 1909 Max Wolf discovered a 'nova' near the spiral nebula M101. Eight years later George Ritchey discovered a 'nova' in NGC 6946. He subsequently found six relatively faint novae in nebulae, two of which were in the Andromeda nebula, but they were very much fainter than S Andromedae. In parallel, Heber Curtis found one nova in galaxy NGC 4527, and two in NGC 4321.

One of the great outstanding issues at the start of the twentieth century was whether all nebulae were associated with the Milky Way, and were thus relatively close, or whether some of them were so-called 'island universes' of stars at much greater distances. If some of them were the latter, the novae seen in them must be incredibly bright, and in 1917 Harlow Shapley calculated in that case that the absolute magnitude of Nova S Andromedae must be about −16, or about ten magnitudes brighter than normal novae seen in the Milky Way. This seemed far too bright, and so Shapley suggested that the nebulae were essentially part of the Milky Way. Ten novae were then found in the Andromeda nebula over the next two years that were ten magnitudes fainter than S Andromedae, leading Curtis to suggest that the Andromeda nebula was an island universe of stars, or galaxy, at some distance from the Milky Way, with two different types of novae of radically different brightness in it, and presumably in other galaxies. Tycho's 'nova' of 1572 and Kepler's of 1604 were very bright novae in the Milky Way, so maybe they were supernovae also.

In 1924 Edwin Hubble discovered Cepheid variables in the Andromeda nebula and other spiral nebulae. This enabled him to estimate their distances, which clearly showed that they were at some distance from the Milky Way, as suggested by Curtis. So some novae, like S Andromedae of 1885, Z Centauri of 1895 and Wolf's 'nova' of 1909, were exceptionally bright. They came to be called supernovae.

Some years later Milton Humason and Walter Baade imaged the spectrum of supernova 1936a in NGC 4273 and found that it had bright emission lines like those of an ordinary nova, but they were much broader, indicating that the star was expanding at about 6,000 km/s. In the same year, William Johnson and Cecilia Payne-Gaposchkin re-examined the original spectrum of the supernova Z Centauri, taken in 1895, and found the same great broadening of the emission lines.

Fritz Zwicky and Josef Johnson had started a systematic search in 1936 of nearby galaxies to find supernovae. Over a three year period they took 1,625 photographs with the 18 inch Palomar Schmidt telescope, and found 12 supernovae. One of these in the galaxy IC 4182 was found to be 100 times brighter than its parent galaxy. The light curve of this supernova showed a decline of three magnitudes in the first month, followed by a more gradual reduction over the next two years.

In 1941 Rudolph Minkowski analysed the first results of this Palomar Supernova Search, and concluded that supernovae were of two types, which he called Types I and II. It was soon found that the light curves of Type I supernovae were very consistent, with a fast rise to maximum, followed by an equally fast decline for two or three weeks. After that the intensity reduced exponentially with a half-life of about 60 days. The supernova in IC 4182 was a typical Type I. The light curves of Type II supernovae were more variable than those of Type I. They typically faded by about 1.5 magnitude and then almost stopped reducing in intensity for about 50 days, before starting to decline again more rapidly. The absolute luminosity of Type I supernovae was typically two magnitudes higher than for Type II.

Minkowski had found that the spectra of Type I supernovae consisted of broad emission lines. These were not identified until 1975 when Robert Kirshner and J. Beverley Oke found that they appeared to be due to the superposition of hundreds of lines of ionised iron. Type I supernovae had no hydrogen lines. They were found in all types of galaxies. Type II supernovae, on the other hand, which were found only in spiral galaxies, had spectra resembling ordinary novae with hydrogen emission lines. Type I supernovae were related to old population II stars, whereas Type II were related to young population I stars. This implied that Type I supernovae occurred in older and less massive stars than those for Type II.

The war, and other research programmes, virtually stopped the Palomar Supernova Search for almost fifteen years but, in 1954, the search was resumed. By 1960, Zwicky had discovered a total of 44 supernovae, ranging in absolute magnitude from about −13 to −18. When the Palomar search was officially closed in 1975 it had resulted in the discovery of 281 supernovae since 1936.

The first 'local' supernova since Kepler's star of 1604 was found by Ian Shelton and Oscar Duhalde in the Large Magellanic Cloud (LMC) on 24 February 1987. As it was much closer than any other supernova detected in recent times, observations of its subsequent development were crucial in checking and modifying theories of supernova explosions.

A few hours after Shelton and Duhalde's discovery, this supernova, SN 1987A, had brightened to visual magnitude 4.5. Subsequent work showed that it had started its rapid brightness increase from magnitude 12 on the previous day. It was also discovered retrospectively that neutrinos had been detected by detectors in Japan, the USA and Russia about three hours before the star had started to increase in brightness.

After its initial rapid rise in intensity, SN 1987A finally achieved magnitude 2.8 in mid May, before beginning a rapid decline. Pre-discovery images showed that its position coincided with that of the blue supergiant Sanduleak −691202 with a surface temperature of 20,000 K and a mass of 20 $M_\odot$. This was the first time that the progenitor star had been found for a supernova, but it was a big surprise as it was thought that red supergiants, rather than blue supergiants, developed into Type II supernovae, which this appeared to be. Ultraviolet observations with the IUE spacecraft 24 hours after the explosion showed an expansion velocity of about 35,000 km/s. Infrared measurements then showed a gas temperature of 5,700 K on 1 March, with a cloud expanding at 18,000 km/s. By the end of March both the gas temperature and the expansion rate had reduced significantly.

Theory predicted that a large amount of nickel 56 should be produced at the high temperatures which occur after the core collapse of massive stars. Nickel 56 would then decay radioactively to cobalt 56 which, in turn, would decay to iron 56. The half lives of the nickel/cobalt and cobalt/iron processes are 6 and 77 days, respectively. The observed light decay curve was found to be consistent with these half-lives. More positive evidence of the production of cobalt 56 was provided by Bouchet and Danziger in 1988 when they detected a cobalt line in the infrared. Analysis indicated that about 0.07 $M_\odot$ of cobalt had been produced.

The decay of cobalt 56 to iron 56 was also expected to produce a great number of gamma rays, some of which would be converted in collisions in the supernova ejecta to X-rays. It was expected that these gamma rays and X-rays would be detected as the ejected gas shell got thinner and allowed them to escape into space. The theory was vindicated when in July 1987 X-rays were discovered by the Ginga spacecraft, and in the following month when gamma rays were first detected by the Solar Maximum Satellite.

Eight months after the supernova explosion, the IUE spacecraft found that light from the explosion had reached a ring of material 1.3 light years in diameter that had surrounded the progenitor star. Then in 1997 the shock wave and high-velocity ejecta from the explosion first reached this ring of material causing it to glow more brightly. Over the next few years the high velocity ejecta reached the whole ring to produce the brightest ring shown in Figure 3.9. Observations of the expansion of this ring indicated that it had been emitted by the progenitor star at least 20,000 years ago. Other ring structures have also been observed, the brightest of which are shown in Figure 3.9. The exact mechanism for their creation is still unclear.

Theories

In the 1930s Subrahmanyan Chandrasekhar had defined the maximum mass for a white dwarf of about 1.4 $M_\odot$. It only became evident gradually, however, that stars could lose a significant amount of mass in the red giant phase, which followed the main sequence phase. This allowed stars with masses sig-

Figure 3.9. The ring structure observed following the explosion of supernova SN 1987A, as imaged by the Hubble Space Telescope in 1994. The brightest ring is that of material that was ejected from the progenitor star at least 20,000 years ago, but the cause of the fainter rings is still not clear. (Courtesy ESA/STScI, HST, NASA.)

nificantly above the Chandrasekhar limit to eventually become white dwarfs.

The red giant mass loss was first examined by Armin Deutsch in 1956. Further work by Ed Ney, Neville Woolf and Andrew Bernat showed that gas haloes surrounding red giants were extensive and massive. It began to appear as though stars of up to 8 $M_\odot$ could lose enough matter to become white dwarfs below the Chandrasekhar limit. More massive stars could not lose enough matter, however, and so would remain above the limit.

Theoretical modelling of possible supernova scenarios has been gradually improved over the last few decades, but even today there is not a consensus on exactly what happens. In addition, not all supernovae fit neatly into Types I and II, and more complex processes than those outlined below occur in some supernovae.

(a) Type I

Type I supernovae, which have no hydrogen lines, are now divided into types Ia, Ib and Ic depending on their spectra. Type Ia contain Si II absorption lines, whilst Type Ib show prominent He lines but not Si II absorption. Type Ic have very weak or no He lines and no Si absorption lines. Types Ib and Ic had lost most of their hydrogen-rich envelopes before they exploded, whilst Type Ic had lost most of its helium-rich region also.

Type Ia supernovae were, like novae, found to occur in close binary systems consisting of a white dwarf and a normal star. The current theory of these Type Ia supernovae was developed mainly in the early 1970s by Craig Wheeler, Carl Hansen and others. The development of the light curve of SN 1972e in NGC 5253 then broadly confirmed the model.

In this model, mass is transferred from the normal star in the binary to the white dwarf, so raising its mass to close to the Chandrasekhar limit. The white dwarf then deflagrates, converting its inner part into nickel 56. The deflagration wave travels outwards, creating lighter elements, finally turning into a detonation which explodes the star. The nickel 56 decays radioactively to cobalt 56 which, in turn, decays to iron 56.

(b) Type II

It is thought that Type II supernovae are produced when a star with a main sequence mass in excess of ~ 8 $M_\odot$ gets to the end of its lifetime. The processes are basically those proposed in the classic B²FH paper published in 1957 by Geoffrey and Margaret Burbidge, Willy Fowler and Fred Hoyle.

When hydrogen in the core of such a heavy star has been converted into helium, the core collapses and the temperature becomes high enough for helium to fuse into carbon and oxygen. This produces more heat, and when the helium has been used up in the core, the carbon is converted to neon and magnesium. Successive processes continue, producing ever higher temperatures and heavier elements, until iron is produced in the core. There are then no further processes possible, and the heat energy is abruptly cut off. At this stage the star consists of shells of gas, with hydrogen and helium in the outer shells, and ever heavier elements as one goes through successive shells towards iron in the centre.

Once the star is no longer producing heat and radiation, nothing can stop it from collapsing and the shells rapidly collapse onto the iron core. There is a limit to this contraction, however, and when the core cannot be compressed any more rebound occurs, setting up a shock wave which progresses outwards through the star to its surface. Neutrinos are produced on core collapse, and they move towards the surface much faster than the shock wave. When the shock wave reaches the surface, we see the sudden increase in light which is the visible signal of a Type II supernova.

Following the explosion, the solid remnant of the original star is normally a neutron star, which is surrounded by a rapidly expanding shell of cooling gas blown off by the shock wave. However, if the original star had a main sequence mass of ~ 25 $M_\odot$, the supernova would result in the formation of a black hole, rather than a neutron star. The precise mass limit for a neutron star production is currently unclear, but it probably depends on the spin rate, amongst other things, of the original star.

Bibliography

Immler, S., Weiler, K., and McCray, R. (eds.), *Supernova 1987A: 20 Years After; Supernovae and Gamma-Ray Bursters*, American Institute of Physics Conference Proceedings, 2007.

Murdin, Paul, and Murdin, Lesley, *Supernovae*, rev. ed., Cambridge University Press, 1985.

Percy, John R., *Understanding Variable Stars*, Cambridge University Press, 2007.

Stephenson, F. Richard, and Green, David A., *Historical Supernovae and their Remnants*, Oxford University Press, 2002.

Stephenson, F. Richard, and Green, David A., *A Reappraisal of Some Proposed Historical Supernovae*, Journal for the History of Astronomy, **36** (2005), 217–229.

Trimble, Virginia, *Supernovae - An Impressionistic View*, The Journal of the American Association of Variable Star Observers, **15**, No. 2, (1986), 181–188.

Weiler, K., (ed.), *Supernovae and Gamma-Ray Bursters*, Lecture Notes in Physics, Springer, 2003.

See also: Supernova remnants.

SUPERNOVA REMNANTS

The Crab nebula

John Bevis first recorded a dim, nebulous object, now called the Crab nebula (see Figure 3.10 for a modern image), in the constellation of Taurus in 1731. Charles Messier saw it in 1758, and recorded it as the first object in his catalogue of nebulae, published in 1781.

Joseph Winlock and Edward Pickering found the green nebula line in the spectrum of the Crab nebula in 1868, showing that the nebula was gaseous. Forty-five years later Vesto Slipher found that all the spectral lines for the Crab were double. He later concluded that this was due to the Doppler effect, and the line separation showed that the nebula was expanding at about 1,000 km/s.

In 1921 John Duncan measured the size of the Crab nebula on photographs taken eleven years apart and found that it was clearly expanding. In the same year Knut Lundmark published a list of novae that had been observed by the Chinese, and pointed out that a new star observed by them in 1054 was near to the current position of the Crab nebula. But the Chinese had apparently recorded the new star as being south-east of ζ Tau whereas the Crab nebula is just over 1° north-west of that star. Seven years later Edwin Hubble concluded that the

Crab Nebula · M1
Hubble Space Telescope · WFPC2

NASA, ESA, and J. Hester (Arizona State University)   STScI-PRC05-37

Figure 3.10. The Crab nebula as imaged by the Hubble Space Telescope. (Courtesy NASA, ESA and J. Hester [Arizona State University].)

expansion of the Crab had been under way for about 900 years. This fitted the timescale of the 1054 explosion remarkably well, and he suggested that the Crab nebula was its remnant. In 1935 Kanda Shigeru had found two Japanese records of sightings of the new star in 1054. Unfortunately these records were not sufficiently clear to be able to solve the discrepancy between the Chinese recorded positions and the position of the Crab nebula. Two years later Nicholas Mayall, using both its angular expansion rate and Doppler velocity, concluded that the Crab nebula was ~1,500 parsec or 4,900 light years away.

Duncan published an up-to-date analysis in 1939 comparing a new image of the nebula with one taken 29 years earlier. He concluded that, assuming the nebula had been expanding uniformly since the initial explosion, the supernova that created it must have occurred in about the year 1172, with an uncertainty of the order of ±50 years. This estimate was modified by Mayall and Jan Oort in 1942 to produce an explosion date of 1138. Then in 1942 Walter Baade deduced a date of 1184. Whichever of these dates was chosen, they were appreciably later than the known 1054 date of the supernova. Mayall and Oort suggested that the gas had accelerated slightly over time, but Baade was more doubtful, suggesting that the deduced acceleration was spurious.

In 1949 Bolton, Stanley and Slee discovered that the Crab nebula was a strong radio source, called Taurus A. It had been known for some time that the nebula consisted mainly of a chaotic web of bright filaments best seen in red light, which have an emission line spectrum like that of planetary nebulae, and a bluish diffuse part with a continuous spectrum.

Iosif Shklovskii suggested in 1953 that the radio emission from the Crab nebula was caused by synchrotron radiation, produced by electrons spiralling in a strong magnetic field. He also suggested that the bluish diffuse part with the continuous spectrum was also caused by synchrotron radiation, and so should be polarised. This was confirmed in the following year by Viktor Dombrovsky and Mikheil Vashakidze. Then in 1957 Mayer, McCullough and Sloanaker found that radio signals were also polarised.

In 1963 Herbert Friedman's group discovered that the Crab nebula was also a strong X-ray source. Friedman speculated that the source of the X-rays was a neutron star, and he and his team took advantage of the occultation of the nebula by the Moon on 7 July 1964 to measure the size of the X-ray source. They found it to be 1 arc minute, or about 2 light years, in diameter, clearly too large to be a stellar object.

David Staelin and Edward Reifenstein discovered a radio pulsar in the Crab nebula in November 1968 with a period of 33 milliseconds. Shortly afterwards Cocke, Disney and Taylor also discovered optical flashes at the same frequency. At about the same time Friedman, Byram, Chubb and Fritz also found X-ray pulses with the same frequency. This X-ray pulsar was found to be very powerful, generating about 20,000 times as much power in X-rays as in radio waves. In 1976 the COS−B spacecraft found that the Crab pulsar was the third most powerful source of gamma rays in the sky, and three years later, the Einstein Observatory spacecraft imaged both the pulsar and the nebula in X-rays.

As a result of this and other work we now know that the Crab pulsar is a neutron star with a surface temperature of about 2 million K. Its pulses, or spin rate, are slowing at the rate of 1.3×10^{-5} s/year, and it is therefore losing rotational energy at a rate equivalent to the radiation from 100,000 Suns. It is this energy that accelerates and lights up the nebula. The Crab nebula is clearly a supernova remnant and it is virtually certain that it is the result of the explosion seen by Chinese and Japanese observers in 1054.

Tycho's and Kepler's supernova remnants

The discovery, in the early decades of the twentieth century, of the link between the Crab nebula and the 1054 supernova caused astronomers to search for the remnants of the Tycho and Kepler supernovae of 1572 and 1604, respectively. Baade tried in 1941 to find the Kepler supernova remnant, but all he could find was a few filaments of gas. Looking for the Tycho

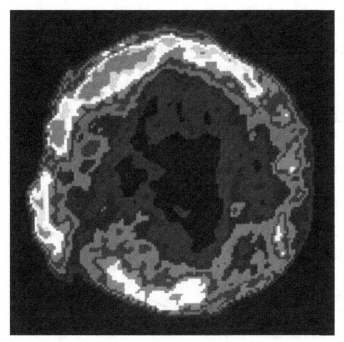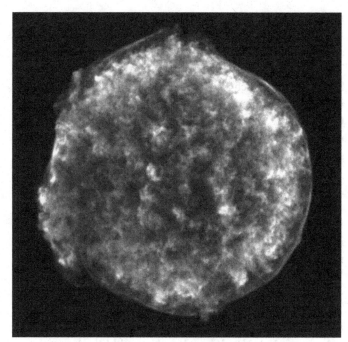

Figure 3.11. The shell-like structure of Tycho's supernova remnant imaged in radio waves (left-hand image) and X-rays (right-hand image). It was found that the highest-energy X-rays, corresponding to a temperature of about 20 million K, were being emitted from the shocked gas of the outer shell. Relatively cooler gas filled the volume inside the shell. (Left-hand image courtesy VLA, NRAO, AUI, NSF, right-hand image courtesy NASA/CXC/Rutgers/J. Warren & J. Hughes et al.)

remnant he could find nothing of interest, then in 1952 a radio source was discovered near the expected place by Robert Hanbury Brown and Cyril Hazard. In 1967 the remnant was also discovered to be a source of X-rays by Friedman, Byram and Chubb.

The structure of the radio emission from the Tycho supernova remnant (SNR) was shown by John Baldwin in 1967 to be that of an almost circular limb-brightened shell (see Figure 3.11, left-hand image). An X-ray image produced by Reid, Becker and Long in 1982 also showed a shell structure (see Figure 3.11, right-hand image). The X-ray shell was coincident with the radio shell, although, unlike the radio shell, the X-ray shell was not complete. Later measurements of the X-ray spectrum showed that the temperature of the shell was many millions of degrees Kelvin. Analysis indicated that the Tycho supernova had been a Type Ia.

Work on Kepler's SNR was difficult because of its southerly declination, which caused problems for northern observatories, and because of its location near the galactic centre, which had many X-ray sources. As a result, although radio emission was first detected by Baldwin and Edge in 1957, the first clear radio image was not produced until eighteen years later by Stephen Gull. It showed an almost circular limb-brightened structure about 3 arcmin in diameter, with an uneven brightening around its circumference. X-ray emission was first detected in 1979 by Bunner and Tuohy and colleagues. The image in X-rays showed a similar structure to that in the radio band, but for some unknown reason the expansion rate in X-rays and in the radio band were not consistent.

Cassiopeia A

In 1942 Grote Reber found a region of the sky in Cassiopeia to be a source of radio waves. This source, now known as Cassiopeia A, was rediscovered by Martin Ryle and F. Graham Smith in 1948. Three years later Graham Smith obtained an accurate estimate of its position using a radio interferometer. This enabled Walter Baade and Rudolf Minkowski to optically identify Cassiopeia A, which is the brightest radio source in the sky, with a very faint nebulosity of about 6 arcmin in extent. Graham Smith, Hanbury Brown, Jennison and Das Gupta helped to confirm this visual correlation when they showed that the radio source had a diameter of about the same size.

Minkowski measured the movement of the nebulosity on photographs taken at optical wavelengths over a number of years, and calculated in 1958 that the supernova that had produced it had exploded about the year 1700. Later measurements by Sidney van den Bergh and Karl Kamper yielded a date of about 1660. Minkowski also estimated that the maximum gas velocity was 7,400 km/s which, given the angular expansion rate that he had measured, enabled him to deduce a distance for Cassiopeia A of about 11,000 light years. At this distance the supernova should have been seen as one of the brightest stars in the sky in about 1660, but no such star had been recorded. So Minkowski suggested that the supernova had been dimmed by a considerable amount of interstellar dust in our line of sight. The only feasible candidate seems to be a sixth magnitude star, 3 Cassiopeiae, which is no longer visible. It had been recorded in 1680 by John Flamsteed, several arcmin from the position of Cassiopeia A. Although the position discrepancy is larger than normally attributed to Flamsteed, it could well have been the supernova.

The Vela supernova remnant

In the 1950s radio astronomers discovered a strong radio source in the constellation of Vela, which was flanked by two other radio sources. Further work showed that these three sources, Vela X, Y and Z, were all part of a very extended source about $4°$ in diameter.

In 1968 a pulsar, PSR 0833-45, that had a period of 89 millisecond, was discovered near the centre of the bright radio source Vela X by Large, Vaughan and Mills. Astronomers had long suspected that the extended Vela source was a supernova remnant, and the discovery of the pulsar confirmed it. Seven years later the pulsar was found to pulse in gamma rays, and in 1977 optical pulses were found. The gamma-ray and optical pulses had two peaks to every one radio peak. The optical, radio and gamma-ray pulses were also found to be out of phase with each other.

The Vela pulsar was found in 1993 to also pulse in X-rays using the Rosat spacecraft. Its period was found to be lengthening by 4×10^{-6} s/year, implying an age of the pulsar, and of the SNR, of 11,000 years. Further work with the Chandra spacecraft showed not only that there were three pulses to every radio pulse, but that the pulse behaviour in low-energy X-rays was very different from that in higher-energy X-rays.

General

As explained above, the realisation that the Crab nebula was a supernova remnant was a gradual event, being the conclusion of a number of investigations carried out over a number of years. Since then supernova remnants have been discovered either as the result of a deliberate search for them from known supernovae, or by the realisation that some objects, not known to be related to supernovae, were actually supernova remnants. In the process, hundreds of supernova remnants have now been found, most of which have not been identified with known supernovae. Like supernovae themselves, their characteristics differ enormously. In the above I have considered the most historically important examples. (The development of the 1987A remnant is covered in the Supernova article.)

Bibliography

Dwarkadas, Vikram, et al., *Supernova Remnants, Pulsars and the Interstellar Medium: Summary of a Workshop held at the University of Sydney, March 1999*, Publications of the Astronomical Society of Australia, **17** (2000), 83–91.

McCray, Richard, and Wang, Zhenru (eds.), *Supernovae and Supernova Remnants: IAU Colloquium 145*, Cambridge University Press, 1996.

Murdin, Paul, and Murdin, Lesley, *Supernovae*, rev. ed., Cambridge University Press, 1985.

Stephenson, F. Richard, and Green, David A., *Historical Supernovae and Their Remnants*, Oxford University Press, 2002.

See also: Supernovae.

WHITE DWARFS

Observations

Friedrich Wilhelm Bessel observed in 1844 that there was an oscillatory motion superimposed on the proper motions of both Sirius and Procyon, which he attributed in both cases to an unseen companion. In 1850 Christian Peters estimated the period of the hypothesised Sirius binary as about 50 years. Then in 1862 Alvan and Alvan Graham Clark observed Sirius B, the companion to Sirius, which was found to have a luminosity of only about 0.0001 that of the main star, Sirius A. From the relative motion of the two stars it was concluded that the mass of Sirius B was about half that of Sirius A, and from the parallax of Sirius, measured in the 1880s, the masses of the two stars were estimated to be about 1 and 2 times $M_\odot$ (mass of the Sun).

There had been some doubt about the oscillatory motion of Procyon, even after it had been confirmed by Johann Heinrich Mädler in 1851. But in 1862 Arthur Auwers analysed observations of Procyon going back to 1750 and confirmed its oscillatory motion. He calculated a period of about 40 years. The companion of Procyon, Procyon B, was finally observed by John M. Schaeberle in 1896. Its mass was about $0.5\ M_\odot$, yet its luminosity was only about 0.00005 that of its companion, Procyon A. So, like Sirius, Procyon's companion had a similar mass to its primary, but had an extremely low luminosity.

At the time it was assumed that as the masses of both Sirius B and Procyon B were similar to those of their companions, they had about the same surface area as their companions. Both dim companions were assumed to have a very dark surface and were called 'dark stars'.

In 1910 Henry Norris Russell found that a white star, 40 Eridani B, did not fit on either the main sequence or giant bands of his colour-magnitude diagram. It had an absolute luminosity of only 0.0025 that of the Sun, whilst having a mass of 0.4 $M_\odot$. In 1914 Walter Adams measured its spectrum and confirmed that it was a very-low-luminosity white star of type A0.

At that time Sirius B was assumed to be a low-luminosity red M-type star, but its colour had not been determined, as it was too close to its very bright companion. But in 1915, spurred on by his success in measuring the spectrum of 40 Eridani B, Adams tried to measure that of Sirius B, and found that it was not a low-luminosity red star but a white A0-type star like 40 Eridani B. A0-type stars were known at the time to have a very high surface brightness, so if the low luminosities of Sirius B and 40 Eridani B were simply due to their physical size, they must be very small and very dense. Alternatively they could be normal-sized stars radiating from a very small part of their surface.

Arthur Eddington showed theoretically in 1924, using the perfect gas laws, that the luminosity of giant stars is determined almost completely by their mass. When he plotted the data for real stars on his mass-luminosity plot, however, he found that both giant stars and main sequence stars lay on the same line, indicating that matter in the main sequence stars also behaved like a perfect gas. But there was one star that did not fit on this line, Sirius B.

Eddington realised that completely ionized or degenerate matter could be packed at much higher densities than matter in normal stars. He calculated a density of about 6×10^4 g/cm^3 for Sirius B, which he described as a 'white dwarf'. He also showed that if it was really this dense there should be a relativistic shift in its spectral lines equivalent to about 20 km/s. Adams confirmed this relativistic shift observationally the following year. So Sirius B was a white dwarf consisting of degenerate matter. Likewise 40 Eridani B and Procyon B were later confirmed to be white dwarfs.

A number of white dwarfs were discovered in the 1930s, and in 1941 Gerard Kuiper published a list of 38 white dwarfs, most, but not all of which, were white. For example, his list included the orange-red white dwarf Wolf 489, which had been previously classified as K5 from its colour, as there were no lines in its spectrum.

Willem J. Luyten undertook a number of proper motion surveys of faint white and blue stars starting in the 1920s, and in 1970 he produced a list of almost 3,000 probable white dwarfs. These and other observations allowed the spectral classification of white dwarfs to be gradually developed from the first classification by Kuiper in 1941. Today all white dwarf spectral types begin with 'D' for 'degenerate' (see Table 3.1), followed by various letters, like O if ionised helium lines are observed, B if non-ionised helium lines are observed, and so on, mimicking, in a way, the spectral classification of normal stars. But whereas the spectra of ordinary stars differ mainly because of temperature differences, white dwarf spectra show real differences in their atmospheric composition.

Theory

Ralph Fowler published a theory of degenerate matter in 1926 in which he confirmed Eddington's hypothesis that a white dwarf is composed largely of degenerate matter, with only its out layer remaining gaseous. In the 1930s Subrahmanyan Chandrasekhar developed Fowler's theory and showed that if a star's mass is greater than $\frac{5.76}{\mu_e^2}$ $M_\odot$, where μ_e is the mean molecular weight of stellar material per electron, it would continue to collapse near the end of its life to below the size of a white dwarf. For most compact stars $\mu_e \approx 2$, giving the maximum mass for a white dwarf of about 1.44 $M_\odot$.

Bibliography

Camenzind, Max, *Compact Objects in Astrophysics: White Dwarfs, Neutron Stars and Black Holes*, Springer, 2007.

Chandrasekhar, Subrahmanyan, *On Stars, Their Evolution and Their Stability*, Nobel lecture, 8 December 1983, (accessed via the Internet).

Eddington, Arthur S., *The Internal Constitution of the Stars*, Cambridge University Press, 1926.

Hearnshaw, J. B., *The Analysis of Starlight; One Hundred and Fifty Years of Astronomical Spectroscopy*, Cambridge University Press, 1986.

Holberg Jay B., *Sirius: Brightest Diamond in the Night Sky*, Springer-Praxis, 2007.

Kaler, James B., *Stars and Their Spectra; An Introduction to the Spectral Sequence*, Cambridge University Press, 1997.

Miller, Arthur I., *Empire of the Stars: Friendship, Obsession and Betrayal in the Quest for Black Holes*, Little, Brown, 2005.

See also: Cataclysmic variables; Neutron stars; Planetary nebulae; Supernovae.

X-RAY BURSTERS

Jonathan Grindlay and John Heise, using the ANS spacecraft, reported the first observations of an X-ray burster in late 1975. (X-ray bursters had previously been discovered using the military Vela spacecraft, but their discovery was not reported until later.) This ANS source, 4U 1820–30, which was in the globular cluster NGC 6624 near the galactic centre, was found to emit sudden (~0.5s), very brief (decay ~8s) bursts of X-rays in the 1 to 30 keV range. Walter Lewin and colleagues discovered three more X-ray bursters in February 1976, all very close to the galactic centre. As more and more X-ray bursters were discovered they were found to be either in Milky Way globular clusters or in the Milky Way disc. Simultaneous optical and X-ray bursts were also observed for some bursters.

Some X-ray bursters (e.g., MXB 1659–29, discovered in 1977) had an almost regular interval between pulses. But the pulse interval for most sources was found to change from hours to days, and sometimes the pulses switched off completely. The strength of the bursts varied for a given source, with the strongest bursts following the longest gaps between bursts. The ratio of soft to hard X-rays increased during bursts which lasted for tens of seconds, indicating that the sources were cooling, typically from 30 million K to 15 million K over that time. Almost all bursters were found to be continuously emitting a strong but variable X-ray flux out of burst mode. Currently about one hundred X-ray bursters are known.

The Einstein Observatory spacecraft showed in the early 1980s that X-ray bursters in globular clusters were located very close to the centre of the clusters. The bursters were thought to be compact binary stars consisting of an ordinary star which lost material to its compact companion. Further work indicated that these cluster X-ray sources had an average mass of ~ 1.5 $M_\odot$ (mass of the Sun), indicating that the compact sources were neutron stars rather than black holes.

Stella, White and Priedhorsky found, using the Exosat spacecraft in the mid-1980s, that the X-ray output of 4U 1820–30 had a 3% amplitude variability with a regular period of 685 s. The modulation did not change frequency over time, indicating that this was the orbital period rather than the spin period of the neutron star. This very short orbital period strongly constrained the size of the low mass X-ray binary system which was thought to consist of a white dwarf losing mass to a 1.5 $M_\odot$ neutron star.

A few months after the discovery of 4U 1820–30, Lewin and colleagues had discovered an X-ray burster which was unlike any other known at the time. Called the Rapid Burster, or MXB 1730–335, it was in globular cluster Liller 1. Burst durations varied from a few seconds to $\sim$10 min, with time intervals between bursts of similar orders of magnitude. At times it produced $\sim$1,000 bursts/day. The energy in the strongest bursts was as much as 1000 times that in the weakest, with, contrary to the situation with normal X-ray bursters, the longest gaps *following* the strongest bursts rather than preceding them. Also, unlike the situation for normal bursters, there was no spectral softening with time. The rapid burster was found to be active for a few weeks every six to eight months.

Normal X-ray bursters were said to exhibit Type I bursts having intervals from hours to days with spectral softening, whilst the rapid burster exhibited Type II bursts having intervals from seconds to minutes with no spectral softening. But, from time to time, the rapid burster also exhibited Type I bursts.

In 1993 Lewin, van Paradijs and Taam proposed the thermonuclear flash model that satisfactorily explains the main features of Type I bursts. In this model hydrogen is accreted onto the surface of a neutron star from its companion star. On the neutron star this hydrogen steadily burns to form helium. Eventually when the density and temperature are high enough, the helium burns explosively to produce an X-ray burst. The longer the period of time between bursts, the more helium is produced on the neutron star's surface and the larger the burst.

Type II bursts are thought to be due to an instability in the accretion disc around the neutron star. Every now and again material from the disc is suddenly released and falls rapidly onto the neutron star where the gravitational energy is released as a Type II burst. After this, the accretion disc needs to replace the material that has been lost to the neutron star. The more material that has been lost, the brighter the X-ray burst, and the longer the time required to replace that material in the accretion disc.

Bibliography

Charles, Philip A., and Seward, Frederick D., *Exploring the X-Ray Universe*, Cambridge University Press, 1995.

Lewin, Walter H. G., and Van der Klis, Michael (eds.), *Compact Stellar X-Ray Sources*, Cambridge University Press, 2006.

Lewin, Walter H. G., and van Paradijs, Jan, *What are X-Ray Bursters?*, Sky and Telescope, May 1979, pp. 446–451.

X-RAY TRANSIENTS

An X-ray source, Cen X-2, was discovered by J. Harries and colleagues on 4 April 1967 where none had existed before. Six days later, Cen X-2 was independently observed by Brin Cooke and colleagues near to its maximum X-ray intensity which was about the same as that of Sco X-1, the brightest X-ray source in the sky. Although the source was still observable on 18 May, it fell below threshold visibility over the next few months and was never seen again. It was the first known X-ray transient.

Soft X-ray transients

The second confirmed X-ray transient was Cen X-4, which was found by Conner, Evans and Belian using the Vela 5A and B spacecraft on 9 July 1969. It reached its peak intensity two days later, of about twice that of Sco X-1. It then started a decline in X-ray intensity, so that it was no longer detectable in late September, having reduced to less than 0.5% of its peak. A second outburst was observed almost ten years later in May 1979 by Lou Kaluzienski and colleagues using the Ariel 5 spacecraft. This enabled its visual counterpart to be identified as a 13th magnitude red star which had increased in intensity by at least six visual magnitudes.

Further analysis showed that the X-ray source was in a low-mass X-ray binary (LMXB) consisting of a neutron star, surrounded by an accretion disc, and a K-type donor star. The binary's period was about 15 hours. As a result of its soft X-ray spectrum, Cen X-4 is now categorised as a soft X-ray transient.

Another soft X-ray transient, A0620-00, was discovered by Martin Elvis and colleagues using Ariel 5 in August 1975. They found that it rapidly increased its X-ray intensity to reach a maximum of about four times that of Sco X-1 about ten days later. It then started a gradual decline. The position of this X-ray transient was found by Forrest Boley and colleagues to coincide with that of the recurrent nova V616 Mon (or Mon 1975), which was then a 12th magnitude blue star, having increased from its normal 18th magnitude. The nova had exhibited a similar outburst in 1917 when it had also reached 12th magnitude.

Spectra taken in 1976 and 1978 after outburst showed that the blue star had reverted to a K-type (reddish-orange) star with emission lines indicative of an accretion disc, which Murdin and colleagues suggested was surrounding an unseen companion. Then in 1983 Jeff McClintock and colleagues found that the orange star's intensity was varying regularly with a period of 7.8 hours. They later measured the Doppler shift in its spectral lines, and in 1986 concluded that the mass of its compact, X-ray emitting companion must be at least 3.4 $M_\odot$ and maybe more than double that.

The maximum theoretical mass of a neutron star is between about 3 and 5 $M_\odot$, depending on assumptions, and any compact object heavier than that is thought to be a black hole. So given the mass estimates for the A0620-00 X-ray source, it could just conceivably be a neutron star, but it was far more likely that it was a black hole. To be absolutely certain that a black hole had been discovered, however, astronomers wanted to find a compact source with a minimum mass estimate of at least 5 $M_\odot$.

Another X-ray transient, GS2023+338, was detected by the Ginga spacecraft in May 1989. Its optical counterpart was identified as a star of visual magnitude 12.8 four days later. Further investigations showed that the star had brightened from about 20th magnitude, and Brian Marsden showed that it was a reappearance of Nova Cygni 1938, or V404 Cygni, a recurrent nova. Then in 1991 Casares, Charles and Naylor discovered periodic variations in the spectrum of V404 Cygni using the William Herschel telescope on La Palma. A late G or early K-type star appeared to be orbiting a dark companion every 6.5 days. They calculated that this dark companion had a mass of at least 6.3 $M_\odot$, and concluded that it was a black hole with a probable mass in the range of 8 to 15 $M_\odot$. Charles Bradley and colleagues concluded in 2008, using observations made by the XMM-Newton spacecraft, that the binary consisted of a $\sim$0.7 $M_\odot$, K-type star orbiting an approximately 12 $M_\odot$ mass black hole.

Currently about one hundred of these soft X-ray transients are known. They are in binary systems in which a small, usually K−M type star overflows its Roche lobe, and is the donor of material to an accretion disc surrounding a black hole or neutron star. The amount of material in the disc gradually increases during quiescence, but when its density exceeds a critical value the disc suddenly collapses onto the compact object, generating X-rays and light. The X-ray intensity increases rapidly over a few days, but then reduces exponentially over a month or two by at least 5 orders of magnitude to often undectable levels in its quiescent state. At outburst the optical intensity increases by about 6 to 8 magnitudes.

Hard X-ray transients

The X-ray source A0538-66 was discovered by Nick White and Geoff Carpenter using the Ariel 5 spacecraft in 1977. Not only was A0538-66 highly variable, but it also had a regular 16 day period between X-ray peaks, indicative of a binary. Johnston, Griffiths and Ward re-analysed HEAO-1 spacecraft data in 1980 to get a more accurate location of the X-ray source, which they found coincided with a variable B-type star, but its optical variability was unusually large for such a star. Later that year Gerry Skinner found that the optical period of this variable star was also 16.6 days, confirming that it was the optical counterpart of A0538-66. Skinner also found that there were times in the preceding fifty years when its optical variability had disappeared.

In 1981 Philip Charles and John Thorstensen measured the optical spectrum of A0538-66. It was that of a B-type star, with hydrogen and helium absorption lines, that was completely transformed during outburst by the addition of many strong, broad emission lines with a P Cygni profile. These showed that the B-type star was surrounded by a shell of gas which, at times, expanded at velocities >3,000 km/s. The Doppler shift of the absorption lines in quiescence showed that A0538-66 was in the Large Magellanic Cloud. This enabled A0538-66's absolute X-ray luminosity to be determined as about 1039 erg/s at its peak, making it the most luminous X-ray stellar source known at the time.

In the following year Gerry Skinner and colleagues detected X-ray pulsations with a period of 69 milliseconds using the Einstein Observatory spacecraft. This showed that one star in the binary system was a rapidly rotating neutron star or pulsar. The observed rate of change of period also showed that A0538-66 was an eccentric binary in which the B-type star and pulsar approached close to each other producing the outbursts. A0538-66 is now categorised as a hard X-ray transient.

The best current model of this hard X-ray transient is of a massive Be-type star which is spinning so fast that it has produced a disc of material surrounding the star. The pulsar in its highly eccentric orbit around the Be-type star passes through this disc once per orbit creating a burst of X-ray energy that heats the disc. This partially depletes the disc, as disc material is accreted onto the surface of the neutron star at its poles. After a number of orbits the material in the disc is not dense enough to overcome the neutron star's magnetosphere, which expands as the disc pressure reduces, and the active period of the binary ends. The optical star continues to eject material into the disc, however, and eventually it accumulates enough material for the process to start again.

Yasuo Tanaka and colleagues discovered another hard X-ray transient on 14 November 1983 using the Tenma spacecraft. It showed erratic short-term variations on timescales down to 100 milliseconds. A few days later James Terrell and William Priedhorsky pointed out that the source was identical with V0332+53, which had first been detected ten years earlier by the military Vela 5B spacecraft. On this previous occasion it was temporarily the second brightest X-ray source in the sky. Following the Japanese re-discovery, Jaap Davelaar and colleagues used the Exosat spacecraft to locate V0332+53's position accurately enough for its visible light counterpart to be detected. This proved to be an O- or B-type star with hydrogen emission and helium absorption lines. Exosat also detected X-ray pulses from V0332+53 every 4.4 seconds with much more pronounced random flickering on timescales as short as 10 milliseconds. The 4.4 second pulses clearly indicated that the source was a neutron star.

Further observations of V0332+53 by Exosat over the next few months showed that the X-ray intensity varied over a period of 34 days, indicating that the source was a binary. The binary orbit was found to be highly elliptical, with the transient X-rays occurring only when the neutron star was closest to its companion. These observations of V0332+53 were crucial as they were the first detection of flickering X-ray emission from a source that was clearly not a black hole. More recent observations have shown that the optical counterpart, BQ Cam, of the neutron star, was a late O-type star with the hydrogen emission caused by a circumstellar accretion disc around it.

A number of these hard X-ray transient systems are now known. They appear to be binaries consisting of a massive Be-type star surrounded by a disc of material, and a neutron star.

Bibliography

Charles, Philip A., and Seward, Frederick D., *Exploring the X-Ray Universe*, Cambridge University Press, 1995.

Lewin, W. H. G., Van Paradijs, J., and Van den Heuvel, E. P. J. (eds.), *X-Ray Binaries*, Cambridge University Press, 1997.

PART 4
GALAXIES AND COSMOLOGY

1 · Milky Way

OVERVIEW – MILKY WAY

Early estimates of size

Various early philosophers had, from time to time, speculated that the Milky Way may consist of numerous distant stars. Galileo was able to show that this was so when he resolved the Milky Way into stars with one of his earliest telescopes in 1609. But the size and structure of the Milky Way were still unclear until the twentieth century.

Thomas Wright, in the mid-eighteenth century, suggested a number of possible configurations for the Milky Way which had religious overtones. In each of these configurations the Sun and stars orbited a gravitational centre which was coincident with the divine centre or seat of God. In one model he placed the Sun and stars in a thin spherical shell surrounded by the region of the damned. In another the Sun and stars were in a flat ring, whilst in another there was an infinite set of concentric shells around the centre. In this last scheme divine punishment was meted out by God moving souls from one shell to a more confined shell.

Immanuel Kant misunderstood Wright's model, as he had read only a summary. Kant latched onto a statement in this summary that compared the Milky Way to the solar system in which the planets orbited the Sun. He did not realise that Wright was suggesting that the stars in the Milky Way were orbiting, at some distance, a divine centre devoid of stars. But Kant, inspired by his erroneous interpretation of Wright's system, proposed that the Milky Way was a disc-shaped system full of stars right to the centre.

Towards the end of the eighteenth century William Herschel undertook a detailed survey of stars near the Sun, assuming that they were equally luminous and equally scattered in space, to a first approximation. He concluded that the Milky Way was a disc, with the Sun near the centre. He deduced a diameter of about 800 times, and a thickness of about 150 times the mean distance of Sirius or Arcturus from the Sun.

Hugo von Seeliger, Karl Schwarzschild and Jacobus Kapteyn all undertook a similar analysis at about the turn of the nineteenth century, finding that the Milky Way was a little larger than Herschel's estimate. The largest was Kapteyn's model which was about 15 kpc in diameter and 2 kpc thick, with the Sun about 2 kpc from the centre.

Star clusters, spiral nebulae and the size of the Milky Way

In the early twentieth century it was generally assumed that open star clusters and globular star clusters were part of the Milky Way, but the size and location of the spiral nebulae were still in doubt.

Lewis Boss had spent many years analysing the proper motions of stars when, in 1908, he found that the stars in the Hyades open star cluster were travelling together in the sky and, apparently, converging on one point. He concluded that this convergence was due to perspective, and that the stars in the cluster were really travelling parallel to each other at the same velocity relative to the Sun. Boss measured their radial velocities and proper motions and deduced the distance of the cluster as 40 parsec. This provided the first clear evidence that stars in open star clusters moved as a group in the Milky Way.

Eighty years earlier, John Herschel had noted that globular clusters were far more prominent in the Sagittarius region of the southern sky than elsewhere. Harlow Shapley pointed out in the early twentieth century that there were also globular clusters in the northern sky. This led him to conclude that they surrounded the Milky Way in a cloud. In 1917 he found, by measuring Cepheid variables in 69 globular clusters, that the centre of the group of globular clusters, which he concluded was the centre of the Milky Way, was ~20 kpc away from the Sun in the direction of galactic longitude $l = 325°$ in Sagittarius. Based on the distances of the most distant globular clusters, he concluded that the Milky Way was ~100 kpc in diameter. Shapley's analysis had suddenly increased the estimated diameter of the Milky Way by about a factor of 10, and had placed the Sun appreciably off-centre. These results were revolutionary when they were announced, and they were not generally accepted for some years.

The first observational evidence that the spiral nebulae may not be part of the Milky Way was published by Vesto Slipher in 1915. He found that they had, in general, very large redshifts, which indicated that they were receding from the Milky Way at very high velocities.

Heber Curtis considered that Shapley's estimate of the size of the Milky Way was far too large, and participated in a debate with Shapley in 1920 on the size of the Milky Way and the location of the spiral nebulae. Shapley favoured a very large Milky Way with the spiral nebulae located within it, whereas Curtis favoured a much smaller Milky Way with the spiral nebulae being island universes, like the Milky Way, some distance outside.

In 1924 Edwin Hubble finally resolved the question of the location of the spiral nebulae when he proved that they were island universes, as argued by Curtis, outside of the Milky Way. But both Shapley and Curtis were wrong in their estimated size of the Milky Way as both had, incorrectly, thought that interstellar absorption would be negligible.

The first conclusive evidence of absorbing interstellar dust was found by Robert Trumpler in 1930. Three years later Joel Stebbins showed that it had a dramatic effect on the size of the Milky Way, reducing Shapley's estimated diameter to ~30 kpc. So its size was intermediate between that of Shapley and that of Curtis.

In 1993 Douglas Lin analysed the orbit of the Large Magellanic Cloud and concluded that the mass of the Milky Way must be $\sim 6 \times 10^{11}$ $M_\odot$ (mass of the Sun), about 4 times the previous estimate. He attributed the extra mass to a halo of dark matter ~200 kpc in diameter. If this is the case, the LMC, which is 55 kpc from the centre of the Milky Way, must be within the halo. More recent results have tended to confirm this general conclusion about the galactic halo and dark matter, although the amount of dark matter is now thought to be somewhat higher.

Early measurements of solar movement

William Herschel discovered in 1783 that the Sun was moving towards the star λ Herculis (263° RA, +26° Dec) by analysing the proper motions of a number of stars near to the Sun. Further measurements by various astronomers in the nineteenth century yielded a similar result.

The first reliable Doppler shifts showing the velocity of the Sun relative to the nearby stars were made by Hermann Vogel and Julius Scheiner around 1890. In 1893 Paul Kempf deduced a solar velocity of 13.0 km/s relative to the nearby stars from Vogel's and Scheiner's measurements. William Campbell published the radial velocities of 280 stars in 1901 and deduced a solar velocity of 19.9 km/s, with a target area or apex of motion of RA 268°, Dec + 25°, very similar to that estimated over a hundred years earlier by William Herschel.

Star streaming and rotation of the Milky Way

In 1871 Hugo Gyldén examined the proper motions of nearby stars and found that in one part of the sky they tended to have a maximum movement across the sky in one direction, while in the opposite part of the sky they moved in the opposite direction. At positions half way between the two there was no general drift. He correctly interpreted this effect as showing that the Milky Way was rotating.

Thirty years later Kapteyn undertook a similar but more extensive analysis. Unlike Gyldén, Kapteyn was able to use both proper motion and radial velocity measurements, giving him a three-dimensional view of stellar movements. Like Gyldén he also found that the stars appeared to be streaming in two different directions, leading him to conclude that there were two clouds of stars moving through one another. Kapteyn's explanation of this so-called 'star streaming' was not generally accepted, however, and Karl Schwarzschild, Arthur Eddington, Herbert Turner, and others devised alternative theories. But none of these was entirely satisfactory.

Most of the stars near the Sun had been found to move with modest speeds, relative to the Sun, of <30 km/s. The directions of motion of these stars had a symmetrical distribution in the plane of the Milky Way. But in 1918 Benjamin Boss found that the direction of motion, relative to the Sun, of the so-called high-velocity stars, whose velocities relative to the Sun exceeded 75 km/s, was asymmetrical, being confined between $l = 140°$ and 340°. Similar results were found independently by Walter Adams and Alfred Joy. This work was greatly expanded over the next few years, separately, by Gustav Strömberg and Jan Oort. It was found that the high velocity stars had an average motion, relative to the Sun, towards $l = 235°$. This implied that the Sun moved, relative to these stars, in exactly the opposite direction, that is towards $l = 55°$.

In 1925 Strömberg also concluded that the Sun moved, relative to the distant globular clusters, towards $l = 70°$ with a velocity of 330 km/s. He also deduced similar results for the movement of the Sun relative to spiral nebulae, which by then were known to be galaxies of stars outside the Milky Way.

Bertil Lindblad provided the theoretical basis for solving the problem of stellar motions in the Milky Way in 1926. He suggested that the Milky Way was composed of a number of subsystems rotating at different velocities about a common axis perpendicular to the plane of the Milky Way. The subsytem that rotated the fastest, at ~300 km/s, was that containing the Sun and local stars. Because it was rotating very fast, this subsystem would be flattened with respect to the plane of the Milky Way. The high-velocity stars were in a subsystem orbiting the Milky Way more slowly, whilst the globular cluster subsystem was orbiting the Milky Way the slowest of all. Because of this slow rotation, the globular cluster subsystem was the least flattened.

In the following year Jan Oort tested Lindblad's theory by analysing the velocities of numerous types of stars, planetary nebulae, and calcium clouds. He found that their movements were consistent with rotation of the Milky Way about a centre at about $l = 325°$, which was perpendicular to the direction of motion of the Sun. This $l = 325°$ was also the direction of the centre of the globular cluster system determined by

Shapley. Oort also showed that when stars have orbits deviating from exact circles, their velocities relative to the Sun showed a preferred direction towards the galactic centre. This was precisely the effect of star streaming that Kapteyn had detected.

Oort concluded that the Milky Way did not rotate like a solid body, as its angular rotation near the edge was slower than near the centre. He deduced that that the centre of the Milky Way was about 6.6 kpc from the Sun, that it had a central attraction equivalent to $\sim 8 \times 10^{10}$ $M_\odot$, and that the Sun orbited the centre once every 150 million years. Oort was concerned that the amount of mass visible in the Milky Way was much less than his calculated 8×10^{10} $M_\odot$, and attributed this effect to interstellar absorption. Interstellar absorption was detected a few years later.

In 1930 John Plaskett and Joseph Pearce studied ionised calcium lines, caused by interstellar gas, in the spectra of O- and B-type stars. As a result they not only confirmed the rotation of the galaxy, but showed that the interstellar gas was approximately uniformly dispersed in the plane of the Milky Way, and was rotating at the same velocity as the stars. They also confirmed the differential rotation of the Milky Way, measuring a 'slip rate' of 17 km/s/kpc.

The spiral arms

Astronomers began to think of the Milky Way as a spiral galaxy in the late 1920s, after the Andromeda nebula had been shown to be a spiral galaxy some distance from the Milky Way. But there was no clear evidence of our galaxy's spiral structure for some time.

Walter Baade and Nicholas Mayall showed in the 1940s that the spiral arms of the Andromeda galaxy could best be shown by O and B-type stars, hydrogen emission nebulae (H II regions), and by clouds of galactic dust. William Morgan, inspired by Baade's early work, tried to find the spiral arms of the Milky Way by detecting as many H II regions as possible and examining the O- and B-type stars within them. By 1951 Morgan, assisted by Stewart Sharpless and Donald Osterbrock, had found the Orion and Perseus arms. There were also suggestions of a third arm, the Sagittarius arm.

Interstellar absorption limits observations of the Milky Way in the visible waveband to those parts within a few kpc of the Sun. In 1951 however, neutral hydrogen was found to emit radio waves at a wavelength of 21 cm with virtually no attenuation. As a result, in the 1950s, Oort and Gart Westerhout in the Netherlands, and Frank Kerr in Australia, used radio telescopes to produce our first view of the overall structure of the Milky Way (see Figure 4.1). This showed that the Sun was on the inner edge of the Orion arm, which was ~ 8 kpc from the centre. The Perseus and Sagittarius arms, on either side of the Orion arm, were also clearly delineated.

There was then a problem of trying to understand how the spiral arms in the Milky Way and other spiral galaxies could retain their shape after many orbits of the centre. Lindblad suggested that the stars were not fixed in the spiral arms. Instead the spiral arms represented regions of enhanced density, or density waves, in the Milky Way, which rotated more slowly than the Milky Way's stars and gas. When gas entered these regions of higher density, the local mass density increased, resulting in new stars being formed. This explains why the stars in the spiral arms tended to be relatively young. Chia-Chiao Lin and Frank Shu developed this density wave theory mathematically in the early 1960s.

The central region

In 1959 Frank Drake concluded from his radio observations of the nucleus of the Milky Way that there was a small, very dense, cluster of stars there. He calculated a density of >1,000 stars/cubic parsec, which was more than a thousand times the density of stars in Sun's vicinity.

X-rays were detected coming from the general direction of the centre of the Milky Way in 1971 by the Uhuru spacecraft. Some years later the Einstein Observatory spacecraft pinpointed the position of the source to within an accuracy of 1 arcminute. The X-rays were also found to vary over a period of about 3 years, indicating that the source was relatively small.

An extremely compact radio source, Sgr A*, was discovered at the centre of the Milky Way in 1974 by Bruce Balick and Robert Brown. Six years later John Lacy and colleagues analysed the movement of gas clouds in the vicinity of Sgr A*, and concluded that there was probably a point-like mass of $\sim 3 \times 10^6$ $M_\odot$ at the centre. Subsequent observations of the movement of individual stars near the galactic centre by Reinhard Genzel, Andrea Ghez and colleagues indicated that most of the unseen mass of $\sim 4 \times 10^6$ $M_\odot$ must be within about 100 AU of the centre.

Recent radio observations of Sgr A* have shown that its diameter is about the same order of magnitude as the Schwarzschild radius for a 4×10^6 $M_\odot$ black hole. These and other results indicate that there is a supermassive black hole at the centre of the Milky Way.

Bibliography

Belkora, Leila, *Minding the Heavens: The Story of Our Discovery of the Milky Way*, Institute of Physics Publishing, 2003.

Bok, Bart J., and Bok, Priscilla F., *The Milky Way*, 3rd ed., Harvard University Press, 1957.

Levy, David H., *The Man Who Sold the Milky Way: A Biography of Bart Bok*, University of Arizona Press, 1993.

Osterbrock, Donald E., *Walter Baade: A Life in Astrophysics*, Princeton University Press, 2001.

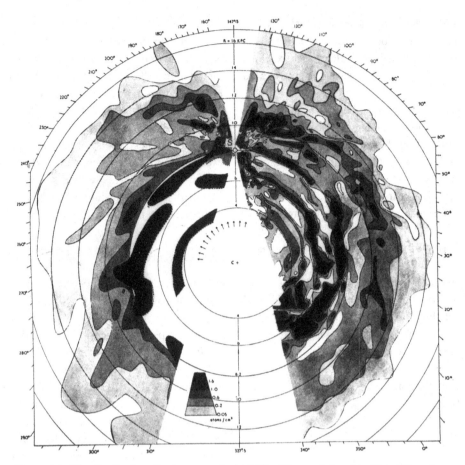

Figure 4.1. The distribution of hydrogen gas in the Milky Way, as deduced by Oort, Westerhout and Kerr, showing the spiral arm structure. The position of the Sun is marked 'S'. (From *Monthly Notices of the Royal Astronomical Society*, Vol. 118 (1958), Plate 6; Courtesy The Royal Astronomical Society.)

Paul, Erich Robert, *The Milky Way Galaxy and Statistical Cosmology 1890–1924*, Cambridge University Press, 1993.

Struve, Otto, and Zebergs, Velta, *Astronomy of the 20th Century*, Macmillan, 1962

See also: Globular clusters.

GLOBULAR CLUSTERS

It took some time before globular clusters were recognised as clusters of stars. For example, the brightest globular cluster, omega Centauri, was listed as a star in Ptolemy's star catalogue of about 150 AD. Edmond Halley recorded it as a nebulous object when he observed it on a journey to St Helena in 1677, but it was not listed as a globular cluster until John Herschel recognised it as such in the 1830s.

M22 in Sagittarius was probably the first globular cluster to be recognised as being something other than a star when Abraham Ihle recorded it as a 'composite nebula' in 1665. M5 was discovered by Gottfried Kirch and his wife Maria Margarethe as a 'nebulous star' in 1702, and M13, the great globular cluster in Hercules, was discovered by Edmond Halley

as a nebulous object in 1714. Charles Messier listed 29 globular clusters (20 of them discovered by him) in his catalogue of nebulous objects in 1781. He listed 28 as 'round nebulae', whilst he had managed to resolve stars in one, M4. This was the first globular cluster in which individual stars were resolved.

Thirty-three globular clusters were known when William Herschel started his comprehensive deep sky survey in 1782. He managed to resolve stars in all of these, whilst discovering a further 37 globular clusters. Herschel introduced the term 'globular cluster' to describe these objects in his second catalogue of deep sky objects in 1789.

John Herschel noticed in the 1830s that globular clusters were far more prominent in the Sagittarius region of the sky than any other, but it was not until the twentieth century that globular clusters were examined in any detail. Karl Bohlin suggested in 1909 that they formed a compact group at the centre of the Milky Way, with the Sun just outside the group. But Harlow Shapley disagreed because, although there were many globular clusters in Sagittarius, there were clusters in the northern sky also. This suggested to him that globulars surrounded the Milky Way in a cloud.

Shapley began his globular cluster investigations in 1914, estimating the distance to the nearest clusters using Cepheid variables. As a result he found that the clusters all had approximately the same absolute size and intensity. This enabled him to estimate the distances of the furthest globular clusters from their apparent sizes and intensities, even though they were too far away for Cepheids to be detected.

Shapley concluded in 1917, after plotting the locations of 69 globular clusters, that the centre of the system of globular clusters was also the centre of the Milky Way. As a result he estimated that the centre of the Milky Way was ~20 kpc from the Sun in the direction of galactic longitude 325° in Sagittarius. He also found that the globular clusters were distributed symmetrically on both sides of the plane of the Milky Way. Distances of the globulars from the Sun ranged from just less than 7 kpc (omega Centauri and 47 Tucanae) to 67 kpc (NGC 7006). The Milky Way system appeared to be ~100 kpc in diameter.

Shapley's dimensions of the Milky Way was disputed for some time, as previous work had indicated that it was only ~10 kpc in diameter. Although Shapley's distance estimates were later found to be of the correct order of magnitude, they were generally too large, particularly for the most distant globulars. This was because he had not taken into account the attenuation of light due to interstellar dust. Then in 1932 Edwin Hubble announced that he had found 140 globular clusters associated with the Andromeda nebula, M31.

Robert Trumpler showed in the 1920s that the Hertzsprung-Russell (H-R) diagram for stars in the Milky Way globular clusters was different from that of stars relatively close to the Sun, as there was no main sequence for stars in globular clusters. In the 1940s Walter Baade found a similar difference between the H-R diagram of stars in the spiral arms of galaxies, on the one hand, and those in the nucleus regions of spiral galaxies and in globular clusters, on the other. He called the former population I and the latter population II stars. A little later it became clear that population I stars have more metals in their atmospheres than population II, indicating that they are much younger. In the case of the Milky Way, population II stars seemed to be as old as the galaxy itself.

In 1975 Woltjer investigated the distribution of globular clusters in the Milky Way. He differentiated between what he called the halo clusters which had an underabundance of the heavier elements or metals, and the disk clusters which had an underabundance that was not nearly as large. It was later found that the halo clusters, which are usually some distance above and below the galactic plane, orbit the galactic centre in orbits with generally high eccentricity and random inclinations. In fact some halo clusters orbit the centre in the opposite sense to the general rotation of the galaxy. The disk clusters, on the other hand, which are within about 2 kpc of the galactic plane, orbit the centre in more circular orbits, all in the same sense.

Today about 160 Milky Way globular clusters have been identified each of which typically contains ~100,000 stars, although omega Centauri, which is the largest known in the Milky Way, has a few million. Globular clusters have also been identified in both Magellanic Clouds and a number of other galaxies. M31 has more than 500 globular clusters, whilst the giant elliptical galaxy M87 has ~10,000.

Bibliography

Ashman, Keith M., and Zepf, Stephen E., *Globular Cluster Systems*, Cambridge University Press, 2008.

Belkora, Leila, *Minding the Heavens:; The Story of Our Discovery of the Milky Way*, Institute of Physics Publishing, 2003.

Kissler-Patig, M. (ed.), *Extragalactic Globular Cluster Systems: Proceedings of the ESO Workshop Held in Garching, 27–30 August, 2002*, Springer, 2003.

Martinez, R. C., Pérez-Fournon, I., and Sánchez, F., *Globular Clusters*, Cambridge University Press, 1999.

Osterbrock, Donald E., *Walter Baade: A Life in Astrophysics*, Princeton University Press, 2001.

Struve, Otto, and Zebergs, Velta, *Astronomy of the 20th Century*, Macmillan, 1962.

See also: Pulsating variables.

INTERSTELLAR ABSORPTION

The question of why the night sky is dark, if the universe is infinite and filled with stars, had been the subject of debate amongst astronomers for some time when Jean Phillipe de Chéseaux made a radical proposal. He suggested in 1744 that it was because there was material between the stars that attenuated their light, but his estimate of the attenuation was far too high. In 1823 Wilhelm Olbers, who has given his name to this paradox about the dark night sky, also suggested that the cause was attenuation. He calculated an attenuation equivalent to ~0.5 magnitude/kiloparsec (mag/kpc), which is very close to today's value. Twenty years later Wilhelm Struve analysed William Herschel's observations and concluded that they showed too few faint stars if, on average, all the stars were of the same intrinsic intensity. Either the Sun was in a region where the stars were exceptionally dense or, as he suggested, there was some material in space attenuating the light from the more distant stars. His estimate of the decrease was equivalent to ~2 mag/kpc.

Meanwhile William Herschel had noticed, in the late eighteenth century, that there were a number of areas in the sky in which he could see virtually no stars, even with his largest telescopes. He interpreted these as regions devoid of stars. His son, John Herschel, also observed dark patches in the southern sky some years later. When Edward Barnard started to photograph the Milky Way in 1889 he discovered a number of black areas.

Barnard published his first photographs of these regions in 1894, still assuming, like the Herschels, that these were regions devoid of stars. Arthur Ranyard disagreed, however, and suggested that they were clouds of absorbing material blotting out the background stars. Unknown to Ranyard, a similar suggestion had already been made by Angelo Secchi in 1853. Until his death in 1923, Barnard oscillated between the two theories of areas empty of stars or regions of absorbing matter.

So there were two basic questions to be answered in the early twentieth century: Was there material in space that attenuated light uniformly throughout the Milky Way? And/or were there dark clouds of material attenuating light just in certain parts of the Milky Way?

Gas

In 1904 Johannes Hartmann discovered the stationary K line of ionised calcium in the spectrum of the binary star δ Orionis, which he attributed to absorption by interstellar gas. The same stationary K line and the stationary H line of calcium was then discovered in the spectra of other binary stars by Vesto Slipher and others. Then in 1919 the same effect was found for sodium in the binaries δ Orionis and β Scorpii by Mary Heger.

Many astronomers thought that these stationary lines were produced by small clouds surrounding the binary stars. But in 1923 John Plaskett showed that the velocities of the calcium absorption lines differed from those of their background stars by up to 50 km/s. So these non-stellar lines were probably caused by interstellar gas permeating the Milky Way. Seven years later, Plaskett and Joseph Pearce showed, by studying ionised calcium lines in the spectra of O and B type stars, that interstellar gas was approximately uniformly dispersed in the plane of the Milky Way, and was rotating at the same velocity as its stars.

Carlyle Beals found in 1936 that interstellar lines were sometimes multiple, which suggested that more than one cloud was between the star and the Earth, and that these clouds were moving with different velocities. Walter Adams measured interstellar absorption lines in hundreds of stars and concluded, in 1949, that the strongest component in the multiplet nearly always corresponded to the lowest velocity, while some of the weaker components showed velocities of up to 100 km/s. Measurements of the absorption lines of neighbouring stars showed the same strongest component of the interstellar lines, but, sometimes, the highly displaced lines were different. This indicated that the high-velocity clouds were generally smaller than the low-velocity clouds and, although the average cloud size was estimated at about 3 parsec, the range of sizes was found to be very large.

Guido Münch discovered in 1956 that the absorption lines for very distant stars near the plane of the Milky Way often showed two strong components, one due to clouds in our spiral arm (the Orion arm), and the other due to clouds in the Perseus arm, which is the next arm out from ours. Otto Struve and others had shown that, although clouds were much scarcer at high galactic latitudes than in the plane of the Milky Way, they were still found there. As the general gas density outside of the clouds was very low at high galactic latitudes, these clouds ought to have dispersed by now. But in 1961 Münch and Harold Zirin showed that this dispersal could have been restrained by a hot, very tenuous galactic corona.

Jan Oort recognised in the early 1940s that, as absorption by interstellar dust reduces with increase in wavelength (see later), observations in the radio waveband would probably be unhindered by dust. So he asked Hendrik van de Hulst in 1944 whether there was a spectral line in the radio waveband that they may be able to detect and use to define the structure of the Milky Way's spiral arms. Although interstellar gas had been found to consist of calcium and sodium, it was thought that it may consist mostly of hydrogen. So van de Hulst studied the likely emission lines of hydrogen and predicted that a hyperfine transition in the ground state of neutral hydrogen should emit radio waves at a wavelength of about 21 cm. He thought that although the probability of emission of this 'forbidden' line was very small, it should be detectable in space, as it was anticipated that there would be a large amount of hydrogen there.

Emission of this 21 cm line of neutral hydrogen was discovered in 1951 by Harold Ewen and Edward Purcell. Jan Oort and colleagues then used Doppler shifts of this line to detect and measure the slow rotation of the thin sheet of neutral hydrogen gas found in the plane of the Milky Way. This showed that the gas in the central few hundred parsecs of the galaxy rotated as if it were a solid body, with an almost constant angular velocity. But outside of this region it had an approximately constant linear velocity of ~250 km/s.

In 1941 Walter Adams reported that interstellar gas consisted not only of atoms, but also of molecules, when he found the lines of methylidyne (CH) and cyanogen (CN) in the spectrum of ζ Ophiuchi. Further discoveries of interstellar molecules had to await the development of infrared and ultraviolet detectors mounted on spacecraft, and of radio telescopes. Then in 1963 Barrett, Meeks and Weinreb discovered the hydroxyl radical, and in 1968 ammonia was found. These were followed by the discovery of water vapour and formaldehyde in 1969, and that of the hydrogen molecule H_2 and of carbon monoxide in 1970. Since then ever more complex molecules have been discovered including formic acid in 1970 and methylamine in 1977, which can react together to form the amino acid, glycine. Over 150 molecules have now been discovered in interstellar gas, a large number of which are organic. In 2008 amino acetonitrile, which is chemically related to glycine, was found.

Some of these molecules were found to have very intense emission lines, for example, water vapour at 1.35 cm and hydroxyl at 18 cm, which were due to maser action. Later work with very high resolution radio interferometers showed that these sources were generally extremely small.

The discovery of interstellar molecules had been a surprise because it was thought that ultraviolet radiation would dissociate all but the most tightly bound molecules. But interstellar dust (see below) was found to shield molecules from this ultraviolet radiation, allowing the molecules to survive. Large quantities of these molecules, thought to be mostly molecular hydrogen, were found to exist in giant molecular clouds that had masses of about 10^4 to 10^6 $M_\odot$. Dust in the densest parts of these clouds produced the dark nebulae previously observed by Edward Barnard. In the 1970s these giant molecular clouds were recognised to be sites of star formation.

Dust

It was known in the early twentieth century that there were far more spiral nebulae around the poles of the Milky Way than near its equator. Heber Curtis suggested, during his debate with Harlow Shapley in 1920, that this was because light from spiral nebulae was being absorbed by a ring of material *surrounding* the Milky Way, as many edge-on spirals seemed to have such an obscuration. But both Shapley and Curtis thought that any absorption effect *within* the Milky Way was small. Kapteyn had earlier considered possible absorption within the Milky Way, but after some vacillation, had come to the same conclusion.

In the late 1920s Robert Trumpler measured the apparent diameters of 100 open star clusters, and the intensity of their stars, and found that the intensities reduced faster with distance than expected, by about 0.7 mag/kpc (photographic), assuming that the clusters were all of the same size. He published his conclusions in 1930, concluding that the Milky Way contained interstellar dust.

Trumpler also showed that stars of the same spectral type were redder, the further away they were. He found that dust absorption reduced with increasing wavelength, such that an average selective absorption (photographic $\lambda = 430$ nm minus visual $\lambda = 550$ nm) was about 0.3 mag/kpc for stars in open star clusters. This appeared to contradict Shapley's observations of ten years earlier that showed no such effect for stars in globular clusters, which were much further away than open star clusters. Open star clusters were generally in the plane of the Milky Way, however, whereas the globular clusters were generally far away from this plane. So Trumpler concluded that this selective absorption was confined to a relatively thin layer of material extending ~100 parsec on either side of the plane of the Milky Way.

Two years later Joel Stebbins showed that those globular clusters in the plane of the Milky Way are redder than those out of the plane due to dust. He also concluded that the estimated diameter of the Milky Way, based on the distances of globular clusters in or near the plane of the Milky Way, needed to be reduced from ~80 kpc to ~30 kpc, as interstellar absorption had given a false idea of distance. In 1938 Alfred Joy studied the radial velocities of Cepheid variables near the plane of the Milky Way, and found that the more distant Cepheids were also fainter than expected due to interstellar absorption.

Albert Hiltner and (separately) John Hall and Alfred Mikesell found in 1949 that the light from a number of stars was plane polarised. They tentatively suggested that this may be due to scattering by elongated particles aligned by magnetic fields of the interstellar medium. Two years later Leverett Davis and Jesse Greenstein calculated the strength of the galactic magnetic field as ~10^{-4} to 10^{-5} gauss from the degree of polarisation.

Bibliography

Longair, Malcolm, *The Cosmic Century: A History of Astrophysics and Cosmology*, Cambridge University Press, 2006.

Paul, Erich Robert, *The Milky Way Galaxy and Statistical Cosmology 1890–1924*, Cambridge University Press, 1993.

Seeley, D., and Berendzen, R., *The Development of Research in Interstellar Absorption, c. 1900–1930*, Journal for the History of Astronomy, 3 (1972), 52–64, and 3 (1972), 75–86.

Struve, Otto, and Zebergs, Velta, *Astronomy of the 20th Century*, Macmillan, 1962.

Whittet, D. C. B., *Dust in the Galactic Environment*, Institute of Physics Publishing 1992.

See also: Globular clusters; Overview – Milky Way.

MAGELLANIC CLOUDS

The Magellanic Clouds have been known since ancient times by people living in the southern hemisphere, like the Aborigines of Australia and the Polynesian seafarers. The clouds are also visible from South America and Equatorial and Southern Africa, and the Large Magellanic Cloud (LMC) is visible from southern Arabia.

The first known written mention of the LMC was as 'al-Bakr', 'the White Ox', by the Arabian astronomer Al-Sufi in his *Book of Fixed Stars* in 964. Both clouds were visible to the first European seafarers who ventured into the southern oceans in the fifteenth and sixteenth centuries. They were initially known as the Cape Clouds or Nebecula Major and Minor, whilst their modern name comes from that of Ferdinand Magellan, who observed them in the first circumnavigation of the globe in the early sixteenth century. Antonio Pigafetta, who sailed with Magellan, recognised that they consisted of stars in his 1524 description of Magellan's voyage.

John Herschel was the first to investigate the composition of the two clouds, listing 919 objects (582 stars, 291 nebulae and 46 star clusters) in the large cloud and 244 objects in the small one in 1847. Twenty years later Cleveland Abbe suggested that the two clouds were galaxies near to, but external to, the Milky Way.

Henrietta Leavitt found a period–luminosity relationship for 25 Cepheid variables in the Small Magellanic Cloud (SMC) in 1912. In the following year Ejnar Hertzsprung was able to normalise this relationship, using a number of relatively close Cepheids, which enabled him to estimate the distance of the SMC for the first time, as about 10 kpc (kiloparsec). But in 1918 Harlow Shapley recalibrated the absolute magnitude scale of Cepheids and estimated the distance of the SMC to be about 30 kpc, with that of the Large Magellanic Cloud a little less.

In 1952 Andrew Thackeray and Adriaan Wesselink found that the first cluster-type Cepheids discovered in the SMC were 1.6 magnitudes fainter than expected. Walter Baade came to a similar conclusion following his observations of spiral galaxies. As a result, Shapley's estimates for the distances of both Magellanic Clouds needed doubling.

J. van Kuilenburg detected a 35° long cloud of neutral hydrogen using a radio telescope in 1972. In the same year Wannier and Wrixon, using higher sensitivity, found that the stream was 60° in length with a systematic variation in its radial velocity of from 400 to 60 km/s. Two years later Mathewson, Cleary and Murray found that the stream extended over 180° of the sky, almost following a great circle. It appeared to include both of the Magellanic Clouds and so was called the Magellanic Stream. In 2000 Brad Gibson and colleagues established the chemical similarity of the stream and the two clouds, both of which are gas rich and metal poor compared to the Milky Way. It is thought that the stream had been created through the interaction of the Milky Way with the Magellanic Clouds early in their history.

Douglas Lin analysed the movement of about 250 stars in the Large Magellanic Cloud and concluded, in 1993, that it was leading the Magellanic Stream and orbiting the Milky Way at about 240 km/s, and approaching the Milky Way at 55 km/s. As a result Lin estimated the mass of the Milky Way to be about 4 times the previous estimate, attributing the extra mass to a halo of dark matter ~200 kpc in diameter. If this was the case, the LMC, which was about 50 kpc from the centre of the Milky Way, must be within the halo.

Bibliography

Chu, You-Hua, et al. (eds.), *New Views of the Magellanic Clouds: Proceedings of the 190th Symposium of the International Astronomical Union Held in Victoria, Canada, 12–17 July 1998*, Astronomical Society of the Pacific, 1999.

Feast, M. W., *The Magellanic Clouds and the Extragalactic Distance Scale*, in Van den Bergh, S., and Pritchet, C. J., *The Extragalactic Distance Scale*, Astronomical Society of the Pacific, 1988.

Humboldt, Alexander von (E. C. Otté, trans.), *Cosmos: A Sketch of the Physical Description of the Universe*, Vols. I–IV, Henry G. Bohn, 1849–52.

Westerlund, Bengt E., *The Magellanic Clouds*, Cambridge University Press, 1997.

2 · Other galaxies and cosmology

ACTIVE GALAXIES

Seyfert galaxies

In 1908 Edward Fath observed very strong emission lines in the nebula NGC 1068. A little later Vesto Slipher produced higher-resolution spectra of the same nebula and concluded that the line widths showed velocities of the order of hundreds of kilometres per second. In 1943 Carl Seyfert examined a number of galaxies, including NGC 1068, that had small, intensely bright, very blue nuclei. He found that their optical spectra were dominated by very intense emission lines of the Balmer series and forbidden lines of ionised elements, such as oxygen and iron, superimposed on a normal G-type spectrum. The lines were broad, indicating velocities of up to 8,500 km/s. In 1955 two of these galaxies (NGC 1068 and NGC 1275), now called Seyfert galaxies, were found to be radio sources.

Four years later Lodewijk Woltjer found that the nuclei of the nearest Seyfert galaxies were unresolved, so they must be less than 100 pc (parsec) in diameter. Assuming a normal mass-to-light ratio, he calculated that the mass of their nuclei was in the range 10^8-10^{10} $M_\odot$ (mass of the Sun). This implied that a significant amount of the galaxy's mass was contained in a volume of less than 100 pc in diameter.

Merle Walker made a detailed examination of one of the brightest Seyfert galaxies, NGC 1068, in 1967. He found a number of clouds, each weighing $\sim 10^6-10^7$ $M_\odot$, being ejected from the central region at velocities of up to 600 km/s. Walker concluded that a new, unexplained energy source was required to explain these violent motions. He also detected similar clouds in Seyfert galaxies NGC 4151 and NGC 7749. A few months later J. Beverley Oke and Wallace Sargent detected moving clouds or filaments of hot (20,000 K) gas, with a total mass of $\sim 2 \times 10^5$ $M_\odot$, within the central 50 pc of NGC 4151. The forbidden lines of iron indicated that the remainder of the nucleus consisted of low-density gas at a temperature of $\sim 10^6$ K.

Geoffrey Burbidge, Margaret Burbidge and Allan Sandage and, independently, Iosif Shklovskii suggested in the mid-1960s that quasars, which had recently been found (see below), seemed to be more violent versions of Seyfert galaxies. Shortly afterwards A. G. Pacholczyk and W. Z. Wisniewski found that Seyfert galaxies, like quasars, were strong emitters of infrared

radiation. As a result, a search was made to see if Seyfert galaxies, like quasars, also showed optical variations. Almost immediately Fitch, Pacholczyk and Weymann found that the intensity of NGC 4151 varied in both visible and ultraviolet wavelengths. Both Seyfert galaxies and quasars are today called 'active galaxies'.

Edward Khachikian and Daniel Weedman concluded in the early 1970s that there were two different types of emission line spectra in Seyfert galaxies. Type 1 Seyferts were defined as those with very broad permitted emission lines with widths of $\sim$5,000 km/s, whilst those with relatively narrow emission lines with widths of $\sim$500 km/s were classified as Type 2 Seyferts. Type 1 Seyferts were also observed to have narrow forbidden lines. It was concluded that the broad permitted emission lines originated close to the source of excitation in high density gas clouds in Type 1 Seyferts, whereas the narrow forbidden lines were emitted in regions of lower density further away from the nucleus.

In 1985 Robert Antonucci and Joseph Miller compared the polarisation spectra of the nucleus of the Seyfert 2 galaxy NGC 1068 with that observed for Seyfert 1 galaxies. As a result they concluded that there is an obscuring torus around the nucleus of Seyfert galaxies. In the case of Type 1 Seyferts we observe the galaxy at a small angle to the axis of the torus. So both the broad-line emitting nuclear region and the narrow-line emitting outer region can be seen. But in the case of NGC 1068, and apparently all other Type 2 Seyferts, we observe the galaxy at a large angle to the axis of the torus, so the nuclear region is obscured and only the narrow-line regions are observed directly. However the broad-line emission can sometimes be detected in the light scattered or reflected by gas and dust in the narrow-line regions. In this case the light is polarised.

Marques dos Santos and J. R. D. Lépine had discovered water vapour maser emission from the Seyfert galaxy NGC 4945 in 1978. Since then similar maser emission has been detected in a number of Seyfert galaxies. For example, in 1995 Naomasa Nakai and colleagues searched for and found high-velocity water maser emission in a number of Seyfert galaxies. The highest-velocity maser emission was detected within the velocity range of $\pm 1,000$ km/s in NGC 4258. They concluded that this was consistent with a molecular disc of diameter 0.26 pc rotating around a black hole of mass 3.6×10^7 $M_\odot$ at the centre (see also *Unified model for active galaxies* below).

Quasars

Thomas Matthews and colleagues found in 1960 that the radio source, 3C 48, appeared to be coincident with a dim blue star which was surrounded by a faint, irregular nebulosity. Matthews and Allan Sandage then found indications of short-term variability in the optical source. In 1961 another radio source, 3C 273, was identified with a faint blue star which also varied in its optical intensity. Harlan Smith and Dorrit Hoffleit found two types of variation in this star, one of about half a magnitude that occurred over a timescale of about ten years or so, and another, much more rapid, variation which appeared to be quite random.

In 1962 Hazard, Mackey and Shimmins found, using lunar occultations, that the radio source 3C 273 was double, with a separation of 20 arcsec. The position of one of these sources correlated with the faint blue star, whilst that of the other coincided with the end of a faint bluish jet coming from the star. Maarten Schmidt found in the following year that the broad emission lines in the star's spectrum were Balmer hydrogen lines with an unexpectedly large redshift of 0.158. This indicated that the source was receding from the solar system at about 0.15 c (the speed of light), and that it was about 470 Mpc (1.5 billion light years) away, assuming that the redshift indicated distance (i.e., was of cosmological origin) and assuming a Hubble constant of 100 km/s/Mpc. If it was at that distance, 3C 273 must be $\sim$100 times as bright in the optical waveband as the Milky Way, assuming that 3C 273 was radiating isotropically.

Jesse Greenstein and Thomas Matthews immediately checked the optical spectrum of the other radio source, 3C 48, and found that it had an even larger redshift of 0.367, indicating a distance of 1.10 Mpc (3.6 billion light years). In addition, the radio brightness of 3C 48 was found to vary substantially during a day, indicating that it was only $\sim$ 1 light day in diameter. Over the next few years many of these quasi-stellar radio sources or 'quasars' were discovered at ever greater distances. For example, 3C 9 had been found as early as 1965 with a redshift of 2.01. Many of these quasars showed similar short period variations in their radio output to 3C 48, indicating their very small physical size. Today quasars are recognised as a class of active galaxies.

Matthews and Sandage had found in 1963 that quasars seemed to have an ultraviolet excess. As a result, Sandage was able to discover a number of possible quasars in 1965 from their ultraviolet excess, rather than from their radio emissions. This new class of radio-quiet object, which were initially called quasi-stellar galaxies, are now called radio-quiet quasars (RQQs) to distinguish them from normal quasars or radio-loud quasars (RLQs). In the event far more RQQs have been found than RLQs.

In 1967 Herbert Friedman and Edward Byram thought that they had detected X-rays from 3C 273. This was confirmed three years later by C. Stuart Bowyer and colleagues using a sounding rocket experiment. In 1979 the Einstein Observatory spacecraft found that the X-ray intensity of 3C 273 was about a million times greater than that of the Milky Way, but it was varying over periods as short as half a day, indicating the very small size of the X-ray source. In the meantime, Andrzej Pacholczyk and Ray Weymann had found that 3C 273 was a strong emitter of infrared radiation. Then in 1978 Boudewijn Swanenburg and colleagues concluded, using the COS – B spacecraft, that 3C 273 was also a source of gamma rays. This was endorsed by Giovanni Bignami and colleagues in 1981. So 3C 273 was a strong source in all wavebands from gamma rays to radio waves.

At first there were doubts as to whether quasar redshifts really were of cosmological origin, in view of their implied enormously large luminosities and their very small size. Fred Hoyle, Geoffrey Burbidge and Halton Arp, in particular, independently suggested that they were, instead, relatively nearby objects. But no quasars had been observed with blueshifts, so it appeared unlikely that the redshifts indicated local velocities, and no other convincing case was put forward to explain them. Then in the 1980s absorption lines, the 'Lyman alpha forest', were discovered in quasar spectra apparently due to intervening matter at different cosmological distances. Nevertheless, Arp continued to argue for many years that some high-redshift quasars appeared to be physically linked to some relatively low-redshift galaxies. But most astronomers dismissed these apparent linkages as simple line-of-sight effects.

Since their first identification in 1963, more than 200,000 quasars have been found so far with redshifts of up to 6.5, most of them from the Sloan Digital Sky Survey. They radiate from X-rays to the far infrared with some, like 3C 273, also being strong gamma-ray and radio sources. Some quasars vary rapidly in the optical waveband and even more rapidly in X-rays, indicating sizes as small as that of the solar system or even less. Most quasars are brightest in their rest-frame ultraviolet, which is observed in the visible or infrared, depending on their redshift. The radio emission of radio-loud quasars seems to be associated with jets of relativistic charged particles coming from the nucleus.

Yakov Zeldovich and Igor Novikov suggested in 1964 that the masses of quasars had to be very large because their luminosities could not be greater than the Eddington limiting luminosity. So for 3C 273, for example, with a luminosity of at least 10^{40} W, the energy source must have a mass of at least 10^9 M$_\odot$. Such a large mass concentration in a volume no larger than $\frac{1}{2}$ a light day across implied the presence of a supermassive black hole. More recent observations and analysis of various quasars have confirmed that they probably contain a central supermassive black hole in the range $10^8 - 10^{10}$ M$_\odot$. So it is thought that quasars are the centres of active galaxies, powered by the accretion of material into supermassive black holes. The

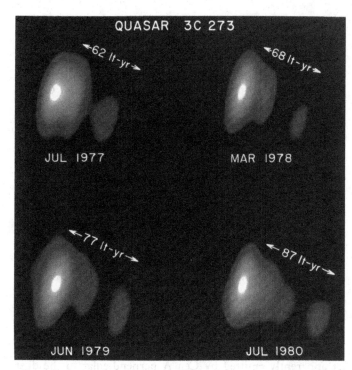

Figure 4.2. Apparent superluminal motion in the jet emitted by the quasar 3C 273. Image produced by the VLBI radio telescopes at Effelsberg, Haystack, Green Bank, Fort Davis, Owens Valley and Hat Creek. (Courtesy California Institute of Technology.)

spectra of quasars showed that they have elements heavier than helium, indicating that their source galaxies have undergone a massive phase of star formation between the time of the Big Bang and their formation. This was a surprise as it had been thought that such processes would not have occurred so early in the universe.

Superluminal sources

In 1969 Jack Gubbay and colleagues found, using a radio interferometer based on radio telescopes in California and Australia, that the expansion of 3C 273 appeared to be superluminal (i.e., faster than the speed of light). Then in 1971 Alan Whitney and colleagues announced the discovery of the apparent expansion of 3C 279 at a velocity of ~10c by using the Goldstone-Haystack interferometer in the USA. A few months later Marshall Cohen and colleagues announced that they had also found apparent superluminal motion for 3C 273 (see Figure 4.2) and 3c for 3C 279. Some years later G. A. Seielstad and colleagues found the same effect in 3C 120 and 3C 345. Then in 1981 R. B. Phillips and Robert Mutel observed superluminal motion in BL Lacertae. Subsequently, superluminal motion was found to be a relatively common effect with quasars and blazars (see below).

A number of possible explanations were put forward for this apparent superluminal motion. One possibility was that the quasars were not at cosmological distances, so the motion

was not superluminal. Another was that there were particles called tachyons with an imaginary mass which could therefore move at superluminal velocity. But the most likely explanation came from the geometry of an explosion that was expanding at velocities close to the speed of light, as described by Martin Rees in 1966. In that case, motion transverse to our line of sight can appear to be superluminal, even though it is not.

Blazars

Blazars, which are another type of active galaxy, are categorised as either *BL Lac objects* or *highly polarized quasars (HPQs)*. They are discussed separately below.

(a) BL Lac objects

BL Lacertae was identified as a variable star by Cuno Hoffmeister in 1929. Then in 1968 John MacLeod and Brian Andrew observed that the radio source VRO 42.22.01 had a peculiar radio spectrum. In the same year John Schmitt identified VRO 42.22.01 with the optical source BL Lac which he found was surrounded by a faint nebulosity. BL Lac's optical emissions were linearly polarised and its optical brightness was found to vary between 12.0 and 15.5 magnitude (mag) during the period April to November 1968 together with rapid variations of ~0.3 mag/day. Its radio emissions varied with a timescale of the order of a month, and were circularly polarised. Amazingly its optical spectrum showed no emission or absorption bands. In 1974 Beverley Oke and James Gunn measured BL Lac's redshift as 0.07 from that of its associated nebulosity which was thought to be a galaxy. Over the next few years similar objects to BL Lac were found, one of which, AP Lib, had, like BL Lac, been thought to be a variable star. Its redshift was found to be 0.049. Then in 1981 Phillips and Mutel found evidence that BL Lac was emitting at least one high-velocity jet.

Optical and infrared luminosities of BL Lac objects were found to vary significantly over the course of days, whereas at shorter wavelengths the variations sometimes occurred in just a few hours. Their linear polarisations also varied rapidly. By the 1980s BL Lac objects were thought to be similar to quasars. Their intensities were found to increase with increasing wavelength from X-rays through optical to reach a maximum in the infrared or microwave bands. Some BL Lacs, now called X-ray loud BL Lacs (XRLs), were found which were brighter than normal BL Lacs, now called radio-loud BL Lacs (RBLs), in the X-ray region, and less bright than RBLs at radio wavelengths. In 1991 the Compton Gamma Ray Observatory spacecraft also found gamma rays coming from three BL Lac objects, supporting the idea that they were similar to quasars.

BL Lac objects were thought for a number of years to be associated only with the nuclei of elliptical galaxies, but in 1991 Ian McHardy and colleagues found that the BL Lac object PKS 1413+135 was at the centre of a spiral galaxy. This was the first indication that relativistic jets, which are associated with active galaxies, may not be limited to the nuclei of ellipticals.

(b) Highly polarized quasars (HPQs)

In 1978 Hervey Stockman and Roger Angel analysed the optical polarisation of 44 quasars and found that all but three had a polarisation of $<1.75\%$. The three with the highest polarisations had values of 3.6%, 4.1% and 13.6%. So in 1981 Richard Moore and Hervey Stockman suggested that a new class of quasars should be instituted of highly polarized quasars (HPQs), which they defined as quasars with a polarisation of $>3\%$. By that time there were four quasars with known values $> 3\%$ and, as the authors noted, each of these had interesting properties in addition to their high polarisation, partly justifying their separate classification. Moore and Stockman then tried to find more quasars that fitted into their new category, and found 10 new HPQs out of more than 200 quasars that they analysed.

HPQs have since been found to be similar to BL Lacs in having their relativistic jet pointing virtually straight at us. But they differ crucially from BL Lacs in having broad and strong emission lines like quasars, rather than the basically featureless optical continuum of BL Lacs. HPQs are also brighter than BL Lacs, being observed at higher redshifts, whilst their observed superluminal motion is also more pronounced than that of BL Lacs.

Radio galaxies

Karl Jansky detected radio waves in 1932 that he thought might be coming from the centre of the Milky Way. Grote Reber undertook a series of observations in the late 1930s and early 1940s and confirmed that the most intense radio signals were coming from the centre of the Milky Way. However Reber also detected a few subsidiary radio peaks, one of which was in the constellation of Cygnus. In 1946, Hey, Parsons and Phillips also discovered radio signals coming from a source in Cygnus, but although their equipment was better than Reber's, its location accuracy was still only $\sim 2°$. Hey and colleagues also thought that they had detected the source's variability, but this detection was later found to be erroneous.

There was much confusion about the increasing number of radio sources that were discovered in the late 1940s because of the poor resolution of the radio telescopes available at that time. There were also problems caused by difficulties in determining the effects of atmospheric absorption on the true location of the radio sources.

(a) Centaurus A

John Bolton announced the discovery of a number of new, discrete radio sources in July 1948, including Centaurus A and Virgo A. About twelve months later, in spite of their rela-

tively inaccurate position estimates, Bolton, Stanley and Slee felt able to identify Cen A with the seventh magnitude galaxy NGC 5128, and Vir A with the galaxy M87, but these identifications were disputed at the time. Three years later Bernard Mills produced a more accurate position estimate of Cen A, confirming its identification with NGC 5128. He concluded that the radio source was $\sim 6'$ in size. John Bolton, on the other hand, found the radio source was nearly $2°$ in diameter. In visible light NGC 5128 appeared to be an almost-circular disc of stars with a dark dust band across its centre. Walter Baade and Rudolph Minkowski suggested that it was the result of an interaction or merger of two galaxies.

C. Stuart Bowyer and colleagues detected X-rays from Cen A in 1969. A few years later Frank Winkler and Alice White found, using the OSO-7 spacecraft, that Cen A had brightened in X-rays by a factor of at least 1.6 over a six day period, indicating that the source was only a few light days in diameter. They also found that Cen A's X-ray intensity had also varied by up to a factor of 14 over the previous few years. In 1975 Victor Blanco and colleagues discovered a short, faint optical jet apparently emitted by Cen A perpendicular to the dust band. Then, four years later, Ethan Schreier and colleagues discovered a similarly oriented, short X-ray jet.

In 1971 William Kunkel and Hale Bradt found a $3'' \times 5''$ hot spot at the centre of Cen A in the near infrared. They concluded that there was a mass of $\sim 1.5 \times 10^9$ M$_\odot$ in the centre of the galaxy with a mass density of $\sim 1.3 \times 10^3$ M$_\odot$/pc^3. Jonathan Grindlay and colleagues detected >300 GeV gamma-rays apparently coming from Cen A over the period 1972–1974 using ground-based Čerenkov detectors. At about the same time Hall, Meegan and Walraven found, using a balloon-borne experiment, that Cen A also emitted gamma rays in the MeV region.

The normal optical image of Cen A shows a ball of stars about $6'$ across with the band of dust cutting across it. More sensitive modern optical images have revealed a main body of the galaxy of $33' \times 27'$, with the long axis of this ellipsoid perpendicular to the dust band. There are also faint streamers of stars extending from either ends of the $33'$ dimension to double the optical width along the ellipsoid's long axis. Similarly ever more sensitive radio observations have showed that the radio source has two huge lobes covering a total of about $8° \times 4°$ or 500×250 kpc. There are two much smaller inner radio lobes near the centre of this massive radio source oriented, like the extended optical source, perpendicular to the dust band.

Cen A is the nearest radio-loud active galaxy. It appears to be a Type 2 Seyfert object in the optical waveband and a BL Lac type at higher energies. Cen A is thought to be the result of the merger of two galaxies, as proposed many years earlier by Baade and Minkowski. The mass of the hypothesised black hole at the centre has been estimated to be $\sim 10^4 - 10^8$ M$_\odot$.

(b) Cygnus A

In 1951 F. Graham Smith was able to measure the position of the radio source Cyg A, which had been detected by Reber (see above), to within 1'. This enabled Baade and Minkowski to locate it optically as an inconspicuous 16th magnitude galaxy which appeared to be double. Its spectrum had broad emission lines indicating gas velocities of ~400 km/s, which caused Baade to conclude that it was two galaxies in collision. However, further investigations showed it to be a single galaxy crossed by a central dust band, like Cen A, but much further away.

Roger Jennison and Mrinal Das Gupta discovered in 1953 that Cyg A's radio emissions come mainly from two diffuse patches on either side of the galaxy. More recent radio images show these two patches as lobes that terminate in bright hot spots at their furthest point from the galaxy. A pair of narrow jets extend about 60 kpc on either side of the galaxy to link it to the lobes. The jets consist of fast electrons which carry energy outwards from the galaxy until they meet the surrounding gas, where their orderly motion is disrupted and where they generate magnetic fields. The electrons then produce radio emissions as they thread their way through the magnetic fields.

Cyg A was first detected as an X-ray source by Riccardo Giacconi and colleagues using the Uhuru spacecraft in 1972. But observations with successive spacecraft have, so far, failed to completely disentangle the X-rays emitted by Cyg A itself from those coming from the intracluster medium surrounding the galaxy.

It is now clear that Cyg A, with its jets, lobes and hot spots, is by far the most powerful radio-loud active galaxy within a redshift of 0.1 of the Milky Way. Its radio emissions are ~10^3 more powerful than those of Cen A which is, in turn, ~10^3 more powerful than a normal galaxy like the Milky Way. Like Cen A, the source of energy in Cyg A is thought to be gas falling into a black hole at the centre of the galaxy.

(c) Virgo A

As mentioned above, Virgo A had been identified as the galaxy M87 in the Virgo cluster of galaxies by Bolton, Stanley and Slee in 1949. But this identification had been disputed for a few years as the position of the radio source was too inaccurate.

In 1918 a small jet had been found extending from M87 by Heber Curtis. Baade and Minkowski observed this jet in 1954, noting its blue colour, by which time the identification of Vir A with M87 had been generally accepted. Two years later Baade found that the jet's optical emission was highly polarised, indicating that it was the source of synchrotron radiation. At about the same time John Baldwin and F. Graham Smith detected a weaker radio halo around M87. Then in 1966 Halton Arp reported the discovery of a counterjet. Thirty years later this faint, short counterjet was observed in some detail in both visual and radio wavebands.

An X-ray source was detected by Byram, Chubb and Friedman in the general direction of M87 in 1966, but the resolution of their equipment was not sufficient to be sure that this galaxy was the source. Six years later Edwin Kellogg and colleagues found that the X-ray source, which had a diameter of ~1°, was clearly centred on M87. At the end of the 1970s the Einstein Observatory spacecraft showed that the bulk of the X-ray emission in the M87 halo was thermal in origin with a temperature of ~3×10^7 K. Spacecraft measurements also confirmed that the ~2 kpc long jet was the source of synchrotron radiation most likely caused by relativistic electrons.

Wallace Sargent and colleagues found evidence of a possible supermassive black hole at the centre of M87 in 1978, based on measurements of gas velocities around the centre of the galaxy. Observations with the Hubble Space Telescope in the 1990s implied a mass density at the centre of the galaxy of ~1×10^5 $M_\odot/pc^3$. Although this mass density was high, it was not above that of stars in dense clusters and therefore did not require the existence of a supermassive black hole. Nevertheless, the presence of the high-velocity jet and other observations indicated that a supermassive black hole may well reside at the centre of M87.

In optical images M87 appears to be a moderate size elliptical galaxy about 40 kpc in diameter, but David Malin's sophisticated image-processing techniques have shown that it has a diameter of ~150 kpc at optical wavelengths. It has since been found to possess ~10,000 globular clusters, which is at least 50 times as many as the Milky Way.

(d) General

Numerous radio galaxies, which are radio-loud active galaxies, have been found since those mentioned above. The host galaxies are usually elliptical. Radio emission is caused by the synchrotron emission of relativistic electrons in a magnetic field. The structure of radio galaxies varies from galaxy to galaxy, but many such objects have twin jets coming from the nucleus which end in lobes, with Cyg A being a particularly clear example. The most usual cause of X-ray emission is thought to be the inverse-Compton process in which relativistic electrons interact with ambient photons.

Unified model for active galaxies

Antonucci and Miller had produced a unified model of Type 1 and Type 2 Seyfert galaxies in 1985 (see *Seyfert galaxies* above) which showed that the observed differences between the two types were largely due to viewing geometry, as there was an obscuring torus around their active nucleus. Four years later Peter Barthel produced a unified model of radio-loud quasars (RLQs) and radio galaxies. He concluded that relativistic beaming is intrinsic to all RLQs, and suggested that both RLQs and radio galaxies have basically the same structure as each other with an obscuring torus. They are seen as RLQs when our line of sight is within about 45° of the relativistic jet, and as radio galaxies when our line of sight is outside that 45° when the torus obscures the nucleus. Blazars are a special case in which we are looking almost exactly along the jet.

This so-called unified model has been developed since Barthel's proposal. In its modern configuration an active galactic nucleus (AGN) consists of a supermassive black hole with a mass of $\sim 10^6 - 10^9$ $M_\odot$, surrounded by an accretion disc. Infalling material heats the accretion disc and the resulting radiation is observed in the blue and ultraviolet parts of an AGN's spectrum. The accretion disc is embedded in a hot rarefied region emitting synchrotron radiation and X-ray radiation. The absorbing torus, which is in the same plane as the accretion disc, is much further from the nucleus. The torus is composed of warm molecular gas and dust. Jets which form from ionising gases are ejected from the nucleus of the AGN perpendicular to the plane of the accretion disc at speeds varying from relative modest speeds to those approaching the speed of light. The broad spectral emission lines are produced by dense fast-moving gas clouds which are roughly randomly distributed around the central core. The narrow lines are produced by low-density slow-moving gas clouds that lie outside the torus.

Radio-loud quasars have relativistic jets ending in radio lobes, whereas radio-quiet quasars have either much slower jets or no jets at all. If the surrounding galaxy provides a great amount of fuel to maintain the AGN, we see the AGN as a quasar; otherwise we see it as a Seyfert galaxy with a less-luminous central source.

Gravitational lensing and Einstein rings

Einstein suggested in 1936 that a star could bend light, or distort space, so that multiple images would be seen of a star if it was behind and along the same line of sight as the gravitational lensing star. If the alignment was perfect the image of the more distant object would appear as a ring. In the following year Fritz Zwicky suggested that multiple imaging would also be observed if the light was bent by a foreground galaxy, rather than a foreground star.

In 1979 Walsh, Carswell and Weymann discovered a twin quasar, 0957 + 561 A and B, both of which had virtually identical redshifts of 1.405 and optical spectra. They speculated that these may be the images of one quasar which had been doubled as the result of a gravitational lens caused by an intervening object. Six months later J. E. Gunn and colleagues found the image of the intervening galaxy at a redshift of 0.4. In 1991 David Roberts and colleagues found, from 11 years of observations with the Very Large Array, that the two sources A and B showed a similar intensity variability to each other, but one was 1.40 years in advance of the other. This enabled the

detailed geometry of the quasar and intervening galaxy system to be determined. As a result the Hubble constant could be estimated, giving a most likely value between 46 and 69 km/s/Mpc.

Roger Lynds and Vahè Petrosian discovered giant luminous arcs in the galaxy clusters Abel 370 and Cl 2244 in 1986. They thought that these arcs were incomplete Einstein rings caused, in both cases, by the imperfect alignment of two galaxies. A little later Jacqueline Hewitt and colleagues found the first complete Einstein ring when they observed the object MG 1131+ 456.

Bibliography

Hey, J. S., *The Evolution of Radio Astronomy*, Science History Publications, 1973.

Kembhavi, Agit K., and Narlikar, Jayant V., *Quasars and Active Galactic Nuclei: An Introduction*, Cambridge University Press, 1999.

Kitchin, Chris, *Galaxies in Turmoil: The Active and Starburst Galaxies and the Black Holes that Drive Them*, Springer-Verlag, 2007.

Longair, Malcolm, *The Cosmic Century: A History of Astrophysics and Cosmology*, Cambridge University Press, 2006.

Page, Thornton, and Page, Lou Williams (eds.), *Beyond the Milky Way: Galaxies, Quasars and The New Cosmology*, Macmillan, 1969.

Peterson, Bradley M., *An Introduction to Active Galactic Nuclei*, Cambridge University Press, 1997.

Schneider, Peter, *Extragalactic Astronomy and Cosmology: An Introduction*, Springer, 2006.

ANDROMEDA NEBULA (M31)

The first known written mention of the Andromeda nebula was as a 'small cloud' by the Arabian astronomer Al-Sufi in his *Book of Fixed Stars* in 964. Simon Marius was the first to observe the Andromeda nebula with a telescope in 1612. Charles Messier observed it in 1764 and catalogued it as number 31 (i.e., Messier or M31) in his nebula catalogue of a few years later. In 1868 William Huggins found that M31 had a continuous spectrum, rather than the line spectrum of a gaseous nebula, indicating that it was composed of stars. In the early twentieth century, however, Vesto Slipher suggested that M31, and similar spiral nebulae, may consist of a central star surrounded by a gaseous nebula which was lit up by the central star. So it was not clear at this time whether M31 was a star system or not.

George Bond detected two dark dust lanes in M31 in 1847. This was the first indication of its spiral arms. Forty years later Isaac Roberts produced the first clear photographs of M31, which showed its basic spiral structure for the first time.

Frank Very estimated the distance of M31 in 1911 as about 500 parsec from a comparison of the brightness of Nova S Andromedae of 1885 with that of a normal nova, Persei 1901. This distance would clearly place M31 in the Milky Way. Six years later Harlow Shapley also concluded that M31 was in the Milky Way when he calculated that, if it was actually outside the Milky Way, the absolute magnitude of Nova S Andromedae would have to be about ten magnitudes brighter than normal novae seen in the Milky Way, which seemed incredible. But ten novae were then found in M31 that were about ten magnitudes fainter than S Andromedae. As a result Heber Curtis suggested that M31 was an 'island universe' of stars, or galaxy, at some distance from the Milky Way, there being two different types of novae of radically different brightness in it and presumably other galaxies.

Shapley and Curtis met in a famous debate on 26 April 1920 to discuss the relative merits of their two different concepts of the structure of the universe. Shapley's universe consisted of a very large Milky Way, which included spiral nebulae like M31. Curtis' Milky Way, on the other hand, was very much smaller, and the spiral nebulae were similar island universes to the Milky Way outside of the Milky Way. The result of the debate was largely inconclusive.

Edwin Hubble was able to measure the distance of the first spiral nebulae in 1924 when he discovered Cepheid variables in M31 and in the Triangulum nebula, M33. As a result he found the distances of both spiral nebulae were about 285 kpc, which showed that they were far away from the Milky Way, as argued by Curtis. Unfortunately Hubble's estimates of the diameters of these two nebulae, given this distance, seemed to indicate that they were considerably smaller than the Milky Way. This naturally concerned those astronomers who thought that all spiral nebulae were probably of roughly the same diameter.

Joel Stebbins showed in 1933 that interstellar dust reduced the estimated diameter of the Milky Way from ~80 kpc to ~30 kpc. In parallel, Hubble and others found that the diameter of M31 was larger than previously thought at ~20 kpc. Then in 1952 Walter Baade found that the distance and hence the diameter estimates of spiral nebulae needed to be doubled following his work on cluster variables. This resulted in a diameter estimate for M31 of ~40 kpc. So M31 was a spiral galaxy larger than the Milky Way.

Robert Hanbury Brown and Cyril Hazard detected weak radio emission from M31 in 1950. Sixteen years later Morton Roberts was able to measure the distribution of neutral hydrogen in M31 using the much higher resolution of the 300 ft (90 m) National Radio Astronomical Observatory telescope. He found that the neutral hydrogen was in a torus whose maximum density was about 10 kpc from the nucleus. Star formation had been completed long ago in the central region, as there was little gas left to form stars. But 10 kpc from the nucleus, where the density of the neutral hydrogen was at its greatest, the galaxy was producing most of its new stars.

In 1987 John Kormendy found evidence for a supermassive black hole at the centre of M31 from the velocity of stars close to the nucleus. Five years later a radio source was discovered at the centre of M31, supporting this black hole theory. Then in 1993 Tod Lauer and colleagues found, using the Hubble Space Telescope, that the nucleus was double. The brighter nucleus, P1, was offset from the centre of M31's bulge, and the fainter nucleus, P2, was at the bulge centre. Two years later King, Stanford and Crane discovered that P2 was much brighter than P1 in the ultraviolet, with the blue light coming from a compact source embedded in P2. In 2005 Ralph Bender and colleagues found that this blue light was coming from a number of A-type stars probably in the form of a disc surrounding a black hole.

Bibliography

Gribbin, John, *Galaxies: A Very Short Introduction*, Oxford University Press, 2008.

Hodge, Paul W., *The Andromeda Galaxy*, Kluwer Academic Publishers, 1992.

Hubble, Edwin, *The Realm of the Nebulae*, Yale University Press, 1936.

Page, Thornton, and Page, Lou Williams (eds.), *Beyond the Milky Way: Galaxies, Quasars, and the New Cosmology*, Macmillan, 1969.

COSMIC RAYS

Shortly after 1896, when Henri Becquerel discovered radioactivity, physicists found that even shielded electrometers seemed to detect some sort of radiation which, if it was Becquerel's type of radioactivity, they should not. At first it was thought that this radiation was caused by a new type of radioactivity, either from material on Earth or from gaseous substances in the air. There were occasional suggestions that the radiation may come from space. But results were inconclusive, as some experiments seemed to show that the intensity reduced with increase in altitude, whereas others showed that it increased. Then in 1911 and 1912 Victor Hess made a number of balloon ascents and found that the intensity began to clearly increase above about 1,500 m up to his maximum altitude of 5,350 m. As a result he concluded that the radiation was coming from space and being attenuated by the Earth's atmosphere. Shortly afterwards Werner Kolhörster confirmed that this increase with altitude continued when he took his balloon to a maximum altitude of 9,300 m.

At first it was thought that the radiation, which was given the name of 'cosmic rays' by Robert Millikan, was ultra-high-energy gamma radiation. But in 1929 Walther Bothe and Werner Kolhörster concluded, using Geiger counters in a coincidence arrangement, that the radiation consisted of highly penetrating charged particles. Shortly afterwards Carl Anderson, using a cloud chamber, found that cosmic rays contain both positive and negative particles in about equal numbers. In 1932 he found that the positive particles appeared to be like electrons, but with a positive charge; they were the positrons or anti-electrons previously predicted by the theorist Paul Dirac. Two years later Patrick Blackett and Giuseppe Occhialini showed that cosmic rays could each produce a shower of up to about 20 positive and negative particles. Some of these particles were electrons and positrons but, in addition, there were some unknown particles.

Carl Anderson and Seth Neddermayer concluded in 1936 that some of these unknown particles had a mass between that of an electron and proton. These particles, which were initially called 'mesotrons' or 'mesons', were thought to be those proposed theoretically the previous year by Hideki Yukawa. But that proved to be incorrect. In fact Yukawa's particle, the 'pi-meson' or 'pion', was discovered by Cecil Powell and colleagues in 1947, whereas the cosmic ray particle, now called the 'mu-meson' or 'muon', was quite different.

Millikan had continued to argue into the early 1930s that cosmic rays that hit the top of the Earth's atmosphere, that is the primary cosmic rays, were ultra-high-energy gamma radiation (i.e., photons). But in 1927 Jacob Clay had detected a latitude effect in the intensity of cosmic rays, which he argued was proof that the primary cosmic rays were corpuscular in nature and were not ultra-high energy photons. Bothe and Kolhörster agreed, but the latitude effect was not confirmed by other observers for a number of years until Arthur Compton's results of the early 1930s. It showed a correlation with geomagnetic rather than geographic latitude.

In 1934 Bruno Rossi observed the almost simultaneous discharge of three Geiger counters spread in a triangular arrangement some distance apart on a horizontal plane. A little later Pierre Auger, unaware of Rossi's result, found a similar effect and researched it in some detail. He concluded that high-energy primary cosmic rays had interacted with atmospheric molecules in the upper atmosphere. This had produced various particles which interacted with the atmosphere at lower levels to produce showers of electrons and other particles. He estimated the energy of cosmic ray primaries, some of which were $> 10^{15}$ eV. In 1937 Homi Bhabha and Walter Heitler explained the production of cosmic ray showers by the cascade production of gamma rays and electron/positron pairs.

In 1948 Pyllis Freier and colleagues found, using experiments on high-altitude balloons, that primary high-energy cosmic rays were mostly protons, with some alpha particles (helium nuclei) and a very small percentage of heavier nuclei. This was the first time that nuclei heavier than helium had been discovered in cosmic rays. Later work showed that the primary high-energy cosmic ray particles were about 85% protons, 12% alpha particles, 2% electrons and positrons, and 1% the nuclei of heavy elements.

Helmut Bradt and Bernard Peters found in 1950 that the abundance of lithium, beryllium and boron in cosmic rays incident on the Earth's upper atmosphere was much higher than the abundance of these elements in the solar system. But the abundance peaks at the carbon, nitrogen and oxygen group in both cosmic rays and in the solar system were very similar. The lithium, beryllium and boron excesses were explained by spallation in the interstellar medium between the sources of cosmic rays and their arrival at the top of the Earth's atmosphere. In this process, carbon, nitrogen and oxygen nuclei in cosmic rays collided with protons and nuclei in the interstellar medium, losing particles in the process, to produce the lithium, beryllium and boron excess observed. Calculations of these changes enabled the composition of the cosmic rays, as they left their sources, to be deduced.

It had long been theorised that high-energy cosmic rays were produced by shock waves in supernova explosions. But it was difficult to detect the sources of these high-energy cosmic rays because their trajectories in the Milky Way were effected by its magnetic field. Calculations showed that cosmic rays with energies $> \sim 10^{15}$ eV cannot be trapped by the Milky Way's magnetic field, but those with energies below this limit can be trapped. If other galaxies behaved in a similar way, then cosmic rays with energies $> \sim 10^{15}$ eV would most likely have come from outside the Milky Way, whilst those with lower energy would have come from Milky Way sources.

In 1961 John Linsley detected a cosmic ray shower with an energy $> 10^{20}$ eV, using a large-scale detector array at Volcano Ranch, New Mexico. This was the first direct evidence of ultra-high energy cosmic rays. This energy was greater than could be produced in supernovae, which led some astronomers to doubt his result. As the energy was greater than the 10^{15} eV cut-off, this ultra-high-energy source was most probably outside the Milky Way. It was later found that the cosmic background radiation would tend to slow down cosmic rays to $< 5 \times 10^{19}$ eV, so any source like Linsley's with an energy $> 10^{20}$ eV, although outside the Milky Way, must be nearer than ~ 100 million light years. In October 1991 a 3.2×10^{20} eV cosmic ray was detected using the Fly's Eye fluorescence array at Dugway, Utah. Then in December 1993 a 2.0×10^{20} eV cosmic ray was detected using the AGASA scintillator array in Japan.

The energy of ultra-high-energy cosmic rays was such that their paths would not be seriously affected by the Milky Way's magnetic field. By 2002 a correlation had been found between the direction of a number of ultra-high-energy cosmic rays and four elliptical galaxies, each of which was thought to have a quiescent supermassive black hole at its centre. A few years later a possible correlation was also found between the direction of four ultra-high-energy cosmic rays and two merging galaxy clusters.

Garcia-Munoz, Mason and Simpson found in 1977, using the IMP-7 and -8 spacecraft, that there was relatively little of the ^{10}Be isotope, compared with other beryllium isotopes, in cosmic rays. The amount of this ^{10}Be isotope, which has a half-life of about 3×10^6 years, indicated that it had taken the cosmic rays about 10 million years to reach Earth. About fifteen years later N. E. Yanasak and colleagues calculated the mean life or confinement times of cosmic rays based on the relative proportions of the isotopes ^{10}Be, ^{26}Al, ^{36}Cl and ^{54}Mn using the Advanced Composition Explorer (ACE) spacecraft. Their times of 21, 22, 25 and 30×10^6 years, respectively, were remarkably consistent.

Bibliography

Dorman, Lev I., *Cosmic Rays in the Earth's Atmosphere and Underground*, Kluwer Academic Publishers, 2004.

Janossy, L., *Cosmic Rays*, Janossy Press, 2007.

Longair, Malcolm, *The Cosmic Century: A History of Astrophysics and Cosmology*, Cambridge University Press, 2006.

Pleijel, H., *The Nobel Prize in Physics 1936: Presentation Speech*, Nobel Prize website.

Shapiro, Maurice M. (ed.), *Composition and Origin of Cosmic Rays: NATO Conference 1982*, Reidel Publishing, 1983.

See also: Ground-based cosmic-ray observatories.

COSMOLOGY: ORIGIN, AGE AND DEVELOPMENT OF THE UNIVERSE

Early theories

Most ancient civilisations had creation myths as people thought that the universe could not have existed forever. Most of these myths tended to envisage a god or gods creating the universe out of something because, as the Roman poet Lucretius pointed out, 'Nothing can be created out of nothing'. In China, for example, the universe was thought to have been created out of a giant egg which contained a sleeping giant. When he awoke he shattered the egg, the lighter part forming the heavens, and the heavier part the Earth.

In Egypt and some other civilisations the universe was thought to have been created out of primeval matter, which in the Egyptians' case was thought to have been water. Water also played a key part in the creation myth of Genesis where, on the second day, 'God divided the waters which were under the firmament from the waters which were above the firmament'. According to Aristotle, the Greek philosopher Thales of Miletus also believed that water was the primeval substance out of which everything else was formed. It is not surprising that water was chosen as the key element by many philosophers and civilisations as it was colourless, odourless and tasteless, and had no shape when left to itself. It could change from ice to water to steam, and for those coastal communities the sea and ocean were seen to be vast. But, more particularly, water was essential for life.

Aristotle, in his philosophy, divided space as that part up to the Moon, which was made of four elements, namely fire, air, water and earth, and that beyond the Moon reaching to the sphere of the fixed stars which was composed of ether. In the seventeenth century René Descartes also concluded that space was filled with massless ether. He explained that the different densities of materials could be explained by different materials having different proportions of massless ether to normal matter. He also suggested that vortices in this ether carried the planets around the Sun. Both Christiaan Huygens and Isaac Newton initially believed in the ether, using its existence to explain the propagation of light in a vacuum.

Newton abandoned the idea of etherial vortices in 1684, and began to construct a model of the solar system based on his theory of gravitation. Newton's solar system would carry on moving like clockwork, provided there were no external influences. But, in the mid-nineteenth century, the laws of thermodynamics led William Thomson (Lord Kelvin) and others to foresee the time when the universe could have wound down completely in what was called the 'heat death of the universe'. Thomson found this conclusion uncomfortable, however, and suggested that it could be avoided if the universe was infinite.

Age of the solar system

The age of the universe, assuming the universe had a beginning, began to be considered in the mid-nineteenth century. Although it was virtually impossible for its age to be determined scientifically at that time, a lower limit could be defined if the age of the Sun, Earth and/or solar system could be determined.

Hermann von Helmholtz calculated in 1854 that the Sun was about 22 million years old, assuming that its energy was being produced by gravitational contraction. Unfortunately fossil evidence later in the nineteenth century indicated that the Earth was considerably older than that. But in 1905 Einstein's special theory of relativity indicated that, if the mass of the Sun could be transformed into energy, it could have a lifetime of many billions of years. This was comfortably more than the age of the Earth, which was by then estimated to be ~1 billion years old from measurements of radioactive decay. But the exact mechanism for this transformation of solar mass into energy was unknown. In 1926 Arthur Eddington outlined a nuclear reaction which would allow the Sun to keep shining for a few ten billion years, whilst consuming only about 1% of its mass. But his scheme was found to be untenable. Then in the late 1930s the proton-proton and carbon-nitrogen-oxygen cycles were developed to successfully explain the generation of heat in both the Sun and stars. This indicated that the Sun was many billions of years old.

In 1947 Arthur Holmes concluded that the Earth was at least 3.35 billion years old, based on radioactive dating. Holmes had earlier suggested using radioactive dating of meteorites to determine the age of the Earth, and in 1956 Claire Patterson did precisely that, concluding that the age of meteorites, and hence of the Earth, was about 4.55 billion years. The age of the Sun has since been determined to be only slightly older.

The universe post-Einstein

Albert Einstein published his general theory of relativity in 1916, and in the following year he used it to examine the structure of the universe. Much to his surprise his equations predicted that the universe was either expanding or contracting. But, as he was convinced that it was basically static, he introduced a term, the cosmological constant Λ, to make his equations predict a static universe. At about the same time Willem de Sitter developed his theory of the universe based on Einstein's equations. But de Sitter's universe was expanding, although it was rather theoretical as it contained no matter.

Alexander Friedmann developed Einstein's equations in the early 1920s and showed that, if the average density ρ of the universe is greater than a critical value ρ_{crit}, gravity will force the universe to collapse on itself. But, if the density is less than this, the universe will continue to expand forever. Einstein dismissed Friedmann's work as pure mathematical speculation, although he later softened his criticism.

In 1927 Georges Lemaître produced his model of the universe, not knowing of Friedmann's work. His universe was similar to Friedmann's, but whereas Friedmann's work was purely mathematical, Lemaître was concerned with the real physical universe. Lemaître's work predicted an expanding universe which he pointed out could provide an explanation of the galactic redshifts then being measured. Unfortunately Lemaître's work was published in a relatively obscure Belgian journal and it was not generally recognised until 1930 when he brought it to the attention of Arthur Eddington.

In the following year, Lemaître suggested that the universe had expanded from what he called the primeval atom. This primeval atom, which consisted of protons, electrons and alpha particles packed together at nuclear densities, was naturally unstable. Lemaître suggested that its disintegration would produce the chemical elements. Einstein finally dropped his cosmological constant in the same year, later referring to its introduction as his biggest blunder.

Age of the universe Part 1

Vesto Slipher had found in 1914 that many spiral nebulae had large redshifts, indicating that they were receding from the Milky Way at high velocities. For some years the true nature of spiral nebulae was unclear, until in 1924 Edwin Hubble was able to determine the distances of the nearest from the variability of their Cepheids. This showed that spiral nebulae were galaxies outside the Milky Way.

Slipher measured the redshifts of 45 spiral galaxies by 1925. They had radial velocities ranging from −300 km/s (for the Andromeda galaxy) to +1,800 km/s. This work was continued by Milton Humason in 1928 using the Mount Wilson 100 inch (2.5 m) telescope to extend observations to more distant galaxies. Hubble found in the following year that the recession velocity or redshift of the galaxies investigated by Humason increased linearly with distance at the rate of about $H_0 = 500$ km/s/megaparsec (km/s/Mpc) (see Figure 4.3). This implied that the universe had been expanding for about 2 billion years, assuming that its expansion rate had been constant. But, unfortunately, this was less than the estimated lifetime of the Sun. Hubble was not aware of Lemaître's work at the time, and suggested that the expansion that he had measured may represent the de Sitter universe.

Big Bang and Steady State theories of the 1940s

In the early 1940s George Gamow began to consider how the heavy elements might have been formed in the early universe. He suggested in 1946 that they might have been produced by neutron capture following the breakup of an original superdense object, somewhat like Lemaître's primeval atom, from which the universe developed. Gamow's theory required the existence of a plentiful supply of neutrons in the early universe, with protons becoming available following neutron beta decay.

Gamow's theory was then developed by his graduate student Ralph Alpher, who was joined a little later by Robert Herman. They showed in 1948 that the universe had to go through a hot, dense phase if the elements were to be produced cosmologically, and went on to show that the early universe was radiation- rather than matter-dominated. They also estimated that this thermal background would have cooled to about 5 K today. Fred Hoyle christened this theory the Big Bang theory in 1949.

It soon became clear that the heavy elements could not have been produced in the Big Bang in anything like the quantities currently observed. But shortly afterwards Hoyle and others showed that they could be produced in stars instead. In addition, there was still a problem with the age of the universe in the Big Bang theory, calculated using galactic redshifts, as it was still less than the age of the solar system. As a result the Big Bang theory was in some trouble in the early 1950s.

In the meantime, Hermann Bondi, Thomas Gold and Fred Hoyle had found a way of solving the inconsistency between the age of the universe and the age of the solar system when they introduced their Steady State theory in 1948. In their theory the universe had existed forever. Matter was created out of nothing at a rate that kept the density of the universe constant as the galaxies receded from each other.

Age of the universe Part 2

Hubble's galaxy distance estimates had to be modified in 1952 when Walter Baade found that there were two different types

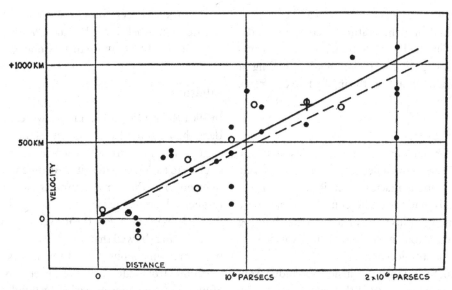

Figure 4.3. The distance-velocity graph, published by Edwin Hubble in 1936, which showed an approximate linear relationship. The slope gave a value for the Hubble constant of about 500 km/s/Mpc. (From *The Realm of the Nebulae*, by E. Hubble, New Haven Press, p. 114; courtesy Yale University Press.)

of Cepheid variables with different period-luminosity relationships. This resulted in the value of the Hubble constant, or H_0, being reduced by 50% to 250 km/s/Mpc, implying that the universe was about 4 billion years old. But this was still too young, as the Earth's age was determined a little later to be about 4.55 billion years.

Hubble had initially estimated the distances of galaxies from the apparent intensities and periods of their Cepheids. But for the more distant galaxies, where this was not possible, he had used the apparent magnitudes of their brightest stars. However, in the 1950s it became clear that some of the objects that Hubble thought were bright supergiants were, in fact, regions of ionised hydrogen or star clusters. Allan Sandage corrected for this in 1958 and reduced the estimated value of H_0 to 75 km/s/Mps. This implied a far more plausible age for the universe of 13 billion years, which was much more than that of the Earth.

Allan Sandage also deduced the ages of various open star clusters in the late 1950s from the shape of their H-R diagrams. They ranged from less than 1 *million* years for cluster NGC 2362 to 5 *billion* years for M67. At about the same time Fred Hoyle deduced an age for M67 of 9.2 billion years, whilst shortly afterwards Allan Sandage showed, using Hoyle's model, that the estimated ages of the globular clusters M3 and M5 were both about 25 billion years. This was significantly more than the 13 billion year estimated age of the universe.

The exact age of the universe derived from a given value of H_0 depends on the values of various cosmological parameters. In a 1961 paper, for example, Sandage quoted an age of the universe of 7.4 billion years for $H_0 = 75$ km/s/Mpc, assuming

$\Lambda = 0$ and a value of the deceleration parameter q_0 of $+1$. (In a closed, elliptical universe $q_0 = +1$, whereas for a Steady State universe $q_0 = -1$. The most probable value was thought to be at the time, according to Sandage, $q_0 = 1 \pm \frac{1}{2}$.) This made the problem, that the age of the universe appeared to be appreciably less than that of some star clusters, even worse. He pointed out that the only way to 'save the timescale', as he put it, was either a Steady State universe or an Einstein-Lemaître model with $\Lambda > 0$, both of which required $q_0 = -1$. Neither model seemed attractive.

The microwave background radiation

In 1962 Riccardo Giacconi and colleagues detected a diffuse source of X-rays isotropically distributed across the sky. The predicted intensity of the X-ray background using the Steady State theory was about two orders of magnitude higher than that observed, causing serious doubts about the Steady State theory. Then three years later Arno Penzias and Robert Wilson accidentally discovered the microwave background radiation at a temperature of about 3.5 K, which was very close to the temperature predicted by Alpher and Herman in 1948 (see above) using the Big Bang theory. Its discovery marked the beginning of the end of the Steady State theory, which had been virtually abandoned by the end of the decade.

Encouraged by the discovery of this microwave or cosmic background radiation, James Peebles and, independently, Zeldovich, Kurt and Syunyaev undertook theoretical work in the late 1960s to explain the development of the early universe after the Big Bang. They analysed the process of recombination of the primordial plasma as the universe expanded and cooled, in an attempt to understand the temperature fluctuations in the cosmic background radiation. They found that the universe was unobservable during this period of recombination, ending at redshifts of ~1,000, due to photon scattering by free electrons.

Dark matter

Fritz Zwicky had measured the velocities of galaxies in the Coma cluster in 1933 and concluded that the mass of this cluster was too small by about a factor of ten if the cluster was to remain gravitationally bound. So he concluded that there was a considerable amount of invisible or dark matter binding the cluster together. But it took many years before this idea of dark matter became generally accepted.

In 1974, Gott, Gunn, Schramm and Tinsley concluded from various observations that the density of the universe today, ρ, is ≤ 0.1 times the critical density ρ_{crit} corresponding to a flat universe, with a most likely value between about 0.05 and 0.09 ρ_{crit}. But, according to the Big Bang theory, any deviation from the critical density would have increased rapidly over time since the Big Bang. So if ρ is $\leq 0.1 \rho_{crit}$ today, it must have been within 10^{-15} of ρ_{crit}, one second after the Big Bang. This led many theorists to conclude that such a tiny deviation from the critical density was unlikely, and that ρ is equal to the critical density now and has always been so. As a result the universe would always have been flat with the density parameter $\Omega = \rho/\rho_{crit} = 1$. In that case $\leq 10\%$ of all matter had been detected, with the remainder apparently being dark matter.

Vera Rubin and Kent Ford had also been studying the rotation of spiral galaxies for a number of years when they concluded in 1975 that there must be more mass in galaxies than detected in their visible discs in order to explain their rotation rates. Linear velocities of stars around their centres initially increased with distance from the centre, but then became almost independent of distance. Rubin, Ford and colleagues continued with their measurements over the next decade or so producing more and more convincing evidence of this effect. At first astronomers were sceptical, but eventually the idea of dark matter in spiral galaxies became generally accepted. The majority of this mass appeared to be in haloes around the galaxies.

Not all galaxies contained dark matter, however. For example, in 1993 Robin Ciardullo and George Jacoby found that the elliptical galaxy NGC 3379 appeared to have no such matter following an analysis of the motion of its planetary nebulae.

In 1986 Amos Yarhil, David Walker and Michael Rowan-Robinson calculated the net gravitational pull on the Milky Way within about 200 Mpc using infrared data. They found that if this volume of the universe was representative of the universe as a whole, then $\Omega = 1 \pm 0.2$, and about 90% of the mass of the universe must be dark matter. Evidence for dark matter has also been found more recently in many clusters of galaxies, like Abell 1689, by dark matter's lensing effect on the light observed from more distant objects.

Inflation

In 1961 Robert Dicke had raised two key questions that the Big Bang theory should answer. Why is the universe so uniform on large scales, and why is the density of the universe apparently so close to the critical density? As time went on these problems, referred to as the horizon problem and the flatness problem, respectively, remained and became much more clearly defined.

Analysis of the size of the universe as a function of time after the Big Bang showed that it was too large for it to have reached a uniform homogenous state. It was always too large for anything to get one from side to another, even travelling with the speed of light. Yet the microwave background radiation measured in the 1960s and 1970s seemed remarkably isotropic.

Alan Guth proposed a major modification to the Big Bang theory in 1980 to solve both these horizon and flatness problems. In his so-called inflation theory he proposed that the universe expanded from about 10^{-34} to 10^{-32} seconds such that its radius increased by a factor of about 10^{43}. So before the inflationary expansion began, the universe had been small enough for it to reach a uniform, homogenous state. This rapid expansion also had the effect of straightening out the geometry of the early universe, driving it towards a flat Euclidean geometry where $\Omega = 1$.

Age of the universe Part 3 and dark energy

There was a continuing controversy over the value of H_0 from the 1970s until the 1990s between Allan Sandage and Gustav Tammann, on the one hand, who favoured a value of about 50 km/s/Mpc, and Gérard de Vaucouleurs and colleagues, on the other, who deduced a value nearer 100 km/s/Mpc. However, by the later 1990s most observational values were in the range 55 to 75 km/s/Mpc, equivalent to an age of the universe from about 17 to 13 billion years for a flat universe.

Then in 1998, much to everyone's surprise, Saul Perlmutter and colleagues and Brian Schmidt and colleagues independently found, after analysing a number of Type Ia supernovae, that the cosmological constant Λ was non-zero and positive at 99% probability. As a result the expansion of the universe was accelerating due to its vacuum or dark energy. For a flat cosmology $\Omega_m + \Omega_\Lambda = 1$, where Ω_m is the density parameter of matter and Ω_Λ is the density parameter of dark energy, Perlmutter and colleagues found $\Omega_m = 0.28$, and an age of the universe of $14.9^{+1.4}_{-1.1}$ billion years. With no such flat cosmology constraint, they deduced an age of the universe of 14.5 ± 1.0 billion years. Schmidt and colleagues deduced an age of the universe of 14.2 ± 1.5 billion years.

Shortly afterwards, Wendy Freedman and colleagues published a value of H_0 of 72 ± 8 km/s/Mpc, following a major investigation using the Hubble Space Telescope. This produced an expansion age of the universe of about 13 ± 1 billion years for a flat universe, if $\Omega_m = 0.3$, and $\Omega_\Lambda = 0.7$. The age was consistent with the best estimate of the age of the oldest globular clusters which had, by this time, been reduced to about 12.5 billion years. The latter followed an extensive analysis of observations with the Hipparcos spacecraft by Eugenio Carretta and others.

Anisotropy in the microwave background radiation

In 1976 David Wilkinson and colleagues found an anisotropy in the microwave or cosmic background radiation due to the motion of the Milky Way, but they could find no anisotropy in the true background radiation itself.

If the cosmic background radiation is a true indication of the early universe, it should have structure, as otherwise it would be difficult to see how the galaxies could then have formed. Eventually in 1992 George Smoot and colleagues found an anisotropy of just one part in 100,000 using the COBE spacecraft.

The WMAP spacecraft, launched in 2001, produced a map of the variation of the cosmic background radiation over the celestial sphere. Gary Hinshaw and colleagues concluded in 2008, following the analysis of five years of WMAP observations, that the cosmic background radiation was consistent with a flat universe dominated by dark matter and dark energy. In particular, the universe is 13.69 ± 0.13 billion years old, consistent with the earlier results of Perlmutter, Schmidt and Freedman, whilst H_0 is $71.9^{+2.6}_{-2.7}$ km/s/Mpc. It consists of just $4.4 \pm 0.3\%$ normal matter, $21.4 \pm 2.7\%$ dark matter, and $74.2 \pm 3.0\%$ dark energy. When the universe became transparent $380,000 \pm 6,000$ years after the Big Bang, it consisted of 12% atoms, 15% photons, 10% neutrinos and 63% dark matter, and virtually no dark energy. Combining these WMAP results with those from the Sloan Digital Sky Survey, the 2dF Galaxy Redshift Survey, and the latest Type Ia Supernova results made only slight changes to the WMAP cosmological parameters, all of which were within the above-quoted errors.

Bibliography

Amendola, Luca, and Tsujikawa, Shinji, *Dark Energy: Theory and Observations*, Cambridge University Press, 2010.

Bertotti, B., Balbinot, R., Bergia, S., and Messina, A. (eds.), *Modern Cosmology in Retrospect*, Cambridge University Press, 1990.

Harrison, Edward, *Cosmology: The Science of the Universe*, 2nd ed., Cambridge University Press, 2000.

Hetherington, Norriss S. (ed.), *Encyclopaedia of Cosmology: Historical, Philosophical and Scientific Foundations of Modern Cosmology*, Garland Publishing, 1993.

Kerszberg, Pierre, *The Invented Universe: Einstein-De Sitter Controversy (1916–17) and the Rise of Relativistic Cosmology*, Clarendon Press, Oxford, 1989.

Kragh, Helge, *Cosmology and Controversy: The Historical Development of Two Theories of the Universe*, Princeton University Press, 1996.

Longair, Malcolm, *The Cosmic Century: A History of Astrophysics and Cosmology*, Cambridge University Press, 2006.

North, J. D., *The Measure of the Universe: A History of Modern Cosmology*, Clarendon Press, Oxford, 1965.

Struve, Otto, and Zebergs, Velta, *Astronomy of the 20th Century*, Macmillan, 1962.

Tropp, E. A., Frenkel, V. Ya., and Chernin, A. D., *Alexander A. Friedman: The Man Who Made the Universe Expand*, Cambridge University Press, 1993.

Weinberg, Steven, *The First Three Minutes: A Modern View of the Origin of the Universe*, Andre Deutsch, 1977.

EXTRATERRESTRIAL LIFE

Early ideas

A number of ancient Greek philosophers, including Leucippus, Democritus and Epicurus, suggested that life existed elsewhere in the universe in what is now called the theory of 'cosmic pluralism' or 'the plurality of worlds'. However, the very influential philosophers Plato and Aristotle opposed the idea, and so the theory of cosmic pluralism was generally abandoned. In fact most Christian authors of the first millennium AD also dismissed the idea. But in the Condemnation of 1277 the Church allowed that God could have created other worlds than the Earth, but whether he had done so was left open to question.

William Vorilong argued in the fifteenth century that God could create an infinity of worlds, and that life could exist on worlds other than the Earth. But it is unclear whether he thought that such life really did exist. However, at about the same time Nicholas of Cusa suggested that life really did exist throughout an infinite universe on the Sun and stars. In the following century Giordano Bruno extended Copernicus' theory of the heliocentric universe when he proposed that there was a limitless number of solar systems in the universe that supported intelligent life. But Bruno's religious ideas fell afoul of the Church and he was burnt at the stake in 1600. In the early seventeenth century Galileo Galilei and René Descartes were more cautious on the subject of the plurality of worlds.

A number of the ancient philosophers had argued that the Moon was inhabited, including Anaxagoras, Xenophanes and Plutarch. Johannes Kepler favoured the idea of life on the Moon in his *Somnium*, published posthumously in 1634, as did John Wilkins a few years later. Then in 1686 Bernard le Bovier de Fontenelle published his popular *Entretiens sur La*

Pluralité des Mondes (Conversations on the Plurality of Worlds) in which he also proposed that the Moon was inhabited. Twelve years later Christiaan Huygens, in his posthumously published *Cosmotheoros*, suggested that there was probably vegetation on the planets of our solar system, and possibly also animals and intelligent beings. But he was not so sure whether the Moon was inhabited, as it appeared to have no water.

In the eighteenth century a number of intellectuals and philosophers, including Alexander Pope, Thomas Gray, and Immanuel Kant, wrote positively about the existence of extraterrestrial life. William Herschel even went so far as to propose that the Sun and stars were inhabited. John Wesley and Leonhard Euler, on the other hand, were in the minority in dismissing the existence of extraterrestrial life. The 'pro' side gained more of an upper hand in the first half of the nineteenth century when some theologians suggested that the belief in extraterrestrial life was not incompatible with Christian doctrine.

William Whewell was initially a pluralist in the mid-nineteenth century, but he had difficulty in reconciling pluralism with his religious beliefs. So by the time he had published his *Plurality of Worlds* anonymously in 1853 he had changed his mind. In this he considered whether conditions for intelligent life were suitable on other planets in our solar system and concluded that they were not, with the possible exception of Mars. Whewell was important as, although his problems with pluralism were based on theology, he used clear, rational scientific arguments to justify his conclusions. His book created a great deal of debate at the time.

Later in the nineteenth century a number of pluralist authors wrote influential books. For example, Camille Flammarion published numerous books, of which his most famous was probably *La pluralité des mondes habités (The extent of inhabited worlds)*, first published in 1862. In this and his subsequent work Flammarion took a relaxed view about the existence of life, assuming that it existed unless he could prove otherwise. As a result he argued that all the planets in the solar system from Mercury to Neptune were most likely inhabited with intelligent life. On the other hand Richard Proctor, in his *Other Worlds than Ours*, first published in 1870, was more cautious. Even so he concluded that Venus, Mars and the moons of Jupiter and Saturn were inhabited, and that Jupiter and Saturn would themselves become inhabited one day when they had cooled down enough (he considered them currently self-luminous). Finally, the theologian Joseph Pohle, in his *Die Sternenwelten und ihre Bewohner (Stellar Worlds and Their Inhabitants)* of 1884, came to broadly similar conclusions to Proctor. But, unlike Proctor, Pohle was heavily influenced by theology, reasoning that, as the universe shows the glory of God, it must be perfect. As a result, it must include numerous intelligent beings dispersed throughout it. Proctor, on the other hand, treated the subject purely scientifically.

The Moon

Astronomers were still speculating on the possible existence of life on the Moon until well into the twentieth century. For example, William Pickering suggested in 1903 that there may be plants, snow and river beds on the Moon. He later speculated that there may be a simple form of vegetation that completed its life cycle in the 14 days of sunlight per month. But as time went on, these and similar ideas were mostly abandoned. However it was still thought possible when the Americans were planning their manned lunar landings in the 1960s that there may be some forms of very elementary life, like bacteria, on the Moon. But none was ever found.

It was thought for many years that the Apollo samples of the lunar surface showed no evidence of water or of organic material. But in 2008 Alberto Saal and colleagues re-examined some of these samples and found evidence of water. If there had been water once, there may have been life.

Kenneth Watson, Bruce Murray and Harrison Brown had pointed out in 1961 that the floor of some polar craters could be in permanent darkness. If this was the case, there might be water ice deposits in such regions left there by cometary impacts. Initial spacecraft and ground-based radar observations produced ambiguous results. But in 2009 water was discovered in a dust plume produced by the impact of a launcher upper stage with a polar crater. So it is possible that some very elementary form of life could have existed on the Moon for a short time in the past, possibly under the surface where it would have been protected from solar ultraviolet radiation and the solar wind.

Mars

As the nineteenth century progressed it seemed as though Mars provided the most suitable environment for extraterrestrial life in the solar system. William Herschel had suggested at the end of the eighteenth century that the reason why its polar caps changed size with the season was because they consisted of ice and snow that melted in summer. His son John Herschel suggested in 1830 that the dark, greenish areas of Mars were seas. Thirty years later Angelo Secchi also concluded that the dark areas were seas, and proposed that the reddish-coloured areas were continents. On the other hand, Emmanuel Liais and others thought that the dark areas, whose size seemed to vary with the seasonal change in the size of the polar caps, were vegetation. So there appeared to be evidence that there was water on Mars, even if it only came from seasonal melting of the polar caps. As a result there might also be some form of plant life.

Giovanni Schiaparelli observed the close oppositions of Mars in 1877, 1879 and 1881. His maps produced from these oppositions showed linear features on the surface, some of which were double, that he called *canali*. English astronomers mis-translated this word to mean canals, which implied that they were produced by intelligent beings. However the correct translation was channels, which Schiaparelli envisaged to be natural features. A number of other astronomers also saw *canali* at subsequent oppositions, including William Pickering in 1892. He noticed that some *canali* crossed some of the dark areas on Mars, showing that these dark areas could not be seas. As a result he concluded that they were areas of vegetation. Two years later Percival Lowell observed numerous *canali*, which he concluded had been dug by intelligent beings to transport water from the melting polar caps to irrigate the dark areas.

Some respected astronomers of the late nineteenth and early twentieth centuries were convinced that the *canali* existed, whether or not they had been made by intelligent beings. On the other hand, some equally reliable, astronomers thought that they were optical illusions. Eventually the controversy died down and the *canali* were generally consigned to history, although a few protagonists still continued to believe in their existence. Nevertheless, the nineteenth-century idea that the dark areas of Mars may be areas of vegetation still persisted.

It had been clear for some time that Mars had an atmosphere. In 1947 Gerard Kuiper found carbon dioxide in this atmosphere, and in 1963 Audouin Dollfus, and Hyron Spinrad and colleagues, working in parallel, found a trace amount of water vapour. Unfortunately for those hoping to find evidence of life, no oxygen was found. Nevertheless, the unambiguous detection of carbon dioxide, which is involved in photosynthesis on Earth, was of great interest.

The Mariner 4 spacecraft relayed 22 images to Earth during a fly-by of Mars in 1965. The relatively small areas of Mars that had been imaged were found to be cratered and apparently dry, looking very much like the surface of the Moon. The dark areas seen previously were not covered with vegetation as had been hoped, but were simply low albedo areas. Mars' surface atmospheric pressure was so low that water could not exist on the surface now. In addition its atmosphere was not thick enough to shield the surface from ultraviolet radiation, which would probably have killed off any micro-organisms long ago. Mariner 4 also found that Mars had no measurable magnetic field, so charged particles emitted by the Sun would impact the surface virtually unimpeded, again being a threat to life. So conditions were highly unfavourable for life on Mars now, and it appeared probable that there had never been any form of life in the past.

The idea of life on Mars was resurrected by Mariner 9 that was put into orbit around Mars in 1971. It revealed a completely different Mars from that of Mariner 4, with very large volcanoes and a giant canyon system, Valles Marineris. There were also sinuous channels in Chryse Planitia that resembled dried-up river beds. Mariner 9 also found evidence that the south polar cap may have a substantial amount of water locked up in it that was partly released during the local summer.

The Viking 1 and 2 spacecraft, which were launched in 1975, were designed to search for life in low lying areas where there may have been water in the past which could have supported life. Initially it was thought that these experiments showed that there may be some sort of very elementary life on Mars. But it is now generally accepted that they did not.

The Viking 1 lander found that Mars' ground-level atmosphere largely consisted of carbon dioxide with just 0.13% oxygen. The Viking orbiter images showed clear signs of flash floods in the Chryse region, and many landslides, some of which seem to have occurred in saturated soil. Dendritic or branching drainage features were also seen resembling terrestrial river systems. The north polar cap was found to be made of water ice in the northern summer, with a covering of carbon dioxide ice in the winter.

In 1996 a team of scientists announced that they had found evidence of early life in a Martian meteorite, ALH 84001, which had been found in Antarctica. It contained elongated, worm-like shapes, which appeared remarkably similar to Earth-based microfossils. Since then, however, this evidence for life on Mars has been largely discounted, although the original researchers are still standing by their original conclusion.

A number of spacecraft launched in the first decade of the twenty-first century found clear evidence of abundant water in Mars' early history. But the planet then appeared to have become colder and dryer over time. Mars Express found that the interior of the south polar cap was almost completely water ice, and buried ice at the south polar cap extended down by as much as 3.5 km in places. This spacecraft also discovered methane in the atmosphere which may have been caused by some form of life or by volcanic activity.

Jupiter's and Saturn's satellites

Images received from the Voyager 1 and 2 spacecraft fly-bys of Jupiter's moon Europa in 1979 showed that it was covered in water ice, with linear markings which may indicate cracks. It was thought that the lower layers of ice may have been melted by tidal and/or radioactive heating, so the surface would consist of ice floes floating on water. Images produced by the Galileo spacecraft, which was put into orbit around Jupiter in 1996, confirmed the ice floe theory, and measurements of Europa's variable magnetic field in 2000 indicated that the water beneath this ice was probably still fluid.

The recent discoveries of life on Earth which was found in oceans in pitch darkness, drawing their energy from planetary heat via volcanic vents, has encouraged the idea that such a sub-surface ocean on Europa could support life. Similarly the Galileo spacecraft also found evidence for sub-surface oceans on both Callisto and Ganymede which could also support life.

José Comas Solá found evidence of a possible atmosphere on Saturn's moon Titan in 1907. This was confirmed in 1944 by

Gerard Kuiper who discovered that its atmosphere contained methane. Then in 1980 the Voyager 1 spacecraft found that the majority constituent of Titan's atmosphere was nitrogen, and that the moon's orange colour was due to photochemical smog. Voyager 1 also found that Titan's surface level atmospheric pressure was higher than that on Earth, and that the temperature of Titan's surface was very close to the triple point of methane. This suggested that methane on its surface could exist in both solid and liquid form. The Cassini-Huygens spacecraft, which arrived at Saturn in 2004, confirmed the main constituents of Titan's atmosphere as being nitrogen and methane, and found evidence of methane rain. It also discovered numerous large lakes of hydrocarbons on the surface, encouraging those who thought that there may be life on Titan.

Recent work on extremophiles on Earth has shown that life can adapt to what appear to be extremely hostile environments. For example, acidophiles have been found that can tolerate acids of a pH of 0 (about as strong as battery acid), and organisms have been found which are resistant to high levels of ionising radiation. Life may flourish in even more challenging environments elsewhere, so it may exist in the atmosphere of Venus, for example, where there are clouds of sulphuric acid.

Exoplanets

In recent years, numerous planets have been found around other stars. But because of the way they have been discovered, most of the planets found so far are at least as large as Jupiter, and orbiting very close to their parent star. As a result they are very hot. In the foreseeable future, however, it is expected that smaller planets will be discovered in more life-friendly environments around their parent stars. The technology is also just becoming available to be able to detect the main gases in the atmospheres of some of these planets. But it will probably take a considerable time before more clear evidence of potential life is found, if our research on the planets and moons of our solar system is any guide.

Search for Extraterrestrial Intelligence (SETI)

The idea of detecting signals transmitted by extraterrestrial civilisations has been around almost since the invention of radio. In 1901 Nikola Tesla erroneously claimed that he had picked up such transmissions. Then in 1924 David Todd persuaded the U.S. military to listen for transmissions from Mars during its opposition of that year. Nothing was found.

The first push to start a programme searching for signals sent by extraterrestrial intelligences originated at the end of the 1950s. In 1959 Guiseppe Cocconi and Philip Morrison published a paper suggesting that such a search should be undertaken of solar-type stars within a distance of 5 parsec of the Sun using the 1.42 GHz frequency (21 cm wavelength)

of neutral hydrogen. In parallel Frank Drake had begun to plan such a search in Project Ozma using the same frequency. In the following year Drake and colleagues began to observe two relatively close solar-type stars with the 26 m Tatel radio telescope at the USA's National Radio Astronomy Observatory (NRAO). They observed for about six hours/day for two months using frequencies covering a range of 360 kHz centred on 1.42 GHz, but no signal was detected. At about the same time in the Soviet Union Iosif Shklovskii and Nicolai Kardashev became interested in the possibility of interstellar communication, and shortly afterwards Kardashev undertook a search for signals. He also drew a blank.

A conference was held at NRAO in late 1961 to discuss interstellar communication. At this conference Drake introduced his famous 'Drake Equation' to calculate the number N of civilisations able to communicate in the Milky Way over interstellar distances at a given time as:

$$N = R^* f_p n_e f_l f_i f_c L$$

where R^* is the rate of star formation in the Milky Way, f_p is the fraction of stars forming planets, n_e is the number of planets per star with environments suitable for life, f_l is the fraction of suitable planets on which life develops, f_i is the fraction of those life-bearing planets on which intelligence evolves, f_c is the fraction of intelligent cultures communicative over interstellar distances and L is the average lifetime of a communicative civilisation. Over the next ten years various astronomers produced their own estimates of the individual probabilities, yielding values of N varying from $\sim 10^3$ to 10^9.

Enrico Fermi had posed the question in 1950, 'If there are extraterrestrials, where are they now?'. This so-called Fermi Paradox received very little attention until 1975 when it was seriously discussed for the first time following articles written independently in the USA and UK. The problem was that extraterrestrial civilisations were likely to be, on average, millions of years in advance of ours. As a result they should have developed the capability to communicate with, visit or even colonise the Milky Way. So why do we see no evidence of them? Discussions on the Fermi Paradox, in the present state of our knowledge can, unfortunately, come to no clear conclusion.

A number of programmes started in America in the first half of the 1970s to search for extraterrestrial intelligence. One used the 43 m and the 90 m NRAO radio telescopes, one by John Kraus and Robert Dixon at Ohio State University used a meridian transit telescope, and one by Drake and Carl Sagan used the 305 m Arecibo dish. Then in 1976 NASA set up a SETI (Search for Extraterrestrial Intelligence) programme office at their Ames Research Center to carry out preparatory work for a SETI programme. Finally in 1990 NASA funded a new High Resolution Microwave Survey (HRMS) programme using two different strategies. One at NASA Ames adopted a targeted approach, observing about 1,000 solar-type stars within about 30 parsec of the Sun using the Arecibo dish. It used a multi-channel spectrum analyser capable of covering millions of channels simultaneously and analysing them for unusual signals. The other at the Jet Propulsion Laboratory undertook systematic sweeps of the sky using 34 m Deep Space Network dishes. Both sets of observations started in 1992, but funding was withdrawn by the U.S. Congress less than a year later. Since then work has continued under private funding in the USA, as well as in Australia to cover the southern sky. So far no unambiguous signals have been detected.

More recently SETI has also used optical telescopes at a number of American observatories looking for bursts of photons. Again no unambiguous signals have been detected.

The origin of life on Earth

(a) Spontaneous generation

Aristotle taught that life was produced either by normal reproduction, or by the internal secretions of various organs, or by spontaneous generation from non-living material. He believed that some fish were spontaneously generated from mud or sand, that fireflies and aphids came from the dew on plants, insects from putrid matter, and mice from moist soil. His idea of spontaneous generation was adopted by Neoplatonic philosophers, by Saint Augustine of Hippo, the early Christian Church, and Thomas Aquinas, amongst others. It was also later accepted in one form or another by many leading philosophers and scientists of the seventeenth and eighteenth centuries, including Isaac Newton, René Descartes and Francis Bacon.

Eventually the idea of spontaneous generation was subjected to experimental tests. For example, in 1668 Francesco Redi undertook an experiment to see if maggots appeared spontaneously in rotting meat, as was generally believed at the time. He placed meat in sealed, open and partly covered containers and found that maggots developed only in meat on which flies were able to lay their eggs. So maggots did not appear spontaneously. A few years later Anton van Leeuwenhoek, the 'father of microscopy', discovered micro-organisms, or 'animalcules' as he called them. In 1745 John Needham's experiment appeared to show that some micro-organisms could appear spontaneously in a vessel of mutton broth after boiling, even though boiling was known to kill living things. Over the next hundred years or so more and more experiments were carried out to examine the phenomenon of spontaneous generation, producing apparently conflicting results. Then in the 1860s Louis Pasteur finally showed that the spontaneous generation of life was a myth when his experiments demonstrated that particles in the air had compromised Needham's experiment. If particles in the unheated air could not get to the boiled broth, the broth stayed clear; as soon as they could, the broth became cloudy. So if life on Earth could not be created spontaneously, where did it come from?

There are now basically two theories on the origin of life on Earth; either it was created on Earth by some type of chemical process, the 'chemical route', or it was produced elsewhere in the universe and came to Earth by some means in a theory called 'panspermia' or 'exogenesis'.

(b) The chemical route

Alexander Oparin and John Haldane independently suggested in the 1920s that the early Earth's atmosphere was hydrogen rich, or 'reducing', and that this encouraged the synthesis of organic molecules if there was an appropriate supply of energy. These molecules would gradually form larger and more complex molecules, which would then form viruses and bacteria and, eventually, more complex forms of life. In 1953 Stanley Miller and Harold Urey undertook laboratory experiments with a reducing atmosphere to try to reproduce such a process. Their chemical apparatus included an atmosphere of water vapour, methane, ammonia and hydrogen, which was thought at the time to be representative of that on the early Earth. They were able to produce four out of the twenty naturally occurring amino acids using electric sparks to simulate lightning. But it was later found that the atmosphere in their experiments was not representative of the early Earth, which is now thought to have been more neutral and to have consisted mainly of water vapour, nitrogen, and carbon dioxide. The production of amino acids in such an atmosphere has been found to be much more difficult.

Although the Miller-Urey experiment may not have been valid for an early Earth environment, something like their simulated environment may have occurred elsewhere in the universe. Unfortunately, more complex experiments that are thought to replicate the correct environment of the early Earth have been unsuccessful so far in producing proteins or nucleic acids like those expected in the development of early life.

Other more radical chemical routes to creating life have also been proposed, for example using hydrogen cyanide to polymerise nucleotides directly, rather than via amino acids. But these more radical solutions have not yet proved successful either.

(c) Panspermia

(i) Meteorites. Carbon was detected in a meteorite that fell in Alais, France in 1806 by Louis Thénard and Louis Vauquelin. Then in the 1850s Friedrich Wöhler detected materials of organic origin in two other meteorites. But in all these cases it was not clear if the organic material was of extraterrestrial origin or was due to Earth-based contamination. In 1864 a large meteorite fell at Orgueil, France that was analysed almost immediately by Stanislas Cloëz and a little later by Marcellin Berthelot, both of whom found that it contained organic material.

Meanwhile Sales Gyron de Montlivant had suggested in 1821 that not only carbon but life had come to Earth on meteorites ejected by lunar volcanoes. Just over forty years later Hermann Richter also suggested that life came to Earth on meteorites, as did William Thomson (Lord Kelvin) in 1871. Thomson's theory was instantly condemned and ridiculed by many eminent scientists and laymen. A little later Hermann von Helmholtz suggested that either meteorites or comets could have brought life to Earth.

Otto Hahn claimed in 1881 that he had discovered minute fossils that resembled terrestrial corals or sponges in semi-transparent meteorite slices that he had examined under a microscope. But by the end of the nineteenth century his fossils were found to be nothing more than crystalline formations.

In the early twentieth century Svante Arrhenius suggested that dormant life was carried through interstellar space by radiation pressure, reviving when it found a suitable planetary environment. Most scientists thought that stellar ultraviolet light would kill such organisms, so the theory found little support at the time. But it was revived in a radically different form by Carl Sagan and others in the early 1960s when microbes were discovered at an altitude of about 40 km in the Earth's atmosphere.

Meanwhile in 1960 Melvin Calvin and colleagues had found complex organic molecules in a carbonaceous chondrite meteorite. One of these molecules was similar to one of the four bases in the DNA molecule. Then Bartholomew Nagy and George Claus, in a claim reminiscent of Otto Hahn, thought that they had discovered structures in two meteorites that resembled fossil algae. As with Hahn the claim was treated with a great deal of scepticism. Then a meteorite that fell near Murchison, Australia in 1969 was found to contain polycyclic aromatic hydrocarbons (PAHs) and 74 amino acids, 55 of which did not occur naturally on Earth. This finally proved that amino acids, which are a key ingredient of life, existed in space.

(ii) Comets. William Huggins had found in the nineteenth century that there were hydrocarbon compounds in comets, which was one reason why Helmholtz thought that comets could bring life to Earth (see above). By the mid-1950s various carbon-, oxygen- and nitrogen-bearing molecules had been found in comets, including C_2, C_3, CH, CN, OH, NH, NH_2, CO^+, CO_2^+ and N_2^+. Karl Wurm and, later, Pol Swings suggested that, as a number of these molecules were chemically very active they were probably the daughter molecules from more stable parent molecules. In 1974 Ulich and Conklin reported the first direct detection of such a parent molecule, CH_3CN (methyl cyanide), in Comet Kohoutek. But this was disputed. The parent molecules HCN (hydrogen cyanide) was detected in Comet Halley in 1986, H_2CO (formaldehyde) in Comet Brorsen-Metcalf in 1989 and CH_3OH (methanol) in Comet Austin in 1990. This showed that comets contained relatively complex organic molecules. Since then numerous other complex organic molecules have been found in comets.

In the late 1970s/early 1980s Fred Hoyle and Chandra Wickramasinghe developed a theory that nucleic acids and proteins were created in interstellar molecular dust clouds. These complex molecules were then assembled into even more complex molecules and possibly bacteria on comets, which ended up on Earth following cometary impact. Their theory, particularly that including bacteria, was treated with scepticism.

(iii) Interstellar medium. In 1941 Walter Adams had found CH and CN in the interstellar medium. Then in 1963 OH was found, and in 1968 NH_3 (ammonia), followed in the next year by water vapour and H_2CO. Over 150 different molecules have now been detected in molecular clouds, a large number of which are organic. In 2008 amino acetonitrile was found, which is chemically related to glycine.

So we now know that some meteorites contain amino acids, and that complex organic molecules exist in comets and the interstellar medium. This gives some support to the panspermia theory, although there is some way to go from the complexity of these amino acids and molecules to that of the simplest forms of life.

Bibliography

Brack, André (ed.), *The Molecular Origins of Life: Assembling Pieces of the Puzzle*, Cambridge University Press, 1998.

Crowe, Michael J., *The Extraterrestrial Life Debate 1750–1900*, Cambridge University Press, 1986.

Dick, Steven J., *Plurality of Worlds; The Origins of the Extraterrestrial Life Debate from Democritus to Kant*, Cambridge University Press, 1982.

Dick, Steven J., *The Biological Universe: The Twentieth Century Extraterrestrial Life Debate and the Limits of Science*, Cambridge University Press, 1996.

Dick, Steven J., *Life on Other Worlds: The 20th-Century Extraterrestrial Life Debate*, Cambridge University Press, 1998.

Dick, Steven J., and Strick, James E., *The Living Universe: NASA and the Development of Astrobiology*, Rutgers University Press, 2005.

Drake, Frank D., *How Can We Detect Radio Transmissions from Distant Planetary Systems?*, Sky and Telescope, January 1960, pp. 140–143.

Grant, Edward, *Planets, Stars, and Orbs: The Medieval Cosmos 1200–1697*, Cambridge University Press, 1994.

GALAXY CLASSIFICATION

William Herschel, followed by his son John, developed a classification system of nebulae in the eighteenth and nineteenth centuries. But at that time it was unclear which of these nebulae were in the Milky Way and which were external galaxies. This was also the case when Max Wolf published his classification system in 1908. The issue was finally solved by Edwin Hubble in 1924, however, when he showed that both the Andromeda and Triangulum spiral nebulae were outside the Milky Way.

In 1926 Hubble devised a classification system for galaxies based on their shape, which he included in his book *The Realm of the Nebulae* published ten years later. His classification system divided galaxies into ellipticals, spirals, barred spirals, and irregulars. Each type was further subdivided depending on their shape: in the case of ellipticals on their degree of ellipticity, and in the case of spirals and barred spirals on the degree of tightness of their spiral arms. About 80% of his sample of 600 nebulae were spirals, 17% were elliptical and the remainder were irregular.

Hubble did not claim that his sequence of galactic types, which he presented in his 'tuning fork diagram', was an evolutionary sequence, but many astronomers considered it as such. They thought that galaxies started life as approximately circular galaxies, becoming more and more elliptical over time, before either developing into spirals or barred spirals, with the arms opening up over time. When the rotational velocities of galaxies were measured it was concluded that this development sequence was unlikely. It also appeared as though the elliptical galaxies had relatively little interstellar matter and a low rate of star formation. If anything the evolution sequence seemed to be the reverse of the Hubble sequence, ending instead of starting with elliptical galaxies.

Hubble continued to improve his classification system until he died in 1953. Since then it has been modified by a number of astronomers like de Vaucouleurs, Holmberg, Sandage and van den Bergh, but its basic structure has been retained. In 1959 van den Bergh not only considered the Hubble types based on shape, but added a parameter based on luminosity. At about the same time William Morgan proposed a galaxy classification scheme that took into account a galaxy's integrated spectrum, as well as its shape and degree of central concentration. Various astronomers have also more recently developed classification schemes for interacting galaxies, starburst galaxies, active galaxies and so on.

Bibliography

Hubble, Edwin, *The Realm of the Nebulae*, Yale University Press, 1936.

Sandage, A., Sandage, M., and Kristian, J., *Galaxies and the Universe*, University of Chicago Press, 1975.

INTERACTING GALAXIES, STARBURST GALAXIES AND THE LARGE-SCALE STRUCTURE OF THE UNIVERSE

Interacting galaxies

William Herschel noticed in the late eighteenth century that nebulae were not uniformly distributed across the sky. His son John confirmed this, noting that there were relatively few nebulae in the plane of the Milky Way, whilst there were clear groupings in Coma Berenices, Virgo, Pisces and Cetus, for example. These observations were made long before it was realised that most nebulae were galaxies outside the Milky Way.

It was thought in the early twentieth century that most galaxies were isolated systems. In the 1920s Knut Lundmark found 200 double and multiple systems amongst 8,000 NGC (New General Catalogue) objects that he examined. But as better telescopes became available more and more examples were found of apparently multiple galaxy systems. Then when it became clear in the late 1920s that redshifts were a measure of distance, some of these groupings were rejected and others found.

In 1936 Edwin Hubble, in his book *The Realm of the Nebulae*, identified a number of galaxies, including the Milky Way, that were members of what he called the Local Group. These were the Magellanic Clouds, which were satellite galaxies of the Milky Way; the Andromeda galaxy (M31) and its two companions, M32 (NGC 221) and M110 (NGC 205); together with M33 and two other galaxies, NGC 6822 and IC (Index Catalogue) 1613.

Erik Holmberg identified over 800 double and multiple galaxy systems in 1937. Three years later he suggested that, as the distances between the galaxies were often relatively small, there should be a number of cases where two or more galaxies were gravitationally attracted to each other. In these cases the galaxies would orbit their common centre of mass, but tidal forces would eventually cause their orbits to shrink, and the galaxies would merge. In 1950 Lyman Spitzer and Walter Baade calculated that the average galaxy in the dense Coma cluster would collide with at least 25 other galaxies during $\sim 3 \times 10^9$ years. They also concluded that these collisions would have little effect on their individual stars. But the interstellar matter in these galaxies would be swept into intergalactic space, provided its density was $\geq \sim 0.1$ hydrogen atoms/cm^3.

Fritz Zwicky produced a survey of galaxy systems in 1956 in which he described a number of cases of galaxies linked to or even colliding with other galaxies. For example the galaxies IC 3481, 3481A and 3483 seemed to be linked together by a connecting bridge of material, as was NGC 5216 and 5218, whilst the Antenna galaxies, NGC 4038 and 4039 (see Figure 4.4), appeared to be colliding. There was a problem with the three IC galaxies, however, as although the redshifts of IC 3481 and 3481A were virtually identical at 7250 km/s, that of IC 3483 was much smaller at 100 km/s. So although the latter appeared to be linked to the other two galaxies by looking at their optical images, this appeared to be just a perspective effect. Subsequently Halton Arp drew attention to a number of these cases of galaxies apparently linked to each other but with radically different redshifts. As a result he argued that these

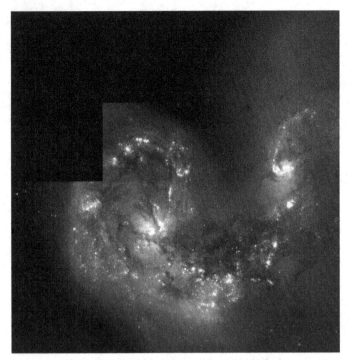

Figure 4.4. A Hubble Space Telescope image of the Antenna galaxies, which appear to be colliding. The cores of the two galaxies are left of centre and near top right; otherwise most of the other bright objects in this image are blue globular clusters. (Courtesy B. Whitmore, STScI and NASA.)

redshifts were not cosmological and the galaxies were linked, but the vast majority of astronomers disagreed.

It is not surprising, in view of the above, that it has proved very difficult to be absolutely certain as to which galaxies belong to a particular group of galaxies. For example even the relatively close Whirlpool galaxy (M51) was confirmed as being connected to NGC 5195 only when they were both imaged at radio wavelengths. Then in 1972 Alar and Juri Toomre's computer simulation of this galaxy pair in a near parabolic intercept successfully explained the bridge and tail structure of M51. In 2000 Heikki Salo and Eija Laurikainen undertook a new simulation and showed that the observed morphology of the pair could be explained if NGC 5195 had passed through the disc plane of M51 about 500 million years ago. However, although such computer simulations are impressive in reproducing many of the observed galactic structures, they can be reproduced with a number of quite different initial conditions. So the models are, at present, by no means definitive.

Starburst galaxies

NGC 7714, otherwise known as Markarian 538, had been known for some time as a distorted spiral galaxy with a small, bright, blue core which had intense but narrow spectral emission lines. In 1977 John Huchra concluded that this and other such galaxies were old galaxies going through a burst of star

formation. In the following year Richard Larson and Beatrice Tinsley suggested that star formation had been triggered in the case of NGC 7714 by interactions with its partner galaxy, NGC 7715. At about this time the International Ultraviolet Explorer (IUE) spacecraft produced an ultraviolet spectrum that indicated that there was a strong gaseous outflow of $\sim$500–1000 km/s from NGC 7714. This so-called starburst galaxy was known in the 1970s to be a strong infrared and weak radio and X-ray emitter. In 1981 Daniel Weedman and colleagues suggested that the starburst activity in its nucleus of $\sim$500 parsec diameter was the result of the explosion of $\sim$1 supernova per year irradiating large numbers of O- and B-type stars.

M82 was known in the 1970s to be one of a number of galaxies that had nuclei with very large infrared luminosities. It appeared to be affected by tidal forces from its companion galaxy M81. Weedman and colleagues concluded that the observations of M82 could, like those of NGC 7715, be explained by the frequent explosions of supernovae. But, unlike the case of NGC 7714, M82 is a very dusty galaxy. So intense ultraviolet emission from the starburst activity in M82's nucleus is heating up its surrounding dust, and so is visible in the infrared rather than in the ultraviolet.

In the mid-1980s the Infrared Astronomical Satellite (IRAS) detected about 75,000 starburst galaxies which emitted well over 90% of their energy in the infrared. Many of these starburst galaxies were also found to have superwinds coming from their nuclei, apparently caused by large numbers of supernova explosions. IRAS also found strong infrared emission from interacting galaxies. Armus, Heckman and Miley found that one particularly powerful, far-infrared galaxy, IRAS 01003−2238, had a large number of Wolf-Rayet stars, but a small number of O-type stars. This indicated that star formation had occurred in this so-called Wolf-Rayet galaxy in a sudden burst $\sim$10 million years ago.

Close encounters between galaxies like NGC 7714 and 7715, and between M82 and M81, are thought to transfer angular momentum between the galaxies. In the galaxy that loses momentum, gas flows towards the centre, increasing its density and facilitating the production of stars.

In the case of galactic collisions, the orbits of the galaxies' gas clouds around their galactic centres is disturbed. This results in these clouds falling towards the centre of the new, merged galaxy to produce a much higher density environment near the centre. Shock waves caused by the collision would also cause the clouds of gas and dust to collapse on themselves and form massive stars. These would then explode as supernovae, producing more shock waves which would help to produce even more stars. If most large galaxies have supermassive black holes at their centres, as many astronomers believe, then a collision between two such galaxies would result in their supermassive black holes merging and possibly the production of an accretion disc. In fact recent results from the Infrared

Space Observatory (ISO) spacecraft and from the European Southern Observatory's (ESO) Very Large Telescope (VLT) have indicated that some starburst galaxies may contain active galactic nuclei.

Large-scale structure of the universe

It gradually became clear in the first half of the twentieth century that galaxies were generally associated gravitationally with other galaxies as groups (like the Local Group) and/or as clusters of galaxies (see above). Groups typically contained a few dozen galaxies within a diameter of up to 2 Mpc (megaparsec), whilst clusters typically contained hundreds or thousands of galaxies within a diameter up to ~10 Mpc. Then in the 1950s Gerard de Vaucouleurs found evidence of the first supercluster, now called the Local or Virgo supercluster, which included the Virgo cluster. Eventually more and more superclusters were found with a large range of sizes varying from about 10 to 100 Mpc.

In 1966 Elihu Boldt and colleagues, using a sounding rocket experiment, detected what appeared to be an extended source of X-rays from the Coma cluster of galaxies. James Felten and colleagues suggested that this could be the result of thermal bremsstrahlung radiation emitted by intergalactic gas at a temperature of $\sim 10^8$ K. Subsequent investigators could find no evidence of this X-ray source until it was rediscovered in 1971 by John Meekins and colleagues, and Herbert Gursky and colleagues in parallel. In the same year Edwin Kellogg, Gursky and colleagues also found similar emission from the Virgo and Perseus clusters with the Uhuru spacecraft. Since then thermal bremsstrahlung radiation has been found to be a normal feature of clusters of galaxies, indicating that they are generally permeated by very hot gas.

Arno Penzias and Robert Wilson had discovered the microwave background radiation in 1965, which was the radiation left over from the Big Bang. Further work by others indicated that this radiation was extremely regular across the sky, supporting the idea, common at the time, that the universe on the largest scales would be essentially unstructured. But in 1981 Robert Kirshner and colleagues found evidence for a large void in Bootes. In the same year Gregory, Thompson and Tifft also found evidence of filamentary structure in the universe when they confirmed the existence of the Perseus-Pisces chain of galaxy clusters at a distance of ~75 Mpc. Then in 1982 Zeldovich, Einasto and Shandarin identified a generally cell-like structure of the local universe of superclusters.

Compared to the size of the universe as a whole, the volume that had been examined by the early 1980s was very small. As a result John Huchra, Margaret Geller and colleagues began an investigation, as part of an extension to the Center for Astrophysics redshift survey, to try to map the universe as far out from the Milky Way as possible. In 1986 they announced their first results for a slice of the universe centred on the Milky Way extending for about 120° in right ascension and 6° in declination, down to a limiting magnitude of 15.5 (equivalent to a distance of ~200 Mpc). In this they found a bubble-like structure, in which the clusters and superclusters of galaxies formed the surface of the bubbles. This created a real problem for the theoreticians who tried to explain it. The largest bubble was ~50 Mpc in diameter, comparable in size to the Bootes void. Three years later Geller and Huchra announced the discovery of a Great Wall of galaxies about 100 Mpc away from the Milky Way. It was ~200 Mpc long, but just 5 Mpc thick.

Alan Dressler and colleagues found in 1987 that the Local Cluster and a number of other nearby clusters of galaxies appeared to be being attracted towards what Dressler called the 'Great Attractor'. The exact nature of this Great Attractor was a subject of much controversy for many years, as there appeared to be something even further away also attracting local clusters and superclusters. In 1996 Renée Kraan-Korteweg and colleagues concluded that the Great Attractor was at or near the Norma Cluster (otherwise called Abell 3627) in a region of space behind the Milky Way as seen from Earth. The further source of attraction has been identified as the Shapley supercluster or Shapley concentration which, at ~200 Mpc, is about three times the distance of the Great Attractor.

Two sky surveys have recently been completed to more completely map the local universe. They are the Two Micron All Sky Survey (2MASS) and the Sloan Digital Sky Survey (SDSS). The 2MASS used two ground-based telescopes, located in the Northern and Southern Hemispheres, to map the whole sky at infrared wavelengths between 1 and 2.2 microns. These wavelengths were chosen to penetrate most of the Milky Way. Between 1997 and 2001 it resolved more than 1.5 million galaxies. Using redshifts from other sources it enabled a map to be created of the local universe out to about 400 Mpc. Meanwhile the SDSS was started in 2000 as a multi-filter imaging and spectroscopic redshift survey to map 25% of the sky and observe the redshifts of ~1 million galaxies. These two surveys between them have provided an incredible amount of detail showing the structure of the local universe. In particular, in 2003 Richard Gott and colleagues announced the discovery of another Great Wall of galaxies, now called the Sloan Great Wall, ~420 Mpc long, ~300 Mpc from the Milky Way.

Meanwhile in 1992 George Smoot and colleagues had found an anisotropy in the microwave background radiation of about 1 part in 100,000 using the COBE spacecraft. This gave an indication of the anisotropy of the universe about 300,000 years after the Big Bang, when this radiation was first visible. This variability has been more accurately mapped using the Wilkinson Microwave Anisotropy Probe (WMAP) which was launched in 2001. Theorists are now trying to explain how this initial variability has been transformed into the bubble-like structure of the universe that we now see closer to the Milky Way.

Bibliography

Charles, Philip A., and Seward, Frederick D., *Exploring the X-Ray Universe*, Cambridge University Press, 1995.

Dubinski, John, *The Great Milky Way and Andromeda Collision*, Sky and Telescope, October 2006, pp. 30–36.

Giacconi, Riccardo, *Secrets of the Hoary Deep: A Personal History of Modern Astronomy*, Johns Hopkins University Press, 2008.

Jarrett, Thomas, *1.5 Million Galaxies Revealed*, Astronomy Magazine, March 2008, pp. 41–48.

Kitchin, Chris, *Galaxies in Turmoil; The Active and Starburst Galaxies and the Black Holes that Drive Them*, Springer-Verlag, 2007.

Nadis, Steve, *What Happens When Galaxies Collide?*, Astronomy Magazine, March 2008, pp. 28–33.

Pendick, Daniel, *Galaxies on Fire*, Astronomy Magazine, March 2008, pp. 64–69.

Struck, Curtis, *Galaxy Collisions: Forging New Worlds from Cosmic Crashes*, Springer-Praxis, 2011.

Struve, Otto, *Galaxies and Their Interactions*, Sky and Telescope, February 1957, pp. 162–166.

Tacconi, L., and Lutz, D. (eds.), *Starburst Galaxies: Near and Far; Proceedings of a Workshop Held at Ringberg Castle, Germany, 10–15 September 2000*, Springer, 2001.

Tucker, Wallace, and Giacconi, Riccardo, *The X-Ray Universe*, Harvard University Press, 1985.

NATURE AND DISTANCE OF SPIRAL NEBULAE

Nebulae had been observed in the sky since the early seventeenth century, and in 1781 Charles Messier published the final version of his catalogue of the brighter nebulae, which he had produced to avoid them being confused with comets. Shortly afterwards William Herschel decided to list all the non-stellar objects that he could see from his observatory just outside London and by 1802 he had catalogued over 2,500 of them.

Thomas Wright had suggested in 1750 that there were other star systems like the Milky Way in the universe. Five years later Immanuel Kant went one step further and suggested that the small nebulous objects which were visible in the telescope were, in fact, these star systems which he called 'island universes'. Originally Herschel agreed with Kant's theory, but his discovery of a planetary nebula, in which a nebula surrounded a central star, caused him to change his mind in 1791. Instead, he concluded that at least some nebulae, like the Orion nebula, consisted of luminous non-stellar material which would eventually condense to form stars.

Herschel could resolve some nebulae into stars, and William Parsons (third Earl of Rosse) found, in the mid-nineteenth century, that he could resolve even more nebulae into stars with his larger telescope. This led some astronomers to believe that if one had a large enough telescope, all nebulae would be resolved into star systems like the Milky Way, with the exception of planetary nebulae, which appeared to be a special case. William Parsons also found that some nebulae had a spiral shape.

This idea of the stellar nature of nebulae received a severe setback in 1864/65 when William Huggins examined the spectra of a number of nebulae, including the one in Orion, and found that they all had bright emission line spectra, indicating that they were gaseous. One of the lines, which was bright green in colour, could not be clearly identified, and was eventually attributed to a new, previously unknown element called nebulum (or nebulium). By 1868 Huggins had examined the spectra of about 70 nebulae, one-third of which were gaseous, but two-thirds, including the Andromeda nebula, gave a continuous spectrum, indicating that they were composed of stars.

The nebulae with continuous spectra tended to have a spiral structure and were thought by some astronomers to be distant star systems like the Milky Way, but others thought that they were gaseous objects in the Milky Way, with their continuous spectra being produced by scattered starlight. In 1889 James Keeler discovered hundreds more spiral nebulae on photographs.

There were many more spiral nebulae away from the plane of the Milky Way than in it, which seemed to indicate that they were connected to the Milky Way. But the existence of some sort of attenuating material in the plane of the Milky Way could explain why objects beyond the Milky Way were not as visible in the plane. In 1904 Johannes Hartmann discovered the spectral signature of interstellar gas, but the general existence of interstellar gas was disputed and was not proved until the mid-1920s. Likewise the existence of significant amounts of interstellar dust was not proved until the early 1930s. So in the first decade of the twentieth century spiral nebulae were still generally thought to be in the Milky Way.

Vesto Slipher discovered in 1912 that the nebulosity in the Pleiades had a stellar spectrum because it was scattering light from its associated stars. This raised the question once more as to whether the nebulae with continuous spectra were composed of stars or gas. Slipher himself suggested that the Andromeda nebula, and similar spiral nebulae, may consist of a central star surrounded by a gaseous nebula which was lit up by the central star. John Reynolds' measurements of the intensity distribution across the Andromeda nebula supported this conclusion. Slipher also found that many spiral nebulae had large red shifts, indicating that they were receding from the Milky Way at very high velocities.

A nova, Nova S Andromedae, had been discovered in 1885 that appeared to be in the bright central regions of the Andromeda nebula. Ten years later a similar nova, Z Centauri, of similar magnitude was discovered apparently in the central region of another nebula, NGC 5253. Annie Cannon found that Z Centauri's spectrum was unlike that of any other nova except S Andromedae, increasing the probability that both novae really were in their nebulae, and were not just cases of chance alignments.

In 1911 Frank Very estimated the distance of the Andromeda nebula as about 500 parsec, assuming that the absolute brightness of Nova S Andromedae was the same as that of Nova Persei 1901. If he was correct this would clearly place the Andromeda nebula in the Milky Way. Five years later Adriaan van Maanen announced that he had detected the rotation of the spiral nebula M101. The angular rotation speed that he detected meant that this nebula must also be in the Milky Way. Then in 1917 Harlow Shapley calculated that, if the Andromeda nebula was outside the Milky Way, the absolute magnitude of Nova S Andromedae must be about -16, or about ten magnitudes brighter than normal novae seen in the Milky Way. This seemed ridiculous to him, and so Shapley suggested that the Andromeda nebula and the other spiral nebulae were part of the Milky Way.

Shapley concluded in about 1918 that the Milky Way was $\sim$100 kpc in diameter, which was an order of magnitude larger than understood at the time, with the Sun $\sim$20 kpc from the centre. He also believed that the Milky Way, which he thought included the spiral nebulae, was also surrounded by a cloud of globular clusters. In short, he thought that the Milky Way was the observable universe.

But ten novae were found in the Andromeda nebula in 1917–18 that were about ten magnitudes fainter than S Andromedae. As a result Heber Curtis suggested that the Andromeda nebula was an 'island universe' of stars, or galaxy, at some distance from the Milky Way, there being two different types of novae of radically different brightness in it, and presumably other galaxies. Curtis disagreed with Shapley's method of estimating the size of the Milky Way based, as it was, on the period-luminosity relationship of Cepheid variables. Instead, he preferred a system based on star counts. As a result he believed that the Milky Way was $\sim$10 kpc in diameter, or about one-tenth that proposed by Shapley, with the Sun about 3 kpc from the centre.

Shapley and Curtis met in a famous debate on 26 April 1920 to discuss the relative merits of their two different concepts of the size of the Milky Way and of the structure of the universe. Shapley's universe consisted of a very large Milky Way, which included the spiral nebulae. Curtis' Milky Way, on the other hand, was very much smaller, and the spiral nebulae were similar island universes to the Milky Way outside of the Milky Way.

Shapley spoke first and raised the following main points:

(i) It was difficult to explain the lack of spiral nebulae in the plane of the Milky Way if they were outside the Milky Way. Curtis suggested that the light from these spiral nebulae was being absorbed by a ring of material surrounding the Milky Way, as many edge-on spirals seemed to have such

an obscuration. (It was later found that Curtis was almost right, but the obscuring material was in the Milky Way, not surrounding it.)

(ii) The absolute luminosity of Nova S Andromedae would have to be much too high if it was in a nebula far outside the Milky Way. Curtis pointed out that the majority of novae in the Andromeda nebula were much dimmer than S Andromedae. So maybe there were two types of novae of radically different intensities, and S Andromeda was exceptionally bright. In that case the Andromeda nebula was an island universe some distance from the Milky Way. (Curtis was later found to be correct, as S Andromeda was found to be a supernova.)

(iii) Vesto Slipher's measurements of the fast recessional velocity of spirals seemed difficult to accept if they were such very large objects as implied by Curtis. Curtis suggested that the wavelength shifts, which had been interpreted as Doppler shifts by Slipher, may be due to another, as yet unknown, property of the spiral nebulae. (It was later found that the recessional velocities were real.)

(iv) Frederick Seares' estimate of the surface brightness of the centre of spiral nebulae, compared with that of the centre of the Milky Way, indicated that the spirals could not be at large distances. (Seares' estimate of the central brightness of the Milky Way was later found to be too low due to interstellar absorption.)

(v) Van Maanen's estimate of the distance of M101, based on its rotation, indicated that it must be in the Milky Way. Curtis argued that the measurements were too difficult to make and too subject to error to be taken at face value. (Curtis was right. The rotation rate was later found to be in error.)

The result of the debate was largely inconclusive at the time, but it was very useful in bringing the evidence on both sides into the open.

The nature of spiral nebulae was finally resolved in 1924 by Edwin Hubble when he discovered Cepheid variables in the Andromeda nebula, M31, and the Triangulum nebula, M33. This enabled him to estimate their distances as about 285 kpc for both nebulae, which showed that they were far away from the Milky Way, as argued by Curtis. Ironically, however, Hubble had used Cepheid variables to determine the distances of the spiral nebulae, a method which Curtis had previously rejected.

Over the next few years Hubble refined his distance estimates slightly, and concluded that the diameter of M31 was about 6.4 kpc, and of M33 was about 4.6 kpc. This compared with a diameter of $\sim$100 kpc for the Milky Way. So the diameters of these two spirals appeared to be less than one-tenth that of the Milky Way. This worried those astronomers who thought that all spiral nebulae were probably of roughly the same size.

Robert Trumpler found the first conclusive evidence of interstellar dust in 1930. Three years later Joel Stebbins showed that this had a dramatic effect on the estimated size of the Milky Way, reducing Shapley's estimated diameter to ~30 kpc. In parallel Hubble found that the diameter of M31 was larger than previously measured on photographic plates which resulted in a new estimated diameter of ~20 kpc.

Finally, in 1952 Walter Baade found that the distance estimates of the spiral nebulae needed to be doubled, following his work on population I and II stars. This resulted in the diameter estimate for M31 being doubled to ~40 kpc, which was now larger than that of the Milky Way, whose estimated diameter had not been affected by Baade's work.

Bibliography

Berendzen, R., Hart, R., and Seeley, D., *Man Discovers the Galaxies*, Science History Publications, 1976.

Hubble, Edwin, *The Realm of the Nebulae*, Yale University Press, 1936.

Shapley, Harlow, and Curtis, Heber D., *The Scale of the Universe*, Bulletin of the National Research Council, **2**, Part 3, May 1921, No. 11, 171–217.

Struve, Otto, and Zebergs, Velta, *Astronomy of the 20th Century*, Macmillan, 1962.

Trimble, Virginia, *The 1920 Shapley-Curtis Discussion: Background, Issues and Aftermath*, Publications of the Astronomical Society of the Pacific, **107** (1995), 1133–1144.

X-RAY BACKGROUND RADIATION

In 1962 Riccardo Giacconi and colleagues detected a diffuse source of X-rays across the sky with a sounding rocket payload sensitive from about 2 to 6 keV. The Steady State theory of the universe predicted an X-ray background, but its predicted intensity was about two orders of magnitude higher than that observed, increasing doubts about the Steady State theory. (The discovery of the microwave background radiation shortly afterwards virtually finished off the Steady State theory, as far as most astronomers were concerned.)

The Uhuru spacecraft measured the 2–20 keV X-ray background in the early 1970s, and found that it was isotropic to better than 2% over a 3° field, providing strong confirmation of the extragalactic nature of this X-ray background. It was unclear, however, if this background was due to many individual faint X-ray sources spread evenly across the sky, or whether it was due to an all-pervading, extragalactic, X-ray-emitting gas or plasma. It was concluded that if it was due to individual sources, they must have a density of at least 1 source/arcmin^2. Giancarlo Setti and Lodewijk Woltjer pointed out that the whole of this X-ray background could be explained by a large number of quasars, if they all had the same X-ray luminosity as that of 3C 273. But there was a problem with the idea of a large number of individual sources, as the spectrum of the X-ray background did not match that of known extragalactic sources.

The HEAO-1 (High Energy Astronomy Observatory-1) spacecraft produced an all-sky X-ray map covering the range 2 to 60 keV in the late 1970s. It showed an approximately isotropic distribution of X-ray emission, once known sources had been eliminated. But the intensity variations from point to point in the sky were greater than expected from normal statistical fluctuations of a smooth background. This implied that the background was not truly isotropic, but was made up of faint unresolved sources that would produce some, or possibly all, of the background radiation. On the other hand the spectrum of the X-ray background in the limited range from 3 to 40 keV seemed to indicate that it was produced by thermal bremsstrahlung emitted by a 5×10^8 K gas. Many astronomers accepted such a conclusion as showing that the X-ray background was really diffuse, and not due to numerous faint sources, even though there were a number of theoretical difficulties connected with such a high temperature.

The X-ray background had also been observed by William Kraushaar and colleagues in a series of sounding rocket experiments over the period from 1972 to 1980. These showed that above 2 keV the background was approximately isotropic. From 0.45 to 1.2 keV the X-ray sky was also approximately isotropic, but with the addition of three large-scale features. However, in the 0.11 to 0.19 keV band, the X-ray distribution across the sky was completely different, with the brightest regions roughly aligned with the galactic poles.

Two extensive deep sky surveys were undertaken in the 0.3 to 3.5 keV range using the Einstein Observatory spacecraft at the end of the 1970s, to see if the X-ray background radiation could be explained by discrete sources. In the deepest survey (DS) about 25% of the X-ray background was resolved into about 100 individual sources in the restricted fields selected. In the medium sensitivity survey (MSS) over larger areas some 800 or so discrete objects were discovered. With successively deeper surveys the percentage of new sources that were clusters of galaxies had reduced from 51% (HEAO-1), to 13% (MSS) to 6% (DS). As a result, clusters of galaxies were not expected to make a major contribution to the total X-ray background radiation. In the case of quasars and active galactic nuclei, however, the percentage increased substantially between the HEAO-1 survey and the deeper Einstein surveys. This caused astronomers to conclude that almost all of the diffuse X-ray background radiation was due to unresolved quasars, at least in the energy range below ~3 keV.

The Rosat spacecraft, launched in 1990, undertook an even longer exposure than that of the Einstein Observatory, totalling about eight days in the energy range from 0.4 to 3.0 keV. It observed through the Lockman hole, where interstellar matter in our galaxy is virtually transparent to X-rays. The inner part of the image resolved about 250 discrete sources, which were shown to account for at least 75% of the background radiation there. Around 80% of the energy from these discrete sources was found to come from quasars or active galactic nuclei. So the vast majority, if not all, of the background radiation at these relatively low X-ray energies appeared to come from discrete sources, most of which were quasars or active galactic nuclei.

Ten years later the Chandra spacecraft's deep survey resolved at least 90% of the X-ray background over the extended range from 0.5 to 10 keV into discrete sources. Chandra also showed that spectrum of all these individual sources when added together matched that of the X-ray background. It also found that the spectrum became harder for the weaker sources, thus explaining the HEAO-1's spectrum which had been erroneously attributed to an ultra-hot gas.

Bibliography

Barcons, X., and Fabian, A. C. (eds.), *The X-Ray Background*, Cambridge University Press, 1992.

Charles, Philip A., and Seward, Frederick D., *Exploring the X-Ray Universe*, Cambridge University Press, 1995.

Giacconi, Riccardo, *Secrets of the Hoary Deep: A Personal History of Modern Astronomy*, Johns Hopkins University Press, 2008.

Hirsh, Richard F., *Glimpsing an Invisible Universe; The Emergence of X-Ray Astronomy*, Cambridge University Press, 1985.

Tucker, Wallace, and Giacconi, Riccardo, *The X-Ray Universe*, Harvard University Press, 1985.

PART 5
GENERAL ASTRONOMICAL TOOLS AND TECHNIQUES
(POST-1600)

General astronomical tools and techniques

ADAPTIVE OPTICS

In 1953 Horace Babcock suggested a method of eliminating the problem of 'seeing' in optical telescopes by using a small mirror consisting of a thin film of oil inserted into the telescope's light path. A control system would sense how atmospheric turbulence was modifying the astronomical image, and would adjust the shape of the oil-film mirror electrostatically to correct the image distortion. This idea was ahead of its time, as such an adaptive optics system could not then be built. But in the early 1970s, unknown to the wider world, the U.S. military built such a system to image Soviet spacecraft in orbit, but with a deformable mirror in place of the oil film.

Meanwhile in the late 1960s David Fried had found that atmospheric turbulence had less effect at infrared wavelengths than at optical wavelengths, so an adaptive optics system should be easier to build at these longer wavelengths. The isoplanatic angle, within which the phase of the wavefront was nearly uniform, was later estimated to be about $2''$ at $0.5\ \mu m$, $10''$ at $2.2\ \mu m$ and $30''$ at $5.0\ \mu m$.

Over the next twenty years or so, until the military systems were declassified, a number of research groups built adaptive optics systems. For example, in 1983 Robert Smithson and colleagues at the Lockheed Research Laboratories built a system using a deformable mirror of nineteen hexagonal segments arranged in a circular array. The orientation of each segment could be adjusted using three piezoelectric pistons or actuators. The light from each mirror segment was focused by one of nineteen small lenses in a Shack-Hartmann array, one lens for each segment, onto a CCD. This enabled the displacement of the light coming from each mirror segment to be determined, enabling correction signals to be applied. In 1983 Smithson and colleagues successfully tested the system by imaging sunspots and solar granulation using the Vacuum Tower Telescope at Sacramento Peak. Initially it achieved an angular resolution of $2''$ to $3''$ in $6''$ seeing conditions, but ten years later this was improved to about $0.33''$ in $1''$ to $3''$ seeing.

A few years later Gerard Rousset, Fritz Merkle and colleagues produced a new adaptive optics system as a prototype for the European Southern Observatory's (ESO's) Very Large Telescope (VLT) to be built on Cerro Paranal in Chile. Called COME-ON, it was similar to Smithson's system, but its deformable mirror had a continuous reflecting surface instead of discrete segments. This continuous surface was bonded to an array of nineteen piezoelectric actuators, each of which changed the local surface curvature when a voltage was applied. Unlike Smithson's system, however, it operated in the infrared, taking advantage of Fried's discovery, and the wavefront changes were determined by observing a bright star close to the target object so that it suffered the same atmospheric distortion.

In 1989 Rousset and Merkle's system was tested on the 1.5 m telescope at the Haute-Provence Observatory in France, where it produced the first diffraction-limited images ever made of a faint astronomical object. It resolved a binary star system with an angular separation of only $0.5''$ at a wavelength of $2.2\ \mu m$, in seeing that varied between $1''$ and $2''$. A second generation 49-element system called ADONIS (Adaptive Optics Near-Infrared System) was later installed on a 3.6 m telescope at ESO's La Silla Observatory in Chile.

These adaptive optics systems used computers to calculate the magnitude of the signals to be applied to each of the actuators attached to the deformable mirror. In the late 1980s, however, François Roddier proposed an alternative adaptive optics system. It used a new type of wavefront detector, called a curvature sensor, that was linked *directly* to a special type of deformable mirror, called a bimorph, which consisted of a sandwich of two sheets of dissimilar piezoelectric materials. No intervening computer was required. In 1995 a thirteen electrode prototype was fitted to the 3.6 m Canada-France-Hawaii telescope, achieving a resolution of $0.15''$ in the near infrared. Eight years later a 185 element system called ALTAIR (ALTitude conjugate Adaptive optics for the InfraRed) was fitted to the 8 m Gemini North telescope, achieving a resolution of $0.06''$ at $1.6\ \mu m$.

One problem with these adaptive optics systems was that they needed a nearby, reasonably bright star to enable the wavefront distortion of the target astronomical object to be determined. This was a particular problem when using wavelengths of less than about $3\ \mu m$. Renaud Foy and Antoine Labeyrie suggested in 1985 that, instead of being limited to observing objects near to a relatively bright star, it may be possible to produce an artificial star adjacent to the target astronomical object using a laser. One solution involved using a sodium laser that could be used to illuminate the layer of meteoritic sodium found at an altitude of about 90 km above the Earth. In 1991, when the military work on adaptive optics was declassified, it was found that the military had been working along the same lines, and had already produced adaptive optics systems to observe military targets using laser guide 'stars'.

The first attempt to use an adaptive optics astronomical system with an artificial guide star was made in 1995 at the Lick Observatory using a pulsed laser facility outside the dome of the 3 m Shane telescope. The tests went so well that, in the following year, the Shane telescope was used to both transmit and receive the laser pulses and also to observe the astronomical objects.

Today adaptive optics are being used all over the world, generally operating in the near infrared between 1 and 5 μm, both with and without artificial guide stars.

Bibliography

Brandner, W., and Kasper, M. (eds.), *Science with Adaptive Optics: Proceedings of the ESO Workshop Held at Garching, Germany 16–19 September 2003*, Springer-Verlag, 2005.

Duffner, Robert W., *The Adaptive Optics Revolution: A History*, University of New Mexico Press, 2009.

Roddier, François (ed.), *Adaptive Optics in Astronomy*, Cambridge University Press, 1999.

Wilson, R. N., *Reflecting Optics II: Manufacture, Testing, Alignment, Modern Techniques*, Springer-Verlag, 1999.

Zirker, J. B., *An Acre of Glass: A History and Forecast of the Telescope*, Johns Hopkins University Press, 2005.

CHARGE-COUPLED DEVICES

Charge-coupled devices, or CCDs, were invented in 1969 by Willard Boyle and George Smith at the Bell Telephone Laboratories in New Jersey. At the time the laboratory was working on semiconductor bubble memories and the CCD, then called a 'charge bubble device', was initially conceived as a memory device. But it rapidly became clear that CCDs would be very useful for both military and astronomical imaging, particularly for space missions, as the images were produced in electronic form ready to be transmitted to ground. Boyle and Smith's initial CCDs were surface-channel devices in which charge was stored and transferred at the semiconductor surface. But in 1973 they also invented the buried-channel CCD in which charge was stored and transferred in the bulk semiconductor below the surface. In 2009 Boyle and Smith were awarded the Nobel Prize for Physics for their work on CCDs.

Fairchild Semiconductor produced a buried-channel CCD that became available commercially as a 100×100 pixel array in 1974. It was the first CCD to be tested by the Jet Propulsion Laboratory (JPL), which found it clearly better than the earlier surface-channel CCDs. Unfortunately the Fairchild device used interline transfer architecture which could not achieve high quantum efficiency. Meanwhile RCA were developing a back-illuminated, frame transfer 512×320 array for television applications. Unfortunately it used surface channel technology. JPL tests showed that the RCA device had poor charge transfer efficiency and high read noise compared with the Fairchild device.

Fairchild and RCA were trying to develop commercial CCDs no larger than 512×320 pixels, which was marginal for many astronomical purposes. It rapidly became clear that, for scientific applications, it would be necessary to work with a commercial organisation to specifically develop devices with all the best attributes then known for such use. As a result JPL worked with Texas Instruments (TI), under contract from NASA, to develop a CCD sensor for scientific applications based on full frame, back-illuminated, buried-channel technology. In 1975 JPL successfully tested a 100×160 back-illuminated TI device. In the following year they tested a 400×400 TI device, which had been designed for the Mariner Jupiter-Uranus mission. Its performance was excellent. Unfortunately the process yield was very low, so TI changed their design from using aluminium gates to polysilicon gates, as originally used by Fairchild and RCA. JPL planned on producing a 800×800 device that could be used on the Jupiter Orbiter and Probe mission (now called Galileo) and possibly on the Large Space Telescope (later called the Hubble Space Telescope) if the CCD was sufficiently cooled.

JPL then produced a 400×400 back-illuminated travelling CCD camera system to demonstrate its capabilities to scientists and astronomers at various observatories. At that time astronomers were basically satisfied with the photographic and electronic systems that were available, but the JPL demonstrations were to change that. The performance of the CCD system was a revelation and, as a result, there was increased support for the use of CCDs in both ground- and space-based applications.

In 1976 Landauer, Hovland and Janesick of JPL and Bradford Smith of the University of Arizona produced the first high-resolution astronomical images with a CCD. These were of Jupiter, Saturn and Uranus using a prototype TI array on the 60 inch (1.5 m) Mount Lemmon telescope in the near infrared. When Smith and his colleagues looked at the 890 nm image of Uranus they were surprised to find, for the first time, evidence of particles high above the methane layer in its atmosphere.

It had been decided in the early 1970s that the SEC Vidicon should be the detector used for the Wide Field Camera of the Hubble Space Telescope (HST). But a few years later NASA were beginning to have second thoughts because of the rapid improvements in CCDs. However, one of the big problems with CCDs of the time was their very poor sensitivity to the ultraviolet, which was a key waveband for the HST. James Westphal suggested that this could be solved by coating the CCD with a fluorescent material called coronene which would convert the incident ultraviolet light to visible light, which the CCD could detect. This proved successful on CCD tests with the 200 inch (5 m) Hale telescope on Mount Palomar.

Westphal proposed using two sets of four CCDs to provide a full image array for the wide field camera of 1,600 × 1,600 pixels operating in either the wide field or planetary camera mode. This camera, which was later called the Wide Field/Planetary Camera (WFPC) 1, was launched on the HST in 1990. It was replaced three years later by the WFPC 2, and in 2009 by the Wide Field Camera (WFC) 3. This WFC 3 had two ultraviolet/visible CCDs, each of 4,096 × 2,048 pixels, and a separate infrared sensitive CCD of 1,024 × 1,024 pixels operating up to 1,700 nm.

Modifications in the design of CCDs over the years have resulted in back-thinned, back-illuminated CCDs now having a quantum efficiency of about 80% to more than 90% over the visual waveband, compared with a few percent for the fastest photographic emulsions. In addition, not only have pixel numbers and physical array sizes increased over the last few decades, but noise levels have also been reduced substantially. So, today, CCDs have taken over from photographic emulsions for astronomical imaging not only in space applications, but also in ground-based telescopes. In addition, there are now CCDs available for detecting X-rays using a thin phosphor screen. This converts the incident X-rays into optical photons, which the CCD detects.

Bibliography

Chromey, Frederick R., *To Measure the Sky: An Introduction to Observational Astronomy*, Cambridge University Press, 2010.

Janesick, James, R., *Scientific Charge-Coupled Devices*, SPIE Publication, 2001.

Smith, Robert W., *The Space Telescope: A Study of NASA, Science, Technology, and Politics*, Cambridge University Press, 1989.

HELIOSEISMOLOGY

Five-minute vertical oscillations of the Sun's photosphere, with amplitudes of about 0.4 km/s, were discovered by Robert Leighton, Robert Noyes and George Simon at the Mount Wilson Observatory in 1961. They were found by measuring the Doppler shifts of various absorption lines at several points on the solar disc. Initially these oscillations appeared to be chaotic, short-lived and a purely local effect, with each 'cell' moving independently of adjacent ones. But subsequent work in the late 1960s and early 1970s by Roger Ulrich and others showed that these local motions were driven by standing acoustic waves excited and trapped in the solar interior. So, as Douglas Gough pointed out, analysis of these photospheric oscillations should yield information about the internal structure of the Sun.

Martin Woodard and Hugh Hudson found in 1981, using the Solar Max spacecraft, that the solar irradiance oscillated

with various discrete periods of around 5 minutes and amplitudes of a few parts per million. Four years later Woodard and Noyes found that the frequency of the five-minute oscillations had reduced from near solar maximum in 1980 to near solar minimum in 1984 by about 0.4 μHz or 1.3 parts in 10^4. Later, Kenneth Libbrecht and Martin Woodard found no such change in frequency for 1.2 mHz (14 minute) vibrations, whilst the shift at 3.6 mHz (4.6 minute) was clear, indicating that the cause of the change in frequency was near the photosphere.

In 1979 Franz-Ludwig Deubner, Roger Ulrich and Edward Rhodes found, using helioseismology, that the Sun did not rotate at the same angular rate throughout its convective layer. In particular they concluded that the convective layer rotated about 80 m/s faster than the surface velocity (of about 2 km/s) about 14,000 km (0.02 $R_\odot$ or solar radii) below the surface. Then in 1986 Thomas Duvall, John Harvey and Martin Pomerantz used a telescope at the South Pole that could provide long-term, uninterrupted coverage. Unlike Deubner and colleagues, they found that the rotation of the convection zone appeared to vary little with depth, sharing the same variation with latitude as observed at the surface. However, in 1989 Timothy Brown and colleagues concluded that there appeared to be a gradual gradient in rotation from the convective zone to the radiative interior. They also found that as depth continued to increase in the radiative interior, the equatorial and polar rotation rates approached a common value intermediate between their surface values.

In 1989 Wojciech Dziembowski, Philip Goode and Kenneth Libbrecht found that there was a sharp radial gradient in the Sun's rotation at the base of the convective zone. This produced an intense shear layer in this so-called tachocline which would create a large amount of turbulence. They further suggested that the tachocline could be the site of the dynamo which drove the sunspot cycle. Two years later Jørgen Christensen-Dalsgaard, Douglas Gough and Michael Thompson found that the base of the convection zone was at a depth of 0.287 ± 0.003 $R_\odot$, using helioseismic measurements of the speed of sound in the solar interior. Then the GONG and BiSON ground-based solar networks and the SOHO spacecraft showed, in the mid-1990s, that the Sun's central radiative region rotated as a rigid body.

The SOHO spacecraft, which had been launched in 1995, measured the vertical motion of the Sun's surface at a million different points once per minute. This enabled the production of high-quality data of a type never seen before of the complex motions of gas in the convective zone just below the Sun's visible surface. Maps of horizontal and vertical motions in this upper convective layer clearly showed vertical convection of up to 1 km/s in approximately evenly spaced columns about 50,000 to 100,000 km apart.

Helioseismology has also enabled observations to be made of subsurface flows in and around sunspots and plages. It has detected subsurface flows of material from the Sun's equator to the poles that move with a velocity of about 20 m/s down to a depth of at least 25,000 km ($\sim$0.04 $R_\odot$). It has enabled images to be made of sunspots on the far side of the Sun using helioseismic holography, and has detected the Sun's reactions to sunquakes induced by solar flares. It has also determined the density and temperature profiles of the Sun, so providing valuable constraints on the Sun's equation of state.

Bibliography

Chaplin, William J., *Music of the Sun: The Story of Helioseismology*, Oneworld Publications, 2006.

Christensen-Dalsgaard, Jørgen, *Helioseismology*, Reviews of Modern Physics, 74, 4 (2002), 1073–1129.

Di Mauro, M. P., *Helioseismology*, Astrophysics and Space Sciences Transactions, 4 (2008), 13–17.

Lang, Kenneth R., *The Sun from Space*, 2nd ed., Springer Verlag, 2008.

See also: Modern optical solar observatories, BiSON and GONG networks.

OPTICAL INTERFEROMETRY

Hippolyte Fizeau explained in 1868 how it was possible to measure the diameter of planetary moons and stars by observing the disappearance of interference fringes seen in the focal plane of a telescope as two apertures, placed in front of the telescope, are gradually moved apart. Experiments in the 1870s to measure the diameters of stars were unsuccessful, but in 1891 Albert Michelson was able to measure the diameters of the four large moons of Jupiter using such an interferometer. Michelson made no attempt to measure the diameters of stars, however, as they were presumed to subtend too small an angle at the Earth, requiring an impossibly large separation between the apertures.

At about the turn of the century, John Gore estimated that Arcturus may have a diameter about the size of the orbit of Venus. Then in 1905 Ejnar Hertzsprung suggested that some stars were giants, and over the next few years further research by Hertzsprung and Henry Norris Russell seemed to confirm this. If some stars were giants, then a Michelson interferometer of feasible proportions might be able to measure them.

George Ellery Hale considered the possibility of attaching an interferometer to the Mount Wilson 100 inch (2.5 m) reflector as it was nearing completion in 1917. He was interested in using it to study double stars, but in 1920 Arthur Eddington produced diameter estimates of giant stars which also appeared to be measurable. Michelson quickly designed a suitable interferometer which was fitted to the 100 inch by Francis Pease and John Anderson later that year. It consisted of a 6 m steel girder placed across the open end of the telescope, with two small

mirrors (instead of apertures), whose separation could be varied from 2 to 5.5 m, mounted near each end of the girder. Light from the target star was reflected off these movable mirrors, along the girder, to two fixed mirrors, which were situated a little to either side of the optical axis of the telescope. These fixed mirrors then reflected the light downwards into the telescope.

Michelson and Pease tried out the interferometer system on the spectroscopic binary Capella to see if it would detect the two components, which it did. Then on 13 December 1920 they first observed the interference fringes disappear for Betelgeuse at a mirror separation of 2.52 m, indicating a star diameter of 0.047″. This gave a stellar diameter of the order of three hundred times the diameter of the Sun, given the expected distance of the star. The diameters of Arcturus and Antares were quickly determined. Pease then designed and used a 50 ft (15 m) interferometer attached to a 36 inch (0.9 m) telescope in the early 1930s, but technical problems caused him to give up in frustration.

Optical interferometry was basically neglected over the next twenty years. But in 1954 Robert Hanbury Brown and Richard Twiss conceived of the concept of the optical intensity interferometer, based on a similar system that Hanbury Brown had devised to operate at radio wavelengths. It used two independently controlled optical telescopes, and had the big advantage over the Michelson interferometer in that only the intensity needed to be detected, not the relative phase of the signals. As a result, unlike a Michelson interferometer, the intensity interferometer was not affected by atmospheric turbulence, although it was practical for only the brightest stars. Hanbury Brown produced a proof-of-concept instrument in 1956, and completed a 188 m diameter system at Narrabri Observatory, Australia, in 1963. It operated from 1963 to 1972 and produced the first accurate measurements of the diameters of 32 main sequence stars.

Michelson interferometry had previously used two beams of light in one telescope because of the difficulty of making and controlling two light paths to an accuracy of better than the wavelength of light. However improvements in mechanical construction and design allowed the basic Michelson system to be extended to a two-telescope system in 1974, when Mike Johnson, Al Betz and Charles Townes used a two-telescope heterodyne interferometer using a CO_2 laser local operator. The interferometer operated at 10 μm with a baseline of 5.5 m. A little later in the same year Antoine Labeyrie extended this technique to the visible waveband using two small telescopes separated by 12 m.

In the meantime, in 1970 a new technique called speckle interferometry had been devised by Labeyrie which used atmospheric fluctuations to produce diffraction limited images with a normal large-aperture telescope. It had the advantage over intensity interferometry in that it could be used to observe relatively faint objects. Then in 1979 a phase-tracking interferometer was devised by Michael Shao and David Staelin making use of computer developments. It compensated for atmospheric turbulence by making continuous phase and amplitude measurements with a 4 msec integration time.

Astronomical optical interferometry has developed substantially in recent years by integrating innovative optical designs with modern computer and mechanical systems. This has resulted in a number of systems being developed in the USA, Europe, Australia and Japan, the best of which have achieved resolutions of ~ 0.2 milliarcsec.

Modern systems include the European Southern Observatory's VLTI (Very Large Telescope Interferometer) and the Keck Interferometer which have become operational using very large telescopes. In addition, more modest arrangements like the Berkeley ISI (Infrared Spatial Interferometer), the CHARA Array (Center for High Angular Resolution Astronomy) on Mount Wilson, the Japanese MIRA (Mitaka optical and InfraRed Array) –I.2, SUSI (Sydney University Stellar Interferometer), and the U.S. Naval Observatory's NPOI (Navy Prototype Optical Interferometer) have been developed. Others such as the GI2T (Grand Interféromètre à 2 Télescopes)/REGAIN (REcombinateur du GrAnd INterferometre), IOTA (Infrared/Optical Telescope Array), and PTI (Palomar Testbed Interferometer) have been closed recently to save money. The history of these interferometers is outlined in a separate article, *Modern optical interferometers*.

Bibliography

Dainty, J. C., *Speckle Interferometry in Astronomy*, in Johnson, Harold L., and Allen, Christine (eds.), *Recent Advances in Observational Astronomy*, Universidad Nacional Autónoma de México, 1981, pp. 95–111.

Hanbury Brown, Robert, *The Intensity Interferometer: Its Application to Astronomy*, Taylor and Francis, 1974.

Labeyrie, A., Lispon, S. G., and Nisenson, P., *An Introduction to Optical Stellar Interferometry*, Cambridge University Press, 2006.

Lawson, Peter R., *Principles of Long Baseline Stellar Interferometry, Course Notes from the 1999 Michelson Summer School, Held 15–19 August, 1999*, NASA, 2000.

Lawson, Peter R., *Optical Interferometry Comes of Age*, Sky and Telescope, May 2003, pp 30–39.

Shao, M., and Colavita, M. M., *Long-Baseline Optical and Infrared Stellar Interferometry*, Annual Review of Astronomy and Astrophysics, 30 (1992), 457–498.

See also: Modern optical interferometers.

PHOTOGRAPHY

Early astronomical photographs were taken using the daguerreotype process invented by Louis Daguerre in 1837.

But not only was this process hazardous, using iodine and mercury vapour, it produced images on metal that could not be copied. It was also relatively insensitive to light, so it was limited to very bright objects like the Sun or Moon. With it John Draper photographed the Moon in 1840, Alexandre-Edmond Becquerel photographed the solar spectrum in 1842 and, three years later, Léon Foucault and Hippolyte Fizeau photographed two large sunspot groups. In 1851 the professional photographer M. Berkowski imaged the solar corona during a total solar eclipse, whilst in the same year George P. Bond exhibited a daguerreotype of the Moon at London's Great Exhibition.

Frederick Scott Archer invented the wet collodion process in 1851, which was safer and much more sensitive to light. It produced images on glass, but exposures were limited to about 15 minutes as the plates had to be exposed when they were still wet. George Bond used it to take photographs of Vega, Mizar and Alcor in the late 1850s, and found that the photographs could be used to measure stellar intensities, as the brighter stars produced larger images.

Warren De La Rue began photographing the Moon in 1852 with exposures of tens of seconds using the wet collodion process. A little later he was asked by the British Association to design an instrument to provide daily photographs of sunspots at the Kew Observatory. As a result he designed an instrument called a photoheliograph with a high-speed shutter which was used at Kew from 1858 to 1872. It was then transferred to the Royal Observatory at Greenwich. In the meantime, in 1860 De La Rue and Angelo Secchi had used photography to prove that prominences moved with the Sun, rather than the Moon, during total solar eclipses. Then in 1872 Henry Draper produced the first photograph of a stellar spectrum, that of Vega. But photography did have a number of limitations, as the photographs taken during the transit of Venus of 1874 showed. These images were not sharp enough to allow accurate measurements to be made and so were virtually useless. As a result photography was not used for serious research during the following 1882 transit.

George Bond had realised in 1849 that the visual and photographic foci of a lens differed because of the different spectral sensitivities of photographic emulsions and the eye. But it was not until 1863 that Lewis Rutherfurd made the first refracting telescope specifically designed for photographic work. Then in the 1870s the wet collodion process was gradually replaced by the far more sensitive and convenient dry photographic plates. In 1880 Henry Draper produced a photograph of the Orion nebula using dry plates, whilst two years later William Huggins imaged emission lines in its spectrum. In 1883 Andrew Common's photograph of the Orion nebula (see Figure 5.1) showed details not clearly seen before. This was the first evidence that astronomical photography could be used not just to provide a permanent record, but also for research purposes.

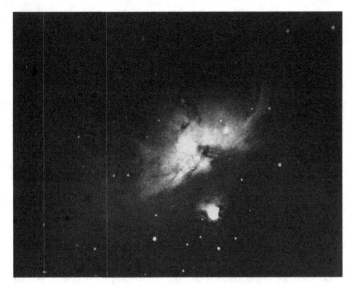

Figure 5.1. The Orion nebula photographed by Common with his 36 inch (0.9 m) reflector in 1883. (From *A Popular History of Astronomy during the Nineteenth Century*, by Agnes Clerke, 1908, Frontispiece.)

Some astronomical photographs produced in the middle decades of the nineteenth century were produced by professional photographers, some by wealthy amateurs, but only a few by professional astronomers, as photography was viewed at the time as just a method of producing impressive-looking images. However, Common's photograph of the Orion nebula and similar photographs of the 1880s was to change this idea and persuade a large part of the professional astronomical community that photography could also be used as a significant research tool.

The first successful photograph of a comet was that of Donati's comet taken in 1858 by William Usherwood, a commercial photographer, using an $f/2.4$ portrait camera with an exposure of only 7 seconds. But on the following night Bond needed 6 minutes to get a photograph of just the comet's inner coma using an $f/15$ refracting telescope. Donati's comet was bright and it was not until the faster dry plates became available in the 1870s that good photographs of normal comets could be produced. Even so Pierre Jules Janssen needed an exposure of 30 seconds in 1881 to photograph Tebbutt's comet with an 11 inch (28 cm) aperture, $f/3$ camera. It was then realised that for extensive objects like comets, unlike point sources like stars, the f ratio of an optical system, rather than its absolute aperture, was the key parameter determining photographic exposure. This was why Usherwood required less exposure to photograph Donati's comet in 1858 with a portrait camera with a fast lens, than Bond who had used an $f/15$ telescope. As a result David Gill, director of the Royal Observatory at the Cape of Good Hope, used a standard $f/4.5$ portrait camera to photograph the Great September Comet of 1882. This not only produced excellent images of the comet, but it also recorded so many stars (as the camera's magnification was low) that it

encouraged many astronomers to use photography to record star fields.

Gill's experience with imaging the Great September Comet convinced him that astronomical photography had a great future. In 1885 he started the mammoth task of photographing the southern sky to enable a star map to be produced covering the region between 19° southern declination and the South Pole. The photographs were analysed at a laboratory set up in the Netherlands by Jacobus Kapteyn. This resulted in the *Cape Photographic Durchmusterung*, or CPD catalogue, which was completed in 1900 and which contained the position and intensity of over 454,000 stars down to 11th photographic magnitude. Gill had received a grant from the Royal Society in 1885 to start the photographic work, but two years later the funding was stopped, following pressure from a number of astronomers who were concerned that photography would supersede meridian instruments. In the event Gill continued the work on his own funding.

The brothers Paul and Prosper Henry used photography in the mid-1880s to complete their mapping of the sky where the Milky Way and the ecliptic crossed, as there were too many stars there to record visually. This led the Paris Observatory director Ernest Mouchez to suggest in 1885 that an all-sky chart should be produced using photographs produced by a number of observatories. Gill and Otto W. Struve (director of Pulkovo Observatory) agreed. As a result an international Astrographic Congress was convened in Paris in 1887 which agreed on the production of the *Carte du Ciel* catalogue of the whole sky by photographing all stars down to magnitude 14. Nineteen nations attended the Congress, and the sky was split into 18 separate sections with one section to be photographed and catalogued per observatory using identical cameras. Unfortunately the production of the plates and cataloguing were much more time-consuming and costly than originally foreseen. As a result, the publication of the catalogue was not completed until 1964.

Photographic astrometry

Friedrich Bessel measured the parallax of 61 Cygni in 1838 as 0.31″, implying a distance of about 10.4 light years (3.2 parsec). His paper announcing this was followed by parallaxes for α Centauri determined by Thomas Henderson, and for Vega by F. G. Wilhelm Struve. Although all these results, which were the result of numerous visual observations, were not particularly accurate, they were of the correct order of magnitude. It was to take another fifty years before parallaxes could be determined photographically.

Charles Pritchard pioneered photographic astrometry in the early 1870s when he used photography to study lunar libration. But it was not until 1886 that he began a successful run of photographic observations of the star 61 Cygni spread over thirteen months to measure its parallax. He produced 330 plates which were subjected to an average of almost 100 measurements per plate. In spite of this, his results were not that accurate, but they showed the way for the future.

A number of astronomers built on Pritchard's pioneering work, of which Frank Schlesinger was probably the most influential. He developed techniques around 1910 that became the standard for the photographic determination of stellar parallaxes using long focus refractors. Schlesinger followed up this work with the design of an instrument to accurately record star positions photographically. This produced a significant improvement in the measured positional accuracies of stars compared with those recorded in the *Carte du Ciel* project.

Photographic photometry

The problem of determining stellar magnitudes from photographic plates was the subject of much research and confusion about the end of the nineteenth century. The Bunsen-Roscoe reciprocity law, which stated that image density, for a given emulsion, depended linearly on the total amount of light striking the emulsion during a timed exposure, was accepted by many. This was the basis of Prosper Henry's proposal for determining stellar magnitudes photographically that had been accepted by the Permanent International Committee (PIC) of the *Carte du Ciel* project in 1889.

But in 1891 the PIC reopened the subject of the determination of stellar magnitudes, because there were doubts that the Bunsen-Roscoe law was correct. Because of this, it was thought that it was safer to measure the size of the stellar images, rather than their density, to determine the magnitude of the stars, as it was considered that the size of the image was proportional to intensity, even if the density was not. Bond had suggested this technique in the 1850s but it was, unfortunately, difficult to decide on the exact image size as there was no clear edge.

After further work, the PIC decided in 1896 that there was no really satisfactory method for determining stellar magnitudes, and so they left it to the individual observatories participating in the *Carte du Ciel* project to devise their own systems. Karl Schwarzschild, in particular, continued his search for a solution to the problem, and found in 1899 that if a photographic plate was placed a little in front or behind the focal plane of the camera or telescope, the stellar images produced are all the same size but of different densities. He initially measured these visually against standard densities, but shortly afterwards Johannes Hartmann produced a visual microphotometer that enabled Schwarzschild to produce much more accurate results.

In 1900 Schwarzschild showed that the Bunsen-Roscoe law was not valid for low-intensity subjects that were subject to long exposures, owing to the reciprocity failure of the emulsion. He proposed, instead, that equal blackening of the emulsion is achieved for long exposures with low-intensity subjects when $I \times t^p$ is the same, where I is the intensity of the incident light, t is the exposure time, and p is a constant whose value is dependent on the type of emulsion used. For the emulsions in use at the time, Schwarzschild estimated a value of p of 0.86.

A more complete overview of photometry, not limited to photographic determinations, is provided in the article entitled *Photometry*.

Waveband sensitivities

In the late nineteenth century, many astronomers still assumed that the density of photographic images was independent of the wavelength of the incident light, although this was not so. The first emulsions had been made more sensitive to blue light with a wavelength of about 480 nm. Green-sensitive plates were first produced in 1884, panchromatic plates in 1904, and infrared-sensitive plates in 1919, with a peak sensitivity at 820 nm. By 1934 the peak sensitivity had been pushed well into the infrared at 1,350 nm.

Originally, because photographic plates were sensitive only to short wavelengths, refractors had to be optimised for these wavelengths if they were to be used for photography. However, this necessity largely disappeared after 1904 with the advent of the panchromatic plate, which had a spectral sensitivity similar to that of the eye. Eventually the requirement to produce large telescopes that could be used visually largely disappeared. So telescopes were then optimised for the spectral sensitivity range of the detectors to be used, whether they be photographic plates, photocells or CCD arrays.

The National Geographic–Palomar Sky Survey

The National Geographic–Palomar Sky Survey was started in 1949 before the *Carte du Ciel* was completed. It involved taking photographs in blue and red light, down to a stellar magnitude of 21 (in blue light), using 14 inch (35 cm) square plates on the Palomar 48/72 inch (1.2/1.8 m) Schmidt. In 1985 a new survey was started, reaching down to magnitude 22.5 in blue light, using modern photographic emulsions on the improved Palomar Schmidt.

Photography has now been largely replaced by CCDs at all the major optical observatories owing to CCDs' high quantum efficiency, simplicity of use and the fact that the images are provided directly in digital form ready to be analysed and enhanced by computers.

Bibliography

Barnard, E. E., *The Development of Photography in Astronomy*, Popular Astronomy, 6 (1898), 425–455.

De Vaucouleurs, Gerard (Wright, R., trans.), *Astronomical Photography: From the Daguerreotype to the Electron Camera*, Faber and Faber, 1961.

Hearnshaw, J. B., *The Measurement of Starlight: Two Centuries of Astronomical Photometry*, Cambridge University Press, 1996.

Hoffleit, Dorrit, *Some Firsts in Astronomical Photography*, Harvard College Observatory, 1950.

King, Henry C., *The History of the Telescope*, Charles Griffin, 1955 (Dover reprint 1979).

Lankford, John, The Impact of Photography on Astronomy, in Gingerich, O. (ed.), *The General History of Astronomy, Vol. 4A: Astrophysics and Twentieth-Century Astronomy to 1950*, Cambridge University Press, 1984.

See also: Photometry.

PHOTOMETRY

Ptolemy's *Almagest* of about 150 AD is the earliest known star catalogue to contain the brightness or magnitude of stars. The magnitudes were specified on a 1 to 6 scale, with 6 being for the faintest visible stars. No significant revision was made to Ptolemy's catalogue for eight hundred years, when a new set of magnitudes were produced by the Arab astronomer Abd al-Rahman al-Sūfi. There was another long gap until Tycho Brahe's star catalogue of 1602 updated Ptolemy's work. This was just before the first astronomical use of the telescope so it, like the previous catalogues, had to rely on naked eye observations.

Galileo noted in 1610 that his telescope showed many more stars than could be observed visually. Subsequently, various astronomers with progressively more and more powerful telescopes observed more and more stars. But there was no agreed way of estimating and recording their intensities until the twentieth century, and even then it was fraught with difficulties (see below).

Early quantitative photometry

Pierre Bouguer was the founder of quantitative photometry when in 1725 he tried to estimate the relative intensity of the Sun compared to that of the full Moon. To do this he observed their images on a screen produced by a concave mirror, and compared these, in turn, with the illumination produced by a candle. His value of 300,000 for the relative intensity of the Sun compared with the Moon was very close to its true value. Fifteen years later Anders Celsius estimated the brightness of stars by noting how many glass filters were required to just make each star invisible.

William Herschel was the first to produce reliable naked-eye estimates of stellar brightness. He foresaw two potential uses of such work: the detection of variable stars and the ability to assess the distribution of stars in space, assuming that a star's apparent magnitude was, on average, an indication of distance. So Herschel undertook a methodical assessment of stellar intensities in the 1780s by estimating the intensity differences of stars in sequences for various areas of the sky. In 1817 he tried to measure the apparent intensities of stars less subjectively by pointing two identical telescopes at different stars, and stopping down one telescope until the images of the two stars appeared to have the same intensity.

John Herschel devised his *astrometer* in the 1830s to measure the brightness of the brightest stars observable from the Cape of Good Hope. He did this by comparing the intensity of stars with an image of the Moon which had been made into a point source by a small lens. Unfortunately his results were inconsistent. This was later attributed to the fact that he had used Euler's incorrect relationship between intensity and phase of the Moon.

At about the same time Karl Steinheil produced a photometer that enabled the images of two defocused stars to be compared side by side. He then reduced the intensity of one image until it appeared to have the same intensity as the other. Steinheil's photometer could be used only for relatively bright stars, however. Ludwig Seidel made the first reliable photometric observations, using a Steinheil photometer, starting in 1844. He made a careful study of the effects of atmospheric absorption with reducing altitude from the zenith. Then in 1862 he produced a photometric catalogue of 208 stars brighter than 3.5 magnitude, corrected for atmospheric absorption, to a relative accuracy of better than 0.10 magnitude.

Friedrich Argelander produced a catalogue of over 300,000 stellar intensities (and positions) in 1859–1862 in the *Bonner Durchmusterung (BD)*. He extrapolated the *Almagest* scale down to 9.5 magnitude using a 3 inch (76 mm) refractor. Stellar intensities were estimated on a memorized scale during a meridian transit, without direct reference to comparison stars. The resulting magnitudes were accurate to about a quarter of a magnitude near the zenith.

In the early 1860s Johann Zöllner designed a much more accurate comparison system than previously available. In this a real star was compared with an artificial star consisting of an illuminated pinhole, and the intensities were equalised using crossed polarising Nicol prisms. Between 1885 and 1905 Gustav Müller and Paul Kempf used such a Zöllner photometer to measure the magnitude of stars seen from Potsdam down to magnitude 7.5. The resulting luminosity catalogue of over 14,000 stars was published as the *Potsdamer Durchmusterung (PD)* progressively between 1894 and 1907. The mean measurement error for each star was only about 0.07 magnitude.

Zöllner's system was difficult to use as the artificial star never looked like a real star. So Edward Pickering designed a meridian instrument in 1879 to simultaneously observe both the object star and a reference star, Polaris, in the same field of view. The intensities were then made equal using a calibrated Nicol prism. With this system he measured all 4,260 naked-eye stars north of $-30°$, publishing his results in the *Harvard Photometry Catalogue (HPC)* in 1884.

A standard magnitude scale

Karl Steinheil and Theodore Fechner independently discovered in the early nineteenth century that the eye sees intensity differences logarithmically, rather than linearly. Then in 1836 Steinheil estimated that the average intensity ratio between any two adjacent stellar magnitudes was about 2.83. In the following year, F. G. Wilhelm Struve measured the ratio as 2.89, followed by S. Stampfer with a ratio of 2.49 and Manuel Johnson who in 1853 produced a figure of 2.36.

In 1856 Norman Pogson proposed that that there should be a standard ratio of 2.512 between any two adjacent stellar magnitudes, that is log (ratio) $= 0.4$, so that a first magnitude star would be one hundred times brighter than a sixth magnitude star. Pickering used this ratio when he started work on the *HPC* in 1879, as did Müller and Kempf at Potsdam working on the *PD*. It was also necessary to define where the zero point of the photometric scale should be. Pickering initially assumed that Polaris had a magnitude of 2.0, which he revised to 2.15 and finally 2.12, whereas Müller and Kempf assumed that the average magnitude of the 144 standard stars they used in the *PD* was 6.0. It was hoped at both observatories that these choices would enable previous catalogues to be used unchanged. But Ernst Zinner found in the 1920s that the *PD* scale was from 0.13 to 0.21 magnitude fainter than the *HPC* scale, indicating the sort of errors contained in even the best photometric catalogues around the end of the nineteenth century.

Further photometric catalogues around the end of the nineteenth century

Edward Pickering extended his *HPC* of 1884, using an improved version of his meridian photometer which included a larger objective to enable fainter stars to be observed. He also changed his reference star from Polaris to λ Ursae Minoris, a 6.4 magnitude star about 1° from the celestial pole, and more suitable for comparison with fainter stars. During the period from 1882 to 1888 he measured over 20,000 stars with magnitudes down to the faintest measured by Argelander for the *BD* catalogue. It had been known for some time that Argelander's catalogue did not follow Pogson's intensity ratio of 2.5 between adjacent magnitudes for the fainter stars. Pickering measured stars in selected regions of the sky to limit the size of his task, and produced a set of correction factors which enabled all of the magnitudes in the BD catalogue to be modified to be consistent with the new photometry scale.

Following completion of Pickering's work with the Harvard meridian photometer, Solon Bailey took it to Arequipa, Peru

in 1889 to undertake a similar survey in the Southern Hemisphere. He measured almost 8,000 stars in selected areas down to magnitude 10. On his return, Pickering re-measured all the stars in his 1884 catalogue, and included all stars visible down to magnitude 7.5. In 1899, and again in 1902, Bailey returned to Peru to complete a similar task there. As a result, in 1908 Pickering and Bailey were able to publish the magnitudes of over 45,000 stars in both hemispheres, which were complete down to magnitude 7.0, with measurements in selected areas down to magnitude 10. This *Revised Harvard Photometry Catalogue* was the standard reference work for many years, mainly because it comprehensively covered both hemispheres.

Starting in 1904 Karl Schwarzschild used photographs to estimate the brightness of a series of stars by measuring the density of their out of focus images (see *Photographic photometry* in the *Photography* article). He compared the density of stellar images using a visual microphotometer designed by Johannes Hartmann. Although his work was very accurate, with mean magnitude errors as low as 0.02 to 0.04, he completed the survey of the sky only between declinations of 0° and 20°. In all, he determined the photographic magnitudes of just over 3,500 stars down to magnitude 7.5, which he listed in the *Göttinger Aktinometrie* published in 1910.

Standard stars and the North Polar Sequence

Pickering had proposed as long ago as 1879 that Polaris should be used as a primary standard to calibrate all visual photometric observations. He also suggested that ten other near-polar stars, with magnitudes ranging from about 6 to 13, should be used as secondary standards. His proposal, which was the origin of the North Polar Sequence (NPS), was not adopted, but in 1907 he resurrected it to provide primary and secondary photographic photometric standards. He proposed to cover the Southern Hemisphere by establishing secondary standards in each of the 48 Harvard Standard Regions that he had previously defined, which would be tied to the NPS.

By 1909 Pickering's idea of using just ten stars in the NPS as standards had been expanded to 47 stars. But by the time that the photometric measurements had been made by Henrietta Leavitt for the NPS in 1912, the NPS had been expanded to include 96 stars with photographic magnitudes ranging from 2.7 to 21. In the following year, this Harvard International System of Photographic Magnitudes was adopted by the International *Carte du Ciel* project. Henrietta Leavitt was still involved in producing more accurate photometric measurements of the standard stars until her death in 1921.

Frederick Seares then spent many years trying to resolve a number of difficulties with the NPS, caused by a lack of clarity on the types of telescopes, wavelength, passbands, and extinction corrections to be used. In 1922 he produced a new set of photographic and photovisual magnitudes for 92 NPS stars, with errors of only a few hundredths of a magnitude. These were accepted by the International Astronomical Union (IAU) as defining the international photometric standard. Unfortunately, over the next thirty years this standard met with numerous problems including the distortion effect of ultraviolet light on photometric measurements, the continuing problem of transferring the standard to other parts of the sky, and so on. So in 1955 the IAU was finally forced to abandon their standard system.

Photometric detectors

The accuracy of the visual photometers of Zöllner, Pickering and their contemporaries of the late nineteenth and early twentieth centuries depended on the accuracy with which the eye could determine if two light sources were of equal intensity. This yielded errors of a few percent. A much more accurate photometer, which eliminated the subjective element, was developed in the early twentieth century by Joel Stebbins and Fay Brown using a selenium detector. They discovered, in particular, that the signal-to-noise ratio of the device improved dramatically at low temperatures and, after further refinements, Stebbins was able, in 1910, to detect the secondary minimum of Algol, which caused a reduction of intensity of only 0.06 magnitude.

Unfortunately, selenium cells were not very stable, but a new type of light-sensitive device also became available, the photoelectric cell, which was more sensitive and stable than the selenium cell. Photoelectric cells were developed by Julius Elster and Hans Geitel at about the turn of the century, and first used in 1912 by Paul Guthnick in Berlin and Hans Rosenberg and Edgar Meyer in Tübingen. However, only Guthnick produced astronomical results. Rosenberg, on the other hand, concentrated on trying to improve the detection of very low currents from these photoelectric cells. Albert Whitford undertook similar work to Rosenberg, then in 1937 he and Gerald Kron were sent a prototype RCA photomultiplier for evaluation. They found it very useful, and mounted it as an autoguider on the Mt Wilson 60 inch (1.5 m) telescope. This was the first time a photomultiplier had been used in astronomical research, and the first time an electronic device had been used as an autoguider.

The photoelectric surface used in the early photoelectric cells, or photocells, was generally potassium hydride. More complex surfaces of caesium and antimony were developed by Paul Görlich in 1938 which had higher sensitivity than potassium, particularly at the blue end of the spectrum. Then two years later RCA produced both photocells and photomultipliers with such a surface. Cooling the photomultiplier tube was also found to reduce the dark current, in a similar way to that discovered by Stebbins and Brown for selenium cells. All

these developments meant that, by 1960, stars of magnitude 23 could be measured to an accuracy of 0.05 magnitude, using the 200 inch (5 m) Palomar telescope.

Subsequent years have seen the development of a number of different types of photometric detectors, including André Lallemand's electronic camera and the image orthicon of the 1930s–1950s, and the vidicon of 1964. But charge-coupled devices (CCDs), which were first available commercially in 1973, have been far the most useful (see *Charge-coupled devices* article).

Colour

Ptolemy listed the colour of six stars in his *Almagest*, but the colour of stars was not generally commented on or recorded until the nineteenth century. William Herschel noted the colours of some visual binaries, but Wilhelm Struve was the first to systematically observe star colours during his observation of double stars starting in 1814. Struve estimated the colours of several thousand stars using a scale of ten colours ranging from blue through bluish white and yellowish white to yellow and red. In 1847 John Herschel published a list of 76 unusually red stars that he had observed from the Cape of Good Hope.

Numerous astronomers recorded the colours of stars in the second half of the nineteenth century, but it was very difficult to agree on a scale of colours. In 1856 Angelo Secchi proposed using the emission lines of a spark spectrum to provide a standard reference, whilst William Smyth produced a reference colour chart in his book of 1864. However, in a paper three years later Sidney Kincaid, discussing the problem of accurate colour estimates, pointing out the difficulty in comparing colour on a colour chart with that of coloured light. He also discussed the differences in colour appreciation between observers (the so-called *personal equation*), as well as atmospheric and instrumental effects. In 1886 William Franks mentioned the effect of atmospheric conditions and the brightness of stars on their apparent colour. In the following year, he suggested using colours in the solar spectrum as a standard, with colour filters based on these colours as secondary standards.

Angelo Secchi had found in 1866 that a star's colour and spectral type seemed to be related. Hermann Vogel and Gustav Müller of Potsdam found the same thing in 1883. They used a much larger sample of over 4,000 stars and a seven-point colour scale going from white to yellow to red. Müller and Kempf estimated the colour of over 14,000 stars in their PD catalogue in 1907 using an extension of this Potsdam colour scale to produce seventeen colour classes. Zinner used this PD colour data in 1926 to re-examine the relationship between colour and spectral type. This showed the rather limited precision in the PD colours which Zinner put down, in part, to the eye's limited colour sensitivity when observing faint stars.

In the early twentieth century, Heinrich Osthoff pointed out that the naked eye could only detect the colour of stars of brighter than about 2.5 magnitude (owing to the fact, discovered by Max Schultze in 1866, that colour is detected by cones in the eye, which are appreciably less sensitive than rods, which are not colour sensitive). Osthoff investigated the effect of star brightness and the closeness of the target star to other stars on the perception of colour, as well as of moonlight, atmospheric dust or haze and the star's altitude. He found that stars appeared redder to older observers. They also appeared redder as a given observer's eye adapted over about the first 45 minutes of observation, whilst eye fatigue resulted in stars appearing bluer over subsequent hours of observation.

Quantitative colour estimates and colour indices

Attempts had been made in the second half of the nineteenth century to produce quantitative estimates of star colours. In 1860 Zöllner used a birefringent rock crystal plate placed between crossed Nicol prisms to change the colour of his artificial star. One of the Nicol prisms was then rotated to match the colour of the star. Using this system he published colorimetric results for 37 stars and planets in 1868. They were expressed as the angle of the Nicol prism to produce a match between the artificial and real stars.

In 1883 Seth Chandler devised another method of producing quantitative colour measurements whilst observing a number of red variable stars. He selected a white comparison star close to his target red star of similar intensity, and then compared their brightness difference with and without a light blue filter attached to his eyepiece.

Photographic emulsions in the late nineteenth century had a blue-sensitive bias compared with the eye. So a colour index could be produced by comparing the magnitudes of stars estimated visually with those estimated photographically. In 1890 Scheiner calculated the differences between the photographic and visual magnitudes of stars in the Pleiades cluster to produce such a colour index. Carl Charlier produced a similar set of results at about the same time. But these results of Scheiner and Charlier were somewhat compromised by the inaccuracies in both their photographic and visual magnitudes. Ten years later Schwarzschild produced the first accurate photographic intensities using his extra-focal method. He found that the colour index based on his photographic magnitudes and on the Potsdam visual magnitudes ranged from about −0.6 for B-type stars to +1.9 for M-type stars. He then calibrated his colour indices using black body flux distributions to obtain stellar temperatures.

In 1901 Max Planck derived theoretically the energy distribution radiated by a black body as a function of wavelength for various temperatures. As a result it was expected that the measurement of the shape of the star's radiation curve with wavelength should give more accurate temperature results than that produced by a simple colour index.

Johannes Wilsing and Julius Scheiner began such an investigation at Potsdam in 1905 using five wavelengths between 448 and 638 nm isolated using a spectroscope. They measured the spectral energy distribution of 109 stars, calibrating their results using an electric oven at known temperature. These observations were then continued by Wilsing and Wilhelm Münch in 1908 using ten passbands. Shortly afterwards Charles Abbot criticised the results for their limited wavelength range and the assumption that stars radiate as black bodies, which may not be correct.

Hans Rosenberg visually measured the density variation along a photographic spectrum between 400 and 500 nm for 70 bright stars in 1914 to derive their colour temperatures. Wilsing criticised Rosenberg for using a very short spectral baseline in the blue region where there are many lines. Ten years later Ralf Sampson used more red-sensitive plates, and used a microdensitometer to measure the density variation along the photographic spectrum to about 660 nm. As a result his colour temperature estimates were more accurate than Rosenberg's.

Near the end of the nineteenth century William Huggins had observed a rapid fall in the intensity of the continuous spectrum of Vega, and of similar stars, beyond the end of the series of Balmer lines in the ultraviolet, that is below about 365 nm. This so-called Balmer jump was also observed by Wright in 1918, showing that stars do not radiate as ideal black bodies. This caused a problem in trying to estimate the temperature of stars based purely on their spectral energy distribution, but the full extent of the problem was not understood at that time. Then in 1926 Ch'ing sung Yü, who was the first to study the Balmer jump in detail, found that the magnitude of the jump varied from star to star, making temperature estimates from the spectral energy distribution of stars even more problematic.

Photographic plates and photoelectric devices were manufactured with an ever-increasing range of spectral sensitivities in the first half of the twentieth century, and so it became essential to standardise on the spectral sensitivity of detectors when determining stellar magnitudes. The International Solar Union had agreed in 1910 that the visual and photographic magnitudes of class A0 stars should be the same between magnitudes 5.5 and 6.5, that is, the colour index of these stars was zero. But this definition was too imprecise for the mid-twentieth century.

So a new system, the UBV system, was published by Harold Johnson and William Morgan in 1953. In this, ultraviolet (U), blue (B) and visible (V) stellar magnitudes were defined using passbands with effective wavelengths of about 350 nm (U), 430 nm (B), and 550 nm (V), the latter being a compromise between the eye's sensitivity in normal and dark-adapted vision. In describing their system, the authors recommended that an RCA 1P21 photomultiplier and aluminised reflecting telescope be used. Six A0 main sequence stars, with apparent magnitudes about 6, were specified as having the same U, B and V magnitudes. Ten stars were then defined as primary standards and 98 as secondary standards. The system was approved in 1955 by the IAU as an official system for three-colour photometry.

Later, Johnson extended the three-colour system to a ten-colour system by adding passbands centred on 700 nm (R), 900 nm (I), 1.25 μm (J), 2.2 μm (K), 3.4 μm (L), 5.0 μm (M) and 10.2 μm (N), the infrared passbands being added to correspond with the transmission windows of the Earth's atmosphere. Johnson's multicolour system was later extended to include passbands centred on 1.65 μm (H) and 21.2 μm (Q).

Bibliography

Hearnshaw, J. B., *The Measurement of Starlight: Two Centuries of Astronomical Photometry*, Cambridge University Press, 1996.

Herrmann, Dieter B. (Krisciunas, K., trans. and rev.), *The History of Astronomy from Herschel to Hertzsprung*, Cambridge University Press, 1984.

King, Henry C., *The History of the Telescope*, Charles Griffin, 1955 (Dover reprint 1979).

Milone, E. F., and Sterken, C. (eds.), *Astronomical Photometry: Past, Present, and Future*, Springer, 2011.

Sterken, Christiaan, and Staubermann, Klaus, *Karl Friedrich Zollner and the Historical Dimension of Astronomical Photometry*, VUB Press, 2000.

Weaver, Harold F., *The Development of Astronomical Photometry*, Popular Astronomy, **54** (1946), 211–230, 287–299, 339–351, 389–404, 451–464, 504–526.

See also: Photography.

RADIOMETRY

Nineteenth century detection devices

William Herschel measured the heat in various parts of the solar spectrum in 1800 using mercury thermometers with blackened bulbs. He found that the temperature progressively increased from a minimum at the blue end of the spectrum to a maximum beyond the red in what is now called the infrared. Herschel did not determine the wavelength of this infrared light, but it is clear from his results that he had detected energy at wavelengths beyond 1 μm. In the following year, Johann Ritter discovered that the Sun also emitted ultraviolet light when he found that silver chloride became darkened when subjected to solar energy beyond the blue end of the spectrum.

Thomas Seebeck discovered the thermoelectric effect in 1822 when he found that a thermocouple consisting of two dissimilar metals deflected a compass needle when one of the two junctions was heated. This was due to an induced voltage

difference and current which were proportional to the temperature difference between the two junctions. This thermocouple effect provided a much more convenient way of measuring heat than with a liquid-in-glass thermometer. In 1831 Macedonio Melloni used a number of thermocouples joined together in what is now called a thermopile to increase sensitivity.

William Crookes designed a radiometer in 1875 that consisted of four thin mica vanes, black on one face and white on the other. They rotated in an evacuated glass tube when illuminated by radiation, the rotational speed depending on the intensity of illumination. A tortional radiometer was also developed consisting of two blackened vanes suspended from a torsion fibre in which one vane was illuminated and the other not. The intensity of the radiation was measured by the deflection of the fibre.

Samuel Langley's bolometer, designed in about 1880, was based on measuring the increase in resistance of a thin strip of blackened platinum as its temperature was increased by incident radiation. The resistance was measured by a galvanometer in a Wheatstone bridge arrangement. Langley's platinum strips were only a few microns thick, producing a higher sensitivity and a much faster response to changes in temperature than possible with thermopiles.

Surface temperature of the Sun

It was impossible in the middle of the nineteenth century to obtain good temperature estimates of the Sun's visible surface, as the two basic laws of the time yielded radically different results. Newton's law, in which luminosity was linearly proportional to surface temperature, produced much higher temperatures than Dulong and Petit's law of 1817, in which radiative energy increased geometrically with increase in temperature.

In 1838 Claude-Servais Pouillet measured the amount of solar heat radiation reaching the Earth's surface with a simple pyrheliometer. This consisted of a small waterbath in a blackened container with a thermometer to measure the temperature rise. He concluded that the Earth was receiving about 1.76 calories/cm²/minute or 1.44 kW/m², which was extremely close to the modern value of 1.37 kW/m². He then went on to calculate that the temperature of the Sun's surface, using Dulong and Petit's law, must be at least 1,730 K.

John Waterston estimated in 1860 that the temperature of the Sun's surface, using Newton's law, was about 7.2 million K. Over the next few decades Vicaire, Violle and others used Dulong and Petit's law to produce solar surface temperature estimates, like Pouillet's, of a few thousand Kelvin, whilst Secchi and Ericsson used Newton's law, like Waterston, to estimate solar surface temperatures of millions of Kelvin.

In 1879 Joseph Stefan found experimentally that the total energy radiated by a perfect emitter, or black body, was proportional to the fourth power of its absolute temperature. This law was not immediately accepted at the time, however, as it was not obvious why it was any more correct than any of a number of other competing laws. But five years later Ludwig Boltzmann gave Stefan's law a theoretical basis. Then in 1893 Wilhelm Wien showed that the wavelength of the maximum energy radiated from a black body was inversely proportional to its temperature. In 1911 Charles Abbot used bolometer measurements at Washington, Mount Wilson and Mount Whitney of solar intensities from 0.30 to 3.0 μm to deduce an effective solar temperature of about 6,200 K (which is roughly correct) using Wien's formula.

Absorption of the Earth's atmosphere

There was a great deal of confusion in the second half of the nineteenth century about how much of the Sun's energy was absorbed by both the solar and Earth's atmospheres. It had been known since the seventeenth century that the Sun was darker near the limb than in the centre. This limb darkening was put down, erroneously as it turned out, to simple absorption in the solar atmosphere. Then in 1871 Sergei Lamansky detected three broad absorption bands in the solar infrared, which he correctly attributed to absorption by the Earth's atmosphere. Four years later Jules Violle studied the absorption of the Earth's atmosphere by comparing sea-level observations of the Sun with those taken from the summit of Mont Blanc (4,800 m altitude). He estimated a solar constant of 2.08 kW/m².

Langley was the first to use a diffraction grating to investigate the solar infrared, and in 1881 he compared bolometer measurements made from the summit of Mount Whitney (3,600 m altitude) with those at sea level. He showed that the maximum of the Sun's thermal radiation was in the orange region, and not in the infrared as previously thought. He also extended the infrared coverage to 2.8 μm, and concluded that there were a number of deep and wide absorption bands in the infrared, and that the absorption of the Earth's atmosphere increased rapidly with decreasing wavelength towards the ultraviolet. He calculated a solar constant of 2.32 kW/m². In the 1890s Langley detected hundreds of absorption lines in the infrared up to 5.3 μm.

Moon and planets

Melloni was the first to clearly detect the heating effects of moonlight from a meteorological station on Mount Vesuvius in 1846. Lawrence Parsons (fourth Earl of Rosse) measured this lunar radiation in about 1870 and concluded that the Moon's surface temperature increased by about 280 K during the lunar day. Otto Boeddicker, Rosse's assistant, later showed that the lunar surface cooled down very rapidly during a lunar eclipse, giving some idea of the nature of the lunar surface. Seth Nicholson and Edison Pettit found in 1927 that the lunar temperature fell from about 340 K to 150 K during such a lunar eclipse. In the same year, Donald Menzel determined the temperature variations of the terrestrial planets which enabled conclusions to be drawn on their possible atmospheres and types of surfaces.

Stellar radiometry

In 1867 William Huggins was the first to detect infrared radiation from stars. He was able, using a thermopile, to detect such radiation from Regulus, Arcturus, Sirius and Pollux. In the following year Edward Stone made similar observations, both astronomers concluding that red stars were stronger infrared emitters relative to their visual brightness than other stars.

Peëtr Lebedev improved the design of thermocouples at the end of the nineteenth century by putting the junctions in an evacuated cell. Herman Pfund constructed such a vacuum thermocouple in 1912 with bismuth alloys which achieved a gain of at least ten because of the vacuum environment. William Coblentz used a similar vacuum thermocouple on the 36 inch (0.9 m) Crossley reflector at the Lick Observatory in 1914 and observed over 100 stars down to visual magnitude 6.7. By making measurements with and without a water cell, which absorbed energy beyond 1.4 μm, he was able to estimate the fraction of the total radiation that stars emitted beyond that limit.

Pettit and Nicholson undertook an extensive series of stellar observations over the period 1921–1927 with a vacuum thermocouple on the Mount Wilson 100 inch (2.5 m) telescope. They observed 124 bright stars and derived a heat index (HI, visual minus radiometric magnitude) and a water cell index (WC, based on water cell absorption). They found, plotting one index against the other for all the stars that they had measured, that the cooler M-type stars diverged progressively from the theoretical black body line. Pettit and Nicholson's results also enabled the diameter of stars to be estimated, knowing their trigonometrical parallaxes and hence their distances. These diameters were found to be similar to those determined by Francis Pease and colleagues using a Michelson interferometer.

Abbot and Lloyd Aldrich attempted in 1922 to measure the spectral energy distribution of stars from the visible into the infrared using a bolometer attached to the Mount Wilson 100 inch. Unfortunately their results were rather inconsistent. In the following year Abbot replaced the bolometer with a tortional radiometer developed and improved by Ernest Nichols. As a result Abbot was able to measure the spectra of nine bright stars up to 2.2 μm, and determine their effective temperatures as ranging from 2,500 K for α Herculis to 16,000 K for Rigel.

Lead sulphide photoconductive cells, which were developed for military applications in the Second World War, had a sensitivity up to about 3.6 μm, with a peak at 2.5 μm. Their sensitivity was about three orders of magnitude better than that of thermocouples. Numerous astronomers used these lead sulphide cells, both cooled and uncooled, immediately after the war to provide infrared spectroscopy of bright stars.

Harold Johnson used an indium antimony photovoltaic device in 1961, cooled down to 77 K by liquid nitrogen, which extended the wavelength coverage to 5 μm. He used this to study the infrared photometry of stars, defining four broad passbands which coincided with the atmospheric windows at 1.3 μm (J), 2.2 μm (K), 3.6 μm (L) and 5.0 μm (M). At about the same time Frank Low developed the gallium-doped germanium bolometer cooled by liquid helium to about 2 K. Low then collaborated with Johnson to produce astronomical observations in the 10.2 μm (N) band in 1963 and in the 21.2 μm (Q) band in 1966.

Infrared sky surveys

The first major infrared sky survey was carried out at 2.2 μm (K band) by Gerry Neugebauer and Robert Leighton starting in 1965 using eight lead sulphide photoconductive cells cooled in liquid nitrogen. They detected about 20,000 infrared sources over a three-year period of which more than 5,000 were catalogued with K band magnitudes greater than 3.0. Most of these catalogued sources could be identified as known stars, but a few dozen sources, with colour temperatures of ~1,000 K, were new. Since that time numerous infrared surveys have been carried out using ground-based observatories, sounding rockets and spacecraft. The latter included IRAS, ISO and the Spitzer Space Telescope.

Cavity radiometers

Radiometers that were available in the early years of the space age were not accurate or stable enough to measure small variations in the heat received even from a source as powerful as the Sun. Also the temperatures recorded by spacecraft in orbit did not match those determined from on-ground testing. It was unclear as to whether this was due to the fact that the solar constant used in ground testing was incorrect, or due to problems with the spacecraft radiometers. So in 1964 Joseph Plamondon and Richard Willson started designing a total irradiance radiometer based on Floyd Haley's successful cavity radiometer which had been used for calibrating line intensities in the ultraviolet.

The detector in a cavity radiometer consisted of an insulated black cavity that was kept in thermal equilibrium. The cavity had two apertures, one for the incident radiation of unknown energy and the other to allow radiation to leave the cavity. The cavity was maintained at a constant temperature by a heater. So the unknown energy of the incident radiation could be determined by subtracting the energy supplied by the heater

from the energy radiated from the cavity, which was a function of its known temperature.

Plamondon and Willson were joined by James Kendall in 1965, and by 1970 Kendall's cavity radiometers were so accurate in the laboratory that they were adopted as international standards. Richard Willson, in the meantime, had been adapting Kendall's instruments for use outside the laboratory, and in 1968 he flew two of his instruments on a high-altitude balloon. By 1973 Willson had achieved measurement accuracies of ±0.2%, and a side-by-side comparison with a thermopile radiometer on a sounding rocket three years later showed the superiority of the cavity radiometer design which was used for future spacecraft.

Microwave radiometry

Radiometry is not limited to the visible and infrared parts of the spectrum. For example in recent years microwave radiometers have been flown on spacecraft to both study the Earth and to detect variations in the cosmic microwave background (CMB). The COBE spacecraft, which was launched in 1989, was the first to examine the CMB. It used three differential microwave radiometers operating at 3.3, 5.7 and 9.5 mm wavelengths (90, 53 and 31.5 GHz). Each radiometer used a pair of horn antennae that measured the temperature difference between two different parts of the sky about 60° apart. This was done using a microwave receiver that rapidly switched between the two horn antennae. Averaged over time it measured temperature anisotropies to better than 10^{-4} K.

Bibliography

Allen, D. A., *Infrared – The New Astronomy*, John Wiley, 1975.

Hearnshaw, J. B., *The Measurement of Starlight: Two Centuries of Astronomical Photometry*, Cambridge University Press, 1996.

Hufbauer, Karl, *Exploring the Sun: Solar Science since Galileo*, Johns Hopkins University Press, 1991.

SPECTROHELIOGRAPH

George Ellery Hale found in 1890 that he could photograph the surface of the Sun in the light of a single wavelength. To do this he made the solar image scan across the first slit of a spectroscope as the photographic plate was moved simultaneously across the second slit. This instrument was only partially successful, however, so he modified it by keeping the solar image and photographic plate stationary while moving the slits. This spectroheliograph, as it was called, was first used to record the solar disc in the light of hydrogen α, but Hale then decided to use the H and K lines of calcium, as photographic plates were more sensitive to the ultraviolet end of the spectrum at that time. These photographs enabled prominences and sunspots to be examined at will across the solar disc when the system was in full operation in 1892. Only much later did Hale learn that the basic idea of the spectroheliograph had been suggested by Jules Janssen as long ago as 1869, and that F. W. Braun had unsuccessfully attempted to make such an instrument in 1872.

Independently, Henri Deslandres invented a similar device in Paris, but it was in operation a little later than Hale's. Deslandres suggested that, as the calcium H and K bands originated in different layers of the Sun, photographs taken at these different wavelengths would image different depths of the Sun's photosphere.

There then followed a controversy between Deslandres and Hale over the priority of invention and first successful operation of a spectroheliograph. Whilst it is now clear that Hale was earlier than Deslandres in inventing and first building such a device, the inventor was really Janssen or Braun, although Hale had been the first to make it work.

Hale subsequently developed a spectrohelioscope over the period 1924 to 1926 in which he replaced the slow-moving slits of the spectroheliograph with fast-oscillating slits, whilst replacing the photographic plate with the eye. Persistence of vision enabled Hale to see a monochromatic image of the Sun, and allowed him to observe rapid changes on its surface.

Bibliography

Gingerich, O. (ed.), *The General History of Astronomy, Vol. 4A: Astrophysics and Twentieth-Century Astronomy to 1950*, Cambridge University Press, 1984.

King, Henry C., *The History of the Telescope*, Charles Griffin, 1955 (Dover reprint 1979).

Wright, Helen, *Explorer of the Universe: A Biography of George Ellery Hale*, American Institute of Physics Press, 1994.

SPECTROSCOPY

Early solar observations and the interpretation of laboratory spectra

William Wollaston noticed in 1802 that the Sun's spectrum, when produced by a glass prism, was crossed by seven dark lines, but he took this discovery no further. Then over the period 1814–1815 Joseph Fraunhofer rediscovered these lines and many more (see Figure 5.2) as part of his attempts to measure the refractive index of optical glass. In total he observed more than 500 lines, the strongest of which he labelled A to H. In 1823 he observed the spectra of the Moon, Mars, Venus, Sirius, Castor, Betelgeuse and a few other stars using a prism spectroscope.

Alexandre-Edmond Becquerel photographed the solar spectrum in 1842, discovering numerous lines in the ultraviolet far beyond the visible limit. It was not completely clear at this time whether the lines observed in the solar spectrum originated on the Sun or not, as some lines seemed to become darker as the Sun approached the Earth's horizon, indicating that they, at least, were produced by the Earth's atmosphere. But by the 1850s most astronomers accepted that the majority of the lines in the solar spectrum did originate on the Sun.

Léon Foucault found in 1849 that the dark D line in the solar spectrum, which was double, coincided with a double bright line in a laboratory arc. He also found that when sunlight was passed through such an arc, the darkness of the D lines increased. Unfortunately Foucault did not speculate on the possible reason for this and other interesting results that he obtained, and his work was largely ignored.

It had been suspected for some time that the spectrum produced by various gases in the laboratory was characteristic of the substance producing it. But the Fraunhofer D line seemed to be present in all laboratory spectra. This led some physicists to speculate that the D line was due to some form of contamination, which was confirmed in the late 1850s. Then in 1859 Julius Plücker found that the position of the Fraunhofer F line coincided with that of a bright hydrogen line which he called Hβ.

In a series of experiments commencing in 1859, Gustav Kirchhoff and Robert Bunsen showed, using very pure samples, that bright spectral lines emitted by a hot gas were unique to each element in the gas. They also found that luminous solids and liquids generally produced a continuous spectrum. If this light was then passed through a gas, the gas would produce either dark absorption lines or bright emission lines, depending on the relative temperature of the gas and of the luminous

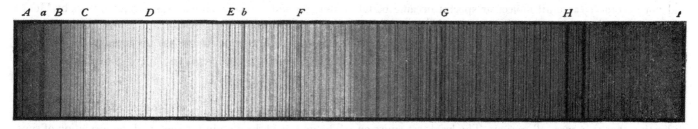

Figure 5.2. Fraunhofer's map of 324 lines in the solar spectrum that he observed over the period 1814–1815. (From *A Short History of Astronomy*, by Arthur Berry, 1898, Figure 97.)

source. This indicated that the Fraunhofer lines in the solar spectrum were the result of absorption in the atmosphere of the Sun that was cooler than the layer beneath it. Kirchhoff also measured the position of thousands of dark Fraunhofer lines from 1859 to 1862 and deduced which elements were producing these lines from a comparison with laboratory spectra. As a result he concluded that iron, calcium, magnesium, sodium, nickel and chromium must be present in the cooler outer layers of the Sun's atmosphere.

Jules Janssen, Norman Lockyer and others observed the spectra of solar prominences during a total solar eclipse in 1868. They found that these spectra consisted of hydrogen lines and a line at about 588 nm which did not appear to correlate with that of any known element. Lockyer hypothesised that the line was due to a new element that he called helium. This was shown to be correct when helium was discovered on Earth by William Ramsay in 1895.

Charles Young and William Harkness found a bright green emission line at about 532 nm in the spectrum of the Sun's corona during a total solar eclipse in 1869. This line was eventually attributed to another new element called coronium. In 1941, however, Bengt Edlén showed that the line was produced, not by a new element, but by highly ionized iron, implying a coronal temperature of over a million degrees K.

Spectra of other celestial bodies

Angelo Secchi produced a system of stellar classification in 1863 based on their spectra (see article *Spectral classification of stars*). Then in the following year William Huggins and William Miller identified hydrogen, sodium, iron, magnesium and calcium in stellar spectra. This showed, for the first time, that stellar atmospheres were composed of the same elements as those found on the Sun and Earth.

In 1864 Giovanni Donati made the first measurements of the chemical composition of a comet by observing the spectrum of 1864 II Tempel. Later observations of comets in the nineteenth century showed that their spectra were continuous when first found, indicating that they were scattering sunlight. But as the comets got closer to the Sun, first hydrocarbon bands appeared, then, when they got even closer, these bands were replaced by various emission lines. The exact hydrocarbon compounds did not become clear until molecular spectra became better understood in the first half of the twentieth century. Molecular carbon, C_2, was identified in the head of a comet just after the turn of the nineteenth century, and by the mid-1950s C_3, CH, CN, OH, NH and NH_2 had been found in the heads of comets.

Huggins examined the spectra of a number of nebulae in 1864 and found that they all had bright emission lines, indicating that they were all gaseous. The brightest emission line was green, which he later attributed to a new element

called nebulum (later 'nebulium'), as its wavelength did not appear to match that of any known element. But in 1928 Ira Bowen showed that the line was not due to a new element but was caused by a rare (forbidden) transition in ionised oxygen.

In the meantime, Huggins had continued observing the spectra of nebulae following the publication of his earlier work on the subject. He found in 1868, with his much larger sample, that although one-third were gaseous, two-thirds, including the Andromeda nebula, had a continuous spectrum, indicating that they consisted of stars. Vesto Slipher measured the spectrum of the Andromeda nebula in 1912 and discovered that it was not only rotating but was approaching us with a velocity of about 300 km/s. Over the next five years he measured the radial velocities of 25 spiral nebulae and found that 21 of them were moving away from us at very high speeds. This was the first indication of the expansion of the universe discovered in 1929 by Edwin Hubble.

The first nova to be studied spectroscopically was Nova T Coronae Borealis of 1866, which Huggins and Miller found had bright hydrogen emission lines. Ten years later Nova Cygni was initially found to have bright hydrogen emission lines, but some months later its spectrum had changed completely to resemble that of a planetary nebula with a single bright green emission line. In 1892 Nova T Aurigae was the first nova to have its spectrum photographed. It showed bright hydrogen, helium and sodium lines, together with a continuous spectrum and dark absorption lines. The Doppler shift of the lines enabled the velocity of its gases to be determined as being up to 600 km/s.

Unfortunately spectroscopic observations of the planets, particularly those of Venus and Mars, were difficult to interpret because of the presence of lines produced by the Earth's atmosphere. But in 1932 Walter Adams and Theodore Dunham found clear evidence of carbon dioxide in the atmosphere of Venus, whilst in 1947 Gerard Kuiper found carbon dioxide in the atmosphere of Mars.

Vesto Slipher undertook a detailed investigation of the spectra of the outer planets from Jupiter to Neptune in the early twentieth century. Although he recorded numerous bands, he had trouble interpreting them. In 1932 Rupert Wildt deduced that a number of the bands in all four planets were due to ammonia and methane. However, subsequent work by Mecke, Dunham, Adel and Slipher showed that some of the lines had been mis-attributed, so there was no ammonia in the atmospheres of Uranus and Neptune.

Theory

The first breakthrough in the theory of the formation of spectral lines came in 1885 when Johann Balmer discovered that the wavelength λ of the four visible lines in the spectrum of hydrogen were given by the formula:

$$\lambda = 364.56 \times \frac{n^2}{n^2 - 4} \text{ nanometers}$$

where $n = 3$, 4, 5 or 6. He predicted that lines would exist at wavelengths given by $n = 7$, etc. Balmer was then told by Eduard Hagenbach that Anders Ångström had already observed a line at 397 nm, consistent with $n = 7$.

Balmer simply presented his numerical series and did not try to explain why it worked. This was not done until 1913 when Niels Bohr outlined his quantum theory of the atom.

Instrumental developments

In 1801 Thomas Young used a simple diffraction grating ruled on glass to demonstrate the wave nature of light. The grating, which had about 500 grooves or lines per inch (200 lines/cm), enabled him to determine that the visible spectrum covered a wavelength range from about 425 to 675 nm.

Fraunhofer experimented with both a prism and a diffraction grating as the dispersing element of his spectroscope in the early nineteenth century. His first prism spectroscope consisted of a 60° prism mounted in front of a small theodolite telescope. Then in 1823 he placed a prism with an apex angle of 37° 40' in front of a 10 cm diameter telescope to produce an objective prism spectroscope. Fraunhofer's first diffraction grating consisted of fine parallel wires spaced at up to 325 wires/inch (130 wires/cm). Then in 1822 he produced a diffraction grating with a spacing of 3,300 lines/inch ruled by a diamond directly onto glass.

In the early 1820s John Draper found that diffraction gratings produced brighter spectra if they were used as reflection gratings and coated with a mercury-tin amalgam. Over the next forty years or so the design and manufacture of gratings improved substantially, so that by the 1870s the best gratings were being made by Lewis Rutherfurd. He produced gratings on polished speculum metal which had typically about 17,300 lines/inch (6,800 lines/cm).

Unlike prism spectroscopes, diffraction gratings can produce an almost linear wavelength scale. In 1868 Ångström had used a diffraction grating spectroscope to image about 1,200 lines in the solar spectrum. The unit of length that he used of 10^{-10} m, which was later called the Ångström unit, was soon accepted as the standard for spectroscopy. It was superseded as a standard by the nanometre (nm) of 10^{-9} m in the second half of the twentieth century. Ångström's best diffraction grating, which he used in transmission, consisted of about 5,000 lines/inch (2,000 lines/cm) ruled on glass.

None of Fraunhofer's spectroscopes used a collimator, relying instead on the slit being some distance from the prism. But in 1839 Jacques Babinet used a collimator in order to achieve a higher resolving power. By about 1860 collimators were a common feature of spectroscopes. Then in 1860 Donati used a cylindrical lens, instead of a slit, to broaden the spectrum and make it brighter, so making it easier to view. He placed this spectroscope near the focus of his telescope objective and used that arrangement to observe stellar spectra.

Hippolyte Fizeau and Léon Foucault had used up to five prisms in their spectroscope in 1848 to increase its resolving power, although this reduced the brightness of the spectrum. Then in 1872 Henry Draper replaced the standard glass prism of his spectrograph with a quartz prism, which transmitted more ultraviolet light. He used this new arrangement with the wet collodion process to produce the first photograph of a stellar spectrum, that of Vega. Four years later Huggins used the new faster dry plates to photograph stellar spectra using quartz lenses and an Iceland spar prism. He trailed the star along the slit to broaden its spectrum, instead of using a cylindrical lens. With this set-up he recorded, for Vega, seven strong lines of hydrogen, five of them in the ultraviolet.

Angelo Secchi had fitted a 12° prism to the front of his objective lens in the 1860s in order to produce bright spectra, albeit of relatively low dispersion. He used this objective prism spectroscope visually. Then in 1885 Edward Pickering placed a large shallow-angle prism in front of his telescope objective to enable stellar spectra to be photographed in quantity. No slits were required as the movement of the stars, with the telescope's drive disconnected, produced widths to the spectra. This system allowed Pickering to photograph and publish the spectra of over 10,000 stars in the *Draper Memorial Catalogue* in 1890. The spectra were about an order of magnitude brighter than with a standard spectrograph, although their dispersions were about an order of magnitude worse. These low dispersions were ideal for stellar classification of large numbers of stars, but not for studying their detailed compositions.

Henry Rowland, on the other hand, was more interested in producing highly detailed spectra. As a result he replaced the standard planar diffraction grating with a concave grating in his spectrograph to greatly increase its dispersion and resolution. No lens was required to focus the spectrum, which meant that it extended further into the ultraviolet, which was normally absorbed by the lens. With this new device Rowland was able in 1888 to photograph 20,000 lines in the solar spectrum in a band 12 m long.

Rowland spent many years perfecting an engraving machine capable of cutting fine grooves of exactly equal spacing. As a result his gratings were the best of the nineteenth century. Not only were they highly precise, but they were up to 6 inches (15 cm) long with 14,400 lines/inch (5,700 lines/cm).

Even though diffraction gratings can produce an almost linear wavelength scale, the diffracted light is dispersed into several diffraction orders, producing relatively dim spectra. As a result diffraction gratings were generally only used for solar work in the nineteenth century. The first successful use for non-stellar work was by James Keeler who in 1890–1891 used the 36 inch (90 cm) Lick refractor and a high quality 14,400 lines/inch Rowland grating. With these outstanding instruments he was able to make accurate wavelength measurements, using the third or fourth orders of diffraction, of very bright stars and emission line nebulae.

Robert Wood and John Anderson pioneered the use of blazed gratings in the early twentieth century. These were able to concentrate much of the spectrum into one diffraction order by the use of carefully shaped diffraction grooves. The blazed grating was particularly useful for photography in the red and near infrared where prisms produced inadequate dispersion. In 1931 John Strong developed a vacuum-evaporation process for aluminising telescope mirrors. He also found that thick aluminium coatings deposited on glass allowed the production of blazed reflection gratings ruled with a sloping groove profile. They concentrated an unprecedented 75% of light into the first-order diffraction spectrum. After the Second World War high-precision blazed gratings became available that, when used with the faster camera objectives then available, produced appreciably brighter spectra.

George Harrison pointed out in 1949 that the resolving power of a blazed grating at a given wavelength depended only on the ruled width of the grating and the angles of illumination and observation, and not specifically on the number of ruled grooves. In addition, the angular dispersion at a given wavelength depended only on the angle of diffraction, and not on the groove spacing. As a result he suggested that what he called an échelle grating could be produced with a large blaze angle, with a groove density just high enough to avoid problems with overlapping diffraction orders. A cross-dispersion element, consisting of a low dispersion grating or prism, could be used to separate these diffraction orders which could run into hundreds. For example he suggested that a 10 inch (25 cm) échelle of just 100 grooves/inch (40 grooves/cm) could have a resolving power of 10^6 and a plate factor of 0.02 nm/mm when used with a lens or mirror of only 100 inch (250 cm) focal length.

In 1951 Keith Pierce, Robert McMath and Orren Mohler were the first to use an échelle for astronomical research. They used a 150×75 mm échelle at the McMath-Hulbert solar observatory with a resolving power of 2.5×10^5 and plate factor of 0.03 nm/mm. The first échelles to be used for stellar spectroscopy were those used by the Crimean Astrophysical Observatory and the Okayama Observatory in Japan in the mid 1960s.

Multi-object spectrographs
Objective prisms had been used by Edward Pickering in the 1880s for the mass production of stellar spectra for stellar classification purposes. They needed relatively short exposure times but had relatively poor dispersion. In 1939 objective prisms were first used with a Schmidt telescope allowing much shorter exposure times. In particular, a 4° prism used with the 24/36 inch (60/90 m) f3.5 Burrell Schmidt allowed spectra of twelfth magnitude stars to be recorded in only 30 minutes.

One of the problems associated with objective prism spectroscopy was the contamination of spectra by background sky illumination. In 1980 Harvey Butcher tried a new approach to producing multi-object spectra without using an objective prism. In this he placed a black anodised aperture plate, with holes in it, in the focal plane of the 4 m Mayall telescope. The positions of the holes corresponded to those of the objects to be studied, whilst the aperture plate eliminated the sky background except at the holes. Butcher then used a grism (consisting of a transmission grating on one surface of a prism) as the dispersing element, after the light had passed through the aperture plate, as a grism can be designed to disperse light in a straight-through optical path. It initially took some time to produce an aperture plate for each set of observations, but an automatic system was soon devised by astronomers using the Canada-France-Hawaii telescope. In that system a CCD image was used to provide image locations to an automatic machine that could produce holes within minutes of the image being produced.

John Hill and colleagues developed an alternative multi-object spectrograph, using the Steward Observatory's 2.3 m telescope, at about the same time as Butcher. They also used an aperture plate like Butcher, but they fixed optical fibres in the holes to transfer the light to a normal spectrograph which could be some distance away. This so-called Medusa system was not as efficient optically as Butcher's due to fibre losses and other optical problems, but it produced good redshift measurements. There were also problems with imperfect positioning of the fibres on the plate, but this was solved by using an automatic fibre positioning system without the aperture plate. This allowed the positions of each of the fibres to be optimised, using computer-controlled robotic arms, for maximum optical output. A prototype was tested on the Hale 200 inch (5 m) telescope in 1981 and a fully operational system, called MX, was in use on the Steward 2.3 m in 1986.

Bibliography

Clerke, Agnes M., *A Popular History of Astronomy during the Nineteenth Century*, 4th ed., Adams and Charles Black, 1902.

Gingerich, O., (ed.), *The General History of Astronomy, Vol. 4A: Astrophysics and Twentieth-Century Astronomy to 1950*, Cambridge University Press, 1984.

Hearnshaw, J. B., *The Analysis of Starlight; One Hundred and Fifty Years of Astronomical Spectroscopy*, Cambridge University Press, 1986.

Hearnshaw, John B., *Astronomical Spectrographs and their History*, Cambridge University Press, 2009.

Hill, John M., *The History of Multiobject Fiber Spectroscopy*, Astronomical Society of the Pacific Conference Series, 3 (1988), 77–92.

King, Henry C., *The History of the Telescope*, Charles Griffin, 1955 (Dover reprint 1979).

Various authors, *Journal of Astronomical History and Heritage*, 13, No. 2, July 2010, pp. 88–148.

See also: Spectral classification of stars.

PART 6
OPTICAL TELESCOPES AND OBSERVATORIES

1 · Overview – Optical telescopes and observatories

MODERN OPTICAL INTERFEROMETERS

SUSI (Sydney University Stellar Interferometer)

Robert Hanbury Brown designed an intensity interferometer (see *Optical interferometry* article) which was completed in 1963 at Narrabri Observatory in Australia. Although this 188 m system, with two 6.5 m diameter light collectors, was insensitive to atmospheric turbulence, it was usable only with stars brighter than 2.5 magnitude. As a result the University of Sydney, which operated the intensity interferometer, decided to produce a Michelson interferometer as a follow-on instrument. Initially a prototype interferometer was built with a relatively short baseline of 11.4 m to iron out a few problems. This was followed by the full-sized instrument, called the Sydney University Stellar Interferometer or SUSI, which was designed to undertake very high angular resolution studies of single stars and close binary systems.

SUSI was designed to operate with baselines ranging from 5 to 640 m, producing resolutions of from 20 to 0.07 mas (milliarcsec) at a wavelength of 450 nm. Only two apertures were used at any one time as it was not intended to produce images. The mirrors were limited to 20 cm in size, with a clear aperture of 14 cm, to reduce problems with atmospheric turbulence. Construction started in 1987 and the instrument was opened in 1991 with the 'blue' system operating in the wavelength range 400 to 540 nm. It was then modified to also allow operation with the 'red' system in the range 500 to 950 nm in 2002.

Berkeley ISI (Infrared Spatial Interferometer)

Mike Johnson, Al Betz and Charles Townes constructed a two-telescope heterodyne Michelson interferometer in 1974 using a CO_2 laser local oscillator. They used the two 81 cm McMath auxiliary telescopes on Kitt Peak with a baseline of 5.5 m operating in the mid-infrared at 10 μm. This was the first mid-infrared interferometer and the first Michelson interferometer to use two telescopes.

In 1988 Townes and colleagues started to use the new dedicated Infrared Spatial Heterodyne Interferometer (ISI) on Mount Wilson, which also operated in the mid-infrared. It initially consisted of two moveable 1.65 m infrared telescopes,

each with its own heterodyne system, with a baseline of 32 m. In 1999 the baseline was increased to 70 m, roughly doubling the resolving power of the system, with a third telescope being added a little later. It had a resolution of about 3 mas at 11 μm with the maximum telescope spacing of 70 m.

I2T (Interféromètre à 2 Télescopes) and later developments

Antoine Labeyrie was, in 1974, the first to observe stellar sources using a two telescope Michelson interferometer in the optical band. His two 26 cm telescopes at the Nice observatory in France were separated by 12 m. This I2T (Interféromètre à 2 Télescopes) system was replaced in 1986 by the GI2T (Grand I2T) which consisted of two 1.5 m telescopes on a baseline extending up to 65 m, operating in the visible and infrared wavebands. It was closed down in 2006.

Mount Wilson Marks I, II and III

Michael Shao and David Staelin designed a 1.5 m baseline stellar interferometer at Mount Wilson in 1979. It was the first interferometer to compensate for atmospheric turbulence by making continuous phase and amplitude measurements. It was succeeded by the Mark II 3.1 m interferometer from 1982 to 1984 which was used as a technology testbed for astrometric measurements. This, in turn, was followed by a Mark III from 1986 to 1992 which had a maximum baseline of 32 m and operated in two spectral channels from about 400 to 600 nm and from about 600 to 900 nm. Like the Mark II, the Mark III was designed to test out new techniques and technologies. Its aim was to be reliable and easy to operate, whilst being capable of extremely accurate astronomical measurements. Lessons learnt in operating the Mark III were applied to the design of the NPOI (see below).

COAST (Cambridge Optical Aperture Synthesis Telescope)

The COAST (Cambridge Optical Aperture Synthesis Telescope) interferometer was designed in the 1980s to provide very-high-resolution images (not just fringes) of a wide variety of stellar sources. These included the envelopes of pre-main sequence stars, circumstellar shells, compact planetary nebulae and close binary systems. It initially consisted of a coherent array of four 40 cm Cassegrain telescopes operating in the red and near infrared from 600 to 2,300 nm using Michelson interferometry. A fifth telescope was added in 1998 and the maximum baseline of the array eventually extended to 100 m to achieve a maximum resolution of 1 mas.

The first stellar fringes were obtained in 1991 and in 1995 it produced the first synthesis image ever produced from separated optical telescopes. This was of the spectroscopic binary Capella at 830 nm. It imaged Betelgeuse shortly afterwards, showing strong limb darkening, and in 1998 detected a periodic modulation in the size of R Leonis.

The COAST team used their experience to assist in designing the Magdalena Ridge Observatory Interferometer (MROI) which is currently (in 2010) being built at an altitude of 3,200 m in the Magdalena mountains of New Mexico. COAST is now mainly used as a testbed for subsystems and components intended for MROI and the VLTI interferometer of the European Southern Observatory (see below).

IOTA (Infrared/Optical Telescope Array)

IOTA (Infrared/Optical Telescope Array) was built by a consortium of American academic institutions in the early 1990s at the Smithsonian Astrophysical Observatory's Whipple Observatory on Mount Hopkins. Its aim was similar to that of COAST, namely to produce high resolution images of astronomical objects using an interferometer array. Scheduled operation started in 1995 with the first two 45 cm telescopes in place on an L-shaped track that allowed spacings of between 5 and 38 m. The interferometer was operational from 550 to 2,200 nm. A third telescope was added in 2000, but the interferometer was closed in 2006 for financial reasons.

PTI (Palomar Testbed Interferometer)

The Palomar Testbed Interferometer (PTI) was designed to act as a testbed for interferometric techniques applicable to the Keck (see below) and other interferometers. These techniques included high-sensitivity infrared interferometry and dual star interferometry with phase referencing for very narrow angle astrometry. It built on experience gained with the Mark III interferometer on Mount Wilson which had been used with active fringe tracking for, amongst other things, wide-angle astrometry in the visible waveband. The PTI was installed at the Palomar Observatory in 1995.

The PTI, which used three 40 cm telescopes, was operated as a two-element interferometer in the near infrared, combining light from two of its three siderostats at any one time. The maximum separation was 110 m to give an absolute accuracy of 1 mas. But compared to nearby stars it could achieve relative accuracies of about 20 microarcsec.

PTI measured the orbits of binary stars and in 2001 measured the distortion of the 3 mas diameter, A-type star, Altair, due to its rapid rotation. This showed that its equatorial diameter was about 14% greater than its polar diameter. The interferometer was closed in 2008.

NPOI (Navy Prototype Optical Interferometer)

The U.S. Naval Observatory's NPOI (Navy Prototype Optical Interferometer) is located at a height of 2,200 m at Anderson Mesa near Flagstaff, Arizona. It was designed for astrometry and imaging, using the experience gained with the Mark III interferometer on Mount Wilson. NPOI became operational in 1996 when it produced its first stellar images (of Mizar A) with three siderostats available, followed a little later by its first observations of stellar limb darkening.

When completed in 2000 the imaging subarray consisted of six movable 50 cm siderostats feeding 12 cm apertures in a Y-shaped array allowing separations of up to 437 m. The astrometric subarray consisted of four fixed 50 cm siderostats feeding 12 cm apertures on the same Y-shaped system as used for imaging, with baselines from 19 to 38 m. Both subarrays used the same receiving system which covered 450 to 850 nm in 32 channels.

NPOI's primary goal was the creation of an astrometric catalogue of stars with a precision of about 2 mas to update the Hipparcos catalogue. This work is still in progress (in 2010). NPOI has also been able to image complex sources as faint as 7 magnitude with a resolution of 0.2 mas.

MIRA (Mitaka optical and InfraRed Array) −I.1 and −I.2 interferometers

The MIRA (Mitaka optical and InfraRed Array) −I.1 was a prototype interferometer built at the Mitaka campus of the National Astronomical Observatory of Japan. This was to enable Japanese astronomers to get used to the operation of a modern interferometer. It consisted of two small telescopes placed on a 4 m baseline, and was operational for five months from 1998 to 1999. It produced fringes for nine stars and allowed experiments to take place in stellar fringe tracking.

The MIRA−I.2 interferometer was then built consisting of two 30 cm siderostats on a 30 m baseline. The aim was to determine stellar diameters with an angular resolution of a few mas.

CHARA (Center for High Angular Resolution Astronomy) array

The CHARA Array on Mount Wilson is operated by Georgia State University's Center for High Angular Resolution Astronomy (CHARA). It is a Y-shaped array of six 1 m telescopes with separations of up to 330 m operating in the visible and near infrared. Construction was begun in 1996 and first fringes obtained with two telescopes three years later. It became fully operational in 2004.

In addition to operating as a scientific instrument, the CHARA Array was also used as a testbed for new technologies. As a result the Paris Observatory installed their FLUOR (Fiber Linked Unit for Optical Recombination) at CHARA. In addition, the University of Michigan, the Côte d'Azur Observatory and astronomers from SUSI in Australia also tested equipment at CHARA.

The CHARA Array produced the first image of a single main sequence star, Altair, in 2007, showing its distorted shape due to its rapid rotation. It also showed that it was hotter at the poles than at the equator. This image in the near infrared had a resolution of better than 1 mas. In the same year the CHARA Array was the first to measure the diameter of an exoplanet directly. The planet, which was orbiting HD 189733 in 2.2 days, was found to have a diameter of 0.38 mas, which equated to about 185,000 km (30% larger than Jupiter) at the star's known distance from Earth. The CHARA Array also provided the first detection of gravity darkening in a rapidly rotating star, Regulus, that is not a member of an eclipsing binary system. Then in 2010 the Array helped to unravel the configuration of ε Aurigae when it detected a dark cloud partially occulting the visible source.

Keck interferometer

The Keck interferometer operates at an altitude of 4,150 m on Mauna Kea, Hawaii. It was initially expected to consist of the two 9.8 m Keck telescopes and six 1.8 m Outrigger telescopes. The deployment of the outrigger telescopes was initially delayed because of environmental and other concerns, then in 2006 finance was withdrawn. So the interferometer currently consists of just the two 9.8 m telescopes separated by 85 m, providing an angular resolution of 5 mas at 2.2 μm and 24 mas at 10 μm. The lack of the outrigger telescopes means that the interferometer cannot be used for interferometric imaging.

The two telescopes of the Keck interferometer were first combined optically in March 2001, when engineering tests produced fringes for a number of stars. Scientific observations started in the following year using adaptive optics on the two telescopes. This enabled the interferometer to measure much dimmer objects than previously possible. The first object to be observed was the T Tauri object DG Tau, which the interferometer found was surrounded by a hot dust disc with an inner edge ∼0.15 AU from the star.

European Southern Observatory's VLTI (Very Large Telescope Interferometer)

The European Southern Observatory's Very Large Telescope Interferometer (VLTI) consists of the four VLT 8.2 m unit telescopes (UTs) and four 1.8 m auxiliary telescopes (ATs) in various combinations. They are located on the top of the 2,640 m Cerro Paranal in Chile's Atacama Desert. The UTs are in fixed locations, whereas the ATs can be located at 30 different locations. The system provides high sensitivity and milliarcsec resolution using groups of two or three telescopes with baselines of up to 200 m. The telescopes are linked together by a complex system of mirrors and delay lines underground.

The basic interferometry system was checked out at 2.2 μm using two 40 cm test telescopes in March 2001, measuring the diameter of α Hydrae as 9.3 ± 0.2 mas. Later that year two 8.2 m UTs were connected 102 m apart to measure the diameter of α Eridani to be 1.92 ± 0.05 mas. The first AT arrived at Paranal Observatory in January 2004, and the first interference fringes were observed between this and the second AT in February 2005.

Bibliography

Baldwin, John E., *Ground-Based Interferometry: The Past Decade and the One to Come*, in Conference Proceedings SPIE, **4838** (2003), 1–8.

Labeyrie, A., Lispon, S. G., and Nisenson, P., *An Introduction to Optical Stellar Interferometry*, Cambridge University Press, 2006.

Lawson, Peter R., *Optical Interferometry Comes of Age*, Sky and Telescope, May 2003, pp. 30–39.

Shao, M., and Colavita, M. M., *Long-Baseline Optical and Infrared Stellar Interferometry*, Annual Review of Astronomy and Astrophysics, **30** (1992), 457–498.

See also: Optical interferometry.

MODERN OPTICAL SOLAR OBSERVATORIES

The development of spectroscopy and the invention of photography in the nineteenth century provided new tools for the investigation of the Sun, resulting in a renewed interest in its observation. Alexandre-Edmond Becquerel photographed the solar spectrum in 1842, although it was not until later in the century that solar and other spectra could be interpreted. Léon Foucault and Hippolyte Fizeau photographed sunspots in 1845 and, six years later, Berkowski imaged the solar corona during a total solar eclipse. But in each of these cases observers were using equipment that was used for other astronomical purposes. There had been no attempt, up to that time, to produce regular, systematic observations of the Sun using equipment specially designed for that purpose. But that was due to change.

At John Herschel's suggestion, a routine daily photographic record of sunspots was made at Kew, near London, starting in 1858. These records were produced using a photoheliograph, designed by Warren De La Rue, which used a stopped-down 3.5 inch (9 cm) refracting telescope. This produced a projected image of the Sun about 4 inches in diameter.

In 1890 George Ellery Hale was the first to successfully make a spectroheliograph which produced an image of the Sun in the light of a single wavelength. But this spectroheliograph was too cumbersome to be fitted to the large telescopes at Chicago's Yerkes Observatory, where he became director in 1892. So, to mount such instruments, he and George Ritchey built a long focus horizontal reflecting telescope at Yerkes with a 30 inch (75 cm) coelostat mirror. In 1903 they built an improved version, called the Snow telescope, and two years later moved this telescope to Mount Wilson in California where the observing conditions were much better. Shortly afterwards Hale built another solar telescope on Mount Wilson, this time in a 60 ft (18 m) high tower to minimise the effect of ground-generated thermals. The 60 ft, which had a coelostat and 12 inch (30 cm) objective, was completed in 1908, followed four years later by a 150 ft (45 m) tower telescope. The 150 ft produced a 17 inch (43 cm) solar image which, although similar in size to that produced by the Snow telescope, was of much higher quality. These three solar telescopes, that is the Snow, 60 ft and 150 ft, were part of the first major solar observatory, the Mount Wilson Solar Observatory, to be built in the twentieth century.

McMath-Hulbert Solar Observatory

The McMath-Hulbert Solar Observatory was founded in 1930 at Lake Angelus, Michigan by Francis McMath, his son Robert, and Henry Hulbert. The observatory began as a private project, but in 1931 it was given to the University of Michigan by its founders. In the following year a new instrument, called a spectroheliokinematograph, was attached to a $10\frac{1}{2}$ inch (27 cm) refractor at the observatory to produce a motion picture of the Sun. Four years later the observatory was expanded by building a 50 ft (15 m) high solar tower and spectroheliograph. Then in 1940 the 75 ft (23 m) McGregor Solar Tower was constructed, and in 1955 a vacuum spectrograph became available. The university ended its support for the observatory in 1979. It has since been decommissioned.

McMath-Pierce Solar Telescope

Leo Goldberg suggested at a conference in 1953 that the National Science Foundation (NSF) should fund a national observatory in the south-western United States to carry out all types of observing, including solar work. As a result, the NSF appointed an advisory panel under the chairmanship of Robert McMath to examine the need and location for Goldberg's suggested observatory. This resulted, a few years later, in the foundation of the Kitt Peak National Observatory and the agreement to build a large solar telescope there.

The selected design of solar telescope consisted of an 80 inch (2.0 m) diameter heliostat mounted 100 ft (30 m) above the ground to minimize atmospheric distortion. This heliostat reflected sunlight down a hollow inclined shaft and underground tunnel onto an image-forming parabolic mirror 60 inches (1.5 m) in diameter some 500 ft (150 m) away. The light was then reflected back to a flat mirror which deflected the light into a room where the solar image could be viewed. This McMath Solar Telescope, which was dedicated in 1962, produced a 34 inch (84 cm) diameter solar image which allowed the study of individual solar granules. Its name was modified to the McMath-Pierce Solar Telescope in 1992, on the thirtieth anniversary of its dedication. This was to honour Keith Pierce who had been responsible for the development of the telescope.

A considerable amount of design effort had been spent in trying to stabilise the air inside the McMath-Pierce to minimise image distortion. Later designers of solar telescopes often solved this problem by evacuating the air from within their telescopes. But these vacuum systems required at least two optical windows which were generally found to be opaque to infrared wavelengths above about 2.5 μm. This did not matter at first as astronomical imagers were available only for the visible waveband. But when infrared imagers became available in the early 1990s, the McMath-Pierce was found to be the only large solar telescope with an imaging capability above this infrared cut-off.

The 60 cm Kitt Peak Vacuum Telescope (KPVT) was constructed at Kitt Peak in 1973 in response to the need for synoptic maps of solar magnetic activity. The telescope also provided valuable support in its first year or so to NASA's Skylab mission. The KPVT was decommissioned in 2002.

Big Bear Solar Observatory

Robert Leighton began an extensive series of site surveys in 1964 to find a suitable location for a Caltech (California Institute of Technology) solar telescope within reasonable distance of their campus at Pasadena, California. This site survey showed that, unlike during the night, all mountains had poor 'seeing' during the day. But almost all sizeable lakes had good seeing, and mountain lakes, in particular, combined good seeing with transparency. As a result it was decided to build the new solar observatory at Big Bear Lake 2,070 m (6,780 ft) above sea level in California's San Bernardino Mountains, as it had the best combination of seeing and transparency. The plan was to build the solar telescope, designed by Harold Zirin, on an island in the lake as the surrounding water would produce much lower thermals than if the observatory had been surrounded by land.

Construction work on the Big Bear Observatory started in 1968 by first building an artificial island and an access causeway on the north side of the lake. The observatory's initial suite of instruments consisted of four telescopes, all mounted in a 42 inch (1.06 m) tube. These were two 10 inch (25 cm) refractors, a 9 inch (23 cm) reflecting coronagraph, and a 16 inch (40 cm) Cassegrain reflector. Both the Cassegrain reflector and the reflecting coronagraph fed through a coudé system to a vertically-mounted spectrograph in a long vertical tube set into the observatory floor.

The suite of observing instruments was later changed to a 65 cm vacuum reflector, a 25 cm vacuum refractor and a 20 cm full disc telescope. In 2007 these were replaced by a new 1.6 m, open frame, clear aperture telescope, and the old dome was replaced by a larger, ventilated dome. An adaptive optics system was then installed, achieving first light in 2010. Two full disc telescopes were also installed nearby in an additional small dome.

Sacramento Peak Solar Observatory

Donald Menzel of Harvard College Observatory asked the U.S. Army Air Forces (after 18 September 1947 the U.S. Air Force) if they would fund a solar observatory to help to study the influence of the Sun on the Earth's ionosphere and atmosphere. At that time the Air Force was interested not only in the effect of the Sun on radio communications, but also on the atmosphere, as that was the environment through which guided missiles and supersonic aircraft would travel. As a result the Air Force gave a contract to the High Altitude Observatory in late 1947 to carry out a site survey for such an observatory. Marcus O'Day, of the Air Force, asked Menzel if a suitable location could be found close to the White Sands Proving Grounds or the Holloman Air Force Base. Although other sites were examined, it was concluded by April 1948 that the 9,200 ft (2,800 m) high Sacramento Peak, overlooking White Sands, was the most suitable. In the following month the Air Force gave a contract to Harvard to design and equip this solar observatory at Sacramento Peak.

A regular programme of solar research was started in 1949 on the site of the new observatory using small outdoor instruments. Then in the following year the first indoor facility, the Grain Bin Dome, was constructed using a small grain silo as the observatory shelter. It housed a 6 inch (15 cm) prominence telescope and a flare patrol camera. The Big Dome facility, now called the Evans Solar Facility, was built shortly afterwards. It contained a 16 inch (40 cm) coronagraph and spectrograph.

Frequent, regular images were produced by the flare patrol camera in the Grain Bin Dome starting in 1950. Then in 1963 these were replaced by more extensive observations from instruments located in the newly built Hilltop Dome. These included whole-Sun images taken every minute in both visible light and the monochromatic light of hydrogen alpha at 656.3 nm.

The Army Corps of Engineers began building the observatory structure of a radically new design of telescope in 1966. It consisted of a vertical tower telescope whose light path was evacuated down to a residual pressure of 0.8 millibar to eliminate image distortion caused by air turbulence. Sunlight entered the 136 ft (41 m) high tower at the top through a 30 inch

(76 cm) fused quartz entrance window. It was then reflected down the evacuated tube by two 44 inch (112 m) diameter fused quartz flat mirrors. One hundred and eighty-eight feet (57 m) below ground level it was reflected by a 64 inch (162 cm) spherical primary mirror back to ground level where it produced a 20 inch (50 cm) diameter image of the Sun. There the light left the evacuated tube through any one of five quartz windows located around the tube, where it entered one of a number of scientific instruments. In 1998 this vacuum tower solar telescope was renamed the Richard B. Dunn Solar Telescope after the name of the original project scientist.

Canary Islands solar telescopes

Solar astronomers from a number of European countries formed the Joint Organization for Solar Observations (JOSO) in 1969 to search for an ideal site for a solar observatory somewhere on the Mediterranean and Atlantic coasts. By 1974 JOSO had reduced the number of possible sites to three, namely Izana at an altitude of 2,380 m (7,250 ft) on the island of Tenerife in the Canary Islands, Roque de los Muchachos at an altitude of 2,300 m (7,600 ft) on La Palma also in the Canaries, and a site in Portugal. The site in Portugal was later eliminated, leaving the two sites in the Canaries as potential sites at the end of the observing campaign in 1979.

In the meantime Spain had built a 25 cm refractor for white light and hydrogen alpha heliography at the Observatorio del Teide at Izana on Tenerife in 1969 (see *Canary Islands optical observatories* article). A few years later, the German Fraunhofer Institute built a 40 cm evacuated Newtonian type solar tower telescope also at El Teide. Then in 1979 an agreement was signed between Spain, on the one hand, and the UK, Sweden and Denmark, on the other, to allow these three countries to build astronomical facilities at the Roque de los Muchachos on La Palma and at Izana on Tenerife. Four years later West Germany became a signatory to the agreement.

A number of solar instruments had already been transferred from Capri to La Palma by Swedish astronomers in 1978, in anticipation of relocating them at the Roque de los Muchachos Observatory. Then in 1985 Sweden upgraded their Roque de los Muchachos facility by building the Swedish Vacuum Solar Telescope (SVST) which used a 50 cm achromatic doublet as the entrance window of their vacuum system. At about the same time the Göttingen Observatory removed their 45 cm solar Gregory-Coudé Telescope (GCT) from their solar observatory at Locarno, Switzerland and rebuilt it at El Teide. (The GCT was dismantled in 2002 to make room for the new 1.5 m GREGOR telescope, which is, in 2010, in the process of being commissioned.) Then in 1987 the Kiepenheuer Institute (prior to 1978 called the Fraunhofer Institute) built a Vacuum Tower Telescope (VTT), also at El Teide, with a primary mirror of 70 cm diameter and a tower height of 33 m. An adaptive optics system was permanently installed on the VTT in 2003, producing an image resolution of about 0.2 arcseconds.

THEMIS (Télescope Héliographique pour l'Etude du Magnétisme et des Instabilités Solaires) was a French/Italian telescope built at El Teide Observatory in the 1990s. It was a 90 cm evacuated Ritchey-Chrétien telescope dedicated to the investigation of solar magnetism. A key element of its design was its ability to image the Sun in up to ten wavelengths simultaneously from about 400 to 1,100 nm. This allowed the surface magnetic fields of the Sun to be determined in three dimensions. THEMIS achieved first light in 1996.

The 45 cm Dutch Open Telescope was built at the Roque de los Muchachos Observatory on La Palma in 1997. As its name implies the telescope was used without any protective cover when observing the Sun. This reduced local thermal irregularities usually caused by a telescope's structure, whilst allowing a free flow of air across the mirror. Finally, the 50 cm Swedish Vacuum Solar Telescope on La Palma was removed in 2000 and replaced with the Swedish 1-m Solar Telescope (SST) which had a fused silica lens of 110 cm diameter and 98 cm clear aperture. The SST produced diffraction-limited images when it saw first light in 2002 with its adaptive optics.

BiSON and GONG networks

Five-minute vertical oscillations of the Sun's photosphere were discovered by Robert Leighton, Robert Noyes and George Simon at the Mount Wilson Observatory in 1961. It was recognised early on that it was essential to observe the Sun 24 hours/day to get a full understanding of these oscillations and avoid false periodicities. So the Birmingham (UK) Solar Oscillations Network (BiSON) was gradually built up from the late 1970s to measure shifts in a chosen solar Fraunhofer line averaged over the whole disc. The first prototype solar instrument, of what was to become the BiSON network, was taken to the Observatorio del Teide in the Canary Islands in 1975. This instrument was gradually developed there over the next few years, leading in 1978 to the discovery of global, acoustic solar oscillations of low degree. This instrument remained at El Teide, whilst improved instruments were located at Haleakala, Hawaii in 1981, Carnarvon, Australia (1985), Sutherland, South Africa (1990), Las Campanas, Chile (1991), and Narrabri, Australia (1992). The station at Haleakala was closed in 1991 and the instrument moved to Mount Wilson, California in the following year.

The BiSON network was followed by the GONG (Global Oscillation Network Group) network that was initiated by the American National Solar Observatory in 1984 and funded by the National Science Foundation. It was calculated that at least six solar telescopes, approximately equally spaced in longitude, were required to allow for weather problems and equipment breakdowns. So ten different sites were tested using a pyrhe-

liometer. In the event, the six chosen sites were at Mauna Loa, Hawaii; Big Bear, California; Cerro Tololo, Chile; El Teide, Canaries; Udaipur, India; and Learmonth, Australia.

The GONG network of six identical, automatic observing stations was inaugurated when its final station was completed in late 1995. These stations monitored the change in wavelength of a single solar absorption line all over the Sun using a Fourier tachometer that had a design based on a Michelson interferometer. The solar disc was initially imaged on to a 256 × 242 pixel detector, which in 2001 was changed to a 1024 × 1024 pixel device. It was initially planned to operate the GONG network for three years, but it was so successful that in 2003 the National Solar Observatory decided to operate it indefinitely.

Bibliography

Chaplin, William J., et al., *BiSON Performance*, Solar Physics, **168** (1996), 1–18.

Dunn, Richard B., *Sacramento Peak's New Solar Telescope*, Sky and Telescope, December 1969, pp. 368–375.

Edmondson, Frank K., *AURA and its US National Observatories*, Cambridge University Press, 1997.

Godoli, Giovanni, *The Joint Organization for Solar Observations (JOSO) and The Large Earth-Based Solar Telescope (LEST) Foundation*, Società Astronomica Italiana, Memorie, **57**, No. 4 (1986), 631–654.

Harvey, J.W., Kennedy, J.R., and Leibacher, J.W., *GONG: To See Inside Our Sun*, Sky and Telescope, November 1987, pp. 470–476.

Kennedy, James R., *GONG: Probing the Sun's Hidden Heart*, Sky and Telescope, October 1996, pp. 20–24.

Kiepenheuer, K. O., *European Site Survey for a Solar Observatory*, Sky and Telescope, August 1974, pp. 84–87.

Kloeppel, James E., *Realm of the Long Eyes: A Brief History of Kitt Peak National Observatory*, Univelt, 1983.

Liebowitz, Ruth Prelowski, *Donald Menzel and the Creation of the Sacramento Peak Observatory*, Journal for the History of Astronomy, **33**, Part 2, No. 111 (2002), 193–211.

Zirin, Harold, *The Big Bear Solar Observatory*, Sky and Telescope, April 1970, pp. 215–219.

REFLECTING TELESCOPES

Early reflecting telescopes

Concave and convex mirrors were known to both classical antiquity and Middle Age Europe, but little is known about their manufacture and use. William Bourne suggested, in his *Inventions and Devices* of 1578, the arrangement of a large lens and concave mirror which would have produced a telescopic effect. But it is thought unlikely that he could have built such a device because of the limitations of sixteenth-century lenses.

James Gregory, in his *Optica Promota* of 1663, proposed a telescope design which he stated would give an erect image, and which, unlike the refractors available at the time, would have little chromatic aberration and no spherical aberration. Gregory's design consisted of a parabolic primary mirror, with a hole through its centre, and a concave elliptical secondary which reflected the light received from the primary back through the central hole to the eyepiece. Unfortunately opticians were unable to make a satisfactory parabolic mirror at the time, so Gregory was unable to build such an instrument.

In 1668 Isaac Newton appears to have been the first person to build a reflecting telescope. Four years later he presented a similar instrument to the Royal Society. It consisted of a $1\frac{1}{2}$ inch (4 cm) spherical primary mirror which reflected light onto an optical flat inclined at 45° to the axis of the telescope tube. This optical flat, in turn, reflected the light from the primary to an eyepiece mounted through the side of the tube. Newton's spherical mirror was made of a speculum metal alloy of copper and tin, which was similar in brightness to silver, with arsenic added to make it even whiter and capable of taking a better polish. His 7 inch (18 cm) long telescope had a maximum magnification of about 40. Its small size disguised the spherical aberration inherent in its design.

Laurent Cassegrain suggested a new design of reflecting telescope in 1672. His design was similar to that of a Gregorian, but he replaced the concave elliptical secondary of the latter with a convex hyperbolic secondary. Also, in the Gregorian the secondary intercepted the light from the primary after it had been brought to a focus, whereas in the Cassegrain the secondary intercepted the light before it was brought to a focus. As a result the Cassegrain would have been shorter than a Gregorian of the same power, and so would have been easier to handle. Newton challenged Cassegrain to make such a telescope, which he was unable to do. In addition, Newton mercilessly criticised Cassegrain's design so much that it quickly passed into relative obscurity.

John Hadley, assisted by his brothers George and Henry, produced a much improved Newtonian design in 1721. Although its 6 inch (15 cm) diameter speculum metal mirror was cast as a spherical mirror, it was polished to an approximately parabolic shape with a focal length of 62 inches (1.6 m). The telescope was mounted onto a very good altazimuth mount of John Hadley's design. Its definition was found to be just as good as that of Huygens' 125 ft (38 m) refractor, whilst Hadley's telescope was substantially easier to use. On the other hand, Huygens' refractor did give brighter images. Hadley made a number of telescopes over time, mostly Newtonians, with a few small Gregorians and Cassegrains.

James Short began making production Gregorian telescopes in 1740 with apertures of up to 18 inches (45 cm) and focal lengths up to 12 ft (3.7 m). They were generally mounted on an altazimuth mount, although Short did occasionally use an equatorial. They had the advantage over Newtonians and Cassegrains in that the images were erect, and so could be used for terrestrial purposes.

William Herschel began making speculum metal telescope mirrors in 1773. He initially made a Gregorian telescope, but found that it was too difficult to keep the mirrors in alignment, and so he produced a Newtonian telescope instead. By 1778 he had produced a number of reflectors, of which the most successful was his 6.2 inch diameter, 7 ft (2.1 m) focus Newtonian, with which he discovered Uranus in March 1781. Two and a half years later he completed a 20 ft (6.1 m) Newtonian reflector with an 18.7 inch (47 cm) diameter mirror on an altazimuth mount. He later modified it to produce a front-view focus in what is now called the Herschelian configuration. To do this he tilted the primary mirror to redirect its focus to the edge of the tube where he placed an eyepiece, without the intervening optical flat. He found the resolution as good as for the Newtonian configuration, but the image was much brighter.

Meanwhile in 1785 Herschel had begun to design a 40 ft (12.2 m) long telescope with a 49 inch (1.2 m) diameter speculum mirror. He completed building the telescope (see Figure 6.1) in August 1789, and on his second night of observations discovered Saturn's sixth satellite, Enceladus. Three weeks later he discovered Saturn's seventh satellite, Mimas. Unfortunately not only did his 40 ft require at least two assistants to operate it, but the weather in England was not conducive to using it to its full potential. The mirror also sagged under its own weight, and took a long time to reach ambient air temperature, often misting over in the process. In addition, the mirror tarnished quickly because of its higher than normal copper content, so requiring frequent repolishing and refiguring. As a result, William Herschel carried out much of his work with the 20 ft. His son John later took this 20 ft to the Cape of Good Hope to survey the southern sky.

Tarnishing of speculum metal mirrors was a problem with all reflecting telescopes of the late eighteenth and early nineteenth centuries. Nevertheless, Herschel's large reflectors did enable him to observe nebulae, double stars and other objects which were not observable in refracting telescopes of the time owing to their limited apertures.

The nineteenth century

William Parsons, the third Earl of Rosse, continued William Herschel's work on nebulae in the mid-nineteenth century. In 1840 he completed a 30 ft (9.1 m), 36 inch (91 cm) aperture Newtonian telescope at Birr Castle in Ireland. Two years later he began work on a 72 inch (1.8 m) speculum metal mirror which he completed in 1845. The resulting 72 inch reflector was slung between two masonry walls each aligned along the

Figure 6.1. William Herschel's 40 ft (12 m) telescope which he built in the grounds of his house at Slough, near London, over the period 1785–1789. The cost of £4,000, which was a considerable figure at the time, was paid for by grants from King George III. (From *A Short History of Astronomy*, by Arthur Berry, 1898, Figure 82.)

meridian. The walls were both 72 ft (22 m) long and 56 ft (17 m) high, built 24 ft (7.3 m) apart. As a result the telescope, which was 56 ft long, was restricted in movement to about ±8° either side of the meridian. With this telescope Lord Rosse discovered that a number of nebulae were spiral in shape, and most, if not all of the brightest nebulae, appeared to be composed of stars.

William Parsons, like William Herschel before him, found that his largest telescope was too unwieldy, and the weather was generally too poor, for it to be used to its full capabilities. So in about 1875, after his death, his son remounted the 36 inch telescope on an equatorial mount. He found that it performed so well in comparison with the 72 inch that the latter was gradually allowed to fall into disrepair.

In the early 1840s James Nasmyth had built a radically new 20 inch (51 cm) diameter reflector. His design was based on a Cassegrain, but he added a 45° inclined flat mirror to intercept the light from the secondary mirror before it reached the

Cassegrain's normal eyepiece position. The inclined flat then directed the light to a hole in the wall of the telescope tube. It then went through the hollow elevation trunnion of his altazimuth mount to the eyepiece, where the image was observed at what is now called the Nasmyth focus. This meant that the eyepiece would always be in the same position, no matter what the elevation of the telescope. The whole telescope and observer's seat were mounted on a turntable, so the observer never had to move as he moved the telescope. Nasmyth completed this telescope in 1842.

Four years later William Lassell was the first to mount a large reflecting telescope on an equatorial mount. This was his 24 inch (61 cm) Newtonian with which he discovered Neptune's largest satellite, Triton. In 1852 he moved this telescope to Malta to take advantage of its better weather and clearer atmosphere. He later built a 48 inch (1.2 m) equatorial Newtonian on Malta. Its tube consisted of an open framework to allow free circulation of air around the mirror and so improve its thermal stability. Lassell made many important observations of the effect of atmospheric conditions on 'seeing', and crucially found that the effect of atmospheric turbulence was more evident in a large telescope than a small one.

Tarnishing of speculum metal mirrors had been a problem with reflectors. But in 1856 a Munich instrument maker, Karl Steinheil, produced a 4 inch (10 cm) diameter reflector with a silver-coated glass mirror, using a process invented by Justus von Liebig. Independently, in the following year Léon Foucault produced a 13 inch (33 cm) diameter reflector also with a silver-coated glass mirror. By 1862 Foucault had produced a reflector with a 31 inch (80 cm) diameter silver-coated mirror, which clearly showed that such a mirror was better than one made of speculum metal. Silver not only had a better reflectivity, but it was far easier, when tarnished, to restore its optical surface to its original condition. A silver coated mirror was also much lighter than a mirror of speculum metal.

The last large reflector to be built with a speculum mirror was the 48 inch (1.2 m) diameter Great Melbourne Telescope that was installed in Australia in 1869. For various reasons, including problems with its mirror, it proved to be a white elephant and never produced any consistent results.

In America, Henry Draper built a 28 inch (71 cm) Cassegrain silver-on-glass reflector which he completed in 1872. He used it to produce a number of excellent photographs which, together with a monograph published by the Smithsonian Institution, encouraged many amateur astronomers to build telescopes with silver-coated glass mirrors. In England, Andrew A. Common installed a 36 inch (90 cm), $f/5.8$ silver-coated mirror in his Newtonian reflector in 1879, which he used mainly as a photographic telescope. A few years later he used his telescope to produce a remarkable photograph of the Orion nebula. It showed details never previously observed either photographically or visually, earning him the Gold Medal of the Royal Astronomical Society. Common later sold this telescope to Edward Crossley who donated it to the Lick Observatory in 1895.

It had become clear in the early 1880s that the f ratio of an optical system, rather than its absolute aperture, was the key in determining photographic exposure of extended objects like comets and nebulae. As a result, in 1904 George Ritchey was able to produce better nebula photographs with his relatively small 24 inch (60 cm) $f/3.9$ reflector than with the 40 inch (102 cm) $f/18.6$ refractor at the University of Chicago's Yerkes observatory.

This was, in many ways, the beginning of the end of the superiority of large refractors in astronomy, as reflectors were perfectly achromatic, whereas refractors were not. In addition, it was easier to make a reflector of short focal length, and hence smaller f ratio, than refractors. With a carefully silvered mirror, reflectors lost very little light, whereas lenses could not transmit all the light incident on them. Reflectors could be made with glass of a lower quality than refractors, and the mirrors could be supported across their whole diameter to make distortion less of a problem with changing telescope orientation. Finally, with most of the weight at the bottom of the tube, the reflector was also easier to balance and, with shorter focal lengths, their observatories could be made smaller.

Mount Wilson and Palomar large reflectors

George Ellery Hale had purchased a glass blank from the St Gobain glassworks in France in 1895 to be used in the construction of a 60 inch (1.5 m) reflecting telescope. But the blank had to be put into storage shortly after it arrived in America as Hale could not raise the money for the telescope's construction and that of its observatory. It took some time for this money to be raised, but in 1904 Hale managed to persuade the Carnegie Institution to pay for the construction of the 60 inch reflector to be built at the new Mount Wilson Solar Observatory in southern California.

When the 60 inch was completed four years later, it was the most powerful telescope in the world. It also set the standard for future large reflectors allowing, as it did, operation in a number of different configurations. For example, it could be used as an $f/5$ Newtonian for photography and low-dispersion spectrography, as a Cassegrain for spectrography and for photography, and as a coudé for high-dispersion spectrography.

The next large reflector to be built at Mount Wilson was the 100 inch (2.5 m). It was mounted in a rectangular steel cradle on an English-type mounting to provide rigidity. Its smooth motion was obtained by relieving the bearings of both pedestals by a mercury flotation system. The mounting's main disadvantage, which was not considered to be too important at the time, was that it could not be used to observe circumpolar stars. This 100 inch telescope, which saw first light in November 1917, was a great success. It was the largest telescope in the world for the next thirty years.

By 1931 it became possible to deposit thin films of aluminium on glass in a high vacuum. These aluminium coatings were a big improvement over silver coatings for reflecting telescopes as they had a much higher reflectivity in the ultraviolet, and they lasted very much longer before needing recoating. By 1933 John Strong had managed to aluminize the secondary mirrors of the 60 and 100 inch reflectors, and by the end of 1935 he had also aluminized their primary mirrors.

The 60 inch and 100 inch telescopes made Mount Wilson the premier observatory in the world from 1908, when the 60 inch was completed, to 1948, when the 200 inch (5.1 m) was inaugurated on Palomar Mountain.

Hale had wanted to build an even larger telescope than the 100 inch when it was completed in 1917, but light pollution was becoming more and more of a problem at Mount Wilson.

So in 1934 Palomar Mountain, near Pasadena, was chosen as the site for this new 200 inch (5.1 m) telescope, following a survey of suitable sites in the south-western United States. In the meantime, General Electric had been contracted in 1928 to provide the mirror blank in fused quartz (pure silica), but their attempts with trial 60 inch discs were unsuccessful. As a result, the contract was cancelled and Corning was contracted to produce the 200 inch blank. Corning used Pyrex (borosilicate) which, although it had a somewhat larger expansion coefficient than fused quartz, it had a lower expansion coefficient than traditional materials and was thought to be quite adequate. A satisfactory blank, which was produced on the second attempt, was transported to Pasadena in 1936.

It was decided to produce a relatively fast $f/3.3$ primary mirror for the 200 inch, as this allowed relatively short exposures for nebula photography. It also enabled the tube to be made shorter, allowing the dome to be made smaller also. The mirror blank was lightened by using an hexagonal pattern of ribs on its back to provide stiffening. A special corrector lens designed by F. E. Ross was fitted to correct for off-axis coma expected at the Newtonian focus. The telescope was also designed to operate as an $f/16$ Cassegrain or $f/30$ coudé.

Grinding of the 200 inch mirror started in 1936, but the work on the telescope was interrupted by the Second World War. So the telescope was not dedicated until 1948, becoming operational the following year. This 200 inch Hale telescope remained the world's largest and most productive optical telescope for many years.

Schmidt telescopes

Photography over large fields was a problem with fast reflectors as they exhibited off-axis coma. In 1930 Bernhard Schmidt of the Hamburg Observatory produced a radical new design which solved the problem. It had a spherical, not parabolic, primary mirror and an aspherical glass plate to correct for its spherical aberration. The corrector plate was smaller than the primary mirror, and placed at the centre of curvature of the primary, thus eliminating coma. The main problem was that the focal plane was curved, so the photographic plate or film placed in that plane also had to be curved to keep the image in focus.

Schmidt made his aspherical glass corrector plate using a stress polishing technique in which he placed a thin optical flat over the end of a tube. The tube was then partially evacuated to distort the flat by a specified amount. The flat was then repolished to make it flat again on one face, and the air let back into the tube, causing the glass to take up a pre-designated shape.

Schmidt's first telescope to his radically new design had a 17 inch (45 cm) diameter $f/1.7$ mirror, and a 14 inch (35 cm) diameter corrector to produce a field of 16° diameter. Many small

Schmidts were then produced, but the first large Schmidt was installed at the Palomar Observatory in 1948. It had a 72 inch (1.8 m) $f/2.4$ primary mirror and a 48 inch (1.22 m) corrector plate, and used curved photographic plates covering an area of 6° × 6°. It was used for the National Geographic–Palomar Sky Survey carried out between 1949 to 1958 which photographed stars and nebulae down to magnitude 21. An almost exact copy of this Palomar or Oschin Schmidt, as it became called, was built in Australia in 1973 at the Siding Spring Observatory.

The change to altazimuth mounts

The 200 inch (5.1 m) Hale telescope's position as the largest optical telescope in the world was lost in 1976 to the 6.0 m (236 inch) Bolshoi Teleskop Alt-azimutalnyi (BTA) or Large Altazimuth Telescope. It was built at the Special Astrophysical Observatory near Zelenchukskaya in the Caucasus. The BTA was the first modern large telescope to have an altazimuth mount, rather than the equatorial of the previous large reflectors. Altazimuth mounts are much smaller and easier to design and build than equatorials. But it is much more difficult to track objects with altazimuth mounts, as they need synchronised movements about two axes, whereas equatorials need to move about just one axis. However by the 1970s computers could be used to control the synchronised movements about the two altazimuth axes. Unfortunately, the BTA's performance was poor because problems with its mirror were compounded by relatively poor seeing conditions at the observatory.

Meanwhile, whilst the BTA 6 m was being constructed, a number of new 4 m-class telescopes were being constructed in the West at a number of new mountain observatories, some of them in the Southern Hemisphere. All of these had equatorial mounts. The telescopes included the almost identical 4.0 m Mayall Telescope on Kitt Peak in Arizona, which was completed in 1973, the 3.9 m Anglo-Australian Telescope (AAT) at Siding Spring in Australia, completed in 1974, and the 4.0 m the Victor M. Blanco telescope at the Cerro Tololo Inter-American Observatory (CTIO) in Chile, also completed in 1974. These were followed by the 3.6 m Canada-France-Hawaii telescope and the 3.8 m United Kingdom Infra-Red Telescope (UKIRT), both completed on Mauna Kea, Hawaii in 1979.

The first large modern optical telescope in the West to use an altazimuth mount was the Multiple Mirror Telescope (MMT), which was completed at the Whipple Observatory on Mount Hopkins, Arizona in 1979. It consisted of six 1.8 m telescopes on a common mount that was designed to bring all six images to a common focus on its central axis. The next large optical telescope on an altazimuth mount was the 4.2 m William Herschel Telescope (WHT), which was completed on La Palma in the Canary Islands in 1987. Since then altazimuth mounts have become the mounts of choice for large optical reflectors.

Glass ceramic mirrors and other innovations

Glass ceramic materials that have essentially a zero coefficient of expansion became available in the 1960s. One of these, Cer-Vit, was put on the market by Owens-Illinois in 1967, and was used for a number of large astronomical mirrors, like the 4 m-class mirrors of the Anglo-Australian Telescope (AAT) and of the Victor M. Blanco telescope at the CTIO. But stability and other problems caused Cer-Vit to be withdrawn from the market in 1978. Designers of reflecting telescopes then tended to choose another glass ceramic, Zerodur, which was produced by the Schott company in Germany.

The AAT, which had become operational in 1974, was one of the first large, ground-based telescopes to take full advantage of the development of CCDs and computer control technology. The observer, situated in a normal office environment, was able to use a computer to automatically slew the telescope to the objects to be studied, automatically taking account of atmospheric refraction, aberration, telescope misalignment and flexure. This AAT also had an automatic focus control.

The MMT, which became operational five years after the AAT, had an even more innovative design. Not only was it the first large telescope to have a number of identical telescopes on one mount, but it was the first major telescope to use a form of active optics. It was planned to continuously monitor the flexure of the structures of the six telescopes using lasers, and correct them in real time to ensure that the six images coincided. Its altazimuth mounting was smaller than the equivalent equatorial, and as each of the telescopes had reasonably fast $f/2.7$ mirrors, the MMT was able to use a smaller 'dome' than normal. In fact the dome was more like a box which was designed to open wide at night to allow the free flow of air around the optics. This box, which also included the computers and control room, rotated with the telescope.

The MMT was able to act like a 4.4 m telescope at about 40% of the capital cost of a normal reflector. But it took quite a number of years to get the telescope to work correctly. In the process the laser system was abandoned and the six images digitally recorded separately. In 1987 it was decided to abandon the MMT concept and replace the six telescopes with a single 6.5 m diameter primary of lightweight construction in the same housing.

The European Southern Observatory's (ESO's) 3.6 m diameter, $f/2.2$ New Technology Telescope (NTT), which was completed at La Silla in Chile in 1989, was one of the first large telescopes to use a Zerodur primary mirror. It was only 24 cm thick. The NTT also used an active optics system to automatically measure and correct the shape of this primary, as well as the position of the secondary, whilst observations were being carried out. The telescope had two Nasmyth foci but, unlike most other reflectors of this size, it had no Cassegrain focus. The building housing the NTT was an improved version of the MMT's housing. It had an hexagonal structure that rotated with the telescope, having large openings and louvres, in front of and behind the telescope, to allow outside air to pass directly over the instrument. This design enabled the telescope to cool down to the ambient air temperature relatively quickly, minimising thermal problems. It was also possible to undertake remote observing from ESO headquarters in Garching, West Germany, using CCD images from the telescope and a satellite link that had been inaugurated in 1987. In addition to normal observing, the NTT was used as a testbed for new technologies to be used in the ESO Very Large Telescope (VLT).

The Apache Point 3.5 m telescope, at $f/1.75$, had an even faster mirror than the NTT. The mirror was the first of two spun-cast by the Steward Observatory Mirror Laboratory (see below) in 1988. This altazimuth telescope was built at an elevation of 2,800 m (9,200 ft) in the Sacramento Mountains of New Mexico. Completed in 1994, it was designed to provide rapid access to a variety of instruments permanently located at different telescope ports accessed by adjustment of a tertiary mirror. Almost all of the observing was undertaken remotely, although decisions about the opening and closing of the dome were taken on-site. Slewing, instrument configuration and exposure management could also be undertaken remotely.

Adaptive optics

In 1983 Robert Smithson and colleagues built a system of adaptive optics using an array of nineteen small lenses to detect distortions of the image wavefront. A deformable mirror with nineteen hexagonal segments then corrected these distortions. Smithson and colleagues successfully tested the system by imaging sunspots and solar granulation.

Gerard Rousset, Fritz Merkle and colleagues produced an adaptive optics system a few years later as a prototype for the ESO's VLT. It had a deformable mirror with a continuous reflecting surface instead of Smithson's discrete segments. The new system was tested in 1989 on the 1.5 m telescope at the Haute-Provence Observatory in France, where it produced the first diffraction-limited images ever made of a faint astronomical object. It resolved a 0.5″ binary star system at infrared wavelengths, in seeing that varied between 1″ and 2″.

Early adaptive optics systems used a computer to calculate the magnitude of the signals to be applied to the deformable mirror. But in the late 1980s François Roddier proposed using a new type of wavefront detector that was linked directly to a different type of deformable mirror, with no intervening computer. In 1995 a prototype was fitted to the 3.6 m

Canada-France-Hawaii telescope, achieving a resolution of 0.15″ in the near infrared.

Adaptive optics systems generally used a reasonably bright star, near to the astronomical object of interest, in order to detect atmospheric wavefront distortions. But it was often impossible to find a suitable nearby star. So Renaud Foy and Antoine Labeyrie suggested that it may be possible to produce an artificial star adjacent to the target astronomical object using a laser. In 1991, when military work on adaptive optics was declassified, it was found that the military had already produced adaptive optics systems to observe military targets using laser guide 'stars'. Subsequently astronomical observations were successfully made with laser guide stars, which resulted in a number of observatories being fitted with more permanent laser guide star systems.

Spun-cast mirrors

As long ago as 1908 R. W. Wood had built a telescope with a rotating container of liquid mercury as its primary mirror, as the rotating liquid naturally formed a parabolic surface. But the telescope suffered from a number of problems, and was limited to looking more or less directly overhead. In the 1980s Roger Angel, of Arizona's Steward Observatory Mirror Laboratory, decided to use the same principle to produce a parabolic mirror blank by rotating molten glass and continuing to rotate it until it set. This had the big advantage over a normal blank as it required much less glass to be ground out to produce the final mirror. In addition, the curvature of the mirror, and hence its f ratio, could be adjusted just by changing the rotational speed of the glass whilst it was still molten.

Angel and John Hill used ordinary borosilicate glass as not only was it cheap, but it had a much lower melting temperature than the ultra-low expansion, glass ceramic alternatives. By 1985 they had produced a 1.8 m, $f/1.0$ mirror blank for the Vatican Advanced Technology Telescope (VATT). Three years later they produced two 3.5m, $f/1.75$ blanks, one for the Astrophysical Research Consortium's (ARC's) telescope at Apache Point, and the other for the Wisconsin, Indiana, Yale and NOAO (National Optical Astronomy Observatory) (WIYN) telescope on Kitt Peak.

The U.S. Air Force contracted the Steward Observatory Mirror Laboratory to develop the Mirror Lab's stress-lapping polishing technique for producing the steep curves for very fast mirrors, and to produce another 3.5 m mirror blank. The first operational mirror to be polished using this technique was the 1.8 m, $f/1.0$ VATT, followed by the Air Force's 3.5 m, $f/1.5$ mirror. Since then the Mirror Lab has produced a 6.5m, $f/1.25$ blank for the MMT, two similar ones for the Magellan project at Las Campanas Observatory in Chile, and two 8.4 m, $f/1.14$ blanks for the Large Binocular Telescope on Mount Graham. In 2005 it also spun-cast the first of seven 8.4 m blanks for the

Giant Magellan Telescope due to be built at the Las Campanas Observatory over the next decade.

Segmented mirrors and the Kecks

The first segmented primary mirror was built as long ago as 1828 by William Parsons (third Earl of Rosse). His 6 inch (15 cm) mirror consisted of a central disc, which could be axially adjusted, surrounded by a torus 1.5 inches (4 cm) wide. Just over a century later Guido Horn d'Arturo, of the University of Bologna, also undertook research into producing segmented mirrors. By 1953 he had completed a 1.8 m diameter primary mirror consisting of 61 hexagonal mirrors with a focal length of 10.4 m. This segmented primary mirror was fixed on a marble slab and used as part of a zenith telescope.

In 1977 Jerry Nelson of the University of California suggested that it should be possible to build a 10 m primary mirror using hexagonal segments. He proposed using very thin segments aligned using active optics, which would not only reduce the mass of the mirror, but would also produce a mirror that could respond more rapidly to changes in its thermal environment. This mirror should be easier to build than one large mirror, and it should be easier to produce one with a very short focal length.

The Keck foundation agreed in 1985 to provide $70 million of the estimated $90 million to construct the 9.8 m telescope (and observatory) which would operate in both the visible and infrared. In the event its $f/1.75$ primary mirror consisted of 36 hexagonal segments, each 7.5 cm (3 inches) thick. Mauna Kea was selected as the site for the new telescope as its very high altitude of 4,150 m (13,600 ft) was ideal for infrared observations.

The Keck Foundation announced in 1991 that it would provide a substantial grant towards the construction of a second 9.8 m telescope, 85 m away from the first Keck, Keck I. The images from both telescopes would be combined optically in an interferometric arrangement. Keck I began scientific observations in 1993 and Keck II three years later, with first fringes of the Keck interferometer in 2001.

Both Keck telescopes have now been fitted with an adaptive optics system which consisted of a 6 inch (15 cm) deformable mirror that changed shape at a frequency of up to 670 Hz.

Metal mirrors

André Couder and Dmitri Maksutov independently investigated the possible use of metal mirrors in reflecting telescopes in the 1930s. Couder showed that metal was theoretically superior to glass because of the former's higher thermal conductivity, whilst Maksutov went so far as to produce a number of metal mirrors, including a stainless steel mirror for a 0.7 m telescope. Maksutov investigated the use of both solid and ribbed metal

Figure 6.2. The Very Large Telescope of the European Southern Observatory situated at an altitude of 2,640 m (8,670 ft) on Cerro Paranal in the Atacama desert. The enclosures for the four 8.2 m telescopes dwarf those for the two 1.8 m Auxiliary Telescopes which were present when this image was taken in 2005. (Courtesy ESO/G. Hüdepohl.)

mirrors, and suggested that copper, bronze and aluminium could be used with chromium coats, or stainless steel could be used directly, all with reflective coats of aluminium.

Little work was done on modern metal mirrors until the 1960s when 36″ (0.9 m) and 50″ (1.3 m) telescopes were installed on Kitt Peak with aluminium primaries. The performance of the 50″ mirror deteriorated rapidly, however, causing it to be eventually replaced by a Cer-Vit mirror. But in 1969 a 1.37 m telescope, with a solid aluminium primary, went into operation in Merate, Italy. Encouraged by its good long-term performance, ESO considered using a nickel-coated aluminium primary for its 3.6 m NTT, but were forced to reject it for timescale reasons. ESO did for a time consider using aluminium as a back-up to glass for the VLT's 8.2 m primary mirrors, but eventually settled on glass. However they did decide to use a nickel-coated beryllium secondary for the VLT, rather than glass, because of its lower mass. With a diameter of 1.1 m the beryllium mirror weighed just 50 kg.

The Very Large Telescope

The European Southern Observatory (ESO) received financial approval for their Very Large Telecope (VLT) project in 1987 which was to consist of four 8.2 m telescopes (see Figure 6.2). These were designed so they could operate individually, or together as an interferometer. ESO planned on using their NTT as a prototype to test out various detailed design solutions for the VLT. For example, the active optics system used on the

NTT was crucial to the VLT with its very thin, large Zerodur mirrors.

In 1990 ESO chose to build the VLT at Cerro Paranal at a height of 2,640 m in Chile's Atacama desert. Two years later Schott spun-cast the first successful 8.2 m diameter, 17.5 cm thick mirror blank. The first of the four VLT telescopes saw first light in 1998. Three years later an adaptive optics system was fitted to one of the 8.2 m telescopes, achieving a resolution of 0.04″ in the infrared.

Meanwhile, a contract had been signed in 1998 for the provision of four 1.8 m auxiliary telescopes (ATs). These were designed to operate with the four 8.2 m unit telescopes (UTs) in various combinations to provide interferometric baselines of up to 200 m. First light with two UTs connected together was achieved in 2001. The first AT arrived at Paranal Observatory in January 2004, and the first interference fringes were observed between this and the second AT about a year later.

Bibliography

Gingerich, O. (ed.,), *The General History of Astronomy Volume 4A; Astrophysics and Twentieth-Century Astronomy to 1950: Part A*, Cambridge University Press, 1984.

King, Henry C., *The History of the Telescope*, Charles Griffin, 1955 (Dover reprint 1979).

Learner, Richard, *The Legacy of the 200-Inch*, Sky and Telescope, April 1986, pp. 349–353.

McCray, W. Patrick, *Giant Telescopes: Astronomical Ambition and the Promise of Technology*, Harvard University Press, 2004.

Osterbrock, Donald E., *Pauper and Prince: Ritchey, Hale and Big American Telescopes*, University of Arizona Press, 1993.

Watson, Fred, *Stargazer: The Life and Times of the Telescope*, Da Capo Press, 2004.

Wilson, R. N., *Reflecting Optics II: Manufacture, Testing, Alignment, Modern Techniques*, Springer-Verlag, 1999.

Wright, Helen, *Explorer of the Universe: A Biography of George Ellery Hale*, American Institute of Physics Press, 1994.

Zirker, J. B., *An Acre of Glass: A History and Forecast of the Telescope*, Johns Hopkins University Press, 2005.

See also: Adaptive optics; Modern optical interferometers.

REFRACTING TELESCOPES

Convex lenses had been used in spectacles to correct for long-sightedness (hypermetropia) since the late thirteenth century, and concave lenses had been used to correct for short-sightedness (myopia) from about two centuries later. But the idea of putting two lenses together to form a telescope appears to date from later still.

Roger Bacon is known to have experimented with lenses in the thirteenth century, and some think that he invented the telescope, although the written evidence for this is very ambiguous. Likewise, cases have been submitted for Leonard Digges, John Dee and William Bourne, all of whom were English, and Giambattista della Porta of Naples as the inventor of the telescope in the sixteenth century, but again the evidence in each case is ambiguous. In fact, the first clear reference to a telescope is in 1608 when Hans Lipperhey or Lippershey, a spectacle maker from Middelburg in the Netherlands, arrived in the Hague with a telescope and submitted a petition on 2 October to the States-General asking for a patent for his invention, which produced a magnification of three.

Two weeks later, Jacob Metius (also known as Jacob Adriaanzoon) of Alkmaar also arrived in the Hague with a telescope and a petition to the States-General asking for a patent. At about the same time the Councillors of Zeeland wrote to the States-General saying that there was another person in Middelburg who had produced a similar instrument to that of Lipperhey. As a result the States-General dismissed Lipperhey's patent application on the grounds that the instrument was too easy to copy and impossible to determine who had invented it.

According to Simon Marius a telescope had been available for sale at the Frankfurt fair in the autumn of 1608, and by the spring of the following year telescopes were also for sale in Paris and Milan. Galileo Galilei first heard of telescopes when visiting Venice in May 1609, and he made one with a magnification of about three diameters immediately on his return to Padua. In July 1609 Thomas Harriot in England produced the first telescope-aided drawing of the Moon, using a telescope with a magnification of about six.

Galileo experimented with the design of his telescope, and made a new instrument with a magnification of eight which he presented to the Venetian senate in August 1609. He then returned to Padua where he tried to improve his design yet again, finally producing an instrument 49 inches (1.25 m) long and of $1\frac{3}{4}$ inch (4.5 cm) aperture that had a magnification of about 30, using lenses that he ground himself.

These early telescopes of Lipperhey, Galileo and others, which were generally made with poor-quality spectacle lenses, consisted of a weak convex objective and a strong concave eyepiece. The lenses were mounted in a metal, cardboard or paper tube only about one inch (2.5 cm) or so in diameter. These telescopes produced an upright image, although they had a very low magnification and a very small field of view.

In 1611 Johannes Kepler proposed an alternative design of telescope in his book *Dioptrice* in which the eyepiece was convex instead of concave. This produced a wider field of view, but at the expense of producing an inverted image which, whilst not being a problem for astronomical work, was a problem for terrestrial use. He suggested that the addition of another convex lens between the objective and the eyepiece would enable the telescope to produce an erect image. But this would have seriously compromised the quality of the image because of the poor quality of the lenses available at the time. Kepler never made such a two- or three-lens telescope, as far as we know. In fact, the first such Keplerian instrument appears to have been made by Christoph Scheiner at Ingolstadt in about 1617.

Anton de Rheita gave details of a four-lens telescope in 1645 that he had been using for astronomical observations. Made in Augsburg by Johannes Wiesel the fourth lens, now called the field lens, was a second lens in the eyepiece placed slightly nearer the objective than the eye lens itself. It increased the telescope's field of view over both the Galilean and Keplerian designs.

The refracting telescopes of the first half of the seventeenth century suffered from both chromatic and spherical aberrations. It was found that both aberrations could be reduced if the surface curvature of the objective lens was made shallow and its focal length long. As a result astronomers built longer and longer telescopes. For example, Christiaan Huygens and his brother Constantyn built a 23 ft (7 m) long telescope of 2.3 inch (5.8 cm) diameter in 1655 that had a magnification of about 100. Some years later Christiaan also found that an eyepiece consisting of two plano-convex lenses, with their flat surfaces facing the eye, produced a better quality image and a larger field of view.

Johannes Hevelius from Danzig built a massive 150 ft (45 m) long telescope in about 1670 that was suspended by a system of ropes and pulleys from a 90 ft (27 m) high mast. Instead of an enclosed tube the objective and eyepiece were connected by long planks of wood joined together to make an L-shape cross section. Although this and similar extremely long telescopes were highly unstable and needed a number of assistants to operate, they were able to produce some useful astronomical observations, but only on dark, windless nights.

One problem with such very long telescopes was the flexibility of their very long tubes or planks, which were used to support the lenses, and the 'give' in the supporting ropes. Huygens sidestepped this problem with his longest telescopes by completely eliminating the tube linking the objective to the eyepiece. Instead he mounted his objective, one of which had a focal length of 210 ft (64 m), in a short metal tube and connected it to a high mast where it could move up and down in a groove. The only physical link between the objective and the eyepiece, in this so-called aerial telescope, was a piece of string which, when taut, was used to approximately align the lenses.

Robert Hooke suggested in 1668 that the light path from a long-focus objective could be folded using plane mirrors between the objective and the eyepiece, so substantially reducing the size of the telescope. In 1682 Boffat of Toulouse suggested fixing the telescope on the ground and using a plane mirror to feed light into it. But the quality of the plane mirrors available at the time was not adequate for such applications.

An English barrister, Chester Moor Hall, largely solved the problem of chromatic aberration in 1733 by making the objective of two lenses. He used a concave element made of flint glass and a convex one of crown glass. On receipt of these elements from the manufacturer, Hall satisfied himself that the resulting composite objective worked. But that is as far as he went. He did not try to patent his design and very few people knew of it. The quality of flint glass available at the time was very poor, however, so the composite objective did not have such a clear advantage over a standard objective as it would have later on.

The optician John Dolland of London also tried to produce an achromatic objective in the 1750s when he stumbled across Hall's work by accident. Following various experiments he successfully made an achromatic doublet in 1758. Five years later his son, Peter, produced an achromatic triplet in which the third element not only reduced the residual chromatic aberration, but also reduced the spherical aberration. His largest telescope with a triple objective had a focal length of 42 inches (1.06 m), an aperture of 3.75 inches (9.5 cm), and a magnification of 150. It was limited to such a relatively small aperture because of the poor quality of the flint glass available This lens apparently produced the best quality image available at the time, being extremely bright with no evident chromatic aberration.

Towards the end of the eighteenth century Pierre Guinand, a Swiss cabinet maker, undertook research into the manufacture of flint glass lens blanks, which were notoriously poor at the time. He examined their constituents and their method of manufacture, which he modified. As a result by 1805 he was able to produce almost flawless blanks which enormously improved the performance of flint and composite flint lenses.

Guinand moved to Benediktbeuern near Munich in 1806 where he was joined by the young Joseph von Fraunhofer. There Fraunhofer improved the furnaces and annealing ovens. He was able to measure the refractive indices of various glasses more accurately than had been possible previously by observing the deviation of spectral lines. His achromatic doublets were also more accurately corrected for both chromatic and spherical aberration than those of the Dollands.

Fraunhofer subsequently produced larger and larger telescope objectives until, in 1819, he made a 9.5 inch (24 cm) diameter objective lens for the observatory at Dorpat in Russia (now Tartu, Estonia). The telescope, which he mounted on a new design of clock-driven equatorial mount, was the largest refractor in the world when it was completed in 1824. Guinand had left Munich in 1814 following disagreements with Fraunhofer. So George Merz, who had worked with Fraunhofer, continued Fraunhofer's work when he died prematurely in 1826. Merz and Franz Joseph Mahler purchased the business in 1839. Mahler died in 1845 but the business, which stayed in the Merz family until 1903, continued to produce high-quality optical work.

The main requirement for telescopes in the early nineteenth century had been to measure the position of stars as accurately as possible and, in the case of double stars, to measure their separation. As a result there was more of an emphasis on the accuracy of telescope mounts than the size of objectives. This to some extent was due to the problem of building large objectives because of limitations in the quality of glass. But by the 1830s glass and optical technology had improved enough for Merz and Mahler to build a 15 inch (38 cm) refractor for the Pulkovo observatory near St Petersburg which they completed in 1839.

Pierre Guinand had died in 1824, but his son Henri moved to Paris where he set up a glass works a few years later at Choisy-le-Roi on the outskirts that became the centre of high-quality optical glass making. During the revolution of 1848 George Bontemps, who worked with Henri Guinand, and others left France and helped Chance Brothers in the UK to improve their own glass making. Guinand's glass works and Chance Brothers were to supply glass blanks for many of the world's largest telescopes in the second half of the nineteenth century.

George Bond had realised in 1849 that the visual and photographic foci of a lens differed because of the different spectral sensitivities of the eye and of photographic emulsions available at the time. Fourteen years later Lewis Rutherfurd made the first refracting telescope specifically designed for photographic work. Subsequently some observatories ordered pairs of telescopes, one for visual and one for photographic work, on a common mount. Other observatories ordered two different objectives that could be interchanged, one corrected for the visual and the other for the photographic wavebands.

In 1861 the American telescope manufacturer Alvan Clark and Sons had started to build an 18.5 inch (47 cm) refractor for the University of Mississippi with glass blanks provided by Chance Brothers. But the American Civil War intervened and the university was unable to pay for the blanks. However, in the meantime, Alvan Graham Clark and his father Alvan had made this 18.5 inch telescope famous in 1862 when they were the first to observe the previously unseen companion of Sirius whilst testing it. In the end the telescope, the largest refractor in the world at the time, was installed in the Dearborn Observatory of the Chicago Astronomical Society in 1866. Chance Brothers also supplied glass blanks to Thomas Cook whose largest telescope, the 25 inch (64 cm) Newall refractor, became the largest refractor in the world when it was finished in 1869.

Henri Guinand's glass works near Paris were, in the 1880s, run by Charles Feil (Pierre Guinand's great-grandson), Mantois and Parra. This factory provided the glass blanks for the 27 inch (69 cm) Vienna telescope built by Howard Grubb of Dublin, which was completed in 1883. It also supplied blanks for the 36 inch (91 cm) Lick telescope built by Alvan Clark and Sons and completed in 1888. This Lick telescope, with an equatorial mount by Warner and Swasey, was built at the first mountaintop observatory in the United States at an altitude of 1,300 m (4,250 ft) on Mount Hamilton. As the lenses got to this size, however, it became more and more difficult to produce acceptable glass blanks without small air bubbles and other flaws. For example it took twenty attempts to produce a satisfactory crown glass blank for this Lick 36 inch.

Mantois, another of Guinand's successors, provided the glass blanks to Alvan Clark for the 40 inch (102 cm) Yerkes refractor. There was a trade-off in designing this lens between making the edge of the lens thick enough to avoid the lens sagging under its own weight, whilst minimising the light loss by not making the lens too thick. This 40 inch Yerkes lens had reached the limit of existing technology. When completed in 1897 it was, and still is, the largest fully steerable refractor in the world.

Bibliography

Bell, Louis, *The Telescope*, McGraw-Hill, 1922 (Dover reprint 1981).

King, Henry C., *The History of the Telescope*, Charles Griffin, 1955 (Dover reprint 1979).

Reeves, Eileen, *Galileo's Glassworks, the Telescope and the Mirror*, Harvard University Press, 2008.

Van Helden, Albert, *The Invention of the Telescope*, Transactions of the American Philosophical Society, **67**, 4 (1977), 5–67.

Van Helden, Albert, *The Telescope in the Seventeenth Century*, Isis, **65** (1974), 38–58.

Warner, Deborah Jean, and Ariail, Robert B., *Alvan Clark & Sons: Artists in Optics*, 2nd ed., Willmann-Bell, 1996.

Watson, Fred, *Stargazer: The Life and Times of the Telescope*, Da Capo Press, 2004.

See also: Photography.

2 · Optical observatories

APACHE POINT OBSERVATORY

The University of Washington found in the late 1970s that they could not get enough access to large optical astronomical telescopes. But when Alex Kane of Ashland, Oregon had died in 1975 he had left an estate worth $250,000 to build a large telescope. It was initially intended to leave this money to Oregon State University but they were not interested in a new large telescope, so the money was left to the University of Washington (UW), instead. The money was not sufficient, on its own, to pay for the facility that the UW required, so Bruce Balick set out in 1978 try to interest other universities in a joint project. Balick's initial ideas for the design of the new optical telescope came from the Kitt Peak Advanced Development Program and from his experience in radio astronomy.

By 1981 a provisional consortium had been assembled of UW, New Mexico State University (NMSU), Washington State University (WSU), and Howard University. The initial telescope design included a 2 m primary mirror, just 6 cm thick, with active optics, on an altazimuth mount. The design team also envisaged a remote observing capability and rapid instrument changes to optimise the use of telescope time. A suitable site was identified at an altitude of 2,800 m (9,200 ft) at Sunspot, New Mexico, near the Sacramento Peak National Solar Observatory.

Over the next few years Howard University had to pull out, but both the University of Chicago (UC) and Princeton University (PU) joined the consortium. By 1984 the size of the telescope had been increased to 3.5 m principally at the request of Princeton. The consortium was legally established, in the same year, as the Astrophysical Research Consortium (ARC). The allocated telescope time, based on the contributions of the various members, was agreed to be 31.25% for both the UW and UC, 15.625% for both the NMSU and PU, and 6.25% for WSU. (The WSU sold its share in the ARC to the University of Colorado in 2001.) It was decided to use the site, renamed Apache Point, on Sacramento Peak, and in 1986 the National Science Foundation (NSF) agreed to grant the consortium a sum of $3.74 m towards its capital costs.

The ARC decided to use a fast $f/1.75$ mirror which the Steward Observatory Mirror Laboratory offered to provide free of charge. This was because the NSF was already funding the laboratory to make spun-cast mirrors, and the laboratory wanted to see how one of its mirrors actually performed in a telescope. However one problem of using an experimental mirror was timescale, and the Steward Observatory's mirror was not completed until 1992, over four years late.

In the meantime the telescope structure and housing had been completed by the end of 1987. So, at that time, the ARC had a telescope and no mirror, but the University of Calgary had a 1.8 m mirror and no telescope. As a result the ARC borrowed the Calgary mirror which they installed in their telescope in 1991 to enable the system to be tested. In the following year the ARC were able to carry out remote operations from the campuses of their member universities with this system.

The ARC telescope was finally completed in 1994 with its 3.5 m mirror. It provided rapid access to a variety of instruments permanently located at different telescope ports and accessed by a movable tertiary mirror. Typically, instruments could be changed in less than 5 minutes, if they were available at one of the telescope ports, or 15 minutes otherwise. Almost all of the observing was undertaken remotely, although decisions about the opening and closing of the dome were taken on-site.

Sloan Digital Sky Survey

The University of Washington and Princeton University had begun to consider building a 2.5 m telescope in the mid 1980s to undertake a redshift survey. Then in 1990 Princeton's Institute for Advanced Study joined the ARC after they had become interested in building such a telescope at Apache Point. In the following year the ARC received a grant from the Alfred P. Sloan Foundation to carry out what came to be called the Sloan Digital Sky Survey (SDSS). A number of new institutions who were interested in the project then joined the ARC or worked with it. So that by 2005 the Advisory Council responsible for the SDSS had members from about 25 different institutions in America, Europe and Asia.

The SDSS project built a dedicated 2.5 m telescope in the mid-1990s that was able to image 1.5 square degrees of the sky at a time. It used a 120 megapixel camera that produced images at five different wavelengths from 350 to 920 nm. In addition, a pair of fibre-fed spectrographs obtained 640 spectra simultaneously over a similar range of wavelengths. The telescope used a drift scanning technique in which the telescope was fixed and the rotation of the Earth was used to record small strips of the sky.

The initial five-year programme, called SDSS-I, observed about a fifth of the sky around the North Galactic Polar Cap down to a limiting magnitude of between about 23 and 25, depending on colour. When completed in 2005 the telescope had imaged 215 million individual objects, obtained the spectra of 675,000 galaxies, 90,000 quasars and 185,000 stars, and had found over 100,000 asteroids. The measured redshifts of distant objects enabled a three-dimensional map of the universe to be constructed over a volume about two orders of magnitude larger than that of previous surveys.

This first programme was followed by SDSS-II which included SEGUE (Sloan Extension for Galactic Understanding and Exploration) which extended observations to lower galactic latitudes, and the Sloan Supernova Survey which undertook repeated imaging of 300 square degrees of sky. Over the course of three 3-month campaigns the latter discovered around 500 Type Ia supernovae. SDSS-II, which lasted from 2005 to 2008, was followed by a series of four new surveys in the SDSS-III programme which is due to finish in 2014. This had involved upgrading the spectrograph with more fibres, new detectors and gratings to effectively double its output.

Bibliography

Finkbeiner, Ann K., *A Grand and Bold Thing: An Extraordinary New Map of the Universe in Ushering a New Era of Discovery*, Simon and Schuster, 2010.

Gunn, James E., et al., *The 2.5 m Telescope of the Sloan Digital Sky Survey*, Astronomical Journal, **131** (2006), 2332–2359.

Peterson, Jim, and Mackie, Glen, *A Brief History of the Astrophysical Research Consortium and the Apache Point Observatory*, Journal of Astronomical History and Heritage, **9** (2006), 109–128.

ARGENTINE NATIONAL OBSERVATORY, CORDOBA

The American astronomer Benjamin Gould began charting stars near the north celestial pole in 1864. Whilst undertaking this work he decided to extend his observations into the southern sky which had been relatively poorly surveyed. So he asked Domingo Sarmiento, then minister of the Argentine republic resident in Washington, if the conditions in Argentina were suitable for such observations. Sarmiento said that they were and told Gould that the Argentine government were prepared to erect and maintain a permanent observatory as a national scientific foundation. As a result Gould ordered from Europe the appropriate instruments for such an undertaking. Sarmiento then became president of Argentina determined to improve its education system, particularly in the natural sciences, which Gould's observatory project fitted very well.

Gould arrived in Argentina in 1870, travelling inland to Cordoba, the site selected for his observatory. Unfortunately the Franco-Prussian war in Europe had held up the delivery of his telescopes, so he started work with his four American assistants to estimate the relative brightness of the brightest stars using naked eye and binocular observations. In the meantime construction of the observatory had begun, with the instruments, which consisted of a 12 inch (30 cm) refractor, a 5 inch (12.5 cm) photographic refractor, and a small meridian circle, arriving a little later. These enabled Gould and his colleagues to compile the *Uranometria Argentina*, published in 1879, which catalogued the positions and magnitudes of stars brighter than seventh magnitude.

Gould, who returned to the United States in 1885, was succeeded as director of the Cordoba Observatory by John Thome, one of his assistants. But in the meantime an observatory that had been set up at La Plata in 1882 had failed to meet its commitment to the international *Carte du Ciel* project, so its region of sky was reassigned to Cordoba. As a result the Cordoba Observatory received a 13 inch (33 cm) photographic refractor in 1902, which was a standard instrument for the project, to produce its photographic plates.

Charles Perrine, the last American to run the Cordoba observatory, took over after Thome's death in 1908. He tried to take the observatory into the twentieth century by closing down the earlier programmes of positional astronomy, and by beginning, instead, programmes of astrophysical and galactic research. He built a 30 inch (76 cm) reflector and tried, unsuccessfully, to build a 60 inch (1.5 m) reflecting telescope. Thome's earlier directorship of the observatory had been hampered by Argentina's economic problems. But Perrine's period as director was even more difficult, particularly towards the end when it was subjected to increasing Argentine nationalism. Perrine survived an assassination attempt in 1931, was demoted in 1933, and was forcibly retired three years later. In the meantime the final volume of the *Cordoba Durchmusterung*, which contained, in total, the positions of over 600,000 stars, had been published in 1932. It was the result of almost sixty years of observations at the Cordoba Observatory.

After Perrine, the control of the observatory was taken over by Argentine scientists, starting with the appointment of Felix Aquilar, who promoted spectroscopic work. He was followed in 1940 by Ernesto Gaviola, a physicist who developed its astrophysical work. Then in 1942 a 60 inch reflector was finally completed at Bosque Alegre, a high altitude site about 30 miles (50 km) from Cordoba.

Galactic astronomy flourished for a time at Cordoba and Bosque Alegre after the end of the Second World War. But in 1955 the Argentine National Observatory became part of the University of Cordoba, where its research was limited to astrometry and the calculation of cometary and asteroid orbits. Its time as a first-class astronomical facility was at an end.

Bibliography

Comstock, George C., *Benjamin Apthorp Gould 1824–1896*, Biographical Memoirs of the National Academy of Sciences, **17** (1922), 153–180.

Evans, David S., *Under Capricorn: A History of Southern Hemisphere Astronomy*, IOP Publishing, 1988.

Perrine, C. D., *The National Observatory of the Argentine Republic*, Publications of the Astronomical Society of the Pacific, **22** (1910), 205–211.

ASIAGO OBSERVATORY

Giovanni Silva, the director of the Padua Observatory, was the first astronomer at the observatory to be interested in astrophysics. In the 1930s he used a tiny 19 cm Dembowski telescope to study the photometry of variable stars, but he hoped to obtain a large reflecting telescope to extend his research. In 1942 he achieved his goal when a 1.22 m reflecting telescope was inaugurated at a new observatory on the 1,050 m high Asiago plateau about 90 km from Padua. It was named after Galileo, who had at one time observed from Padua. This Galileo telescope was, at the time, the largest telescope in Europe.

Unfortunately the inauguration of this telescope could not have come at a worse time as it was in the middle of the Second World War. As a result, research did not really begin at the Asiago Observatory until some years later. In 1958 the observatory obtained a 40/50 cm Schmidt which detected numerous supernovae and many variable stars. Nine years later the observatory also inaugurated a 67/92 cm Schmidt to extend this research to include quasars, RR Lyrae, Wolf-Rayet and flare stars.

Light pollution was becoming more and more of a problem at Asiago in the 1960s, so when approval was given to build a 1.82 m reflector it was decided to build this at an altitude of 1,350 m on Cima Ekar, a mountain ridge a few kilometres from Asiago. Copernicus had once studied at Padua, and the telescope was inaugurated in 1973, which was the 500th anniversary of his birth. As a result the telescope, which is still the largest telescope in Italy, was named the Copernicus telescope. In the 1990s both Schmidt telescopes were moved from the original Asiago observatory to the Cima Ekar site.

Atmospheric conditions were not good enough to justify building an even larger telescope in Italy than the 1.8 m Copernicus telescope. So the largest current Italian optical telescope, the 3.6 m diameter Telescopio Nazionale Galileo, was built on La Palma, in the Canary Islands. It saw first light in 1998.

Bibliography

Barbieri, C., Rosino, L., and Stagni, R., *The 72-Inch 'Copernicus Telescope'*, Sky and Telescope, May 1974, pp. 298–300.

Capaccioli, Massimo, and Buson, Lucio, *The Observatories of Padova/Asiago*, Memorie della Società Astronomia Italiana, **60** (1989), 427–458.

AUSTRALIAN OPTICAL OBSERVATORIES

Early years

Thomas Brisbane, the Governor of New South Wales and an amateur astronomer, set up the first permanent observatory in Australia at Paramatta, near Sydney, in 1822. The instruments included a $3\frac{3}{4}$ inch (9.5 cm) transit instrument, a 24 inch (60 cm) mural circle, and a 16 inch (40 cm) repeating circle. Observations made at this observatory over the next four years resulted in the publication of the Brisbane catalogue of over 7,000 stars.

Brisbane returned to the United Kingdom in 1826 leaving behind his astronomical instruments which he sold to the government. The observatory was used only intermittently over the next twenty years before it was closed in 1847.

The New South Wales government subsequently decided to build an astronomical observatory at Sydney harbour for navigational purposes. It was opened in 1858. At one stage in the early 1860s William Scott, the first observatory director, tried to persuade the exceptional amateur, John Tebbutt, to become the next observatory director when Scott retired, but he declined. In 1861 the Sydney observatory acquired a $7\frac{1}{4}$ inch (18.5 cm) Mertz refractor, and about twelve years later it added a $11\frac{1}{2}$ inch (29 cm) Schröder refractor for observing double stars. In 1887 the director of the observatory, Henry Russell, agreed to cover the area of the sky from 52° to 64° south declination for the international *Carte du Ciel* photographic project. The Sydney task, which eventually resulted in the publication of 53 volumes of results, took many years to complete, being completed only in 1971.

A Southern Telescope Committee had been set up in 1852 jointly by the British Association and the UK Royal Society to consider the location of a large reflector in the Southern Hemisphere. In 1865 the government of Victoria agreed to fund the telescope which was installed near Melbourne four years later. This Great Melbourne Telescope, as it became known, had a 48 inch (1.2 m) diameter speculum mirror. This was to be its Achilles heel as no-one at Melbourne could successfully repolish the mirror when it became tarnished, which it did rather rapidly. The dusty atmosphere at Melbourne also compromised the telescope's use, so this impressive looking instrument rapidly fell into disuse.

Mount Stromlo

The Australian spectroscopist W. Geoffrey Duffield suggested at a meeting of the International Solar Union at Oxford in 1905 that a solar observatory be set up in Australia to complement the one in England. The idea received widespread support from both British and Australian astronomers, but the First World War intervened before this solar observatory could be established. In the event, the Commonwealth Solar Observatory was finally built at Mount Stromlo near Canberra over the period 1925–1926 with Duffield as its first director.

A little earlier the amateur astronomer John Reynolds, who was later to become president of the Royal Astronomical Society, had offered Duffield a 30 inch (75 cm) reflecting telescope with equatorial mounting for this new observatory. This Reynolds telescope was installed on Mount Stromlo in the 1930s, but technical problems over the next few years compromised its performance. Meanwhile in 1931 a vertical solar telescope with an 18 inch (45 cm) coelostat and 12 inch (30 cm) objective had also been built on Mount Stromlo.

The Reynolds telescope was used in the early 1950s by Gerard de Vaucouleurs, after some modifications, to undertake a photographic survey of relatively bright southern galaxies. Then over the period 1969–1972 John Hart made further substantial modifications to the instrument to improve its use for spectroscopic observations.

In the meantime, Richard v. d. R. Woolley, who was later to become the UK Astronomer Royal, had been appointed director of the Commonwealth Solar Observatory in 1939. Woolley wanted to change the work of the observatory from solar to stellar research, and in 1945 he refurbished the Great Melbourne Reflector. He changed the mirror to Pyrex, reducing the focal length in the process, and added an electric drive. Woolley also hoped to acquire a 74 inch (1.9 m) reflector, similar to that planned for the Radcliffe Observatory in Pretoria, but delays due to the Second World War meant that the 74 inch was not installed until 1953. Unfortunately it was found that the primary mirror suffered from astigmatism. So it was eventually returned to the manufacturers for refiguring when a spare mirror was available. It was reinstalled in 1961.

The Commonwealth Solar Observatory was renamed the Commonwealth Observatory in 1945. It was subsequently renamed the Mount Stromlo Observatory in 1957 when control was transferred from the Department of the Interior to the new Australian National University in Canberra. Meanwhile in 1953 a 26 inch (65 cm) photographic refractor had been moved from the Yale-Columbia southern station in South Africa to Mount Stromlo because of the better observing conditions in Australia. The telescope was given to the Australian National University in 1963 when the Yale-Columbia southern station was moved to Argentina. In addition, the Swedish University of Uppsala had, at Woolley's invitation, located its new 20/26 inch (50/66 cm) Schmidt telescope on Mount Stromlo in 1957.

Siding Spring

Light pollution near Canberra was becoming a serious problem after the Second World War, and this led to the search for a new observing site to act as a field station of the Mount Stromlo Observatory. Eventually in 1962 it was decided to locate this field station at an altitude of 1,150 m (3,800 ft) at Siding Spring in New South Wales, about 270 miles (430 km) north of Mount Stromlo. By 1964 a 40 inch (1.0 m) and a 16 inch (40 cm) Boller and Chivens reflector had been installed at Siding Spring to be used mostly for photometry and photography. The Mount Stromlo Observatory and its Siding Spring field station were then renamed the Mount Stromlo and Siding Spring Observatory. Then in 1981 the Uppsala Schmidt was moved to Siding Spring to escape the increasing light pollution at Mount Stromlo.

Meanwhile, the UK Royal Society and the Australian Academy of Sciences had been discussing the provision of a jointly financed major optical telescope in Australia. In 1967 the Australian government suggested that the instrument, which should be built at Siding Spring, should have a design based on that of the Kitt Peak 158 inch (4.0 m). In the event this was agreed and the resulting 154 inch (3.9 m) Anglo-Australian Telescope (AAT) saw first light in 1974 with its very low expansion Cer-Vit primary mirror.

The AAT was one of the first telescopes to take full advantage of the development of computer control technology. Its computer slewed the telescope to the objects to be studied, automatically taking account of atmospheric refraction, telescope misalignment and flexure. At the time these were very much state-of-the art developments.

In the meantime it had been agreed that a wide-field Schmidt telescope should be built at Siding Spring to extend the Palomar Schmidt photographic sky survey from −30° declination to the south celestial pole. This 48/72 inch (1.2/1.8 m) UK Schmidt was completed in record time and installed at Siding Spring in 1973 with a design based on that of the Palomar Schmidt. Initially the UK Schmidt facility was treated as an outstation of the Royal Observatory, Edinburgh, and it wasn't until 1988 that it officially became part of the Anglo-Australian Observatory.

Don Mathewson, the acting director of the Mount Stromlo and Siding Spring Observatory, suggested in 1978 that the observatory should build a 3 metre, new technology telescope. Shortly afterwards he became aware that Owens-Illinois had a 95 inch (2.4 m) Cer-Vit mirror blank in stock, so he compromised on the size of the new telescope and purchased this mirror blank. He then had it cut in half horizontally and sold one half for almost the same amount of money as he had paid for the complete blank. (This half was later used in the Hiltner Telescope on Kitt Peak.) The finished 2.3 m mirror of the Advanced Technology Telescope (ATT), as it was called, was relatively thin, so a specially designed support system was built consisting of an air bed underneath the mirror and a mercury collar around it.

This ATT, which was built at Siding Spring, had an altazimuth, computer-controlled mount, fast primary mirror and rotating building, like that of the Multi-Mirror Telescope. These design solutions allowed the observatory building to be much smaller than for a traditional equatorially mounted instrument. The telescope was completed in 1986.

An infrared photometer-spectrometer became available for the Anglo-Australian Telescope in 1979 allowing observations to be made at infrared wavelengths. Twenty years later the so-called 2dF system was built for the AAT that allowed the spectra of 400 objects to be obtained simultaneously by using robotically positioned optical fibres. This allowed the production of the 2dF Galaxy Redshift Survey, which was completed in 2002, that measured the position of over 220,000 galaxies within about 750 megaparsec of the Milky Way.

The United Kingdom joined the European Southern Observatory in 2002 and, as a result, gradually reduced its funding of the Anglo-Australian Observatory. Finally in 2010 UK funding came to an end and the observatory became 100% financed by Australia as the Australian Astronomical Observatory, with an unchanged abbreviation, AAO.

Bushfires on Mount Stromlo

A major bushfire, started by a lightning strike, engulfed the Stromlo Observatory in 1952. The workshop, two storage buildings, machine tools and records of the Great Melbourne Telescope were destroyed, but all the major facilities survived. Unfortunately the same could not be said of the bushfire of 2003. All the old domes and all the old telescopes, except for an old 6 inch (15 cm) refractor, were destroyed. The workshop was also destroyed which contained the almost-completed NIFS (Near-Infrared Integral-Field Spectrograph) for the Gemini North telescope in Hawaii. Fortunately, all the major new telescopes had been built at Siding Spring, which was not affected. In addition, two office buildings at Mount Stromlo survived, which contained all the observatory's computerized archives. A new NIFS was later delivered to Hawaii.

Bibliography

Evans, David S., *Under Capricorn: A History of Southern Hemisphere Astronomy*, Adam Hilger, 1988.

Frame, Tom, and Faulkner, Don, *Stromlo: An Australian Observatory*, Allen and Unwin, 2003.

Gascoigne, S.C.B., Proust, K.M., and Robins, M.O., *The Creation of the Anglo–Australian Observatory*, Cambridge University Press, 1990.

Haynes, Raymond, et al., *Explorers of the Southern Sky: A History of Australian Astronomy*, Cambridge University Press, 1996.

See also: Modern optical interferometers.

BERLIN AND POTSDAM OBSERVATORIES

Towards the end of the seventeenth century the Electress Sophie Charlotte of Brandenburg, advised by Gottfried Leibniz, suggested that an Academy of Sciences and astronomical observatory should be founded in Berlin along the lines of those that had already been established in Paris and London. Then on 18 February 1700 (old date) the Protestant Estates of Germany adopted the Gregorian Calendar. This was followed a few months later, by the establishment of an observatory in Berlin by Frederick III, Elector of Brandenburg. Two months later the Brandenburg Science Society was founded, which was later to become the Prussian Academy of Sciences.

Gottfried Kirch was appointed the first director of the Berlin Observatory in 1700. But Frederick expected the observatory to be financed by sales of a national astronomical calendar, which proved to be very low. As a result, construction of the observatory in Dorotheenstrasse was very slow, and Kirch died before building could be completed in 1711. Funding got even worse over subsequent decades, and the observatory eventually fell into disuse. But in 1764 Johann Bernoulli was appointed director and asked by King Frederick II to revive the observatory. Four years later Bernoulli purchased the observatory's first significant instrument, a large wall quadrant. Johann Bode joined the observatory in 1772 where he was put in charge of the annual publication of the *Berliner Astronomisches Jahrbuch*, the first edition of which was published in 1774. Bode, who took over from Bernoulli in 1787, was director until 1825.

The observatory was in need of a radical overhaul by the time that Johann Encke took over from Bode as director. In fact Encke had been promised that a new observatory would be built when he was appointed. But it needed the intervention of Alexander von Humboldt with King Frederick William III to make that promise turn into reality. The cornerstone was laid in 1832, and the observatory was finished three years later, situated just inside the city walls. Its main instrument was a twin of the $9\frac{1}{2}$ inch (24 cm) Fraunhofer Dorpat refractor. In 1846 Johann Galle and Heinrich d'Arrest discovered Neptune with this telescope.

Encke had undertaken important work on comets and asteroids at the observatory. To facilitate this he had arranged for the production of a number of new star charts, one of which had helped in the discovery of Neptune. He was followed as director by Wilhelm Foerster in 1865. Foerster was more of an organiser and populariser of astronomy than a research astronomer. He was a sponsor or inaugurator of a number of new organisations and, in particular, in the early 1870s he was instrumental in founding the Potsdam Astrophysical Observatory about 30 km (19 miles) from the centre of Berlin. Foerster continually tried to keep the instruments at the Berlin Observatory up to date. For example, in 1868 he purchased a new 8 inch (20 cm) meridian circle that provided excellent service both at Berlin and, subsequently, when the Berlin Observatory was moved to Babelsberg in the early twentieth century (see below). In 1879 a new transit instrument was built, designed by Foerster. Ten years later Karl Küstner detected that the latitude of the Berlin observatory was varying using this transit instrument. This was eventually found to be the result of the Earth's spin axis moving relative to the surface of the Earth.

It had become clear by 1904, when Karl von Struve took over from Foerster, that the observatory would have to be moved because of the expansion of Berlin and its associated dust and light pollution. The $9\frac{1}{2}$ inch refractor was also, by this time, showing its age, so it was replaced by a 12 inch (30 cm) refractor with a Zeiss objective and Repsold base.

Paul Guthnick undertook a site survey in 1906 to find a suitable location for the new observatory. As a result a site was found in the Royal Park of Babelsberg, near Potsdam. The foundations were built for the new observatory in 1911, and the move from Berlin was started two years later. New instruments were even delivered to the new observatory during the First World War. Its main instrument was a 65 cm ($25\frac{1}{2}$ inch) Zeiss refractor installed in 1915, accompanied ten years later by a 1.2 m (48 inch) Zeiss reflector. The latter was moved to the Crimean Astrophysical Observatory by the Soviet Union as part of reparations after the Second World War.

Potsdam Astrophysical Observatory

The solar observer Gustav Spörer was the first to suggest that a special institute should be established in Germany to investigate the Sun. In 1871 Foerster, at the suggestion of K. H. Schellbach, the former teacher of the Crown Prince Frederick Wilhelm, wrote a paper on the establishment of a solar observatory. He emphasised the influence of the Sun on the Earth, but did not mention the use of spectral analysis that had been developed by Gustav Kirchhoff and others over the preceding decade. After reviewing Foerster's paper, the Academy of Sciences recommended that the work of the proposed observatory should include the whole of the emerging field of astrophysics. As a result the Potsdam Astrophysical Observatory was founded in 1874, with Spörer and the astronomical spectroscopist Hermann Carl Vogel as astronomical observers. It had been hoped that Kirchhoff would accept the post of director, but he declined. So initially the observatory was managed by a board of directors, but in 1882 Vogel was appointed the sole director, a position he retained until his death in 1907.

There were no permanent buildings when Spörer started working for the observatory in 1874, so he carried out his sunspot observations from the tower of a former military orphanage in Potsdam. Construction of the observatory was begun in 1876 on the Telegrafenberg hill just south of Potsdam, and the main building and instruments were finished three years later. Initial instruments included a 30 cm (12 inch)

refractor and a 16 cm ($6\frac{1}{4}$ inch) heliograph. In 1889 a double refractor was installed with a 33 cm (13 inch) diameter photographic objective, and a 24 cm ($9\frac{1}{2}$ inch) diameter visual objective, to enable the observatory to participate in the international *Carte du Ciel* project. Ten years later an even larger double refractor was completed with 80 cm ($31\frac{1}{2}$ inch) photographic and 50 cm ($19\frac{1}{2}$ inch) visual objectives.

The early work at Potsdam resulted in the publication of a number of important star surveys and catalogues. For example in the early 1880s Vogel and Gustav Müller produced a spectroscopic survey of stars. Then in the late 1890s Vogel and Johannes Wilsing determined the spectral classification of over 500 stars in the northern sky. About the same time Müller and Paul Kempf used a Zöllner photometer to measure the intensity of stars down to magnitude 7.5. This resulted in the publication, around the turn of the century, of a luminosity catalogue of over 14,000 stars in the *Potsdamer Durchmusterung*.

Karl Schwarzschild was appointed director of the Potsdam Astrophysical Observatory two years after Vogel's death in 1907. Schwarzschild continued his previous work on radiative heat transfer in stellar atmospheres and started work on spectroscopy. He unsuccessfully tried to detect the gravitational redshift in the solar spectrum predicted by Einstein's general theory of relativity. During the First World War Schwarzschild was sent to the Russian front where he wrote his papers on relativity, including that on what is now called the Schwarzschild radius. Unfortunately he died at Potsdam in 1916 after contracting a rare skin disease at the front. Ejnar Hertzsprung also joined the observatory at the same time as Schwarzschild where he undertook valuable work on Cepheid variables. As a result, in 1913 he was able to determine the distance of the Small Magellanic Cloud from the observation of its Cepheids.

In 1917 it was suggested that a solar tower telescope should be built with the specific aim of detecting the gravitational redshift in the solar spectrum as predicted by Einstein's general theory of relativity. Two years later the British total solar eclipse expedition detected the deflection of starlight as it passed close to the Sun, as also predicted by Einstein's general theory. This provided an additional impetus to the proposed German solar tower telescope project.

Construction of this Einstein Tower solar telescope, as it was called, was completed in 1924 at the Potsdam Observatory. The 20 m (65 ft) high tower contained a solar telescope with a coelostat, using two 85 cm (33 inch) mirrors, and a 60 cm (24 inch) Zeiss objective to produce a 14 cm ($5\frac{1}{2}$ inch) diameter solar image. Although the gravitational redshift was not detected, the solar telescope was used for general solar observations until it was heavily damaged during the Second World War. It was fully restored for its 75th anniversary in 1999, after which it was used to detect variations in the solar magnetic field.

After the Second World War the Potsdam Astrophysical Observatory and the Babelsberg Observatory were situated in the Russian sector of Germany, later to become East Germany. As a result research there stagnated and some instruments were removed as part of war reparations. The main development in East Germany was the founding of the Karl Schwarzschild Observatory in 1960 in the Tautenburg Forest where a 2 m optical telescope was installed (see article on *Karl Schwarzschild Observatory, Tautenburg*). On the other hand, after the war the West Germans built the majority of their optical telescopes abroad, at Calar Alto in Spain, in the Canary Islands, and at the European Southern Observatory in Chile (see separate articles on *Calar Alto*, *Canary Islands optical observatories*, and *European Southern Observatory*).

The East Germans had, in the meanwhile, created the Central Institute for Astrophysics of their Academy of Sciences in 1969. It included the Potsdam Astrophysical Observatory, the Babelsberg Observatory, the Karl Schwarzschild Observatory and, later, the Einstein Tower Solar Observatory. Shortly after the reunification of Germany in 1989, this Central Institute for Astrophysics was dissolved and replaced by the Astrophysical Institute Potsdam with responsibility for all but the Karl Schwarzschild Observatory.

Bibliography

Dick, Julius (Friese, C. E., trans.), *The 250th Anniversary of the Berlin Observatory*, Popular Astronomy, 59 (1951), 524–535.

Dick, Wolfgang R., and Fritze, Klaus, *300 Jahre Astronomie in Berlin und Potsdam: Eine Sammlung von Aufsätzen aus Alaß des Gründungsjubiläums der Berliner Sternwarte*, Verlag Harri Deutsch, 2000.

Gingerich, Owen (ed.), *The General History of Astronomy, Vol. 4A; Astrophysics and Twentieth-Century Astronomy to 1950*, Cambridge University Press, 1984.

Hearnshaw, J. B., *The Analysis of Starlight: One Hundred and Fifty Years of Astronomical Spectroscopy*, Cambridge University Press, 1986.

See also: Calar Alto; Canary Islands optical observatories; European Southern Observatory; Karl Schwarzschild Observatory, Tautenburg.

BTA-6 OPTICAL TELESCOPE, ZELENCHUKSKAYA

Astronomers at the Pulkovo Observatory near Leningrad (now St Petersburg) examined, in the 1950s, the possibility of building a very large telescope for deep sky observing. They envisaged one that was larger than the 200 inch (5.1 m) Hale telescope on Mount Palomar, and so would be the largest optical telescope in the world. Initially the Pulkovo design team under Dmitrii Maksutov considered an equatorially mounted instrument, like that of all other large telescopes of the period. But an equatorial mounting is asymmetrical, causing loading and flexure problems with such a large instrument, whereas an altazimuth design is symmetrical. So the decision was made to change to an altazimuth design instead.

This change to an altazimuth design resulted in a lighter and cheaper structure. But it also involved rotating the telescope simultaneously about two perpendicular axes at nonuniform rates to track objects. In addition, the field of view would rotate during an observation, and that needed to be compensated. These were serious problems at the time in the early days of computers. But the Soviet designers solved them, although the telescope could not be used to observe a small region centred on the zenith where the telescope drive rates were too large.

The project to build this 6 metre (236 inch) diameter instrument, the so-called BTA-6 or Bolshoi Teleskop Altazimutalnyi (Large Altazimuth Telescope), was approved in 1960 by the Soviet Academy of Sciences. It was decided that the BTA-6 should be situated near the proposed RATAN-600 radio telescope, and a number of expeditions were sent to various regions of the USSR, whilst the BTA-6 telescope was being designed, to find a suitable location. Eventually it was decided to locate both telescopes at what was to become the Special Astrophysical Observatory. This was at a height of 2,080 m (6,820 ft) in the North Caucasus near Zelenchukskaya.

Bagrat Ioannisiani, who was a consultant at Pulkovo, was responsible for the design of the BTA-6. He was also section head at the Leningrad Optical Equipment Works where the telescope was built. The 6 m glass blank, which was made of a Pyrex-type low-expansion borosilicate glass, was cast at the Optical Glass Works near Moscow. Although the 6 m, $f/4$ telescope was completed in 1976, the original mirror had significant surface defects that had to be covered with black cloth. This mirror was replaced with an identical but undamaged one in 1978.

Very little in the way of results were published in the West in the BTA-6's first decade or so, leading to rumours about the telescope's poor quality. Eventually it was disclosed that observing conditions at the site were rather poor, often caused by the turbulent downwind wake of the nearby higher Caucasus peaks. The weather conditions also tended to be rather variable. In fact, better observing conditions had been found during the original site search, but these were mostly in Central Asia, which was considered to be too remote. There were also problems caused by thermal inertia of the mirror and the enormous dome, which could have been much reduced if the telescope had been designed with a significantly shorter focal length.

Bibliography

Ioannisiani, Bagrat K., *The Soviet 6-Meter Altazimuth Reflector*, Sky and Telescope, November 1977, pp. 356–362.

Ioannisiani, B. K., et al., *The Zelenchuk 6 m Telescope (BTA) of the USSR Academy of Sciences, Proceedings of the Sixty-seventh Colloquium, Zelenchukskaya, USSR, September 8–10, 1981*, Reidel Publishing, 1982, pp. 3–10.

Keel, William C., *Galaxies Through a Red Giant*, Sky and Telescope, June 1992, pp. 626–632.

BYURAKAN ASTROPHYSICAL OBSERVATORY

The Byurakan Astrophysical Observatory was founded in 1946 at the instigation of Viktor Ambartsumian. It was located at an altitude of 1,405 m (4,610 ft) on the slopes of Mount Aragatz in Armenia. Over its first decade the observatory acquired a number of modest-sized instruments, the largest of which was a 21/21 inch (53/53 cm) Schmidt. During these early years Ambartsumian discovered stellar associations, and proposed theories of star formation in stellar clusters and associations, and theories of activity in active galactic nuclei (AGNs). Then in 1960 a 40/52 inch (102/132 cm) Schmidt was installed at the observatory. Beniamin Markarian and colleagues used this to undertake a survey of galaxies with ultraviolet excesses, now called Markarian galaxies. By 1980 he and his colleagues had discovered about 1,500 such galaxies in the First Byurakan Survey.

The largest telescope at the observatory, the ZTA-2.6 (Armenian Reflecting Telescope-2.6 m) reflector, became operational in 1976. This 102 inch telescope was built by the Leningrad Optical-Mechanical Association. It was an updated version of the same-sized G. A. Shajn reflector that had been operating for fifteen years at the Crimean Astrophysical Observatory. In 1984 the ZTA-2.6's main mirror was replaced by one made of ultra-low-expansion Sitall glass.

The Byurakan Astrophysical Observatory suffered badly from the break-up of the Soviet Union in 1991. Many of the staff left, but in the mid-1990s France helped the observatory equip the 2.6 m with new instruments. A number of summer schools were held at Byurakan starting in 1995 which helped to re-energise the observatory. The International Astronomical Union (IAU) held a symposium on active galaxies at Byurakan in 1998, followed a few years later by an IAU Colloquium on AGN surveys, devoted to the memory of Markarian. Then over the period 2002 to 2005 the plates of the First Byurakan Survey were digitised in collaboration with the University of Rome and Cornell University.

Bibliography

Khachikian, Edward Ye, and Weedman, Daniel W., *The Byurakan Observatory in Soviet Armenia*, Sky and Telescope, April 1971, pp. 217–219.

Mickaelian, A. M., *The Byurakan Observatory*, 5th ed., Edit Print, Yerevan, 2008.

CALAR ALTO

Optical astronomy in West Germany was beginning to suffer in the early years after the Second World War because of the limited size of their telescopes. The climatic conditions in West Germany were also not conducive to high-quality optical observing. As a result a site survey was begun in 1968 to try to find a more suitable site in the Northern Hemisphere on which to build a new astronomical observatory. A number of sites were surveyed in Greece and southern Spain until one was chosen by the Max Planck Institute for Astronomy at Heidelberg, the new centre of astronomical research in West Germany, who were responsible for the project. The chosen site was at an altitude of 2,160 m on Calar Alto in the mountains of southern Spain near Grenada. An agreement was then signed between the West German and Spanish governments covering the construction and operation of such an observatory, with the Spanish government agreeing to restrict the use of artificial lights within about 5 km of the observatory.

Construction of the Calar Alto Observatory started in 1973. The first permanent telescope, a 1.2 m reflector, was completed shortly afterwards by the Carl Zeiss company, and by the end of the decade a 2.2 m had also been installed. The Madrid Observatory had also set up an independently operated 1.5 m reflector. In 1980 the 80/120 cm Schmidt was moved from the Hamburg Observatory to Calar Alto, and in 1985 a new 3.5 m reflector was completed at the observatory. This 3.5 m was the last telescope of that size to be built anywhere that was equatorially mounted.

No other large telescopes have been built at Calar Alto, as more recent German funding has been concentrated on building and equipping telescopes at the European Southern Observatory sites at La Silla and Cerro Paranal in Chile.

Bibliography

Robinson, Leif J., *The German-Spanish Connection*, Sky and Telescope, April 1980, pp. 279–286.

CANADIAN OPTICAL OBSERVATORIES

The Dominion Observatory and the Dominion Astrophysical Observatory

The early explorers of Canada naturally used the stars to navigate on both land and sea. Then in 1666 the Collège de Québec began to teach astronomy to pilots who worked on the St Lawrence River. A number of small astronomical observatories were set up in Canada in the mid-nineteenth century, but it was not until 1905 that the first substantial astronomical observatory was founded, the Dominion Observatory in Ottawa.

Initially the Dominion Observatory had a 15 inch (40 cm) refractor and a horizontal solar telescope, which John Plaskett and colleagues used to observe spectroscopic binaries and solar rotation. But the small size of the refractor was soon found to be too limiting, and in 1913 the government approved the construction of a 72 inch (1.8 m) reflector. Following a site survey it was found that the astronomical 'seeing' and thermal temperature range were better at a site near Victoria, British Columbia, than at the Dominion Observatory in Ottawa. So the Victoria site was chosen for the 72 inch telescope at what was to become the Dominion Astrophysical Observatory (DAO).

The crown glass mirror blank for the 72 inch was produced by the St Gobain glass works in France and was shipped out in July 1914, a few days before the outbreak of the First World War. The telescope mounting and dome were completed by late 1916, but problems with grinding the mirror resulted in the completion of the telescope being delayed until May 1918. Nevertheless, this was still a remarkable achievement considering that the First World War was still ongoing at that date.

During the subsequent interwar years research at the DAO concentrated on radial velocities, spectroscopic parallaxes and research into the interstellar medium. The 72 inch, later called the Plaskett Telescope, was the only large telescope at the DAO until a 48 inch (1.2 m) reflector was installed there in 1961. Then in 1974 the crown glass primary of the 72 inch was replaced by a Cer-Vit mirror.

Meanwhile the Dominion Observatory in Ottawa had been closed in 1970. Then by 1975 the national astronomy infrastructure in Canada had been modified so that all federal astronomy, including that at the DAO and the equivalent radio astronomy observatories, became the responsibility of the National Research Council's Herzberg's Institute of Astrophysics.

David Dunlap Observatory

Clarence Chant, one-time president of the Royal Astronomical Society of Canada, had begun to investigate the possibility of building a new observatory for the University of Toronto before the First World War. But funding was a problem until 1921 when he managed to interest a mining executive, David Dunlap, in his project. Unfortunately Dunlap died three years later, but Chant persuaded his widow to fund an observatory in memory of her husband. As a result Chant ordered a 74 inch (1.9 m) instrument which, when completed in 1935, was the second largest telescope in the world. Like the Victoria reflector it was also used for spectroscopic studies.

Light pollution became more and more of a problem at the David Dunlap Observatory in the 1960s and, for a time, consideration was given to moving the 74 inch to a new site. But it was eventually decided that the funds would be better employed in building a new 60 cm reflector at a dark site in the southern hemisphere. An agreement was then reached with the Carnegie Institution of Washington to locate this telescope at their Las Campanas site in Chile. As a result the University of Toronto established their Southern Observatory there to accommodate the 60 cm. This telescope, which saw first light in 1971, was later named the Helen Sawyer Hogg Telescope after the well-known Canadian astronomer.

The University of Toronto's Southern Observatory was closed in 1997 and its telescope, including the building and dome, were given to the National Observatory of Argentina who moved it, shortly afterwards, to their observatory at El Leoncito. The Argentineans operated and maintained the telescope whilst reserving 25% of observing time for the University of Toronto. The University of Toronto also sold the site of their David Dunlap Observatory on the outskirts of Toronto in 2008.

Other university observatories

More modest facilities than the David Dunlap 74 inch have been built over the years at university level. These included a 48 inch (1.2 m) reflector that was built at the University of Western Ontario's Elginfield Observatory in 1969, and a 64 inch (1.6 m) that was built at the University of Montreal's Mont Mégantic Observatory, Quebec, ten years later. In addition, a 1.5 m telescope was built at the Rothney Astrophysical Observatory of the University of Calgary optimised for the infrared. It became operational in 1987.

The University of Calgary also received a 1.8 m mirror that they had ordered from the University of Arizona's Mirror Laboratory. Unfortunately at that time they did not have a completed telescope in which to fit the mirror. The Apache Point Observatory, on the other hand, had the opposite problem of a telescope but no mirror. So Apache Point borrowed the Calgary mirror in the late 1980s whilst waiting for their own, larger mirror from Arizona. Eventually the Calgary mirror was returned to the Rothney Observatory in 1993, and first light was achieved for their 1.8 m telescope three years later.

International collaboration

In the 1960s Canadian astronomers had examined the possibility of building a 150 inch (3.8 m) telescope, to be called the Queen Elizabeth II telescope, on Mount Kobau, British Columbia. But the project was cancelled in 1968. A few years later the 3.6 m Canada-France-Hawaii (CFH) telescope was approved instead, which was completed on Mauna Kea in 1979. Canada also became a member of the international Gemini project, which consisted of two 8 m telescopes, one in each of the Earth's hemispheres, that saw first light in 1999 and 2002. So, as with most other countries, the largest telescopes available to Canada around the end of the twentieth century were the result of international collaboration.

Liquid mirror telescopes

In 1908 the physicist Robert Wood of Johns Hopkins University used a rotating 20 inch (50 cm) mirror of liquid mercury as the primary mirror of a reflecting telescope. With this he was able to resolve stars with separations as low as 2.3 arcseconds. Although such mirrors were very much cheaper to make than conventional mirrors, they could only be used to observe the zenith, and so Wood's idea was not followed up for many years. But in the early 1980s Ermanno Borra and colleagues of Laval University (LU), Quebec, built and successfully tested a 1 m diameter, $f/1.6$ liquid mercury mirror. This was followed by a collaboration between Paul Hickson and colleagues of the University of British Columbia (UBC) and Borra and his team to build a 2.7 m $f/1.9$ Liquid Mirror Telescope (LMT). By 1994 Hickson and his team had used a drift scanning CCD system with this telescope to produce astronomical images limited only by atmospheric turbulence.

The Institut d'Astrophysique de Paris then joined the UBC/LU team which was hoping to build a zenith-pointing telescope using a rotating liquid-metal primary mirror of at least 5 metres diameter. This was the origin of the Large Zenith Telescope (LZT) project which has since included astronomers from the State University of New York at Stony Brook and Columbia University. In the event they built a 6 m diameter telescope whose construction was completed in 2003. Problems were then experienced with the rotation speed of the mirror, which seemed to be affected by local gusts of wind, and with the large velocity of the outer parts of the mirror, which caused ripples on the surface of the mercury. Both problems have subsequently been resolved.

Bibliography

Gibson, Brad K., *Liquid Mirror Telescopes: History*, Journal of the Royal Astronomical Society of Canada, **85**, No. 4 (1991), 158–171.

Hodgson, J. H., *The Heavens Above and the Earth Beneath: A History of the Dominion Observatories*, Geological Survey of Canada, 1989.

Jarrell, R. A., *The Cold Light of Dawn: A History of Canadian Astronomy*, University of Toronto Press, 1988.

Plaskett, J. S., *The Dominion Astrophysical Observatory*, Publications of the Astronomical Society of the Pacific, **39** (1927), 88–96.

Plaskett, J. S., *The History of Astronomy in British Columbia*, Journal of the Royal Astronomical Society of Canada, **77** (1983), 108–120.

See also: Mauna Kea Observatory; National New Technology Telescope and Gemini.

CANARY ISLANDS OPTICAL OBSERVATORIES

In 1856 the Astronomer Royal for Scotland, Charles Piazzi Smyth, went to the island of Tenerife in the Canary Islands, off the western coast of Africa, to make astronomical observations from high altitude. There he set up two observing stations, one on the summit of Guajara at an altitude of 2,700 m (8,900 ft), and the other at Alta Vista at an altitude of 3,250 m (10,700 ft) just below the summit of El Teide. His observations showed the great advantage of observing at these high altitudes. But this did not persuade anyone to build a large observatory at high altitude until the Lick Observatory was built on the 1,300 m (4,250 ft) Mount Hamilton in 1880s.

But returning to astronomy in the Canary Islands, Jean Mascart of the Paris Observatory photographed Halley's comet from there in 1910, and suggested that an international astronomical observatory should be established on Mount Guajara on Tenerife. Following his suggestion, discussions started between the Spanish, French and German governments, but they were discontinued following the start of the First World War and were never restarted.

Astronomical interest in the Canaries was finally reignited when astronomers observed a total solar eclipse from there in 1959. Shortly afterwards the Spanish set up the Observatorio del Teide (now part of the European Northern Observatory) at Izana, on Tenerife, at an altitude of 2,380 m (7,250 ft). The first telescope was a 30 cm French photopolarimeter installed there in 1964 to study the zodiacal light, followed by a Spanish 25 cm heliographic telescope installed in 1969, and a 40 cm West German evacuated tower solar telescope in 1972. The largest telescope at El Teide was the UK's 1.5 m Infrared Flux Collector (IRFC). This was built in 1971 as the prototype for the 3.8 m United Kingdom Infrared Telescope that was completed on Mauna Kea eight years later. (The IRFC was transferred to Spanish control in 1983. Two years later its name was changed to the Carlos Sanchez Telescope.)

Although telescopes continued to be built at El Teide, the largest telescopes in the Canaries have been installed since the early 1980s on the nearby island of La Palma, rather than Tenerife. This is because of La Palma's better observing environment.

The Greenwich Royal Observatory had been gradually moved to Herstmonceux Castle in Sussex during the 1950s to escape from the light pollution of London. The move was completed in 1958, and the new 98 inch (2.5 m) Isaac Newton Telescope (INT) was completed there in 1967. Unfortunately the seeing and weather conditions in England were not good enough for a large telescope. So the UK surveyed a number of sites in the North Atlantic area in the early 1970s and compared them with Mauna Kea on Hawaii to find a suitable location to build its northern hemisphere observatory. This was to complement the UK's shared Southern Hemisphere observatory in Australia. The UK eventually chose the site at the Roque de los Muchachos, at an altitude of 2,300 m (7,600 ft), on La Palma. Although this observatory was not as high as that on

Mauna Kea, it was much closer geographically to the UK, while being generally above the normal cloud ceiling of about 1,500 m.

The Canary Islands were part of Spain, and so agreement was required from the Spanish authorities to construct an observatory on La Palma. Eventually an agreement was reached in 1979 between Spain, on the one hand, and the UK, Sweden and Denmark, on the other, to allow the latter three countries to construct an observatory (also part of the European Northern Observatory) at the Roque de los Muchachos on La Palma. In return for giving their permission, Spain would receive 20% of observing time on all telescopes.

Under the terms of this agreement the UK were to build three main telescopes. These were the 4.2 m William Herschel Telescope (WHT), the 2.5 m INT, and a 1.0 m astrometric reflector (later called the Jacobus Kapteyn Telescope or JKT). Sweden was to build a vacuum solar tower telescope with a 60 cm heliostat feeding a 44 cm Cassegrain, and a general purpose 60 cm reflector, whilst Denmark was to provide an automatic transit circle.

The JKT saw first light at the Roque de los Muchachos observatory in 1984. The INT was moved from Herstmonceux to this new observatory, where it also saw first light in 1984 with a new 101 inch (2.6 m) mirror. But the largest telescope on La Palma was, for many years, the 4.2 m (165 inch) WHT. This had been designed to complement the 3.9 m Anglo-Australian Telescope in the southern hemisphere which had been completed in 1974. Initially conceived as a 4.5 m telescope, the 4.2 m WHT was almost cancelled in 1979 because of escalating costs. But a radical redesign, and the agreement of the Netherlands to contribute financially, enabled the project to be given the go-ahead two years later.

The WHT was built by Grubb Parsons of Newcastle, UK as their last telescope, using a Cer-Vit mirror blank which was one of four produced ten years earlier by Owens-Illinois. (The other three had been used on the Anglo-Australian Telescope, the Canada-France-Hawaii Telescope, and the Blanco Telescope on Cerro Tololo.) The WHT, which saw first light in June 1987, was one of the first large modern telescopes to use an altazimuth mount. At that time it was the third largest optical telescope in the world.

In 1988 the Spanish authorities passed the Sky Law to control light pollution in the Canaries. Then in the following year the Nordic Optical Telescope (NOT), jointly owned by Sweden, Finland, Denmark and Norway, came into operation on La Palma. It had a form of active optics which allowed the use of the thinnest single 2.5 m primary mirror up to that time which was only 18 cm (7 inches) thick. The telescope complemented the Danish 1.5 m instrument located at the European Southern Observatory (ESO). About ten years later an Italian telescope was also built on La Palma. This was the 3.6 m Telescopio Nazionale Galileo, whose design was based on that of ESO's New Technology Telescope at La Silla.

Planning for the construction of the largest telescope, the Gran Telescopio Canarias, on La Palma was begun in 1987 following the inauguration of the WHT. At that time Spain and the United Kingdom considered building a telescope with a monolithic 8 m mirror, but in 1990 the UK decided to join the American-led Gemini project instead. Five years later the Spanish abandoned the design with a monolithic mirror and decided to build a telescope with a 10 m diameter segmented mirror instead, like that of the two Keck telescopes on Mauna Kea. The Canarian Government set up the Gran Telescopio de Canarias company, in which the Spanish and Canarian Governments each had a 50% share, to build the telescope. Then in 2001 both Mexico and the University of Florida agreed to contribute 5% each to the costs of the telescope in return for each using 5% of the telescope's observing time. This 10.4 m Gran Telescopio Canarias saw first light in 2007 with 12 of its 36 Schott Zerodur mirror segments in place. All 36 mirror segments had been fitted by 2009.

The original Swedish solar telescope, which had been transferred from Capri to La Palma in the late 1970s, was replaced by a new Swedish Vacuum Solar Telescope in 1985. Fifteen years later this was in its turn removed and replaced with the Swedish 1 m Solar Telescope, which had a 110 cm diameter lens of fused silica, with a clear aperture of 98 cm. It was a vacuum telescope with adaptive optics which allowed it to produce diffraction limited images in May 2002.

Over the years the countries involved in the various Canary Island telescopes have changed. For example, in 1997 Iceland became the fifth member of the Nordic Optical Telescope Scientific Association (NOTSA), which ran the NOT. In addition, the financial contribution of various counties to the telescope projects has changed. So in the early twenty-first century, for example, the UK contribution to the Isaac Newton Group of telescopes, which comprised the WHT, INT and JKT, was reduced and balanced by both efficiency savings and additional contributions from the UK's partners.

Bibliography

Boksenberg, A., *The William Herschel Telescope*, Vistas in Astronomy, Pergamon Press, **28** (1985), 531–553.

Heck, André (ed.), *Organizations and Strategies in Astronomy*; *Vol. 5*, Kluwer Academic Publishers, 2004.

Jones, Anthony W., *The Canary Islands – An Astronomer's Experiment*, Sky and Telescope, September 1981, pp. 199–201.

Smith, David H., *Approaching First Light on La Palma*, Sky and Telescope, March 1984, pp. 214–217.

See also: Modern optical solar observatories, Canary Islands solar telescopes.

CARNEGIE SOUTHERN OBSERVATORY (LAS CAMPANAS)

An advisory committee set up by the Carnegie Institution of Washington recommended in 1902 that an observatory should be set up in the Southern Hemisphere as there were, at that time, ten times as many observatories in the Northern Hemisphere as in the Southern. This suggestion was endorsed by George Ellery Hale, Walter Adams, and others in the 1920s and 1930s, but the collapse of the United States stock market in 1929, and the ensuing depression, meant that such an idea would have to be put on ice.

Merle Tuve, of the Carnegie Institution, Ira Bowen, and others resurrected the idea of a Southern Hemisphere observatory in 1962. Then in the following year the Carnegie Trustees approved funds for a site survey. It was, at the time, envisaged that the observatory would include a copy of the 200 inch (5.1 m) Hale reflector, a 48 inch (1.2 m) Schmidt telescope, and a few smaller instruments.

Horace Babcock supervised the site testing for this proposed Carnegie Southern Observatory (CARSO) in New Zealand, Australia and Chile, starting in 1963, and concluded that it should be located in the Chilean Andes. Eventually the decision was made in 1968 to set up the observatory at an altitude of between 2,250 and 2,520 m (7,400 to 8,250 ft) on the Las Campanas mountain ridge in Chile. Construction started there in the following year on its first telescope, a 1.0 m (40 inch) reflector, that was dedicated in 1971 and named after the former Carnegie astronomer, Henrietta Swope, who helped to finance it.

Funding could not be found at that time for the originally envisaged 5 m telescope. So, as a compromise, it was decided to build a 2.5 m (100 inch), with possibly a larger telescope later. Finance of the 2.5 m, which had an exceptionally wide field of view of 2.1° diameter, was provided to a large extent by Mr and Mrs Crawford Greenewalt. The telescope, which had a primary mirror of Corning fused silica, was completed in 1976, and named the Irénée du Pont telescope after Mrs Greenewalt's father.

The Carnegie Institution decided to withdraw financial support from the Mount Wilson Observatory in California in 1985, and to concentrate their financial resources in astronomy on the Las Campanas Observatory. In the following year the Carnegie Institution, plus the University of Arizona and the Johns Hopkins University, agreed to build an 8 m telescope at the Las Campanas Observatory. Johns Hopkins were forced to withdraw from this so-called Magellan Project in 1991 for financial reasons, but the other two partners decided to proceed with a 6.5 m, $f/1.25$ telescope instead. This telescope, which came to be called the Walter Baade Telescope, had a spun-cast borosilicate glass mirror produced in 1994 by the University of Arizona's Mirror Laboratory. In the meantime construction of the telescope had started on-site in 1993. The mirror arrived at the observatory in 1999, where it was coated with aluminium. First light for the telescope occurred in September 2000.

Harvard University had, in the meantime, joined the Magellan Project in 1995, and the University of Michigan and the Massachusetts Institute of Technology joined in the following year. This enabled the construction of a twin telescope at Las Campanas about 60 m from the first. First light occurred for this second telescope, now named after Landon T. Clay, a Harvard graduate and Boston philanthropist, two years after that of the first.

Bibliography

Babcock, Horace, *First Tests of the Irénée du Pont Telescope*, Sky and Telescope, August 1977, pp. 90–94.

Edmondson, Frank K., *AURA and its US National Observatories*, Cambridge University Press, 1997.

Hammond, Allen L., *Big Astronomy in Chile: The Southern Observatories Come of Age*, Science, **198** (1977), 1235–1239.

CHINESE OPTICAL OBSERVATORIES

The decline of astronomy in China which had accelerated in the fourteenth century was not really rectified until the second half of the twentieth century. Prior to then, however, a number of small, foreign-owned observatories had been founded in the late nineteenth century.

Xu Jiahui Observatory had been established in Shanghai in 1872 by French Jesuits. It later began transmitting time signals primarily for the use of French warships. Then in 1900 the Sheshan Observatory was founded some 40 km (25 miles) from Shanghai by a French Catholic mission. There they installed a 40 cm (16 inch) double astrograph. Both observatories were taken over by the new Chinese communist government in 1950, with that at Xu Jiahui becoming the headquarters of the national time service. In 1962 the two observatories were amalgamated to become the Shanghai Astronomical Observatory. Then in the 1980s a 1.56 m (61 inch) reflecting telescope was installed at Sheshan.

The Quigdao observing station was another foreign-owned observing station. It had been built by Germany in 1898 on Mount Guanxiang, but was taken over by the Chinese government in 1924. That year the observatory provided the first Chinese time service and it began the study of sunspots a year later.

The Chinese Astronomical Society was founded in Peking (Beijing) in 1922. Then in 1928 the Institute of Astronomy was founded in Nanking (Nanjing), following the transfer of the national capital from Peking to Nanking by the new Chinese

Nationalist government. The main task of the institute was, initially, to establish the Purple Mountain Observatory (PMO) on the outskirts of Nanking, where operations began in 1934. A 60 cm (24 inch) reflector was built there.

After the outbreak of the Sino-Japanese war in 1937, equipment and research workers of the PMO were moved from Nanking to the outskirts of Kunming, in Yunnan Province. But even here work gradually came to a standstill. In 1946, after the war, the PMO returned to their Nanking site, leaving Kunming as a solar observing station. In 1964 the Chinese built their own 60 cm (24 inch) Schmidt at the main PMO. Then, eight years later, the Kunming station was extended to become the Yunnan Observatory. It was equipped with a chromospheric telescope and a 40 cm (16 inch) solar coelostat capable of a one arcsecond resolution.

The Chinese Cultural Revolution in the late 1960s/early 1970s had a massive effect on all intellectual activity in China, as all universities were closed and astronomical research all but ceased. In fact China's main professional astronomical journal, *Acta Astronomica Sinica*, was not published for eight years.

Work on the modern Peking Astronomical Observatory had been begun in 1958. It had four field stations, Xinglong (for stellar astronomy), Sha-ho (solar physics), Mi-yun (radio astronomy) and Tientsin (latitude variation work). The initial construction of the Xinglong field station, built at an altitude of 960 m (3,150 ft) in the Yanshan Mountains about 130 km (80 miles) from Beijing, was completed in 1968. Its main instrument in the early days was a 60 cm/90 cm (24 inch/36 inch) Schmidt telescope built by Zeiss Jena.

Meanwhile in 1966 the Nanking Astronomical Instruments Factory had been founded to produce most of China's professional optical equipment. One of its key early tasks was to build a 2.16 m reflector for the Xinglong observatory using a primary mirror blank produced in the Soviet Union. But the telescope took much longer than expected to complete as it was by far the largest telescope to have been constructed at that time in China. It was completed in 1989.

The Xinglong Observatory now includes the Guoshoujing Telescope (GSJT), which was formally called LAMOST (Large Area Multi-Object Spectroscopic Telescope). It is a 4.8 m/6.1 m Schmidt telescope with its optical axis fixed in the meridian plane. Construction of the GSJT was completed in 2008. The Xinglong Observatory also has a 1.26 m infrared telescope, and three reflectors with aperture in the range 85 to 65 cm, whilst its 60/90 cm Schmidt is still operational.

Bibliography

Goldberg, Leo, *American Astronomers Visit China-1 and -II*, Sky and Telescope, October 1978, pp. 279–282, and November 1978, pp. 383–387.

Guoxiang, Ai, *Chinese Astronomy in the Century Transition*, in Yongxiang, Lu (ed.), *Science Progress in China*, Science Press, 2003.

Miley, G. K., *Astronomy in China Today*, Sky and Telescope, March 1974, pp. 148–152.

CRIMEAN ASTROPHYSICAL OBSERVATORY

N. S. Mal'tsov, a wealthy amateur astronomer, built his own astronomical observatory in the early years of the twentieth century. It was at an altitude of 360 m (1,200 ft) on Mount Koshka virtually on the coast of the Crimea near Simeiz. In 1908 he donated his observatory to the Pulkovo Observatory of St Petersburg, who operated it as an outstation. At the time the Simeiz observatory included a Zeiss double astrograph, with a 12 cm (5 inch) objective, and a heliograph. Pulkovo used the astrograph for astrophotography to both search for asteroids and comets and to extend their study of variable stars.

In 1912 a 1.0 m (40 inch) reflector was ordered from Grubbs of Dublin for the Simeiz observatory. But it was not delivered until after the First World War, the Russian Revolution and the Russian Civil War. When it was installed in 1925 it was the largest reflector in Europe. Grigorij Shajn used it to study spectroscopic binaries, gaseous nebulae and stellar rotation. The Simeiz observatory was completely destroyed in the Second World War.

After the Second World War the Simeiz observatory was restored but not as an outstation of the Pulkovo Observatory. It became, instead, the independent Crimean Astrophysical Observatory (CrAO) of the USSR Academy of Sciences with Shajn as its director. Under his direction a new optical observatory was built at a more suitable site. This was at an altitude of 560 m (1,850 ft) at Nauchny, further away from the coast. A series of instruments was moved there which had been taken from Germany as war reparations. These were a 40 cm (16 inch) double astrograph that was installed in 1949, a coronagraph that was installed in 1950, and, two years later, the 1.2 m (48 inch) Babelsberg reflector which was used for spectroscopic work. A new 50 cm (20 inch) reflector was located at the CrAO in 1953 and, in the following year, a 1 m diameter solar tower telescope. Then the 2.6 m (102 inch) Shajn reflector was built at the CrAO in 1961. It was, at the time, the largest optical telescope in Europe, and was used for galactic research, particularly that of active galaxies and their nuclei.

Following the break-up of the Soviet Union in 1991, the CrAO came under the control of Ukraine's Ministry of Science and Technology. Today the CrAO is able to observe over a wide range of frequencies, with the site at Nauchny devoted to optical and gamma ray observations and that at Simeiz devoted to radio research.

Bibliography

Artemenko, T. G., Balyshev, M. A., and Vavilova, I. B., *The Struve Dynasty in the History of Astronomy in Ukraine*, Kinematics and Physics of Celestial Bodies, 25 (2009), 153–167.

Heck, André (ed.), *Organizations and Strategies in Astronomy; Vol. 7*, Springer, 2006.

Kulikovskii, P. G., *250 Years of the Academy of Sciences and Astronomy*, Soviet Astronomy, 18 (1974), 131–135.

EUROPEAN SOUTHERN OBSERVATORY

In 1953 Walter Baade suggested to Jan Oort that European astronomers should establish a large joint European observatory in the Southern Hemisphere, as most of the largest telescopes at that time were in the Northern Hemisphere. In addition, the Magellanic Clouds and the central region of the Milky Way were best observed from well south of the Earth's equator. Baade suggested that the observatory's main instruments should be a 120 inch (3 m) reflector similar to that of the Lick Observatory and a 48 inch (1.2 m) Schmidt like that on Mount Palomar. Using existing designs should enable the European versions to be built more quickly and cheaply than if the telescopes had to be designed from scratch. Baade's idea was discussed shortly afterwards by a group of European astronomers who had gathered at Groningen, in the Netherlands, for a conference on galactic research. At the time South Africa was the envisaged location of the observatory in view of its known good observational conditions and the fact that a number of European countries already owned or used observational facilities there. In the following year a committee was established, which included representatives of Belgium, France, Great Britain, the Netherlands, Sweden, and West Germany, to examine the project's feasibility. Subsequently Great Britain withdrew from the project in favour of establishing an Anglo-Australian observatory in Australia.

The possibility of setting up the proposed European Southern Observatory (ESO) was given a major boost in October 1959 when the Ford Foundation agreed to provide a grant of $1.0 million towards the $5 million estimated capital cost. This was not only very useful funding in its own right, but it also demonstrated confidence in the project and helped to persuade France to participate in it. Subsequently, the ESO Convention was signed by five Member States (Belgium, France, the Netherlands, Sweden, and West Germany) in 1962, and ratified by each between 1963 and 1967. In addition, Denmark joined ESO in 1967, bringing the total number of Member States at the end of that year to six.

In the meantime, site testing in South Africa in the early 1960s had shown that there were no suitable sites there. But information from the American AURA (Association of Universities for Research in Astronomy) site surveys, conducted more or less in parallel, had shown that there were excellent observing conditions in the Andes. As a result, in November 1962 AURA decided to locate their Inter-American Observatory at Cerro Tololo in the Chilean Andes. Eighteen months later the ESO Council also decided to locate their European Southern Observatory 2,400 m (7,900 ft) above sea level at La Silla in the Andes, just over 100 km from Cerro Tololo. ESO also sought to control the local environment around their observatory by purchasing 625 km^2 of the surrounding land.

ESO headquarters was built in the late 1960s at Santiago, Chile on land donated by the Chilean Government next to the United Nations' South American Headquarters. In the mid-1970s this building was sold and the ESO headquarters moved to Garching, West Germany, where its administrative headquarters was already located. This move from Chile was partly to satisfy the West German government's desire to have a major European scientific centre on their soil, and partly because of economic problems in Chile.

The initial programme of telescope construction at La Silla included a 1.0 m (40 inch) telescope designed primarily for photoelectric photometry (later known as the Photometric Telescope), a 1.0 m telescope designed primarily for spectrographic work, and the Grand Prism Objectif (GPO) radial-velocity astrograph, which was a copy of the GPO installed at the Haute-Provence Observatory in France. The designs of the photometric and spectroscopic telescopes were delegated to Dutch and French astronomers, respectively, because of their different expertise and interests.

The 1.0 m photometric telescope, which operated from the ultraviolet to the infrared, was the first permanent telescope to be built on La Silla. It was mounted in a temporary observatory building in 1966 and moved to its permanent observatory two years later. The optics, which used low-expansion Schott glass for the primary mirror, were made by Jenoptic in East Germany.

As mentioned above, the spectrographic telescope was expected to have a 1.0 m aperture. But French astronomers suggested in 1961 that it would be more cost-effective to purchase a duplicate of the 1.5 m spectrographic telescope then being designed for their Haute-Provence Observatory. This suggestion was accepted, but with the addition of a Cassegrain focus to the Haute Provence design. The telescope was delivered to La Silla in 1968.

The GPO consisted of twin 40 cm photographic and visual refracting telescopes on a common mount. The photographic telescope's objective lens was covered by an objective prism to enable it to measure stellar radial velocities. The GPO had initially been used in South Africa for site testing but, after refurbishment, it was installed on La Silla in 1968.

Walter Baade had suggested in 1953 that the proposed southern European observatory should have a 1.2/1.8 m (48″/72″) Schmidt like that on Mount Palomar. In the following year a 0.8/1.2 m (32″/48″) Schmidt had been installed at the Hamburg observatory. But, unlike the Palomar Schmidt, the Hamburg Schmidt had an objective prism to produce spectra. European astronomers also wanted their proposed Schmidt to have an objective prism, which meant that they had to compromise on its size to make the telescope affordable. As a result, ESO decided in 1962 to build a 1.0/1.6 m which, although somewhat smaller than the Palomar Schmidt, would produce the same plate scale of 67 arcsec/mm.

Otto Heckmann, from 1962 the first director of the ESO, had been the director of the Hamburg Observatory when the Hamburg Schmidt had been built. Naturally he took a very strong interest in the ESO Schmidt, and suggested that the firm of Heidenreich and Harbeck, which had been responsible for the mechanical parts of the Hamburg Schmidt, should be given the same tasks for the ESO Schmidt, as he was very happy with their work. But the ESO Committee wanted an in-house group to be responsible for developing all major instrumentation. As a compromise an attempt was made to involve both internal engineers and Heidenreich and Harbeck, but this did not work. So in 1965 Heckmann signed a design contract with W. Strewinski, who had led the mechanical design team at Heidenreich and Harbeck, but who had left in the meantime to set up his own design bureau. The contract covered both the ESO Schmidt and the ESO's planned 3.6 m reflector.

Strewinski was late with his Schmidt mechanical drawings, partly because he was also working on the 3.6 m reflector and did not have enough staff. Heidenreich and Harbeck, who were still responsible for manufacture, were also late in making the Schmidt's mechanical parts. So it was not until the end of 1971 that the Schmidt actually arrived on La Silla and was installed in its dome which had been empty there for three years.

Walter Baade had also suggested in 1953 that the proposed ESO observatory should have a 3 m (120 inch) telescope similar to that at the Lick Observatory. But it had since been found that the observer's prime focus cage was really too small in the Lick. As a result, in 1961 Charles Fehrenbach, the director of the Haute-Provence Observatory, and Otto Heckmann suggested that the diameter of the planned European telescope should be increased to 3.5 m. ESO ordered a fused silica mirror blank from Corning in 1965 for the reflector's primary mirror. But when the blank was delivered two years later it was found that it had a useable diameter of 3.6 m, rather than the 3.5 m expected. The mirror, after polishing, was formally accepted by ESO in February 1972.

In the meantime, ESO had found that their main problem with the 3.6 m was not with the primary mirror, as was usually the case with large reflectors, but with the telescope's mechanical parts. It had become clear in the late 1960s that ESO's chosen external design bureau, run by Strewinski, was under-resourced, creating continuing delays to their schedule of both the 3.6 m and the Schmidt. So in 1969 ESO turned to CERN (Conseil Européen pour la Recherche Nucléaire) for assistance. This resulted in the creation of a joint ESO/CERN Technical Project (TP) Division to oversee the design and construction of the 3.6 m telescope, with Strewinski's design bureau being asked to concentrate on completing the Schmidt. In 1972 work had progressed sufficiently with the 3.6 m to allow ESO to place construction contracts for the building, dome and telescope structure. Finally the 3.6 m telescope saw first light in November 1976 to complete the observatory's initial building programme.

Italy and Switzerland joined ESO in 1982 and provided much-needed funding to enable a new 3.6 m reflector, called the New Technology Telescope (NTT), to be built. This NTT incorporated a number of new design features, including a very thin Zerodur primary mirror with active optics, an altazimuth mount, and a housing, based on that of the Multiple Mirror Telescope (MMT), which rotated with the telescope. The altazimuth mount enabled the telescope's housing to be much smaller than for a conventional equatorial mount like that of the existing 3.6 m. In addition, the idea of having the housing rotate with the telescope also reduced its size. First light of the NTT was achieved in early 1989. In addition to normal observing, this telescope was used as a testbed for new technologies to be used in the ESO Very Large Telescope (VLT, see below).

In 1984 the Max Planck Gesellschaft provided their 2.2 m reflector on permanent loan to ESO. Three years later ESO inaugurated the world's first routine system for the remote control of telescopes between continents when they linked La Silla to ESO Headquarters in Garching via an Intelsat communications satellite. Astronomers were then able to remotely control and receive images from the 2.2 m telescope. This remote control facility was later extended to include the NTT, allowing transmission of its first images to Garching in 1989.

National telescopes on La Silla

The very favourable viewing conditions on La Silla became an attraction to many national authorities. The first proposal to place a national telescope at the ESO site was made in the mid-1960s by the director of the Bochum Observatory in West Germany. The idea was approved by the ESO Council in 1966, but it took three years to agree on the conditions under which the 60 cm Bochum telescope would be located there. In the event, ESO astronomers were allocated 30% of the observing time, in return for ESO allowing the telescope to be located on La Silla and allowing the use of ESO facilities. In the meantime the Bochum telescope had been installed on La Silla in September 1968, a year before the conditions of its location and use were formally approved.

The next national telescope to be installed on La Silla was the 50 cm photometric telescope from Copenhagen Observatory in Denmark. It was initially intended to be located there only temporarily and, as a result, it was put in the building which had been used temporarily for the ESO 1 m photometric telescope. But, by the time that this Copenhagen 50 m was installed on the mountain in February 1969, ESO had agreed to let it stay there on a more permanent basis. It was moved into its permanent dome two years later.

Since then a series of national telescopes have been located on La Silla, including the 90 cm Dutch Telescope that was moved from South Africa and re-erected at La Silla in 1978, and the Danish 1.5 m that became operational in 1979. The Danish 1.5 m is still operational, as are the Swiss 1.2 m Leonhard Euler Telescope of the Geneva Observatory, the Italian 0.6 m Rapid Eye Mount (REM) Telescope, the French TAROT (Téléscope à Action Rapide pour les Objects Transitoires), and the Belgian/Swiss 0.6 m TRAPPIST (TRAnsiting Planets and Planetesimals Small Telescope).

Both the REM and TAROT telescopes were designed to observe the afterglow of gamma-ray bursts. Both telescopes quickly slewed automatically to the celestial coordinates of a gamma-ray burst after receiving location data from satellite observatories. REM produced images in the infrared, whilst TAROT produced sub-arcsecond positions of the burst. TRAPPIST was devoted to the detection of exoplanets and the study of comets.

Cerro Paranal

ESO's Very Large Telescope (VLT) project, which consists of four 8.2 m telescopes, was approved by ESO's eight member states in December 1987. It was envisaged that these telescopes could operate either individually or together as an interferometer. Three years later ESO chose to build the VLT at Cerro Paranal at a height of 2,640 m in Chile's Atacama desert. This new site was chosen as it has more than twice as many nights of exceptional seeing as La Silla.

The 8.2 m VLT primary mirrors provided by Schott were, like the primary of the NTT, thin meniscus mirrors requiring the addition of active optics. The first of the four VLT telescopes saw first light in 1998, and the Paranal Observatory was dedicated the following year when the four 8.2 m telescopes were named Antu (Sun), Kueyen (Moon), Melipal (Southern Cross), and Yepun (Venus) from the Mapuche language of a native Chilean people. An adaptive optics system was fitted to one of the 8.2 m telescopes in late 2001, achieving a resolution of $0.04''$ at 1.2 μm.

In the meantime, a contract had been signed in 1998 for the provision of four 1.8 m auxiliary telescopes (ATs) to complete the VLT interferometer. This interferometer consisted of the four 8.2 m unit telescopes (UTs) plus these four 1.8 m ATs in various combinations to provide baselines of up to 200 m. The telescopes were linked together underground by a complex system of mirrors and delay lines. First light with just two

UTs connected together was achieved in 2001. The first AT arrived at Paranal Observatory in January 2004, and the first interference fringes were observed between this and a second AT about a year later. The final two ATs became operational in 2006.

Portugal joined ESO in 2001, followed by the UK (in 2002), Finland (2004), Spain (2006), Czech Republic (2007) and Austria (2008).

The UK provided the 4.1 m VISTA (Visible and Infrared Survey Telescope for Astronomy) to ESO as part of its joining fee. VISTA, which had been conceived and developed by a consortium of eighteen UK universities, was a wide-field (1.65°), infrared survey telescope built near the VLT on Cerro Paranal and completed in 2009. Its primary mirror, which was made of Schott glass, was polished by LZOS in Russia. Its 2,048 × 2,048 pixel detector, which was cooled to 70 K, was made of mercury cadmium telluride chips, rather than silicon, to provide infrared coverage up to 2.5 μm.

Bibliography

Blaauw, Adriaan, *ESO's Early History: The European Southern Observatory from Concept to Reality*, European Southern Observatory, 1991.

Edmondson, Frank K., *AURA and its US National Observatories*, Cambridge University Press, 1997.

Fehrenbach, Charles, *Des Hommes, Des Télescopes, Des Étoiles*, 2nd ed., Vuibert, 2007.

Heck, André (ed.), *Organizations and Strategies in Astronomy*; *Vols. 2 and 7*, Kluwer Academic Publishers, 2001, and Springer, 2006.

West, Richard M., *Europe's Astronomy Machine*, Sky and Telescope, May 1988, pp. 471–481.

See also: Adaptive optics; Modern optical interferometers.

GREENWICH OBSERVATORY

One of the key problems of the seventeenth century for a maritime country like Britain was that of determining the longitude of ships at sea. A number of possible solutions had been proposed over the years, of which the most promising seemed to be one that was brought to the attention of King Charles II in 1674 by Le Sieur de St Pierre. This involved observing the movement of the Moon relative to the stars and planets. Charles II set up a committee to examine this solution but, as John Flamsteed advised the committee, the orbit of the Moon was not known with sufficient accuracy at the time for the method to work. As a result Sir Jonas Moore, the Surveyor General at the Ordnance Office, persuaded the king in 1675 that an astronomical observatory should be built to produce observations of the required accuracy. The Ordnance Office was tasked with building the observatory at Greenwich, just outside London, and Flamsteed was appointed the 'astronomical observator' or Astronomer Royal, as we now know the position, and director of the observatory. The observatory building, sometimes called 'Flamsteed House', was designed by Sir Christopher Wren and was completed in the following year. Moore paid for the observatory's key instruments, which included a mural quadrant designed by Robert Hooke, a 7 ft (2.1 m) iron sextant, and two highly accurate clocks. The iron sextant was used until 1689 when a 7 ft radius mural arc was completed, paid for by Flamsteed from his own personal funds.

Flamsteed and his assistants made numerous observations of stars, planets, comets and sunspots. Flamsteed was very reticent about publishing these results, as they had been carried out with instruments that he considered his own property. Some results were published in 1712, but the complete authorised version was not published until 1725, six years after his death. This was contained in the three-volume *Historia Coelestis Britannica*, which contained much else in addition to Flamsteed's own observations. His star catalogue, also included in this publication, gave the position of over 2,900 stars. It was a great achievement, providing a significant improvement over all previous catalogues.

Edmond Halley was appointed the second Astronomer Royal in 1720, but Flamsteed's widow had removed all his instruments from the observatory. So Halley was the director of an observatory with no instruments. Nevertheless he was provided with funding to purchase some new ones and, in the following year, he set up a meridian transit instrument in a small new building. This was followed four years later by an 8 ft (2.4 m) wall-mounted iron quadrant built by George Graham, which was installed in a new quadrant room. Halley's discovery of the secular acceleration of the Moon, his work on the transits of Mercury and Venus, and his work on comets largely predated his appointment. But his observations of the Moon at the observatory showed that it was often possible to determine the longitude at sea within about 70 miles at the equator. This was not accurate enough for mariners, and sometimes the error was much larger than this, but his work was at least a step in the right direction.

James Bradley took over as Astronomer Royal on Halley's death in 1742. Prior to his appointment he had discovered stellar aberration, which was due to the finite velocity of light, and had detected the nutation of the Earth's axis. He (correctly) suggested in 1748 that the latter effect, which had an amplitude of $\pm 9''$ with a period of about 18 years, was caused by the gravitational attraction of the Moon. Bradley, after taking up his appointment, ordered an 8 ft brass mural quadrant and an 8 ft transit instrument. These were installed in a new observatory building in 1750, where he also sited his own $12\frac{1}{2}$ ft (3.8 m) zenith sector. The transit instrument defined the Greenwich meridian until 1816, when it was replaced by a new 10 ft (3 m) transit instrument. Bradley's measurements, which were highly accurate, were still being used to determine the proper motion of stars at the end of the nineteenth century.

Nevil Maskelyne, who was appointed the fifth Astronomer Royal in 1765, was responsible in the following year for the publication of *The Nautical Almanac and Astronomical Ephemeris*. This included daily tables of the positions of the Sun, Moon and planets, as well as tables of angular distances of the Moon from the Sun and ten stars. These enabled finding longitude at sea to within $\frac{1}{2}°$. The publication later became the standard almanac for mariners worldwide.

John Pond became Astronomer Royal in 1811, and seven years later the Admiralty became responsible for the Royal Observatory. Pond instituted the first public time signal in England in 1833, when a time ball was dropped down a mast on Flamsteed House every day at 1 o'clock precisely.

The next Astronomer Royal, George Airy, was in post from 1835 to 1881. During this time he paid meticulous attention to the design of new instruments, striving for the highest possible accuracy in making and analysing the astrometric observations used for navigation and time keeping. He also used the newly created railway and international telegraph systems for the widespread distribution of accurate time signals. In 1850 Airy installed a new transit circle, which was then used to define the position of the Greenwich meridian, a little to the east of the previous two transit instruments. Then in 1884, at the Washington Conference, this meridian through the Airy transit circle was accepted as the international zero of longitude, and Greenwich Mean Time was accepted as the international time standard.

Airy concentrated on meridional astronomy, which was routine and rather uninteresting. However, he did accept that some of his astronomers wanted to undertake other astronomical work. So in 1859 he acquired a 12.8 inch (32 cm) Merz visual refractor, which encouraged the development of astrophysical studies at the Greenwich observatory. Long-term photographic observations of the Sun were begun in 1873.

William Christie, Airy's successor, equipped the observatory with new telescopes and equipment for astrophysical observations. These included a 28 inch (71 cm) visual refractor installed in 1893, followed shortly afterwards by a 26 inch (66 cm) photographic refractor and a 30 inch (76 cm) photographic reflector on a twin mounting. The observatory also acquired a 13 inch (33 cm) photographic refractor in order to participate in the international *Carte du Ciel* project.

Frank Dyson's period of Astronomer Royal straddled the First World War, during which he kept all essential work going even though the staff had other commitments to wartime work. After the war he helped to organise the joint Greenwich—Cambridge expedition to observe the 1919 total solar eclipse, during which observations were made that supported the predictions of Einstein's general theory of relativity. Shortly before Dyson's retirement, William Yapp offered to fund a telescope in recognition of Dyson's work. This was the 36 inch (91 cm) Yapp reflector which was dedicated in 1934.

The move from Greenwich

Seeing conditions at Greenwich had gradually deteriorated in the nineteenth century, mostly caused by industrial and domestic pollution. The problem had then accelerated in the early twentieth century with the building in 1906 of a large generating station nearby, and with the increasing use of electric lights. Suggestions were made as early as the turn of the nineteenth century that the observatory should be moved to a more suitable site, but it wasn't until 1938 that a move was seriously considered. Matters were then put on hold because of the Second World War, but in 1944 the Admiralty approved the move in principle, and a short list of five sites was produced. Two years later it was agreed to move the observatory to Herstmonceux Castle in rural Sussex. But the move was not completed until 1958.

The new observatory, called The Royal Greenwich Observatory at Herstmonceux, acquired the 98 inch (2.5 m) Isaac Newton Telescope (INT) which was completed in 1967. But the seeing at Herstmonceux was not good by international standards. So the UK surveyed a number of possible sites in the North Atlantic area in the early 1970s to find a suitable location to build a new observatory. They eventually chose one at the Roque de los Muchachos, at an altitude of 2,300 m (7,600 ft), on La Palma in the Canary Islands.

Under the terms of the agreement with the Spanish authorities on the new observatory, the UK was to build three main telescopes there. These were the 4.2 m William Herschel Telescope (WHT), the 2.5 m INT and a 1.0 m astrometric reflector (later called the Jacobus Kapteyn Telescope or JKT). The JKT saw first light at the Roque de los Muchachos observatory in 1984. The INT was moved from Herstmonceux to this new observatory, where it also saw first light in 1984 with a new 101 inch (2.6 m) mirror. But the largest telescope on La Palma was, for many years, the 4.2 m (165 inch) WHT which saw first light in June 1987. At that time it was the third largest optical telescope in the world.

In the meantime the administration of the Royal Greenwich Observatory (RGO) had been transferred in 1965 from the Admiralty to what was to become known as the Science and Engineering Research Council (SERC). In 1972 the positions of Astronomer Royal and director of the observatory were separated. Then in 1986 SERC decided that the RGO would be moved from Herstmonceux to near the Institute of Astronomy at Cambridge. This move was completed in 1990. Finally eight years later the RGO was closed, with just a few staff being retained for work on La Palma, at the Royal Observatory, Edinburgh, and at the Nautical Almanac Office (NAO) which was moved to the Rutherford Appleton Laboratory. In 2006 the NAO was moved to the UK Hydrographic Office at Taunton, Somerset.

Bibliography

Forbes, Eric G., Meadows, A. J., and Howse, Derek. *Greenwich Observatory: The Royal Observatory at Greenwich and Herstmonceux (1675–1975), Vol. 1 by* Eric G. Forbes, *Origins and Early History (1675–1835); Vol. 2 by* A. J. Meadows, *Recent History (1836–1975); Vol. 3 by* Derek Howse, *Buildings and Instruments,* Taylor and Francis, 1975.

Gingerich, Owen (ed.), *The General History of Astronomy, Vol. 4A: Astrophysics and Twentieth-Century Astronomy to 1950,* Cambridge University Press, 1984.

Krisciunas, Kevin, *Astronomical Centers of the World,* Cambridge University Press, 1988.

Maunder, Edward Walter, *The Royal Observatory Greenwich; A Glance at its History and Work,* Religious Tract Society, 1900 (BiblioBazaar republished 2010).

Ronan, Colin A., *Their Majesties' Astronomers,* Bodley Head, 1967.

Willmoth, Frances, (ed.), *Flamsteed's Stars: New Perspectives on the Life and Work of the First Astronomer Royal, 1646–1719,* Boydell Press, 1997.

HARVARD COLLEGE OBSERVATORY

In 1839 Harvard College persuaded the well-known Boston clockmaker and amateur astronomer William Cranch Bond to take the post of Astronomical Observer to the University. However, not only was this position unpaid, but Bond also had to bring his own instruments which he set up in a special cupola mounted on the roof of Dana House.

The cupola was not really very satisfactory, and Bond's instruments were relatively modest for an institutional observatory, particularly compared with those in Europe. But public interest in astronomy in America, and in the Boston area in particular, was much increased by the very bright comet of 1843. As a result, public subscriptions were collected to provide Harvard with a state-of-the-art telescope in a new observatory. As a result Bond was finally paid a salary and the new observatory was completed on the outskirts of Cambridge, Massachusetts, in 1846. In the following year it was equipped with a twin of the 15 inch (38 cm) refractor of the Pulkovo Observatory which was, at that time, the largest refractor in the world.

In 1848 W. C. Bond and his son George Phillips Bond discovered Saturn's eighth satellite, Hyperion, with the 15 inch refractor, followed two years later by their discovery of Saturn's dusky C ring. Both Bonds also experimented with astrophotography at the Harvard College Observatory. G. P. Bond exhibited his daguerreotype of the Moon in 1851 at London's Great Exhibition. Six years later he also published photographs of Alcor and the double star Mizar using the relatively new wet collodion process. He suggested that the relative intensities of stars could be determined from the physical size of their photographic images for a standard photographic exposure. G. P. Bond succeeded his father as director of the observatory in 1859. But his tenure was cut short by his death in 1865 at the early age of thirty-nine.

Harvard College Observatory was bedevilled by lack of money in its early decades. Nevertheless Joseph Winlock, who was its third director from 1866 to 1875, was able to purchase a spectroscope and new meridian circle shortly after taking up his appointment. Winlock supplemented the observatory's limited income by providing a time service to the railways and to jewellers to set their watches. He made thousands of double star measurements with the 15 inch refractor during his directorship. He also continued with experiments in photography, and began a spectroscopic and photometric programme. Winlock was succeeded by Edward C. Pickering who was director from 1877 to 1919, determined to extend Winlock's work in astrophysics.

Pickering's appointment initiated a period of rapid growth in the observatory's staff and equipment, as he started a series of major projects. But before he could begin these, he needed to raise additional funds. At that time the observatory received no funding from the university. It relied, instead, on the relatively modest amount of money raised from the sale of its publications, the provision of the time service organised by Winlock, and earnings from investments. Within a year Pickering had managed to secure additional public donations to increase the observatory's annual income by about 35%.

Pickering wanted to determine quantitatively the visual magnitude of all naked-eye stars visible from Harvard. To do this he designed a meridian instrument in 1879 to measure a star's brightness relative to that of a standard reference star, Polaris, by using a calibrated Nicol prism to equalise the brightness of the two stars. With this meridian photometer he measured the intensities of over 4,000 stars, publishing his results in the *Harvard Photometry Catalogue* of 1884. Pickering also used a modified version of his photometer to enable fainter stars to be observed, and asked Solon Bailey to go to Arequipa, Peru, to measure the intensities of the southern stars. The results were published in the *Revised Harvard Photometry Catalogue* in 1908 which contained the magnitudes of over 45,000 stars. It was the standard reference work for many years.

Pickering had outlined plans in 1883 for a major measurement and analysis programme of stellar spectra, and an extension of the Harvard plate collection of astronomical photographs to include a complete photographic map of the sky. To undertake these programmes he needed additional sources of funds. Over the period from 1886 to 1889 Pickering managed to obtain four major sources of funding for Harvard, including more than $300,000 in 1886 from the Robert Treat Paine Fund. He used the income from this for routine operational costs. In the same year he obtained another large sum from Henry Draper's widow to carry out a long-range programme, as a memorial to her husband, of imaging and classifying the spectra of a large number of stars.

Pickering placed a shallow prism over the objective of the 8 inch (20 cm) Bache photographic refractor, which had been paid for by the National Academy of Sciences, to enable stellar spectra to be photographed in quantity. This allowed Williamina Fleming and Pickering to photograph and analyse the spectra of over 10,000 stars using his Harvard spectral classification system. The results were published in 1890 in the *Draper Memorial Catalogue*. This Harvard classification system, which was progressively modified over the years mainly by Annie Cannon and Pickering, was adopted as a provisional standard by the International Solar Union in 1913. Nine years later it was also accepted by the International Astronomical Union as the official system. Eventually the *Henry Draper Catalogue*, which was published over the period 1918–1924 by Cannon and Pickering, included the spectral classification of about 225,000 stars.

A bequest of about $240,000 had become available to Harvard from Uriah A. Boyden's estate in 1887 to build a high-altitude observatory. Pickering used this to establish the Harvard Boyden outstation at an altitude of about 2,450 m (8,040 ft) at Arequipa, Peru. Two years later Catherine W. Bruce gave Harvard $50,000 to enable the observatory to purchase a 24 inch (61 cm) refractor to photograph the whole sky. Pickering planned to undertake this project at Harvard and at Arequipa, instead of joining the international *Carte du Ciel* project, which he heavily criticised. In the event the first photographic map of the entire sky, which was published in 1903, was produced using 2.5 inch (6.5 cm) lenses at Harvard and Arequipa.

The Bache 8 inch refractor, which had first been taken to the Boyden station in 1889, was routinely used to photograph large areas of the sky, so considerably extending the Harvard plate collection. The Bruce 24 inch refractor, which was first set up in Peru in 1896, was used to produce a complete map of the southern sky. In 1899 William Pickering, Edward's brother, discovered Phoebe, the ninth satellite of Saturn, on plates taken with the Bruce 24 inch. But by far the most important work undertaken with these plates was the work on Cepheid variables done by Henrietta Leavitt in which she discovered the period-luminosity relationship by observing Cepheids in the Small Magellanic Cloud.

So by the early twentieth century Edward Pickering had managed to transform the Harvard College Observatory from a small, poorly equipped observatory to the foremost astrophysical observatory in the United States, with a high-altitude outstation in Peru.

Harlow Shapley was the next director of the observatory, serving from 1921 to 1952. In his first two decades at Harvard he studied variable stars in the Milky Way, studied the Magellanic

Clouds, produced a catalogue of the brighter galaxies, and undertook a major survey of the distribution of faint galaxies in the southern hemisphere. In 1927 he moved the Boyden station at Arequipa to Mazelspoort near Bloemfontein in South Africa. He also installed a 60 inch (1.5 m) reflector at Mazelspoort, which Harvard had purchased from the Andrew A. Common estate. In the 1930s Shapley discovered the first dwarf galaxies of the local group from photographic plates taken in South Africa.

Shapley introduced a graduate programme in astronomy at Harvard in the 1920s, and, in the following decade, he encouraged foreign astronomers to work for a year or two at Harvard. Later in the 1930s he led a countrywide effort to rescue Jewish scientists and their families from totalitarian regimes in Europe and bring them to the United States.

The Smithsonian Astrophysical Observatory (SAO) moved its headquarters to Cambridge, Massachusetts in 1955. Eighteen years later the Harvard College Observatory and the SAO merged to form the Harvard-Smithsonian Astrophysical Observatory. In common with many other traditional observatories, their best ground-based observational facilities are now jointly owned at very remote sites with other institutions.

Bibliography

Bailey, Solon I., *The History and Work of Harvard Observatory 1839 to 1927*, McGraw-Hill, 1931.

Bok, Bart J., *Harlow Shapley, 1885–1972*, Biographical Memoirs of the National Academy of Sciences, **49** (1978), 241–291.

Hearnshaw, J. B., *The Analysis of Starlight: One Hundred and Fifty Years of Astronomical Spectroscopy*, Cambridge University Press, 1986.

Jones, Bessie Zaban, and Boyd, Lyle Gifford, *The Harvard College Observatory: The First Four Directorships, 1839–1919*, Harvard University Press, 1971.

Stephens, Carlene E., *Astronomy as Public Utility: The Bond Years at the Harvard College Observatory*, Journal for the History of Astronomy, **21** (1990), 21–35.

HAUTE-PROVENCE OBSERVATORY

In the 1930s French astronomers began to look for a new site for an astrophysical observatory. After five years testing with an 0.8 m reflector it was concluded that a site in the foothills of the French Alps, about 100 km north of Marseilles, would be suitable as it benefited from a Mediterranean climate. As a result the Haute-Provence Observatory was built there on a 650 m high plateau, near the village of St Michel and about 10 km from the town of Forcalquier. Irène Joliot-Curie, the French Minister for Scientific Research, approved the decision to build the observatory in 1936. Construction started two years later and the first astronomical observations were made from

the observatory in 1943 with a 1.20 m reflector. However, little work was done until after the end of the Second World War two years later.

The 0.80 m reflector, which had been used in 1932 for the initial site survey, was installed at the observatory in 1945. Four years later it was the first telescope to be used by foreign astronomers, Geoffrey and Margaret Burbidge. Then in 1958 the observatory's largest reflector, a 1.93 m, was installed at the observatory. This telescope had been made by the British firm Grubb Parsons using a primary mirror that had been cast in France in 1939. A 60 cm reflector was also installed in 1958 to undertake photoelectric photometry. The 60 cm remained at the observatory until the late 1970s when it was moved to the Nice Observatory. In the meantime a 1.52 m reflector had been operating at the Haute-Provence Observatory from 1967.

Foreign-owned or jointly owned telescopes have also been installed at the Haute-Provence Observatory. For example the Geneva Observatory located its 1.0 m telescope there in 1966, and a 62/90 cm Schmidt, jointly owned by the Haute-Provence Observatory and the University of Liège, saw first light in 1970.

In more recent times the Haute-Provence observatory has had to change from undertaking frontier-level research to undertaking large-scale surveys. Frontier research is now generally the prerogative of observatories at much higher altitudes with much larger telescopes. But observing time at these better facilities is at a premium, making large-scale surveys problematical for them, thus providing a niche for observatories like Haute-Provence.

One such survey carried out at the Haute-Provence Observatory yielded dramatic results in 1995, a year after Michel Mayor and Didier Queloz had begun a programme to try to detect planets by measuring variations in stellar spectra using the 1.93 m telescope. They found the first clear evidence of a planet around another star, other than a pulsar. This discovery opened the floodgates, so that by now hundreds of exoplanets have been discovered by astronomers worldwide.

Bibliography

Dufay, Jean, and Fehrenbach, Charles, *France's Haute-Provence Observatory*, Sky and Telescope, July 1961, pp. 4–9.

Heck, André (ed.), *The Multinational History of the Strasbourg Astronomical Observatory*, Springer, 2005.

See also: Adaptive optics.

KARL SCHWARZSCHILD OBSERVATORY, TAUTENBURG

Germany lost a number of astronomical telescopes as a result of the Second World War. Some were destroyed accidentally by enemy action, and some, including the 1.2 m Babelsberg reflector, were dismantled by the Soviet Union and confiscated as war reparations. So, in the war's aftermath, the question was raised in Germany: Which telescopes should be replaced with more modern instruments? Also, should German astronomers continue their previous spectrographic and photometric research using normal reflecting telescopes, or should they concentrate on galactic and extragalactic astronomy using Schmidt telescopes? In 1948 Hans Kienle, the director of the Potsdam Astrophysical Observatory, suggested both could be achieved by building a universal reflecting telescope that could be operated both as a conventional telescope and as a Schmidt. This was the basic idea behind what came to be called the 2 m Alfred Jensch Teleskop.

This new 2 m reflector was built at the new Karl Schwarzschild Observatory at an altitude of just 350 m about 10 km from Jena in East Germany's Tautenburg Forest. The telescope, which was completed in 1960, could be operated in a Nasmyth configuration, a coudé configuration, or as a 1.34/2.00 m Schmidt camera. Built by the nearby Carl Zeiss of Jena, it is still the largest Schmidt camera in the world, and the largest telescope in Germany. A second corrector plate was also produced which was prismatically ground to produce objective spectra. (This was the first time that such a prismatic Schmidt corrector plate had been made anywhere.) Unfortunately, the operation of the 2 m telescope was compromised by its relatively low observatory altitude and by its proximity to towns and villages, in spite of attempts by the authorities to control locally generated light pollution.

In 1969 the observatory became part of the Central Institute for Astrophysics of the Academy of Sciences of the German Democratic Republic, but after the reunification of Germany it was operated as a Thuringian State Observatory. Its main instrument is still the 2 m.

Bibliography

Van den Bergh, Sidney, *Visiting Germany's Largest Telescope*, Sky and Telescope, May 1964, pp. 268–272.

KITT PEAK AND CERRO TOLOLO OBSERVATORIES

The foundation of AURA

There was no tradition of publicly funded observatories in the United States before the Second World War, unlike in many European countries. As a result, many American astronomers had no access to the best facilities, like those on Mount Wilson, as their use was limited to members of the observatory or university who operated them. But in 1940 Otto Struve suggested setting up a collaborative observatory between a large number of astronomical institutions to provide wider access to first class facilities. But his suggestion had to be put on hold until after the war.

The Second World War saw a turning point in the funding of technology by the United States government, as centralised government funding was used, in many cases for the first time, for all types of scientific research. This continued after the war in many areas, but naturally at a lower level. In 1949 the National Research Council (NRC) asked the American Astronomical Society (AAS) to consider how the proposed National Science Foundation (NSF), which was to be established in the following year, could best support astronomical research in the United States. In response, the AAS suggested in 1951 that the NSF should provide something of the order of $200k–$400k annually for new capital facilities.

Thirty-five astronomers attended a conference at Flagstaff, Arizona, in 1953 called by the NSF to discuss photoelectric photometry research and to consider possible new facilities. At this conference Leo Goldberg, of the University of Michigan, suggested that a National Observatory should be funded by the NSF. It should have both a 100 to 120 inch (2.5 to 3.0 m) telescope, and a 36 inch (0.9 m) telescope. In addition, he suggested that the observatory, which could carry out all types of observing, including photometric, spectroscopic, photographic and solar work, should be located in the south-western United States where the weather cycle was out of phase with that of California. As a result, the NSF appointed an advisory panel under the chairmanship of Robert R. McMath, the president of the AAS, to examine the need and location for Goldberg's suggested National Astronomical Observatory (NAO).

McMath's panel endorsed the idea of a National Astronomical Observatory at its meeting in November 1954, and recommended that Aden B. Meinel, of the Yerkes Observatory, should undertake a search for suitable sites for both stellar and solar work. They further recommended that a 30 inch (0.8 m) telescope should be installed as soon as possible, followed by an 80 inch (2.0 m), but that a decision on a very large telescope should be deferred. It was later agreed that Helmut Abt should assist Meinel in site testing, and that a 36 inch (0.9 m) should be built instead of a 30 inch.

Meinel and Abt's site testing of potential sites in the south-western United States narrowed the choice down to sites on Kitt Peak and Hualapai, both in Arizona, with most of the selection criteria favouring Kitt Peak. As a result in March 1958 the Scientific Committee overseeing the site surveys unanimously voted for Kitt Peak as the site for the National Astronomical Observatory. Unfortunately, things were complicated by the fact that the 6,875 ft (2,100 m) high Kitt Peak was on a Papago Indian reservation. But permission was eventually obtained from the Tribal Council to use the site, provided the land was used only for astronomical and related research.

In the meantime, institutional plans had been developed for the observatory. This resulted in seven universities, namely

California, Chicago, Harvard, Indiana, Michigan, Ohio State and Wisconsin, becoming founder members of the Association of Universities for Research in Astronomy (AURA) in October 1957. The National Science Foundation agreed to provide funds to AURA for observatory buildings and capital equipment. Two months later the AURA Board appointed Aden Meinel as the first Observatory Director. A year later the observatory was named the Kitt Peak National Observatory (KPNO).

The first artificial satellite, Sputnik 1, had been successfully launched on 4 October 1957 whilst these plans for the National Astronomical Observatory were being finalised. This Sputnik 1 launch, followed only a month later by that of the much heavier Sputnik 2, created a degree of panic in the United States, because of their military significance. This caused Congress to agree on a rapid increase in funds for both military and civil projects, including substantial extra funding for the NSF. This allowed preliminary work to start on the large solar telescope, previously proposed by McMath, which had been deferred due to lack of funds.

National Observatory telescopes on Kitt Peak

Planning for the first medium-sized telescope to be built at the National Astronomical Observatory had been started in 1955, the same year that Meinel and Abt had started their site survey. In the event this 36 inch (0.9 m) telescope was completed on Kitt Peak in March 1960. Its Kanigen-coated, aluminium primary mirror was replaced in 1963, and the original mirror was used in a second 36 inch telescope that was completed in 1966. The mirror of the first instrument was again replaced in both 1966 and 1970. In the latter case its new mirror was made of Cer-Vit, a new very-low-expansion material.

The next reflector to be built at Kitt Peak was an 84 inch (2.1 m) that used a Pyrex blank purchased from Corning. Aden Meinel's first design had been rejected by AURA as it was thought to be too innovative. Nevertheless he was still able to incorporate some non-standard concepts in the telescope that was eventually built. For example, most reflectors at that time used a standard Cassegrain arrangement with a parabolic mirror, whereas Meinel adopted a Ritchey–Chrétien design using a hyperbolic primary. This had the advantage of producing less off-axis distortion, and so made it more suitable for wider field photography. The telescope was ready for use by visiting astronomers in September 1964.

In 1958 the NSF, when it agreed to fund the large solar telescope, had insisted that it must be located on the same site as the National Astronomical Observatory to minimise its construction and operating costs. The selected design consisted of an 80 inch (2.0 m) diameter heliostat that reflected the Sun's light onto an image-forming parabolic mirror 60 inches (1.5 m) in diameter some 500 ft (150 m) away along the polar axis of the telescope. The light was then reflected back to a 48 inch (1.2 m) diameter flat mirror, which deflected the light into an observing room, where the solar image could be viewed. The heliostat was mounted 100 ft (30 m) above the ground, to minimize atmospheric distortion, with about the last 300 ft (90 m) of the 500 ft light path, between the heliostat and the 60 inch mirror, being located in a tunnel underground.

This Kitt Peak telescope was named the McMath (later McMath-Pierce) solar telescope at its dedication in November 1962. The solar image was 34 inches (84 cm) in diameter, allowing the study of individual solar granules, which were about 0.4 mm in diameter on the image. The 70 ft (20 m) solar spectrum was used, amongst other things, to confirm the existence of fluorine and chlorine on the Sun.

The 60 and 48 inch mirrors used in the initial construction of this solar telescope had been produced by General Electric as 60 inch quartz blanks as part of their contract to produce the 200 inch main Palomar mirror. They had remained unused since the 1930s, so Ira S. Bowen at Palomar agreed to give them to McMath for use in a solar telescope. One of these mirrors was then cut down to 48 inch diameter and used in the Kitt Peak solar telescope until 1965, when it was replaced by a 48 inch beryllium mirror. Another of the 60 inch mirrors was used in the solar telescope until it was replaced in 1966.

McMath suggested in early 1961 that AURA should start planning for an even larger astronomical reflector than the 84 inch, with a mirror of about 150 inches (3.8 m) in diameter. This was accepted in principle by AURA, and preliminary work was undertaken later that year. So by January 1962 there was already an outline design available. It was suggested that the telescope should be about 150 ft (45 m) above ground level to improve the seeing. After a number of problems in the bid phase for the primary mirror blank, the Corning bid was disallowed, and the contract was awarded to General Electric. Their 158 inch (4.0 m) fused quartz mirror blank was delivered to Tucson, Arizona about three years later in October 1967 where its surface was ground and polished to the required shape in Kitt Peak's optical shop.

The cylinder, observatory structure and dome of the 158 inch were completed by the end of 1970. First light was achieved on 27 February 1973, and the telescope was formally dedicated a few months later. It was officially named after Nicholas U. Mayall, who had been director of KPNO from 1960 until he retired in 1971. At the time it was the second largest telescope in the United States, behind the 200 inch Hale reflector.

Non-NAO telescopes on Kitt Peak

First light of the 158 inch (4.0 m) saw the end of the main construction phase of the Kitt Peak National Observatory. But this was by no means the end of the construction of optical telescopes on the mountain as the KPNO telescopes were joined by those of a number of other American institutions.

The first such telescope was the 36 inch (0.9 m) Steward Observatory reflector of the University of Arizona, which had originally been built on its Tucson campus in 1922. Whilst at Tucson it had been used in 1955 to show the Moon to representatives of the Papago Tribal Council, to obtain permission to construct the National Astronomical Observatory on Kitt Peak. But light and atmospheric pollution at Tucson had reduced the usefulness of this instrument, and so in 1962 it was moved to Kitt Peak.

Three years later, the NSF provided a grant to the University of Arizona to build a 90 inch (2.3 m) reflector for its Steward Observatory on Kitt Peak. The State of Arizona donated money to pay for the building and dome. General Electric were chosen to supply the mirror blank of fused quartz, whilst its surface was figured at the University of Arizona's optical shop in Tucson, and aluminised at the Lick Observatory. At first light in 1968 it was the fourth largest optical telescope in the USA. It was officially named the Bok telescope in April 1996, in honour of Bart Bok, who had been director of the Steward Observatory from 1966 to 1969.

In 1967 the NSF agreed to partially fund a new 52 inch (1.3 m) reflector at the University of Michigan's Portage Lake Observatory, near the university's Ann Arbor campus. This was to enable easy access for both undergraduate and graduate students who were to be its major users. The telescope had a Cer-Vit primary blank provided by Owens-Illinois.

Unfortunately, Portage Lake proved unsuitable for the 52 inch, as not only was the weather too cloudy, there was also a growing problem of light pollution. So in 1975 it was moved to Kitt Peak and placed in the McGraw-Hill Observatory, which had been purpose-built for the telescope with money provided by McGraw-Hill and the Alfred P. Sloan Foundation. Operation of the observatory was undertaken by a consortium consisting of the University of Michigan, Dartmouth College and MIT, or MDM, all of whom provided instrumentation.

As a result of these developments, there was by 1975 a plethora of optical telescopes on Kitt Peak. KPNO had the 158 inch (4.0 m) Mayall, an 84 inch (2.1 m), and a number of smaller reflectors. There was also the McMath solar telescope, plus a vacuum solar telescope, which had been built to support the Skylab space mission. In addition, two other observatories had been set up there, namely the Steward Observatory with its 90 inch (2.3 m) and 36 inch (0.9 m) reflectors, and the McGraw-Hill Observatory with its 52 inch (1.3 m) reflector.

Shortly afterwards Case Western Reserve University (CWRU) moved their 24/36 inch (0.6/0.9 m) Burrell Schmidt telescope to the mountain in 1979. It had originally been located at CWRU's Warner and Swasey Observatory at Cleveland, Ohio in 1941. But light pollution had caused it to be moved 30 miles away to Montville, Ohio in 1957. Now it had to be moved again. Its primary mirror was refigured and a new corrector plate installed to provide much sharper images than previously.

The MDM consortium decided in 1982 to extend their collaboration on Kitt Peak by building a 95 inch (2.4 m) reflector there. The design was based on that of the 91 inch (2.3 m) University of Wyoming instrument that had been installed a few years earlier on Jelm mountain. The Cer-Vit primary mirror was cast by Owens–Illinois as a 50 cm thick blank that was cut in half horizontally; one went to the MDM Observatory, and the other was used in the Advanced Technology Telescope at Siding Spring, Australia. The MDM telescope, which was dedicated in 1989, was named the Hiltner telescope, after William A. Hiltner, chairman and professor emeritus of Michigan's astronomy department, who had overseen its development.

National funding of the National Optical Astronomy Observatory (NOAO), which consisted of KPNO, the Cerro Tololo Inter-American Observatory (see later), and the Sacramento Peak Observatory, was progressively reduced in real terms in the 1980s. As a result, by 1988 the NOAO director, Sidney Wolff, was faced with the prospect of closing down some of the KPNO facilities. To assist her in making a decision, she wrote a letter to about 2,000 astronomers, asking for their suggestions. She saw two basic options. Either the NOAO could devote most of its resources to preserving and operating existing telescopes, or it could close these down and spend most of its money on developing major new facilities. In the first case, it could keep open as many of the existing telescopes as possible, so serving the maximum number of astronomers, or it could concentrate its resources on those facilities that were unique as far as the American astronomical community was concerned. These latter facilities were mostly on Cerro Tololo in Chile (see next section), which had access to the far southern sky.

In response to Sidney Wolff's letter, the American astronomical community overwhelmingly opted for keeping open existing telescopes, at the expense of curtailing the development of new facilities. As a result, NOAO closed their Advanced Development Program, and stopped supporting Roger Angel's spinning oven laboratory in Tucson. They also agreed to close one of their two 36 inch (0.9 m) telescopes. Maintenance was also deferred.

One possible solution to KPNO's financial problems was to undertake more joint programmes with universities, and to sell off or privatise some of their existing telescopes. These solutions were not ideal, as they restricted access of the general American astronomical community to these facilities, but it was considered to be a compromise worth investigating.

In fact, the first significant example of a public–private partnership at Kitt Peak had been started in the late 1980s. This was the 3.5 m (138 inch) WIYN telescope of the universities of Wisconsin, Indiana and Yale, and the NOAO.

The WIYN telescope was one of a new breed of lightweight, compact, medium-sized telescopes of the late 1980s and early 1990s, which had an altazimuth mount, instead of the usual equatorial. This made the telescope mounting system far simpler and compact. The WIYN's highly reflective octagonal 'dome' was designed to minimise solar heating during the day, and optimise air flow over the telescope at night to minimise atmospheric distortions. In addition, the telescope used active optics to control distortions of the primary mirror. There was also a highly sophisticated thermal control system for the primary mirror, which allowed individual parts to be cooled by passing thermally conditioned air through parts of its honeycomb substrate. The WIYN telescope was formally dedicated in October 1994.

Interestingly, the 4.0 m Mayall, which had achieved first light some twenty years before the 3.5 m WIYN, had a 15 ton mirror compared with the WIYN's 2 tons. In addition, the cost of the Mayall was over double that of the WIYN, in real terms. And yet the WIYN outperformed the Mayall by a large margin, with a much lower wavefront error.

It had been hoped, when KPNO was founded in the late 1950s, that it would serve as the top-quality optical observatory for those American astronomers who, at that time, did not have access to top-quality instruments at a premier observing site. In this it has been very successful. But over the last 50 years other optical observatories have been developed outside of the continental United States that have been found to have better 'seeing' than Kitt Peak. So it is to these that today's best telescopes are generally directed.

Cerro Tololo Inter-American Observatory

In 1958 Federico Rutllant, director of the National Observatory of the University of Chile, asked whether the NSF and AURA were interested in building and operating an astronomical observatory in Chile as a joint project with the University of Chile. AURA felt that they would not be able to support such an idea at the time, as they were in the process of setting up their observatory on Kitt Peak. On the other hand, Gerard Kuiper and William Hiltner, of the Chicago Yerkes Observatory, were interested.

At that time, the collaborative agreement between the Universities of Texas and Chicago to run the McDonald Observatory was coming to an end, and Kuiper and Hiltner both saw the benefits of an expansion of the Chicago/Texas cooperation into the Southern Hemisphere. Hiltner contacted the United States Air Force Geophysics Research Department (GRD) to ask if they were interested in such a facility. They responded positively, and this eventually led to an agreement between the universities of Chicago and Texas to set up a joint observatory in Chile with the University of Chile, with funding provided by the Air Force GRD. Formal approval of Air Force funding

was received in December 1959 for the provision of a 60 inch (1.5 m) reflector.

In the meantime, Kuiper had asked Jurgen Stock to commence site testing in Chile. Kuiper also tried once again to interest AURA in the project, as he felt that the project, because of its international aspects, needed a broader political base as well as additional funding. The NSF and AURA showed more interest this time, and the AURA Scientific Committee studied the design of the proposed 60 inch in March 1960. As a result they suggested a number of design changes. Two weeks later, Vice-President Warren Johnson of the University of Chicago proposed that AURA be asked to take over the running of the project, although the universities of Chicago and Texas still wished to be involved. This AURA agreed to do. The Air Force agreed to provide a grant for the design of the telescope mounting, purchase of the 60 inch mirror blank, and general planning, whilst the NSF provided money for preliminary work in FY 1962.

From his site surveys, Stock had finally settled on two alternative sites in the lower Andes, Cerro Tololo and La Peineta. La Peineta had more hours of clear sky than Cerro Tololo, but it was subject to higher winds. In addition, Cerro Tololo was on private land that could be purchased, whereas La Peineta was on government-owned land that could not. As a result, in late 1962 the 2,200 m altitude Cerro Tololo was chosen as the site for the AURA southern observatory, which became known as the Cerro Tololo Inter-American Observatory (CTIO). Cerro Tololo was near to the southern part of the Atacama desert, making it ideal for infrared observing. The AURA Executive Committee also decided that the observatory headquarters would be located at La Serena not far away on the Pacific coast.

Work started on flattening the top of the mountain in January 1964. Mayall suggested that this flattening should be limited to providing enough room for just the 60 inch and for a possible 150 inch (3.8 m) later. Fortunately for future users of the site, Stock ignored this advice and produced a flat area 190 × 110 yards (210 × 120 m) in size, which was enough for the location of a few more telescopes.

The NSF had approved funding in 1963 for a 36 inch (0.9 m) CTIO reflector, the first medium-sized telescope to be erected on the mountain. It used the primary mirror that had been removed from the first 36 inch telescope on Kitt Peak in 1966. First light was achieved at CTIO in the following year.

The 60 inch reflector, which had first light in November 1967, had a modified Ritchey–Chrétien design with a Pyrex primary mirror. This had been polished at the KPNO optical shop at Tucson. The telescope's mounting, which had been made by the Westinghouse Electric Corporation, was of a similar design to the 36 inch.

Once work was under way in 1961 on designing the KPNO 150 inch (3.8 m) telescope, attention turned to providing a similar instrument at AURA's southern observatory. The AURA Scientific Committee debated the relative merits of a large Schmidt telescope and a 150 inch reflector, and continued to prefer the latter. But there were financial problems which caused the proposed CTIO 150 inch to be deleted from the FY 1967 budget. Then in January 1967 the Ford Foundation agreed to fund half of the cost of the 150 inch, on condition that the NSF provided matching funds.

This was excellent timing, as AURA was just about to invite bids for the construction of the dome and observatory building for the KPNO 150 inch, and so they were able to add a second dome to the bid invitation. The mirror blank for the KPNO 150 inch had been ordered some time ago, however, so there was no chance of doubling up on this order for Cerro Tololo. But the Anglo-Australian Observatory and the French National Observatory wanted to order similar-sized blanks for their new telescopes. So AURA joined with them to produce a multiple-purchase bid invitation. General Electric, Corning and Owens-Illinois submitted bids, with the latter winning. Owens-Illinois proposed using Cer-Vit at about half the price of fused silica.

The Cerro Tololo building and dome were completed in 1972, and the telescope mounting was shipped to Chile in the following year. Grinding and polishing of the 158 inch (4.0 m) Cer-Vit mirror blank had started in early 1972 at the KPNO optical shop at Tucson. After completion the mirror was sent to Chile, where it arrived in September 1974. Once on the mountain, the mirror was coated with aluminium and installed in its mount. Shortly afterwards the first photographs were taken at the telescope's prime focus by Victor Blanco, the observatory director. It is now named after him as the Victor M. Blanco Telescope.

Non-NAO telescopes on Cerro Tololo

The University of Michigan had located its 24/36 inch (0.6/0.9 m) f/3.5 Curtis Schmidt, a twin of the Burrell Schmidt, at its Portage Lake Observatory in 1950. But in 1967 the Curtis Schmidt was moved to Cerro Tololo, where it was put in a new observatory building. The telescope was initially installed on CTIO on a 10 year loan basis, with the University of Michigan allocated one-third of its observatory time. But in 1975 the loan was extended by 25 years.

As mentioned earlier, financial pressure had caused NOAO to concentrate its resources in later years on maintaining and operating the largest telescopes at Kitt Peak, and closing or handing over operational responsibility of the smaller telescopes to university consortia. NOAO adopted a similar approach on Cerro Tololo, with operation of the 36 inch (0.9 m) and 60 inch (1.5 m) being handed over to the SMARTS (Small and Moderate Aperture Research Telescope System) consortium in 2003. This consortium consisted of the uni-

versities of Delaware, Fisk, Georgia State, Northern Arizona, Ohio State, SUNY-Stony Brook, Vanderbilt and Yale, plus the American Museum of Natural History, CTIO, NOAO and the Space Telescope Science Institute. In 1973 Yale University moved their 40 in (1.0 m) Boller and Chivens reflector from Bethany, Connecticut to Cerro Tololo. Since 2004 this telescope has also been operated by the SMARTS consortium.

A 1.3 m (50 inch) reflector was built on Cerro Tololo in the late 1990s specifically optimised for the infrared. It was used for a 2 micron all-sky survey starting in 1998. This Cerro Tololo 2MASS telescope, as it was called, covered the southern part of the sky, and an identical telescope, located on Mt Hopkins, Arizona, covered the northern part.

The infrared detectors of this 2MASS telescope were cooled by liquid nitrogen to minimise thermal noise. But the most unusual part of the telescope was the Cassegrain secondary mirror support, which moved during each 1.3 second exposure. This was to compensate for the telescope movement as it smoothly and continuously scanned across the sky at about 1 arcmin/sec. After each exposure, the Cassegrain returned to its starting position, ready for the next 1.3 second exposure. All the operations were automatic. In February 2003, after the 2MASS survey had been completed, this telescope was also handed over to the SMARTS consortium to operate.

The CTIO observatory was very popular for some time after it was completed in the 1970s, as it was one of the few astronomical observatories located at altitude in the Southern Hemisphere with a good range of telescopes. But the flattened top of the mountain was completely full by that time, so the 8.1 m Gemini South and the 4.1 m SOAR (Southern Astrophysical Research) telescopes were built on the nearby Cerro Pachón around the turn of the twentieth century.

Bibliography

Blanco, Victor M., *The Inter-American Observatory in Chile*, Sky and Telescope, February 1968, pp. 72–76.

Edmondson, Frank K., *AURA and its US National Observatories*, Cambridge University Press, 1997.

Evans, David S., *Under Capricorn: A History of Southern Hemisphere Astronomy*, Institute of Physics Publishing, 1988.

Goldberg, Leo, *The Founding of Kitt Peak*, Sky and Telescope, March 1983, pp. 228–232.

Kloeppel, James E., *Realm of the Long Eyes: A Brief History of Kitt Peak National Observatory*, Univelt, 1983.

Krisciunas, Kevin, *Astronomical Centers of the World*, Cambridge University Press, 1988.

McCray, W. Patrick, *Giant Telescopes: Astronomical Ambition and the Promise of Technology*, Harvard University Press, 2004.

Meinel, Aden B., *The National Observatory at Kitt Peak*, Sky and Telescope, August 1958, pp. 493–499.

Figure 6.3. A drawing of the 36 inch (90 cm) Lick refractor made shortly after it was completed in 1888.

LICK OBSERVATORY

In 1876 James Lick left $700,000 in his will to the University of California to enable them to build the largest telescope in the world in a new observatory. Lick himself chose the site of the observatory as the 1,300 m (4,250 ft) Mount Hamilton about 20 miles (32 km) from San Jose, California. This followed the advice given him by George Davidson, the President of the California Academy of Sciences, who suggested that Lick's telescope should be located on the top of a mountain where the quality of the air was better than at sea level. As a result, the Lick Observatory became the first major research observatory to be located on a mountain top. Its excellent observing conditions, when compared to those of sea-level sites, subsequently persuaded other institutions to locate their new, large telescopes at high altitude.

The top of Mount Hamilton was levelled in 1880, and over the next few years a number of small telescopes were installed there, including a 12 inch (30 cm) Clark refractor and a 6 inch (15 cm) Repsold meridian circle. Finally Lick's major telescope, the 36 inch (0.9 m) refractor (see Figure 6.3), was

installed there in 1888. Built by Alvan Clark & Sons, with an equatorial mount by Warner and Swasey, it was, at the time, the largest telescope in the world.

The Lick bequest not only provided enough money for building and equipping the original observatory, but there was also some money left over to help with its annual costs. The State of California also provided some annual funding, but even then the observatory was underfunded. As a result it had to rely on private funding to help pay for both new equipment and staff. So it was with some relief that Edward Crossley, an Englishman, donated his 36 inch reflector, which had been made by Andrew Common, to the observatory in 1895. When it was installed there the following year it was the largest reflecting telescope in America.

The first director of the observatory, Edward S. Holden, was successful in persuading wealthy individuals to support the observatory. He was also successful in recruiting a number of excellent observers including Edward Barnard, S. W. Burnham, James Keeler and John Schaeberle. Keeler left the Lick Observatory in 1891 to become director of the Allegheny Observatory. Personality problems with Holden also led to the departure of Burnham in 1892 and Barnard three years later. In addition, Holden antagonised other important local figures, resulting in his being forced out in 1897. Keeler then returned to the Lick Observatory to become its new director.

Keeler used the Crossley reflector to pioneer nebular photography, but he died early just three years later. He was followed as director by William Campbell, who remained in the post for thirty years, for the last seven of which he was also president of the University of California. In 1933 the Crossley reflector was the first large reflector to have the silver coating of its primary mirror replaced by aluminium, which had a longer lifetime and a much higher reflectivity in the ultraviolet.

Campbell, who had originally assisted Keeler in his spectroscopic work in the early 1890s, was interested, when he became director, in producing a complete catalogue of stellar radial velocities. This required extending observations to the Southern Hemisphere. As a result he established a southern station of the Lick Observatory at Cerro San Cristobal, just outside Santiago, Chile. There the Lick Observatory set up a 36 inch Brashear reflector in 1903 that was paid for by Darius Mills, an early Lick trustee. Although it was initially intended to operate the observing station in Chile for two years, it remained there until 1929 when it was sold to the Catholic University of Santiago.

It became clear in the 1930s that the Lick Observatory facilities needed substantial improvement if its astronomers were to remain at the forefront of astronomical research. But the Great Depression in the United States and the Second World War delayed the approval and construction of any large new telescopes. In the meantime, during the war the Lick Observatory had considered building a number of possible reflectors with sizes ranging from about 80 to 120 inches (2.0 to 3.0 m). C. Donald Shane was appointed director of the Lick Observatory in 1945 and, in the following year, the University of California submitted a request to the State of California legislature for funding of a 120 inch. The state legislators approved it later in the year.

Unfortunately post-war inflation was higher than anticipated, and so savings were required to meet the approved budget. As a result, the Lick Observatory decided to use a 120 inch plane mirror that was already available at Caltech as the blank for their primary mirror. (It had originally been produced in 1933 by Corning as a test blank for the Palomar 200 inch.) They also decided to figure it in house, and to eliminate the telescope's Cassegrain focus. But problems with figuring the mirror and with other elements of the telescope meant that the telescope was not completed until 1959, a year after Shane had retired as observatory director. It was, at the time, the second largest telescope in the world. In 1977 the University of California decided to call it the Shane telescope.

Anna Nickel, a resident of San Francisco, donated a large part of her estate to the Lick Observatory in the late 1970s. As a result Lick Observatory staff were able to build a 40 inch (1.0 m) telescope. To save money, the telescope was designed and built in house, and located in the dome that had once held the 12 inch Clark refractor. A number of surplus parts were used, including a 40 inch Pyrex mirror that had originally been acquired for the Crossley reflector. The Nickel 40 inch Reflector, as it is called, was built with the same focal ratio as the 120 inch Shane, so that instruments could be shared between the two telescopes. Prototype instruments could also be tested on the 40 inch first, rather than using valuable 120 inch time.

The Lick Observatory had been interested in having an observatory in the Southern Hemisphere even after they had sold their southern station in Chile in 1929. But it was not until the early 1960s that they, along with the University of California's other astronomy departments at Berkeley, Los Angeles and San Diego, began to look seriously at building such an observatory in an all-university effort led by the Lick. They initially investigated setting up a collaborative observatory in Australia, but this fell through when the Australians agreed to set up the Anglo-Australian observatory with the British. With that avenue shut off the Lick considered building a new observatory in the United States away from the lights of San Jose, but they were not able to acquire funding.

Finally in the late 1970s the University of California and Caltech set up a joint project to build a telescope with a considerably larger mirror than the Mount Palomar 200 inch (5.1 m). This eventually resulted in the construction of the two 9.8 m Keck telescopes on Mauna Kea, Hawaii, which saw first light in the mid-1990s. So the main University of California observatory was no longer on Mount Hamilton but on Mauna Kea.

Bibliography

Osterbrock, Donald E., *James E. Keeler: Pioneer American Astrophysicist and the Early Development of American Astrophysics*, Cambridge University Press, 1984.

Osterbrock, Donald E., *The Rise and Fall of Edward S. Holden*, Journal for the History of Astronomy, **15** (1984), 81–127 and 151–176.

Osterbrock, D. E., Gustafson, J. R., and Unruh, W. J. S., *Eye on the Sky: Lick Observatory's First Century*, University of California Press, 1988.

Wright, Helen, *James Lick's Monument: The Saga of Captain Richard Floyd and the Building of the Lick Observatory*, Cambridge University Press, 1987.

Wright, W. H., *William Wallace Campbell 1862–1938*, Biographical Memoirs of the National Academy of Sciences, **25** (1949), 35–74.

MAUNA KEA OBSERVATORY

A tsunami devastated the town of Hilo on Hawaii in May 1960, causing serious problems for the local economy. As a result, ideas were put forward to bring much needed money to the island, one of which involved setting up an astronomical observatory which would take advantage of its unpolluted skies and high mountains. Mitsuo Akiyama of the Hawaii Chamber of Commerce sent letters to many American and Japanese universities in June 1963 outlining the possibility of setting up such a facility on either of the two large extinct volcanoes, Mauna Kea and Mauna Loa. Gerard Kuiper, of the University of Arizona, had already had his assistant Alika Herring undertake a testing programme at Haleakala on the island of Maui in Hawaii, funded by NASA and the United States Department of Defense. This indicated that Haleakala's altitude of 3,000 m (10,000 ft) was not high enough above the normal cloud tops in the area. Mauna Kea, on the other hand, is 4,200 m (13,800 ft) high, so Kuiper drew up a proposal in January 1964 to investigate Mauna Kea as a possible site for an observatory.

Progress then accelerated. An agreement was signed between the University of Hawaii and the University of Arizona for the establishment of a 12 ft (3.7 m) dome and test telescope on Mauna Kea. A rough road was built up Mauna Kea in April and May 1964, testing of the viewing conditions commenced on 13 June and the dedication of the Mauna Kea Observatory took place on 20 July. At this stage the only telescope on the mountain was the 12½ inch (32 cm) reflector being used by Herring for site testing, which showed that observing conditions were excellent. The air was very dry, so the site should be ideal for both visible and infrared observations.

Kuiper then submitted a proposal to NASA to build a 60 inch (1.5 m) reflector on Mauna Kea, while the University of Hawaii submitted a proposal a few months later to build an 84 inch (2.1 m). Much to Kuiper's annoyance, the University of

Hawaii's proposal was accepted by NASA even though they had no astronomy department at the time. NASA's choice was because the observatory was to be managed locally, and, in addition, it would provide a larger telescope. In the end, the University of Hawaii built a slightly larger telescope of 88 inches (2.2 m) that was completed in 1970. Twenty-two years later David Jewitt and Jane Luu discovered the first Trans-Neptunian Object, 1992 QB$_1$, using this 2.2 m telescope.

Canada-France-Hawaii (CFH) Telescope

French astronomers had been looking at potential sites for a large telescope in the late 1960s when John Jefferies, professor of physics at the University of Hawaii, suggested that they consider installing it on Mauna Kea. The French ordered the telescope's 3.6 m Cer-Vit primary mirror blank from Owens-Illinois in 1970, and in the following year they decided to site their telescope on Mauna Kea. At about the same time Canadian astronomers had also been looking at potential sites for a large telescope and, on hearing of the French decision, suggested to them that they join forces and share the cost. This resulted in an agreement between Canada, France and the University of Hawaii in 1973 in which it was agreed that a 3.6 m Canada-France-Hawaii (CFH) telescope would be built. Canada and France would share the costs of building the telescope 50/50, whilst the University of Hawaii would provide the land and the midlevel facilities at Hale Pohaku. (These facilities, at about 2,800 m altitude, were required for acclimatisation because of the excessive height of the mountain.) Canada, France and the University of Hawaii agreed to share the operating costs 45/45/10. The University of Hawaii would get 15% of the telescope time, with the remainder being shared equally between Canada and France.

Construction of the CFH telescope began in 1974, and was completed five years later. The telescope, which had originally been built for photography, had an unusually wide field-of-view of 1° at prime focus. In 2003 the MegaPrime imager was installed to cover most of this field-of-view with a 18,400 × 18,400 pixel CCD camera.

NASA InfraRed Telescope Facility (IRTF)

NASA decided in the early 1970s that they needed a large infrared telescope to support their Voyager planetary mission. James Westphal looked at a number of sites in the United States, Mexico and Chile and concluded that Mauna Kea was the most suitable. Unfortunately, there was still bad feeling between the University of Arizona and the University of Hawaii following NASA's decision to fund the latter's 2.2 m telescope. In addition, the University of Arizona had a strong infrared group and wanted NASA's proposed infrared telescope built on Mount Lemmon in Arizona. In the event NASA decided

to locate their InfraRed Telescope Facility (IRTF) on Mauna Kea, and in 1974 awarded a contract to the University of Hawaii for its design and construction.

In 1975 the IRTF's 3 m primary mirror blank had to be written off following the appearance of a crack after cutting the Cassegrain hole at the Kitt Peak National Observatory. Another mirror blank had to be purchased and NASA took advantage of the delay to re-examine the telescope's design. This resulted in a major redesign of its mounting system.

The NASA IRTF facility (see Figure 6.4) was completed in 1979, the same year as the CFH telescope.

The United Kingdom Infrared Telescope (UKIRT)

The United Kingdom's InfraRed Telescope (UKIRT) was the third telescope to be completed on Mauna Kea in 1979. Originally called the Infrared Flux Collector, it was the brainchild of Jim Rig of Imperial College, London and Gordon Carpenter of the Royal Observatory, Edinburgh who looked into the possibility of building a 4 m infrared telescope in the early 1970s. But they were heavily constrained by lack of money, so they decided to use a very thin mirror, supported both axially and radially, together with the smallest dome possible.

UKIRT's design was derived from that of the 1.5 m Infrared Flux Collector (now called the Carlos Sanchez Telescope) built on Tenerife in the Canaries in 1972. When completed in 1979 the UKIRT had cost only $5 million, which was very cheap for a 3.8 m infrared telescope. It was jointly operated with the submillimetre James Clerk Maxwell Telescope (JCMT) when it was completed on Mauna Kea in 1986.

In 2009, following a re-evaluation of UK priorities in ground-based astronomical observatories, it was decided to undertake a phased withdrawal from UKIRT starting in 2011.

The Kecks

Light pollution was beginning to cause a serious problem in the 1970s at the University of California's Lick Observatory on Mount Hamilton. One possible solution was to build a 3 m diameter telescope at a better site. But Joe Wampler suggested that it might be easier to obtain funding for something more revolutionary like a 10 m diameter telescope. As a result the university's astronomy faculty set up a committee in 1977 to consider the various options.

Jerry Nelson was asked to join the committee and look into the problems of building a 10 m telescope. In the course of his investigation he compared the relative benefits and problems of building such a large telescope with either a monolithic mirror or a segmented mirror, and concluded that the segmented mirror was the better option. In particular he suggested that it should be possible to build a 10 m primary mirror using very thin hexagonal segments aligned using active optics. It should

Figure 6.4. The domes of the two 10 m Keck telescopes on Mauna Kea are centrally located in this photograph. The enclosure for the 8.3 m Subaru telescope is on the left and the dome for the NASA 3 m Infrared Telescope Facility (IRTF) on the right. (Courtesy Sasquatch/Wikimedia Commons.)

also be easier to build a segmented mirror, rather than a monolithic mirror, with a very short focal length. This would then enable a relatively small observatory building to be used.

Nelson, assisted by Jacob Lubliner, Terry Mast and others, came up with a more detailed design over the next few years. The f/1.75 mirror would consist of 36 hexagonal segments, each 1.8 m (6 ft) across and 7.5 cm (3 inches) thick. The alignment of the individual segments would be monitored and adjusted continuously using a computer-controlled system. The segments would each be polished using a stress polishing technique similar to that used by Bernhard Schmidt to produce his corrector plates in the 1930s.

Caltech joined the project in 1979, and shortly afterwards a site selection committee was set up chaired by Robert Kraft of the Lick Observatory. They recommended Mauna Kea as the preferred location as it was then known to be an excellent site for both optical and infrared astronomy. In January 1985 the Keck Foundation agreed to provide $70 million of the estimated $90 million to construct the telescope and observatory. Shortly afterwards the University of California and Caltech founded the California Association for Research in Astronomy (CARA) to design, build and operate the telescope.

The Keck telescope saw first light in November 1990, with nine of its 36 Zerodur mirror segments in place, and in April 1992 the last of its mirror segments was installed. In the meantime, in 1991 the Keck Foundation had agreed to provide a grant towards the cost of a second 10 m telescope, Keck II,

85 m away from Keck I. The images from both telescopes would be combined optically in an interferometric arrangement.

Keck I began scientific observations in 1993 and Keck II three years later. The two telescopes of the Keck interferometer were first combined optically in March 2001, when fringes were observed for a number of stars. Scientific observations with the interferometer started the following year.

Subaru

In the mid-1980s Japanese astronomers looked into the possibility of building a 7.5 m telescope on Mauna Kea optimised for the infrared. The National Astronomical Observatory of Japan, representing many Japanese universities, was organised in 1988 to oversee this project, which was then called the Japan National Large Telescope. A potential site had already been identified on Mauna Kea. But the telescope's precise location was determined by building a scale model of the mountain's summit and testing it in a wind tunnel. In addition, the design of the telescope's enclosure was determined using computer simulations and wind tunnel tests. The result was a cylindrical enclosure, air conditioned in the day, with fans to circulate the air at night.

Initially the telescope was envisaged as being 7.5 m diameter, but when funding was approved in 1991 its size had been increased to 8 m. Shortly afterwards the telescope's name was changed to Subaru (Japanese for the Pleiades) and its planned size was further increased to 8.3 m. Its primary mirror was of monolithic construction using Corning's Ultra Low Expansion (ULE) glass, just 20 cm thick, and supported by over 250 computer-controlled actuators. The mirror was made of glass segments fused together into a single blank. The blank was then reheated and allowed to slump onto a convex mould of the appropriate shape.

Construction of this Subaru telescope was started on Mauna Kea in 1992. The mirror blank was completed two years later, with first light for the telescope in December 1998. Subaru was available for scientific observations in the following year.

Mauna Kea is now the site of not just optical and infrared telescopes, but also of submillimetre and radio telescopes. It is, in addition, home to the Gemini North telescope. (The Gemini and more submillimetre telescopes are covered in separate articles.)

Bibliography

Jefferies, John T., and Sinton, William M., *Progress at Mauna Kea Observatory*, Sky and Telescope, September 1968, pp. 140–145.

Krisciunas, Kevin, *Astronomical Centers of the World*, Cambridge University Press, 1988.

Morrison, David, and Jefferies, John T., *Hawaii's Mauna Kea Observatory Today*, Sky and Telescope, December 1972, pp. 361–365.

Parker, Barry, *Stairway to the Stars: The Story of the World's Largest Observatory*, Perseus Publishing, 1994.

Zirker, J. B., *An Acre of Glass: A History and Forecast of the Telescope*, Johns Hopkins University Press, 2005.

See also: Modern optical interferometers; National New Technology Telescope and Gemini; Submillimetre radio telescopes on Mauna Kea.

McDONALD OBSERVATORY

In 1926 the Texan banker and amateur astronomer W. J. McDonald died and left the bulk of his $1.26 million estate to the University of Texas specifically to enable them to build an astronomical observatory. His will was contested by distant relatives, but the university still received just over $800,000 in an out-of-court settlement in 1929. Unfortunately the university had no working astronomers and no idea how to plan and build an observatory with their unexpected legacy. But, unknown to the University of Texas, at about the same time Otto Struve, of the University of Chicago's Yerkes Observatory, had exactly the opposite problem. The Yerkes Observatory had a number of excellent astronomers, but their best telescopes were no longer good enough for Struve's cutting edge research. So he tried to persuade his university to build a new large telescope, but it had no money.

The president of the University of Texas, Harry Benedict, sought advice from a number of observatories as to how to plan and build an astronomical observatory, and, in the process, contact was made with Chicago's Yerkes Observatory. It was almost inevitable, in view of their complementary interests, that the two universities would agree to build and operate a joint observatory. Struve visited the University of Texas in 1932 to discuss possible collaboration. This resulted in a formal agreement to set up a joint observatory under a director who was also the director of the Yerkes Observatory, which was Struve himself. The collaboration agreement was specified to last thirty years. The University of Texas was required to build and equip this McDonald Observatory, as it was to be called, at a cost not to exceed $375,000, and the University of Chicago was to pay for most of its operating expenses.

Christian Elvey and T. G. Mehlen undertook an extensive survey of potential observatory sites in Texas in the summer of 1932. As a result Struve selected the 6,790 ft (2,070 m) Mount Locke in the Davis Mountains of western Texas. The owner of the land, Violet Locke McIvor, made a gift of it to the university. Then, in the following year, the University of Texas also acquired the deeds of a nearby peak, called Mount Fowlkes, for possible future use.

The University of Texas decided to buy a telescope of about 80 inch (2.03 m) diameter, which was about as large as they could afford. In 1933 Warner and Swasey were given the contract for building this telescope, including the mount and dome, using a Pyrex glass primary mirror. The idea of giving the same contract for the mirror and the mechanical side of the telescope to the same firm was unusual at that time for a large telescope. But the university decided on this approach as they believed that it would minimise costs. Grinding and polishing of the primary mirror took four years, which was much longer than expected. As a result the observatory was not dedicated until 1939. But when it was, the 82 inch (2.08 m) telescope, later called the Otto Struve Telescope, was the second largest telescope in the world. With it in the first decade W. Albert Hiltner discovered that starlight was polarised by the interstellar medium, and Bengt Strömgren perfected his method for photoelectric classification of stellar spectra.

A new agreement was made between the two universities when the initial agreement expired in 1962. According to this new arrangement, the McDonald Observatory would become an autonomous facility of the University of Texas, although its use was still shared with the University of Chicago. Harlan J. Smith was the first director under this new regime.

At about the same time NASA contacted the University of Texas to discuss the possibility of the university undertaking planetary observations in support of NASA's planned spacecraft missions. NASA had contacted the University of Texas because of previous work at the McDonald Observatory on the Moon and planets, and because of the observatory's southern location, making it ideal for planetary observations. This NASA approach resulted two years later in their agreeing to finance a new large telescope at the McDonald Observatory, whilst the university paid for the telescope's building. This was the origin of the 107 inch (2.72 m) telescope, later called the Harlan J. Smith Telescope, which had a primary mirror of fused silica from Corning. A hole of 30 inches (76 cm) diameter was cut for the Cassegrain focus, and the resulting spare piece of glass used to provide a 30 inch reflector for the observatory. When completed in 1968 the 107 inch was the third largest telescope in the world.

A new McDonald Observatory employee, who suffered a mental breakdown, fired seven bullets at the 107 inch primary mirror in 1970. The mirror did not crack, so the craters that had been produced by the impacts were simply bored out and painted black to minimise any scattered light. This produced the equivalent of a 106 inch (2.69 m) mirror.

In the 1980s astronomers at the University of Texas looked into the possibility of building an even larger telescope at the McDonald Observatory optimised for spectroscopy. This became the Hobby-Eberly Telescope (HET, previously called the Spectroscopic Survey Telescope) which had an 11×9.8 m primary mirror composed of 91 hexagonal segments made of Schott Zerodur glass. Built on the 6,660 ft (2,030 m) Mount Fowlkes, the telescope was named after Bill Hobby, a previous Lieutenant-Governor of Texas, and Robert E. Eberly, a Pennsylvania State benefactor. Unlike most other large optical telescopes, the primary mirror was not a paraboloid but spherical, requiring a complex corrector system to bring images into focus. This reduced the HET's effective aperture to that of a circular 9.2 m mirror. To further save costs the HET was fixed in elevation but movable in azimuth, with its detectors following celestial objects by moving on a rail system at the telescope's $f/1.2$ primary focus. It was a joint enterprise of the universities of Texas, Pennsylvania State and Stanford, plus universities in Munich and Göttingen in Germany. Dedicated in October 1997, the HET became operational two years later.

Bibliography

Evans, David S., and Mulholland, J. Derral, *Big and Bright: A History of the McDonald Observatory*, University of Texas Press, 1986.

Osterbrock, Donald E., *Yerkes Observatory 1892–1950: The Birth, Near Death, and Resurrection of a Scientific Research Institution*, University of Chicago Press, 1997.

Smith, Harlan J., *McDonald Observatory's 107-Inch Reflector*, Sky and Telescope, December 1968, pp. 360–367.

Struve, Otto, *The Birth of McDonald Observatory*, Sky and Telescope, December 1962, pp. 316–320.

MOUNT GRAHAM INTERNATIONAL OPTICAL OBSERVATORY

Astronomers at the University of Arizona started planning in the early 1980s for an international observatory on the 10,700 ft (3,250 m) Mount Graham, about 75 miles (120 km) from Tucson, Arizona. Mount Graham was chosen because of its low light pollution, low atmospheric water vapour, ease of access and excellent seeing, which was only marginally worse than that on Mauna Kea. The planned telescopes included the 1.8 m Vatican Advanced Technology Telescope (VATT), a submillimetre telescope (later called the Heinrich Hertz Submillimeter Telescope — see separate article), and a binocular telescope, then called the Columbus Project, with two mirrors on a common mount.

The University of Arizona's mirror laboratory spun-cast the 1.8 m, $f/1.0$ mirror blank for the VATT in 1985. In the same year construction work was also expected to start on the new observatory on Mount Graham, but environmental concerns over the mountain's protected red squirrels and its other sensitive habitats led to a series of delays. It appeared as though the impasse had been broken in October 1988 when university-sponsored special legislation was passed by the United States Congress to allow the three planned telescopes to be built without Forest Service approval. This special legislation allowed the university to build four more telescopes, if the first three were found to have had a minimal impact on the red squirrel. But in March 1990 a federal judge ordered a temporary halt to the observatory construction. Although this temporary injunction was overturned by the U.S. Court of Appeals two months later, other potential legal problems had also arisen. In the event the legal problems associated with building the VATT on Mount Graham were solved, and the telescope, which was designed to operate in the visible and infrared, was dedicated in September 1993, achieving first light in the following year. Two major donors subscribed to its costs, Fred A. Lennon and Thomas J. Bannan. As a result the telescope was named the Alice P. Lennon Telescope, after Fred Lennon's wife, and its associated astrophysics facility was named after Thomas Bannan.

But litigation was still ongoing to try to stop the construction of the binocular telescope of the Columbus Project, now called the Large Binocular Telescope (LBT). As a result this construction was not resumed until July 1996, when a federal judge rejected the last case against it. In the following year, the first of the two 8.4 m, $f/1.14$ blanks for the LBT was spun by the University of Arizona's Mirror Laboratory, followed by the second three years later. The LBT saw first light in 2005 with just one mirror installed, and with both mirrors installed two years later. By then the consortium required to finance, build and operate the LBT had increased from the initial one of the University of Arizona and Ohio State University twenty years earlier to include seven universities in the

United States, five in Germany, plus the Instituto Nazionale di Astrofisica.

Bibliography

Heck, André (ed.), *Organizations and Strategies in Astronomy: Vol. 4*, Kluwer Academic Publishers, 2003.

Istock, Conrad A., and Hoffman, Robert S. (eds.), *Storm Over a Mountain Island: Conservation Biology and the Mt. Graham Affair*, University of Arizona Press, 1995.

See also: Heinrich Hertz Submillimeter Telescope.

MOUNT HOPKINS' WHIPPLE OBSERVATORY AND THE MMT

The Smithsonian Astrophysical Observatory (SAO) set up an observatory in the late 1960s on Mount Hopkins, about 35 miles (55 km) from Tucson, Arizona. It initially included a laser and Baker-Nunn camera for satellite range-finding and tracking, and a 10 m diameter optical reflector for gamma-ray astronomy. These detectors were mounted on a 7,600 to 7,800 ft (2,320 to 2,380 m) high ridge, leaving the 8,590 ft (2,620 m) mountain summit for the construction of a large optical telescope. Then in 1969 a 60 inch (1.5 m) telescope, to be used for photoelectric spectrophotometry, was built on the ridge. It was named after Carlton W. Tillinghast, a Smithsonian administrator who died in 1969 at the age of 36.

Whilst the 60 inch was being constructed, Fred Whipple, the director of the SAO, led a team to investigate the potential designs of a large optical telescope. As the SAO were considering their various design options, Gerard Kuiper, of the University of Arizona's Lunar and Planetary Laboratory (LPL), suggested that they consider a Multiple Mirror Telescope (MMT) concept that Frank Low and colleagues were investigating at the LPL. Aden Meinel, also of the University of Arizona, suggested to Whipple that he could obtain some surplus lightweight mirror blanks from the U.S. Air Force. Eventually Low and colleagues' MMT concept was accepted as the basis of the design of the new large telescope.

The basic idea of a multiple mirror telescope, in an MMT-like configuration, had originally been published in 1930 by the Irish scientist Edward Synge. Twenty years later Yrjö Väisälä, the director of the Turku Observatory in Finland, had also built a multiple mirror telescope model with six 32 cm diameter mirrors arranged in a ring, with a seventh mirror in the centre. But he did not build an operating instrument. These ideas were unknown to Frank Low and colleagues when they started their design of an MMT.

Whilst the design concept of the MMT was being developed, the Smithsonian had begun to investigate possible locations for their proposed large telescope. Three potential sites

were seriously considered, all in southern Arizona, namely Kitt Peak, Mount Lemmon, and Mount Hopkins. Kitt Peak was rejected because of its relatively low altitude. Mount Lemmon was the highest peak, and so should be the best for infrared observations, but it was only 16 miles (25 km) from Tucson and was already suffering from light pollution. So Mount Hopkins was chosen.

The MMT was eventually adopted as a joint project of the SAO and the University of Arizona in 1971. When completed eight years later it was the first large modern optical telescope in the West to use an altazimuth mount and the first large telescope to use a form of active optics. The telescope consisted of six 1.8 m telescopes on a common mount that was designed to bring all six images to a common focus on its central axis. It was planned to continuously monitor the flexure of the structures of the six telescopes using lasers, and to correct them in real time to ensure that the six images coincided. Its altazimuth mounting was smaller than the equivalent equatorial, and as each of the telescopes had reasonably fast $f/2.7$ mirrors, the MMT was able to use a smaller 'dome' or observatory building than normal. This building rotated with the telescope.

The net result of all these innovations was that the MMT, which was dedicated in 1979, was able to act like a 4.4 m telescope at about 40% of the capital cost of a normal reflector. On the other hand, it took quite a number of years to get the telescope to work correctly. In the process the laser system had to be abandoned and the six images digitally recorded.

In 1987 it was finally decided to abandon the MMT concept altogether as by then it was clear that it was possible to build very large monolithic mirrors, which had been very problematic when the MMT had been originally designed. It was found to be possible to replace the six telescopes of the MMT by a single 6.5 m diameter lightweight monolithic mirror in the same observatory building. This new telescope would have twice the light-gathering power of the MMT and, with suitable design, its angular field of view could be increased by a factor of 15. In the event the $f/1.25$, 6.5 m mirror blank was spun-cast by the University of Arizona's Mirror Laboratory in 1992, and the modified telescope, which had the very wide field of view of 1°, saw first light in the year 2000.

In 1996 a 1.3 m (50 inch) equatorially-mounted reflector had also been built on Mount Hopkins, specifically optimised for the infrared. It was used for a 2 micron all-sky survey, led by the University of Massachusetts. This 2MASS telescope, as it was called, covered the northern part of the sky, and an identical telescope, located on Cerro Tololo, Chile, covered the southern part. The Cassegrain secondary mirror support of both telescopes moved during each 1.3 second exposure to compensate for the telescope movement as it smoothly and continuously scanned the sky at about 1 arcmin/sec. After each exposure, the secondary mirror returned to its starting position, ready for the next exposure. The 2MASS survey was completed in 2001 and the resulting All Sky Catalog was published two years later.

Bibliography

Beckers, Jacques M., and Ulich, Bobby L., *The Multiple Mirror Telescope*, in Burbidge, G., and Hewitt, A. (eds.), *Annual Reviews Monograph; Telescopes for the 1980s*, Annual Reviews Inc., 1981, pp. 63–128.

Carleton, N. P., and Hoffman, W. F., *The Multiple-Mirror Telescope*, Physics Today, **31** (1978), 30–37.

Skrutskie, Michael, *2MASS: Unveiling the Infrared Universe*, Sky and Telescope, July 2001, pp. 34–42.

Weekes, Trevor C. (ed.), *The MMT and the Future of Ground-Based Astronomy*, Smithsonian Astrophysical Observatory Special Report No. 385, 1979.

See also: Ground-based gamma-ray observatories.

MOUNT WILSON OBSERVATORY

William Hale, George Ellery Hale's father, had agreed in 1895 to pay for a 60 inch (1.5 m) glass blank from the St Gobain glassworks in France. It was his contribution to a 60 inch reflector which G. E. Hale hoped could be built at the University of Chicago's Yerkes Observatory, where he was director. At that time there were no funds available from the university or anyone else to complete and operate the telescope. But the younger Hale reasoned that having the mirror would help to generate enough enthusiasm to get someone or some organisation to provide the necessary funds.

The blank arrived from France two years later and George Ritchey began grinding it. Shortly afterwards, just before he died, William Hale agreed to give the mirror to the University of Chicago on condition that they pay for the telescope's structure and observatory, together with its operating expenses. But the partially ground mirror had to be put into storage in 1898 as money was not forthcoming from the university or any other source to build and operate the telescope.

In the following year, Hale discovered that Andrew Carnegie had decided to donate $10 million to fund the Carnegie Institution, dedicated to scientific research and development. Hale wrote to John Billings, who was on the institution's Executive Committee, hoping to interest him in his 60 inch reflector project. This resulted in Hale being invited to serve on the institution's Advisory Committee on Astronomy which, in 1903, asked William J. Hussey of the Lick Observatory to look for a site for a 'southern and solar observatory'. As a result of his site surveys, Hussey was particularly impressed by Mount Wilson and Palomar Mountain, both in southern California. But, although Palomar had the better viewing, he recommended the 1,740 m (5,700 ft) high Mount Wilson, because

of its much better accessibility. As a result, Hale urged the Carnegie Institution via its Advisory and Executive Committees to build a high altitude solar observatory on Mount Wilson where his planned 60 inch reflector could be mounted.

Whilst waiting for the Carnegie Institution's decision Hale spent the winter of 1903–4 undertaking astronomical observations on Mount Wilson, including solar observations using a 12 inch (30 cm) coelostat sent from Yerkes. Finally in December 1904 the Carnegie Institution decided to establish the Mount Wilson Solar Observatory (after 1919 the Mount Wilson Observatory) with Hale as its first director.

Hale first moved the newly completed, horizontally mounted Snow solar telescope from Yerkes to Mount Wilson. He then built a solar telescope on Mount Wilson in a 60 ft (18 m) high tower, using the experience gained with the Snow telescope. The 60 ft was completed in 1908, followed four years later by a 150 ft (45 m) tower telescope. The 150 ft produced a 17 inch (43 cm) solar image, compared with the 5 inch (13 cm) image obtained with the smaller telescope.

In the meantime the partially ground 60 inch mirror had stayed in storage until 1904 when Hale obtained funding from the Carnegie Institution for the 60 inch reflector to be built at the Mount Wilson Solar Observatory. Ritchey then moved to Mount Wilson to finish the grinding of the mirror. When the 60 inch telescope saw first light in December 1908 it was the most powerful telescope in the world.

The 60 inch was to set the standard for most large reflectors of the future allowing operation in a number of different configurations. As a Newtonian it operated at $f/5$ for photography and low dispersion spectrography, as a modified Cassegrain it operated at $f/16$ for spectrography and $f/20$ for photography, and as a coudé it was used at $f/30$ for high dispersion spectrography.

With the manufacture of the 60 inch well in hand, Hale had started planning the construction of an even larger instrument and, in 1906, persuaded John D. Hooker, a Los Angeles businessman, to pay for its 100 inch (2.5 m) primary mirror. Five years later the Carnegie Institution agreed to pay for its mounting and observatory.

Meanwhile, in 1908 when the $4\frac{1}{2}$ ton glass blank arrived from the St Gobain glassworks for the 100 inch it was found to have a large number of small subsurface air bubbles and to have partially lost its rigidity. Subsequent attempts by St Gobain to produce a better blank failed. Fortunately, tests carried out by Walter Adams had shown that the mirror could probably be figured from the original disc without cutting into the bubbles. But Hale had to persuade Ritchey to start grinding the mirror, because Ritchey was convinced that the risk of cutting into the bubbles was too great. Ritchey started the long grinding process in 1910, and, four years later, he had finished grinding the mirror to a spherical shape. W. L. Kinney then took over the job of parabolising it, which was successfully completed in

1916. In the meantime, the St Gobain glass works had been destroyed by fire, and so a replacement mirror was out of the question in the event of problems. In the event the 100 inch telescope, which saw first light in November 1917, was a great success. It was the largest telescope in the world for the next thirty years.

The names of staff working at Mount Wilson in the early years were impressive. Walter Adams, Ferdinand Ellerman and Francis Pease had come with Hale and Ritchey from Yerkes. Arthur King had joined in 1908, Harold Babcock and Frederick Seares in 1909, Adriaan van Maanen in 1912, Harlow Shapley in 1914, John Anderson in 1916, and Edwin Hubble and Paul Merrill in 1919. The observatory, which was well funded in its early days, had a well-stocked laboratory where new instruments were continually being developed. It was a Mecca for astrophysicists in the early twentieth century.

The 60 and 100 inch telescopes and the solar telescopes made Mount Wilson the premier observatory in the world from 1908, when the 60 inch was completed, to 1948, when the 200 inch (5.1 m) was inaugurated on Palomar Mountain. With the 60 inch, Walter Adams and Arnold Kohlschütter developed the technique of spectroscopic parallax to measure the distance of distant stars, and Harlow Shapley photographed Cepheid variables in globular clusters and estimated the size of the Milky Way. With the 100 inch Hooker reflector, Edwin Hubble proved that spiral nebulae were not part of the Milky Way and estimated the expansion rate of the universe, Walter Baade discovered Population I and II stars in the Andromeda nebula, Michelson, Pease and Anderson made the first measurement of stellar diameter, and Horace Babcock discovered the first evidence of stellar magnetic fields.

Light pollution and photochemical smog from Los Angeles and the surrounding area resulted in the Carnegie Institution's decision to withdraw financial support from the Mount Wilson Observatory in 1985, and redirect their financial resources to Mount Wilson's sister observatory at Las Campanas in Chile. As a result the 100 inch was mothballed, and the Carnegie Institution handed over the Mount Wilson Observatory to the non-profit-making Mount Wilson Institute.

Albert Michelson had designed a stellar interferometer for the Mount Wilson Observatory in 1920 that consisted of a 6 m long steel girder and two movable mirrors which he mounted across the open end of the 100 inch telescope. This enabled him to measure the diameter of a number of red giants. Little more interferometric work was done on Mount Wilson until 1979 when Michael Shao and David Staelin designed a 1.5 m baseline interferometer that was the first interferometer to compensate for atmospheric turbulence. A 3.1 m Mark II was operational from 1982 to 1984, followed by a 32 m Mark III from 1986 to 1992.

In 1988 Charles Townes and colleagues achieved first fringes with the University of California at Berkeley's ISI (Infrared Spatial Heterodyne Interferometer) on Mount Wilson. It consisted, at that time, of two movable 1.65 m infrared telescopes with a baseline of 32 m. In more recent years the CHARA (Center for High Angular Resolution Astronomy) Array of Georgia State University was also built on Mount Wilson. It consisted of six 1 m telescopes, with separations of up to 330 m, that operated in the visible and infrared.

All three solar telescopes of the historic Mount Wilson Observatory have now been updated, and the 100 inch has been reactivated and fitted with an adaptive optics system. So the three solar telescopes and the 60 inch and 100 inch telescopes of the original observatory are still in use, alongside the up-to-date ISI and CHARA interferometric arrays. The Snow solar telescope is used for educational purposes, and the 60 inch is available for public observations. But the other two solar telescopes and the 100 inch are still used for astronomical research.

Bibliography

Adams, W. S., *Early Days at Mount Wilson*, Publications of the Astronomical Society of the Pacific, **59** (1947), 213–231 and 285–304.

Adams, W. S., *The Founding of the Mount Wilson Observatory*, Publications of the Astronomical Society of the Pacific, **66** (1954), 267–303.

Bartusiak, Marcia, *The Day We Found the Universe*, Pantheon Books, 2009.

Christianson, Gale E., *Edwin Hubble: Mariner of the Nebulae*, Farrar, Straus and Giroux, 1995.

Gingerich, O. (ed.), *The General History of Astronomy Volume 4A; Astrophysics and Twentieth-Century Astronomy to 1950: Part A*, Cambridge University Press, 1984.

King, Henry C., *The History of the Telescope*, Charles Griffin, 1955 (Dover reprint 1979).

Krisciunas, Kevin, *Astronomical Centers of the World*, Cambridge University Press, 1988.

Osterbrock, Donald E., *Pauper and Prince: Ritchey, Hale and Big American Telescopes*, University of Arizona Press, 1993.

Sandage, Allan, *Centennial History of the Carnegie Institution of Washington: Vol. 1, The Mount Wilson Observatory: Breaking the Code of Cosmic Evolution*, Cambridge University Press, 2004.

Wright, Helen, *Explorer of the Universe: A Biography of George Ellery Hale*, American Institute of Physics Press, 1994.

See also: Optical interferometry; Modern optical interferometers.

NATIONAL NEW TECHNOLOGY TELESCOPE AND GEMINI

Various studies were carried out in the late 1970s/early 1980s into a proposed National New Technology Telescope (NNTT) which was envisaged as being appreciably larger than the 200 inch (5.1 m) Palomar reflector. By 1982 the choice of designs had been reduced to two; either a larger Multiple Mirror Telescope (MMT) consisting of four 7.5 m mirrors, or a telescope with a single segmented 15 m mirror. In the following year Geoffrey Burbidge, the director of the Kitt Peak National Observatory (KPNO), which is now part of NOAO (National Optical Astronomy Observatory), invited Robert Gehrz of the University of Wyoming to chair a committee to select a winning design from these two design alternatives. In July 1984 the committee recommended, by a small margin of votes, that the larger-size MMT should be the NNTT.

The existing MMT was partially owned by the University of Arizona, and it was assumed that the mirrors of the larger MMT would also be moulded in the University of Arizona's rotating furnace. Some astronomers thought that this looked as though the University of Arizona was taking over the new national telescope, in competition with Caltech's Keck telescope whose funding had just been approved. At about the same time Ohio State University and the University of Arizona announced that they were considering building a Large Binocular Telescope (then called the Columbus Project) with two 8 m telescopes on a common mount. So many people now asked whether there was a real requirement for the NNTT with both the Keck and this new Large Binocular Telescope on the stocks.

Assuming the NNTT proceeded, there was also the question of whether it should be located on Mauna Kea, Hawaii, in spite of its distance from the USA mainland, or whether there was an equivalent site on the mainland. At this time the University of Arizona was planning to build its next generation of astronomical facilities, including the Large Binocular Telescope, on the 3,250 m (10,700 ft) high Mount Graham in Arizona. So they favoured putting the NNTT on Mount Graham, whereas John Jefferies, NOAO's new director, favoured Mauna Kea, a location he had helped to develop. Eventually in early 1987, after three years of site surveys, NOAO announced that they had chosen Mauna Kea as their preferred site for the NNTT.

It became clear by September 1987 that this large NNTT was not affordable. In the meantime Goetz Oertel, the new president of AURA (Association of Universities for Research in Astronomy), NOAO's parent body, had set up a committee to examine the future role of NOAO. This committee recommended in late 1987 that NOAO build two 8 metre class telescopes, one in each hemisphere, instead of the NNTT, which was left as a long-term goal. Subsequently in September 1989 AURA submitted its proposal for two 8 m telescopes to the National Science Foundation (NSF).

The idea of scaling scale back the American national large telescope project first from 25 m, which it had been in 1980, to 15 m two years later, and then to two 8 m telescopes, together

with other cost-saving ideas, caused many staff to resign from NOAO. Not only was the 10 m Keck under construction on Mauna Kea by this time, but the Carnegie Institution of Washington was planning an 8 m telescope of their own, and the University of Arizona and others had already started designing the Large Binocular Telescope. But, worse than that for American prestige, the European Southern Observatory (ESO) were planning to build the Very Large Telescope, consisting of an array of four 8 m telescopes in Chile. And the Japanese planned to build an 8 m telescope on Mauna Kea. Nevertheless, two 8 m telescopes were better than nothing, so Goetz Oertel and Sidney Wolff, NOAO's new director, set out to convince both American astronomers and Congress that this was the national project to support.

The AURA proposal to the NSF was based on using the University of Arizona's spun honeycomb mirrors developed by Roger Angel, as it was thought that these mirrors would provide better image quality than a segmented mirror like the Keck. Adaptive optics were also incorporated into the design from the start, rather than retrospectively as with many other telescopes. But the capital cost of the two telescopes, which was estimated at about $144 million, was considerable.

In October 1990 the United States Congress agreed to fund no more than half of the capital budget for the two telescopes, which by then had reached an estimated $176 million. Congress also limited the American contribution to $88 million, even if the final capital cost was more than twice this amount. The remainder of the capital costs was expected to come from international collaborators. This was quite possible as informal meetings had already taken place the previous year between American, British and Canadian scientists to explore international collaboration on the project. A provisional financial split between the USA, UK and Canada of 50/25/25 was assumed on this two-telescope project, which was to be called Gemini.

The National Academy of Sciences published its influential Bahcall Report in April 1991 into the future of astronomy in the USA. It recommended that an 8 m telescope should be built on Mauna Kea optimised to operate in the infrared between 2 and 10 μm. In addition, an 8 m telescope should be built in the Southern Hemisphere optimised for optical and near ultraviolet wavelengths. No mention was made of international collaboration, as Bahcall, in particular, thought that they should both be American telescopes.

Bahcall was very well aware that many people would say that, as the 10 m Keck telescope taking shape on Mauna Kea would provide American astronomers with a world-class facility, there was no need to provide another telescope to do the same thing at the same observatory. But, as he said in his report, 'The proposed national infrared-optimized 8 m telescope will differ from the private 10 m Keck telescope in three ways: (1) it will be available to all U.S. astronomers, whereas the Keck telescope will be available to only about 3 percent of the national astronomical community; (2) it will be the only large telescope in the world optimized for performance in the infrared; and (3) it will use adaptive optics to achieve the maximum possible spatial resolution in the near-infrared.'

In the event Bahcall's report was overtaken by events, as the twin 8 m telescope project was already getting under way as an international project, as requested by Congress, when the Bahcall Report was published. America was to contribute 50% of the total cost, with the other 50% divided equally between the UK and Canada.

The early days of this Gemini project were full of problems, however, most of them people-oriented. Many of the Americans in the project team resented the fact that Congress had insisted that Gemini be an international project, when they saw it as their project. After all, Gemini was the result of work done at the NOAO over many years. This problem was not helped by the fact that the Gemini project was based at NOAO's headquarters at Tucson, Arizona.

Matters were made worse in June 1991 when Canada announced that it was pulling out of Gemini, although a little later it said that it would participate at 15 %, rather than the 25 % initially agreed. Then in early 1993 Chile, Argentina and Brazil agreed to participate with contributions of 5/2.5/2.5%, respectively, to make up the missing 10%. At about the same time the 2,700 m high Cerro Pachón in Chile was chosen as the site for Gemini's southern telescope. Then in 1998 the Australian Research Council agreed to make a financial contribution, which gave Gemini additional funding stability. This was just as well as, by then, it had become clear that the budget for instrumentation was far too small.

In the early 1990s the choice of supplier for the two primary mirrors was reopened because of concerns that the University of Arizona's Mirror Laboratory might be unable to meet the required timescale. There were then basically three design options available, namely spun honeycomb mirrors produced by the University of Arizona, thin meniscus mirrors of the sort produced by Schott in Germany and Corning in the United States, and segmented mirrors like those used in the Keck telescope.

At the time, all three mirror types were being used in the design of various large telescopes. For example, the Keck telescope had been dedicated in November 1991. Then in February 1992 Schott had successfully made the first 8.2 m diameter thin meniscus mirror blank for ESO's Very Large Telescope. Two months later the University of Arizona had produced the 6.5 m mirror blank for the MMT on Mount Hopkins. In addition, a little earlier Corning had been chosen to produce the 8.3 m meniscus mirror blank for the Japanese Subaru telescope. So when, in May 1992, Gemini's management issued a formal request for quotations for their 8.1 m mirrors, all three competing technologies were available.

Gemini sent out their request for quotations to Corning, Schott and the University of Arizona in May 1992. In reply, Corning bid for two thin meniscus mirror blanks, made of Ultra Low Expansion (ULE) glass, whereas the University of Arizona bid for two blanks made of their normal borosilicate glass. In September 1992 the Gemini Board decided to award the contract to Corning to supply one mirror blank, with an option for a second.

The Gemini mirror decision caused a number of disgruntled American astronomers to join forces and push for various changes to the project. Not only were Peter Strittmatter and Roger Angel of the University of Arizona concerned about their failure to win the mirror contract, but John Bahcall felt that the design of the Mauna Kea telescope had not been properly optimised for infrared astronomy. In this he was backed up by Frank Low, who was a respected infrared astronomer at the University of Arizona. Bahcall had also never accepted that the Mauna Kea telescope was an international instrument, and in this he was supported by Donald Hall, director of the University of Hawaii's Institute of Astronomy. Unfortunately for their arguments, all these people, with the exception of Bahcall, had institutional reasons to press their point of view. What was clearly required was an independent review of the project. As a result, in December 1992 the NSF set up an independent committee to review Gemini chaired by James Houck of Cornell University, who was a respected infrared scientist.

The Houck Committee condemned the mirror choice as too risky. The main problem that the committee highlighted was that of supporting the very thin meniscus mirror. It was true that the ESO and Japan had opted for a similar-sized thin meniscus mirror in their new large telescopes, but their imaging requirements were considered to be less onerous than Gemini, and their telescopes had not yet been completed. The Gemini Board addressed the Houck Committee's concerns, and said that all the technical issues raised would be considered at Gemini's Preliminary Design Review (PDR) in late 1993. In the event this PDR showed that the existing telescope design was sound.

As with all thin primary mirrors, the shape of the Gemini primary mirrors was maintained by an active optics system, in this case using 120 actuators per mirror. The coating plants at both the north and south observatories would be able to coat the mirrors with either silver, for infrared observations, or aluminium.

Corning produced their mirror blank in the same way as they had produced the 8.3 m blank for the Japanese Subaru telescope, fusing together glass segments into a single blank. After fusing, this blank was rough ground and then heated over a mould, until it sagged and took up a meniscus shape with the required radius of curvature.

Corning finished its first 8.1 m mirror blank for Gemini in October 1995 in the USA, and then shipped it to a company near Paris to be polished. Polishing was completed in January 1998, and the 20 cm thick primary mirror installed in the Mauna Kea or Gemini North telescope about one year later. First light was achieved in February 1999. The Gemini project then used an adaptive optics camera on loan from the University of Hawaii in May 1999 to achieve star images with Gemini North of just 0.08 arcsec in diameter at 2.2 μm. The second primary mirror, which was destined for Gemini South on Cerro Pachón in Chile, followed the Mauna Kea mirror about two years later.

Gemini North was dedicated on 25 June 1999 and Gemini South on 18 January 2002. At another ceremony on 13 November 2002 the Gemini North telescope was named after Frederick C. Gillett, a pioneer in infrared astronomy and associate project scientist on Gemini, who had died prematurely in 2001.

Bibliography

Barr, L. D., *The U.S. National Large Telescope Project: The NNTT and Others*, in Ulrich, Marie-Helene (ed.), *Proceedings of a ESO Conference on Very Large Telescopes and Their Instrumentation, Held in Garching, March 21–24*, 1988, pp 73–92.

McCray, W. Patrick, *Giant Telescopes: Astronomical Ambition and the Promise of Technology*, Harvard University Press, 2004.

Robinson, Leif J., and Murray, Jack, *The Gemini Project: Twins in Trouble?*, Sky and Telescope, May 1993, pp. 26–32; letters in response. Sky and Telescope, September 1993, pp. 6–7 and 70.

Zirker, J. B., *An Acre of Glass: A History and Forecast of the Telescope*, Johns Hopkins University Press, 2005.

PALOMAR MOUNTAIN OBSERVATORY

The 200 inch (5.1 m) Hale Telescope

George Ellery Hale wanted to build an even larger telescope when the 100 inch (2.5 m) was completed on Mount Wilson in 1917. His initial idea was to build a 300 inch (7.6 m) reflector, but practical considerations forced him to reduce it to a 200 inch (5.1 m). Wickliffe Rose of the Rockefeller Foundation's International Education Board was very interested in Hale's proposed telescope, and in 1928 persuaded the Rockefeller trustees to pay its estimated construction cost of $6 million. Running costs were to be covered by an endowment provided by Henry Robinson, a Los Angeles Banker. The Rockefeller Foundation gave the money to the California Institute of Technology (Caltech) who were to build and operate the telescope, with the cooperation of the Carnegie Institution and the Mount Wilson Observatory who had gained experience with the 100 inch Hooker reflector.

Light pollution was already becoming a problem at Mount Wilson in the late 1920s. So a number of alternative sites in the south-western United States were tested for weather conditions and seeing and, in 1934, Palomar Mountain was chosen as the site for the new observatory. Construction of a road to the observatory was begun the following year.

Hale and Francis Pease decided to make the mirror an $f/3.3$ because this allowed relatively short exposures for nebula photography. It also enabled the tube to be made relatively short, allowing the dome to be made smaller also. The mirror blank was lightened considerably by having an hexagonal pattern of ribs on its back to provide stiffening. General Electric were contracted in 1928 to provide the mirror blank but their attempts to produce 60 inch (1.5 m) test pieces were unsuccessful. So their contract was cancelled in 1931, and Corning was contracted to produce the 200 inch blank. After many exploratory trials, a perfect 200 inch disc was cast in December 1934 on the second attempt, after several cores had broken loose with the first disc. The successful blank, which took a year to cool, was transported to Pasadena by train in April 1936, taking two weeks to cross the country.

Grinding of the mirror was then started in the Pasadena optical shop. Two years later Hale died, then the Second World War interrupted work, and so it was not until November 1947 that the mirror was transported to the new observatory on Palomar Mountain. The telescope (see Figure 6.5) was finally dedicated to Hale on 3 June 1948 by Vannevar Bush, president of the Carnegie Institution, and Lee Du Bridge, president of Caltech. It was commissioned for regular scheduled observations in November 1949.

The 200 inch Hale telescope remained the largest and most productive optical telescope in the world for many years. In the 1950s, observations with it confirmed the linear relationship between redshift and distance to greater distances than previously possible. It also showed that the Andromeda nebula was twice as far away as previously thought, and that the value of the Hubble constant needed substantial revision. The Hale telescope also played an important role in the discovery of quasars in the early 1960s.

Schmidt telescopes

The first telescope to be built at the Palomar Observatory was not the 200 inch, but a Schmidt telescope. It had a 26 inch (66 cm) $f/2$ primary mirror and a 18 inch (46 cm) corrector plate. This was followed by the much larger Palomar Schmidt whose construction was started in 1938 but, because of the war, was not finished until 1948. It had a 72 inch (1.8 m) $f/2.4$ primary mirror and a 48 inch (1.22 m) corrector plate. This Schmidt was used for the National Geographic—Palomar Observatory Sky Survey (POSS I) which was carried out between 1949 and 1958 on both blue- and red-sensitive plates recording stars, galaxies and other objects down to magnitude 21.

Figure 6.5. The 200 inch (5.1 m) Hale reflector which was completed on Palomar Mountain in the late 1940s. This image shows the observer sitting in the prime focus cage. (Courtesy California Institute of Technology.)

A new corrector plate and an automatic guider were installed on the Palomar Schmidt in the early 1980s, and a new sky survey (POSS II) started in 1985 using blue-, red- and near infrared-sensitive plates. It was completed in 2000. In the meantime, in 1986 this Palomar Schmidt telescope had been renamed the Samuel Oschin Telescope following a donation of $1 million by the Samuel Oschin Family Foundation to help to finance astronomical research at Palomar.

In 2001 the Jet Propulsion Laboratory's NEAT (Near Earth Asteroid Tracker) CCD camera was installed on the Oschin Schmidt, followed two years later by Yale University's QUEST (QUasar Equatorial Survey Team) which was a 161 megapixel CCD camera.

Over its lifetime the Palomar/Oschin Schmidt, whilst imaging millions of stars, has discovered thousands of supernovae and many thousands of asteroids. Its 1982 Quick V Survey was used to produce the northern half of the Hubble Space Telescope's first Guide Star Catalog, published in 1989, which included the position and intensity of about 19 million stars. More recently the Oschin Schmidt has discovered the dwarf planet Eris, the large classical Kuiper belt object Quaoar, the detached object Sedna and the plutino Orcus.

The Palomar Interferometer

The designers of the Palomar Testbed Interferometer (PTI), which was installed at the Palomar Observatory in 1995, used the experience gained with the Mark III interferometer on Mount Wilson. The PTI combined light from any two of its three 40 cm telescopes to act as a two element interferometer in the near infrared. Its maximum separation of 110 m enabled an absolute accuracy of 1 milliarcsecond (mas) to be achieved, but compared to nearby stars it could achieve relative accuracies of about 20 microarcsec. In 2001 the PTI was able to measure the distortion of Altair, which was only 3 mas in diameter, due to its rapid rotation. The instrument was closed down in 2008.

Bibliography

Gingerich, O. (ed.), *The General History of Astronomy Volume 4A; Astrophysics and Twentieth-Century Astronomy to 1950: Part A*, Cambridge University Press, 1984.

Hale, G. E., *The Possibilities of Large Telescopes*, Harper's, **156** (1928), 639–646.

Hale, G. E., *Building the 200-Inch Telescope*, Harper's, **159** (1929), 720–732.

King, Henry C., *The History of the Telescope*, Charles Griffin, 1955 (Dover reprint 1979).

Kuiper, Gerard P., and Middlehurst, Barbara M. (eds.), *Stars and Stellar Systems, Vol. I, Telescopes*, University of Chicago Press, 1960.

Osterbrock, Donald E., *Pauper and Prince; Ritchey, Hale and Big American Telescopes*, University of Arizona Press, 1993.

Sandage, Allan, *The First 50 Years at Palomar: 1949–1999 The Early Years of Stellar Evolution, Cosmology, and High-Energy Astrophysics*,

Annual Reviews of Astronomy and Astrophysics, **37** (1999), 445–486.

Wright, Helen, *The Great Palomar Telescope*, Faber and Faber, 1953.

Wright, Helen, *Explorer of the Universe: A Biography of George Ellery Hale*, American Institute of Physics Press, 1994.

PARIS AND MEUDON OBSERVATORIES

Louis XIV's Royal Observatory was founded in 1667, just one year after the formation of the French Academy of Sciences. It was formed at the instigation of Adrien Auzout and Jean-Baptiste Colbert and was located just south of Paris on the Paris meridian.

Christiaan Huygens, who was a founding member of the French Academy of Sciences, had come to Paris in 1666. Three years later, Gian Domenico or Jean Dominique Cassini was invited to Paris to supervise the construction of the new observatory which had been designed by the respected architect Claude Perrault. In 1671 Cassini became the observatory's first director, which was an unofficial position at the time. He argued with Perrault on the impracticability of his observatory design, but he was only able to get it slightly modified. This impractical design, together with the self-importance shown by the French court of the time, resulted in the observatory sometimes being called 'l'Observatoire de parade'.

The early instrumentation consisted of long- and very-long-focus aerial telescopes to observe solar system objects, and mural sectors and quadrants with micrometers for positional astronomy. From the observatory J. D. Cassini discovered the gap in Saturn's rings, now known as the Cassini division, and discovered four of Saturn's moons, the last two with Campani telescopes of 100 and 136 ft (30 and 42 m) focal lengths. Around the same time Ole Römer, a Danish astronomer working in Paris, concluded that the velocity of light was finite by timing the eclipses of Jupiter's moons. Initially the Royal Observatory was administered by the Academy of Sciences, but in 1795, after the French Revolution, its administration was transferred to the newly created Bureau des Longitudes.

During the seventeenth and eighteenth centuries the observatory undertook a great deal of work in geodesy and cartography. In the 1790s, for example, a survey undertaken along the French meridian resulted in an accurate determination of the length of the metre, which was defined as one ten-millionth of a quarter of the meridian.

The Paris Observatory was, by the early nineteenth century, poorly equipped with instruments. So in 1822 a 11 cm (4.3 inch) aperture mural circle was installed and fitted with six microscopes. A 15 cm (6 inch) transit instrument was installed in the next decade, and a 12 cm (5 inch) meridian circle in 1843. The observatory director François Arago ordered a 38 cm (15 inch) diameter equatorial refractor. But this was not installed until two years after his death in 1853.

Urbain Le Verrier had fallen out with Arago towards the end of his directorship. As a result Le Verrier let it be known in government circles that he thought the work of the observatory needed a radical overhaul. As he hoped he was appointed observatory director after Arago's death, with authority to make the changes required.

Le Verrier introduced a more structured method of working, in place of the previous free-for-all. But he went too far, treating the astronomers as his assistants to do as they were told. Naturally this did not go down too well, and they resigned as a body in 1870, causing his removal. But his successor Charles-Eugène Delaunay died two years later, which resulted in a much chastened Le Verrier being reinstated. Le Verrier had obtained a large transit circle in 1863 and fifteen years later his successor, Admiral Ernest Mouchez, installed a state-of-the art meridian instrument. He also set up a photographic laboratory, began time determinations and produced astrometric catalogues.

Mouchez was impressed by the excellent quality of the astrophotographs taken at the observatory by the brothers Paul and Prosper Henry, which showed an enormous number of very faint stars. As a result, in 1887 Mouchez arranged an international astrophotography congress in Paris to discuss a programme for photographing the whole sky. All the observatories that participated in this *Carte du Ciel* project were to use the same standard design of telescope as that developed by the Henry brothers, which had a focal length of 3.4 m (11.2 ft) and aperture of 33 cm (13 inches). Guiding was undertaken with an attached 24.5 cm (9.6 inch) refractor of the same focal length. Unfortunately this *Carte du Ciel* project turned out to be far more onerous than originally foreseen, so it was not completed until 1964.

Admiral Mouchez had, many years earlier, ordered two coudé equatorial instruments for the Paris Observatory, the first of which was installed in 1883. The second, which was installed seven years later, had interchangeable visual and photographic objectives of 60 cm (24 inch) aperture and 18 m (59 ft) focal length. Maurice Loewy (director from 1896 to 1907) and Pierre Puiseux used the instrument to produce a photographic atlas of the Moon using 10,000 plates. When it was completed in 1910 it was probably the best atlas of the Moon available for next fifty years.

Amalgamation of Paris and Meudon observatories

The Meudon Observatory had been founded in 1875 just outside Paris as one of the world's first centres for astrophysics research. Jules Janssen, its first director, was authorised to use the old royal estate of Meudon, but its old Château Neuf was in a serious state of disrepair, and it took nearly twenty years to restore it. The observatory's main instrument was the second largest refractor in the world when it was completed by 1891. This was a double refractor consisting of a 83 cm (32.6 inch) visual objective and 62 cm (24.4 inch) photographic objective of 16 m (52.5 ft) focal length. At about the same time a 1 m (39 inch) reflector was installed at Meudon followed in 1898 by a spectroheliograph built by Henri Deslandres to photograph the outer regions of the Sun. In 1906 Deslandres installed a second spectroheliograph and, shortly afterwards, succeeded Janssen as observatory director. Deslandres also installed siderostats and started a long series of solar observations. Eugène Antoniadi also started a series of observations of Mars in 1909 with the 33 inch double refractor, from which he concluded that the canals were an optical illusion.

Deslandres negotiated a merger between the Meudon Observatory and the Paris Observatory in 1926, with Meudon becoming the Astrophysics Section of the Paris Observatory. Deslandres became the joint observatory's first director. At about the same time Bernard Lyot developed a number of new instruments at Meudon, including his coronagraph which he later used at the Pic du Midi Observatory.

After the Second World War, André Danjon directed the reconstruction of the Paris Observatory and the astrophysics section at Meudon. Under his direction, the Nançay Radio Astronomy Station became part of the Paris Observatory. A radio astronomy group had also been set up at Meudon after the war, from where the partial solar eclipses of 1949, 1952 and 1954 were observed. In 1961 a 60/40 cm (24/16 inch) Schmidt telescope was installed at Meudon, followed a few years later by a 36 m (118 ft) solar tower telescope with a 85 cm (33 inch) coelostat. Then in 1972 a 10 m (33 ft) ultraviolet vacuum spectrograph became operational.

Bibliography

Alder, Ken, *The Measure of All Things: The Seven-Year Odyssey that Transformed the World*, Little, Brown, 2002.

Aubin, David, *The Fading Star of the Paris Observatory in the Nineteenth Century: Astronomers' Urban Culture of Circulation and Observation*, OSIRIS, 18 (2003), 79–100.

Débarbat, S., Grillot, S., Lévy, J., *Observatoire de Paris – Son histoire (1667–1963)*, Observatoire de Paris, 1984.

Gingerich, Owen (ed.), *The General History of Astronomy, Vol. 4A: Astrophysics and Twentieth-Century Astronomy to 1950*, Cambridge University Press, 1984.

Wolf, C., *Histoire de l'Observatoire de Paris de sa création à 1793*, Gauthier-Villars, Paris, 1902.

See also: Early French radio astronomy.

PIC DU MIDI OBSERVATORY

The first plans for an observatory on the 2,890 m (9,480 ft) Pic du Midi in the French Pyrenées were drawn up in 1866 by a learned society in the nearby town of Bagnères de Bigorre.

Although these plans came to nothing, two members of the society, Charles de Nansouty and Célestin Vaussenat, persevered, and in 1878 observatory construction was begun at the summit. Later, as a result of financial problems, the observatory was donated to the French government on condition that it settle the observatory's debts.

Initially the main function of the observatory was as a meteorological station, and it was not until 1891 that regular astronomical observations were begun. These were mainly of the Sun. Then in 1903 the observatory became affiliated with the Toulouse University Observatory. A few years later Benjamin Baillaud, the director of that observatory, built a twin telescope consisting of a 20 inch (50 cm) reflector and a 9 inch (23 cm) refractor at the Pic du Midi. But the observatory was little used until 1920 because access was very difficult for two-thirds of the year. Then a new director, Camille Dauzère, was appointed, but he turned the observatory into a geophysics establishment, with astronomy taking a back seat.

Solar facilities

Bernard Lyot came to the Pic du Midi observatory in 1930 to observe the Sun with his newly invented coronagraph. He returned a number of times over the next few years, because the exceptional viewing conditions enabled him to produce high-quality solar and planetary observations. Pierre Auger also studied cosmic rays from the Pic du Midi in the 1930s. In fact cosmic ray showers were recorded there continuously for twenty years.

The observatory did not become fully used until after the Second World War because of difficulty of access. In 1949 a cable railway was opened to transport supplies, but this was limited to the summer months. Then, three years later, a cable car system was installed allowing year-round access for the first time.

In 1954 Jean Rösch began to use a 23 cm refractor to photograph solar granulations. This research was later continued with a 38 cm refractor enclosed in a special turret dome. Then in 1975 its 38 inch objective was replaced by a 50 cm that enabled fine structure of the solar surface to be detected down to 0.25 arcsecond resolution.

High-resolution lunar, planetary and other work

Jules Baillaud, Benjamin's son, who had become director of the Pic du Midi observatory in 1937, was determined to improve his father's 20 inch reflector. He eventually replaced it with a 24 inch (60 cm) photographic lens from the Paris Observatory which, because its focal length was three times that of the reflector, required the light to be folded by two plane mirrors. This 'lunette Baillaud', as it was called, was used by Lyot in 1945 to produce some lunar photographs of exceptional quality. It was later used for both lunar and planetary research until 1964 when it was joined by a 106 cm reflector. This 106 cm was funded by NASA to extend the work of the 60 cm which had been used to take detailed photographs of the Moon in preparation for the Apollo Moon landings. Both telescopes took full advantage of the exceptional observing conditions, with the 106 cm sometimes producing images with a resolution of 0.15 arcseconds. The 60 cm was dismantled in 1970.

The excellent performance of the 106 cm persuaded the observatory to obtain a 2 m telescope, now known as the Bernard Lyot telescope, which was installed on the mountain in the summer of 1979. It was equipped more recently with a high-resolution spectrograph, a stellar polarimeter and an infrared camera, as its mission has been changed from planetary research to stellar and galactic research. This 2 m is still the largest optical telescope in France.

Suggestions had been made in the early 1990s that the Pic du Midi Observatory would have to be closed to allow funding to be diverted to the European Southern Observatory in Chile and the Haute-Provence Observatory in France. But protests from many French organisations, together with international support, resulted in the French government, the European Commission and others funding modifications to the observatory and its access to allow tourists to visit it as well as astronomers. It was first opened to tourists in July 2000.

Bibliography

Davoust, Emmanuel, *A Hundred Years of Science at the Pic du Midi Observatory*, in Novak, Giles, and Dandsberg, Randall H. (eds.), *Astrophysics from Antarctica: Proceedings of the ASP Summer Scientific Symposium Conference, Chicago, 1997*, Astronomical Society of the Pacific, 1998.

Rösch, Jean, and Dragesco, Jean, *The French Quest for High Resolution*, Sky and Telescope, January 1980, pp. 6–13.

PULKOVO OPTICAL OBSERVATORY

The St Petersburg Academy of Sciences, which was established in 1724 by the order of Czar Peter I, included a school and university to train Russian scientists. One of the academy's tasks was to produce the first detailed atlas of the Russian Empire. As a result astronomical observers were sent on expeditions all over the empire to determine the location of places in geographical coordinates. The basic techniques for this work were developed, and the training of observers was carried out in an observatory located in a tower in the centre of the Academy of Sciences building. This was built at St Petersburg on Vasil'evskii Island in the Neva River.

The Frenchman Joseph Delisle, who was invited to St Petersburg in 1726 as the first director, stayed in this post until 1747 when the observatory building burnt down. It was rebuilt and re-equipped with the same type of simple instruments. Then in 1761 the Russian government purchased an 8 foot (2.4 m) radius mural quadrant, and a 5 foot (1.5 m) focal length transit instrument, both from England, to enable more accurate observations to be produced. But the location of the observatory next to a river in the centre of the city was clearly not optimum. In addition, the location of the instruments on an upper floor did not provide the necessary stability. Better facilities were clearly required, although nothing much changed until the early 1830s.

In the meantime, Dorpat University in Livonia (now Tartu University in Estonia) had been restored 165 miles (265 km) south-west of St Petersburg by Czar Aleksander I. Friedrich Georg Wilhelm Struve, who obtained his doctorate at Dorpat University in 1813, taught courses there in astronomy and geodesy. He became director of the Dorpat observatory in 1820 and ordered a Fraunhofer 24 cm (9$\frac{1}{2}$ inch) achromatic refractor, which, when it arrived in 1824, was the largest refractor in the world. Struve used this to measure double stars and determine the annual parallax of Vega.

In 1827 the St Petersburg Academy of Sciences asked Georg Parrot, a former professor at Dorpat, to produce a plan for a major new Russian observatory. By then it was clear that the old observatory at St Petersburg was located in the wrong place, but the czar still wanted the new observatory to be located as close as possible to St Petersburg, as that was his capital city. Eventually it was decided in 1834 that the best location was on a hill near the village of Pulkovo, about 18 km (11 miles) south of the city. The new observatory would be devoted to astronomy, geodesy and hydrography. It was dedicated in 1839.

F. G. Wilhelm Struve, who was appointed director of the Pulkovo Observatory, installed a number of excellent instruments, including the 15 inch (38 cm) Merz and Mahler refractor in 1839 which was, at the time, the largest in the world. He also obtained an Ertel transit instrument and vertical circle, a Repsold meridian circle and transit instrument, a Merz and Mahler heliometer, and a number of smaller instruments. These superb instruments, along with extremely careful observations and analysis by Struve and others, yielded astronomical results of exceptional precision. This resulted in highly accurate estimates of aberration, precession and nutation. As a result the star catalogues produced at Pulkovo contained the most highly accurate positional measurements available at the time by concentrating on a relatively small number of relatively bright stars.

As far as geodesy and cartography were concerned, Struve was also involved in a project to determine the arc of the meridian at 25° 20′ between the Danube and the Artic Sea, and in an international project to measure the arc of the 64th parallel from England to the Urals.

F. G. Wilhelm Struve was followed by his son, Otto W. Struve, who was director of the observatory from 1862 to 1889. He continued his father's work on positional astronomy. He also purchased the 30 inch (75 cm) Repsold refractor, with a lens by Alvan Clark, which was installed at the observatory in 1885. Again, for a time, it was the largest refractor in the world.

Otto W. Struve had tried in the 1860s to observe the spectrum of aurorae, although his main interest was measuring double stars. In 1872 Bernhard Hasselberg, a Swede, was the first astrophysicist to work at Pulkovo. He set up an astrophysical laboratory where he carried out experiments to try to understand stellar spectra, and he also introduced photographic methods to the observatory. His astrophysical work was extended by Aristarkh Belopol'skii, an expert in stellar spectroscopy and solar research, who arrived in 1888, and Fëdor Bredikhin who was appointed director of the observatory in 1890. Shortly afterwards a double astrograph, of the type used for the *Carte du Ciel* project, was installed at Pulkovo. It had a 33 cm (13 inch) diameter objective for photography and a 24.5 cm (9.6 inch) refractor for guiding.

A large solar Littrow spectrograph, with a coelostat, was purchased from England in 1922, and two years later Belopol'skii began regular observations of the Sun's rotation. In the following year a new fundamental determination of longitude between Pulkovo and Greenwich was completed. Then in 1927 the observatory obtained a new Zeiss astrograph for zonal observations. This enabled the photography of the northern polar stars to be completed.

A southern outstation of the Pulkovo Observatory had been set up at Odessa in the Ukraine in 1898 where a new transit instrument and vertical circle were installed. Just over ten years later this southern outstation was moved to the nearby naval observatory at Nikolaev. Meanwhile, N. S. Mal'tsov, a wealthy amateur, had donated his observatory near Simeiz in the Crimea to the Pulkovo Observatory, who also operated it as an outstation. Not only did the Ukraine and the Crimea have many more clear nights than St Petersburg, these new outstations enabled the Pulkovo Observatory to extend their observations further into the southern sky. In 1925 a 1.0 m (40 inch) reflector was installed at the Simeiz Observatory. It was, at the time, the largest reflector in Europe.

The Pulkovo Observatory was destroyed during the Second World War. But the observatory's smaller instruments and the most important manuscripts from the extensive Pulkovo library had previously been removed and put into safe storage. The larger instruments could not be easily moved, so their optics were removed and also stored. After the end of the war, the Pulkovo Observatory was rebuilt to almost its original design, being completed in 1954. The surviving old instruments were repaired and modernised, and a number of new instruments obtained, including a 65 cm (26 inch) refractor that had been removed from Germany as part of war reparations.

The Simeiz Observatory was also completely destroyed in the Second World War. Subsequently the observatory was restored, not as an outstation of the Pulkovo Observatory, but as the independent Crimean Astrophysical Observatory (see separate article).

In 1948 a mountain station was established by the Pulkovo Observatory for solar research near Kislovodsk in the Caucasus, followed about ten years later by a laboratory built at Blagoveshchensk for measuring variations in latitude. In 1954 S. E. Khaykin set up a new department of radio astronomy alongside the optical observatory at Pulkovo (see *Early Soviet radio astronomy* article). And shortly afterwards Pulkovo designers were also involved in the design of the 6 m (236 inch) BTA-6 or Bolshoi Teleskop Alt-azimutalnyi (Large Altazimuth Telescope) (see separate article).

Bibliography

Gingerich, O. (ed.), *The General History of Astronomy, Vol. 4A: Astrophysics and Twentieth-Century Astronomy to 1950*, Cambridge University Press, 1984.

Gurshtein, Alexander A., and Ivanov, Constantin V., *Science Feasts While the Public Starves: A Note on the Reconstruction of the Pulkovo Observatory After World War II*, Journal for the History of Astronomy, **26** (1995), 363–368.

Krisciunas, K., *A Short History of the Pulkovo Observatory*, Vistas in Astronomy, **22** (1978), 27–37.

Krisciunas, K., *The End of Pulkovo Observatory's Reign as the 'Astronomical Capital of the World'*, Quarterly Journal of the Royal Astronomical Society, **25** (1984), 301–305.

Krisciunas, Kevin, *Astronomical Centers of the World*, Cambridge University Press, 1988.

Kulikovskii, P. G., *250 Years of the Academy of Sciences and Astronomy*, Soviet Astronomy, **18** (1974), 131–135.

See also: Crimean Astrophysical Observatory.

SOAR

The University of North Carolina and Columbia University looked into the possibility in the late 1980s of building a new world-class, 4 m-class telescope in the Southern Hemisphere. The universities thought that they may be able to find sufficient funds to build the telescope, but not to pay for its infrastructure and annual operating costs. On the other hand, the Cerro Tololo Inter-American Observatory (CTIO), which already had telescopes on Cerro Tololo in the Chilean Andes, would find it easier to fund its infrastructure and operating costs rather than its capital costs. As a result the three organisations decided to get together and try to build such a 4 m telescope, now called the Southern Astrophysical Research (SOAR) Telescope. A site survey, carried out in 1989, found a suitable site on the 2,700 m Cerro Pachón in Chile, not far from the CTIO.

Over the years since SOAR was first conceived, the project's partners have gradually changed, with Columbia University pulling out, and Brazil's Ministry of Science and Michigan State University joining the team. The telescope site was levelled in 1998 and major contracts let the following year for the mirror blanks, the active optics system for the thin primary mirror, the telescope mount, and the building and dome.

As built the SOAR primary mirror was a 4.2 m diameter, 10 cm thin meniscus mirror made of Corning's Ultra Low Expansion (ULE) material. Corning produced the mirror blank by fusing together glass segments. After fusing, the blank was rough ground and then heated over a mould until it sagged and took up a meniscus shape with the primary mirror's required radius of curvature. The blank was then sent to Raytheon in Danbury, Connecticut, for grinding and polishing. On completion, it was despatched in December 2003 via New York and the Panama Canal to Chile, arriving on Cerro Pachón about a month later. There it was aluminised using the Gemini South vacuum chamber.

SOAR was designed to operate between the atmospheric cut-off at 320 nm up to about 5 μm in the near infrared. Its secondary mirror, like its primary, was made of ULE glass. An instrument support box (ISB) at each Nasmyth focus, on the altazimuth mount's altitude axis, could carry a cluster of three heavy instruments, whilst lighter instruments could be mounted at two folded Cassegrain ports. A tertiary mirror rotated to select the focal station to be used, whilst beam steering optics in each ISB directed light to the chosen instrument. This allowed the observer to switch between instruments in a few minutes. The tertiary mirror was also designed to vibrate in two axes at up to 50 Hz to partially compensate for atmospheric turbulence.

SOAR's mirrors were installed in February 2004, and the telescope was dedicated in April of that year. But early scientific observations were severely compromised by a serious problem with the primary mirror's lateral support system. New lateral supports were installed in 2006, so enabling the telescope to become fully operational.

Bibliography

Carney, Bruce W., *The SOAR Telescope*, ASP Conference Series, **55** (1994), 45–55.

SOUTH AFRICAN OBSERVATORIES

The Cape observatory

The British Admiralty had founded the Royal Observatory at the Cape of Good Hope in 1820 to extend the work of the Greenwich Observatory to the Southern Hemisphere. But it was underfunded and periodically forgotten by its patrons. Thomas Maclear became director of the Cape observatory in

1834, where he remained for over thirty-five years. Meanwhile John Herschel began his four-year survey of the southern sky also in 1834, from a location near the Cape observatory, to match his father's northern survey. Herschel gave Maclear much moral and other support both when Herschel was in South Africa and for many years after he returned. Although Maclear and his assistants carried out numerous astrometric observations over the decades, their reductions were largely uncompleted when he retired in 1870. These old observations were reduced by Edward Stone, who was the next observatory director, to produce the Cape Catalogue of over 12,000 stars.

A new, more positive era began at the Cape in 1879, when David Gill was appointed Her Majesty's Astronomer at the Cape. When Gill took over, the observatory was disorganised and run down, but by the time of his retirement in 1907 it had become an internationally respected observatory. In particular, Gill's work on stellar photography in the 1880s had helped to persuade others of the enormous advantages of this new technique. His *Cape Photographic Durchmusterung (CPD)* star catalogue, which used plates taken at the Cape, was second to none when it was completed in 1900.

Previously, in the 1880s Gill, assisted by the American W. L. Elkin, had used a heliometer to measure the parallax of nine stars. He later acquired a 7 inch Repsold and Söhne heliometer with which he measured the solar parallax using observations of asteroids. His final value of the solar parallax was adopted internationally as the definitive value for almost fifty years.

In 1894 Frank McClean offered the Admiralty a 24 inch (60 cm) photographic refractor which was coupled to an 18 inch (45 cm) visual refractor on the same mount. This so-called Victoria telescope became operational at the Cape seven years later where it became the observatory's workhorse for many years.

Harold Spencer Jones, who was appointed director of the Cape observatory in 1924, revived the observatory which had, once again, become somewhat disorganised. He recruited a number of good South African astronomers, and initiated the *Cape Photographic Catalogue* of the southern sky which contained precise stellar positions and proper motions all the way to the south celestial pole. The final section of the catalogue, which included stellar magnitudes and colours, was not finished until 1968.

John Jackson, who took over from Spencer Jones in 1933, concentrated almost entirely on projects initiated by his predecessor. Richard Stoy, who became Jackson's chief assistant in 1935, concentrated on trying to improve the photometry of southern stars, which was quite deficient at that time. He continued this work when he became observatory director himself in 1950, extending it into the ultraviolet using an 18 inch (45 cm) reflector, and in 1964 using the 40 inch (1.0 m) Elizabeth reflector.

The Radcliffe Observatory, Pretoria

The Radcliffe Observatory had been founded at Oxford in 1771, but in the late 1920s Sir William Morris offered to purchase the site of the observatory to expand the Radcliffe Infirmary, which was grossly overcrowded. The Observatory were keen to set up an observatory in a more suitable climate, and readily agreed to the sale. So in 1929, the British astronomer William Steavenson went to Pretoria to find a suitable site but legal problems delayed setting up the Radcliffe Observatory there until 1938. A 74 inch (1.9 m) reflector was ordered for the new observatory from Grubb Parsons in the UK, with a Corning Pyrex primary mirror. But the Second World War started before the telescope could be completed. As a result the 74 inch did not become operational in Pretoria until 1948, when it was the largest telescope in the Southern Hemisphere. In 1951 an agreement was signed that allocated one-third of the observing time on the 74 inch to observers from the Cape Observatory.

The Union Observatory, Leiden Southern Station and Yale-Columbia Southern Station, Johannesburg

David Gill, Jacobus Kapteyn and Jöns Backlund, who was the director of the Pulkovo Observatory in Russia, persuaded the Transvaal government in 1905 to establish an observatory just outside Johannesburg. It became known as the Union Observatory when the Union of South Africa was established in 1910. A 26 inch (65 cm) visual refractor was installed in 1925 for double star work.

The Leiden Observatory in the Netherlands installed a twin 16 inch (40 cm) photographic refractor, funded by the Rockefeller Foundation, at the Union Observatory in 1938 to observe variable stars. The Leiden equipment was moved in the mid-1950s to a darker location at Hartebeespoort further from Johannesburg, and a new 36 inch (0.9 m) Flux Collector telescope installed there at what was called the Leiden Southern Station. However the 36 inch was eventually moved to the European Southern Observatory at La Silla, Chile in 1978, owing to increasing light pollution at the South African site.

From 1925 Yale Observatory also operated a 26 inch (65 cm) photographic refractor for measuring stellar parallaxes at Witwatersrand University, Johannesburg. Twenty years later Columbia University joined Yale to produce the Yale-Columbia Southern Station. By 1953 66,000 plates had been taken with the 26 inch, producing the parallaxes of over 1,700 southern stars.

The Boyden Station and the Lamont-Hussey Observatory, Bloemfontein

In 1927 the Boyden station of the Harvard College Observatory was moved from Arequipa, Peru, to Mazelspoort, a few miles north of Bloemfontein, South Africa. A 60 inch (1.5 m) Fecker reflector was installed there accompanied by a number of smaller telescopes. The 60 inch mirror was refigured in 1933 and its telescope mounting replaced, Cer-Vit optics were installed in the 1960s, and in 2001 a new control system and other improvements were carried out. The Boyden Observatory, now run by the University of the Free State, is still operational.

William Hussey of the University of Michigan had persuaded the industrialist R. P. Lamont in about 1910 to provide funding for an observatory in the Southern Hemisphere, to allow him to extend his observations of double stars. Hussey obtained a 27 inch (70 cm) lens and, in 1924 went to South Africa to find a suitable site for what was to be called the Lamont-Hussey Observatory. He chose Naval Hill, Bloemfontein, but he died before the observatory could be established in 1928. E. C. Slipher observed Mars from there during its close opposition of 1939, but the observatory was only used intermittently until it was finally closed in 1974.

Observatory status mid-twentieth century

So in 1950 South Africa had four main observatories or observatory groups at or near the Cape, Pretoria, Johannesburg and Bloemfontein. These observatories had equipment belonging to the Greenwich and Radcliffe observatories in the UK, Yale, Harvard and Michigan observatories in the USA, and Leiden observatory in the Netherlands. In 1953 the Yale station was closed and moved to Mount Stromlo, Australia, because of deteriorating observing conditions at Johannesburg.

In 1960 the Cape Observatory was merged with the Royal Greenwich Observatory. Then in the following year the Union Observatory was renamed the Republic Observatory, following South Africa's expulsion from the British Commonwealth. In 1965 the British Science Research Council (SRC) took over control of the merged Cape Observatory from the Admiralty, putting it squarely in the scientific arena. Unfortunately at that stage much of its equipment was beginning to look outdated, compared with that proposed for other southern hemisphere observatories. In addition, both the Cape and Radcliffe observatories were suffering from light pollution. One solution considered was to build a new observatory at a new dark site, replacing both the Cape and Radcliffe observatories. Three possible sites were identified near the town of Sutherland, but the idea was not taken any further.

However, the idea of a common observatory was not dead, and in 1970 the British SRC and its equivalent institution in South Africa agreed to build a new, joint astronomical observatory at a dark site. This new South African Astronomical Observatory would involve amalgamating the British-owned Cape and Radcliffe observatories and the South African-owned Republic Observatory.

The South African Astronomical Observatory

As built this South African Astronomical Observatory (SAAO) specialised in photometry and spectroscopy. It had a base station in Cape Town, where some instruments were retained, and an observing station at an altitude of about 5,700 ft (1,750 m) near Sutherland. The 40 inch Elizabeth telescope was moved from the Cape Observatory to Sutherland, and a 20 inch (50 cm) reflector was moved to Sutherland from the Republic Observatory, where it had been in operation since 1967. The new observatory was dedicated in 1973. Shortly afterwards the 74 inch Radcliffe telescope was also moved to Sutherland, where it was used for spectroscopy and infrared photometry.

Astronomy in South Africa suffered during the country's relative isolation from the international community during its apartheid regime. But international relations gradually recovered after Nelson Mandela's release from prison in 1990.

The Infrared Survey Facility, consisting of a 1.4 m telescope and three-colour infrared camera, became operational in 2000. It was a joint project of the SAAO and Nagoya University, Japan. In the meantime, in 1998 the South African government had agreed to pay a significant proportion of the costs of the proposed Southern African Large Telescope (SALT). This telescope, with a diameter of about 10 m, used a design based on that of the Hobby-Eberly Telescope at the McDonald Observatory in the United States, modified for the difference in latitude of the two observatories. Funding was eventually provided by international partners from the United States, Germany, Poland, India, the UK and New Zealand, as well as South Africa. Construction has now been completed and the telescope is currently (2010) being commissioned.

Bibliography

Buckley, David, et al., *Africa's Giant Eye: Building the Southern African Large Telescope*, SALT Foundation, 2005.

Evans, David S., *Under Capricorn: A History of Southern Hemisphere Astronomy*, Institute of Physics Publishing, 1988.

Gill, David, *A History and Description of the Royal Observatory, Cape of Good Hope*, Stationary Office, London, 1913.

Heck, André, (ed.), *Organizations and Strategies in Astronomy: Vols. 3 and 5*, Kluwer Academic Publishers, 2002 and 2004.

Olowin, R. P., *South Africa's Newest Observatory*, Sky and Telescope, June 1973, pp. 347–350.

Warner, Brian, *Astronomers at the Royal Observatory Cape of Good Hope*, Balkema, 1979.

TOKYO ASTRONOMICAL OBSERVATORY

A small astronomical observatory was set up for students at the Hongo campus of Tokyo University in 1878. Ten years later this observatory was moved to Azabu, in another district of Tokyo, where it formed the basis of the Tokyo Astronomical Observatory. In the early years the Tokyo Astronomical Observatory's equipment was limited to a meridian circle and two rather modest refractors. The observatory produced almanacs, undertook routine solar observations and provided a time service.

In about 1920 the Tokyo Astronomical Observatory was moved to Mitaka, further out from the centre of Tokyo, because the rapid expansion of the city had made the earlier site unsuitable. A 20 cm (8 inch) meridian circle was installed there in 1926, followed a few years later by a 65 cm (26 inch) Zeiss refractor and a 48 cm (19 inch) solar tower telescope. Observational work was limited at the time to positional measurements of solar system objects, photometry of variable stars, and solar studies. Then in 1949 radio astronomy started at Mitaka (see *Japanese terrestrial radio telescopes* article).

By the late 1940s it was clear that the Mitaka site was unsatisfactory for deep sky observing because of the continuing expansion of the city, so a number of new branch observatories were established. These were the Norikura Solar Observatory in 1949, Okayama Astrophysical Observatory (1960), Dodaira Observatory (1962), Nobeyama Radio Observatory (1970) (see *Japanese terrestrial radio telescopes* article), and Kiso Observatory (1974).

The Norikura Solar Observatory was built at an altitude of 2,880 m (9,440 ft) on Mount Norikura. Coronal surveys began in 1951 using a 10 cm (4 inch) coronagraph, then in 1971 a 25 cm (10 inch) coronagraph was built there by Nikon. Another 10 cm coronagraph was installed twenty years later with a CCD camera and image processing unit.

The Science Council of Japan decided in 1953 to obtain some modern optical telescopes for Japanese observatories. Following financial approval, it was decided to import a 188 cm (74 inch) Grubb Parsons reflector from England, and to ask Nikon to produce two 91 cm (36 inch) reflectors. These 91 cm reflectors were appreciably larger than any previous domestically built telescopes. Following a two-year site survey it was agreed to locate the 188 cm and one of the 91 cm telescopes at an altitude of 372 m (1,220 ft) on Mount Chikurinji in the Okayama prefecture, and to locate the other 91 cm not far from Tokyo at an altitude of 880 m (2,900 ft) on Mount Dodaira.

The 188 cm Grubb Parsons reflector, which was installed at the Okayama Astrophysical Observatory in 1961, was similar in design to the telescopes built by Grubb Parsons for observatories in Canada, South Africa, and Australia. This large Okayama reflector was used by all Japanese astronomers, not just those of Tokyo Astronomical Observatory, which encouraged the development of Japanese observational astrophysicists in the 1960s. In addition to the 91 cm (36 inch) Nikon reflector, the observatory also had a 65 cm solar telescope.

Work began at the Dodaira Observatory in 1962 with the installation of the other 91 cm Nikon reflector. A satellite tracking facility was also built at the observatory using laser reflections, as well as a 50 cm Schmidt which was used for comet work. A 20 cm (8 inch) polar tube telescope was installed for studying the detailed motion of the celestial pole resulting from precession, nutation and aberration. The Dodaira Observatory was closed down in 2000.

Kiso Observatory was the last of the five observatories to be established as branches of Tokyo Astronomical Observatory. It was built at an altitude of 1,120 m (3,670 ft) on Mount Ontake and dedicated in 1974. Its main instrument was a 105 cm/150 cm (41 inch/59 inch) Schmidt. The observatory started to develop a CCD camera for the Schmidt in 1987 which resulted in a 1024 × 1024 pixel camera, followed by a 2048 × 2048 camera in 1997. A near infrared camera was also developed in parallel.

The Tokyo Astronomical Observatory was amalgamated with the International Latitude Observatory of Mizusawa (established 1899) and part of the Research Institute of Atmospherics at Nagoya University (established 1949) to form the National Astronomical Observatory of Japan (NAOJ) in 1988. The NAOJ subsequently developed the Japan National Large Telescope, later to be called the Subaru telescope. This 8.3 m reflector saw first light on Mauna Kea in late 1998 (see *Mauna Kea Observatory* article).

Bibliography

Kozai, Yoshihide, *A Flowering of Japanese Astronomy*, Sky and Telescope, June 1988, pp. 590–594.

Moriyama, Fumio, *Tokyo Observatory Today*, Sky and Telescope, November 1975, pp. 276–281.

See also: Japanese terrestrial radio telescopes; Mauna Kea Observatory.

TYCHO BRAHE'S OBSERVATORIES

Tycho Brahe's interest in astronomy was stimulated by the solar eclipse of 1560. Shortly afterwards he found that the conjunction of Jupiter and Saturn, as predicted by Erasmus Reinhold's *Prutenic Tables*, was several days in error. This encouraged Tycho to try to measure the positions of the heavenly bodies much more accurately than had previous observers to enable more accurate predictions to be made. Then in 1572 Tycho observed a bright new object which appeared suddenly in the sky. He could detect no parallax, so it appeared to be a new star. The object gradually decreased in brightness and changed colour, and then in 1574 it finally disappeared.

The appearance of this new star, now called Tycho's supernova, persuaded Tycho Brahe to found an astronomical observatory to observe and measure the heavens to unprecedented accuracy. His search for a site came to the attention of King Frederick II of Denmark who decided to offer him the small island of Hven near Copenhagen on which to build his observatory. The king created him the lord of the manor, which provided Tycho with a regular income, and also offered to pay for the construction of the observatory, which began in 1576. This observatory, which Tycho called Uraniborg, was the last significant observatory to be built in the West which did not include a telescope.

The observatory of Uraniborg was basically a castle or large manor house (see Figure 6.6) with rooms for astronomical instruments. There were also rooms for students, and there was an alchemical laboratory in the basement. The astronomical instruments included various quadrants, parallactic instruments, sextants, and equatorial armillaries, totalling over a dozen large, high-quality instruments. In fact there were so many large instruments, with typical sizes ranging from 1.5 to 3 m (5 to 10 ft), that Tycho had to build another observatory, called Stjerneborg, in 1584 to take further instruments. Some of the instruments at Uraniborg suffered from wind vibrations, so Tycho built Stjerneborg largely underground. Stjerneborg, which was near Uraniborg, contained some of Tycho's best instruments including a great equatorial armillary which was about 3 m in diameter.

To improve the accuracy of his measurements Tycho devised a special, non-optical, sight. He also produced a system of transversal dots to allow accurate interpolation between markings on the arcs. As a result his large mural quadrant, which he completed in 1582, produced relative measurements to within about 10 arcseconds. For various reasons, however, his absolute accuracies were only about 2 to 5 arcminutes.

Tycho gave many details of his buildings and instruments in his *Astronomiae Instauratae Mechanica* of 1598. His main task at Hven was to catalogue the positions of the stars, and to make observations of the Sun, Moon and planets to try to understand their motions. His star catalogue, which contained the positions of about one thousand stars, was published only after his death.

Tycho Brahe did not treat his tenants at Hven very well, and his arrogant attitude also alienated a number of Danish noblemen. Then in 1588 Tycho's patron, Frederick II, died, to be replaced by Christian IV, who did not have the same interest in astronomy. As a result Tycho Brahe was forced to leave Hven and then Denmark in 1597, taking his moveable instruments with him. He died four years later in Prague. Subsequently Tycho's observatories fell into disuse and by about 1620 all his instruments had been lost or destroyed.

Bibliography

Christianson, John Robert, *On Tycho's Island: Tycho Brahe and His Assistants, 1570–1601*, Cambridge University Press, 2000.

Krisciunas, Kevin, *Astronomical Centers of the World*, Cambridge University Press, 1988.

Thoren, Victor E., *The Lord of Uraniborg: A Biography of Tycho Brahe*, Cambridge University Press, 1990.

ORTHOGRAPHIA

PRAECIPVAE DOMVS   ARCIS VRANIBVRGI IN
INSVLA PORTHMI DANICI   HVÆNNA, Aſtronomiæ inſtaura-
dægratiâ circa annum 1580. àTYCHONE BRAHE
exædi-   ficatæ.

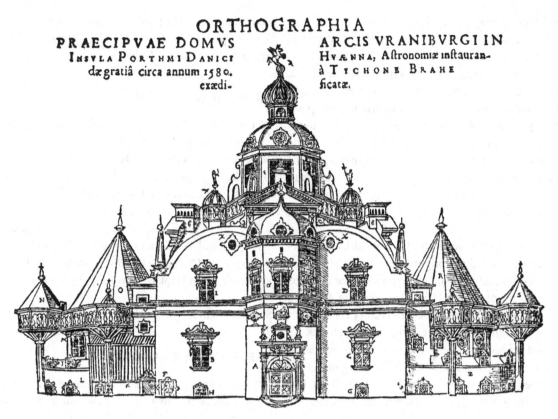

Figure 6.6. Tycho Brahe's Uraniborg observatory in about 1580. The building included an alchemical laboratory and aviary. (From *A Short History of Astronomy*, by Arthur Berry, 1898, Figure 51.)

UNITED STATES NAVAL OBSERVATORY

The reticence of American administrators to spend public funds on science delayed the setting up of a national astronomical observatory until 1830. And even then it was done rather quietly as part of the Depot of Charts and Instruments in the Department of the Navy. Astronomical observations were initially used to calibrate chronometers used for determining longitude, but in 1842 Congress voted funds for a permanent building to house an astronomical observatory in Washington. Its main purpose was to accurately measure the positions of the Sun, Moon, planets and stars for use in preparing the National Almanac, and to give the exact time of day for all those who needed it. The observatory could also be used for general scientific purposes, as long as that did not compromise its primary rôle.

The Depot of Charts and Instruments acquired a small meridian circle and portable achromatic refractor in 1838. Six years later, when the new building of the Naval Observatory was completed, it was the best-equipped astronomical observatory in the United States. It contained numerous instruments including a 9.6 inch (24.4 cm) achromatic refractor, a 5.3 inch (13.5 cm) aperture meridian transit, and a prime vertical transit of 4.9 inch (12.4 cm) aperture. In 1845 the observatory installed a time ball which was dropped at noon precisely to allow ships in the Potomac River to set their chronometers before putting out to sea.

The Naval Observatory's first director, Lieutenant Matthew Maury, observed in January 1846 that Biela's comet had broken into two pieces. In the following year Sears Walker, an assistant at the observatory, analysed pre-discovery observations of Neptune by Michel de Lalande. He found that Neptune was in a very different orbit from the orbit that had been predicted by Le Verrier, which had been used to facilitate its discovery. So its discovery had been somewhat fortuitous.

In 1865 the observatory obtained a 8.5 inch (21.6 cm) aperture meridian circle. This was followed eight years later by the 26 inch (66 cm) Great Equatorial refractor made by Alvan Clark and Sons, which was, at the time, the largest refractor in the world. In 1877 Asaph Hall discovered Phobos and Deimos, the two small satellites of Mars, with this 26 inch. The observatory sponsored expeditions to observe total solar eclipses, and was the headquarters of the commission coordinating the American expeditions to observe the transits of Venus of 1874 and 1882.

Simon Newcomb, who had served as professor of mathematics at the observatory from 1861 to 1877, realised that the lunar tables that had been published in 1857 by Peter Hansen were in error. Newcomb found, when he visited Paris in 1870, that the French had accurate records going back almost one hundred years before those used by Hansen in producing his tables. Newcomb revised Hansen's tables using these earlier observations. His modified tables were later used in all of the major ephemerides of the world.

The Naval Observatory suffered from fogs, as it was near the river, and from the ever increasing city traffic in Washington. So in 1893 it was moved to higher ground in Georgetown Heights, well outside the city, a few miles from its original site. After the move the observatory acquired a new 6 inch (15 cm) transit circle which was to become its key instrument for positional astronomy. The move to the new site was also used to re-evaluate the observatory's programmes, which resulted in the addition of a new programme involving the daily monitoring of solar activity with a photoheliograph. But the observatory did not expand its work into photographic, photometric or spectrometric research, leaving it, in the eyes of many astronomers, stuck in the past.

So the observatory continued its traditional work of observing solar system bodies and the brighter stars into the twentieth century. It continued to maintain a time service and was responsible for the preparation and publication of the *American Ephemeris and Nautical Almanac*. The observatory also continued to sponsor expeditions to observe total solar eclipses. In 1914 it acquired a photographic zenith tube which was used for determining the variation in latitude and the constant of aberration. Other observatories in the United States at this time were mainly devoted to astrophysical observations. So the Naval Observatory fulfilled a valuable rôle by undertaking the more traditional astronomical observations, whilst taking advantage of its more accurate facilities.

In 1934 the Naval Observatory began to use the photographic zenith tube to determine time. Then in the following year George Ritchey completed an aplanatic 40 inch (1.0 m) reflecting telescope, which was to be the last large telescope to be built at the Washington site. In 1955 the 40 inch was moved to a dark sky site near Flagstaff, Arizona. Eight years later a 61 inch (155 cm) astrometric reflector was also built at the Flagstaff site. The two telescopes were used to determine the relative positions, brightness, colours and spectral types of stars. In 1978 James Christy discovered Charon, the first-known satellite of Pluto, using the 61 inch.

The old 26 inch refractor was used visually over the period 1961 to 1990 at the Washington site to measure double stars in a programme which was largely unaffected by the urban environment of the observatory. Since 1990 double stars have been observed with the 26 inch using speckle interferometry and CCDs.

The measurement of time has changed over the last century. Quartz crystal clocks, which began to replace pendulum clocks, showed, for the first time, the irregularity in the Earth's rotation. As a result Ephemeris time, which was based on the orbital period of the Earth around the Sun, was adopted by the International Astronomical Union (IAU) in 1952, replacing Earth-spin time as a new uniform time scale. Six years later the U.S. Naval Observatory and the UK's National Physical Laboratory published the results of joint experiments that defined the relationship between atomic clock time and Ephemeris

time. Then in 1976 two new timescales known as dynamical time, which took into account relativistic effects, were adopted by the IAU to replace Ephemeris time.

Bibliography

Dick, Steven, J., *John Quincy Adams, the Smithsonian Bequest, and the Origins of the U.S. Naval Observatory*, Journal for the History of Astronomy, **22** (1991), 31–44.

Dick, Steven J., *Sky and Ocean Joined: The U.S. Naval Observatory 1830–2000*, Cambridge University Press, 2002.

Norse, J. E., *Memoir of the Founding and Progress of the United States Naval Observatory*, Astronomical and Meteorological Observations made at the U.S. Naval Observatory, 11, d1–d52.

Weber, G. A., *The Naval Observatory: Its History, Activities and Organization*, Johns Hopkins University Press, 1926.

See also: Modern optical interferometers.

YERKES OBSERVATORY

Edward Spence, a trustee of the University of Southern California, concluded in 1887 that the university should have an even larger telescope than the 36 inch (0.9 m) Lick, then nearing completion on Mount Hamilton as the largest telescope in the world. Spence pledged $50,000 and other potential investors made various promises. The trustees then commissioned Alvan Clark and Sons to make a 40 inch (1.0 m) diameter lens for this new telescope. But, before the lens could be completed, a number of the potential investors, including Spence, went bankrupt. This left Alvan Clark and Sons with two partially ground glass disks and a large bill from the French glass disk supplier.

George Ellery Hale became aware of these partially ground disks in the summer of 1892. He had recently been appointed Associate Professor of Astral Physics and Director of the Observatory of the new University of Chicago. So, along with William Harper, the president of the university, he visited the wealthy streetcar and property tycoon Charles T. Yerkes and persuaded him to pay not only for completing the 40 inch lens, but also for its complete telescope and observatory. In the following year the university chose the site of this new observatory, to be called the Yerkes Observatory, at Williams Bay, Wisconsin, some 80 miles (140 km) from Chicago.

Hale planned the observatory building to have spectroscopic laboratories, photographic darkrooms and small auxiliary telescopes. His aim was to compare the spectra of the Sun and stars and so determine their physical condition. When it was completed in 1897 the 40 inch Yerkes telescope was the largest refractor in the world. In fact it still is.

Hale had also looked into the possibility of building a large reflector whilst the 40 inch refractor was still being built. In fact he had gone so far as to persuade his father, William Hale, in

1895 to purchase a 60 inch (1.5 m) diameter mirror blank from France, even though there was no money at Yerkes at the time to build a complete telescope. So, on receipt in America, the blank was partially ground and then put into storage. Eventually it was used in the construction of a 60 inch reflector on Mount Wilson paid for by the Carnegie Institution.

In the meantime, George Ritchey had built a 24 inch (0.6 m) reflector at Yerkes, at his own expense. He began using it for nebular photography in 1901. Ritchey and Hale were convinced, at that time, that reflectors, not refractors, would be the key telescopes of the future. They were impressed, in particular, when they found that photographs taken at the 24 inch reflector's prime focus compared very favourably with the best photographs taken with the 40 inch refractor.

Hale was a very experienced observer of the Sun, but he found the 40 inch Yerkes refractor too cumbersome to operate with his large spectroheliograph attached. As a result he and Ritchey built a long focus horizontal reflecting telescope at Yerkes with a 30 inch (0.75 m) coelostat mirror for capturing the Sun. Unfortunately this solar telescope was completely destroyed by fire in 1902. So, in the following year they built an improved version using money donated by Helen Snow. Unfortunately the seeing was rather poor at Yerkes, as a result this Snow solar telescope was also moved, like the 60 inch mirror blank, to Mount Wilson in early 1905.

So in the early years of the twentieth century the Yerkes Observatory had had some key equipment transferred to Mount Wilson. The Carnegie Institution then agreed to fund this new observatory, which became known as the Mount Wilson Solar Observatory, with Hale as its first director. Ritchey moved with Hale from Yerkes to Mount Wilson, as did Walter Adams, Ferdinand Ellerman and Francis Pease. This left a much-depleted staff at Yerkes, together with just two large telescopes, the 40 inch refractor and the 24 inch reflector. After Carnegie's grant, the Mount Wilson Solar Observatory became a stand-alone facility completely separate from Yerkes.

Edwin Frost took over from Hale as the director of the Yerkes Observatory in 1904, staying in that position until he retired in 1932. Unfortunately the observational facilities at Yerkes were not substantially improved during his tenure, even though the University of Chicago continued to produce first-rate astronomers such as Edwin Hubble, Otto Struve and William Morgan. Struve took over from Frost as director in 1932.

Although the University of Chicago had excellent astronomers when Frost retired, it had no money to build a new large telescope. But the University of Texas had the opposite problem. It had money to build a telescope, because of a legacy left to it by W. J. McDonald, but no astronomers. As a result the two universities got together and agreed to build and operate a joint observatory. Under the terms of this agreement, which was specified to last thirty years, the University of Texas would build and equip this McDonald Observatory and the University of Chicago would pay for most of its operating expenses.

A site was found for the McDonald Observatory in 1932 on the 6,790 ft (2,070 m) Mount Locke in western Texas. It was then agreed to build a reflector of about 80 inch (2.03 m) diameter, later called the Otto Struve Telescope, at this observatory. When the telescope, which was actually 82 inch diameter, was dedicated in 1939 it was the second largest telescope in the world.

A new agreement was made between the two universities when the initial agreement expired in 1962. According to this new arrangement, the McDonald Observatory would become an autonomous facility of the University of Texas, although its use was still shared with the University of Chicago. However, by this time the 82 inch was a relatively modest instrument. Shortly afterwards, a 40 inch reflecting telescope was installed at Yerkes in the dome that had contained Ritchey's 24 inch reflector.

In the meantime, in 1957 the University of Chicago had become a founder member of AURA (Association of Universities for Research in Astronomy) which then built large telescopes on Kitt Peak and Cerro Tololo. So, although the university still had some excellent astronomers like Subrahmanyan Chandrasekhar, Gerard Kuiper and Albert Hiltner, the Yerkes observational facilities had gradually become less and less important to the university as the twentieth century progressed. As a result in 2005 the university considered selling the Yerkes Observatory and its site to a property development company. This naturally led to an outcry, and the university backed off, appointing a committee to consider the various options for the observatory and site. It published its report towards the end of 2007.

Bibliography

Krisciunas, Kevin, *Otto Struve, 1897–1963*, Biographical Memoirs of the National Academy of Sciences, **61** (1992), 351–387.

Osterbrock, Donald E., *Yerkes Observatory 1892–1950: The Birth, Near Death, and Resurrection of a Scientific Research Institution*, The University of Chicago Press, 1997.

Struve, Otto, *The Story of an Observatory: The Fiftieth Anniversary of the Yerkes Observatory*, Popular Astronomy, **55** (1947), 227–244, and 283–294.

Wright, Helen, *Explorer of the Universe; A Biography of George Ellery Hale*, American Institute of Physics, 1994.

See also: Kitt Peak and Cerro Tololo Observatories; McDonald Observatory.

PART 7
RADIO TELESCOPES, OBSERVATORIES AND RADAR

1 · Overview — Radio telescopes and observatories

The first radio device that clearly pinpointed a source of radio waves from outside the Earth's atmosphere was built in the early 1930s by Karl Jansky of the Bell Telephone Laboratories. It was a 30 m long antenna array that rotated horizontally every 20 minutes. Jansky actually designed it to study the cause of interference at 20 MHz (or wavelength of 15 m) on the newly opened trans-Atlantic radio link. In 1932 he found that the noise was coming from the Milky Way.

Five years later Grote Reber built the first antenna specifically designed to detect radio sources in the universe. It was a 10 m diameter parabolic dish which he operated over the frequency range from 160 MHz to 3.3 GHz (wavelength 1.9 m to 9 cm). The dish, which he constructed in his garden in Wheaton, Illinois, was a meridian transit type that could be moved in declination but not in right ascension.

Radio and radar technology improved enormously during the Second World War, and at the end of hostilities there was a large amount of surplus military equipment available. Some of this was used at universities to produce radio telescopes. The main developments in this immediate post-war period took place in Australia and the United Kingdom.

The biggest problem facing these early radio astronomers was that of accurately locating astronomical sources as the long wavelengths of radio waves required very large antennae to produce even modest resolutions. The obvious solution was to use interferometric techniques, like those used optically to measure the diameter of stars.

The first radio interferometer was the cliff interferometer designed by Joe Pawsey, Lindsay McCready and Ruby Payne-Scott and built near Sydney, Australia in 1945. Its configuration, which was analogous to that of the Lloyd's mirror optical arrangement of the nineteenth century, consisted of a 200 MHz wartime radar antenna on a cliff overlooking the sea. The source position could then be determined by measuring the interference pattern between the beam received directly from the source, and that reflected from the sea. At about the same time Martin Ryle and Derek Vonberg built the first Michelson radio interferometer. This device, which was built at Cambridge, UK, used two antennae each consisting of five dipoles mounted over wire netting. A few years later Ryle and F. Graham Smith built another interferometer consisting of two 7.5 m diameter parabolic German radar antennae 280 m apart, operating mostly at 214 MHz. This produced much-improved position estimates of what, at the time, were called radio stars.

Martin Ryle developed the idea of aperture synthesis in the early 1950s. This required just two antennae to be positioned at two locations in a two-dimensional array to sample the signals from the radio source. The positions of the two antennae were then changed a number of times, and the radio image produced by a computer which combined all the observations. The first such aperture synthesis instrument was built by John Blythe at Cambridge in 1954. The main constraints on the performance of such early aperture synthesis systems was the limited capability of early computers.

At about the same time another type of interferometer, the Mills Cross unfilled aperture system, was built near Sydney, Australia, by Bernard Mills and Alec Little. It consisted of two 460 m linear arrays arranged at right angles to each other. This cross arrangement, which operated at 85 MHz, enabled high resolution to be achieved in two directions at once.

Although interferometers were the only practical way to resolve radio sources, large dishes were still required to detect faint sources and to provide spectroscopy of radio sources.

The first large, movable radio dish to be built was the 75 m (250 ft) diameter parabolic dish at Jodrell Bank, UK, which was designed to operate down to the wavelength of neutral hydrogen at 21 cm (1.4 GHz). A new, more accurate reflecting surface was installed in 1970–71 to allow the telescope to operate down to 6 cm (5 GHz). This surface was replaced yet again in 2001–03 by another surface that expanded the useable frequency range by a factor of four, and provided an increase in sensitivity at 5 GHz by a factor of five. Even now, over fifty years after 'first light', this antenna, now called the Lovell telescope, is the third largest fully steerable radio telescope dish in the world.

The American National Radio Astronomy Observatory

The American National Radio Astronomy Observatory (NRAO) was founded in 1956 at Green Bank, West Virginia, and was provided in 1959 with its first major operational radio telescope. This was the 26 m (85 ft) diameter Howard Tatel telescope that could be used down to 3.75 cm wavelengths (8 GHz). In the mid-1960s two other 26 m radio telescopes were built to the same design at Green Bank which, with the Tatel telescope, formed the NRAO Green Bank Interferometer (GBI) with a maximum baseline of 2.7 km. In addition, to operating as an instrument in its own right, the GBI was also used as a prototype for the Very Large Array (VLA).

In the meantime, a 90 m (300 ft) transit radio telescope had been built at Green Bank in the early 1960s. It was covered with aluminium mesh, so making it cheaper to build, but consequently it was only able to operate down to 21 centimetres. Three years later a 43 m (140 ft) fully steerable equatorially mounted dish, with a continuous surface, was completed at the NRAO. It could operate down to 1 cm.

The 90 m transit telescope collapsed in 1988 due to metal fatigue, and was replaced by a 100 × 110 m altazimuth dish, the Robert C. Byrd Green Bank Telescope (GBT), which was formally dedicated in 2000. It has a computer-controlled active surface that allows operation down to about 3 mm (100 GHz). It is currently (2010) the largest fully steerable radio telescope dish in the world.

The move to millimetre and sub-millimetre wavelengths

The NRAO built an 11 m (36 ft) millimetre-wave telescope on Kitt Peak in 1967, because high altitude is important for very short radio wavelengths. In 1983 the telescope's reflecting surface and surface support structure were replaced. This new 12 m dish was designed to cover all atmospheric windows from 68 to 180 GHz (4.4 to 1.7 mm wavelength), and in 2000 the University of Arizona took over its operation. The University of Arizona also operates the Heinrich Hertz SubMillimeter Telescope (SMT) that was built on Mount Graham in the early 1990s. It can operate at up to 1,000 GHz (0.3 mm wavelength).

Three other large submillimetre dishes had been built in the 1980s. A 30 m telescope saw 'first light' in 1984 at the Pico Veleta Observatory at an altitude of about 2,900 m in the Spanish Sierra Nevada. This was followed about three years later by the 15 m James Clerk Maxwell Telescope (JCMT), now jointly funded by the UK, Canada and the Netherlands, and the 10.4 m Caltech submillimetre dish of the Caltech Submillimeter Observatory (CSO), both at an altitude of about 4,100 m on Mauna Kea. The JCMT and CSO, both of which can operate at higher frequencies than the Pico Veleta instrument, are both protected from inclement weather, whereas the latter is not. Unique amongst such millimetre/submillimetre telescopes, a Gore-Tex membrane covers the JCMT's dish to enable it to routinely observe the Sun. The JCMT and CSO have, from time to time, been operated together as the first astronomical submillimetre interferometer.

The SubMillimeter Array (SMA) was conceived in 1983 at the Smithsonian Astrophysics Laboratory as part of an initiative to achieve high resolution in various wavebands. It was initially planned to install six 6 m telescopes near the JCMT and CSO on Mauna Kea, so they could all be linked together. Then in 1996 the Academia Sinica Institute of Astronomy and Astrophysics of Taiwan joined the project, and agreed to add two more telescopes. The SMA was formally dedicated in 2003. It operates from 180 to 900 GHz (1.7 to 0.3 mm), producing maximum resolutions of from 0.5 to 0.1 arcsec.

Caltech built a millimetre-wave interferometer in the 1980s at Owens Valley, California. It operated at a frequency between that of their single submillimetre dish on Mauna Kea and the Very Long Baseline Array (VLBA) which generally operated in the centimetre band. The Caltech millimetre-wave interferometer consisted of six 10.4 m dishes. In 2004 it was decided to merge the Owens Valley millimetre array with that of the Berkeley-Illinois-Maryland Association (BIMA), which consisted of nine 6.1 m telescopes. This new array, called CARMA (Combined Array for Research in Millimeter-wave Astronomy), began routine operations at the new high-altitude Cedar Flat location in 2007. The array operates at wavelengths of 7, 3 and 1.3 mm.

Currently the European Organization for Astronomical Research in the Southern Hemisphere, the United States National Science Foundation (NSF), and similar organisations in Canada, Japan and Taiwan are involved in constructing the Atacama Large Millimeter/submillimeter Array (ALMA) at an altitude of about 5,000 m on the Llano de Chajnantor in Chile. The first three 12 m antennae operated together for the first time at the end of 2009. When completed ALMA will have at least fifty 12 m and twelve 7 m diameter antennae operating at millimetre and submillimetre wavelengths, allowing imaging in all atmospheric windows from 3.5 mm to 300 μm (84 to 950 GHz). Array configurations will have a maximum baseline of 16 km, producing a maximum spatial resolution of 5 milliarcseconds (mas).

Very large transit instruments

William Gordon's original concept of what was to become the Arecibo radio telescope was of a radar system using a large, fixed parabolic dish designed to study the Earth's ionosphere. But a parabolic dish would have severely constrained its potential use in other areas of research that required the telescope to point to different positions in the sky. As a result a spherical antenna was designed with a movable radar feed suspended above the antenna.

Cornell University were contracted by the Department of Defense in 1959 to build this Arecibo radio telescope in a natural depression in Puerto Rico. It was initially built to undertake ionospheric and planetary research using a powerful UHF radar, with radio astronomy as a secondary use. Its 300 m (1,000 ft) diameter fixed dish was made of wire mesh designed to operate at wavelengths greater than 50 cm. The mesh was replaced in the early 1970s by perforated aluminium panels, enabling it to operate at 6 cm and above. The telescope has been used for a wide number of research projects, including imaging the Moon and planets using radar, tracking near-Earth asteroids, detecting and observing pulsars, and the Search for Extraterrestrial Intelligence (SETI).

In the mid-1960s the USSR Academy of Sciences also decided to build a very large transit telescope. This was to become RATAN (Radio Astronomical Telescope of the Academy of Sciences)-600, which was built near Zelenchukskaya in the northern Caucasus. It consisted of a 576 m diameter ring of 7.4 m high panels, which operated between 8 mm and 30 cm. It had about one-sixth of the collecting area of the Arecibo radio telescope, with a best angular resolution of 2 arcsec. First light with a partially completed instrument was in 1974. Construction was completed in 1977.

The large Parkes and Effelsberg dishes

The 64 m (210 ft) Parkes radio telescope in Australia was the brainchild of Edward George 'Taffy' Bowen, who wanted to take Australian radio astronomical research beyond what was possible in the late 1940s and early 1950s using essentially war surplus equipment.

Construction of this Parkes altazimuth radio telescope took just over two years with operations beginning in 1961. The completed 64 m dish had a central zone made of welded steel plate, with the remainder being wire mesh. Its minimum operating wavelength was 10 cm. Just over twenty-five years later the Parkes antenna became part of the Australia Telescope, which was a large radio interferometric array. After several upgrades, including making more of the dish from solid sheets, the Parkes telescope can now operate at frequencies up to 43 GHz (wavelength 7 mm). It has currently discovered more than half of all known pulsars.

In the mid-1960s West Germany decided to build a 100 m diameter altazimuth telescope at Effelsberg, near Bonn, to operate at centimetre wavelengths. The design of the Effelsberg dish, which was based on a self-compensating elastic structure, was unique at the time. It enabled the shape of the dish to still accurately match a true paraboloid as its vertical orientation was changed. When construction was completed in 1972 it was the largest fully steerable dish in the world.

Initial operations of the Effelsberg antenna were at 10.6 GHz (3 cm). Since then the alignment of the solid panels has been improved, and the wire mesh outer panels have been changed to allow the telescope to operate at up to 86 GHz (3.5 mm). Even now, forty years after 'first light', this antenna is the second largest, fully steerable, radio telescope dish in the world. It is now linked to European and other radio telescopes in the European Very Long Baseline Interferometer (VLBI) Network, managed from Dwingeloo in the Netherlands.

The VLA and very long baseline interferometers

Construction of the 27-dish Very Large Array (VLA) aperture synthesis telescope was started in 1974. It was located at an altitude of 2,200 m near Socorro, New Mexico, and within three years the first six 25 m dishes were operational. The Y-shaped array was completed in 1981, allowing the simulation of apertures of up to 36 km. It initially operated at wavelengths of between 21 and 1.3 cm, with an angular resolution of between 1.4 and 0.08 arcsec. Readings were obtained from one pair of dishes at a time, which resulted in over 350 pairs of measurements. Sophisticated self-calibration procedures were then undertaken with a large computer to produce spectacular radio images of extended objects.

Longer baseline radio interferometers were developed in the 1960s to try to improve the spatial resolution of radio sources. In 1966 radio telescopes at Jodrell Bank and Defford in the UK, separated by 127 km, were linked by a triple-hop microwave radio link giving a resolution of about 0.025 arcsec at 6 cm. It was found difficult to preserve the phases of the signals using coaxial cable or microwave links over much larger distances, however, and in 1967 a new technique was developed using atomic clocks. In 1967 Canadian astronomers were the first to use atomic clocks and data recorded on tape to extend the baseline to about 3,000 km. This produced a resolution of 0.06 arcseconds at a wavelength of 67 centimetres. By 1970 NASA had extended the baseline from Goldstone, USA to Canberra, Australia, a distance of more than 10,000 km, and achieved a resolution of better than 1 mas at 13 cm.

Jodrell Bank astronomers began to extend their long baseline interferometer in 1975 by building a permanent network of radio telescopes in the UK. Called MERLIN (Multi-Element Radio Linked Interferometer Network), it consisted of the 75 m Lovell telescope and six 25 m dishes with a maximum separation of 134 km. When it became fully operational in 1981 MERLIN had a better angular resolution (0.04 arcsec at 6 cm) and dynamic range than the VLA, although the VLA had greater sensitivity.

Five major radio astronomy institutes created the European VLBI Network (EVN) in 1980. Unfortunately the informal nature of this network produced many problems. So in 1993 the Joint Institute for VLBI in Europe, JIVE, was formed to provide a dedicated, state-of-the-art, EVN correlator which became available in 2000. Initial observational data was recorded at the individual observatories on magnetic tape, and delivered in that form to the central network correlator. The EVN sites have recently been linked together using fibre optics networks to enable the observational data to be transferred in real time at up to 1 Gbps.

The United States also decided to build a radio interferometer system in the early 1980s, called the Very Long Baseline Array (VLBA), which extended 8,000 km east-west from the Virgin Islands to Hawaii, and 4,000 km north-south. Ten identical 25 m diameter dishes were installed over the period 1986 to 1993, when the system was completed. It operated over a frequency range of from 325 MHz to 43 GHz (90 cm to 7 mm), and achieved resolutions of 22 to 0.17 mas. The system was controlled from an operational centre near Socorro, New Mexico, where tapes were received with the recorded data. Four other sites, Arecibo, Green Bank, the VLA, and Effelsberg, were brought online from time to time to complete what was called the High Sensitivity Array.

One way of improving the resolution of ground-based radio telescope arrays even more is to have the array linked to one or more radio telescopes in space. In 1986 tests were carried out linking a TDRSS (Tracking and Data Relay Satellite System) spacecraft with ground-based radio telescopes in Japan and Australia. Then in 1997 Japan launched the HALCA (Highly Advanced Laboratory for Communications and Astronomy) radio astronomy spacecraft, which was linked with ground-based radio telescopes in the United States, Europe, Australia, China, South Africa and Japan. A resolution of 0.2 mas was achieved in the 5 GHz frequency band.

So in just 50 years radio astronomy has gone from resolutions of the order of a degree (in 1948) to resolutions of 0.2 mas, an improvement of ~20 million times.

Bibliography

Christiansen, W. N., and Högbom, J. A., *Radiotelescopes*, Cambridge University Press, 1969.

Hey, J. S., *The Evolution of Radio Astronomy*, Science History Publications, 1973.

Lovell, Bernard, *The Jodrell Bank Telescopes*, Oxford University Press, 1985.

Malphrus, Benjamin K., *The History of Radio Astronomy and the National Radio Astronomy Observatory: Evolution Toward Big Science*, Krieger Publishing Company, 1996.

Marx, Siegfried, and Pfau, Werner, *Observatories of the World*, Blandford Press, 1982.

Orchiston, Wayne (ed.), *The New Astronomy: Opening the Electromagnetic Window and Expanding our view of Planet Earth*, Springer, 2005.

Robertson, Peter, *Beyond Southern Skies: Radio Astronomy and the Parkes Telescope*, Cambridge University Press, 1992.

Ryle, Martin, *Radio Telescopes of Large Resolving Power*, Nobel Lecture 1974, Nobel website.

Smith, F. Graham, *Radio Astronomy*, Pelican Books, 1960.

Sullivan, W. T. (ed.), *The Early Years of Radio Astronomy: Reflections Fifty Years after Jansky's Discovery*, Cambridge University Press, 1984.

Sullivan, Woodruff T., *Cosmic Noise: A History of Early Radio Astronomy*, Cambridge University Press, 2009.

Verschuur, Gerrit L., *The Invisible Universe Revealed: The Story of Radio Astronomy*, 2nd ed., Springer-Verlag, 2007.

2 · Early radio astronomy and observatories

KARL JANSKY'S RADIO ASTRONOMY

Karl Jansky, of the Bell Telephone Laboratories, designed a radio receiver in 1930 to study the cause of interference at a wavelength of about 15 m (20 MHz frequency) used for transatlantic communications. It consisted of a 30 m long, 4 m high antenna array mounted on a frame that rotated horizontally every 20 minutes. Nicknamed 'the merry-go-round', the array was made from brass pipes and timber, and was mounted on the wheels taken from an old Model T Ford.

At first he thought that the signals that he had detected must be coming from the Sun, but he soon realised that they peaked once every sidereal day rather than once every solar day, so the Sun could not be the source. Jansky spent a year observing and analysing the signals, and concluded that they were coming from the Milky Way at about 18 hours R.A., 10° Dec., the declination estimate having a large uncertainty. He assumed that they had a non-stellar origin in the Milky Way, as he could detect no radio waves coming from our nearest star, the Sun. He suggested two possible origins, either the centre of the Milky Way in Sagittarius, or the direction of space towards which the Sun is moving in Hercules.

He first published his findings in the December 1932 issue of *Proceedings of the Institute of Radio Engineers*, and presented them in April 1933 to a meeting of the International Union for Radio Science. His discovery was reported by the *New York Times* of 5 May 1933, and in the 8 July 1933 edition of *Nature*.

In spite of this and more publicity, Jansky's results were largely ignored by the astronomical community as astronomers were basically optically oriented at the time and regarded radio research with a degree of incomprehension. Jansky proposed building a dish-type radio telescope to further examine the signals, but Bell Labs asked him to move on to other work. This was the time of the Great Depression and Jansky was even concerned that he, like 20% of his colleagues, might be made redundant. So his 'radio astronomy' work basically stopped at the end of 1933.

Bibliography

Sullivan, W. T. (ed.), *The Early Years of Radio Astronomy: Reflections Fifty Years after Jansky's Discovery*, Cambridge University Press, 1984.

Sullivan, Woodruff T., *Cosmic Noise: A History of Early Radio Astronomy*, Cambridge University Press, 2009.

Verschuur, Gerrit L., *The Invisible Universe Revealed: The Story of Radio Astronomy*, 2nd ed., Springer-Verlag, 2007.

GROTE REBER'S RADIO ASTRONOMY

Grote Reber, a radio engineer, read Karl Jansky's papers of the early 1930s and decided to try to detect the cosmic radio signals that Jansky had detected, but at higher frequencies. Reber reasoned that the signals should be more intense at higher frequencies if they followed the black body radiation curve. He also knew that the resolving power of a telescope increased with frequency.

Reber decided, as a result, to build a 31 ft (9.5 m) diameter parabolic dish of sheet metal in his garden at Wheaton, Illinois, in 1937, which could operate above 1 GHz, compared with the 20 MHz used by Jansky. The dish was a meridian transit type that could be moved in declination, but not in right ascension, relying on the rotation of the Earth to accomplish the latter. His first observations at 3.3 GHz and 910 MHz (wavelengths of 9 and 33 cm) were unsuccessful, indicating that the source was not radiating as a black body. But in early 1939 he finally detected the signals at 160 MHz (187 cm wavelength).

Reber spent the next six years mapping this sky noise at 160 and 480 MHz. The most intense signals were coming from the centre of the Milky Way (one of Jansky's suggested sources), but there were other areas of high intensity that did not appear to coincide with any obvious bright optical sources. Reber suggested that the radio emission could be caused by bremsstrahlung produced by the deceleration of electrons in the centre and spiral arms of the Milky Way, whilst Karl Kiepenheuer suggested, in 1950, that they could be synchrotron radiation caused by electrons spiralling in a general galactic magnetic field. Unsöld, Ryle, Oort and Westerhout, on the other hand, thought that the radio sources were really stars that were radio-bright but optically dim. Three of Reber's sources have now been recognised as the supernova remnants Cas A and Vel X, and the active galaxy Cyg A.

Bibliography

Hey, J. S., *The Evolution of Radio Astronomy*, Science History Publications, 1973.

Orchiston, Wayne (ed.), *The New Astronomy: Opening the Electromagnetic Window and Expanding our View of Planet Earth*, Springer, 2005.

Sullivan, W. T. (ed.), *The Early Years of Radio Astronomy: Reflections Fifty Years after Jansky's Discovery*, Cambridge University Press, 1984.

Sullivan, Woodruff T., *Cosmic Noise: A History of Early Radio Astronomy*, Cambridge University Press, 2009.

JAMES HEY AND THE SUN

Thomas Edison had tried to detect radio signals from the Sun in 1890, but without success. However, in the Second World War the British Army were investigating the jamming of their radar when they picked up interference at a wavelength of between 4 and 8 m which lasted from 27 to 28 February 1942. James Hey, who was asked to investigate, realised that the interference was due to radio emissions from the Sun as, at the same time, a large solar flare had been observed which was associated with a large sunspot that was then on the Sun's meridian. Four months later, George Southworth, working at Bell Laboratories, independently discovered radio emissions from the Sun, and Grote Reber detected them in the following year.

Radio experts of the time were astonished that these solar radio emissions had not been detected previously. It had been known for some time that there was a correlation between sunspots and disturbances in the Earth's magnetic field, which appeared to be caused by the Sun emitting some sort of elementary particles. In the 1930s Howard Dellinger had concluded that problems of short-wave radio reception on Earth were due to some sort of ionising solar radiation interfering with the Earth's ionosphere. But no-one until Hey had clearly determined that the Sun was emitting radio waves.

Bibliography

Hey, J. S., *The Evolution of Radio Astronomy*, Science History Publications, 1973.

Orchiston, Wayne (ed.), *The New Astronomy: Opening the Electromagnetic Window and Expanding our View of Planet Earth*, Springer, 2005.

Sullivan, Woodruff T., *Cosmic Noise: A History of Early Radio Astronomy*, Cambridge University Press, 2009.

EARLY AUSTRALIAN RADIO ASTRONOMY

Radio astronomy started in Australia immediately after the end of the Second World War in 1945. In October of that year Joe Pawsey, Ruby Payne-Scott and Lindsay McCready, of the CSIR's (Council for Scientific and Industrial Research) Sydney's Radiophysics Laboratory, used a Royal Australian Air Force radar antenna located at Collaroy, near Sydney, to study the Sun. The antenna consisted of 32 half-wave dipoles operating at 200 MHz (1.5 m wavelength). Much to their surprise they found that the Sun was emitting radio waves during solar minimum, and that the temperature of parts of the Sun appeared to be of the order of a million K. This result was greeted with incredulity, although, unknown to Pawsey and colleagues, Grote Reber had come to a similar conclusion in 1944.

A few months later, Pawsey's team used a similar antenna to observe the Sun from the cliffs at Dover Heights, also near Sydney. The source position could then be determined by measuring the interference pattern between the direct beam received from the source and that reflected from the sea. A few days after starting work, a very large sunspot started crossing the Sun's disc causing widespread disruption to radio communications. The Dover Heights interferometer clearly showed that the sunspot was the cause of the radio interference.

John Bolton, Bruce Slee and Gordon Stanley built new antennae for sea interferometry at Dover Heights in late 1946/early 1947. Bolton and Stanley used these to investigate other possible radio sources than the Sun, and almost immediately in June 1947 detected the strong source in Cygnus that had been previously observed by Hey, Parsons and Phillips. Hey had suspected that the Cygnus source, now called Cygnus A, was relatively small. Bolton and Stanley provided an upper limit for its size when they found that they were unable to resolve it with their interferometer, which had a resolution of 5 arcminutes.

By July of the following year Bolton had announced the detection of six new sources; the three most obvious he named Taurus A, Virgo A and Centaurus A. To improve the measurements of source positions, Bolton and Stanley then built a sea interferometer on the west coast of New Zealand facing westwards. This complemented the Dover Heights system, which was operated by Slee, that pointed eastwards. Using this pair of interferometers Bolton, Stanley and Slee were able to deduce the position of four sources to within 7 arcminutes. In particular they found that the position of Taurus A seemed to correlate with that of the Crab nebula.

New research facilities were built in Australia subsequent to this early work on the Sun and other cosmic sources. In 1949 Bernard Mills set up a two-element interferometer at Badgery's Creek, 40 miles from Sydney, to produce accurate position measurements of cosmic sources. He later built a three-element interferometer operating with antenna spacings of 60 m and 270 m, and in 1952 published a list of 77 sources.

Bolton, Richard McGee and colleagues built an 80 ft (24 m) diameter fixed dish antenna at Dover Heights in 1953 resting directly in a large dish-shaped depression in the ground. They used this so-called 'hole-in-the ground antenna' at 400 MHz, and were able for the first time to locate the radio source at the centre of the Milky Way, Sagittarius A.

Chris Christiansen had built an interferometer in 1951 to study the Sun at Potts Hill. It initially consisted of 32 six foot (1.8 m) dishes in a 200 m line along one wall of this reservoir near Sydney. He later extended it with 16 paraboloids in a line at right angles to the first, to form an L-shape on two sides of the reservoir. Inspired by Christiansen's interferometer, Mills, Alec Little and colleagues built the so-called Mills Cross unfilled aperture system at Fleurs, which was completed in 1954. It consisted of two 460 m linear arrays of 250 half-wave dipoles each, operating at 85 MHz, arranged at right angles to each other. This cross arrangement enabled high resolution to be achieved in two directions at once. It was used to produce a survey of galactic and extragalactic sources. A few years later Christiansen built the Chris-Cross, also at Fleurs, which consisted of 64 six metre paraboloids on equatorial mounts extending 380 m in two orthogonal directions that could operate up to 1.4 GHz. It was used to study the Sun. Then in 1960 Mills and Christiansen joined the University of Sydney, and their research projects at Fleurs were closed down.

Bibliography

Haynes, Raymond, Haynes, Roslynn, Malin, David, and McGee, Richard, *Explorers of the Southern Sky: A History of Australian Astronomy*, Cambridge University Press, 1996.

Orchiston, Wayne (ed.), *The New Astronomy: Opening the Electromagnetic Window and Expanding our View of Planet Earth*, Springer, 2005.

Robertson, Peter, *Beyond Southern Skies: Radio Astronomy and the Parkes Telescope*, Cambridge University Press, 1992.

Sullivan, W. T. (ed.), *The Early Years of Radio Astronomy: Reflections Fifty Years after Jansky's Discovery*, Cambridge University Press, 1984.

Sullivan, Woodruff T., *Cosmic Noise: A History of Early Radio Astronomy*, Cambridge University Press, 2009.

Various articles in *Journal of Astronomical History and Heritage*, in The Slee Celebration Issues, June 2005, December 2005, June 2006; Christiansen Memorial Issues, November 2008 and March 2009, also March 2010.

See also: Parkes Radio Telescope.

EARLY CAMBRIDGE RADIO ASTRONOMY

Martin Ryle and Derek Vonberg built the first Michelson radio interferometer in 1946. This device at Cambridge, UK, used two antennae each consisting of five dipoles mounted over wire netting. With this they were able to show that short-duration radio bursts from the Sun were often circularly polarised and came from discrete areas on the Sun's disc, and not from the disc as a whole.

Two years later Ryle and F. Graham Smith used another Michelson interferometer at Cambridge to study Cyg A, which Hey, Parsons and Phillips had found in 1946 appeared to fluctuate over a period of a few seconds. Ryle and Graham Smith's interferometer consisted of two groups of four Yagi antennae spaced 500 m apart, operating at 80 MHz (3.75 m wavelength). On their first night of observations they not only observed Cyg A, but found another radio source, this time in the constellation of Cassiopeia. This turned out to be Cas A, which had been discovered some years earlier by Reber. In addition, Ryle and Graham Smith also found evidence of other, weaker sources.

In order to detect these weaker sources Ryle, Smith and Bruce Elsmore constructed yet another interferometer. They used a phase-switching receiver, invented by Ryle, which enabled small radio sources to be discriminated from large ones, which were rejected. This enabled Ryle, Smith and Elsmore to produce the first Cambridge (or 1C) catalogue of 50 sources published in 1950.

Ryle and Graham Smith also built an interferometer consisting of two 7.5 m diameter parabolic German radar antennae 280 m apart, operating mostly at 214 MHz. This was to try to produce accurate position estimates of radio sources to enable their optical equivalents to be found. In 1951 Smith managed to reduce the positional errors of bright radio sources to 1 arcmin, and this enabled Walter Baade and Rudolph Minkowski to use the 200 inch (5 m) Palomar reflector to find the optical sources of Cas A and Cyg A. The first was identified as a faint dispersed nebula which had fast-moving gas filaments. But the second intense source was attributed to an inconspicuous 16th magnitude galaxy with a redshift of about 15,000 km/s. This and similar discoveries over the next few years finally showed that many radio sources were not radio stars, as originally thought, but were galaxies at some distance from the Milky Way.

The number of radio sources detected at Cambridge, Sydney and elsewhere increased steadily in the early 1950s. In 1950, 50 northern sources were listed in the Cambridge 1C catalogue, and 22 southern sources were listed by Gordon Stanley and Bruce Slee of the Sydney Radiophysics Laboratory in Australia. But the second Cambridge Catalogue, which was published in 1955, listed 1,936 sources. This was many more than listed by the Sydney group at the time, causing a major disagreement between the two groups. Minkowski, acting as unofficial mediator, concluded that many of the Cambridge sources were spurious. Ryle disagreed, but it eventually became clear that the beamwidth of the Cambridge interferometer was too large to cope with the high density of sources, resulting in source confusion. The Cambridge group then modified their interferometer to operate at a higher frequency. This improved the resolution and reduced the source confusion, so that when the third Cambridge Catalogue was published in 1959 it contained only a quarter of the sources of the second catalogue.

In the early 1950s Ryle developed the idea of aperture synthesis, for which he received the Nobel Prize in 1974. In principle, this required just two antennae to be positioned in two locations of a two-dimensional ground-based array to sample the amplitude and phase of the signals received from a source. The two antennae were then placed in various other positions in the two-dimensional array, and the image of the radio sky was produced by a computer which carried out the Fourier inversion involved in combining all the observations. The first such aperture synthesis instrument capable of mapping an arbitrary distribution of sources was built by John Blythe at Cambridge in 1954. This consisted of 48 full-wave dipoles operating at 38 MHz arranged in a line, plus a single dipole that was moved step by step in a perpendicular direction.

Blythe's antenna was the prototype for the 178 MHz interferometer that was used at Cambridge to produce the almost 5,000 source, 4C catalogue in the mid-1960s. The interferometer consisted of a 440 m long, fixed east-west array of antennae representing slices of a cylindrical paraboloid, and a moveable array which was a 58 m section of an exactly similar cylindrical paraboloid on 305 m of north-south rails. This interferometer was built at the new Mullard Observatory at Lord's Bridge outside Cambridge. The main constraints on the performance of this and other early aperture synthesis systems was the limited capability of the early computers required to perform the Fourier inversion required.

Bibliography

Robertson, Peter, *Beyond Southern Skies: Radio Astronomy and the Parkes Telescope*, Cambridge University Press, 1992.

Ryle, Martin, *Radio Telescopes of Large Resolving Power*, Nobel Lecture 1974, Nobel website.

Sullivan, W. T. (ed.), *The Early Years of Radio Astronomy: Reflections Fifty Years after Jansky's Discovery*, Cambridge University Press, 1984.

Sullivan, Woodruff T., *Cosmic Noise: A History of Early Radio Astronomy*, Cambridge University Press, 2009.

LATER CAMBRIDGE RADIO TELESCOPES

Antony Hewish and colleagues designed and built a radio telescope at Cambridge in the mid-1960s to detect scintillating radio sources, which were assumed to be quasars. This telescope, which was designed to detect rapidly fluctuating signals at 3.7 m wavelength, consisted of an array of permanently fixed wires connected to over 1,000 posts spread out over an area of about 2 hectares (5 acres). Just after this telescope was

completed in 1967, Jocelyn Bell discovered a source with it that showed a regular pulse of 1.34 seconds. A little later she discovered three more sources that also had regular short-duration pulses. These were the first pulsars to be discovered, for which Antony Hewish was awarded the Nobel Prize for Physics in 1974.

Three interferometers were built at the Mullard Radio Astronomy Observatory at Cambridge in the 1960s and early 1970s. These were the One Mile telescope built in 1962–1964, followed by the Half Mile telescope, and finally the Five Kilometre telescope, which was completed in 1971. Each of these consisted of one line of dishes, aligned east-west, which used the rotation of the Earth to provide a two dimensional scan using Earth-rotation aperture synthesis.

The One Mile telescope consisted of three 18 m dishes, two of which were fixed about 750 m apart with the third mounted on an 800 m length of rails, so providing a maximum separation of up to 1,550 m. The telescope was designed both to detect more distant sources and to determine the structure of individual sources. Once the One Mile telescope was in operation, John Shakeshaft and John Baldwin produced plans for a Half Mile telescope mounted on the same rails as the One Mile. The Half Mile used four 9 m dishes, and was designed to map the distribution and velocity of neutral hydrogen at 21 cm in nearby galaxies.

Martin Ryle completed the basic design of the Five Kilometre telescope in 1967. It consisted of eight 13 m dishes, four fixed at intervals of 1.2 km and four on a 1.2 km rail track, to provide a maximum baseline of 4.8 km. The telescope was designed as a survey instrument to detect the faintest sources, without ambiguity, and to provide accurate coordinates to facilitate optical identification. It could be used down to 2 cm wavelength (15 GHz frequency), with a source resolution of about 1 arcsec. This enabled detailed maps to be produced of the complex structure of radio emission from distant galaxies and quasars. Its name was later changed to the Ryle Telescope.

In the last few years the eight 13 m Ryle antennae have been moved to form a two-dimensional array, called the Large Array. Another array, the so-called Small Array, consisting of ten 3.7 m antennae, is situated within a small, 4.5 m high metal enclosure. The two arrays constitute the Arcminute Microkelvin Imager designed to survey clusters of galaxies.

Bibliography

Ryle, Martin, *Radio Telescopes of Large Resolving Power*, Nobel Lecture 1974, Nobel website.

Sullivan, W. T. (ed.), *The Early Years of Radio Astronomy; Reflections Fifty Years after Jansky's Discovery*, Cambridge University Press, 1984.

EARLY JODRELL BANK RADIO ASTRONOMY

Immediately after the Second World War Bernard Lovell had tried to detect radar echoes from cosmic ray showers using an army 4.2 m mobile radar with a Yagi antenna. He was unsuccessful. Then in December 1945 he moved the equipment to Manchester University's horticultural site at Jodrell Bank, to avoid interference from sources in the city. There he was able to detect transient echoes which turned out to be due to meteors, rather than cosmic rays. By October 1946, Lovell, C. J. Banwell and John Clegg were detecting up to 168 echoes per minute during the Giacobinid meteor shower.

Lovell was still determined to detect cosmic ray showers, however, so he and Clegg decided to build as large a parabola as possible to aid detection. This was the 218 ft (66 m) diameter fixed wire dish, built in 1947, which was supported just above the ground by a number of posts. The feed was placed on the top of a 126 ft (38 m) high pole, pivoted at the base. Again the search for cosmic ray showers was unsuccessful. But Lovell decided to use the dish as a transit telescope to search for cosmic radio sources instead, as its beamwidth was much smaller, and its collecting area was much larger, than any other comparable antenna of the time.

In 1950 Robert Hanbury Brown and Cyril Hazard used this 218 ft to show that the Andromeda nebula was a radio source. They also undertook a survey of the sky between 38°N and 68°N declinations, finding a concentration of the brighter radio sources around the plane of the Milky Way. This implied that some, at least, of these sources must be members of our galaxy. Bernard Mills in Australia had already come to the same conclusion, whilst Martin Ryle at Cambridge disagreed. It later became evident that Ryle was mistaken, however, because of unforeseen problems in interpreting the results of the Cambridge interferometer.

Hanbury Brown, Henry Palmer and Richard Thompson built an interferometer at Jodrell Bank with a baseline of 20 km using a small movable array and the 218 ft. They showed in 1956 that three of the sources were unresolvable even using their longest baseline, implying an angular diameter of less than 12 arcsec. They also had a very high brightness temperature of at least 2×10^7 K at 158 MHz. Four years later Rudolph Minkowski of the Mount Wilson and Palomar Observatories identified one of these sources, 3C 295, as a peculiar galaxy with a redshift of 0.46. Between 1958 and 1961 384 sources were measured using baselines up to 115 km. Seven of these sources, one of which was 3C 48, were still unresolved, implying diameters of less than 3 arcsec. Shortly afterwards 3C 48 was found by Jesse Greenstein in America to have a redshift of 0.37. It was the second quasar to be discovered.

Bibliography

Hanbury Brown, R., *Boffin: A Personal Story of the Early Days of Radar, Radio Astronomy and Quantum Optics*, Adam Hilger, 1991.

Hey, J. S., *The Evolution of Radio Astronomy*, Science History Publications, 1973.

Lovell, Bernard, *Astronomer by Chance*, Oxford University Press, 1992.

Sullivan, W. T. (ed.), *The Early Years of Radio Astronomy: Reflections Fifty Years after Jansky's Discovery*, Cambridge University Press, 1984.

Sullivan, Woodruff T., *Cosmic Noise: A History of Early Radio Astronomy*, Cambridge University Press, 2009.

See also: Later Jodrell Bank radio telescopes; MERLIN.

LATER JODRELL BANK RADIO TELESCOPES

The initial success of the 218 ft fixed dish at Jodrell Bank (see *Early Jodrell Bank radio astronomy* article) led Bernard Lovell in 1948 to suggest to Patrick Blackett, the Professor of Physics at Manchester, that a fully steerable dish should be built of at least the same diameter. The initial design of a 300 ft dish proved to be too expensive, so Lovell and colleagues settled on a 250 ft (76 m) dish instead. This dish design had a surface of 2 inch (5 cm) wire mesh, which would enable it to be used down to about 1 metre wavelength (frequency 300 MHz). Then in 1951 the 21 cm emission line of neutral interstellar hydrogen was discovered by Harold Ewen and Edward Purcell in the United States. As a result Lovell asked for the design of the 250 ft dish to be modified to enable it to work at 21 cm (1.4 GHz). Eventually this resulted in the surface being built with solid steel sheets.

It is always very difficult to estimate the cost of a civil engineering project which is completely different from that of any other, and the Jodrell Bank dish was no exception. Added to that was the inexperience of anyone at Manchester University managing such a complex engineering project. The net result was a telescope which, on completion, cost about an order of magnitude more than its original estimate. This resulted in not only Lovell being in trouble with the government funding authorities, but the authorities were also unwilling to meet the increase in cost.

The 250 ft radio telescope (see Figure 7.1) was finished in 1957 just as the first artificial Earth satellite, Sputnik 1, was launched by the Soviet Union. Lovell used the telescope to track the satellite and its carrier rocket using radar. Then, shortly afterwards, the Americans asked Jodrell Bank to track some of their spacecraft. So Lovell and the Jodrell Bank dish was all in the news, and eventually in 1960 the money problem was solved, assisted by a public subscription and a donation from the Nuffield Foundation.

Figure 7.1. The 250 ft (76 m) diameter Lovell telescope at Jodrell Bank as seen in 2007 following modifications to improve its performance at 6 cm. (Courtesy Mike Peel; Jodrell Bank Centre for Astrophysics, University of Manchester/Wikimedia Commons.)

Cyril Hazard used 250 ft telescope to observe the occultation of the radio source 3C 212 by the Moon in 1960, to determine 3C 212's celestial coordinates to a much higher accuracy than previously possible. Two years later he used the same technique with the 210 ft (64 m) Parkes radio telescope in Australia to determine the structure of 3C 273. In the following year it was found to be a quasar; the first to be discovered.

The 250 ft radio telescope had been renamed the Mark I telescope in about 1961. Fatigue cracks were discovered in its elevation drive system in 1967, and shortly afterwards a programme of repair and improvement was started that ended in 1971, when it was renamed the Mark IA. As part of this programme, a more accurate bowl surface was constructed above the old surface to enable the telescope to be used down to 6 cm (5 GHz). In 1980 the Mark IA was used as part of the MER-LIN array, and in 1987 it was renamed the Lovell Telescope. The 1971 surface of the dish was replaced by another surface in 2001–2003 to expand the useable frequency range and provided an increase in sensitivity at 6 cm by a factor of five.

In early 1960 Bernard Lovell started to consider the design of a much larger radio telescope at Jodrell Bank. Later called the Mark IV, he envisaged its dish as having a maximum dimension of 1,500 ft (450 m), maybe substantially more, with a height limitation of 500 ft (150 m). In discussions with Charles Husband, who was the engineer responsible for the design of the 250 ft, he suggested building a prototype. This was to become the Mark II, which had an altazimuth elliptical dish of 125 ft × 83 ft (38 m × 25 m). He also envisaged a portable version of the Mark II, which was to become the Mark III.

The Mark II was completed at Jodrell Bank in 1964 with a solid surface to operate down to 6 cm. This surface was upgraded in 1987 to operate down to 1.4 cm. The Mark III, on the other hand, which was built at Wardle about 24 km from Jodrell Bank over the period 1964–1966, was to form an interferometer with the Jodrell Bank telescopes. Its surface was made of wire mesh that limited its operation down to about 21 cm. Although it was made to be transportable, the Mark III was never moved and it was decommissioned in 1996. The Mark IV was never built. Neither were the planned Mark V or Mark VA telescopes of 400 ft (122 m) or 375 ft (114 m) diameters, respectively.

Currently (2010) there are three active research radio telescopes at Jodrell Bank: the Lovell Telescope, the Mark II, and a 42 ft (13 m) which had been built in 1982.

Bibliography

Agar, Jon, *Science and Spectacle: The Work of Jodrell Bank in Post-War British Culture*, Harwood Academic Publishers, 1998.

Hey, J. S., *The Evolution of Radio Astronomy*, Science History Publications, 1973.

Lovell, Bernard, *Astronomer by Chance*, Oxford University Press, 1992.

Lovell, Bernard, *The Jodrell Bank Telescopes*, Oxford University Press, 1985.

Lovell, Bernard, *Voice of the Universe: Building the Jodrell Bank Telescope*, rev. ed., Praeger Publishers, 1987.

See also: Early Jodrell Bank radio astronomy; MERLIN.

MERLIN

Long-baseline radio interferometers were developed in the 1960s in the UK to try to improve the spatial resolution of radio sources. In 1965 the Mark I radio telescope at Jodrell Bank (see *Later Jodrell Bank radio telescopes* article) and a 25 m radio telescope 127 km away at Defford were linked by a triple-hop microwave radio link. A number of sources were found to be smaller than 0.1 arcsec in at least one dimension at a wavelength of 21 cm (frequency 1.4 GHz). In the following year the Mark I was replaced in the interferometer by the Mark II at Jodrell Bank and observations were made at 11 cm (2.7 GHz) and 6 cm (5 GHz), to produce resolutions of 0.05 and 0.025 arcsec, respectively. Again a number of sources were still unresolved.

The Mark III 38 × 25 m radio telescope had been completed at Wardle in 1966 about 24 km from Jodrell Bank. It was designed to be used, like the Defford 25 m, to form an interferometer with the Jodrell Bank instruments. Six years later Henry Palmer suggested adding more telescopes to this system to improve the observation of small-scale structure of extended objects.

Phase 1 of this Multi-Telescope Radio Link Interferometer (MTRLI) consisted of the Mark IA or Mark II at Jodrell Bank, together with the 25 m telescope at Defford, the Mark III at Wardle, and a new 25 m radio telescope installed at Knockin, 68 km from Jodrell Bank. This Knockin instrument, which was purchased from E-Systems Inc., was based on the 25 m radio telescopes being built by them for the VLA in America. The Knockin 25 m was constructed in 1976 and the microwave links for the system were installed two years later. Phase 1 of the MTRLI was completed in January 1980.

Phase 2 of the MTRLI was started by the construction of two identical 25 m radio telescopes to that at Knockin in 1979. One was at Pickmere 11 km NW of Jodrell Bank and the other at Darnhall 18 km SW of Jodrell Bank. They were linked into the existing array of radio telescopes in December 1980. When the MTRLI became fully operational in the following year the system, with a maximum separation of 134 km, had a better angular resolution (0.04 arcsec at 6 cm) and dynamic range than the VLA, although the VLA had greater sensitivity. The MTRLI system became known in 1981, unofficially, as MERLIN (Multi-Element Radio Linked Interferometer Network).

The surface of the Jodrell Bank Mark II was upgraded in 1987 to allow it to operate with the three E-Systems telescopes at 22 GHz. Then in 1991 a 32 m radio telescope was built at Cambridge and added to the system, increasing the maximum baseline to 217 km. The electronics of the existing MERLIN telescopes were also improved, and in 1997 the first dual frequency observations were made at 5 and 22 GHz. Then in 2009 the microwave links were replaced by fibre optic links increasing the bandwidth of the system from 30 MHz to 4 GHz.

Bibliography

Burke, Bernard F., and Graham-Smith, Francis, *An Introduction to Radio Astronomy*, 3rd ed., Cambridge University Press, 2010.

Lovell, Bernard, *Out of the Zenith: Jodrell Bank, 1957–1970*, Oxford University Press, 1973.

Lovell, Bernard, *The Jodrell Bank Telescopes*, Oxford University Press, 1985.

See also: Later Jodrell Bank radio telescopes.

EARLY SOVIET RADIO ASTRONOMY

Vitaly Ginzburg, Iosif Shklovskii and David Martyn independently concluded in 1946, based on theoretical considerations, that the metre wavelength radiation emitted by the Sun must be coming from the corona rather than the photosphere or chromosphere. This was proved in the following year by a Soviet expedition to Brazil led by S. E. Khaykin to observe a total solar eclipse. It found that radio emission at 1.5 m wavelength did not completely disappear at totality.

A number of radio telescopes were operational in the Soviet Union in the late 1940s, including a captured 7.5 m Würzburg German radar antenna which was installed at the Lebedev Physical Institute's (LPI) Crimean station. Its normal mesh surface was covered with aluminium sheets to enable it to be used at 10 cm wavelength (frequency 3 GHz). Other radio telescopes were used as sea interferometers overlooking the Black Sea. In 1951 V. V. Vitkevich, also of LPI, observed the occultation of the Crab's radio source by the Sun's corona and found, as a result, that the corona was much larger than previously thought.

P. D. Kalachev and colleagues designed a fixed 31 cm diameter dish formed of concrete and covered with a metal film at the LPI's Katsiveli station in the Crimea. In 1957 this was used to obtain the first two-dimensional radio image of the Sun at 3 cm. In the same year, A. D. Kuz'min and V. A. Udal'tsov used this antenna to detect polarisation of the Crab radio emission at 10 cm, which confirmed that it was caused by the synchrotron process.

In the meantime a group of radio astronomers at Gorky State University, led by V. S. Troitsky, had undertaken radio observations of the Moon to try to determine the physical properties of its surface. They found that its mean brightness temperature increased with wavelength, which they interpreted to show that there was a constant heat flow from the lunar interior. They also concluded that the surface was not covered by the deep layer of dust proposed by some researchers.

In 1954 Khaykin had set up a new department of radio astronomy alongside the optical observatory at Pulkovo to facilitate the interaction between the two different astronomical disciplines. Khaykin's team designed the Variable Profile Antenna (VPA), which became operational in the centimetre waveband two years later. It consisted of 90 tiltable, flat steel plates, each 1.5 m wide × 3 m high, arranged in the arc of a circle. The beamwidth was just 1 arcmin at 3 cm wavelength in an east-west direction. Later this was reduced to 15 arcsec at 8 mm.

A new LPI radio astronomy observatory was established in 1956 at Pushchino in the Moscow region with the start of construction of a fully steerable 22 m parabolic radio telescope, the RT-22. Operational three years later, it could be used down to 8 mm wavelength (40 GHz) with an angular resolution of 2 arcmin. In this it outperformed any such instrument in the West at the time. An improved RT-22 was built at the Crimean astrophysical observatory at Simeiz, becoming operational in 1966.

In the meantime construction of another radio telescope, the DKR-1000, had been started at Pushchino. It consisted of two 1000 × 40 m parabolic cylinders operating in the metre waveband, and was a development of the Mills Cross radio telescope in Australia. The east-west arm became operational in 1964, with the north-south arm a few years later. Unlike the Mills Cross, however, it allowed simultaneous observations at a relatively wide range of operating frequencies.

In 1959 N. S. Kardashev analysed hydrogen line emission in transitions between levels with large quantum numbers, and concluded that those between adjacent levels should be detectable in the far infrared to decimetre wavebands. Five years later R. L. Sorochenko and E. V. Borodzich detected the 3 cm line resulting from transitions between levels 91 and 90 in the spectrum of the Omega nebula using the LPI RT-22. At about the same time A. F. Dravskikh and Z. V. Dravskikh also detected the line at Pulkovo resulting in the transition from level 101 to level 100.

Construction of a very large transit telescope was started in 1968 near Zelenchukskaya in the northern Caucasus. This was the RATAN (Radio Astronomical Telescope of the Academy of Sciences)-600, which was a development of the design of the Pulkovo VPA. The RATAN-600 consisted of a 576 m diameter ring of 7.4 m high panels, which operated between 8 mm and 30 cm, with a best angular resolution of 2 arcsec. Inside the ring was a flat 400 m long, 8.3 m high reflector and a number of secondary feeds, enabling the telescope to be operated in a number of different modes. One of these modes used the whole ring, but others used various parts. First light was in 1974 and it was completed in 1977 for manual settings. Since then the RATAN-600 has been progressively modified to make its adjustments and limited tracking automatic.

Bibliography

Dagkesamanskii, R. D., *The Pushchino Radio Astronomy Observatory of the P N Lebedev Physical Institute Astro Space Center: Yesterday, Today and Tomorrow*, Physics-Uspekhi, **52** (2009), 1159–1167.

Strelnitski, Vladimir S., *The Early Post-War History of Soviet Radio Astronomy*, Journal for the History of Astronomy, **26** (1995), 349–362.

Sullivan, W. T. (ed.), *The Early Years of Radio Astronomy: Reflections Fifty Years after Jansky's Discovery*, Cambridge University Press, 1984.

Sullivan, Woodruff T., *Cosmic Noise: A History of Early Radio Astronomy*, Cambridge University Press, 2009.

EARLY FRENCH RADIO ASTRONOMY

Radio astronomy took some time after the end of the Second World War to become established in France. This was hardly surprising as French radio research had barely existed during the war, and the knowledge of developments elsewhere had been virtually non-existent during the German occupation. But there was one notable exception in that Yves Rocard, a member of the French resistance and Professor of Physics at the Sorbonne, had been secretly flown to Britain in 1943 where he worked on radar. After the war, in 1945 Rocard was appointed Director of the Physics Laboratory at the École Normale Supérieure (ENS) where, in the following year, he set up a radio astronomy group.

As in a number of other countries, early radio astronomy in France used captured, altazimuth-mounted, Würzburg radar antennae to get started. The ENS group mounted a 3 m Würzburg dish on the roof of the physics laboratory and used it at 1.2 GHz (25 cm wavelength) for solar research. They also erected a 7.5 m Würzburg dish on railway lines at Marcoussis, near Paris, and used it at 158 MHz to observe the 1949 partial solar eclipse. This occultation by the Moon was used to compensate for the dish's limited resolution in observing the Sun. ENS later erected another 7.5 m Würzburg at Marcoussis on a concrete block several hundred metres away from the first dish.

In the meantime the Institute of Astrophysics (IA) in Paris had erected another 7.5 m Würzburg at the Meudon Observatory on the outskirts of Paris, which they used to observe the 1949, 1952 and 1954 partial solar eclipses.

The first 7.5 m dish at Marcoussis was used between 1954 and 1956 to observe the plane of the Milky Way at 900 MHz. It detected a number of discrete sources which included Sag A, Cyg A, Cyg X, Cas A and Tau A, along with a number of new sources. The ENS group of Denisse, Lequeux and Le Roux also observed the North Polar Spur, and initially concluded that the brightness temperature of the sky was less than 3 K. This is interesting in view of Arno Penzias and Robert Wilson's 1965 discovery of the microwave background radiation at a temperature of 3.5 K. But in 1957 the ENS group of Delannoy, Denisse, Le Roux and Morlet had modified their conclusion to say that the brightness temperature was not greater than 20 K.

In the mid-1950s a number of small interferometers were built in France mainly to observe the Sun. The Arsac instrument, consisting of four 1.1 m dishes and operating at 9.4 GHz, was one of the first. Its antennae were located at positions 0, 1, 4 and 6 (in arbitrary units) to provide six spatial frequencies of equal amplitude without redundancy. Although ingenious in design, this instrument was difficult to use in practice. A far more useful instrument, built at about the same time, consisted of two equatorially mounted 2 m dishes. They had a separation of 60 m and were used to undertake aperture synthesis work at 9.4 GHz.

Construction of a large Nançay interferometer had been started in 1953 to be used for solar research. It consisted of thirty-two 5 m antennae operating at 169 MHz on a 1.6 km east-west baseline, and, in 1959, eight 10 m antenna were added aligned north-south. In the same year the IA built a transit interferometer at the Saint Michel Observatory operating at 300 MHz. It had two large fixed cylindrical-parabolic antennae 1.1 km apart and was used for galactic and extragalactic research.

In the meantime, the two 7.5 m Würzburg antennae had been transferred from Marcoussis to Nançay where they were fitted on equatorial mounts. They were then mounted on two railway tracks, one running 1,500 m east-west and the other 400 m north-south. This interferometer, which was designed to measure the fine structure of the most intense radio sources, operated at 1,415 MHz (21 cm) and had a resolution of better than 1 arcmin. It became operational in 1959 and not only provided valuable information on the structure of various bright radio sources, but found a number of sources, later found to be quasars, that were still unresolved at the highest possible resolutions.

It had been hoped in 1955 to build an interferometer at Nançay consisting of two 25 m dishes on railway tracks to observe the 21 cm line of interstellar hydrogen. But the design was completely changed in 1956 to a transit instrument based on Kraus' Ohio radio telescope. It is not completely clear why the change was made, although it was claimed at the time that the subsoil was not strong enough to support such a large mobile antenna.

When fully operational in 1967 this large Nançay radio telescope consisted of a 300 m wide, 35 m high mirror that was part of a 560 m radius sphere, and a tiltable plane mirror 200 m wide and 40 m high facing it 460 m away. The surfaces of both mirrors were made of 12.5 mm metallic mesh to operate at 21 cm, with a minimum operating wavelength of 9 cm. The idea behind this design was to produce an instrument with a large surface area at minimum cost, whilst accepting the limitation that it was a transit instrument that could be used only for about 30 minutes on either side of the meridian. There were a number of other operational limitations which had not been appreciated when the design was originally produced, and some of these were rectified only in the late 1990s. Although the designers achieved their goal of producing a large radio telescope at relatively low cost per unit area, the Parkes dish that was completed a little earlier in Australia was only slightly more expensive per unit area. And the Parkes dish was fully steerable and could operate at a lower wavelength.

Bibliography

Sullivan, W. T. (ed.), *The Early Years of Radio Astronomy: Reflections Fifty Years after Jansky's Discovery*, Cambridge University Press, 1984.

Various authors, Highlighting the History of French Radio Astronomy, in Journal of Astronomical History and Heritage, Vol. 10, No. 1, March 2007, pp. 11–19; Vol. 10, No. 3, November 2007, pp. 221–245; Vol. 12, No. 3, November 2009, pp. 175–188; Vol. 13, No. 1, March 2010, pp. 29–42; Vol. 14, No. 1, March 2011, pp. 57–77.

EARLY DUTCH RADIO TELESCOPES

The first significant radio telescope in the Netherlands used a 7.5 m (25 ft) ex-German Würzburg radar antenna located at the Dutch PTT's telecommunications station at Kootwijk. Work began in 1948, then three years later C. A. (Lex) Muller and Jan Oort used this antenna to detect the 21 cm (1.4 GHz) emission line of interstellar neutral hydrogen, just a few months after its discovery by Harold Ewen and Edward Purcell in the USA. Over the next few years Hendrik van de Hulst, Muller and Oort observed the distribution of neutral hydrogen in the Milky Way which they published in map form in 1954. This clearly showed the Orion and Perseus arms of the Milky Way's spiral structure.

In the meantime, a contract had been placed in 1951 to design a 25 m (82 ft) diameter altazimuth radio telescope. It was eventually built at Dwingeloo at the edge of the State Forest, where radio interference could be controlled. When regular observations were started in 1956 it was, for a short time, the largest, fully steerable, radio telescope dish in the world.

Oort began to consider the design of the next generation Dutch radio telescope just as the Dwingeloo dish was becoming operational. This initially resulted in a joint Dutch/Belgian programme to produce an advanced version of a Mills Cross antenna, but it proved too expensive, and a much cheaper design was adopted. Belgium eventually pulled out of the programme in 1967, leaving the Dutch to continue alone. In the event they built the Westerbork Synthesis Radio Telescope on the site of the old Westerbork transit camp, which had been used by the Germans to accommodate Jews in the Second World War.

The Westerbork radio telescope, which became operational in 1970, initially consisted of ten fixed 25 m, equatorially mounted dishes and two movable dishes on a 1.6 km baseline. Although initially designed to use down to 21 cm, the dishes were found to be accurate enough to operate down to 6 cm. Later, two additional, movable dishes were built to extend the baseline to 3 km.

Bibliography

Orchiston, Wayne (ed.), *The New Astronomy: Opening the Electromagnetic Window and Expanding our View of Planet Earth*, Springer, 2005.

Raimond, Ernst, and Genee, René, *The Westerbork Observatory, Continuing Adventure in Radio Astronomy*, Kluwer Academic Publishers, 1996.

Sullivan, Woodruff T., *Cosmic Noise: A History of Early Radio Astronomy*, Cambridge University Press, 2009.

Van Woerden, Hugo, and Strom, Richard G., *The Beginnings of Radio Astronomy in the Netherlands*, Journal of Astronomical History and Heritage, Vol. 9, No. 1, June 2006, pp. 3–20.

EARLY AMERICAN RADIO ASTRONOMY

Radio astronomy took some time to take off in America after the pioneering observations of Karl Jansky in 1932 and of Grote Reber in the late 1930s and early 1940s. This was in spite of the tremendous advances in radio technology achieved in the Second World War, and the subsequent pioneering results in radio astronomy achieved in the UK, Australia and elsewhere.

The U.S. Army Signal Corps detected radar signals from the Moon in 1946 as a test of their system to detect incoming missiles from the Soviet Union. A little later the U.S. Naval Research Laboratory used parabolic antennae up to 10 ft (3 m) diameter to undertake radio astronomy observations of the Sun between 8.5 mm and 9.4 cm wavelengths (35 to 3.2 GHz). But it was not until the 1950s that significant radio astronomy was undertaken in the United States. Because of the number of institutions involved, this is outlined in separate articles covering the period from 1950 to the present.

Bibliography

Hey, J. S., *The Evolution of Radio Astronomy*, Science History Publications, 1973.

Sullivan, Woodruff T., *Cosmic Noise: A History of Early Radio Astronomy*, Cambridge University Press, 2009.

See also: Grote Reber's radio astronomy, Karl Jansky's radio astronomy.

3 · Later radio observatories

ALGONQUIN RADIO OBSERVATORY

Arthur Covington started radio frequency observations of the Sun from the Ottawa Radio Field Station of the Canadian National Research Council (NRC) in mid-1946. He used a 4 ft (1.2 m) parabolic antenna at 10.7 cm wavelength (2.8 GHz). Later that year he used the occultation of the Sun by the Moon during a partial solar eclipse to show that radio emission from the Sun could be divided into two components, one from the Sun as a whole and one from sunspots. Routine daily observations of the Sun were begun the following year, but interference from other equipment at the field station caused Covington to move his equipment in 1948 about 8 km to what became known as the Goth Hill radio observatory. Although this solved a number of problems, increasing problems with radar and radio interference in the Ottawa area in the 1950s resulted in another new site having to be found for radio astronomy. This was at Lake Traverse in Algonquin Park, about 200 km from Ottawa. In 1962 this Algonquin Radio Observatory became Canada's national radio observatory.

Covington, the first director of the Algonquin Radio Observatory, installed a 6 ft parabolic dish there in 1960. Six years later, a new radio telescope was completed consisting of an array of thirty-two 10 ft parabolic dishes arranged over a common 700 ft (215 m) long waveguide. Meanwhile construction had started of a 150 ft (46 m) fully steerable radio telescope dish which was completed in 1966. It was designed to study galactic and extragalactic objects down to 2 cm wavelength (15 GHz frequency). The central 37 m diameter of the 46 m dish was made of solid metal plates with the remainder consisting of wire mesh.

Shortly after completion, the 46 m was used in conjunction with the 26 m dish at the Dominion Radio Astrophysical Observatory about 3,000 km away at Penticton to create the world's first successful very long baseline interferometer (VLBI). The observations from both radio telescopes was recorded on magnetic tapes, and the time signals recorded using atomic clocks.

The NRC looked into improving the performance of the 46 m in the 1980s to operate in the millimetre waveband. But they decided, instead, to participate in the James Clerk Maxwell Telescope (JCMT) on Mauna Kea and close down the Algon-

quin Radio Observatory, which they did in 1991. The 46 m dish is now used for geodesic work.

Bibliography

Covington, Arthur E., *Algonquin Radio Observatory, Lake Traverse, Ont., Canada*, in Solar Physics, 9 (1969), 241–245.
Covington, Arthur E., *Beginnings of Solar Radio Astronomy in Canada*, in Sullivan, W. T. (ed.), *The Early Years of Radio Astronomy; Reflections Fifty Years after Jansky's Discovery*, Cambridge University Press, 1984.

ARECIBO RADIO TELESCOPE

In the mid-1950s William Gordon of Cornell University suggested a new way of studying electrons in the Earth's ionosphere by bombarding them with high-energy radar signals. He then tried to obtain funding for a new facility to undertake this work by pointing out its potential use for both radar astronomy and radio astronomy. Gordon initially tried to obtain money from the National Science Foundation (NSF), but the money was far more than it could afford. So he lobbied and obtained funding from the Department of Defense (DoD) instead who could see the possibilities of tracking ion trails produced by missile exhaust systems using the proposed new facility.

In the event Cornell University were contracted by the DoD in 1959 to build this Arecibo radio telescope, or DoD's Ionospheric Research Facility as it was called at the time, in a natural depression in Puerto Rico. This location 18° north of the equator was chosen because of its good visibility of solar system objects. The dish was also expected to be used as a radio telescope to observe more distant celestial objects.

The requirement was to build as large a dish as possible, so the dish was stationary and built directly onto the ground. Its surface was spherical, rather than the normal paraboloid, which enabled it to use a movable radar feed to observe objects up to ±20° from the zenith.

The Arecibo 300 m (1,000 ft) diameter fixed dish was initially made of wire mesh and was designed to operate at wavelengths greater than about 50 cm. The telescope became operational in 1963 with a radar transmitter that could produce a pulsed power of 2.5 megawatts or continuous power of 100 kilowatts. In 1969

Figure 7.2. The 1,000 ft (300 m) diameter Arecibo radio telescope that covers an area of about 18 acres (7 hectares) in a natural depression in Puerto Rico. The surface of the dish consists of about 40,000 6 ft × 3 ft (1.8 × 0.9 m) perforated panels, with a 900 ton feed structure suspended about 450 ft (140 m) above it. (Courtesy NAIC—Arecibo Observatory, a facility of the NSF.)

the National Science Foundation took over the facility from the Department of Defense. At about the same time the feed system was modified (see Figure 7.2) to allow observations up to ±24° from the zenith, which enabled it to observe Cygnus A, Cygnus X and M31, the Andromeda galaxy. In the early 1970s the wire mesh dish surface was replaced by perforated aluminium panels, enabling it to operate down to 6 cm (5 GHz frequency). Then in the mid-1990s a 15 m high ground screen was installed around the dish perimeter to reduce noise, and a Gregorian secondary antenna system was installed in a new radome to correct the main dish's spherical aberration.

The telescope has been used for a wide number of research projects, including imaging the Moon, planets and asteroids using radar, tracking near-Earth asteroids, detecting and observing pulsars, and the Search for Extraterrestrial Intelligence (SETI). In particular, in 1964 Gordon Pettengill and Rolf Dyce used the Arecibo radio telescope to determine the rotation period of Mercury. Four years later Richard Lovelace and colleagues discovered the periodicity of the Crab Pulsar using Arecibo. Likewise, in 1974 Joe Taylor and Russell Hulse discovered the first binary pulsar, PSR 1913+16, and eight years later Don Backer and colleagues discovered the first millisecond pulsar, PSR 1937+214, using Arecibo.

Funding of the Arecibo Observatory has been gradually reduced over the first decade of the twenty-first century. Then in 2011 the National Science Foundation selected SRI International, along with a number of collaborator institutions, to run Arecibo.

Bibliography

Butrica, Andrew J., *To See the Unseen: A History of Planetary Radar Astronomy*, NASA SP−4218, 1996.

Federer, Charles A., *Some Current Programs at Arecibo − I and − II*, Sky and Telescope, July 1964, pp. 4–8, and August 1964, pp. 73–77.

Shawcross, William E., *Arecibo Observatory Today − I and − II*, Sky and Telescope, April 1972, pp. 214–217 and 228, and May 1972, pp. 293–295.

CULGOORA, FLEURS AND THE AUSTRALIA TELESCOPE

The Radiophysics Laboratory in Australia started to consider, in the late 1950s, what new radio telescopes would be needed in the next decade, in addition to the Parkes radio telescope. Two instruments were considered. One was a metre wavelength radioheliograph which had been proposed by Paul Wild to study radio emission from the Sun. The other, proposed by Bernard Mills, was the SuperCross to continue the work of the Mills Cross at Fleurs in detecting and cataloguing galactic and extragalactic sources. In 1959 the Radiophysics Laboratory chose the radioheliograph. At this stage, however, its design was still only conjectural and it was some time before it was finalised as being a circular array.

The radioheliograph, which was designed and built by Radiophysics staff, was completed in 1967 at Culgoora, about 550 km from Sydney. It consisted of ninety-six 45 ft (14 m) diameter dishes arranged in a circle of 3 km in diameter. These dishes were surfaced with 2 inch (5 cm) galvanised mesh to operate at 80 MHz (3.75 m wavelength). The array produced an image of the Sun every second in two polarisations. This radioheliograph enabled the Radiophysics group to remain at the forefront of solar radio research for another decade or so. But the facility was eventually closed down in 1984 as it had come to the end of its useful life.

With the rejection of the SuperCross idea, Mills left the Radiophysics Laboratory in 1960. He joined the University of Sydney where Harry Messel, head of the physics department, had decided to set up a small astronomy group. Money was eventually found by Messel and Mills for the SuperCross which was then built at Molonglo, near Canberra, by staff and students of the physics department. This so-called Molonglo Cross, which was completed in 1967, was a transit radio telescope consisting of two 1.6 km long arms, each in the form of a parabolic cylindrical reflector 12 m wide. Its operational frequency was 408 MHz (73 cm). One arm could be tilted mechanically and the other could be phased to enable a scan of the southern sky to be produced. The Molonglo Cross detected many new pulsars and enabled a catalogue to be produced of more than 12,000 strong radio sources.

The Molonglo Cross had finally broken the monopoly of the Radiophysics Laboratory on radio astronomy in Australia. In 1978 the north-south arm was closed down, and the east-west arm was modified over the next three years to produce what was called the Molonglo Observatory Synthesis Telescope (MOST) operating at 843 MHz (36 cm).

Chris Christiansen also left the Radiophysics Laboratory in 1960 and joined the University of Sydney. At that time Radiophysics planned to demolish his unused Chris-Cross at Fleurs which consisted of thirty-two 6 m parabolic dishes on each of two 380 m arms. However, Radiophysics agreed to transfer the ownership of the Fleurs site to the University of Sydney. Subsequently Christiansen added two 14 m parabolic dishes to each arm, extending the baseline to 786 m and significantly increasing the telescope's resolving power at its operational frequency of 1.4 GHz (21 cm). This so-called Fleurs Synthesis Telescope (FST) was used to observe the structure of large radio galaxies, supernova remnants and emission nebulae. In 1984 the east-west arm was extended to 1,585 m by the addition of two more 14 m paraboloids. The FST was closed down in 1988, on the same day as the inauguration of the Australia Telescope.

The Australia Telescope

It had become increasingly clear in the 1970s that Australia was gradually slipping further and further behind other countries in the development of modern astronomical facilities. One way of addressing this problem was to build a state-of-the-art aperture synthesis radio telescope, as there was no such instrument available at that time in the Southern Hemisphere. It would also take advantage of Australia's previous expertise in building and operating radio interferometers. So in the late 1970s Australian radio astronomers considered a number of possible designs and settled on one based on the Parkes 64 m antenna as the prime instrument, together with a number of new, smaller, movable dishes mounted on rails. But the request for funding of this so-called Australian Synthesis Telescope (AST) was rejected in 1981.

This rejection, although a disappointment, was not entirely unexpected. So it took very little time for an alternative scheme to be proposed. It was based, not on the Parkes antenna, but on a new radio telescope array to be built at the Culgoora site of the radioheliograph that was already destined to close. The idea was to link this new array to a new radio telescope at Siding Spring as well as to the Parkes dish. This solution had the advantage of accommodating the diverse objectives of many different astronomy groups in Australia. It would also provide an impressive example of modern Australian design and engineering expertise, as well as involving industrial contractors in a number of different localities. Imaginatively called the Australia Telescope, it was proposed as a Bicentenary project celebrating, in 1988, the bicentenary of the first British settlement in Australia.

The Australia Telescope, which was approved in 1982, consisted of the Compact Array at Culgoora, plus a new radio telescope at Mopra, near Siding Spring, about 120 km south of Culgoora, and the 64 m Parkes antenna to provide a total baseline of about 320 km. The Compact Array consisted of six 22 m dishes on an east-west baseline of 6 km. Five of the dishes were mounted on a 3 km railway track, with the sixth being stationary 3 km from one end of the track. The 22 m dish at Mopra was identical in design to the Culgoora dishes.

Since completion other radio telescopes around Australia have, from time to time, been connected to the Australia Telescope to provide a very long baseline interferometer (VLBI) with baselines of up to 3,000 km. They have included the NASA Deep Space Network dish at Tidbinbilla, and antennae at Hobart, Ceduna and Perth.

About ten years after becoming operational, the Compact Array was modified to add a 200 m long north spur perpendicular to the existing 6 km baseline. The outer surface area of the inner five dishes was also improved and new receivers added to enable observations down to 3 mm (100 GHz). This improvement work was completed in 2004.

Bibliography

Haynes, Raymond, et al., *Explorers of the Southern Sky: A History of Australian Astronomy*, Cambridge University Press, 1996.

Orchiston, Wayne (ed.), *The New Astronomy: Opening the Electromagnetic Window and Expanding our View of Planet Earth*, Springer, 2005.

Robertson, Peter, *Beyond Southern Skies: Radio Astronomy and the Parkes Telescope*, Cambridge University Press, 1992.

See also: Early Australian radio astronomy; Parkes Radio Telescope.

DOMINION RADIO ASTROPHYSICAL OBSERVATORY

The Dominion Radio Astrophysical Observatory (DRAO) was inaugurated in 1960 near Penticton in Canada. Its first radio telescope to become operational had a 84 ft (26 m) diameter parabolic dish designed to operate at the 21 cm (1.4 GHz) line of neutral interstellar hydrogen. In 1967 this telescope was used in conjunction with the 150 ft (46 m) radio telescope of the Algonquin Radio Observatory, about 3,000 km away, to produce the world's first successful very long baseline interferometer (VLBI). The observations from both radio telescopes was recorded on magnetic tapes, and the time signals recorded using atomic clocks. Following modifications, the Penticton 26 m can now be operated at various frequencies from 408 MHz to 8.4 GHz.

A 6 ft (1.8 m) dish, like that at the Algonquin Radio Observatory, was installed at Penticton in 1965 to complement the solar observations obtained at Algonquin. Penticton took over Algonquin's responsibility for solar monitoring in 1991 when the latter observatory was closed down.

A low-frequency radio telescope, consisting of a T-shaped array covering an area of 65,000 m^2, was also operational at the DRAO over the period 1965–1969. This was a period of solar minimum which was an optimum for observing at these low frequencies. Its main task was to produce a map of radio sources in the Milky Way, as well as observing a number of specific sources, particularly quasars, in more detail.

In the late 1960s a two-element interferometer was built at Penticton consisting of two 8.5 m dishes operating at 1.4 GHz over a baseline of 600 m. This interferometer was later extended to have seven such dishes operating at 1.4 GHz and 408 MHz over the same baseline. It produced high-resolution imaging of discrete sources and mapping of the galactic plane over the period 1995 to 2005.

Bibliography

Landecker, T. L., et al., *The Synthesis Telescope at the Dominion Radio Astrophysical Observatory*, Astronomy and Astrophysics Supplement Series, **145** (2000), 509–524.

EFFELSBERG RADIO TELESCOPE

West Germany was excluded from the development of radio astronomy in the crucial five years immediately after the end of the Second World War as restrictions on undertaking all forms of radio research were not lifted until 1950. The country then started basic radio astronomy research, but it was not until the early 1960s that large projects could be funded. In 1966 a decision was made to build a 100 m diameter telescope at Effelsberg, near Bonn, under the control of the new Max Planck Institute for Radio Astronomy. This altazimuth telescope was built in a valley in the Eifel mountains to help shield it from terrestrial radio sources. The inner 60 m diameter of the dish was constructed of high-accuracy aluminium panels, the ring between 60 and 80 m was made of aluminium frame panels, whilst outside that the dish was made of wire mesh. When construction was completed in 1972 it was the largest fully steerable dish in the world.

The design of the Effelsberg dish, which was based on a self-compensating elastic structure, was unique at the time. So although its surface could be distorted by up to 10 cm due to gravity, its shape still matched a true paraboloid within about 0.4 mm. As the angle of the dish to the vertical was changed, the position of the receiver was simply moved to match the change in focus position. Initial operations were at 10.6 GHz (3 cm), but since then the alignment of the solid panels has been improved, and the wire mesh outer panels have been changed, allowing the telescope to operate at up to 86 GHz (3.5 mm). A new secondary reflector was installed in 2006 with active optics to improve system performance. Even now, forty years after its initial construction, this antenna is the second largest fully steerable radio telescope dish in the world.

Bibliography

Baars, Jacob W. M., *The Parabolic Reflector Antenna in Radio Astronomy and Communication*, Springer, 2007.

Hachenberg, O., *The New Bonn 100-Meter Radio Telescope*, Sky and Telescope, December 1970, pp. 338–343.

Heck, André (ed.), *Organizations and Strategies in Astronomy: Vol. 6*, Springer, 2006.

Wielebinski, R., *The New Era of Large Paraboloid Antennas: The Life of Prof. Dr. Otto Hachenberg*, Advances in Radio Science, **1** (2003), 321–324.

Wielebinski, R., Junkes, N., and Grahl, B. H., *The Effelsberg 100 m Radio Telescope: Construction and Forty Years of Radio Astronomy*, Journal of Astronomical History and Heritage, Vol. 14, No. 1, March 2011, pp. 3–21.

EUROPEAN VLBI NETWORK

The European VLBI (Very Long Baseline Interferometry) Network (EVN) was formed in 1980 by a consortium of five major radio astronomy institutes in Europe. But it subsequently became clear that this rather informal arrangement was the cause of many problems. This included the lack of a state-of-the-art correlator to integrate all the observations. So in 1993 the Joint Institute for VLBI in Europe, JIVE, was formed by the European Consortium for VLBI with a centre at Dwingeloo in the Netherlands. JIVE's main task was to provide, operate and develop the much-needed EVN correlator that became operational in 2000. JIVE also assisted EVN in coordinating observations. The membership of JIVE has now been expanded to include organisations in China, France, Germany, Italy, Netherlands, Spain, Sweden, and the UK.

Observational data was initially recorded at the individual EVN observatories on high-capacity magnetic tapes, together with very accurate timing information. These tapes were then delivered to the central network correlator for analysis. But the EVN sites have recently been linked together using fibre optics networks in the so-called electronic European VLBI network (e-EVN). This transferred observational data in real time at up to 1 Gbps. Similar electronic networks have also enabled e-EVN to link with other radio telescopes in real time all over the world. In early 2009 a demonstration of this global e-VLBI system capability was organised using 17 telescopes in Asia, Australia, Europe, North America and South America, 12 of which operated simultaneously using real-time correlation.

Miguel Perez-Torres and colleagues used e-EVN in 2008 to image the central 150 parsec of the starburst galaxy IC 694 at 5 GHz (6 cm wavelength). This showed a rich cluster of 26 compact radio-emitting sources, most of which appear to be young radio supernovae or supernova remnants. Gabriele Giovannini and colleagues also used the e-EVN in 2009 and 2010 at 5 GHz to examine the structure and velocity of M87's jet. They found that part of the jet, about 1 arcsec from the core, was moving with an apparent velocity of 2.7c. This was thought to be equivalent to a real velocity of 0.94c.

Bibliography

Mantovani, Franco, and Kus, Andrzei (eds.), *The Role of VLBI in Astrophysics, Astrometry and Geodesy*, Kluwer Academic Publishers, 2004.

Pérez-Torres, M. A., et al., *An Extremely Prolific Supernova Factory in the Buried Nucleus of the Starburst Galaxy IC 694*, Astronomy and Astrophysics, **507** (2009), L17–L20.

Szomoru, Arpad, *e-EVN Progress; 3 Years of EXPReS*, Science and Technology of Long Baseline Real-Time Interferometry; The 8th International e-VLBI Workshop – EXPReS09, Madrid, Spain, June 22–26, 2009.

FIVE COLLEGE RADIO ASTRONOMY OBSERVATORY

The Five College Radio Astronomy Observatory (FCRAO), located near Amherst, Massachusetts, was founded in 1969

by the state-supported University of Massachusetts and four small private colleges (Amherst, Hampshire, Mount Holyoke and Smith). Its initial radio telescopes, which were the idea of Richard Huguenin, were basically smaller versions of the Arecibo radio telescope. They were four 120 ft (37 m) diameter fixed, spherical dishes with movable feeds, used for observing pulsars at metre wavelengths. This pulsar array was decommissioned in 1981.

In the meantime a decision had been made in 1974 to purchase a 45 ft (14 m) diameter millimetre-wave radio telescope from the Electronic Space Systems Corporation (ESSCO). The telescope, which was completed two years later, was enclosed in a radome to protect it from the weather. Although it was located at low altitude, it could operate at millimetre wavelengths on cold winter days when the water vapour was largely frozen out of the atmosphere. This radio telescope was used to study the physics and chemistry of interstellar clouds, circumstellar envelopes, planetary atmospheres and comets. It was, in 1987, the first to detect phosphorous nitride in the interstellar medium.

The 14 m millimetre-wave telescope was becoming noncompetitive internationally in the late 1980s, compared with other, more modern millimetre-wave instruments. As a result, FCRAO radio astronomers began to consider ways of replacing it with a 50 m diameter, state-of-the-art, millimetre-wave dish that was at least as performant as the 45 m Nobeyama Radio Observatory instrument in Japan. After many iterations this became the 50 m diameter Mexican-American Large Millimeter Telescope (LMT) which was built on the 4,600 m Sierra Negra in Mexico. A joint project of the Instituto Nacional de Astrofisica, Optica y Electronica in Puebla and the University of Massachusetts, it has so far (2010) achieved first light at centimetre but not millimetre wavelengths.

Bibliography

Arny, T., and Valeriani, G., *The Five College Radio Astronomy Observatory*, Sky and Telescope, June 1977, pp. 431–435.

Irvine, William M., Reflections on the Growth of Astronomy at the University of Massachusetts and the Five College Astronomy Department, http://www.astro.umass.edu/history.pdf.

HEINRICH HERTZ SUBMILLIMETER TELESCOPE

The Max-Planck-Institut für Radioastronomie (MPIfR) looked into the possibility of building a submillimetre-wave radio telescope in the late 1970s. But they eventually gave priority to working with France to build the IRAM (Institut de Radioastronomie Millimétrique) millimetre-wave observatory on the Plateau de Bure in the Alps. This location was not thought to be suitable for submillimetre observations.

In 1982 MPIfR started to look once more at providing a submillimetre telescope, this time in cooperation with the

University of Arizona. The plan was to build a 10 m diameter submillimeter telescope (SMT) on the summit of Mount Lemmon in Arizona. The telescope, which would operate down to the atmospheric cut-off at about 0.3 mm, would be provided by MPIfR, whilst the enclosure would be built by the University of Arizona. The telescope instruments would be developed jointly. A little later it was decided to locate the SMT on Mount Graham as it was about 500 m higher than Mount Lemmon and was more suitable for submillimetre observations.

Work was due to start on Mount Graham in 1985 but environmental concerns over its protected red squirrel habitat led to a frustrating delay. As a result the site work was not started until 1990. Nevertheless, the telescope was dedicated as the Heinrich Hertz Telescope (HHT) just three years later, with first light taking place in the following year. The surface of the dish was made of sheets of carbon fibre reinforced plastic (CFRP) bonded to an aluminium honeycomb core. The support structure was made of CFRP tubes connected to joints of invar steel to produce a low thermal expansion system of low weight and very high stiffness. The HHT or SMT (it is called both) is operated and maintained by the Arizona Radio Observatory. One of the HHT's main fields of research has been that of interstellar molecular clouds and the early stages of star formation.

Bibliography

Baars, J. W. M., and Martin, R. N., *The Heinrich Hertz Telescope: A New Instrument for Submillimeter-Wavelength Astronomy*, Reviews in Modern Astronomy, 9 (1996), 111–126.

Istock, Conrad A., and Hoffman, Robert S. (eds.), *Storm Over a Mountain Island: Conservation Biology and the Mt. Graham Affair*, University of Arizona Press, 1995.

See also: Mount Graham International Optical Observatory.

INDIAN RADIO TELESCOPES

Although India was a relatively poor country in the 1940s it was anxious to develop its scientific base after independence in 1947. At that time radio astronomy was still a very new science, and so it did not figure in India's plans. But in 1955 the CSRIO (Commonwealth Scientific and Industrial Research Organisation) in Australia agreed to donate thirty-two 6 ft (1.8 m) redundant antennae from Potts Hill to India under the Colombo Plan. After a number of delays, it was agreed to use the 32 antennae to produce a grating-type interferometer at Kalyan, near Bombay (now called Mumbai). This interferometric array, which was completed in 1965, consisted of 24 antennae on a 630 m east-west baseline and 8 on a 256 m north-south baseline. It was used to observe the Sun at 610 MHz (wavelength 49 cm).

Over the years radio astronomy in the developed world has progressed from observations at these relatively long wavelengths to observations at ever shorter wavelengths in order to obtain higher resolution. The technology of these developed countries has allowed this transition to take place. But in a country like India, where labour costs were low, it made more sense to build very large antennae operating at longer wavelengths, where the mechanical tolerances were lower.

As a result, the next large radio telescope to be built in India operated at long wavelengths and was very labour intensive to build. This telescope, which was completed in 1970, was built at Ootacamund in southern India at an altitude of 2,100 m. It consisted of a 530 m long, 30 m wide parabolic cylinder, the 'surface' of which consisted of one thousand one hundred 530 m long, thin, stainless steel wires running parallel to each other. They were supported on 24 steerable parabolic frames. The telescope was built with its long axis along a north-south hill with a slope at the same angle, 11.35°, as the latitude of the station. This meant that the telescope's long axis was parallel to the Earth's rotational axis, giving it an equatorial mount. As a result the telescope was able to track radio sources within 35° of the celestial equator for up to 9.5 hours per day by a simple mechanical rotation of the array. The telescope, which operated at 327 MHz (91.7 cm), could also be steered in the north-south direction by introducing a phase delay along the dipole array.

This so-called Ooty Radio Telescope (ORT) observed the lunar occultation of a large number of relatively faint radio sources like distant radio galaxies and quasars. These occultations provided accurate position estimates of the sources and showed their structure to arcsecond resolution. In addition, observations of interplanetary scintillation provided information on the compact structure of radio sources. Then in the mid-1980s seven small parabolic cylindrical antennae were combined with the ORT to form the Ooty Synthesis Radio Telescope.

India built one of the few large low-frequency radio telescopes of the world at Gauribidanur, about 80 km north of Bangalore. When it was completed in the early 1980s it consisted of a T-shaped array 1.38 km east-west, 445 m north-south, of 1,000 broad-band dipole antennae about 1.5 m above a 60,000 m² reflecting screen. The array, which operated at 34.5 MHz (9 m), could be pointed to any direction along the meridian with declination range of −45° to +75°. It was used to observe the Sun and large areas of diffuse radio emission like those associated with clusters of galaxies.

The Giant Metrewave Radio Telescope (GMRT), which was approved in 1987, was built near Pune in west central India. It consisted of thirty 45 m fully steerable altazimuth parabolic dishes. About half of these were located in the central 1 km², with the remainder being on three 14 km long arms arranged in an approximate "Y" formation. It became fully operational in 2000 and could be operated between 40 and 1,430 MHz (7.5 m to 21 cm). The GMRT has been used to observe an enormous variety of objects from the Sun and planets to pulsars, supernova remnants, interstellar hydrogen clouds, cosmic masers and quasars. It has even observed the afterglow of a gamma-ray burst.

A radioheliograph was built at the Gauribidanur radio observatory in the 1990s to study the Sun's corona between about 0.2 to 0.8 solar radii of the Sun's surface. When completed in 1998 it operated between 40 and 150 MHz (7.5 to 2 m), and consisted of 192 log periodic dipoles arranged in a T-shape 1.2 km east-west and 0.45 km north-south. This radioheliograph array was used to study coronal mass ejections, coronal holes and active regions, amongst other things.

Bibliography

Ananthakrishnan, S., *The Giant Metrewave Radio Telescope (GMRT): Salient Features and Recent Results*, 29th International Cosmic Ray Conference, Pune, **10** (2005), 125–136.

Swarup, Govind, *The Ooty Synthesis Radio Telescope: First Results*, Journal of Astrophysics and Astronomy, **5** (1984), 139–148.

Swarup, Govind, *From Potts Hill (Australia) to Pune (India): The Journey of a Radio Astronomer*, Journal of Astronomical History and Heritage, **9**, No. 1, June 2006, pp. 21–33.

Swarup, Govind, *Reminiscences Regarding Professor W. N. Christiansen*, Journal of Astronomical History and Heritage, **11**, No. 3, November 2008, pp. 194–202.

IRAM RADIO TELESCOPES

Millimetre-wave radio astronomy took an important step forward in 1969 with the detection of the 2.6 mm (115 GHz frequency) spectral line of carbon monoxide by the new 11 m millimetre radio telescope of the American National Radio Astronomy Observatory (NRAO) on Kitt Peak. This was quickly followed by the detection by the same telescope of the lines of other molecules in the 2 to 4 mm (150 to 75 GHz) range.

Inspired by this work, astronomers in France and Germany looked at various possibilities of building their own millimetre-wave radio telescope. This was to lead in 1979 to the founding of IRAM (Institut de Radioastronomie Millimétrique) by the French CNRS (Centre National de la Recherche Scientifique) and the German MPG (Max Planck Gesellschaft). Initially the Spanish IGN (Instituto Geográfico Nacional) was an associate member, becoming a full member in 1990.

Initial design work for a large millimetre-wave radio telescope had been started at the Max-Planck-Institut für Radioastronomie (MPIfR) in 1975. The main requirement was to design as large a dish as finances would allow that could be used down to about 1 mm (300 GHz) wavelength. After the

Figure 7.3. The 30 m diameter IRAM Millimeter Radio Telescope on Pico Veleta in the Spanish Sierra Nevada. (Courtesy JuanJaén/Wikimedia Commons/Flickr.)

formation of IRAM it was decided that the telescope should be of 30 m diameter and the MPIfR was contracted to be responsible for its construction. Subsequently the detailed design and construction of what was to be called the Millimeter Radio Telescope (MRT) was undertaken at an altitude of 2,900 m on Pico Veleta in the Spanish Sierra Nevada (see Figure 7.3) by the same industrial consortium that had designed the Effelsberg 100 m radio telescope.

The designers used a self-compensating design, similar to that used for the Effelsberg dish, to minimise the effect of gravitational and wind loads. They also minimised thermal effects on the MRT by designing a cavity underneath the 30 m dish that was thermally controlled by fans. In addition, the dish's aluminium panels were coated with a heat-reflecting paint, and thermal insulation sheets were installed between these panels and the steel support structure. Active thermal control was also used elsewhere on the telescope structure.

Tests of the MRT were begun in 1984 at a wavelength of 3.5 mm (80 GHz), and the telescope became operational the following year, operating down to 1.3 mm. Later improvements enabled it to be used down to the atmospheric window at

0.87 mm (350 GHz). The telescope's excellent thermal design also enabled it to be used in daylight hours.

IRAM had also tried to find a reasonably flat mountain site, with a relatively dry atmosphere, on which to build a millimetre-wave interferometer to complement the MRT. They eventually settled on the Plateau de Bure at an altitude of 2,600 m in the French Alps, whose atmosphere had a precipitable water vapour content of less than 2 mm in dry winter weather. Construction work started in 1983 on this interferometer which was designed to operate down to 0.8 mm. It was built with three 15 m diameter dishes that could be placed at any of 26 stations on a 'T'–shaped railway track stretching 160 m north-south and 288 m east-west. The first fringes were obtained with two dishes in late 1988 and with three in mid-1989. Since then the number of dishes has been increased to six, and both arms of the 'T' have been progressively increased in size to produce a maximum baseline of 760 m.

This interferometer has, amongst other things, been used to examine the dust and gas in the coma of comets, and to measure the wind and thermal profile of Mars' atmosphere. It has also been used to undertake a systematic study of proplyds, and to detect the distribution of dust and gas in distant galaxies.

Bibliography

Baars, J. W. M., *The 30 Meter Millimeterwavelength Radiotelescope*, Mitteilungen der Astronomischen Gesellschaft, **54** (1981), 61–69.

Baars, J.W.M., et al., *The IRAM 30-m Millimeter Radio Telescope on Pico Veleta, Spain*, Astronomy and Astrophysics, **175** (1987), 319–326.

Guilloteau, S., et al., *The IRAM Interferometer on Plateau de Bure*, Astronomy and Astrophysics, **262** (1992), 624–633.

JAPANESE TERRESTRIAL RADIO TELESCOPES

Radio astronomy started in Japan a little later than in some other countries after the Second World War owing to its damaged infrastructure. But in 1949 radio astronomy research started at both the Tokyo Astronomical Observatory at Mitaka and at Nagoya University's Atmospheric Research Institute at Toyokawa.

One of the first instruments to be built at Mitaka was a 5 m by 4.5 m dipole array to observe the Sun at 200 MHz (1.5 m wavelength). In 1953 a 10 m (33 ft) parabolic antenna was also built at Mitaka to study the Sun at 200 MHz and 3 GHz (10 cm). Meanwhile a 2.5 m dish had been built at Toyokawa in 1951 to observe solar emissions at 4 GHz. Two years later a grating interferometer consisting of five 1.5 m dishes was built at Toyokawa, also operating at 4 GHz, for solar observations. Over the 1950s and 1960s the radio astronomy facilities at both these and other sites were expanded. In particular, by 1970 a 6 m paraboloid had been built at Mitaka to operate at frequencies

from 70 to 100 GHz (4.3 to 3 mm). This enabled Japan to enter the relatively new field of interstellar molecular research.

Building of the Nobeyama Observatory was started in 1970 on a 1,350 m high plateau at the foot of the Yatsugatake Mountains. Its 160 MHz solar interferometer consisted of an east-west array of nine 6 m and two 8 m parabolic antennae extending over a baseline of 2.2 km. In addition, there was a north-south line of six 6 m dishes. There was also a 17 GHz interferometer at Nobeyama, which was designed to measure the brightness distribution across the solar disc. It consisted of twelve 1.2 m dishes aligned east-west. These two interferometers, which became operational shortly after the observatory was opened, were closed down in 1989 and 1992, respectively. In the meantime, in 1978 the Nobeyama Radio Observatory had been designated the Japanese national radio observatory specialising in microwave observations.

A fully steerable 45 m parabolic dish was built at the Nobeyama Radio Observatory around the end of the 1970s that could operate at wavelengths of just a few millimetres. This was to extend the observations of interstellar molecules that had been started some years earlier by the 6 m paraboloid at Mitaka. The 45 m became operational in 1982 and, at that time, it was the largest radio telescope in the world operating at such short wavelengths. The panels of the main dish were made of a 10 cm thick honeycomb sandwiched between sheets of carbon fibre reinforced plastic. A cavity underneath that main dish was thermally controlled by fans to try to maintain the temperature of the dish supporting structure constant, and so reduce its thermal distortion. The dish had a similar self-compensating system to that of the Effelsberg antenna in West Germany. Distortions due to gravity, as the Japanese dish was moved from the zenith, were compensated by moving the secondary reflector. This Japanese facility has been used to study molecular clouds to aid our understanding of interstellar chemistry, the formation of stars and the structure of galaxies.

A second major instrument at Nobeyama, which was built in the early 1980s, was the Nobeyama Millimeter Array which initially consisted of five 10 m dishes operating down to 2.6 mm (115 GHz). The dishes were arranged on two baselines of just over 500 m each, one east-west with the other intersecting it at about 60°. Later a sixth 10 m dish was added and the frequency coverage extended to 365 GHz (0.8 mm). This millimeter array was used, amongst other things, to study the fine structure of interstellar clouds.

A radioheliograph was constructed at Nobeyama in the early 1990s which consisted of eighty-four 80 cm diameter parabolic dishes operating at 17 and 34 GHz (1.75 and 0.9 cm). They were arranged in a T-shaped configuration with an east-west baseline of 500 m and a north-south baseline of 220 m. The radioheliograph had a temporal resolution of 0.1 s, and a spatial resolution of 5 arcsec at 34 GHz. Its main purpose was to observe solar flares.

Various other radio telescopes have been built in Japan in the last two decades or so, a number of which have been used for geodesy and Earth rotation studies. For example, in 1988 a 34 m radio telescope was built at Kashima as part of the geodetic Western Pacific VLBI (Very Long Baseline Interferometer) network project. After that project was finished it was used partly for astronomical purposes.

The VERA (VLBI Exploration of Radio Astronomy) network consisted of four 20 m dual beam radio antennae built over the period 2001–2002. They were located at Mizusawa, Iriki, Ogasawara and Ishigaki. This VERA network was designed to measure the coordinates and proper motion of galactic masers with a 10 microarcsec accuracy. Today (2010) these antennae are being integrated into the East Asia VLBI network which will consist of about 20 radio telescopes from Japan, China and the Republic of Korea. One of its main tasks will be to improve our knowledge of galactic structures, particularly that of the Milky Way.

Bibliography

Akabane, Kenji, et al., *The Nobeyama Radio Observatory*, Sky and Telescope, December 1983, pp. 495–499.

Nakajima, Hiroshi, *A New 17-GHz Solar Radio Interferometer at Nobeyama*, Publications of the Astronomical Society of Japan, **32** (1980), 639–650.

Tanaka, Haruo, *Development of Solar Radio Astronomy in Japan up until 1960*, in Sullivan, W. T. (ed.), *The Early Years of Radio Astronomy: Reflections Fifty Years after Jansky's Discovery*, Cambridge University Press, 1984.

See also: HALCA.

JPL/NASA'S DEEP SPACE NETWORK

The Advanced Research Projects Agency (ARPA) of the U.S. Department of Defense ordered three 26 m (85 ft) parabolic antennae via the Jet Propulsion Laboratory (JPL) to support the early American spacecraft programme. One of these was to be used for communicating with the Pioneer spacecraft en route to the Moon, whilst the other two were for later programmes. In the event the first antenna, designated HA–DEC (for HourAngle–Declination, describing its tracking mode), was built at a radio quiet site at Goldstone, California in 1958. In the following year a second 26 m antenna (not one of the above three ordered), called AZ–EL (for Altazimuth–Elevation) was built nearby to participate with HA–DEC in the ECHO balloon spacecraft passive communications experiment. Signals were transmitted from the east coast via Echo and detected by the HA–DEC antenna at the so-called Pioneer site. They were then transmitted back to the east coast by the AZ–EL antenna at the Echo site. As a result the HA–DEC antenna had a receiver, and the AZ–EL had a transmitter.

These two 26 m antennae were used in 1961 to measure the distance of Venus by bistatic radar. The AZ–EL antenna transmitted the signal to Venus, and the HA–DEC antenna, fitted with a maser receiver for detecting weak signals, received it back after reflection by the planet. In the following year Roland Carpenter concluded, from analysing the received signals, that the rotation of Venus was retrograde.

NASA wanted to locate the two extra 26 m antennae about 120° longitude apart, 120° longitude from Goldstone. As a result one was built near Woomera, Australia in 1960, and the second was built near Johannesburg, South Africa in the following year. The three antennae made up what was called the Deep Space Instrumentation Facility or DSIF, which was renamed the Deep Space Network (DSN) in 1963. Meanwhile, in 1962 the AZ–EL antenna had been moved from the Echo site to a new site, the so-called Venus site a few miles away, where it was used as a communications test facility. It was replaced at the Echo site by a new antenna, similar to the Pioneer one, which was added to the DSIF spacecraft communications network.

It became clear in the early 1960s that there were not enough antennae in this four-antenna network to cover all the spacecraft that NASA were expecting to launch over the next few years. So NASA decided to augment this network with two more 26 m antennae, one at Robledo in Spain and the other at Tidbinbilla in Australia. These were both operational in 1965.

Meanwhile in 1960 NASA had also concluded that larger antennae were required to cover deep space missions. This resulted in the construction of a 64 m (210 ft) antenna whose design was based on that of the Parkes radio telescope in Australia. This new 64 m, which was built at the so-called Mars site at Goldstone, a few miles from the Pioneer and Echo sites, detected signals from the Mariner 4 Mars spacecraft in 1966. Since then the DSN ground stations at Goldstone, Robledo and Tidbinbilla have been expanded considerably, each now having at least one 26 m, two 34 m, and one 70 m antenna.

Over the years the DSN antennae have been used, in addition to their primary role of spacecraft monitoring and control, for both radio astronomy and solar system radar research. In 1967 Robertson, Gubbay, Moffet and colleagues used the DSN antennae at Goldstone, Woomera and Tidbinbilla to carry out very long baseline interferometry. This yielded resolutions in the milliarcsecond (mas) region. Another early use of DSN antennae was in the study of pulsars and in the study of radio emissions from Jupiter. In 1969 Reichley and Downs detected a sudden change in the period of the Vela pulsar using the Goldstone 64 m. Then in 1971 Alan Whitney and colleagues detected superluminal motion in 3C 279 using an interferometer consisting of the 64 m Goldstone and the 37 m Haystack antennae.

Bibliography

Imbriale, William A., *Large Antennas of the Deep Space Network*, John Wiley and Sons, 2003.

Mudgway, Douglas J., *Uplink–Downlink: A History of the Deep Space Network*, NASA SP-2001-4227, 2001.

Mudgway, Douglas J., *Big Dish: Building America's Deep Space Connection to the Planets*, University Press of Florida, 2005.

MIT LINCOLN LABORATORY, HAYSTACK

Lincoln Laboratory began the construction of a Ballistic Missile Early Warning System radar at Millstone Hill in Westford, Massachusetts, in 1956. Its 26 m (84 ft) diameter dish antenna began operating at 440 MHz (wavelength 68 cm) in October 1957, just in time to track the first Sputnik spacecraft. The Millstone dish was used for both military and civilian purposes, including ionospheric research, and radar studies of the aurora, meteors and the Moon. In 1959 it also detected Venus by radar.

The Millstone Hill antenna was upgraded in 1962 to operate at 1.32 GHz (23 cm). In parallel Lincoln Laboratory had started the design and construction of a 37 m (120 ft) diameter radar dish that could operate up to 16 GHz (1.9 cm), and later up to 35 GHz (8.5 mm). It was protected by a radome which also considerably reduced the total construction costs by eliminating wind load on the main dish, so enabling it to be made much lighter. On completion in 1964 this 37 m so-called Haystack antenna took over from the nearby Millstone antenna as the Lincoln Laboratory's planetary radar facility. The Haystack dish was also used as a ground terminal for space communications, as a radar to track spacecraft, and as a radio telescope.

In parallel with the construction of the 37 m Haystack radar antenna, Lincoln Laboratory also built an 18 m (60 ft) radome-protected radar for use with the controversial Project West Ford. In this project about 500 million thin copper wires were launched into orbit in 1963 to enable messages to be transmitted from coast to coast across the United States. At about the same time the laboratory also built a fixed, vertically pointing, 67 m (220 ft) antenna which was used for ionospheric research.

In 1970, at the suggestion of Lincoln Laboratory, the Haystack 37 m dish was transferred to the civilian control of a number of Northeast universities under the umbrella of the Northeast Radio Observatory Corporation (NEROC). Although still used as a radio telescope, use of this facility for planetary radar reduced over the next few years and ceased completely in 1974. In 1992 a deformable subreflector was installed to help compensate for gravitational distortions of the main dish as its angle was changed to the vertical. Today the 37 m dish operates part time as a radio telescope up to 115 GHz (2.6 mm), and part time as an element of the U.S. Military Space Surveillance Network.

The MIT Haystack Observatory today operates the 18 m Project West Ford antenna which is mainly used for geodetic research, the 67 m zenith antenna, and a 46 m (140 ft) radar antenna that had been built at the Sagamore Hill Air Force facility at Hamilton, Massachusetts, in 1963. It was moved to Millstone Hill in 1978 where, along with the 67 m, it has been undertaking ionospheric research.

Bibliography

Anonymous, *Haystack Radar-Radio Telescope Dedicated*, Sky and Telescope, December 1964, pp. 349–351.

Barvainis, Richard, and Salah, Joseph E., Upgrade Capabilities of the Haystack Radio Telescope, in Clemens, D. P, and Barvainis, R. (eds.), *Clouds, Cores and Low Mass Stars*, Astronomical Society of the Pacific Conference Series, 65 (1994), 411–418.

Butrica, Andrew J., *To See the Unseen: A History of Planetary Radar Astronomy*, NASA SP−4218, 1996.

OHIO TRANSIT RADIO TELESCOPE

A transit radio telescope of a completely new configuration was designed by John Kraus at Ohio State–Ohio Wesleyan Radio Observatory in the 1950s. It consisted of a tiltable flat reflector of wires stretched on a 80 m long by 30 m high metal frame, which reflected radio waves from the source on to a fixed section of a parabolic reflector 110 m long by 20 m high. From there they were reflected to the feeds at the focus. The ground between the two reflectors was covered with aluminium sheet to reduce thermal radiation from the ground and simplify the design of the feeds at frequencies below 1 GHz. The telescope was operational between 20 MHz and 2 GHz (wavelengths of 15 m and 15 cm), having a beamwidth of about 11 by 40 arcmin at 1.4 GHz (wavelength 21 cm). It could track sources for about 30 minutes near the meridian by using movable feeds.

This Ohio reflector, which took almost seven years to build by university technicians and students, was completed in 1962 in the grounds of Ohio Wesleyan University's Perkins Observatory. Its success encouraged French radio astronomers to design a larger, improved version at Nançay, which became fully operational five years later.

The Perkins radio telescope, otherwise called 'Big Ear', did not initially use the whole width of the parabolic section as funds had run out before the flat section could be completed. In 1970 this flat section was extended from 80 m to its full width of 100 m.

One of the first tasks of the Ohio reflector in 1963 had been to produce the most detailed radio map of the Andromeda galaxy available at that time. This was followed by a thorough survey of the sky from declinations 63° N to 36° S, mainly at 1.4 GHz. Sixty percent of the sources detected were new. Then in 1973 it was first used by the SETI (Search for Extraterrestrial

Intelligence) project, which continued to use it for a number of years until the telescope was demolished in 1997.

Bibliography

Kraus, John, *Big Ear Two: Listening for Other-Worlds*, Cygnus-Quasar Books, 1994.

Kraus, John D., *The Large Radio Telescope of Ohio State University*, Sky and Telescope, July 1963, pp 12–16.

Shuch, H. Paul (ed.), *Searching for Extraterrestrial Intelligence: SETI Past, Present and Future*, Springer, 2011.

OWENS VALLEY RADIO OBSERVATORY

The California Institute of Technology (Caltech) built an interferometer in the mid-1950s at Owens Valley, near Big Pine, California. Owens Valley was chosen as it was shielded from man-made interference by nearby mountains. The interferometer consisted of two 90 ft (27 m) equatorial dishes with a maximum separation of 1,600 ft (500 m) on an east-west baseline. It went into operation in 1957 and was the first operational radio interferometer in the United States. During the 1960s a north-south track was added, to form an 'L' with the original baseline.

A 130 ft (40 m) altazimuth dish was built in 1965–1967 about one kilometre east of the two 90 ft telescopes to extend the size of the interferometer. This 130 ft dish, which could also be operated independently, was used at wavelengths from about 1 m (300 MHz) to 2 cm (15 GHz). It has been used for many research projects including a recent one to study the microwave background radiation.

Three 1.8 m antennae were added to the two 27 m dishes by 1997 when management of this 27 m/1.8 m interferometer, called the Owens Valley Solar Array, was transferred to the New Jersey Institute of Technology. Two more 1.8 m antennae were added by 2004. One of the main tasks of the Solar Array has been to observe the development of magnetic fields on the Sun that were associated with solar flares.

The millimetre array

Caltech installed a 10.4 m dish-type radio telescope at Owens Valley in 1978 that could operate at millimetre wavelengths and, six years later, two more 10.4 m dishes were added to form an interferometer. The antennae were movable along a T-shaped railway track 100 m north to south and 200 m east-west. In 1989 the receivers were changed to reduce the operating wavelength to 1.4 mm (210 GHz). Three more 10.4 m dishes were later added to form a six-dish millimetre array.

In 2004 it was decided to merge this Owens Valley millimetre array with that of the Berkeley-Illinois-Maryland Association (BIMA), which consisted of nine 6.1 m telescopes. This new array, called CARMA (Combined Array for Research in Millimeter-wave Astronomy), began routine operations at the nearby high altitude Cedar Flat location in 2007. This was about 1,000 m higher than Owens Valley, and so had typically about 50% less atmospheric water vapour. Then in 2008 the University of Chicago joined CARMA with the addition of eight 3.5 m antennae.

CARMA operated at wavelengths of 7, 3 and 1.3 mm and was used to detect and image radio emission from molecules and dust. This has enabled research to be undertaken into the formation, evolution and dynamics of galaxies, the formation of stars and any associated planetary systems, and the composition of planetary atmospheres.

PARKES RADIO TELESCOPE

Australia had developed a thriving radio astronomy community after the Second World War using essentially war surplus equipment (see *Early Australian radio astronomy* article). But by 1950 it was becoming evident that, if Australia wanted to retain a leading role in radio astronomy, it would have to build a much more performant telescope than any it had built so far. As a result Edward George 'Taffy' Bowen, head of the Radiophysics Laboratory, proposed that a large parabolic radio telescope should be built, similar in size to that already under construction in the UK at Jodrell Bank, but operating at a higher frequency. He raised finance from the Carnegie Corporation and the Rockefeller Foundation, which was matched by the Australian government. He was determined to learn from the problems experienced with the construction of the 75 m Jodrell Bank dish. He also sought advice from the highly respected design engineer Barnes Wallis as a design consultant.

The location for this large radio telescope was chosen to be a valley floor surrounded by low foothills about 20 km from the small town of Parkes, and 400 km over the mountains from Sydney. The Parkes town council even agreed to restrict industrial development within 10 km of the telescope, in order to keep electromagnetic interference to a minimum. With a minimum temperature of 4°C, icing of the dish could be ignored, and the site was also found to have very low wind speeds.

The main requirement of the telescope was that it could operate down to 21 cm (1.4 GHz), preferably lower. An altazimuth dish of 64 m (210 ft) diameter was selected following a size/cost tradeoff. It was also decided to restrict the movement of the dish to a maximum angle of 60° to the vertical to simplify the design and reduce costs. This would still allow the telescope to reach as far north as +27° declination, which was good enough considering the number of telescopes in the Northern Hemisphere covering northerly declinations. Design work was started at the end of 1955 and the construction of the telescope was completed six years later.

Figure 7.4. The 210 ft (64 m) diameter Parkes radio telescope in Australia which has been operational since 1961. It was used by NASA to cover the Apollo landings, and has been very successful in the discovery and study of pulsars. (Courtesy Ian Sutton/Wikimedia Commons/Flickr.)

The completed 64 m dish had a 17 m diameter central zone made of welded steel plate, with the remainder being wire mesh. The telescope's minimum operating wavelength was 10 cm. In 1962 Hazard, Mackey and Shimmins used the Parkes antenna to observe the occultation of the source 3C 273 by the Moon. They found that the source was double, with a separation of 20 arcsec. This enabled Maarten Schmidt to identify its optical equivalent as a faint blue star and the end of a faint bluish jet coming from the star. The 'star' was found to have an unexpectedly large redshift of 0.158. It was the first quasar to be discovered.

Twenty-five years later the Parkes antenna became part of the Australia Telescope, a large radio interferometric array. After several upgrades, including making more of the dish from solid sheets (see Figure 7.4), the Parkes telescope can now operate at up to 43 GHz (7 mm). In 2003 it discovered the first known double pulsar system. It has currently discovered more than half of all known pulsars.

Bibliography

Goddard, D. E., and Milne, D. K. (eds.), *Parkes: Thirty Years of Radio Astronomy*, CSIRO Publishing, 1994.

Haynes, Raymond, et al., *Explorers of the Southern Sky: A History of Australian Astronomy*, Cambridge University Press, 1996.

Robertson, Peter, *Beyond Southern Skies: Radio Astronomy and the Parkes Telescope*, Cambridge University Press, 1992. ·

See also: Early Australian radio astronomy; Culgoora, Fleurs and the Australia Telescope.

SOUTH POLE RADIO TELESCOPES

The South Pole, at an altitude of 2,850 m, is very cold, resulting in an exceptionally low amount of water vapour in its atmosphere, particularly in winter when the average precipitable amount is just 0.3 mm of water. As a result it is the ideal place, from a scientific point of view, to make microwave and infrared astronomical observations. On the other hand its environment is exceptionally harsh for both people and equipment, and it is a considerable distance from civilisation. So astronomical observation campaigns have been undertaken only relatively recently.

A French experiment, EMILIE (Emission Millimétrique), was the first to make submillimetre astronomical observations from the South Pole during the summer of 1984–85. It measured diffuse radiation from the Milky Way using a single pixel bolometer, operating at a wavelength of 0.9 mm, which was fed by a 45 cm off-axis mirror.

Over the next few years a number of astronomers set up various equipment at the pole to try to measure the anisotropy of the cosmic microwave background (CMB). Although no consistent anisotropy was detected, the observations were encouraging. But an enormous amount of effort was required to ship the equipment to the South Pole where it was operated only through the short, relatively moist summer season, before it all had to be dismantled. Clearly a permanent facility was required.

ASTRO (Antarctic Submillimeter Telescope and Remote Observatory), which was started in 1989, was the first radio telescope to operate all year round at the South Pole. It was a general purpose 1.7 m diameter radio telescope that could be used between 0.2 and 1.3 mm. Two years later ASTRO became part of the newly founded United States National Science Foundation's (NSF's) Center for Astrophysical Research in Antarctica (CARA).

CARA assembled a number of telescopes, some of which, like the earlier instruments, also undertook CMB research. By 1997 they had given the first indication that the spectrum of spatial anisotropy in the CMB was consistent with a flat cosmology. More precise results were provided by DASI (Degree Angular Scale Interferometer) starting in 2000. DASI was a compact interferometer consisting of an array of thirteen 20 m diameter horns on a fully steerable mount. It operated from 26 to 36 GHz (1.15 to 0.85 cm wavelength) and provided baselines from 25 to 121 cm.

Viper, which began observations in 1998, consisted of a 2.1 m off-axis radio telescope designed to measure low-contrast millimetre-wave sources. It was used with a variety of instruments including ACBAR (Arcminute Cosmology Bolometer Array Receiver) which was operational on Viper from 2001 to 2005. It made high quality maps of the Sunyaev-Zel'dovich effect (SZE) in nearby clusters of galaxies and made measurements of CMB anisotropy on a range of scales.

The success of these various facilities persuaded the NSF to fund a much larger radio telescope at the South Pole. This is the South Pole Telescope, a 10 m off-axis radio telescope that was equipped with a state-of-the-art 966 element bolometer array receiver. The telescope, which was installed in 2006–2007, has so far observed at frequencies from about 100 to 220 GHz (3 to 1.3 mm). Its primary goal is to detect clusters of galaxies using the SZE, covering thousands of square degrees at 1.3 arcminute resolution. Its dish surface is accurate enough to enable it to also operate at submillimetre wavelengths.

Bibliography

Ruhl, J. E., et al., *The South Pole Telescope*, Proceedings of SPIE, **5498** (2004), 11–29.

Wilson, Gregory, *Astronomy at the South Pole*, Optics and Photonics News, January 2004, pp. 34–39.

Wilson, Robert W., and Stark, Antony A., *Cosmology from Antarctica*, Proceedings of the Smithsonian at the Poles Symposium 2007, Smithsonian Institution Scholarly Press, pp. 359–367, 2009.

SUBMILLIMETRE RADIO TELESCOPES ON MAUNA KEA

Radio astronomy had begun in the 1930s and 1940s by observing at relatively long radio wavelengths as it was relatively easy to construct antennae capable of such observations. But such radio telescopes had poor angular resolution, so large interferometers had to be built to resolve this problem. In the 1960s, however, it became possible to build parabolic radio antennae that could operate at millimetre wavelengths with improved resolution. And every time the lower wavelength limit was breached, more interesting data was received. So engineers and astronomers began to consider whether it was possible to build a reasonably sized radio dish that could operate at submillimetre wavelengths down to the atmospheric cut-off at about 0.3 mm. This would enable observations to be made of many important molecular transitions in molecular clouds, which would facilitate a detailed study of the physical structure of these clouds. In addition, the shorter wavelengths would increase the resolutions achievable for a given size of antenna.

James Clerk Maxwell Telescope

The United Kingdom's Science Research Council debated these issues and concluded in 1975 that it should be possible to build a 15 m diameter dish to observe down to 0.75 mm. It was initially decided to build this telescope on La Palma. Then the Dutch joined the project in 1981, and the decision was made to build this facility on Mauna Kea, because of the very low water vapour content of its atmosphere. Construction started in 1983, and the telescope, now called the James Clerk Maxwell Telescope (JCMT), saw first light on Mauna Kea just over three years later. The facility was funded by the United Kingdom, Canada (from 1987) and the Netherlands.

The JCMT dish, which was on an altazimuth mount, consisted of over 250 individually adjustable panels. It was protected from the elements by a transparent membrane attached to a co-rotating dome. The original detector, a continuum single pixel bolometer, was replaced by the Submillimetre Common-User Bolometer Array (SCUBA) in 1996 to detect interstellar dust grains. It found voids in the distribution of dust grains around nearby stars that may have been caused by newly formed planets. Observations with SCUBA, which was optimised for observations at 0.45 and 0.85 mm, also showed bright knots in molecular clouds where stars were being born. In addition, submillimetre spectral lines were observed using heterodyne receivers which provided information on the temperature, density and velocity of interstellar gas.

Caltech Submillimeter Observatory

In 1982 the very influential Field Report of the United States National Academy of Sciences had recommended the construction of a 10 m aperture submillimetre telescope at a high, dry site. This was to become the 10.4 m diameter Caltech Submillimeter Observatory (CSO) which was built shortly afterwards very close to the JCMT on Mauna Kea. Caltech's aim was to build the most accurate dish possible at about 10 m aperture, whereas the JCMT designers built a larger dish, whilst accepting a compromise in dish accuracy. As a result, although the Caltech dish had only half the area of the JCMT, it was able to operate down to shorter wavelengths of about 0.35 mm. The 10.4 m millimetre-wave Caltech dishes at the Owens Valley radio observatory were effectively prototypes of the CSO dishes. In the event the CSO was dedicated in late 1986, two weeks before the JCMT saw first light.

The CSO dish was made of hexagonal segments individually machined to a high accuracy. The initial detectors were heterodyne receivers and a single-channel bolometer. The latter was later replaced by SHARC (Submillimeter High Angular Resolution Camera), a 24 element imaging bolometer array. The CSO has, amongst other things, improved our understanding of the chemistry of the interstellar medium, improved our knowledge of the development of red giant stars, mapped molecular gas and dust in galaxies and helped to determine the volatile constituents of comets.

The CSO undertook short-baseline interferometry with the JCMT about 160 m away to produce a best angular resolution of about 0.5 arcseconds at 345 GHz (0.9 mm wavelength). Experience gained with these observations assisted the design of the subsequent Submillimeter Array (see below). Current plans call for the decommissioning of the CSO in 2016 as observations start with more advanced instruments.

Submillimeter Array

When Irwin Shapiro took over as director of the Harvard-Smithsonian Center for Astrophysics in 1983 he decided to try to extend the development of high-angular-resolution instruments across as wide a range of wavelengths as possible. In the submillimetre waveband this was to result in the construction of the Submillimeter Array (SMA). Mauna Kea was selected in 1992 as the preferred site for this SMA which was expected to operate down to the atmospheric cut-off at about 0.3 mm. It was initially planned to install six 6 m telescopes near the JCMT and CSO on Mauna Kea so they could be linked together. Then in 1996 the Academia Sinica Institute of Astronomy and Astrophysics of Taiwan joined the project, and agreed to add two more telescopes, increasing the number of possible baselines from 15 to 28.

The first prototype telescope was built at the MIT Haystack Observatory in 1996, and two copies of a second prototype operated there two years later as an interferometer. The production American telescopes were also assembled at Haystack for testing before being disassembled and shipped out to Mauna Kea, where the SMA was formally dedicated in 2003. There were 24 pads for locating the eight individual dishes to provide baselines varying from 9 to 500 m. The array operated between 180 to 900 GHz (1.7 to 0.3 mm), producing maximum resolutions of from 0.5 to 0.1 arcsec.

Bibliography

Holland, W. S., et al., *SCUBA: A Common-User Submillimetre Camera Operating on the James Clerk Maxwell Telescope*, Monthly Notices of the Royal Astronomical Society, 303 (1999), 659–672.

Moran, James M., *The Submillimeter Array*, ASP Conference Series, 356 (2006), 45–57.

Smith, David H., *The Submillimeter Giants*, Sky and Telescope, August 1985, pp. 119–123.

Steiger, Walter R., *A Brief History of the Caltech Submillimeter Observatory*, http://www.cso.caltech.edu/cso_history/CSO_History/index.html.

See also: Owens Valley Radio Observatory.

UNITED STATES NATIONAL RADIO ASTRONOMY OBSERVATORY

There was quite a degree of soul-searching in the United States in the early 1950s as it was becoming clear that, unless a bold approach was taken, it would continue to fall further and further behind other countries, particularly the United Kingdom and Australia, in radio astronomy. As a result, a number of American organisations had either started to develop radio telescopes or were considering doing so. Then in January 1954 a conference took place to discuss the state of radio astronomy, jointly sponsored by the United States National Science Foundation (NSF), Caltech and the Department of Terrestrial Magnetism. This conference, which included radio astronomers from the UK, Australia, the Netherlands and Canada, helped to persuade many of the American participants that a new approach was required to the provision of radio astronomy facilities in the United States.

Subsequent to the conference, the NSF established a subpanel chaired by Merle Tuve to investigate the future role of the agency in radio astronomy. In parallel Lloyd Berkner, the head of AUI (Associated Universities Incorporated), set up an ad hoc group on cooperative radio astronomy covering a wide range of American interests and personnel. As it happened, Tuve was generally in favour of encouraging individual researchers, whilst Berkner was more of a 'big science' man.

This produced the inevitable clash of ideas, although by 1955 a draft programme had been proposed by AUI to the NSF for a national radio astronomy organisation. The radio telescopes proposed included a 600 ft (180 m) completely steerable dish, a 250 to 300 ft (75 to 90 m) dish, a 140 ft dish of the highest possible precision, and two 84 or 60 ft (26 or 18 m) dishes for interferometry. It was suggested that the first instrument to be built should be the 140 ft, although this was later delayed owing to technical problems. It was also agreed that further studies were required on the possible 600 ft dish. Shortly afterwards AUI became aware of the U.S. Navy's 600 ft project, which was later to be abandoned owing to severe technical problems and cost overruns (see *United States Naval Research Laboratory radio telescopes* article).

In 1955 the NSF accepted the need for national radio astronomy facilities, in spite of reservations from Tuve, and decided to set up a National Radio Astronomy Observatory (NRAO) at Green Bank, West Virginia. This site was chosen as it was subjected to exceptionally low radio interference partly as a result of its mountainous surroundings. The NSF then decided in 1956 to award AUI a contract to be the management institution for this new observatory.

The first major radio telescope at Green Bank, which was operational in 1959, was the 85 ft (26 m) diameter Howard Tatel telescope which could be used simultaneously at 21 and 3.75 cm wavelengths (1.4 and 8 GHz). Its structure was basically identical to that of the University of Michigan's 85 ft telescope. The Green Bank instrument was used shortly after completion to determine the positions of 30 sources to an accuracy of ~10 arcsec. Several sky surveys were then begun, including one by David Heeschen which took advantage of the telescope's ability to perform simultaneous or successive observations at 0.44, 1.2, 1.4 and 8 GHz (68, 25, 21, and 3.75 cm). Planets were also observed and the telescope was used, somewhat controversially, to search for signals emitted by extraterrestrial civilisations.

In the mid-1960s two other 85 ft radio telescopes were built at Green Bank to the same design which, with the Tatel telescope, formed the NRAO Green Bank Interferometer (GBI) with a maximum baseline of 2.7 km. In addition to operating as a research instrument in its own right, the GBI was also used as a prototype instrument for the Very Large Array (VLA), whose design studies had started in 1961. A 42 ft (13 m) telescope, which was located 35 km away, was designed to operate with one of the GBI 85 ft instruments in 1969 using a microwave link. A few years later the 42 ft was replaced by a 45 ft (14 m) instrument to form a four-element interferometer. This had a resolution of 0.2 arcsec at 3.75 cm wavelength which allowed observation of fine structure in radio sources. The GBI was closed down in 2000 for financial reasons.

In 1959 the NRAO began preliminary design work on the design of a relatively inexpensive 300 ft (90 m) transit instrument, adjustable in declination and capable of operating at 21 cm. The dish was made of 0.62 inch (16 mm) aluminium mesh, which was relatively inexpensive and lightweight. The design and construction phase of this very large telescope lasted just 23 months, being completed in 1962, and costing only $850,000. The surface panels were replaced in 1970, allowing the telescope to be operated down to 6 cm (5 GHz). Various other improvements were installed over the years increasing the telescope's sensitivity by a factor of ~100.

The 300 ft radio telescope was a very important survey instrument observing hundreds of sources, primarily extragalactic objects, for the first time. It discovered numerous pulsars, and was used extensively for observations of neutral hydrogen in the Milky Way and beyond. It was also used for planetary observations. The telescope far outlived its expected lifetime, and finally collapsed in 1988 due to metal fatigue. It was replaced by a 100 × 110 m altazimuth dish (see Figure 7.5), the Robert C. Byrd Green Bank Telescope (GBT), which was formally dedicated in 2000.

Unlike the case of similar radio telescopes, the feed of the GBT was not supported by a symmetrical array of legs over the centre of the dish. Instead it was carried by a structure that projected upwards from one edge of the dish. This produced an unblocked aperture that gave the dish a higher efficiency, lower sidelobes and reduced pick-up of ground radiation. The GBT had a computer-controlled active surface that allowed operation down to about 3 mm (100 GHz). It is currently (2010) the largest fully steerable radio telescope dish in the world.

The concept of a 140 ft (43 m) fully steerable radio telescope had been investigated by an advisory committee in the mid-1950s before the NRAO was founded. The question of whether the mounting should be equatorial or altazimuth was the subject of much disagreement, but eventually an equatorial mount was selected in spite of the structural problems that were expected to be involved. In the event these problems were found to be even worse than expected.

Construction of the foundations of the 140 ft started in 1958, but the telescope, capable of operating down to 1 cm (30 GHz), was not completed until 1965. Nine years later it was converted from a prime focus to a Cassegrain configuration in which the feed horns pointed upwards towards the subreflector, rather than downwards towards the ground which radiated noise. The subreflector was of a nutating design which pointed the beam alternately to the object being observed and to the sky beside it. This significantly increased the effective sensitivity whilst minimising atmospheric effects. Then in 1979 a deformable subreflector was installed to compensate for dish distortions that were produced as the angle of the dish to the vertical was changed.

The 140 ft has been particularly useful in galactic astronomy and in molecular spectroscopy. For example, it was used in 1965 by Höglund and Mezger to make the first unambiguous

Figure 7.5. The 100 × 110 m Robert C. Byrd Green Bank Telescope (GBT) that replaced the 300 ft (90 m) dish, following the latter's collapse in 1988. Unlike most such dishes, the GBT has an off-axis feed to eliminate dish obstruction. The shape of the dish surface is adjusted by over 2,000 actuators to compensate for gravitational and other distortions and allow it to operate down to about 3 mm. (Courtesy NRAO/AUI.)

detection of a recombination line of ionised hydrogen. This was the N110 to N109 line at 5.009 GHz which they observed in M17, Orion and a number of other galactic HII regions. Then in 1968 Gerrit Verschuur used the 140 ft to make the first unambiguous detection of the Zeeman effect in neutral hydrogen regions, which clearly proved the existence of interstellar magnetic fields.

The NRAO built an 11 m (36 ft) millimetre-wave telescope at an altitude of about 6,200 ft (1,900 m) on Kitt Peak in 1967, because of the relatively low water vapour content of its atmosphere at high altitudes. It was enclosed in a dome which rotated with the telescope. Although the telescope failed to meet its original specifications, it was highly successful in detecting numerous different complex molecules in the interstellar medium. In 1983 the telescope's reflecting surface and surface support structure were replaced by a new 12 m dish designed to cover all atmospheric windows from 68 to 180 GHz (4.4 to 1.7 mm wavelength). The University of Arizona took over its operation in 2000.

The NRAO was also responsible for managing the VLA and the VLBA, which are covered in separate articles.

Bibliography

Drake, F. D., *A Review of the History, Present Status, and Course of American Radio Astronomy*, Proceedings of the National Academy of Sciences, **49** (1963), 759–766.

Hey, J. S., *The Evolution of Radio Astronomy*, Science History Publications, 1973.

Lockman, F. J., Ghigo, F. D., and Balser, D. S. (eds.), *But It Was Fun: The First Forty Years of Radio Astronomy at Green Bank*, National Radio Astronomy Observatory, 2007.

Malphrus, Benjamin, *The History of Radio Astronomy and the National Radio Astronomy Observatory: Evolution Toward Big Science*, Krieger Publishing, 1996.

See also: United States Naval Research Laboratory radio telescopes, Very Large Array, Very Long Baseline Array.

UNITED STATES NAVAL RESEARCH LABORATORY RADIO TELESCOPES

Just after the Second World War John Hagan and Fred Haddock of the U.S. Naval Research Laboratory (NRL) used parabolic antennae up to 3 m diameter to undertake radio astronomy observations of the Sun between 8.5 mm and 9.4 cm wavelengths (frequencies of 35 and 3.2 GHz respectively).

NRL then purchased a 50 ft (15 m) parabolic dish made of machined aluminium to enable it to be used at wavelengths as low as 1 cm. Its beamwidth was only 3 arcminutes, which was very good for the time. The dish was mounted on the laboratory roof at Washington, using a modified naval gun mounting which operated in the altazimuth mode. This radio telescope, which was the first large reflector to operate down to centimetre wavelengths, became operational in 1951. Two years later NRL made the first detection and measurement of interstellar ionised hydrogen clouds as discrete radio sources. Then Mayer, McCullough and Sloanaker detected 3.15 cm radio emissions from Venus and concluded, in 1956, that its surface temperature was very high at about 600 K. At about the same time NRL also established surface temperatures for Mars and Jupiter.

In the mid-1950s NRL were the leading radio astronomy group in the United States. Subsequently they built an 85 ft (26 m) equatorially-mounted radio telescope at Maryland Point which became available in 1957, and another 85 ft which became available at Maryland Point in 1965. The latter operated down to about 1 cm wavelength.

NRL were also involved in the U.S. Navy's 600 ft (180 m) radio telescope project whose construction had started at Sugar Grove, West Virginia, in the late 1950s. It was to be fully steerable and to operate down to 21 cm. In 1958 the estimated cost was $79 million. It was finally abandoned in 1962 when $96 million had been spent, and when its estimated cost had increased to more than $200 million. NRL also built a 60 ft (18 m) paraboloid at Sugar Grove in 1957 as a prototype for the 600 ft, and a 150 ft (46 m) constructed partly from materials left over from the 600 ft project. Sugar Grove is a military installation and these, as well as other antennae located there, have been used for military purposes.

As time went on, NRL stopped building its own state-of-the-art radio telescopes and began using U.S. national facilities, particularly those at Green Bank. The second NRL 85 ft radio telescope was also later used with other radio telescopes worldwide as part of the Very Large Baseline Interferometry (VLBI) network.

Bibliography

Hey, J. S., *The Evolution of Radio Astronomy*, Science History Publications, 1973.

Malphrus, Benjamin K., *The History of Radio Astronomy and the National Radio Astronomy Observatory: Evolution Toward Big Science*, Krieger Publishing Company, 1996.

McClain, Edward, *A High-Precision 85-Foot Radio Telescope*, Sky and Telescope, July 1966, pp. 4–6.

VERMILION RIVER OBSERVATORY, UNIVERSITY OF ILLINOIS

There was considerable confusion in the mid-1950s about the existence or otherwise of a number of cosmological radio sources as the Cambridge and Sydney radio source catalogues disagreed where their celestial coverage overlapped. At about this time the University of Illinois decided to build a radio telescope to help to resolve the discrepancy. This was the 400 ft (120 m) meridian transit telescope whose overall design was the responsibility of George Swenson.

The telescope consisted of a parabolic cylindrical reflector 600 ft (180 m) long by 400 ft (120 m) wide, aligned north-south, and made of wire mesh placed directly on the ground. Its downward-looking feed elements were mounted on a horizontal beam carried on 165 ft (50 m) high towers at the linear focus of the reflector. The telescope operated at about 610 MHz (49 cm wavelength) and was designed to reduce the confusion of radio sources by having low beam sidelobes. Although it was a stationary instrument, it could detect sources at various declinations by adjusting the phase of its feed elements.

The construction, which was begun in 1959, was finally completed three years later. By that time the differences between the Cambridge and Sydney catalogues had been largely resolved. But by the time that the 600 ft telescope was closed down in 1970 it had detected over 1,000 discrete radio sources beyond the Milky Way. It had also mapped many ionised hydrogen regions of the Milky Way and detected two previously unknown supernova remnants.

The University of Illinois had also planned in the late 1960s to build an interferometer consisting of three 120 ft (37 m) parabolic antennae. The first antenna, operating at 18 cm (1.67 GHz) and 49 cm (0.61 GHz), was built in-house, using staff and students to keep costs to a minimum. But during construction it became clear that funding would not be provided for the other two antennae. So, instead, the one antenna was used as an element in a very large baseline interferometer (VLBI) with other radio telescopes worldwide.

The Vermilion River Observatory was closed down in 1981.

Bibliography

McVittie, G. C., *Illinois' Vermilion River Observatory*, Sky and Telescope, December 1962, pp. 322–325.

Schriefer, Arno H., Yang Kwang-Shi, and Swenson, George W., Jr., *The Illinois' 120-Foot Radio Telescope*, Sky and Telescope, March 1971, pp. 132–138.

Swenson, George W., *The Illinois 400-Foot Radio Telescope*, IEEE Antennas and Propagation Society Newsletter, 28 (1986), 12–16.

VERY LARGE ARRAY

Studies were begun in the United States in the early 1960s into how to improve the resolution and sensitivity of radio telescopes, whilst making it easier to produce accurate maps of radio sources. This resulted in 1967 in a proposal from the National Radio Astronomy Observatory to build a very large radio telescope consisting of thirty-six 25 m (82 ft) diameter paraboloids in a Y-shaped array. This basic concept was endorsed in 1972 by the influential Whitford report of the National Research Council which put the array as its top-priority astronomy project for the next ten years. The project was approved by Congress in the same year.

The requirements placed on the site for what became the Very Large Array (VLA) aperture synthesis radio telescope were quite constraining. It needed to be at high altitude to minimise atmospheric effects. It also needed to be at least 36 km in diameter, relatively flat and at low latitude. Furthermore it should be remote from large population centres to minimise man-made interference, whilst being reasonably accessible for visiting observers. The best site meeting all these criteria was at an altitude of 2,200 m (7,000 ft) on the Plains of San Augustin near Socorro, New Mexico. Construction of the 27 antenna VLA was started there in 1974. Within three years the first six 25 m dishes were operational, and the completed Y-shaped array, with arms about 21 km length, was completed in 1981, allowing simulation of apertures of up to 36 km.

The 27 antennae were mounted on rails so that the configurations could be changed from the tightest, in which all antennae were within 650 m of the centre, to the widest using the full length of the rails. The smallest array gave the highest sensitivity and the largest array the highest resolution. The VLA initially operated at wavelengths of between 21 and 1.3 cm (frequencies 1.43 to 23 GHz). Readings were obtained from one pair of dishes at a time, which resulted in 361 pairs of measurements. Well-known sources were observed typically every 30 minutes for calibration purposes.

The range of available observational wavelengths has now been increased to 400−0.7 cm (75 MHz−43 GHz) to produce maximum resolutions of from 24 to 0.04 arcsec, respectively. The computer system has also been progressively modernised over the years so that sophisticated self-calibration procedures can now be undertaken and high-resolution images produced.

The VLA has been used to produce spectacular radio images of extended objects including protoplanetary discs, supernova remnants, radio galaxies and the central region of the Milky Way. Other observations have included those of the Sun, interstellar hydrogen, astrophysical masers, pulsars and quasars.

The Expanded VLA project was begun in 2001 to provide major improvements in the VLA's sensitivity, frequency coverage, bandwidth and spectroscopic capability. Operation of the partially-completed array began in 2011, with completion expected at the end of 2012.

Bibliography

Heeschen, David S., *The Very Large Array*, Sky and Telescope, June 1975, pp. 344–351.

Heeschen, D. S., The Very Large Array, in Burbidge, G., and Hewitt, A. (eds.), *Annual Reviews Monograph: Telescopes for the 1980s*, Annual Reviews Inc., 1981.

Malphrus, Benjamin K., *The History of Radio Astronomy and the National Astronomy Observatory: Evolution Toward Big Science*, Krieger Publishing Company, 1996.

Perley, R. A., et al., *The Expanded Very Large Array - A New Telescope for New Science*, Astrophysical Journal, 739 (2011), L1−L6.

Tucker, Wallace, and Tucker, Karen, *The Cosmic Inquirers: Modern Telescopes and Their Makers*, Harvard University Press, 1986.

VERY LONG BASELINE ARRAY

The National Radio Astronomy Observatory began planning in 1975 for a permanent array of ten radio telescopes distributed across the United States that would be exclusively devoted to very long baseline interferometry (VLBI). This resulted in the Very Long Baseline Array (VLBA) which was a key project recommended in the influential Field Report of the National Academy of Sciences in 1982. The VLBA was planned to consist of ten 25 m (82 ft) diameter parabolic antennae located at various places around the USA, including Hawaii, to give an angular resolution of 0.3 milliarcsecond (mas). It was approved by Congress in 1984.

The first of the 25 m dishes to be commissioned was at Pie Town, New Mexico, in 1988. In the following year this dish was linked to the 25 m VLBA dish at Kitt Peak, Arizona, to make the VLBA's first coordinated observation of an astronomical source. This was the radio galaxy 3C 84 which was observed at a wavelength of 6 cm (a frequency of 5 GHz). It had initially been expected that construction of the VLBA would take four years, but financial constraints meant that it was not completed until 1993, some nine years after project approval.

When completed, the VLBA extended 8,000 km east-west from the Virgin Islands to Hawaii, and 4,000 km north-south. It operated at ten frequencies from 325 MHz (90 cm wavelength) to 43 GHz (7 mm), and achieved resolutions of 22 to 0.17 mas. The system was controlled from a joint VLA-VLBA operations centre near Socorro, New Mexico where tapes were received with the recorded data. Non-VLBA radio telescopes at Arecibo, Green Bank, the Very Large Array, and Effelsberg (in Germany), were brought online from time to time to work with the VLBA to complete what was called the High Sensitivity Array.

The VLBA has observed the motion of masers near the centre of galaxies, including the Milky Way, enabling the masses of their possible black holes to be determined. It has measured the motion of gas in the circumstellar envelope of evolved stars, and has showed the evolution of magnetic field structures in active galaxy jets using high resolution polarimetry.

Bibliography

Gordon, Mark A., *VBLA - A Continent-Size Radio Telescope*, Sky and Telescope, June 1985, pp. 487–490.

Kellermann, K. I., and Thompson, A. R., *The Very Long Baseline Array*, Science, **229** (1985), 123–130.

Ulvestad, James, and Goss, Miller, *Radio Astronomy's Resolution Machine: The Very Long Baseline Array*, Sky and Telescope, December 1999, pp. 36–46.

PART 8
OTHER GROUND-BASED OBSERVATORIES

Other ground-based observatories

GRAVITY WAVE DETECTORS AND OBSERVATORIES

Gravity waves, predicted by Einstein's general theory of relativity, should be produced as ripples in space-time when large masses move rapidly. As a result they should be observable in very close neutron star or black hole binary systems, supernovae or other stellar collapses, rapid pulsars, and in newly forming or evolving quasars.

Joseph Weber was the first to try to detect gravity waves in the 1960s by attempting to measure vibrations in a large cylindrical bar of aluminium alloy mounted in a vacuum chamber. He looked for simultaneous vibrations in two bars located far apart, with one bar at the University of Maryland's College Park campus and the other about 1,000 km away at the Argonne National Laboratory. In 1969 he reported that he had detected statistically significant simultaneous excitations in the two bars. In the following year he also concluded that these excitations were more evident when the detectors were observing the centre of the Milky Way.

Weber's conclusions were highly controversial at the time, as his detectors were thought to be too insensitive to detect the extremely weak signals expected. Nevertheless a number of laboratories built similar apparatus, some of which had their bars cooled to liquid helium temperatures to reduce thermal noise. A number of observers reported coincidence observations, some of these apparently connected with the galactic centre. But there were also a number of negative results, making it unclear as to whether gravity waves had really been detected.

In parallel with these experiments with Weber-type bars, a number of research groups had unsuccessfully tried to detect gravity waves using a Michelson laser interferometer with evacuated orthogonal arms. In 1975 the Max Planck Institute for Astrophysics in Munich started with a prototype with 3 m arms. Five years later the University of Glasgow started operating a 10 m prototype, whilst in 1983 the Max Planck Institute of Quantum Optics in Garching produced a 30 m instrument. At about the same time the California Institute of Technology built a 40 m system.

Meanwhile in 1974 Joseph Taylor and Russell Hulse had detected the first binary pulsar (PSR 1913+16). It appeared to be a 1.4 solar mass neutron star which was in orbit around its common centre of mass with another neutron star of about the same mass. By 1979 Taylor and his colleagues had found that the orbit's size and period were shrinking at the rate expected by the emission of gravity waves. This was the first clear evidence of their existence.

So Einstein seemed to be correct and gravity waves appeared to exist. But Earth-bound detectors had, at that time, been unable to convincingly detect them reaching Earth. As a result a number of investigators began to design more sensitive equipment. For example, in the 1980s a joint Caltech/Massachusetts Institute of Technology (MIT) team began to examine the possibility of building interferometers with 4 km arms at sites in California and New England.

Eventually both a 4 km and 2 km Caltech—MIT gravity wave detector was built around the turn of the century at Hanford, Washington, and a 4 km device was built at Livingston, Louisiana. These constituted the Laser Interferometer Gravitational—Wave Observatory or LIGO. No gravity wave events had been unambiguously detected by 2008 when a major upgrade was begun to give more than a factor of ten increase in sensitivity.

Other gravity wave observatories also became operational at about the turn of the twentieth century. TAMA, near Tokyo, was a 300 m instrument with the first of a new generation of detectors. It became operational in 1999, followed by GEO, a 600 m German/UK instrument built at Hanover, and VIRGO, an Italian/French 3 km instrument at Pisa. It is estimated that operating all these instruments together would result in gravitational wave sources being located to within about $1/2°$. But so far (2010) no such gravitational waves have been unambiguously detected on Earth.

Bibliography

Bartusiak, Marcia, *Einstein's Unfinished Symphony: Listening to the Sounds of Space-Time*, Joseph Henry Press, 2000.

Ciufolini, I., et al. (eds.), *Gravitational Waves*, Institute of Physics Publishing, 2001.

Fidecaro, Francesco, and Ciufolini, Ignazio (eds.), *Gravitational Waves: Sources and Detectors: Proceedings of the International Conference, Cascina, Italy, 19–23 March, 1996*, World Scientific Publishing, 1997.

Sanders, Gary H., and Beckett, David, *LIGO: An Antenna Tuned to the Songs of Gravity*, Sky and Telescope, October 2000, pp. 40–48.

Trimble, Virginia, *Gravity Waves: A Progress Report*, Sky and Telescope, October 1987, pp. 364–369.

GROUND-BASED COSMIC-RAY OBSERVATORIES

The main methods of detecting cosmic ray air showers on Earth
involved observing either the Čerenkov radiation or the lumi-
nescence (fluorescence, as it is now termed) produced in the
Earth's atmosphere by secondary particles. Alternatively, the
secondary particles could be detected directly using ground-
based particle detectors such as Geiger counters, scintillation
counters and water-Čerenkov detectors.

Particle detectors had been used to detect cosmic rays
for some time when Patrick Blackett suggested in 1949 that
Čerenkov radiation produced in the Earth's atmosphere by
cosmic ray showers should also be detectable. A few years
later Bill Galbraith and John Jelley at Harwell, UK, detected
this atmospheric Čerenkov radiation using a simple detector consisting of a photomultiplier tube (PMT) at the focus
of a 25 cm (10 inch) parabolic mirror. They found that the
pulses of Čerenkov radiation that they observed correlated
with cosmic radiation detected by a nearby Geiger counter
array.

The first attempts to detect cosmic ray showers by air fluorescence were made by Kenneth Greisen and colleagues at
Cornell University in the mid-1960s. But their detector was
not sensitive enough, and it was not until 1976 that physicists
from the University of Utah were able to detect fluorescence
from cosmic-ray air showers. They used three prototype detectors at Volcano Ranch, New Mexico, each of which consisted of
an array of PMTs at the focus of a 1.8 m (6 ft) diameter mirror.
Their prototype fluorescence detectors observed air showers
that correlated with those detected by a nearby array of particle
detectors.

It was clear that it would be impossible to determine the
initial source of all but the highest-energy cosmic rays because
their tracks in space would be distorted by magnetic fields in
the Milky Way and elsewhere. But this should be less of a
problem for cosmic rays with energies above about 10^{19} eV.
As a result, many early cosmic ray observatories attempted to
determine the direction of these ultra-high-energy cosmic rays
(UHECRs) as they hit the Earth's atmosphere, to enable their
cosmic sources to be found.

Then in 1966, shortly after the discovery of the cosmic
microwave background (CMB) radiation, Kenneth Greisen
and, independently, Georgiy Zatsepin and Vadim Kuz'min
proposed what was called the GZK effect. They suggested
that the interaction between this CMB and very-high-energy
protons in cosmic rays should result in the energy spectrum
of cosmic rays steepening above about 5×10^{19} eV. This
assumed that the sources of these very-high-energy protons
were distributed uniformly throughout the universe. As a
result it became important to not only determine the original

direction of UHECRs, as previously, but also measure their
energy spectrum to confirm the GZK effect.

Although atmospheric-Čerenkov detectors were useful for
detecting medium energy cosmic rays, they were not the preferred method for detecting UHECRs. For these UHECRs, so
called water-Čerenkov arrays, which detected Čerenkov radiation produced by particle impacts in water, fluorescence arrays,
and particle detector arrays, particularly of scintillators, were
more suitable.

The earliest arrays to detect UHECRs were the MIT scintillator array at Volcano Ranch, New Mexico, which was in
operation from 1959 to 1963; the world's first water-Čerenkov
array of detectors at Haverah Park, UK, operational from 1968
to 1987; the SUGAR (Sydney University Giant Airshower
Recorder) scintillator array in Australia, operational from 1968
to 1979; and the scintillator and atmospheric Čerenkov array at
Yakutsk, Soviet Union, operational since 1970. These detected
cosmic rays above 10^{19} eV and gave the first indications that
there might be cosmic rays with energies above 10^{20} eV, which
was well above the predicted GZK steepening or GZK cut-off,
as it was generally called. Subsequent arrays included the Fly's
Eye fluorescence array, operational at Dugway, Utah, from
1981 to 1992, which detected a 3.2×10^{20} eV cosmic ray in
October 1991, and its extension HiRes. The AGASA (Akeno
Giant Air Shower Array) scintillator array of the University
of Tokyo, which started operation in 1991, also detected trans-
GZK particles above 10^{20} eV.

The detector arrays available in the 1990s were not large
enough to detect enough UHECRs to be able either to determine what type of objects emitted them or to determine their
energy spectrum above the GZK cut-off. As a result a major
international project was begun in the mid-1990s to build what
came to be known as the Pierre Auger Observatory to detect
large numbers of UHECRs.

The Pierre Auger Observatory, situated on Pampa Amarilla
near Malargüe, Argentina, was the result of an international
collaboration of 19 countries (Argentina, Armenia, Australia,
Bolivia, Brazil, China, Czech Republic, France, Germany,
Greece, Italy, Mexico, Poland, Russia, Slovenia, Spain, United
Kingdom, USA and Vietnam). It consisted of 1,600 water-
Čerenkov detectors, each filled with 12 tonnes of water,
deployed on a 1.5 km hexagonal grid which covered an area of
about 3,000 km². In addition, there were six fluorescence telescopes located on each of four hills overlooking the array. Each
fluorescence telescope had a 11 m² mirror which reflected light
onto a camera with 440 PMTs. Data collection started in 2004
with 125 water-Čerenkov detectors and 6 fluorescence telescopes operational. The array was completed five years later.

Bibliography

Dorman, Lev I., *Cosmic Rays in the Earth's Atmosphere and Underground*, Kluwer Academic Publishers, 2004.

Prouza, Michael, *Studies of the Ultra-High Energy Cosmic Ray Composition at the Pierre Auger Observatory*, Europhysics Conference on High Energy Physics, July 16–22, 2009, Krakow, Poland, http://pos.sissa.it/archive/conferences/084/106/EPS-HEP%202009_106.pdf .

Watson, Alan A., *Catching the Highest Energy Cosmic Rays*, Astronomy and Geophysics, 50, April 2009, pp. 2.20–2.27.

Wilkinson, Christopher R., *The Application of High Precision Timing in the High Resolution Fly's Eye Cosmic Ray Detector*, PhD Thesis, University of Adelaide, http://www.cosmic-ray.org/thesis/wilkinthesis.pdf.

See also: Cosmic rays.

GROUND-BASED GAMMA-RAY OBSERVATORIES

Patrick Blackett suggested in 1949 that about 0.01% of night-sky photons were the result of Čerenkov light emitted by cosmic rays and their secondary particles as they traversed the Earth's atmosphere. A few years later Blackett brought this prediction to the attention of Bill Galbraith and John Jelley at Harwell, UK, who, within a week, had built a rudimentary detector consisting of a photomultiplier tube (PMT) at the focus of a 25 cm (10 inch) parabolic mirror. They obtained pulses from the night sky about every two minutes which correlated with the cosmic radiation detected by the nearby Harwell Geiger counter array.

It was clear that the tracks of all but the highest-energy charged cosmic-ray particles would be effected by magnetic fields in the Milky Way, so it would be impossible to determine the original source in space of normal-energy cosmic-ray particles. But this would not be the case for gamma rays, whose secondaries would also produce Čerenkov radiation. Fortunately Čerenkov radiation preserved much of the original direction of the secondary particles. So observing this Čerenkov radiation produced by these secondary particles should enable the original direction of the gamma-ray primary to be determined. In addition, the intensity of the Čerenkov light was proportional to the total number of secondary particles, and hence the energy of the primary gamma ray. Finally, as the lateral spread of Čerenkov radiation at ground level from a gamma-ray air shower was only ∼100 m, only a relatively small detector array would be required to detect very-high-energy gamma rays.

The Lebedev Research Institute in the Crimea produced a simple atmospheric Čerenkov telescope (ACT) using searchlight mirrors in the early 1960s to try to detect very-high-energy gamma rays. They were hoping to detect such gamma rays from supernova remnants, like the Crab nebula, and from radio galaxies. But, although many air showers were detected, no individual gamma-ray sources could be isolated from the cosmic-ray background.

The first major purpose-built optical reflector for gamma-ray astronomy was the Smithsonian's 10 m (33 ft) parabolic reflector built in 1968 at an altitude of 7,700 ft (2,350 m) on Mount Hopkins, Arizona. This Whipple Observatory reflector consisted of 248 small hexagonal mirrors that focused the Čerenkov light onto a PMT at their focus. One of the problems with this and other Čerenkov detectors was how to discriminate Čerenkov emission produced by gamma-ray air showers from that produced by the more numerous cosmic-ray showers. A number of techniques were proposed. In particular A. M. Hillas suggested in 1985 that the images produced by gamma-ray showers would in most cases be significantly different from those produced by cosmic rays. The Whipple 10 m telescope had had a 37 pixel camera fitted a few years earlier, and this finally enabled Trevor Weekes to detect the first clear source of gamma rays, namely the Crab nebula in 1989. Their energies exceeded 0.7 TeV. This detection was followed three years later, using the Whipple Observatory, by the first detection of an extragalactic source, Markarian 421. The intensity of the TeV gamma rays emitted by this blazar were found to vary, like in other wavebands, in just two days.

The Crab detection provided a welcome boost to ground-based gamma-ray observing which had been in the doldrums for many years. Such ground-based imaging atmospheric Čerenkov telescopes (IACTs) as the Whipple 10 m had a natural advantage over their space-based counterparts in that, by their very size, they could have much greater sensitivity. They were also able to detect much higher-energy gamma rays via Čerenkov radiation.

As a result a number of new ground-based gamma-ray telescopes were then built, including HEGRA (High Energy Gamma Ray Astronomy) on La Palma, which was the result of a Spanish/German/Armenian collaboration. The first prototype telescope was installed in 1992, and the full five IACT telescope array was completed in 1998. All of these five telescopes were identical, each having a 3.4 m diameter mirror and a 271 pixel camera of PMTs. Four of the telescopes were located at the corners of a square of side 100 m, with one in the middle, to provide stereoscopic imaging. This array improved both the discrimination of background cosmic-ray showers, and the energy resolution of the incident gamma rays. HEGRA achieved an angular resolution of about 0.1° for observations above 0.5 TeV.

A multinational team built the HESS (High Energy Stereoscopic System) array of IACTs in the Khomas Highland of Namibia, using the experience gained with HEGRA. Namibia was chosen as not only did the centre of the Milky Way pass almost through the zenith, but the sky in its highlands was also very clear. Its first 12 m telescope was operational in 2002 and a four-telescope array just two years later.

In parallel the Australian/Japanese CANGAROO (Collaboration of Australia and Nippon [Japan] for a GAmma Ray Observatory in the Outback) consortium had built a 3.8 m diameter gamma ray telescope in the early 1990s. They then built a 10 m around the end of the decade, and CANGAROO III, consisting of four 10 m IACTs, which became operational in 2004. Like HESS these were especially useful for observing sources near the centre of the Milky Way.

HESS and CANGAROO were both considered third-generation atmospheric Čerenkov arrays. Two other third-generation arrays were MAGIC (Major Atmospheric Gamma-ray Imaging Čerenkov Telescope) and VERITAS (Very Energetic Radiation Imaging Telescope Array System). The 17 m diameter European MAGIC array was, like its predecessor HEGRA, also built on La Palma. It began routine operations in 2004, with a second MAGIC telescope (MAGIC 2) starting operations five years later 85 m from the first. It was sensitive to gamma rays with energies between 50 GeV and 50 TeV. On the other hand the VERITAS gamma-ray array was funded by American, Canadian, Irish and United Kingdom organisations. It consisted of four 12 m IACTs at the 1,300 m (4,250 ft) high basecamp of the Whipple Observatory in Arizona. VERITAS, which became operational in 2007, operated from 100 GeV to 30 TeV.

Two other basic configurations of ground-based gamma ray observatories have been built in the recent past: one like CELESTE (Čerenkov Low Energy Sampling and Timing Experiment), see below, used the mirrors of existing solar power facilities, and the other, like Milagro (also outlined below), used particle detectors.

The CELESTE system was designed in the late 1990s to fill the energy gap from about 30 to 300 GeV that then existed between space-based and ground-based gamma-ray observatories. It used 40 heliostats of the Themis solar power facility in the Pyrenees and was operational from 1999 to 2004. A similar system called STACEE (Solar Tower Atmospheric Čerenkov Effect Experiment) used the solar thermal test facility in New Mexico starting also in 1999. Amongst other things both facilities provided valuable new information on the Crab nebula and on the blazars Markarian 421 and 501.

If the energy of a gamma ray is high enough, enough particles from the subsequent air shower will survive to ground level to be detected. Unlike IACTs, such particle detector arrays can be used in daylight, but they have poor angular resolution and greater difficulty in rejecting particles from cosmic-ray showers. The Milagro telescope, which was built at an altitude of 2,600 m near Los Alamos, New Mexico in the late 1990s, was such a detector. It had an energy response from 110 GeV to 100 MeV.

The Milagro telescope consisted of a central water tank of about 4,800 m^2 (1.2 acre) area, which was covered in a light-tight barrier, surrounded by an outrigger array of much smaller tanks. The main tank had about 750 PMTs arranged in two layers, one at a depth of 1.4 m and the other at 6 m. This arrangement enabled particles produced in gamma-ray showers to be distinguished from those produced in cosmic-ray showers. Milagro was the first observatory to make measurements of the Milky Way's diffuse emission in the TeV energy band.

Bibliography

Jelley, John V., and Weekes, Trevor C., *Ground-Based Gamma-Ray Astronomy*, Sky and Telescope, September 1995, pp. 20–24.

Ribó, Marc, *TeV Gamma-Ray Astrophysics, Chinese Journal of Astronomy and Astrophysics*, 8 (2008), Supplement, 98–108.

Weekes, T. C., *Very High Energy Gamma-Ray Astronomy*, Institute of Physics Publishing, 2003.

Weekes, Trevor C., *The Atmospheric Cherenkov Imaging Technique for Very High Energy Gamma-ray Astronomy*, 2005, preprint (astr-ph/0508253).

NEUTRINO OBSERVATORIES

Homestake observatory

Raymond Davis of the Brookhaven National Laboratory devised a facility in the early 1960s to detect neutrinos that were thought to be emitted by the Sun. It consisted of an 85,000 gallon (380,000 litre) tank of dry-cleaning fluid, tetrachloroethylene, which was to detect the neutrinos by measuring the amount of argon produced as they interacted with the fluid. The tank was large, as the number of neutrinos expected to be detected was very small (only about one in every three days with this tank), and it was placed deep underground in the Homestake gold mine in South Dakota to reduce spurious signals from energetic cosmic rays.

The facility, which was commissioned in 1967, appeared to detect fewer neutrinos than expected. As a result it was modified in 1970 to allow more accurate results to be obtained. These confirmed that the number of neutrinos was only about a third of that expected. Something was wrong either with the experiment or with the theories explaining how the Sun generates its heat. Alternatively, as Bruno Pontecorvo had suggested in the late 1960s, some of the electron neutrinos emitted by the Sun may be transformed to muon neutrinos en route to the Earth which would not be detected by Davis' equipment. Later the existence of tau neutrinos was postulated which gave another possible transformation route for solar electron neutrinos.

Kamioka observatory

The construction of Kamioka Underground Observatory, which was built to find whether protons decay, was completed in Japan in 1983. Its detector, Kamiokande (Kamioka Nucleon Decay Experiment), was a tank about 16 m (52 ft) high and 16 m wide which contained 3,000 tons (3 million litres) of pure water and about 1,000 photomultiplier tubes (PMTs). The detector was modified in 1985, to become Kamiokande-II, to allow it to detect solar neutrinos.

Kamiokande-II detected neutrinos from the supernova SN 1987A in the Large Magellanic Cloud, and detected solar neutrinos for the first time in 1988. Unlike the Homestake detector, experimenters using Kamiokande-II were able to detect the direction of the neutrinos, and show for the first time that the Sun was clearly a source. It also confirmed Davis' result that the number of solar electron neutrinos was less than theory predicted.

Kamiokande-II was, however, unable to achieve its primary goal to detect proton decay. So it was replaced by a much larger detector, Super-Kamiokande, which started operations in 1996. Two years later it provided the first evidence for neutrino oscillations from one type to another, showing for the first time that neutrinos may not be massless particles. The evidence was indirect but, if confirmed, it would solve the solar neutrino problem, as one sort of neutrino could change to another en route from the centre of the Sun to the Earth, to give spurious results in the earlier neutrino detectors.

Super-Kamiokande was seriously damaged in an accident in 2001 when about half of its PMTs imploded. The PMT design was modified and the broken tubes replaced to produce SuperKamiokande-III in 2006.

Sudbury neutrino observatory

Meanwhile in 1984 Herb Chen had suggested using heavy water as a detector for solar neutrinos as, unlike previous detectors, it was sensitive to all types of neutrino. Such a detector was built about 6,800 ft (2,000 m) underground at Sudbury, Canada at the Sudbury Neutrino Observatory which became operational in 1999. The detector consisted of 1,000 tons of heavy water surrounded by almost 10,000 PMTs. In 2001 it found that the total number of all types of neutrino was as predicted by nuclear physics. So Pontecorvo had been correct as the neutrinos were changing type en route to the Earth.

Other neutrino observatories

Neutrino observatories have also been built to try to detect neutrinos emitted by ultra-energetic astronomical sources such as supernovae (like SN 1987A) and gamma ray bursts.

One of the most interesting of these 'cosmological' neutrino arrays was AMANDA (Antarctic Muon and Neutrino Detec-tor Array) built at the South Pole as the result of a suggestion by Francis Halzen and John Learned in the late 1980s. Its detector initially consisted a number of kilometre-long strings each lowered into a separate hole drilled in the ice, with 20 PMTs attached to the bottom 200 m of each string. Each hole was then allowed to freeze, with the ice acting in place of the fluid in the earlier detectors. The idea was to use the Earth to screen out background noise, with the detector array detect-ing neutrinos from the Northern, rather than the Southern, Hemisphere. AMANDA, which was first operated in 1994, was expanded as AMANDA-II to consist of over 600 PMTs mounted on 19 separate strings spread out in a rough circle of diameter 200 m. It detected hundreds of neutrinos, with an energy in excess of 50 GeV, coming from all over the north-ern sky. The neutrinos did not appear to cluster in an astro-nomical significant way. In 2005 AMANDA-II became part of its successor project IceCube before being decommissioned in 2009.

The intention of IceCube was to build an array similar to AMANDA but on a much larger scale, with the detector array frozen into about 1 cubic kilometre of ice. This was to be achieved by drilling 86 holes, each 2,450 m (8,040 ft) deep, covering a hexagon of surface with sides 800 m long. The plan was to put a string in each hole with the bottom 1,000 m being instrumented by 60 PMTs. Construction was completed in December 2010.

Bibliography

Abe, K., et al., *Solar Neutrino Results in Super-Kamiokande-III*, Phys-ical Review D, **83** (2011), 052010.

Anonymous, *The Sudbury Neutrino Observatory - Canada's Eye on the Universe*, CERN Courier, December 4, 2001.

Bahcall, John N., and Davis, Raymond, *Solar Neutrinos: A Scientific Puzzle*, Science, **191** (1976), 264–267.

Semeniuk, Ivan, *Astronomy and the New Neutrino*, Sky and Telescope, September 2004, pp. 42–48.

Suzuki, Y., and Totsuka, Y., (eds.), Neutrino Physics and Astro-physics, Proceedings of the XVIII International Conference on Neutrino Physics and Astrophysics, Takayama, Japan, 4–9 June, 1998. Elsevier Science, 1999.

PART 9
SOLAR SYSTEM EXPLORATION SPACECRAFT

1 · Overview – Solar system exploration spacecraft

The first Earth-orbiting spacecraft

The Space Age began on 4 October 1957 when the Soviet Union launched the 85 kg Sputnik 1 spacecraft into a low Earth orbit as part of their contribution to the International Geophysical Year (IGY). This 'year' was due to last from 1 July 1957 to the end of 1958. Although the Russians had promised to launch a spacecraft as part of the IGY, their launch of Sputnik 1 was not announced beforehand, and the fact that they had managed to launch a spacecraft before the Americans was a big surprise to everyone in the West. In addition, Sputnik 1's mass, although modest, was far higher than that of the intended American spacecraft. The potential military significance of this Russian achievement was not lost on the American public nor on the West in general.

The launch of Sputnik 1 was followed by that of the 500 kg Sputnik 2 a month later. This clearly showed that Russian rockets were powerful enough, when used as intercontinental ballistic missiles, to launch a nuclear warhead against the USA. As a result on 25 November 1957 the U.S. Senate set up a committee chaired by Lyndon B. Johnson, the Senate majority leader, to review the whole range of America's defence and space programmes. This was to result, in the following year, in the creation of the National Aeronautics and Space Administration (NASA) to define and run the civilian American space programme.

In the meantime, America's first attempt to launch an Earth-orbiting spacecraft had turned into a disaster on 6 December 1957 when the U.S. Navy's Vanguard launcher exploded. But on 1 February 1958 (Universal Time) Wernher von Braun's U.S. Army team successfully launched the Jet Propulsion Laboratory's (JPL's) Explorer 1 spacecraft. They used a modified Jupiter C intermediate range ballistic missile with a modified Redstone as first stage. The instruments on Explorer 1 included a Geiger counter developed by James Van Allen of Iowa State University.

The launch of Explorer 2 failed, but that of Explorer 3 was a success with a payload almost identical to that of Explorer 1. On 1 May 1958 Van Allen announced, using the combined results of these two successful spacecraft, that he had discovered a belt of elementary particles, later called the Van Allen belt, around the Earth. At about the same time the first successful Vanguard spacecraft discovered that the Earth was pear-shaped. These were the first purely scientific discoveries of the Space Age. But the provision of scientific results was not the primary motivation of either the Russian or American space programmes. Once the space race had started, both superpowers became more and more interested in prestige than in scientific results, demonstrating, as they saw it, the advantages of their own political systems, which they were then trying to export to countries in the unaligned third world.

Early lunar spacecraft

In the meantime, three weeks after the launch of Sputnik 1, and before the American's first successful spacecraft launch, William Pickering, the director of JPL, had pushed for the approval of a plan to send a spacecraft to the Moon. Although his proposed plan was not approved, on 27 March 1958 the secretary of defense, Neil McElroy, approved America's first lunar programme of five Pioneer spacecraft. In the event only Pioneer 4 managed to get as far as the Moon, missing the Moon by about 60,000 km, somewhat more than planned, in March 1959. In the meantime, the Russians had managed to fly-by the Moon with Luna (or Lunik) 1 at a distance of just 5,960 km in January 1959. Then in September of that year the Russian Luna 2 was the first spacecraft to impact the Moon. This was followed by Luna 3, which was launched on 4 October 1959, precisely two years after the launch of Sputnik 1. Luna 3 was the first spacecraft to take photographs of the far side of the Moon.

In all there had been twelve launches of lunar probes in the first two years of the Space Age, seven Russian and five American, only three of which (Lunas 1, 2 and 3) were really successful. Engineers in both countries had learnt a great deal about how to build, launch and control spacecraft in orbit. But the scientific results had been rather modest. The configuration of the Van Allen belts had been better understood, the Moon had been found to have no measurable magnetic field, and the far side of the Moon had been found to have fewer mare regions and more craters than the near side. In addition, Luna 2 had been the first spacecraft to measure the solar plasma away from the influence of the Earth's magnetic field. The next ten lunar launch attempts up to the end of 1962, eight American and two Russian, were all failures.

The first missions to Mars and Venus

In the meantime, the first planetary probe, Korabl 4, had been launched by the USSR on 10 October 1960. It and its twin, which was launched a few days later, were intended to reach Mars, but their launchers failed. In the following year the Russians were also the first to try to reach Venus, but these two missions failed also.

The American plans to reach Venus had changed over the years. In November 1958 they had planned to build two 170 kg Pioneer spacecraft to be launched during the June 1959 Venus launch window. But in January 1959 the Russians had successfully launched Luna 1 to the Moon. So NASA decided to redesignate their so-called 'heavy' Pioneers as lunar spacecraft to catch up with the Russians. (Subsequently the launch of these heavy Pioneers failed.) Having missed the 1959 Venus launch window NASA decided in 1960 to develop a series of spacecraft for launches in the next three Venus launch windows of mid-1962, early 1964 and late 1965. These spacecraft were given the generic name of Mariner; the 1,000 kg Mariner A was to fly-by Venus, Mariner B was to eject a probe to crash-land or soft-land on Venus or Mars, and the smaller Mariner R, which was based on the lunar Ranger spacecraft, was to fly-by Venus. Over the next two years NASA kept changing the target of their spacecraft from Venus to Mars and back again. In the event they launched two spacecraft called Mariners 1 and 2, based on the Mariner R design, towards Venus in the summer of 1962.

The launch of Mariner 1 was a failure but that of its twin, the 204 kg Mariner 2, was a success. When it flew past Venus at a distance of about 35,000 km on 14 December 1962 it was the first successful planetary spacecraft from either the USA or USSR. Mariner 2 measured the solar wind en route to the planet. At Venus it found that the planet's surface temperature appeared to be about 450°C (about 725 K) and its surface atmospheric pressure about 20 bars.

Mariner 2 was a big psychological boost to the American space programme as, for the first time, they had managed to upstage the Russians by building the first spacecraft to successfully fly past another planet. In the same year three Soviet attempts to reach Venus and two to reach Mars had all failed. So far the Americans had not tried to reach Mars.

Ranger, Lunar Orbiter, Apollo and the Russian response

The Americans were beginning to realise, after their first unsuccessful attempts to reach the Moon, that a more systematic approach was required. This resulted in a programme of five Ranger spacecraft being approved in 1960 for launch in 1961–62. Two of these spacecraft were designed to traverse a highly eccentric orbit with an apogee beyond the Moon's

orbit, whilst the other three were each to include a capsule to land on the Moon. However, before any of these could be launched, in May 1961 President John F. Kennedy committed the United States to a programme to land a man on the Moon and return him safely to Earth. As a result, the Ranger spacecraft were redesigned to support the manned programme. New Ranger spacecraft were also approved with a mission to take high-resolution images of the Moon as they crashed onto its surface.

In the event, all of the original five Ranger spacecraft failed. But three of the four new Ranger missions were successful, imaging craters on the Moon down to about 1 m diameter just before impact. They indicated that the surface dust, which had been a potential concern to the manned landing programme, was not very deep.

Ranger was followed by the Surveyor programme of soft landers to take images from the surface of the Moon, and Lunar Orbiter to provide high-resolution images from orbit to help NASA to locate suitable places for the astronauts to land. Ranger, Surveyor and Lunar Orbiter were followed by the Apollo manned lunar programme, culminating on 20 July 1969 in the first manned landing on the Moon.

Apollo and its predecessor programmes provided some interesting scientific information on the Moon. For example, the lunar material brought back by Apollo 11 from the Sea of Tranquillity (Mare Tranquillitatis) was found to be generally of igneous basalt, indicating that the Moon's dark maria were of volcanic origin. Breccias (impact rocks) were also found indicating that the lunar craters were the result of meteorite impacts. Apollo 15 found a sample of the original lunar crust which was about 4.5 billion years old, and very small green-coloured glass spheres embedded in some lunar rocks. These glass spheres seem to have been brought to the surface from deep in the Moon by fire fountains. But the main purpose of the Apollo programme had been political, proving to the world that America had overtaken the Soviet Union in space and was the world's leading technical superpower.

The Russians had, in the meantime, followed up their early Luna missions by being the first to soft-land a capsule on the Moon with Luna 9 in February 1966. Two months later Luna 10 had been the first spacecraft to orbit the Moon, but their Luna programme had then been overtaken by the Americans. Nevertheless the Russians continued their programme, concentrating on bringing small amounts of lunar soil back to Earth with automatic sample-return missions. The last one, Luna 24, retrieved core samples in 1976 at depths of up to 2.3 m from the Mare Crisium.

Later missions to the Moon

The hectic American and Russian lunar spacecraft programmes of the 1960s and early 1970s was followed by a lull lasting

Table 9.1. *Selected solar system spacecraft missions*

(A) Selected lunar spacecraft missions

Spacecraft	Country or institution	Launch date	Mission	Comments
Luna 1	USSR	2 Jan. 1959	Impact	First to fly-by Moon.
Luna 2	USSR	12 Sept. 1959	Impact	First to impact Moon.
Luna 3	USSR	4 Oct. 1959	Fly-by	First to photograph Moon's far side.
Ranger 7	USA	28 July 1964	Impact	First to photograph Moon during impact sequence.
Luna 9	USSR	31 Jan. 1966	Lander	First Moon soft landing. Transmitted images from lunar surface.
Luna 10	USSR	31 Mar. 1966	Orbiter	First to orbit Moon.
Surveyor 1	USA	30 May 1966	Lander	Designed to find safe landing areas for Apollo manned spacecraft.
Luna Orbiter 1	USA	10 Aug. 1966	Orbiter	Imaged Moon to find landing sites for Apollo astronauts.
Apollo 8	USA	21 Dec. 1968	Manned	First manned spacecraft to orbit Moon.
Apollo 11	USA	16 July 1969	Manned	First manned lunar landing.
Luna 16	USSR	12 Sept. 1970	Sample return	First automatic sample return.
Luna 17	USSR	10 Nov. 1970	Rover	First automatic lunar rover to land on Moon.
Apollo 15	USA	26 July 1971	Manned	First manned lunar rover.
Apollo 17	USA	7 Dec. 1972	Manned	First geologist to land on Moon.
Clementine	USA	25 Jan. 1994	Orbiter	Produced first lunar compositional and topographical map.
Lunar Prospector	USA	7 Jan. 1998	Orbiter	Impacted polar region of Moon at end of life to try to detect water ice. None found.
Chandrayaan-1	India	22 Oct. 2008	Orbiter and Impactor	Detected evidence of water ice in permanently shadowed craters at lunar north pole.
Lunar Reconnaissance Orbiter/LCROSS	USA	18 June 2009	Orbiter and Impactor	Detected first clear evidence of water ice on Moon.

almost twenty years in launching spacecraft to the Moon. This was broken by the American Clementine spacecraft, which was launched in 1994. Subsequently a series of modest-sized spacecraft was sent to the Moon by America, the European Space Agency (ESA), Japan, China and India. These were a mixture of proof-of-concept technological missions, and others to try to detect water, in the form of ice, in the Moon's permanently shadowed polar craters. Some of these missions provided circumstantial evidence of such water. But water was finally unambiguously detected on the Moon by the NASA Lunar Reconnaissance Orbiter/ Lunar Craft Observation and Sensing Satellite or LRO/LCROSS mission in 2009.

Table 9.1A lists a selection of the more important lunar spacecraft.

Mars

We left American and Russian attempts to reach Mars and Venus with NASA's Mariner 2 which flew past Venus in 1962 as the world's first successful planetary mission.

Following this success, NASA decided to send their next planetary spacecraft to Mars. They prepared two identical spacecraft, Mariners 3 and 4, with a television camera as their main payload. The launch of Mariner 3 failed, but Mariner 4 successfully flew past Mars in July 1965, the first spacecraft to do so. It showed a desolate-looking, crater-covered landscape, much like the Moon, with a surface atmospheric pressure of just 4–7 millibars.

Four years later Mariners 6 and 7 found that, although Mars' atmosphere was almost 100% carbon dioxide, it contained detectable amounts of water vapour. Liquid water could clearly not exist on the surface today because of Mars' low atmospheric pressure. But astronomers were anxious to find if there was any evidence that significant amounts of water had existed on the surface in the past, as this could have helped life to form and develop on Mars.

In 1971 NASA's Mariner 9, which was the first spacecraft to orbit Mars, showed a much more interesting landscape than the previous Mariners. It imaged large volcanoes, a giant canyon system and sinuous channels that looked very much like dried-

Table 9.1. *(cont.)*

(B) Selected Mars spacecraft missions

Spacecraft	Country or institution	Launch date	Arrival date	Comments
Mariner 4	USA	28 Nov. 1964	15 July 1965	First to fly-by Mars. Imaged cratered surface. Surface atmospheric pressure found to be very low.
Mariner 9	USA	30 May 1971	14 Nov. 1971	First to orbit Mars. Imaged Olympus Mons and Valles Marineris.
Viking 1	USA	20 Aug. 1975	19 June 1976	Orbiter and landing module. Latter contained miniature biological laboratory to search for life. No clear evidence found.
Viking 2	USA	9 Sept. 1975	7 Aug. 1976	Repeat of Viking 1's mission.
Mars Global Surveyor	USA	7 Nov. 1996	12 Sept. 1997	Orbiter included high-resolution camera and laser altimeter. Detected surface changes.
Mars Pathfinder	USA	4 Dec. 1996	4 July 1997	Minimum-cost mission. Included small Sojourner surface rover.
Mars Odyssey	USA	7 Apr. 2001	24 Oct. 2001	Orbiter. Found evidence of water in Mars' topsoil.
Mars Express	ESA	2 June 2003	25 Dec. 2003	Orbiter. Found further evidence of subsurface water on Mars. Detected atmospheric methane.
Mars Exploration Rover-Spirit	USA	10 June 2003	3 Jan. 2004	Surface rover. Found evidence of previous surface water.
Mars Exploration Rover-Opportunity	USA	8 July 2003	25 Jan. 2004	Surface rover. Found evidence of previous surface water.
Mars Reconnaissance Orbiter	USA	12 Aug. 2005	10 Mar. 2006	Orbiter. Found evidence of subsurface water ice at mid-latitudes.
Phoenix Mars Lander	USA	4 Aug. 2007	25 May 2008	Lander first to directly detect and image surface water ice.

up riverbeds. The Russian Mars 2 and 3 spacecraft, which arrived at Mars at about the same time as Mariner 9, produced very limited results. In fact, compared with the Russian Venus spacecraft (see below), their Mars spacecraft programme was only a limited success.

Ever since the true nature of Mars had been understood as a planet, mankind had wondered if it was inhabited by living things. However, by the time that the Space Age started it was clear that, if Mars had any life on its surface, it must be very rudimentary. NASA had managed to orbit Mars with Mariner 9, so the next task was to build a soft-lander to search for signs of life. This was the mission of the twin Viking spacecraft, launched in 1975, each of which consisted of a lander and an orbiter. Both landers were highly successful, but their complex biological laboratory payloads produced conflicting evidence for signs of life. At first some of the results appeared to be positive, but as time went on doubts began to increase. So today it is generally accepted that Viking did not find any evidence for life on Mars. On the other hand, the Viking images from orbit showed clear signs of flash floods, that may have been made by water in the distant past, as well as dendritic features resembling terrestrial river systems.

Following Viking, there was a gap of almost twenty years until NASA restarted their Mars exploration programme with the Mars Observer spacecraft. This failed just three days before it was due to enter Mars orbit in August 1993. Mars Global Surveyor (MGS), which was launched in 1996, was a partial replacement for Mars Observer. Instead of just using on-board thrusters, MGS used the novel technique of aerobraking in Mars' atmosphere to reach its final orbit, so minimising the use of on-board propellant.

Since MGS a number of NASA spacecraft have visited Mars, some of them with small surface rover vehicles, to study the landscape and rock types. NASA hoped to find clear evidence for water, now or in the past, on the surface. Then in 2003 the NASA orbiters were joined in orbit by ESA's Mars Express spacecraft. These spacecraft provided much circumstantial evidence of water on the surface of Mars in the past, and at present just under the surface. However, NASA's Phoenix Mars Lander, which landed on Mars in 2008, was the first to detect unambiguous evidence of water away from the polar ice caps when it uncovered water ice just beneath the surface.

Table 9.1B lists a selection of the more important Mars spacecraft.

Venus and Mercury

Returning now to the exploration of Venus. Mariner 2 had successfully visited Venus in 1962, and Mariners 3 and 4 had been sent to Mars two years later. Although the launch of Mariner 3 had failed, Mariner 4 had been successful. So NASA decided that Mariner 5, which had been built as a back-up Mars spacecraft to Mariners 3 and 4, should be modified and sent to Venus instead. In the event, it flew past Venus at an altitude of about 4,000 km in October 1967, just one day after the Soviet Union's first successful planetary spacecraft, Venera 4. Mariner 5 detected a surface temperature of at least 430°C (about 700 K) and a surface atmospheric pressure of between 75 and 100 bars.

Unlike Mariner 5, the Russian Venera 4 had a landing capsule with a parachute system to slow its descent through Venus' atmosphere. When the capsule stopped transmitting it had measured an atmospheric pressure of 20 bars and a temperature of 270°C. The Russians maintained that the capsule had reached the surface, but later analysis showed that it had failed at an altitude of about 25 km.

The Russians were more successful in 1970 with their Venera 7 capsule which still transmitted a feeble signal for 23 minutes after crashing onto the surface at about 70 km/h. It was the first artificial object to transmit a signal from the surface of any planet, measuring a ground-level atmospheric temperature of about 470°C. Five years later, landers from Venera 9 and 10 were the first spacecraft to take photographs from the surface of Venus. The landers relayed their data back to Earth via their main spacecraft, which were the first spacecraft to orbit Venus. In 1978 the Venera 11 and 12 landers found that some of the clouds on Venus were composed of sulphuric acid. The Venera 13 and 14 landers of 1982 both included a drilling and analysis system which determined that the surface at both sites was made of basalt. Finally Venera 15 and 16, which were launched in 1983 as the final spacecraft of the Soviet Venera series, did not carry landers. They were replaced by a synthetic aperture radar (SAR) and radar altimeter to see through Venus' thick clouds. They produced a surface resolution of about 2 km and a height resolution of 50 m. The SAR images showed several ten to forty metre diameter craters in the Lakshmi highland region, but few craters were seen in the lowland regions suggesting that they may be relatively young lava plains.

We left the American Venus exploration programme after Mariner 5 in 1967. The next American spacecraft to visit Venus was NASA's Mariner 10 which was launched in 1973 as the first spacecraft to visit both Venus and Mercury. It flew within about 6,000 km of Venus in February 1974, and then undertook three successive fly-bys of Mercury at six monthly intervals, observing the same half of Mercury's surface on each fly-by. It was found to be covered with craters, the largest of which was the 1,350 km diameter Caloris basin.

NASA's next two Venus spacecraft were the Pioneer-Venus Orbiter (PVO) and the Pioneer-Venus Multiprobe (PVM) launched in May and August 1978, respectively. The main instrument on the PVO was a radar altimeter to provide the first detailed map of Venus with a maximum surface resolution of 20 km and altitude resolution of 100 m. (This was five years before the launch of the Russian Venera 15 and 16 spacecraft with their SARs and radar altimeters.) The PVO mapped over 90% of the surface over the next two years, showing the 11,500 m high Maxwell Montes, the 2,000 m deep Diana Chasma, and a number of shield volcanoes. The PVM consisted of four atmospheric probes, none of which was designed to survive surface impact. In the event they arrived at Venus two weeks before Venera 11 and 12 and found, like them, that Venus' clouds were composed mainly of sulphuric acid.

The two Russian Vega spacecraft, launched in 1984, were designed to fly-by Venus on their way to observe Halley's comet. As they flew past the planet they each released a balloon and landing module. The balloon from Vega 1, which continued to operate for two days, drifted about a quarter of the way around the planet at 7° N until its batteries went flat. The balloon from Vega 2 undertook a similar mission at 7° S. Both balloons were tracked by numerous antennae on Earth, including those of NASA's Deep Space Network, which also received their observational data. This provided valuable data on Venus' atmospheric dynamics.

NASA's Magellan spacecraft, launched in 1989, had just two experiments on board: a SAR and a radar altimeter designed to have maximum horizontal and vertical resolutions of about 120 and 10 m, respectively. Most of the impact craters imaged by Magellan appeared fresh, indicated that weathering on Venus was a slow process, and the lack of any large craters also suggested that the surface was relatively young. Lava flows on Venus appeared to have been of radically different viscosities, with very long channels being produced by extremely fluid lava and large pancake-shaped volcanic domes being produced by very viscous lava.

Venus Express, the first ESA spacecraft to visit Venus, was a low-cost mission based on the reuse of their Mars Express spacecraft bus. The Venus spacecraft, which was launched in 2005 by a Russian rocket, was designed to observe the structure of the planet's atmosphere and clouds from a polar orbit. It was, for the first time, able to image both the horizontal and vertical structures of the double-eyed vortices over both poles.

The NASA MESSENGER (MErcury Surface Space ENvironment GEochemistry and Ranging) spacecraft, launched in 2004, was the first spacecraft to visit Mercury since Mariner 10 in the mid-1970s. To save propellant, its orbit included one fly-by of Earth, two of Venus and three of Mercury. It was then finally put into a 12 hour near polar orbit around Mercury

Table 9.1. *(cont.)*

(C) Selected Venus and Mercury spacecraft missions

Spacecraft	Country or institution	Launch date	Arrival date at Venus	Comments
Mariner 2	USA	27 Aug. 1962	14 Dec. 1962	First successful planetary fly-by. Detected high surface temperature and high surface atmospheric pressure on Venus.
Venera 4	USSR	12 June 1967	18 Oct. 1967	First to descend through Venus' atmosphere.
Venera 7	USSR	17 Aug. 1970	15 Dec.1970	First to land and transmit data from a planet's surface.
Mariner 10	USA	3 Nov. 1973	5 Feb. 1974	First to fly-by Venus and Mercury. Flew-by Mercury on 29 March 1974, 21 September 1974 and 16 March 1975.
Venera 9	USSR	8 June 1975	22 Oct. 1975	Orbiter plus landing capsule. Landing capsule took first photographs of Venus' surface.
Pioneer-Venus Orbiter	USA	20 May 1978	4 Dec. 1978	First spacecraft to produce radar map of Venus.
Pioneer-Venus Multiprobe	USA	8 Aug. 1978	9 Dec.1978	Four atmospheric probes to help understand dynamics of Venus' atmosphere. Found clouds mostly of sulphuric acid.
Venera 13	USSR	30 Oct. 1981	1 Mar.1982	Fly-by spacecraft plus landing capsule. Capsule had drill to enable Venus' surface to be analysed.
Venera 15	USSR	2 June 1983	10 Oct. 1983	Orbiter with Synthetic Aperture Radar (SAR) and radar altimeter. No lander.
Vega 1	USSR	15 Dec. 1984	11 June 1985	Spacecraft released balloon and landing module during Venus fly-by. Then undertook fly-by of Comet Halley in 1986.
Vega 2	USSR	21 Dec. 1984	15 June 1985	Repeat of Vega 1's mission.
Magellan	USA	4 May 1989	10 Aug.1990	Radar mapper with SAR and radar altimeter.
MESSENGER	NASA	3 Aug. 2004	24 Oct. 2006 5 June 2007	Two fly-bys of Venus en-route to eventually be the first spacecraft, on 18 March 2011, to orbit Mercury.
Venus Express	ESA	9 Nov. 2005	11 April 2006	Undertook global atmospheric measurements.

in March 2011. For the first time, MESSENGER found clear evidence of volcanic structures along the margins of the Caloris basin and elsewhere on Mercury, indicating that at least some of Mercury's planes are of volcanic origin.

Table 9.1C lists a selection of the more important spacecraft to visit Venus and Mercury.

The outer planets

(a) Pioneer

In 1963 NASA had contracted TRW (Thompson-Ramo-Wooldridge) to build a series of Pioneer spacecraft (eventually numbered 6 to 9) to investigate the solar wind. It became clear shortly afterwards that modified versions of these spacecraft could also be used for a Jupiter intercept mission. Detailed design studies were undertaken and in 1969 NASA decided to launch Pioneers 10 and 11 in 1972 and 1973 to fly

past Jupiter. If Pioneer 10 failed, Pioneer 11 would repeat its Jupiter intercept mission. But if Pioneer 10 succeeded, NASA would consider redirecting Pioneer 11 to fly by Saturn after its Jupiter intercept. After the Voyager programme was approved in 1972 (see below), the Pioneers were considered as not only important missions in their own right, but as pathfinder missions to assist the later more complex Voyager missions.

Pioneer 10 was launched to Jupiter by an Atlas-Centaur in March 1972, followed 13 months later, at the next Jupiter launch window, by Pioneer 11. To get to Jupiter both spacecraft had to fly through the asteroid belt, with an unknown risk of collision. In the event Pioneer 10 survived and successfully flew past Jupiter in December 1973, just 21 months after launch. After evaluating the Pioneer 10 results, NASA decided that it would adjust Pioneer 11's intercept trajectory with Jupiter to allow it to fly past Saturn in September 1979.

The Pioneers' imaging system produced images of Jupiter up to four times sharper than those of the best Earth-based telescopes. The spacecraft confirmed that Jupiter emitted more heat than it received from the Sun, and made the first measurement of the amount of helium in Jupiter's atmosphere. They also provided much valuable information on Jupiter's magnetic field, magnetosphere and radiation belts. Pioneer 11 provided similar information for Saturn and found a new ring and new satellite.

(b) Voyager

In the early years of the Space Age it appeared as though it would require a massive launcher to send a spacecraft to visit the outer planets of the solar system. But it was possible, in theory, to use the gravity of one planet to accelerate and redirect a spacecraft on to the next, and so reduce the size of launcher required. In the mid-1960s Michael Minovich showed this was feasible. Then in 1965 Gary Flandro found that the four major planets of the outer solar system (Jupiter, Saturn, Uranus and Neptune) were perfectly aligned for an optimum gravity-assist trajectory if launch took place in the period 1976–80. Such an alignment would occur only once every 175 years, so there was pressure on NASA to implement a so-called 'Grand Tour' mission with a launch in this period.

After examining many options, NASA finally approved a simplified version of the original Grand Tour spacecraft with two spacecraft based on the Mariner design. Voyager 1 would be launched in 1977 to fly past Jupiter (in 1979) and Saturn (1980). Voyager 2 would also be launched in 1977 to fly past Jupiter (in 1979), Saturn (1981), and possibly Uranus (1986) and Neptune (1989). All spacecraft functions, except for trajectory changes, were to be controlled automatically on board, as the round-trip time for signals from Earth to Neptune, for example, would be more than eight hours.

A key element of the Voyager 1 mission was a close intercept of Titan, Saturn's largest satellite, which had a dense atmosphere. This intercept was considered to be so important that, if it failed, Voyager 2 would be retargeted for a close fly-by of Titan, even though this would prevent it from visiting Uranus and Neptune.

The Voyager 2 spacecraft was launched on 20 August 1977 to fly past Jupiter in July 1979. Voyager 1 was launched two weeks later on a slightly different trajectory so that it would overtake Voyager 2 en route to Jupiter. Voyager 1 crossed Jupiter's bow shock in February 1979 as it approached the planet, and recrossed the bow shock and Jupiter's magnetopause several times during the next few days as their positions responded to the varying intensity of the solar wind. Voyager 1 then passed through the Jupiter system imaging both the planet and a number of its satellites. Probably the most interesting satellite images were those of Io, which was found to have a

very young surface, with numerous volcanic caldera and rivers of lava. Then, three days after closest approach to Io, a volcanic eruption was seen on its edge. Eventually it was found that there had been eight volcanoes active during Voyager 1's closest approach.

When Voyager 2 flew past Jupiter a few months later one of these volcanoes had become dormant, but two more were found. Then in the following year Voyager 1 was successful in its intercept of Saturn and Titan. As a result, NASA announced in January 1981 that Voyager 2 would fly-by Uranus and Neptune after its Saturn intercept.

It is not feasible here to go into detail of what happened during the Voyager 1 and 2 planetary intercepts, all of which were highly successful. These intercepts are covered in the *Voyager* article in the 'Individual solar system exploration spacecraft' section following this 'Overview'. Nor is it feasible here to outline what was discovered at each of the four planets by both Voyager spacecraft, as this was extensive. These discoveries are covered in the individual *Jupiter*, *Saturn*, *Uranus*, and *Neptune* articles in Part 2.

(c) Galileo

The idea of sending a spacecraft to orbit Jupiter and aiming a probe to descend into its atmosphere had been mooted in the late 1960s. But the Voyager 1 and 2 Grand Tour spacecraft had been approved instead. However, in 1977 the Jupiter Orbiter and Probe mission, now called Galileo, was approved.

NASA had originally planned to launch Galileo from the space shuttle in 1982 using the Inertial Upper Stage (IUS). But development problems with the IUS forced NASA to change to the more hazardous Centaur for a launch in 1985. Although the basic Centaur was already developed, and had an impressive success record, it needed modifications to enable it to be launched by the shuttle. But further delays with the Galileo programme and the Space Shuttle Challenger accident in January 1986 eventually forced NASA to return to the IUS.

The Galileo spacecraft, which was launched from the Space Shuttle Atlantis in 1989, had a detachable probe to enter Jupiter's atmosphere. To reach Jupiter the spacecraft and attached probe undertook fly-bys of Venus (in 1990) and Earth (in 1990 and 1992), which provided it with the required additional velocity. But this extended itinerary meant that it would take 74 months to reach Jupiter, compared with just 18 months for Voyager 1. En route, however, Galileo was able to fly past the asteroids Gaspra in 1991, and Ida in 1993. As a result the spacecraft produced the first close-up views of an asteroid, and discovered the first known satellite, Dactyl, of an asteroid, Ida.

Unfortunately, Galileo's high-gain antenna failed to fully deploy en route to Jupiter, which resulted in NASA having to modify the spacecraft's on-board software and improve its

Table 9.1. *(cont.)*

(D) Missions to the outer planets

Mission timelines

Spacecraft	Launched	Jupiter fly-by	Saturn fly-by	Uranus fly-by	Neptune fly-by
Pioneer 10	3 Mar. 1972	4 Dec. 1973			
Pioneer 11	6 Apr. 1973	3 Dec. 1974	1 Sept. 1979		
Voyager 1	5 Sept. 1977	5 Mar. 1979	12 Nov. 1980		
Voyager 2	20 Aug. 1977	9 July 1979	26 Aug. 1981	24 Jan. 1986	25 Aug. 1989

Spacecraft	Launched	Gaspra fly-by	Ida fly-by	Jupiter atmospheric probe released	Atmospheric probe and orbiter arrive at Jupiter
Galileo	18 Oct. 1989	29 Oct. 1991	28 Aug. 1993	13 July 1995	7 Dec. 1995

Spacecraft	Launched	Jupiter fly-by	Saturn orbit insertion	Huygens probe released	Huygens probe lands on Titan
Cassini-Huygens	15 Oct. 1997	30 Dec. 2000	1 July 2004	25 Dec. 2004	14 Jan. 2005

ground receiving stations to help compensate for the much weaker signals received on Earth.

The Jupiter entry probe was released from the main Galileo spacecraft in July 1995, and five months later it entered Jupiter's atmosphere at high velocity. About an hour later the main spacecraft was put into orbit around Jupiter to, amongst other things, act as a communications relay between the probe and Earth. The probe started transmitting atmospheric data when it reached the 0.35 bar level in Jupiter's atmosphere, and it continued transmitting until the atmospheric pressure reached 23 bars about 160 km further into the atmosphere.

The probe provided much valuable information on the structure and composition of Jupiter's cloud layers (see *Galileo* article). The Galileo orbiter found that Ganymede had an induced component to its magnetic field, implying that there was a salty ocean tens of kilometres thick beneath its icy surface. Evidence for similar sub-surface salty oceans was also found for Callisto and Europa. Galileo's images of the surface of Europa showed numerous rafts of ice, some of which were tilted, indicating that they had been subjected to lateral compressional pressure at some stage in the past.

(d) Cassini-Huygens

NASA's Outer Planet Working Group had suggested in 1969 that the agency should undertake a comprehensive programme of missions to the outer planets including a Saturn orbiter and probe to be launched in 1978. Voyager 1 and 2 were approved as a first step in such a comprehensive programme. Then in 1982 the United States National Academy of Sciences and the Euro-

pean Science Foundation suggested that a Saturn Orbiter and Titan Probe mission be undertaken. Two years later NASA and ESA agreed on such a joint mission with NASA responsible for the Saturn Orbiter (later called Cassini) and ESA responsible for the Titan Probe (called Huygens). The resulting Cassini-Huygens spacecraft was eventually launched in October 1997.

En-route to Saturn, Cassini-Huygens undertook fly-bys of Venus (in 1998 and 1999), Earth (1999) and Jupiter (2000) in order to gain enough energy to reach Saturn in 2004. Whilst the spacecraft was still some distance from Saturn it was realised that there would be a problem of pulse synchronisation between the Huygens transmitter and the Cassini receiver when Huygens was descending through Titan's atmosphere. This was solved by changing the relative velocity of the two spacecraft after separation by releasing Huygens seven weeks later than originally planned. In the event Huygens landed on Titan on 14 January 2005, the first spacecraft to land on a planetary satellite.

The Huygens probe found methane and nitrogen in Titan's stratosphere, and imaged dendritic channels on Titan's surface during its descent. The Cassini orbiter radar imaged what appeared to be large lakes on Titan. Then in 2007 one of these lakes was found to contain liquid hydrocarbons, including ethane. Cassini also discovered plumes of water ice ejected into one of Saturn's rings from cracks in Enceladus' surface near its south pole. In March 2008 Cassini flew through one of these plumes and detected water, carbon dioxide and hydrocarbons.

Table 9.1D lists the missions to the outer planets.

The Sun and solar wind

Early observations of the solar wind outside the influence of the Earth's magnetic field were made by spacecraft en route to the Moon and planets. For example, in 1959 Luna 2 was the first to detect the solar wind in interplanetary space and, three years later, the Venus spacecraft Mariner 2 was the first to measure the solar wind's velocity in interplanetary space. It generally ranged from about 400 to 700 km/s, whilst occasionally exceeding 1,200 km/s.

In the meantime, the first Orbiting Solar Observatory (OSO) spacecraft had been proposed in 1959 as a one-off, low-budget, Earth-orbiting solar observatory to observe the Sun in the ultraviolet, X-ray and gamma-ray wavebands. Other instruments on this 210 kg OSO-1 spacecraft measured the elementary particles of the solar wind. The spacecraft, which was launched in 1962, observed more than 75 flares and subflares, and monitored the inner Van Allen belt. But by the time it had been launched, NASA had decided to transform this single OSO-1 spacecraft mission into an ongoing programme of similar spacecraft. In the event eight out of nine OSO spacecraft were successfully launched over the period to 1975.

OSO-3 measured the spectrum of a solar flare in 1967 which indicated an extremely high local temperature. It also detected about 600 cosmic gamma rays during its sixteen month lifetime. A later spacecraft, OSO-7, discovered clouds of protons leaving the Sun in coronal mass ejections (CMEs) in 1971. In the following year it discovered gamma-ray emission from a solar flare with a main line at 0.5 MeV produced by electron/positron annihilation.

Whilst the OSO series of spacecraft were observing the Sun, the IMP (Interplanetary Monitoring Platform) series of spacecraft were investigating the Earth's magnetosphere and the near-Earth solar wind. The first of these, IMP-1, was launched in 1963 into a highly eccentric orbit that allowed it to spend a varying amount of time in interplanetary space beyond the Earth's bow shock. It provided the first clear-cut evidence of the Earth's bow shock, and discovered the neutral sheet at the centre of the Earth's geomagnetic tail.

NASA had begun the Apollo Applications Program in 1965 by examining the use of Apollo hardware for extended manned Earth-orbiting and lunar surface missions. But successive budget cuts over the next few years reduced the programme to just one Earth-orbiting manned space observatory called Skylab. Its main astronomical mission was to study the Sun in the X-ray, ultraviolet and visible wavebands. Skylab's instruments were much better in performance than those of earlier, unmanned spacecraft, as most of the Skylab instruments used film, rather than electronic imaging. In addition, they were built with much less stringent mass and power constraints.

Unfortunately, an explosion during Skylab's launch in May 1973 caused serious problems. But emergency repairs by astronauts enabled Skylab to be used, albeit with a much reduced power supply. Skylab detected six low-latitude coronal holes in X-rays, most of which lasted a number of months. It also detected hundreds of X-ray bright points, and its ultraviolet images showed large, bright, active regions in the Sun's chromosphere.

In 1959 the United States had offered to launch scientific experiments or complete spacecraft of other countries free of charge, provided the experiment results were shared with the Americans. A number of countries took up this offer including the United Kingdom, Italy and France. The Helios spacecraft programme was a later bilateral programme with West Germany providing the spacecraft bus and the United States two Titan-Centaur launch vehicles. Both countries provided experiments.

The two Helios spacecraft, which were launched just over a year apart in the mid-1970s, were placed into orbits around the Sun. Both spacecraft flew closer to the Sun than any previous man-made objects with perihelia of about 0.3 AU, and aphelia at the Earth's orbit. They confirmed that there was both a slow and a fast solar wind, and that the fast wind was being emitted by coronal holes.

It was very difficult to observe how the Earth's magnetosphere reacted to changes in the solar wind in three dimensions using a series of radically different spacecraft operating in a relatively uncoordinated way. It would clearly be much better if a number of similar spacecraft could be launched at essentially the same time and operated in a coordinated programme. This was the thinking behind the NASA/ESA International Sun-Earth Explorer (ISEE) mission which was designed to study the effect of the solar wind on the Earth's magnetosphere, and separate out spatial and temporal variations using three spacecraft. The designs of the ISEE-1 and -3 spacecraft, which were provided by NASA, were based on those of the IMP spacecraft, whilst the ISEE-2 spacecraft was provided by ESA.

ISEE-1 and -2 were launched together in October 1977 into a highly elliptical orbit which, depending on the time of year, took them outside the Earth's magnetosphere for up to 75% of the orbit's period. Their separation could be varied. ISEE-3 was launched about a year later and placed into a halo orbit around the L_1 Lagrangian point upstream in the solar wind between the Sun and Earth. These spacecraft provided a great deal of information on the magnetic structure of the geomagnetic tail, and on the plasma flow near Earth and further out.

The Solar Max (or Solar Maximum Mission, SMM) spacecraft was designed to observe the Sun, and solar flares in particular, during the 1980–1981 solar maximum. Its payload included the first telescope designed to image the Sun in high-energy X-rays, and two non-solar instruments to detect gamma-ray bursts. An important feature of the spacecraft was the on-board coordination of instrumental operations in response to solar flares.

Solar Max lost its fine-pointing capability less than a year after its launch in February 1980, reducing it to a non-imaging mission. It was repaired in April 1984 by astronauts from the Space Shuttle Challenger, making it the first spacecraft to be retrieved, repaired, and redeployed in orbit. Shortly after launch Solar Max showed that the solar constant decreased when two large sunspot groups crossed the central solar meridian. Over the next few years, data from Solar Max and from the Nimbus 7 remote sensing spacecraft showed that the solar constant went through a minimum at solar minimum.

All the work on the Sun so far had been undertaken by spacecraft very close to the ecliptic. So no-one knew how the solar wind, particularly that emitted from high solar latitudes, behaved out of the ecliptic. As a result NASA and ESRO (ESA's predecessor) agreed in 1974 to undertake a joint study into what was to become the Out-of-Ecliptic programme to observe the Sun and solar wind out of the ecliptic. The name of this mission was later changed to the International Solar Polar Mission, and finally to Ulysses.

Initially the idea was to launch two spacecraft to fly approximately over opposite poles of the Sun, using Jupiter's gravity to swing them out of the ecliptic. One spacecraft would be provided by NASA and one by ESA, with American and European experiments on both. Then in 1981 NASA unilaterally cancelled their spacecraft, leaving a number of European experimenters without a spacecraft platform. This and other problems created a distinct chill in ESA/NASA relations. But eventually these were patched up and the Ulysses spacecraft was launched by the Space Shuttle Discovery in October 1990.

The Ulysses spacecraft consisted of an ESA bus with five European and five American experiments. These were to investigate, amongst other things, the solar wind, the solar and heliospheric magnetic field, solar emissions, and cosmic rays. Ulysses found that, around solar minimum, the slow solar wind was constrained to equatorial regions. But the fast solar wind, which originated from the Sun's polar and other coronal holes, was found to fill about 70% of the heliosphere.

Japan had launched their first solar spacecraft, Taiyo, in 1975. But their Solar-A, or Yohkoh as it was later called, was their first to make a major contribution to solar research. Yohkoh was designed to study high energy emissions from the Sun, and solar flares in particular. Its payload consisted of X-ray telescopes and X-ray and gamma-ray spectrometers. Scientists from the United States and the UK contributed to the payload.

Observations from the Yohkoh spacecraft, which was launched in August 1991, covered almost a complete solar cycle. Shortly after launch it showed that hard X-rays were emitted from the apex of flaring loops, probably as a result of magnetic reconnection. In addition, Yohkoh detected X-ray jets, which were also thought to have been caused by magnetic reconnection, moving outwards in the outer solar corona.

The Solar and Heliospheric Observatory (SOHO) had originally been proposed as an ESA-only programme in 1982, but over the next few years it became clear that its potential costs were too high. As a result, ESA approached NASA to see if they were interested in working on a joint programme, which they were. Subsequently the spacecraft was also simplified in order to get it within the agreed financial envelope. The spacecraft's main objectives were to study the solar interior using helioseismology, study the potential heating mechanisms of the solar corona and find out more about the origin and method of acceleration of the solar wind.

SOHO was launched in late 1995 and inserted into an L1 halo orbit two months later. It provided the first images of the structure of the Sun below its surface, and showed that the decrease in angular velocity with latitude, which had been observed at the surface, extended through the convective zone. At the base of this zone, about one-third of the way into the Sun, was the interface with the more orderly radiative interior. SOHO also showed that strong converging downflows stabilised the structure of sunspots, which were found to be relatively shallow.

SOHO found evidence of the upward transfer of magnetic energy from the Sun's surface to the corona though a magnetic carpet in which the observed magnetic flux was found to be highly mobile. This upward transfer of energy was thought to make a major contribution to coronal heating. SOHO also found that the number of CMEs varied with the solar cycle, and imaged over 2,000 mostly Sun-grazing comets.

The International Solar Terrestrial Physics (ISTP) programme had been formed in the 1980s to improve our understanding of the generation and flow of energy from the Sun through the Earth's space environment. This ISTP programme had been created by NASA, ESA and ISAS of Japan as the union of NASA's Global Geospace Science (GGS) programme and the ESA/NASA Solar Terrestrial Programme. The aim was to carry out several coordinated solar-terrestrial missions during the 1990s. Missions in the ISTP programme included Geotail (ISAS/NASA programme, spacecraft launched 1992), WIND (NASA, 1994), SOHO (ESA/NASA, 1995), POLAR (NASA, 1996) and Cluster (ESA, 2000). Since then a number of spacecraft have been launched, covering the same area of research, ranging from the 290 kg RHESSI (launched 2002), which was the sixth of NASA's Small Explorer (SMEX) spacecraft, through the 900 kg Japanese Hinode spacecraft (launched 2006 as a follow on to Yohkoh), to the 3,100 kg American Solar Dynamics Observatory (launched 2010).

These and similar spacecraft have successfully added to our detailed knowledge of the Sun and of its interaction with the Earth's environment. For example, Geotail provided evidence of magnetic reconnection in the geomagnetic tail. The POLAR spacecraft detected Alfvén waves propagating along magnetic field lines from the geomagnetic tail towards the Earth. These waves were found to carry enough energy to create aurorae during magnetospheric substorms. RHESSI confirmed that solar

Table 9.1. *(cont.)*

(E) Selected Sun and solar wind missions

Spacecraft	Main country or institution	Launch dates	Comments
OSO 1–8	USA	1962–1975	Observed Sun and celestial sphere in all wavebands from gamma rays to the visible. Also observed solar wind.
IMP-1	USA	27 Nov. 1963	Measured solar wind and Earth's magnetosphere.
Skylab	USA	14 May 1973	Manned observatory. Observed Sun in X-ray, ultraviolet and visible wavebands.
Helios 1 and 2	West Germany	10 Dec. 1974 and 15 Jan. 1976	Solar orbiting spacecraft. Studied Sun and interplanetary medium within 0.3 AU of Sun.
ISEE-1, -2 and -3	USA/ESA	22 Oct. 1977 and 12 Aug. 1978	Observed solar wind and solar flares. ISEE-3, renamed ICE, subsequently intercepted Comet Giacobini-Zinner.
Solar Max	USA	14 Feb. 1980	Observed Sun, and particularly solar flares, during solar maximum.
Ulysses	ESA	6 Oct. 1990	Measured solar wind and Sun from out-of-ecliptic orbit. Flew over both poles of Sun.
Yohkoh	Japan	30 Aug. 1991	Observed solar flares and other high energy emissions from Sun.
WIND	USA	1 Nov. 1994	Made a number of substantial orbit changes. Measured upstream interplanetary medium.
SOHO	ESA	2 Dec. 1995	Measured internal structure and dynamics of Sun, using helioseismology. Studied potential coronal heating mechanisms and observed solar wind.
RHESSI	USA	5 Feb. 2002	Imaged solar flares simultaneously in X-ray and gamma-ray wavebands with high resolution spectroscopy.
Hinode	Japan	22 Sept. 2006	Observed magnetic activity and plasma outflows on Sun.
STEREO A and B	USA	26 Oct. 2006	Provided stereoscopic imaging of coronal loops, polar coronal jets and coronal mass ejections.
Solar Dynamics Observatory	USA	11 Feb. 2010	Measured Sun's variable radiative, particulate and magnetic plasma emissions.

flares were caused by magnetic reconnection, and the STEREO spacecraft imaged CMEs spectroscopically, in an attempt to understand their origin and development in detail. In addition, Hinode found that the chromosphere was permeated by Alfvén waves, which were sufficiently strong to accelerate the solar wind and possibly to heat the quiet corona.

Table 9.1E lists a selection of the more important Sun and solar wind spacecraft missions.

Asteroids

As mentioned above, the Galileo spacecraft had flown past Gaspra (951) in 1991 on its way to Jupiter, producing the first close-up images of an asteroid. It was found to be an irregular $18 \times 11 \times 9$ km in size, with dozens of small craters on its surface. Crater counts indicated an age of the order of a few hundred million years. Two years later Galileo imaged the 54 km long Ida (243) which was found to have a small spherical companion, about one kilometre in diameter, now called Dactyl. The discovery of this satellite allowed the mass and density of Ida to be determined.

The NEAR (Near Earth Asteroid Rendezvous) Shoemaker spacecraft, which was the first spacecraft to orbit an asteroid, was the first of NASA's 'Faster, Better, Cheaper' Discovery spacecraft to be launched. It included a multispectral camera, a laser altimeter, and an X-ray/gamma ray spectrometer.

Launched in February 1996, NEAR Shoemaker flew past the asteroid Mathilde (253) in mid-1997 whilst the spacecraft was en route to intercept the asteroid Eros (433). Crater counts on the 59×47 km Mathilde indicated that it was at least 2 billion years old. Spacecraft orbital deviations around closest approach showed that Mathilde's density was very low, indicating that it was extremely porous.

It had originally been planned to put NEAR Shoemaker into orbit around Eros in January 1999, but spacecraft problems caused NASA to abandon this attempt and have the spacecraft

Table 9.1. *(cont.)*

(F) Timelines of asteroid and comet spacecraft intercepts

Asteroid intercepts (asteroid names in italics)

(1)	Galileo spacecraft launched 18 October 1989. En route to Jupiter it:		
	Flew past *Gaspra*	29 October 1991	distance 1,600 km
	Flew past *Ida* and Dactyl	28 August 1993	distance 2,400 km
(2)	NEAR Shoemaker spacecraft launched 17 February 1996 to orbit *Eros*. It:		
	Flew past *Mathilde*	27 June 1997	distance 1,200 km
	Flew past *Eros*	23 December 1998	distance 3,800 km
	Entered *Eros* obit	14 February 2000	
	Landed on *Eros*	12 February 2001	
(3)	Stardust spacecraft launched 7 February 1999. En route to Comet Wild 2 it:		
	Flew past *Annefrank*	2 November 2002	distance 3,300 km
(4)	Hayabusa spacecraft, launched 9 May 2003, sample return mission to *Itokawa*. It:		
	Arrived at *Itokawa*	12 September 2005	
	Landed on *Itokawa*	19 November 2005	
	Return capsule landed on Earth	13 June 2010	
(5)	Rosetta spacecraft, launched 2 March 2004. En route to Comet Churyumov-Gerasimenko it:		
	Flew past *Steins*	5 September 2008	distance 800 km

Comets (comet names in italics)

(1) *Comet Halley* spacecraft intercepts undertaken in 1986

Spacecraft	Country or Institution	Fly-by date in 1986	Fly-by distance (km)	Other comets visited	Fly-by date	Fly-by distance (km)
Vega 1	USSR	6 March	8,890			
Suisei	Japan	8 March	151,000			
Vega 2	USSR	9 March	8,030			
Sakigake	Japan	11 March	6,990,000			
Giotto	ESA	14 March	596	*Grigg–Skjellerup*	10 July 1992	200
ICE	USA	25 March	28,100,000	*Giacobini–Zinner*	11 Sept. 1985	7,800

(2)	Deep Space 1 technology spacecraft launched 24 October 1998. It:		
	Flew past *Comet Borrelly*	22 September 2001	distance of 2,200 km
(3)	Stardust spacecraft, launched 7 February 1999, as a sample return mission to *Comet Wild 2*. It:		
	Undertook closest approach and sample collection at *Comet Wild 2*	2 Jan. 2004	dist. 240 km
	Return capsule landed on Earth	15 Jan. 2006	
	Main spacecraft undertook fly-by of *Comet Tempel 1*	15 Feb. 2011	dist. 180 km
(4)	Deep Impact spacecraft, launched 12 January 2005, to impact *Comet Tempel 1*:		
	Its impactor impacted *Comet Tempel 1*	4 July 2005	
	Main spacecraft undertook fly-by of *Comet Hartley 2*	24 Nov. 2010	dist. 700 km

fly-by Eros instead. NASA then put NEAR Shoemaker into a heliocentric orbit that closely matched that of Eros, allowing them to have another attempt to orbit the asteroid in February 2000. This was successful. NASA then decided to attempt an unplanned landing on Eros which they successfully achieved a year later.

The Japanese Hayabusa spacecraft was designed to return samples from the surface of an asteroid to Earth. Launched in 2003, it used ion engines to slowly approach its target, Itokawa (25143). When Hayabusa arrived it was planned to briefly land on the Itokawa's surface. It was also planned to fire small projectiles onto the asteroid's surface and collect part of the resulting spray, returning it to Earth in a small re-entry capsule.

Although the main Hayabusa spacecraft did land, it was not known whether the projectiles had been fired or whether the spacecraft had collected any samples. But in spite of further problems Hayabusa's return capsule successfully landed on Earth in 2010. Preliminary results indicated that it contained numerous extremely small dust particles of extraterrestrial origin.

Table 9.1F includes those spacecraft to visit asteroids.

Comets

In 1985 Halley's comet was expected to make its first visit to the inner solar system since the start of the space age. As a result four space agencies planned to send spacecraft to intercept the comet: two spacecraft (Vega 1 and 2) from Intercosmos in the Soviet Union, two (Suisei and Sakigake) from ISAS in Japan, one (Giotto) from the European Space Agency, and one (International Cometary Explorer or ICE) from NASA. The expected fly-by distances ranged from <1,000 km for Giotto, to <10,000 km for Vega 1 and 2, to 150,000 and 7 million km for Suisei and Sakigake, to about 28 million km for ICE. All the participating agencies aimed for an encounter on the comet's outbound track in March 1986 when its 18° inclined orbit crossed the ecliptic.

An informal Inter-Agency Consultative Group was established with representatives from each agency to coordinate the various spacecraft missions. In the event this worked extremely well, allowing each agency to optimise their intercepts with assistance, as required, from the others.

The spacecraft intercepts showed that Halley's comet's nucleus was about $16 \times 8 \times 8$ km in size with what looked like craters on its surface. Bright jets could clearly be seen streaming towards the Sun. Measured water vapour and dust production rates indicated that Halley's comet would be able to survive about 1,000 more orbits of the Sun before it ran out of material.

Twenty years later NASA's Stardust spacecraft was the first spacecraft to capture cometary material and return it safely to Earth after it had flown by Comet Wild 2 in 2004. This comet had been chosen as it had only been diverted into the inner solar system by a recent close encounter with Jupiter. As a result it was expected that it would be almost in its original primordial state.

Stardust was designed to collect cometary dust particles with blocks of ultra-low-density aerogel foam in a collector tray. This foam was expected to slow down and capture the particles without damaging them during the spacecraft's high-speed fly-by. The tray was contained in a capsule that was to be released from the main spacecraft on its return to the vicinity of the Earth.

In the event, samples recovered from the capsule, which landed in the Utah desert in 2006, showed that Comet Wild 2 consisted of some material which had been formed at very high temperatures, and some which had been formed at very low temperatures. The amino acid glycine was also identified, which was the first time that an amino acid had been found in a comet.

The next cometary mission, Deep Impact, was designed to impact the nucleus of the Comet Tempel 1 and penetrate its crust to expose the relatively fresh material underneath. Launched in January 2005, the Deep Impact spacecraft arrived at Tempel 1 six months later. It released its impactor about 24 hours before impact. The main Deep Impact spacecraft, as well as other space- and Earth-based observatories, observed the impact that took place in July 2005. Individual species in the ejecta included water, water ice and, as in the case of Comet Wild 2, some that required high temperatures to form and some that required low temperatures, indicating that they did not all originate from the same region of the solar system. Unfortunately, the Deep Impact spacecraft could not image the crater produced by its impactor probe as it was obscured by the ejecta plume. So NASA decided to redirect the Stardust spacecraft to image the crater, which it did successfully in February 2011.

Table 9.1F includes those spacecraft that visited comets.

Bibliography

Leverington, David, *New Cosmic Horizons: Space Astronomy from the V2 to the Hubble Space Telescope*, Cambridge University Press, 2000.

2 · Individual solar system exploration spacecraft

ACE

The Advanced Composition Explorer (ACE) was a NASA mission to study the elemental and isotopic composition of accelerated nuclei from 100 eV solar wind particles to 500 MeV galactic cosmic rays, with a charge and mass resolution much better than previously achieved. Particle sources included the solar wind, coronal mass ejections, impulsive solar flares, local interstellar medium (LISM), and galactic matter. The LISM was sampled by detecting pick-up ions and anomalous cosmic rays, and galactic matter by detecting galactic cosmic rays.

Although it was first conceived in 1983, the ACE programme was not begun until eight years later. Then in August 1997 the 785 kg ACE spacecraft was launched into an orbit around the L1 libration point on the Sun-Earth line about 1.5 million km sunward from Earth. From this orbit it was able to give about one hour's warning of a geomagnetic storm. The spacecraft's target lifetime was five years to cover a significant part of a solar cycle. In the event, it was still operating over ten years after launch.

In March 1998 ACE and Ulysses at 1 and 5 AU from the Sun, respectively, provided the first solar wind observations of the same coronal mass ejection, which was also a magnetic cloud, at two very different distances from the Sun. These two spacecraft showed that the average temperature had fallen by a factor of 5 and the density by a factor of 50 between their two locations. The temperature decrease was less than expected for a non-interacting gas, allowing an estimate to be made of the interaction between solar wind particles.

In May 1999, ACE and a number of other spacecraft recorded a dramatic event when the solar wind almost disappeared. Its density was only a few percent of normal. As a result the bow shock was much further away from Earth than normal. A similar but less dramatic event had also occurred the previous month.

Analysis of positively charged helium pick-up ions indicated that the interstellar material surrounding our solar system had a velocity of around 25 km/s and a plasma temperature of about 7,500 K. ACE also found evidence that cosmic rays were produced when supernovae accelerated interstellar material, not directly by the supernova material itself.

Bibliography

Russell, C. T., Mewaldt, R. A., and von Rosenvinge, T. T. (eds.), *The Advanced Composition Explorer Mission*, Kluwer Academic Publishers, 1999.

Various articles, Space Science Reviews, 86 (1998), Nos. 1–4.

APOLLO

President Kennedy had asked his vice president, Lyndon B. Johnson, in early 1961 to recommend a high-profile space programme to demonstrate America's technological and military supremacy to the outside world. The result, announced on 25 May 1961, was a commitment to 'achieving the goal, before this decade is out, of landing a man on the Moon and returning him safely to Earth'. This was to be the Apollo programme. It was clearly a political programme – any scientific returns being a bonus. Prior to the manned landings, the Surveyor spacecraft were to make soft, unmanned landings to determine the strength and condition of the lunar surface, and the Lunar Orbiters were to image the surface of the Moon from orbit to find suitable landing sites for the manned Apollo landers.

In the meantime NASA needed to develop a man-in-space capability, beginning with Alan Shepard's suborbital Mercury flight of 5 May 1961, just twenty days before president Kennedy's announcement. The suborbital Mercury flights were followed by orbital Mercury flights, and by two-man Gemini flights, before the first manned Apollo mission, Apollo 7, reached Earth orbit in October 1968. At the end of that year NASA launched Apollo 8, the first manned flight around the Moon. This was followed by one more Earth-orbiting and one more lunar-orbiting Apollo mission, before NASA launched Apollo 11 to the Moon on 16 July 1969. Four days later Neil Armstrong and Buzz Aldrin landed their Apollo 11 Lunar Excursion Module in the Mare Tranquillitatis (Sea of Tranquillity). It was the first Apollo mission to land on the Moon.

During a moonwalk of about $2\frac{1}{2}$ hours, Armstrong and Aldrin collected 22 kg of Moon rocks which they brought back to Earth. They also left a seismometer and laser reflectors behind on the Moon, the seismometer to measure moonquakes and the reflectors to measure the Earth-Moon distance with high precision. The lunar material that they brought back was generally igneous basalt, which confirmed previous indications that the Moon's dark maria were of volcanic origin. These basalts implied an age for the Tranquillity lavas of 3.65 billion years. Breccias (impact rocks) were also found, many of which pre-dated the Tranquillity lavas. There appeared to be no trace of water in any of the rocks, nor was there any organic material, which was a disappointment for those who were hoping that the Moon may have previously been able to support simple forms of life.

The basalt samples returned from the next mission, Apollo 12, which landed in the Oceanus Procellarum (Ocean of Storms), were similar to those of Apollo 11. But the Apollo 12 samples were about 500 million years younger, showing that the maria did not all form at the same time. This indicated that the Moon's interior was not uniform in composition.

The next mission, Apollo 13, failed to land on the Moon following a major emergency on the flight out from Earth. An oxygen tank exploded inside the service module, damaging other vital spacecraft components. NASA, ably assisted by numerous other ground-based engineers, managed to instruct the astronauts James Lovell, John Swigert and Fred Haise how to adapt various parts of their spacecraft to enable them to survive until they could land on Earth. This they successfully did six days after launch.

Apollo 14, the first successful mission after Apollo 12, was the first manned mission to visit the lunar uplands. It landed at Fra Mauro, a region of hills and craters on the edge of the Mare Imbrium (Sea of Showers). Here the astronauts found breccias that indicated that the impact that had created the Imbrium basin, and the adjacent Fra Mauro hills, had occurred about 3.85 billion years ago. Some basalt samples showed evidence of partial melting of the primitive crust.

Apollo 15 landed near the Hadley Rille at the foot of the lunar Apennines in July 1971. NASA were hoping that the astronauts would find samples of the original lunar crust in the form of anorthosite, which would have risen to the surface of the Moon whilst it was still molten. This mission included the first lunar roving vehicle to land on the Moon, which allowed the astronauts to explore further afield from their lander. They found what came to be known as the Genesis Rock, which was, at about 4.5 billion years old, an example of the original lunar crust. They also brought back samples of the lunar regolith containing coloured pyroclastic glass beads that had, apparently, been brought to the lunar surface from hundreds of kilometres down by fire fountains. On the surface of the beads were volatile elements which had been produced by volcanic vapours during the eruptions. In 2008 Alberto Saal and colleagues re-examined these volcanic beads and found evidence of water.

NASA landed Apollo 16 in the Descartes highland region, expecting to find that part of this region was of volcanic origin. In the event, most of the rocks that the astronauts brought back were breccias, with a few anorthosite remnants of the ancient crust. There were no examples of volcanic rock. This indicated the limits of undertaking geology from photographs.

Harrison Schmitt, who flew on Apollo 17, was the only geologist to land on the Moon during the Apollo programme. The area selected for this last Apollo landing, in December 1972, was the Taurus-Littrow valley on the edge of the Mare Serenitatis (Sea of Serenity), chosen because of its unusual very dark surface with what looked like very small volcanic craters nearby. The astronauts brought back 110 kg of moonrocks from moonwalks that totalled about 22 hours. Probably the most interesting discovery was made on the edge of a 100 m diameter impact crater where Schmitt discovered a layer of orange soil under the surface dust. This orange soil was found to include very small beads of glass, like those from Apollo 15, about 3.7 billion years old. This material, which had originated deep inside the Moon, had been brought to the surface by the impact that had produced the crater 19 million years ago.

Magnetic measurements made by Ian Garrick-Bethell and colleagues in 2008 of a 4.2 billion year old rock returned by Apollo 17 indicated that the Moon had once had a global magnetic field. This implied that it once had a liquid core and dynamo.

Later Apollo missions were cancelled to save money for other programmes, particularly the space shuttle. Remaining Apollo hardware was used in the Skylab programme and in the Apollo-Soyuz Test Project of the mid-1970s.

Bibliography

Beattie, Donald A., *Taking Science to the Moon: Lunar Experiments and the Apollo Program*, Johns Hopkins University Press, 2001.

Chaikin, Andrew, *A Man on the Moon: The Voyages of the Apollo Astronauts*, Penguin Books, 1995.

Compton, William David, *Where No Man Has Gone Before: A History of NASA's Apollo Lunar Expeditions*, NASA, 1989.

Mellberg, William F., *Moon Missions: Mankind's First Voyages to Another World*, Plymouth Press, 1997.

Wilhelms, Don E., *To a Rocky Moon; A Geologist's History of Lunar Exploration*, University of Arizona Press, 1993.

CASSINI—HUYGENS

Cassini—Huygens (see Figure 9.1) was a joint spacecraft programme to Saturn of NASA, the European Space Agency (ESA) and the Italian Space Agency. The Cassini spacecraft orbited Saturn and carried and released the Huygens probe to parachute into Titan's atmosphere. They were the first spacecraft to undertake these types of missions.

In 1969 NASA's Outer Planet Working Group had proposed an Outer Planets programme that included two 'Grand Tour' spacecraft, to be followed by a Saturn orbiter and probe to be launched in 1978—81. The Voyager 1 and 2 'Grand Tour'

Figure 9.1. The 6.8 m (22 ft) high Cassini-Huygens spacecraft with the 4 m (13 ft) diameter high gain antenna on top. The Huygens entry probe is seen beneath its conical heat shield at the side of the main Cassini spacecraft. (Courtesy NASA/JPL.)

spacecraft were approved in 1972. Ten years later the United States National Academy of Sciences and the European Science Foundation recommended that a Saturn Orbiter and Titan Probe mission should be undertaken. NASA and ESA then carried out a joint study, and in 1984 it was agreed that such a programme should be undertaken with NASA responsible for the Saturn Orbiter (later called Cassini) and ESA should be responsible for the Titan Probe (later called Huygens).

The Cassini orbiter's main mission, once it had released the Huygens probe, was to determine the structure and dynamical behaviour of Saturn's rings and of its magnetosphere, observe the larger of Saturn's satellites, particularly Titan, during numerous fly-bys, and study the dynamical behaviour of Saturn's clouds. In the meantime the Huygens probe would descend through Titan's atmosphere and land on its surface, relaying information back to Earth via the Cassini orbiter.

The 5,600 kg Cassini–Huygens was launched by a Titan IVB/Centaur launch vehicle on 15 October 1997. Because of

launcher constraints, the spacecraft undertook fly-bys of Venus (in 1998 and 1999), Earth (1999) and Jupiter (2000) in order to generate enough energy to reach Saturn in July 2004. During its Jupiter fly-by it undertook coordinated observations with the Galileo spacecraft that had been in orbit around the planet since 1995.

En route to Saturn it was realised that the receiver on Cassini was not operating correctly to enable it to satisfactorily receive signals from the Huygens probe as it descended through Titan's atmosphere. There was a problem of pulse synchronisation between the Huygens transmitter and the Cassini receiver. This was solved by changing the relative velocity of the two spacecraft after they had been separated, which required changing the spacecraft trajectories both before and after separation. So the Huygens probe was released on 25 December 2004, seven weeks later than originally planned, landing on Titan on 14 January 2005.

The Huygens probe found methane and nitrogen in Titan's stratosphere, with the methane concentration increasing as the probe descended. Clouds of methane were detected at an altitude of about 20 km. The probe discovered dendritic channels during its descent, and imaged large pebbles, probably of dirty water ice, at its landing site, which had an ambient temperature of 93.7 K. Heat generated by the probe caused bursts of methane gas to be emitted by Titan's surface.

The Cassini orbiter radar imaged what appeared to be large lakes on Titan. In 2007 one of these lakes, Ontario Lacus, was found to contain liquid hydrocarbons, of which ethane was positively identified. The ethane was thought to be in a liquid solution with methane, other hydrocarbons and nitrogen.

Cassini found that Enceladus had a thin, asymmetric atmosphere, most prevalent in the south polar region, that consisted mostly of water vapour. Shortly afterwards the spacecraft discovered plumes of water ice that was being ejected into Saturn's E ring from surface cracks or "tiger stripes" near Enceladus' south pole. Although the average temperature of the south polar region was 85 K, that in the tiger stripes was up to about 180 K. In March 2008 Cassini flew through a plume and detected water, carbon dioxide and hydrocarbons.

The Cassini spacecraft spent a number of hours in the shadow of Saturn in September 2006 and was able to make images of the ring system in silhouette. These revealed a tenuous ring between the F and G rings, associated with Janus and Epimetheus, and another associated with Pallene, a satellite that had been discovered by Cassini in 2004. Jets were also seen to be emitted by Enceladus replenishing the densest part of the E ring.

Cassini also discovered a number of new satellites of Saturn, two of which, Methone and Anthe, were associated with ring arcs which seemed to be constrained by resonances with Mimas. Anthe was found to move backwards and forwards along its arc, which was about 20° longitude in length.

Bibliography

Brown, R. H., Lebreton, J-P, and Waite, J. H. (eds.), *Titan from Cassini-Huygens*, Springer, 2009.

Dougherty, M. K., Esposito, L. W., and Krimigis, S. M. (eds.), *Saturn from Cassini-Huygens*, Springer, 2009.

Godwin, Robert, and Whitfield, Steve, *Deep Space: The NASA Mission Reports*, Apogee Books, 2005.

Harland, David M., *Cassini at Saturn: Huygens Results*, Praxis Publishing, 2007.

Spilker, Linda J. (ed.), *Passage to a Ringed World: The Cassini-Huygens Mission to Saturn and Titan*, NASA SP−533, 1997.

See also: Saturn.

CHANDRAYAAN−1

The Chandrayaan−1 spacecraft, which was launched in October 2008, was India's first lunar space mission. The spacecraft consisted of a lunar orbiter and a detachable probe that was planned to impact the Moon. It was hoped that the debris ejected by the impactor would confirm that there was water on the Moon, as had been implied by measurements made by the Lunar Prospector spacecraft ten years earlier. In addition to its impact mission, Chandrayaan−1 was designed to map both the Moon's surface at 5 metre resolution and its mineralogical composition.

The Chandrayaan−1 spacecraft mission had been announced during Indian Prime Minister Atal Vajpayee's Independence Day speech in August 2003. The launcher and spacecraft were built in India, but foreign scientists were invited to provide spacecraft experiments alongside Indian researchers. In the event, five Indian experiments were flown, together with four from Europe and two from the United States.

Chandrayaan−1 was launched by the Indian Space Research Organisation (ISRO) from their Satish Dhawan Space Centre into a geosynchronous transfer orbit. The spacecraft's apogee was then gradually increased by a series of five firings of its onboard liquid-fuelled rocket motor. The last firing put the spacecraft into a lunar transfer orbit. Subsequent on-board rocket firings put the spacecraft into a circular polar orbit around the Moon at an altitude of 100 km. Two days later the lunar impact probe was released from the main spacecraft, hitting the lunar surface near the crater Shackleton at the south pole. An instrument on-board this impact probe detected water in its vapour phase in the tenuous lunar atmosphere over a large range of latitudes (40° N to 90° S) and altitudes (from 98 km to the surface).

In September 2009 it was announced that an imaging spectrometer on-board the main spacecraft had detected the presence of low concentrations of OH/H_2O species in the top few millimetres of soil over significant areas of the Moon's surface. Then in the following year it was announced that bistatic radar observations from Chandrayaan−1 showed the presence of water ice at least 2 m thick in 40 permanently shadowed craters near the lunar north pole.

Bibliography

Anand, Mahesh, *Lunar Water: A Brief Review*, Earth Moon Planets, **107** (2010), 65–73.

CLEMENTINE

Clementine was a joint Ballistic Missile Defense Organisation/NASA spacecraft project to test new sensor technologies in space for both military and civilian applications. To do this Clementine was to be put into orbit around the Moon for about $2\frac{1}{2}$ months, followed by a fly-by of the asteroid Geographos.

Unusually for a lunar spacecraft, Clementine was launched by a surplus Titan intercontinental ballistic missile from the Vandenberg Air Force Base in California, instead of a NASA launcher from Cape Canaveral. On 20 February 1994 the 450 kg spacecraft entered a highly eccentric polar orbit around the Moon. From there it imaged the surface in eleven different narrow wavebands from 415 nm to 2.78 μm, and measured the Moon's topography with a laser altimeter having a resolution of about 40 m.

Clementine showed, for the first time, the full extent of the 2,500 km diameter South Pole-Aitken basin on the far side of the Moon, which it found was about 12 km deep. The spacecraft also provided much more accurate data on gravitational anomalies and, in particular, of mascons under the maria. It found that the centre of mass of the Moon was displaced from the Moon's geometric centre by 2 km in the direction of the Earth.

Clementine's excellent spectral coverage enabled geologists to produce a mineralogical map of the Moon, confirming, for example, that anorthosite is the dominant constituent of the lunar highlands. Measurements undertaken to try to detect evidence of ice in permanently shadowed craters at the south pole were inconclusive.

The spacecraft failed shortly after leaving lunar orbit en route to the asteroid Geographos.

Bibliography

Bussey, Ben, and Spudis, Paul, *The Clementine Atlas of the Moon*, Cambridge University Press, 2004.

Various authors, *Science*, **266** (1994), 1835–1862.

CLUSTER

Cluster was a European Space Agency (ESA) mission to study the Earth's magnetosphere, particularly the magnetotail and polar cusp. (The polar cusp is where particles that originated in the solar wind followed the converging field lines in Earth's polar regions down to the ionosphere.)

The Cluster programme consisted of four identical spacecraft to characterise the magnetosphere in three dimensions, allowing them to differentiate between spatial and temporal features. It was approved in 1986 as a joint programme with the SOHO solar observatory spacecraft, on condition that the total cost of both missions was reduced by 50%. To save money, ESA decided to launch all four Cluster spacecraft, each weighing 1.2 tons, one on top of another on an Ariane 5 qualification flight. Unfortunately, the launch in June 1996 was a failure.

After the failure, ESA decided to launch four new Cluster spacecraft in pairs on two Russian Soyuz launchers. These launches took place successfully in July and August 2000 from Baikonur in Kazakhstan, from which they achieved their operational $20,000 \times 120,000$ km polar orbits, with a period of 57 hours. The spacecraft were flown in a tetrahedral formation with spacecraft to spacecraft distances that could be varied from 600 to 20,000 km throughout their mission. As the orbits were inertially fixed, they bisected the Earth's magnetotail at apogee in about August each year, and passed through the northern cusp about 6 months later.

The Cluster spacecraft determined the characteristics of the magnetotail current sheet, which separates the northern and southern parts of the magnetotail, for the first time. They measured the bow shock thickness, and discovered vortices of solar wind particles that appeared to be tunnelling their way into the magnetosphere. Magnetic reconnection was observed at small-scale boundaries in the turbulent plasma of the magnetopause. This was the first time that magnetic reconnection had been observed in any turbulent plasma, the results having far-reaching significance in the fields of astronomy and plasma physics in general.

On 27 December 2004, radiation from a starquake on a neutron star was recorded by the Cluster and Double Star spacecraft, apparently giving the first observational evidence of cracks in a neutron star's crust. According to Steven Schwartz and colleagues, it indicated that the fracture size on the SGR 1806–20 neutron star or magnetar was about 5 km, which is significant on a body of only a few tens of kilometres diameter.

Bibliography

Credland, J., and Schmidt, R., *The Resurrection of the Cluster Scientific Mission*, ESA Bulletin, No. 91, August 1997, pp. 5–10.

Krige, J., Russo, A., and Sabesta, L., *A History of the European Space Agency, 1958–1987: Volume II, The Story of ESA, 1973–1987*, ESA SP 125, 2000.

Paschmann, G., et al. (eds.), *Outer Magnetospheric Boundaries: Cluster Results*, Springer, 2005.

Taylor, M. A., et al., *The Legendary Cluster Quartet: Celebrating Ten Years Flying in Formation*, ESA Bulletin, No. 145, February 2011, pp. 46–57.

Walsh, A. P., et al., 10 Years of the Cluster Mission, *Astronomy and Geophysics*, **51** (2010), 5.33–5.36.

DEEP IMPACT

Deep Impact was the first spacecraft mission to include an impactor to collide with a cometary nucleus, that of Tempel 1, and a main, fly-by spacecraft to observe the resulting crater and plume. The plume was also observed by other spacecraft and Earth-based observatories.

The main spacecraft included two imagers and an infrared spectrometer, whilst the 372 kg impactor included a 128 kg pure copper cratering mass and a targeting sensor. The impactor, which impacted the comet's nucleus at about 10 km/s, was designed to penetrate the comet's crust and expose the relatively fresh material underneath.

Deep Impact, which was launched in January 2005, arrived at Tempel 1 six months later. The impactor was attached to the main spacecraft until it was released about 24 hours before impact. The main spacecraft then used its control jets to ensure that it would fly past the comet at a distance of about 500 km. It was still about 10,000 km from the comet when impact occurred, flying past the comet about 16 minutes later. In 2007 the main Deep Impact spacecraft was retargeted to fly by Comet Hartley 2 in November 2010.

Comet Tempel 1 was found to have an irregular 7.6×4.9 km nucleus. The ejecta, which contained $\sim 10^7$ kg of material, included water, water ice, and some species that required high temperatures to form and others that required low temperatures. This indicated that the material had not all originated in the same region of the solar system. The ejecta obscured the impact crater which could not be seen with the fly-by spacecraft. But NASA later redirected their Stardust spacecraft to Tempel 1 to image the crater, which it did successfully in February 2011. It was about 150 m in diameter, slightly larger than expected.

Deep Impact's fly-by of Comet Hartley 2 in 2010, at a distance of 700 km, showed an active nucleus with numerous jets. These were driven by the heating of subsurface frozen carbon dioxide, not frozen water. The nucleus itself was peanut-shaped, about 2 km long and 0.4 km wide at its narrowest point, looking in many ways like the rubble-pile asteroid Itokawa. Its waist, which looked much smoother than the rest of the nucleus, appeared to be composed of fine-grained material that had drifted to fill the gravitationally lowest areas.

Bibliography

Various articles, Science, **310** (2005), 258–283.

Warner, Elizabeth, and Redfern, Greg, *Deep Impact: Our First Look Inside a Comet*, Sky and Telescope, June 2005, pp. 40–45.

DOUBLE STAR

Double Star was China's first spacecraft programme to study the Earth's magnetosphere. It consisted of two 660 kg spacecraft, TC (Tan Ce or Explorer) –1 and –2, both designed and launched by the Chinese National Space Administration (CNSA). Half of the experiments on board were provided by Chinese and half by European research establishments. The two spacecraft operated simultaneously in an extended constellation with the four Cluster spacecraft of the European Space Agency (ESA) which had been launched a few years earlier.

In 1992 the Chinese Academy of Sciences had agreed to collaborate with ESA on their Cluster mission. Then, seven years later, ESA was invited by CNSA to collaborate on their complementary Double Star programme.

TC-1, the equatorial spacecraft, was launched in December 2003 into a 570 × 79,000 km orbit inclined at 28.5° to the equator. This enabled it to study the Earth's magnetotail, where particles are accelerated towards the Earth's magnetic poles by magnetic reconnection processes. TC-2, the polar spacecraft, was launched seven months later into a 700 × 39,000 km polar orbit to study the physical processes taking place over the magnetic poles.

On 26 September 2005 the Double Star and Cluster spacecraft detected three consecutive substorms over a period of just two hours. Magnetic reconnection was found to occur in the magnetotail much closer to Earth than usual, and was almost co-located with the current disruption process. In turn NASA's IMAGE spacecraft imaged auroral brightenings just a few tens of seconds later. These, and similar observations, have provided a much better understanding of the role of magnetic reconnection and current disruption in causing magnetospheric substorms, and their associated auroral effects.

Bibliography

Laakso, H., Taylor, M., and Escoubet, C. P. (eds.), *The Cluster Active Archive: Studying the Earth's Space Plasma Environment*, Astrophysics and Space Science Proceedings, Part 1, Springer, 2010.

Yongxiang, Lu (ed.), *Science Progress in China*, Science Press, Beijing, 2003.

EARLY EXPLORER MAGNETOSPHERIC SPACECRAFT

The Explorer program had its origins in the U.S. Army Ballistic Missile Agency (ABMA) 1954 proposal, led by Wernher von Braun, to launch a very small scientific spacecraft into orbit using a modified Redstone rocket. ABMA suggested that this could be part of the U.S. contribution to the International Geophysical Year. In 1955 the proposal was turned down in favour of the Naval Research Laboratory's Project Vanguard.

Project Vanguard soon ran into trouble, and it was not ready to launch when the Soviet Union launched Sputniks 1 and 2 in late 1957. The Soviet Union's launch of these spacecraft created a mixture of surprise and panic amongst the American public when they realised their potential military significance. This public pressure helped to persuade President Eisenhower to authorize ABMA to launch its spacecraft as soon as possible. This spacecraft was the 14 kg Explorer 1 which was launched successfully from Cape Canaveral on 31 January 1958 into a 360 × 2,550 km orbit inclined at 33° to the equator. Its instruments included a Geiger counter developed by James Van Allen. There was no on-board data storage, so data could be obtained only when the spacecraft was within visibility of a ground station, which made it difficult to separate altitude effects from longitude/latitude effects. The particle counts appeared to saturate at about 800 km before reducing to zero at higher altitudes.

The payload on Explorer 3 was almost identical to that on Explorer 1, but it included a tape recorder. This enabled Van Allen to make sense of his results from both Explorers 1 and 3. On 1 May 1958 he announced the discovery of a belt of elementary particles around Earth, soon called the inner Van Allen belt. Explorer 4, launched on 26 July, mapped this inner belt for 2.5 months and, unrecognised at the time, also detected part of the outer Van Allen belt.

Pioneer 3, launched on 6 December 1958, discovered the outer Van Allen belt at an altitude of about 15,000 km during its abortive attempt to reach the Moon. Shortly afterwards particle tracks on photographic emulsions carried on a Thor-Able rocket proved, for the first time, that most of the particle radiation in the inner Van Allen belt was due to protons, with energies of at least 75 MeV.

The 64 kg Explorer 6, launched in August 1959, showed that both Van Allen belts were more complex and variable than previously realised. In March 1961, Explorer 10, which had been launched into a highly eccentric orbit with an apogee of 240,000 km, obtained the first indications of the tail-like structure of the magnetopause around local midnight. Then in 1962 Explorer 12 crossed the magnetopause a number of times at distances from Earth of between 8 and 12 Earth radii. It detected electrons with energies of about 4 MeV in the outer Van Allen belt.

Bibliography

Bille, Matt, and Lichock, Erika, *The First Space Race*, Texas A & M University Press, 2004.

Newell, Homer E., *Beyond the Atmosphere: Early Years of Space Science*, NASA SP-4211, 1980.

Stern, David P., *Reviews of Geophysics*, **34** (1996), 1–31.

Van Allen, James A., *Origins of Magnetospheric Physics*, University of Iowa Press, 2004.

EARLY SOUNDING ROCKET EXPERIMENTS

Sounding rocket experiments were undertaken in the United States after the Second World War to, amongst other things, measure the structure and composition of the Earth's upper atmosphere and how it was influenced by solar emissions. The early sounding rockets included the WAC Corporal, which had its first successful flight in October 1945; the V2, which had its first successful flight at White Sands, New Mexico, in May 1946; and the Aerobee, which was first launched the following year. The V2s, which had been captured from Nazi Germany at the end of the war, were able to carry payloads of 900 kg to heights of about 160 km. This was far better than the WAC Corporal (11 kg to 80 km) and the Aerobee (70 kg to 130 km), but the number of V2s available was strictly limited.

On 10 October 1946, Richard Tousey's Naval Research Laboratory (NRL) group, using a V2 experiment, managed to photograph the ultraviolet spectrum of the Sun down to 230 nm for the first time. Two years later, Robert Burnight of NRL, using a V2, found the first indication that the Sun may be emitting X-rays. Much clearer evidence was found by Herbert Friedman, also of NRL, using a V2 in September 1949. He found that the counting rate of his X-ray detectors was higher when they pointed towards the Sun. Then, using a Viking sounding rocket in December 1952, Friedman showed that the X-ray intensity was sufficient to sustain the E region in the Earth's ionosphere.

In 1952 William Rense's group at the University of Colorado, using an Aerobee, produced the first solar spectrum showing Lyman-α emission at 121.6 nm. Four years later they obtained the first photograph of the Sun in Lyman-α. In the meantime, Friedman's group had found that the majority of solar X-ray radiation in the 0.8 to 2.0 nm range was due to the solar corona. Then in 1956 they discovered that solar flares produce very-high-energy X-rays.

Bibliography

DeVorkin, David H., *Science with a Vengeance*, Springer-Verlag, 1993.
Friedman, Herbert, *The Astronomer's Universe*, Norton, 1990.

FAST

The Fast Auroral Snapshot (FAST) spacecraft was the second of NASA's Small Explorer (SMEX) missions. Weighing just 191 kg, it was designed to measure fields and particles in the auroral zones at about 65° magnetic latitude, and so help to determine how particles are accelerated to create the aurora. The spacecraft was launched in 1996 into an 83° inclined 350 × 4,175 km orbit that crossed the auroral zones four times per orbit. High-resolution data obtained during these auroral cross-

ings were relayed back to Earth in slower time during the remainder of the orbit.

Prior to launch it was known that the region where particles are accelerated is at an altitude of about 2,000–10,000 km above high latitude regions, whilst the visible aurora is emitted at much lower altitudes of about 100–250 km.

FAST discovered that the plasma inside the particle acceleration region was hot, not cold as expected. It also found that parallel electric fields accelerated electrons downwards to produce the visible aurora. Electrons were also accelerated in the opposite direction to create a non-visible inverse aurora, thus showing how the auroral electric current returned to space. The inverse aurora was not visible as the atmospheric density was too low by the time the upwards accelerating electrons had reached sufficient velocity.

FAST showed that the power source for the Earth's auroral kilometric radiation (AKR) was the electron-cyclotron maser powered by parallel electric fields. This AKR is the most powerful naturally-occurring radio emission radiated by the Earth into space.

Bibliography

Burch, J. L. (ed.), *Magnetospheric Imaging: The IMAGE Prime Mission*, Kluwer Academic Publishers, 2003.
Pfaff, Robert F. Jr. (ed.), *The FAST Mission*, Kluwer Academic Publishers, 2001.

GALILEO

In 1969 NASA's Outer Planet Working Group had proposed a comprehensive Outer Planets programme that included two 'Grand Tour' spacecraft, plus a Jupiter orbiter to be launched in 1976–78, followed by a Jupiter atmospheric multiprobe two years later. In the event the Voyager 1 and 2 'Grand Tour' spacecraft were approved in 1972, and the Jupiter Orbiter and Probe mission (now called Galileo) was approved as a one-spacecraft mission five years later.

NASA originally planned to launch the Galileo spacecraft from the space shuttle in 1982 using the Inertial Upper Stage (IUS) to provide the required additional velocity. But problems with the IUS forced NASA to change to the liquid-hydrogen-fuelled Centaur for a launch in 1985. Further programme delays and the Space Shuttle Challenger accident in January 1986, followed by a safety review, eventually forced NASA to return to the IUS.

The Galileo Jupiter orbiter and attached entry probe were launched from the Space Shuttle Atlantis in October 1989. It was the first orbiter and probe to be launched to Jupiter. Because of launcher constraints, Galileo then undertook fly-bys of Venus (in 1990) and Earth (in 1990 and 1992) in order to generate enough velocity to reach Jupiter in December 1995.

This resulted in a trip time to Jupiter of 74 months, compared with just 21 months for Pioneer 10 and 18 months for Voyager 1. However, NASA took advantage of this extended itinerary to send Galileo to fly by the asteroids Gaspra in October 1991 and Ida in August 1993. This resulted in the first close-up views of an asteroid.

Galileo was expected to deploy its high-gain antenna in April 1991. Unfortunately it failed to fully deploy, in spite of numerous attempts to free it. This resulted in NASA having to modify Galileo's on-board software and improve its ground receiving stations to partly compensate.

The Galileo spacecraft found that Gaspra was $18 \times 11 \times 9$ km in size, with a crater count that indicated a surface age of a few hundred million years. It also found that the 56 km long Ida had a 1.5 km diameter satellite, now called Dactyl. There had been some earlier suspicions that a few asteroids may have satellites, but this was the first to be confirmed.

The Galileo entry probe was released from the main Galileo spacecraft in July 1995, entering Jupiter's atmosphere some five months later at a velocity of 170,000 km/h. The probe started transmitting atmospheric data when it reached the 0.35 bar level in Jupiter's atmosphere, and continued transmitting until the atmospheric pressure had reached 23 bar, some 160 km further into the planet. The main Galileo spacecraft was put into orbit around Jupiter an hour or so after the probe had entered the atmosphere.

The probe, which had started to measure Jupiter's particle environment about 3 hours before atmospheric entry, found two particle belts, one inside and one outside of Jupiter's main ring. The outer belt was that previously discovered by the Pioneer spacecraft, but the inner one had not been seen before. Somewhat surprisingly, energetic helium ions were found in this inner belt.

It had been thought, before the Galileo mission, that there were three cloud layers on Jupiter, with the lower one consisting of water ice or possibly water droplets at about the 8 bar level. In the event the Galileo probe detected the top cloud layer at about 0.5 bar. The next cloud layer was between 0.8 and 1.3 bar, followed by a vertically thin layer at 1.6 bar. But, surprisingly, the expected cloud layer of water ice or droplets did not appear to exist. This is thought to be because the probe had entered Jupiter's atmosphere in an infrared hot spot where the atmosphere was exceptionally dry.

Ganymede was the first of the Galilean satellites to be observed close-up by the main Galileo spacecraft after it had been put into orbit around Jupiter. The spacecraft's first pass was at a distance of just 850 km above the surface, some 70 times closer than during the Voyager fly-bys. The generally dark surface of the 3,000 km diameter Galileo Regio was found to be littered with craters of all sizes, with crater counts indicating a basic surface age of about 4.2 billion years. But its surface had clearly been substantially modified since its forma-

tion apparently caused by a mixture of tectonics and cryovolcanism. Spectroscopic analysis showed that the bright regions on Ganymede were rich in water ice with patches of carbon dioxide ice.

Surprisingly, Ganymede was found to posses a magnetic field with an induced component, implying that the satellite has a salty ocean tens of kilometres thick, about 170 km below the surface. Galileo also detected a magnetosphere along with a thin ionosphere. It also found clear evidence of a tenuous atmosphere of atomic hydrogen, with ionised hydrogen streaming away at high speed.

Callisto was the second of Jupiter's satellites to be imaged by the Galileo orbiter. Voyager had shown that this, the outermost of Jupiter's Galilean satellites, had a dark surface almost completely saturated with craters. But to everyone's surprise, Galileo showed that there was an almost complete lack of smaller craters less than about one kilometre diameter. Callisto seemed to be covered with a dark material that created a very smooth surface, quite unlike that seen anywhere else in the solar system. This dark material, which could be several metres thick, might well be covering the small craters.

Galileo detected hydrogen atoms escaping from Callisto's surface. It later found evidence of oxygen probably as both a frost and as a gas in Callisto's very thin atmosphere, which appeared to consist of hydrogen, oxygen and carbon dioxide. Galileo also found that Callisto had a variable magnetic field caused by Jupiter's magnetic field inducing subsurface electric currents. Calculations showed that these currents could be in a salty ocean 10 to 20 km deep about 200 km beneath the surface.

Europa was the Galileo orbiter's next target. The spacecraft found that the area of chaotic terrain, now called Conamara, consisted of numerous polygonal rafts of ice. Some of the rafts were tilted, as if they had been subjected to lateral compressional pressure at some stage. Some were seen to carry segments of ridges, and reconstructions have shown that about half of the original surface was missing, most likely because of localised melting.

Galileo spacecraft measurements indicated that Europa, like Ganymede, had a differentiated structure, the radius of the core in Europa's case being about half the total radius. Galileo also found that Europa had an induced magnetic field, probably generated in a 100 km deep salty ocean that may be only a few kilometres below Europa's surface.

Galileo undertook a number of long distance observations of Io before its first close fly-by in October 1999. During one of these earlier observations it detected a plume from the Pillan Patera volcano. This produced a 400 km diameter, dark, pyroclastic blanket that covered part of the reddish halo around the volcano Pele. Unlike most other plume deposits on Io, however, which were white, yellow or red, Pillan's deposit was grey, resembling that of Babbar Patera nearby. Its temperature indicated that it was probably due to silicate volcanism.

The Galileo spacecraft made the first in situ measurements of electrons flowing up and down Io's flux tubes. It found that the energy dumped into each of Jupiter's polar regions was enough to produce Jupiter's aurorae. Galileo also undertook a number of ring observations and found that Jupiter's gossamer ring had two basic components, one inward of Amalthea, and one connected with Thebe. Dust from both of these satellites appeared to have supplied the gossamer ring.

NASA deliberately caused the Galileo orbiter to impact Jupiter on 21 September 2003, to protect Europa from possible contamination in the event of a future impact.

Bibliography

Bagenal, F., Dowling, T., and McKinnon, W. (eds.), *Jupiter: The Planet, Satellites and Magnetosphere*, Cambridge University Press, 2004.

Godwin, Robert, and Whitfield, Steve, *Deep Space, The NASA Mission Reports*, Apogee Books, 2005.

Harland, David M., *Jupiter Odyssey: The Story of NASA's Galileo Mission*, Springer-Praxis, 2000.

Meltzer, Michael, *Mission to Jupiter: A History of the Galileo Project*, NASA SP−2007−4231, 2007.

See also: Jupiter.

GENESIS

Genesis was a spacecraft mission designed to collect solar wind samples and return them to Earth. It was expected to determine elemental abundances, with critical isotope ratios being measured to about 1%. The most important elements were considered to be oxygen, nitrogen and the noble gases.

Genesis' collector arrays consisted of a series of ultra-pure low-density wafers of fifteen different materials including, aluminium, gold, silicon, germanium and sapphire to retain different elements in the solar wind. The spacecraft also included ion and electron monitors that detected changes in the solar wind, which was characterised as fast, slow and coronal mass ejections. Depending on which type of wind was detected, the spacecraft automatically deployed the appropriate collector array to collect samples.

In addition to these collector arrays, Genesis included a solar-wind concentrator, which concentrated the solar wind onto an assembly of small tiles. The concentrator was exposed to the solar wind for the duration of the collection mission.

It was important that the solar wind samples were not affected by the Earth's magnetosphere. So the 640 kg Genesis spacecraft was launched in August 2001 into a halo orbit around the L1 Lagrangian point, on the Sun-Earth line 1.5 million km sunward of Earth, some distance in front of the Earth's bow shock. Genesis collected samples from Decem-ber 2001 to April 2004, and then returned to the vicinity of Earth, where it released a landing capsule which landed on Earth on 8 September 2004. The return vehicle was to have been snatched in mid-air by a helicopter, but a fault with the landing capsule caused it to crash land into the Utah desert at about 190 mph (310 km/h). The impact fractured the capsule, creating a contamination problem, although it was still found possible to analyse the collector arrays.

Genesis found that the proportion of oxygen 16 to oxygen 17 and 18 in the solar wind, and presumably in the Sun, was similar to that found in some meteorites, suggesting that these were the values in the original solar nebula. But the ratio is higher than that on the Earth and Moon, which are identical, or on Mars or other meteorites. The reason for these differences is currently (2010) unclear.

Bibliography

Krot, Alexander N., et al., *Oxygen Isotopic Composition of the Sun and Mean Oxygen Isotopic Composition of the Protosolar Silicate Dust: Evidence from Refractory Inclusions*, Astrophysical Journal, 713 (2010), 1159–1166.

Russell, C. T. (ed.), *The Genesis Mission*, Kluwer Academic Publishers, 2003.

Ryschkewitsch, Michael, et al., *Genesis Mishap Investigation Board Report*, NASA, 2005.

GEOTAIL

Geotail was a joint Japanese/United States programme to investigate the plasma, electric and magnetic field characteristics of the distant and near geomagnetic tail, and to determine the role of this tail in substorm phenomena. Japan provided the spacecraft bus and the majority of the scientific instruments, whilst the United States provided the launcher and the remainder of the experiments. Launched in 1992, the 1,000 kg Geotail was the first spacecraft to be launched in the International Solar Terrestrial Physics (ISTP) programme. Other spacecraft involved in the ISTP included NASA's WIND and POLAR, and the European Space Agency's SOHO and Cluster.

Geotail spent just over two years investigating the distant region of the geomagnetic tail between about 80 and 210 R_E (Earth radii), using the Moon to adjust its orbit. The orbit was changed in late 1994 to investigate the near-Earth region of the geomagnetic tail, with an apogee of 50 R_E for four months, before it was reduced to 30 R_E. The spacecraft undertook numerous crossings of the magnetopause and bow shock. Sometimes it skimmed along the dayside magnetopause, so that numerous encounters were made with the magnetopause and the low-latitude boundary layer. The apogee of the lowest near-Earth orbit was in the neutral sheet around the December solstice.

The spacecraft measured the convective motion of plasma and magnetic field lines in the geomagnetic tail in both normal and disturbed periods, and observed the changes in flow direction at the onset of the substorm expansion phase. Geotail provided evidence of magnetic reconnection in the geomagnetic tail. In April 1995 it detected intense bursts of energetic particles at about 13 R_E downwind in the magnetosphere flowing earthwards at up to 2,000 km/s, or about twice the velocity of typical plasma flows. They coincided with the onset of a magnetic substorm at Earth. In May 1999 Geotail and a number of other spacecraft recorded a dramatic event when the solar wind almost disappeared. Its density was very low and its velocity only about half its normal velocity. As a result, Geotail observed that the Earth's bow shock was over 5 times further away from Earth than normal.

Bibliography

Burch, J. L. (ed.), *Magnetospheric Imaging: The IMAGE Prime Mission*, Kluwer Academic Publishers, 2003.

Nishida, A., Baker, D. N., and Cowley, S. W. H. (eds.), *New Perspectives on the Earth's Magnetotail*, American Geophysical Union, 1998.

Schindler, Karl, *Physics of Space Plasma Activity*, Cambridge University Press, 2007.

HALLEY'S COMET INTERCEPTS

Halley's comet has an orbit that takes it into the inner solar system past the Earth's orbit once every 76 years. The first of these visits since the start of the space age was due in 1985–86, when spacecraft technology was sufficiently mature to plan for a spacecraft intercept. The structure and composition of a cometary nucleus was still unknown at that time. In addition, Halley's comet was an active comet with a long tail when it passed close to the Sun, so making it a very interesting object to visit. On the other hand, this level of activity would put extra requirements on any spacecraft that was designed to go very close to the nucleus.

NASA plans in the mid-1970s to intercept Halley's comet were formulated at a time of peak funding for the space shuttle, and when a number of expensive space missions were being considered like the Hubble Space Telescope, the Galileo mission to Jupiter, and the Ulysses solar mission. To make matters worse, astronomers were also hoping to send a Halley spacecraft to fly with the comet for a number of months, which was a complex mission. As programme approval was not forthcoming, NASA considered, instead, sending a spacecraft to fly-by Halley's comet where it would release a small probe to fly closer to the comet. NASA suggested that the European Space Agency (ESA) could provide the probe, although at that time ESA were considering their own Halley comet intercept, later

called Giotto. Then in 1981 NASA's programme was cancelled for financial reasons.

NASA still wanted to send a spacecraft to observe a comet, however, but this now had to be done at minimum cost. So they decided to use their ISEE-3 spacecraft, renamed ICE (International Cometary Explorer), that was already in orbit. They planned to send this spacecraft to fly through the tail of the comet Giacobini-Zinner in September 1985, at a distance of about 7,800 km from its nucleus, and then use it to observe Halley's comet from a distance of about 28 million km.

Intercosmos (USSR) and ISAS (Japan) were also planning to send spacecraft to fly past Halley's comet. Intercosmos planned to send their two spacecraft, Vega 1 and 2, to Halley after they had flown past Venus, whereas ISAS planned to send two spacecraft, called Suisei and Sakigake, dedicated to the Halley mission. Intercosmos planned to send their two Vega spacecraft to within about 10,000 km of the Halley nucleus, whereas ISAS planned to send Suisei and Sakigake to within about 150,000 and 7 million km, respectively. ESA, on the other hand, were planning to send their Giotto spacecraft to within about 1,000 km of the nucleus. All the participating agencies decided to aim for an encounter on the comet's outbound track around 10 March 1986, when its 18° inclined orbit crossed the ecliptic.

It was clearly important to coordinate these missions from ESA, Intercosmos, ISAS and NASA. So an informal Inter-Agency Consultative Group was established with representatives from each agency. One of the problems discussed by the group was how to improve the targeting of the Giotto spacecraft, which was expected to pass closest to Halley's nucleus. Unfortunately, the position of the nucleus was not known at that time with sufficient accuracy to permit such a close fly-by. But the Russian Vega spacecraft were due to fly-by Halley a few days before the Giotto intercept. So NASA agreed to use their Deep Space Network to accurately locate the Russian Vega spacecraft as they flew past the comet's nucleus. This would allow ESA to retarget Giotto at the last minute for its close intercept.

Halley's comet orbits the Sun in a retrograde sense, which meant that the spacecraft would intercept the comet, and its associated dust, with an exceptionally high velocity of about 70 km/s. This was a potential problem for the Giotto spacecraft as it was to pass closest to Halley's nucleus, and so pass through a dense dust environment. As a result, Giotto was protected by a two-layer bumper shield.

Early on 6 March 1986, Vega 1 crossed the comet's bow shock about 1.1 million kilometres from the nucleus. Then at a distance of 320,000 km, Vega 1 detected its first impact by dust particles. It eventually produced a number of images of the nucleus that appeared to have a surface temperature of about 330 K. The shape of the nucleus imaged by Vega 2 looked quite different from that imaged by Vega 1, indicating that the nucleus had rotated between the two images.

For some months the Japanese Suisei spacecraft's ultraviolet telescope had been observing the hydrogen corona that extended some 10 million km from the comet, and found that it tended to brighten at a frequency of 52.9 hours. This was similar to the rotation period of the nucleus deduced in 1985 from enhanced copies of photographs taken during Halley's last visit to the vicinity of the Sun in 1910. Suisei showed that the 400 km/s solar wind reached a minimum of 60 km/s at the spacecraft's closest approach. Sakigake, the other Japanese spacecraft, detected long-period plasma waves induced by pick-up ions from the comet.

The Giotto spacecraft crossed Halley's bow shock on 13 March, 1.3 million kilometres from the nucleus. The first dust particles struck the spacecraft whilst it was still some 290,000 km from its target. About one minute before closest approach Giotto crossed the ionopause 4,700 km from the nucleus, the only spacecraft to do so, where the interplanetary field fell to zero. Then 7.6 seconds before closest approach Giotto was hit by a large dust particle that started the spacecraft nutating, causing an intermittent signal to be received on Earth.

Giotto found that inside the ionopause there was a stream of neutral molecules and cold ions flowing away from the nucleus at about 1 km/s. These neutral molecules were the agent that dragged dust particles away with them to form the coma or head of the comet. Bright jets could be seen streaming towards the Sun from the $16 \times 8 \times 8$ km nucleus in both the Vega and Giotto images. The Vegas and Giotto found that the nucleus had a density of about 0.2 g/cm^3 and its inactive part an albedo of only 0.03.

Halley's inner coma was found by the Vegas and Giotto to consist mainly of water, with smaller amounts of carbon monoxide, carbon dioxide, methane, ammonia and polymerised formaldehyde. These spacecraft found that the dust particles consisted of carbon, hydrogen, oxygen and nitrogen, and simple compounds of these elements, or of mineral-forming elements such as silicon, calcium, iron and sodium. Water vapour and dust production rates were about 20 tons/s and 5 tons/s, respectively, but even at these rates there is still enough material in the nucleus to allow about 1,000 more orbits of the Sun.

Bibliography

Brandt, John C., and Chapman, Robert D., *Introduction to Comets*, 2nd ed., Cambridge University Press, 2004.

Calder, Nigel, *Giotto to the Comets*, Presswork, 1992.

Leverington, David, *New Cosmic Horizons: Space Astronomy from the V2 to the Hubble Space Telescope*, Cambridge University Press, 2000.

Reinhard, R., and Battrick, B. (eds.), *Space Missions to Halley's Comet*, ESA SP−1066, 1986, and *The Giotto Mission − Its Scientific Investigations*, ESA SP−1077, 1986.

Reinhard, Rüdeger, *The Halley Encounters*, in Beatty, J. Kelly, and Chaikin, Andrew (eds.), *The New Solar System*, 3rd ed., Cambridge University Press, 1990.

Various authors, Comet Halley Supplement, Nature, **321** (1986), 259–366.

HAYABUSA

Hayabusa was the first spacecraft designed to return samples from the surface of an asteroid. Launched in 2003, it used ion engines to slowly approach its target asteroid, Itokawa (25143), which had been discovered in 1998. When Hayabusa arrived in 2005 it remained close to Itokawa in a station-keeping heliocentric orbit, surveying the asteroid and looking for a suitable place to land. It was also planned to fire small projectiles into the surface and collect part of the resulting spray, returning it to Earth in a small re-entry capsule. The spacecraft was also expected to release a very small hopper called Minerva to land on Itokawa. But problems with the release sequence caused Minerva to drift into space.

There were a number of problems with the main Hayabusa spacecraft, so, although it did land, it is not known whether the projectiles were fired or whether the spacecraft collected any samples. Further problems delayed the return to Earth from 2007 to 2010 when its sample return capsule successfully landed in Australia. Initial inspection showed that it had captured numerous extremely small dust particles of extraterrestrial origin.

Hayabusa found Itokawa to be only $535 \times 294 \times 209$ m in size with a density of 1.90 g/cm^3. This indicated that it was basically a rubble pile with a microporosity of 40%. Itokawa's surface had no craters, but it was generally very rough with many large boulders. The remaining 20% of its surface was smooth, being covered with pebbles and dust.

Bibliography

Beatty, J. Kelly, *The Falcon's Wild Flight*, Sky and Telescope, September 2006, pp. 34–38.

Various articles, Special Issue: Hayabusa at Asteroid Itokawa, Science, **312** (2006), 1327–1353.

HELIOS

Helios was a West German/United States two spacecraft programme designed to study the Sun and the interplanetary medium. The spacecraft investigated the solar wind, magnetic and electric fields, cosmic rays and micrometeorites. Helios 1 was launched on 10 December 1974 and Helios 2 just over a year later. Both spacecraft flew closer to the Sun than any previous man-made objects with perihelia about 0.3 AU from the

Sun and aphelia at the Earth's orbit. The orbital period was about 190 days.

The Helios programme was conceived in the mid-1960s as part of an attempt by the United States to improve their technological ties with Western Europe, provided any technology transfer had little or no military applications. Helios grew out of this American initiative, with West Germany providing the spacecraft bus and the United States two Titan-Centaur launch vehicles. Germany also provided seven of the ten experiments, and the United States the remainder.

Helios 1 operated over a complete solar cycle of about eleven years, whereas Helios 2 operated for just over four years before succumbing to the intense heat around perihelion. The spacecraft confirmed that there were both a slow and a fast solar wind. The slow solar wind, with an average velocity of about 350 km/s, showed much more variability than the fast solar wind, that had an average velocity of about twice as much. The fast wind was emitted by coronal holes, whereas the slow wind appeared to be coming from magnetic regions that had become temporarily open.

The Helios spacecraft detected significant differences in the two solar winds. For example, protons had higher temperatures than electrons in the fast solar wind, whereas the opposite was the case in the slow wind.

Bibliography

Kraemer, Robert S., *Beyond the Moon: A Golden Age of Planetary Exploration 1971–1978*, Smithsonian Institution Press, 2000.

Krige, J., and Russo, A., *A History of the European Space Agency 1958–1987*, Vol. **I**, European Space Agency SP-1235, 2000.

Porsche, Herbert (ed.), *10 Years HELIOS*, Wenschow Franzis Druck, 1984.

HINODE

Hinode (Japanese for "sunrise"), known as Solar-B before launch, was launched on 22 September (UT) 2006 from the Uchinoura Space Centre in Japan into a 600 km Sun-synchronous, polar orbit. It was designed to investigate magnetic activity on the Sun, including methods of magnetic field generation, transport and dissipation. Of particular interest were the mechanisms responsible for heating the corona and for the production of solar flares and coronal mass ejections. Hinode was a Japanese-led mission with United States, United Kingdom, European Space Agency and Norway support.

Launched as a follow-up to Yohkoh (Solar-A), the 900 kg Hinode included three payload instruments; a 50 cm diameter solar optical telescope, a high resolution grazing incidence X-ray telescope, and an extreme ultraviolet (EUV) imaging spectrometer. The solar optical telescope had an unprecedented resolution of 0.2 arcsec over the 480 to 650 nm range. The soft X-ray telescope had a spatial resolution of 2 arcsec and time cadence of 2 seconds. It was designed to produce images of the corona at different temperatures. The EUV imaging spectrometer, which operated over the 17 to 29 nm range, was designed to measure the flow of hot gases in the solar atmosphere to an accuracy of about 2 km/s.

Hinode identified the source of the low speed solar wind for the first time. The spacecraft found that it was emitted from around the base of bright coronal loops. Hinode also discovered that the chromosphere was permeated by Alfvén waves, with periods of several minutes, which were sufficiently strong to accelerate the solar wind, and possibly to heat the quiet corona. In addition, Hinode detected Alfvén waves in the corona, and provided new information on how magnetic field lines may cross and reconnect to cause solar flares.

Bibliography

Publications of the Astronomical Society of Japan Special Issue, **59**, SP3, 30 November 2007.

Science Magazine Hinode Special Issue, **318**, 7 December 2007.

Shibata, K., Nagata, S., and Sakwai, T., New Solar Physics with Solar-B Mission, *Astronomical Society of the Pacific Conference Series*, **369**, (2007), 1–593.

Various articles, Solar Physics, **243**, No. 1 (2007), 3–92, and **249**, No. 2 (2008), 167–279.

IMAGE

The Imager for Magnetopause-to-Aurora Global Exploration (IMAGE) spacecraft was the first spacecraft mission dedicated to imaging the Earth's magnetosphere. It was also the first of NASA's Medium Class Explorer Missions (MIDEX). The 490 kg spacecraft was launched in 2000 into a 1,000 × 46,000 km polar orbit whose apogee precessed from 40° N latitude to 90° N over the first year, reaching 40° N again about a year later. It operated satisfactorily for over 5 years.

The imagers on board IMAGE included ultraviolet instruments to image auroral emissions excited by precipitating protons and electrons, auroral band emissions from molecular nitrogen, and emissions from singly ionised helium in the Earth's plasmasphere (the inner magnetosphere). There was also a neutral atom instrument to image energetic neutral atom (ENA) emissions between 10 eV and 500 KeV, and a radio imager that produced soundings of the magnetopause, plasmapause (the outer edge of the plasmasphere) and ionosphere.

The images from this innovative spacecraft produced a wealth of new information. For example, they showed, for the first time, fast-moving neutral particles in the solar wind. Some of these neutral particles were thought to have been emitted by coronal mass ejections, whereas some others were charged particles that had been neutralised by collisions en route. IMAGE also produced the first global images of the proton aurora, and showed its relationship with the electron aurora. The IMAGE and SAMPEX spacecraft showed that the shrinkage of the slot region between the inner and outer Van Allen belts, which is usually devoid of high-energy electrons, was directly related to erosion of the plasmasphere.

Bibliography

Burch, J. L., (ed.), *Magnetospheric Imaging: The IMAGE Prime Mission*, Kluwer Academic Publishers, 2003.

Various articles, Space Science Reviews, **91** (2000), Nos. 1–2, and **109** (2003), Nos. 1–4.

IMP-1

Interplanetary Monitoring Platform (IMP)-1 was the first and most important of a series of ten Explorer-type spacecraft launched between 1963 and 1973 to investigate the spatial and temporal variability of the Earth's magnetosphere and the near-Earth solar wind.

The 62 kg IMP-1, otherwise known as Explorer 18, was to have been launched into a highly eccentric, six-day orbit with an apogee of 278,000 km on the sunward side of the Earth. But a launcher malfunction on 27 November 1963 placed the spacecraft into a 198,000 × 190 km orbit. On board were three magnetometers, four cosmic ray experiments, and four solar wind particle experiments. The achieved orbit enabled IMP-1 to spend about 75% of each orbital period in interplanetary space beyond the Earth's bow shock. But by mid-February 1964 its time in interplanetary space on each orbit had reduced to zero.

IMP-1, which was launched around solar minimum, discovered that the magnetic field lines in interplanetary space made an angle of about 45° to the Sun-Earth line as predicted by Eugene Parker in 1958. But, surprisingly, the magnetic field changed direction every few days by about 180°. Norman Ness and John Wilcox then found, by comparing the IMP-1 results with those of the Sun at Mount Wilson, that the interplanetary magnetic field's polarity was correlated with that of the Sun's photosphere at the Sun's equator with a time lag of about 4.5 days. This implied an average solar wind velocity of about 380 km/s, which was consistent with that measured by IMP-1's plasma trap. So the interplanetary magnetic field was an extension of that of the Sun.

Ness and Wilcox analysed the results of almost three solar rotations and found that the abrupt, 180° changes in field direction occurred at about the same place relative to the Sun's surface on each revolution, except during a solar storm. This led them to conclude that the Sun-generated interplanetary magnetic field was composed of organised sectors of alternately inward- and outward-pointing magnetic fields.

IMP-1 found that the geomagnetic tail on the night side of the Earth appeared to extend well beyond the spacecraft's apogee of 32 R_E (Earth radii). It also discovered a narrow region at local midnight where the geomagnetic field fell to almost zero. This so-called neutral sheet was sandwiched between a region to its north, where the geomagnetic field lines were pointing towards the Earth, and a region to its south, where the field lines pointed in the opposite direction.

The spacecraft's instruments provided the first clear-cut evidence for the Earth's bow shock which it crossed many times at about 13 to 25 R_E from Earth, depending on the bow shock's position relative to the Sun-Earth line. Inside the magnetopause, at about 10 to 14 R_E, the geomagnetic field was well ordered. But between the magnetopause and bow shock it was quite turbulent. Outside the bow shock it was relatively steady.

Bibliography

Hess, Wilmot N., *The Radiation Belt and Magnetosphere*, Blaisdell Publishing Company, 1968.

Hufbauer, Karl, *Exploring the Sun: Solar Science since Galileo*, Johns Hopkins University Press, 1991.

Watts, Raymond N. Jr., *Interplanetary Monitoring Platform*, Sky and Telescope, January 1964, pp. 27–28.

ISEE

International Sun-Earth Explorer (ISEE) was a three-spacecraft NASA/European Space Agency (ESA) programme designed to study the effect of the solar wind on the Earth's magnetosphere. It was the first mission to be able to observe in detail how the magnetosphere reacted to changes in the solar wind, and be able to separate out spatial and temporal variations. The designs of the ISEE-1 and -3 spacecraft, weighing 340 kg and 470 kg, respectively, were based on those of the IMP (Interplanetary Monitoring Platform) spacecraft. They were provided by NASA, whereas the 160 kg ISEE-2 spacecraft was provided by ESA.

ISEE-1 and -2 were launched together on 22 October 1977 into a highly elliptical orbit with an apogee of 23 R_E (Earth radii), or 138,000 km altitude, and a perigee of 290 km. Depending on time of year, this orbit took the spacecraft outside of the Earth's magnetosphere for up to 75% of their orbital period. The spacecraft separation could be varied. ISEE-3 was launched almost one year later and placed into a halo orbit around the L_1 Lagrangian point, about 235 R_E upstream between the Earth and Sun. The solar wind that passed this point arrived in the vicinity of the Earth about one hour later, where it could be observed by ISEE-1 and -2.

In 1982 ISEE-3 was moved from its position near the Lagrangian point to undertake a series of transits of the geomagnetic tail, or geotail, the furthest of which was 236 R_E from Earth. Renamed ICE (International Cometary Explorer), the spacecraft was later retargeted to pass through the plasma tail of comet Giacobini-Zinner in 1985, about 7,800 km from the comet's nucleus. In the following year, ICE was used to observe Comet Halley from a distance of about 28 million km.

Previous spacecraft had found that there were two distinct lobes in the Earth's geotail, with steady earthward and anti-earthward directed magnetic fields separated by a neutral sheet embedded in a sheet of hot plasma. But these previous space-craft had generally been limited to observing the geotail within about 60 R_E of Earth. ISEE-3 found that the diameter of the Earth's geotail increased by about 30% from its near-Earth value out to a distance of about 120 R_E. After that its diameter remained constant with distance. ISEE-3 also found that the basic magnetic structure of the tail continued virtually unchanged into the distant tail. It found that, whilst the near-Earth plasma flow was mostly earthward, there was a clear outward flow in the distant plasma sheet which appeared to accelerate as it got beyond about 80 R_E from Earth.

Bibliography

Durney, A.C., *The International Sun-Earth Explorer (ISEE) Mission*, ESA Bulletin, No. 12, February 1978, pp. 12–18.

Knott, K., Durney, A., and Ogilvie, K. (eds.), *Advances in Magnetospheric Physics with GEOS-1 and ISEE. Proceedings of the Thirteenth ESLAB Symposium, Innsbruck, June 1978*, reprinted from Space Science Reviews, 22:4–6 and 23:1, Springer, 1979.

Wenzel, K.-P., *Earth's Distant Geomagnetic Tail Explored by ISEE-3 Spacecraft*, ESA Bulletin, No. 37, February 1984, pp. 46–50.

LUNA

Luna, or Lunik, was a series of unmanned spacecraft launched to the Moon by the Soviet Union between 1958 and 1976. The first three launches failed, but the first successful launch, on 2 January 1959, resulted in the 360 kg Luna 1 (see Figure 9.2) becoming the first spacecraft to reach the vicinity of the Moon. It detected a zone of intense radiation about 30,000 km from Earth, and flew past the Moon at a distance of about 6,000 km 34 hours after launch. Eight months later, Luna 2 was the first spacecraft to impact the Moon, followed three weeks later by Luna 3, which became the first spacecraft to photograph the Moon's far side. Although poor by later standards, these photographs showed that there were fewer maria on the far side than on the near side of the Moon.

The next series of Luna spacecraft were designed to soft-land on the Moon. They each weighed about 1,400 kg, and included a 100 kg landing capsule complete with camera. In 1963 Luna 4 missed the Moon by about 8,500 km because of a problem with its trajectory-correction manoeuvre. Then in

Figure 9.2. Luna 1, the first successful lunar spacecraft, launched by the Soviet Union in January 1959. It passed within about 6,000 km of the Moon, but the Soviet Union were deliberately vague at the time about its exact mission. It was not until two years later that they admitted that they had intended it to crash-land on the Moon. (Courtesy NASA.)

1965, Luna 5 impacted the lunar surface when its retrorocket failed. Luna 6, 7, and 8 also failed, but on 3 February 1966, the landing capsule of Luna 9 successfully landed on the Moon. Early the next day it became the first spacecraft to transmit images from the lunar surface, showing a relatively dust-free, rock-strewn surface apparently of volcanic origin.

Two months later, Luna 10 became the first spacecraft to orbit the Moon. It detected a weak magnetic field that did not change with altitude, indicating that the spacecraft was probably measuring the interplanetary magnetic field, rather than that of the Moon itself. Luna 13, which landed on the Moon in December 1966, included an explosively driven probe. This found that the Moon's soil had the load-bearing characteristics of terrestrial soil.

Luna 16 undertook the first successful automatic sample-return mission, returning 101 g of material from Mare Fecunditatis in September 1970. The sample was mainly of basaltic composition, but it also included spherical glazed particles of impact origin. Two months later, Luna 17 landed an eight-wheeled vehicle, called Lunokhod, on the Moon. It travelled over the lunar surface controlled by a team of drivers on Earth. Over the course of 11 months, Lunokhod travelled a total of 10.5 km in the Mare Imbrium, taking thousands of images and performing numerous soil analyses. These confirmed that the lunar surface, at least in the maria regions, was composed mostly of basalt. Luna 20 samples, which were returned to Earth from a highland region in 1972, had a high concentration of anorthosite material like those returned by Apollo 16 and 17, which had also landed in a highland region. Luna 24, on the other hand, which landed in the Mare Crisium four years later, retrieved core samples from depths of up to 2.3 m that contained basalt of about 3.2 billion years old.

Bibliography

Harford, James, *Korolev: How One Man Masterminded the Soviet Drive to Beat America to the Moon*, John Wiley and Sons, 1997.

Leverington, David, *New Cosmic Horizons: Space Astronomy from the V2 to the Hubble Space Telescope*, Cambridge University Press, 2000.

Reeves, Robert, *The Superpower Space Race: An Explosive Rivalry through the Solar System*, Plenum Press, 1994.

LUNAR ORBITER

Lunar Orbiter was a spacecraft programme designed to image possible landing sites for the manned Apollo spacecraft. It was also to monitor micrometeorites and particle radiation in the Moon's vicinity to show that the environment was safe. NASA launched five Lunar Orbiters between August 1966 and August 1967, all of which operated successfully in lunar orbit.

The Surveyor Lunar Orbiter programme had been conceived in 1959 as a purely scientific programme consisting of a main spacecraft to orbit the Moon, and a smaller detachable probe that would soft-land. In the following year, NASA decided that the soft-lander and lunar orbiting spacecraft should be autonomous spacecraft, based on a common bus design, and launched separately. But on 25 May 1961 President Kennedy committed America to landing an astronaut on the Moon before the end of the decade. As a result NASA started to add Apollo requirements to this common bus design.

Before long it became clear that the various, often competing, requirements being placed on the common bus concept made it unviable. So the orbiter part of the mission was cancelled, and only the Surveyor soft-landing design retained. But the Apollo project still needed high-quality images of possible landing sites. As a result, six months later the Lunar Orbiter project was begun.

Orbiters 1, 2, and 3 were put into equatorial lunar orbits. Their performance was so good, however, that NASA had, by April 1967, settled on eight potential landing sites for Apollo. As a result, NASA re-designated Orbiters 4 and 5 as scientific spacecraft, launching both into polar orbits. This resulted in almost complete image coverage of both sides of the Moon to at least medium resolution.

Although Lunar Orbiter was not a scientific programme, slight deviations in the orbit of Orbiter 1 indicated that there were mass concentrations or mascons beneath some of the Moon's maria. Image analysis showed large differences in crater counts between different parts of the Moon. This indicated either that these were areas of radically different age, or that many craters were of volcanic origin. A few fresh-looking craters were found indicating that, whatever the cause, the surface of the Moon was changing, albeit very slowly.

Bibliography

Byers, Bruce K., *Destination Moon: A History of the Lunar Orbiter Program*, NASA TM X–3487, 1977.

Kosofsky, L.J., and El-Baz, Farouk, *The Moon as Viewed by Lunar Orbiter*, NASA SP–200, 1970.

Nicks, Oran W., *Far Travelers: The Exploring Machines*, NASA SP–480, 1985.

LUNAR PROSPECTOR

Lunar Prospector was a small, lunar-orbiting spacecraft launched in 1998 as a follow-on mission to Clementine. Ten years earlier, Lunar Prospector had been started as a private initiative to show that lunar exploration could be undertaken quickly and inexpensively. But private programme funding for the whole programme was not forthcoming, so it was eventually undertaken as part of NASA's 'faster, better, cheaper' Discovery Program.

The aim of the Lunar Prospector mission was to map the Moon's surface composition and measure its magnetic and gravity fields at high resolution. In the process it was hoped that the spacecraft would determine whether there was water ice in the permanently shadowed craters at the Moon's poles. The spacecraft was launched on 7 January 1998 and was placed into its operational near-circular, 100 km altitude, polar orbit nine days later. Near the end of the year, its orbital altitude was reduced to provide high-resolution data.

Lunar Prospector had six experiments on board, including a neutron spectrometer to detect neutrons that are emitted from the Moon when cosmic rays impact the lunar surface. It showed a reduction in epithermal neutrons, but no decrease in fast neutrons, at both poles. This is consistent with deposits of water ice, covered with a thin layer of regolith, in the permanently shaded craters at the poles. Lunar Prospector also indicated that KREEP-rich material had been excavated from the Moon's crust/mantle boundary by the Imbrium impact, and that there was a relatively strong magnetic field at the antipodes of the Mare Imbrium and Mare Serenitatis. Gravity mapping identified a number of new mascons, and indicated that the Moon had a 650 km diameter iron rich core.

The mission ended on 31 July 1999 when the spacecraft was targeted to impact the Moon in a permanently shadowed area of a polar crater, hoping to liberate water vapour. But no such water vapour was detected by Earth-based telescopes.

Bibliography

Binder, Alan B., *Lunar Prospector: Against All Odds*, Ken Press, 2005.
Various authors, *Science*, 281 (1998), 1475–1500.

LUNAR RECONNAISSANCE ORBITER/LCROSS

Lunar Reconnaissance Orbiter/Lunar Craft Observation and Sensing Satellite, or LRO/LCROSS, was a NASA mission to prepare the way for astronauts to return to the Moon. Its main purpose was to produce a comprehensive atlas of the Moon's features and resources to assist with the design of an inhabitable lunar outpost. In the process it was hoped that the spacecraft would resolve the question, once and for all, as to whether there was water ice in permanently shadowed craters near the Moon's poles.

NASA's new Moon strategy, of which LRO/LCROSS was the first mission, was determined in 2004, aided by a newly published National Research Council report, *New Frontiers in the Solar System: An Integrated Exploration Strategy*. It was initially intended to launch the Lunar Reconnaissance Orbiter (LRO) spacecraft with a Delta II, but in 2005 NASA changed the launch vehicle to a more powerful Atlas V, which allowed the addition of an extra payload. As a result it was decided to retarget the spent Centaur upper stage of the Atlas to impact the Moon in a shadowed crater near the Moon's poles. In addition a second spacecraft, LCROSS, would be added to the launch vehicle's payload. This LCROSS spacecraft would follow the Centaur at a distance of about 600 km and fly through its impact plume, before crashing on the Moon about 4 minutes after the Centaur impact. It was expected that the Centaur impact would be observed not only by LCROSS and the lunar-orbiting LRO, but by Earth-orbiting spacecraft and Earth-based telescopes.

In the event, the Centaur crashed into the Cabeus crater near the south pole of the Moon in October 2009. None of the Earth-based telescopes nor Earth-orbiting spacecraft observed the impact as it, apparently, produced much less debris than expected. But two instruments on LCROSS found clear evidence of ~100 kg of water vapour and water ice in that part of the plume in their field of view. The LRO/LCROSS mission also found that the impact released ammonia, methane, hydrogen, carbon dioxide, carbon monoxide, sodium and, most surprisingly, quite large amounts of mercury.

One of the most surprising observations from the orbiting LRO spacecraft was that parts of the Moon's polar regions had temperatures as low as 35 K. This was far lower than expected and the lowest surface temperature yet detected in the solar system.

Bibliography

Anand, Mahesh, *Lunar Water: A Brief Review*, Earth Moon Planets, **107** (2010), 65–73.
Beatty, J. Kelly, *NASA Slams the Moon*, Sky and Telescope, February 2010, pp. 28–33.
Redfern, Greg, *Lunar Fireworks*, Sky and Telescope, June 2009, pp. 20–25.
Various articles, Science, **330** (2010), 463–486.

MAGELLAN

The Magellan spacecraft (see Figure 9.3), launched to Venus in May 1989, was based on NASA's cancelled VOIR (Venus Orbiting Imaging Radar) Venus spacecraft of the previous decade. VOIR was to have included a high-resolution synthetic aperture radar (SAR), similar to that flown on the Earth-orbiting Seasat, instead of the simple radar altimeter flown on Pioneer-Venus. However, budget cuts caused NASA to reduce the new mission's scope and insist on maximum re-use of Voyager and Galileo components. This resulted in the new smaller spacecraft called Magellan. It was the first planetary spacecraft to be launched by a space shuttle.

The Challenger Space Shuttle disaster of January 1986 caused an automatic delay in Magellan's launch date, along with those of all other planned shuttle launches. But in Magellan's case it also caused NASA to replace the cryogenic, liquid-fuelled Centaur rocket, which was to be used to boost Magellan from Earth orbit, by the safer solid-fuelled Inertial Upper Stage (IUS). But the IUS was not as powerful as the Centaur, and as a result the travel time to reach Venus was also increased from 6 to 15 months.

Magellan had just two experiments on board: a SAR and a radar altimeter designed to have horizontal and vertical resolutions of about 120 and 10 m, respectively, at periapsis. When Magellan finally arrived at Venus in August 1990, it was put into a 290 × 8,030 km altitude near-polar orbit. The SAR antenna operated for 37 minutes per orbit storing data on an on-board tape recorder, before the same antenna was used to transmit data to Earth. This required the antenna to point towards Earth and then back to Venus on each orbit.

The first mapping cycle lasted 243 days during which Venus had spun once below the spacecraft's orbit. In all, three complete mapping cycles were completed, allowing mapping of about 98 percent of the surface. The periapsis was then reduced to 185 km to allow gravitational mapping around periapsis by measuring the effect of local gravity concentrations on the spacecraft's orbit. Finally, the apoapsis was reduced by

Figure 9.3. The Magellan spacecraft, mounted on top of the Inertial Upper Stage, during their deployment from the space shuttle. Magellan's main antenna at the top doubled as both its SAR antenna and as the high-gain antenna for communications with Earth. (Courtesy NASA.)

aerobraking and the orbit circularised to allow gravitational mapping at higher latitudes.

Most of the 900 impact craters imaged by Magellan appeared fresh, indicating that weathering on Venus was a slow process. The lack of any craters larger than about 275 km diameter, and the relative paucity of double-ringed craters, indicated that the surface was relatively young. Crater counts implied that most of the surface appeared to be about only 400 million years old, with no features older than about 900 million years.

Venus showed clear signs of tectonic activity in the highland regions, although there was no global network of faults like there was on Earth. There appeared to have been no obvious plate tectonic activity over the last few hundred million years. The plains of Venus, which covered most of its surface, showed complex patterns indicating that they were volcanic in nature.

They also had thousands of individual shield volcanoes, so it was a surprise when Magellan's gravitational measurements indicated that the rigid lithosphere appeared to be at least 30 km thick.

Venus' lava flows indicated that the lava on Venus was sometimes extremely fluid and at others very viscous. The fluid lava had produced channels over 6,000 km long that look more like rivers on Earth, whereas the viscous lava had produced large pancake-shaped volcanic domes, about 20 to 60 km in diameter and about 100 to 750 m high, unlike anything seen on Earth.

Bibliography

Butrica, Andrew J., *To See the Unseen: A History of Planetary Radar Astronomy*, NASA SP−4218, 1996.

Cooper, Henry S. F., Jr., *The Evening Star: Venus Observed*, Johns Hopkins University Press, 1994.

Roth, Ladislav E., and Wall, Stephen D., *The Face of Venus: The Magellan Radar-Mapping Mission*, NASA SP−520, 1995.

MARINERS TO MARS

Encouraged by their success with Mariner 2 to Venus in 1962, the Americans had decided to pass up their next opportunity to visit Venus and send spacecraft to Mars instead. These twin 260 kg Mariner 3 and 4 spacecraft were the first U.S. spacecraft to visit Mars. Their main payload was a 200 line television camera to image the surface as the spacecraft flew past the planet at an altitude of about 10,000 km. It was also planned to measure the density profile of Mars' atmosphere by measuring the effect on Mariner's radio signals as the spacecraft swung behind Mars as seen from Earth. The launch of Mariner 3 failed, but NASA successfully launched Mariner 4 (see Figure 9.4) just 23 days later on 28 November 1964.

For many years, astronomers knew that Mars' white polar caps reduced in size as local summer approached, which, based on the temperature estimates, indicated the presence of thin water ice. In addition, there were dark areas on Mars that also changed their appearance with season, indicating that they might be covered by some form of vegetation. In 1964 a NASA-sponsored summer school concluded that the surface atmospheric pressure on Mars would be in the range of 10 to 80 millibars, and that the atmosphere would be composed mostly of nitrogen, as on Earth, with carbon dioxide as a minor constituent. So, at the start of spacecraft observations, hopes were high that evidence of some form of elementary life would be found.

Mariner 4 dramatically changed this view of the red planet. The first photographs showed a dead-looking, crater-covered surface, much like the Moon, with a surface atmospheric pressure of just 4 to 7 millibars. This low pressure implied, based on Earth-based spectroscopic observations, that most of the

Figure 9.4. Mariner 4, the first spacecraft to successfully fly-by Mars, in its orbital configuration. The dish-shaped, high-gain antenna, designed to communicate with Earth, is seen on the top of the spacecraft, whilst the camera, which was to photograph the Martian surface, is at the bottom. The deployed solar arrays were each about 2 metres long. (Courtesy NASA/JPL/Caltech.)

atmosphere must be carbon dioxide. The spacecraft showed no evidence that liquid water had ever existed on the Martian surface, and it could certainly not exist there now with such a tenuous atmosphere. In addition, solar ultraviolet radiation would have killed any micro-organisms long ago in such a thin atmosphere. Mariner 4 also found no measurable magnetic field on Mars, indicating that it had no molten magnetic core. The lack of a significant magnetic field would allow charged particles emitted by the Sun to impact the surface, which would also be a threat to any life. So conditions appeared to be highly unfavourable for life now, and probably in the past.

But some astronomers pointed out that Mariner 4 had only imaged about 1% of the surface at a resolution of 3 km. Some of the planet may look different from that imaged by Mariner 4, and evidence of the previous existence of water may be seen when the resolution was significantly improved. In addition, the atmosphere might have been significantly denser in the past which would have allowed significant amounts of water to have existed on the surface. So although Mariner 4's observations made the existence of life unlikely now, they did not rule out the possibility of life in the past.

The 410 kg Mariners 6 and 7 had much better cameras than Mariner 4, and data transmission rates some 2,000 times higher. In addition, the new Mariners both had spectrometers and radiometers to study the atmosphere and polar caps. Mariner 6 was launched on 24 February 1969. Just one month later, Mariner 7 was also on its way to Mars.

Mariner 6 flew over the Martian equator in July 1969 at an altitude of 3,400 km. It measured a surface temperature that varied typically from about 15°C during the day to −75°C at night, suggesting that the surface was covered in a highly insulating material such as dust. It imaged more craters and found that the atmosphere was almost 100% carbon dioxide. Its twin spacecraft, Mariner 7, generally confirmed these Mariner 6 results, although it also found that the Hellas plain was apparently featureless. It detected minute amounts of water vapour in the atmosphere and dust and carbon dioxide crystals in the clouds. The surface of the south polar ice cap appeared to be covered in frozen carbon dioxide in the form of ice or snow.

Mariner 8 and 9 were the first U.S. spacecraft designed to orbit Mars. They each weighed about 1,000 kg, of which 440 kg was propellant. Unfortunately the Mariner 8 spacecraft was lost during launch, but the launch of Mariner 9 was a success on 30 May 1971.

A planet-wide dust storm was raging when Mariner 9 arrived at Mars on 14 November 1971. The spacecraft was then put into orbit around Mars. After two days, this orbit was trimmed, as planned, to a 1,400 × 17,100 km orbit inclined at about 65° to Mars' equator. This allowed repeat imaging every 17 days to enable the study of transient features such as seasonal colour changes and the seasonal advance and retreat of the polar caps.

When routine imaging of Mars started on 2 January it revealed a radically different Mars from that previously observed. The planet was seen to have massive volcanoes, the largest of which was the 25,000 m (80,000 ft) high Olympus Mons, and a giant canyon system. The Hellas plain was seen to be the floor of an enormous impact basin. Other images showed sinuous channels in Chryse Planitia that looked like riverbeds. But it was clear that water could not exist in liquid form on the surface of Mars today, because of the low atmospheric pressure.

Mariner 9 found water vapour over both Martian polar caps when it was their local summer. This indicated that there may be a substantial amount of water locked up in these polar caps which was partially released during summer. This discovery created great excitement, even though the previous Mariners had shown that the Martian surface was now apparently dry and cratered. If there was some water vapour in the atmosphere now, maybe there had been liquid water on the surface earlier in the planet's history.

From Earth, astronomers had seen parts of Mars change colour with the season, indicating that there may be plant life on its surface. However, Mariner 9 showed that the colour change was due to seasonal winds alternately covering and uncovering the darker substrate with lighter dust.

Bibliography

Collins, Stewart A., *The Mariner 6 and 7 Pictures of Mars*, NASA SP−263, 1971.

Ezell, Edward C., and Ezell, Linda N., *On Mars: Exploration of the Red Planet 1958–1978*, NASA SP−4212, 1984.

Godwin, Robert (ed.), *Mars: The NASA Mission Reports*, Apogee Books, 2000.

Hartmann, William K., and Raper, Odell, *The New Mars: The Discoveries of Mariner 9*, NASA SP−337, 1974.

Mariner 9 Team, *Mars as Viewed by Mariner 9*, NASA SP−329, 1976.

MARINERS TO VENUS AND MERCURY

In 1960 NASA planned to develop a series of Mariner spacecraft to be launched to Venus starting in mid-1962. The first two spacecraft, Mariner A and B, were large, weighing about 1,000 kg each. The third spacecraft, Mariner R, was smaller, based on the lunar Ranger design. Over the next year, it became clear that the launcher for the heavy Mariners would not be ready in time, so NASA decided, instead, to launch two Mariner R spacecraft. These spacecraft, called Mariner 1 and 2, were launched to Venus on a fly-by mission in 1962.

The launch of Mariner 1 failed. But the 204 kg Mariner 2 (see Figure 9.5) was launched successfully just over a month later in August 1962. Mariner 2, which was the first spacecraft to successfully intercept a planet, flew past Venus on 14 December at a distance of 35,000 km. It measured a Venus surface temperature of about 450°C and a surface atmospheric pressure of about 20 bar. The cloud tops, with temperatures of about −40°C, were about 70 km above the surface. Venus had no measurable magnetic field and no discernible radiation belts.

The next American Venus probe, Mariner 5, had been a backup for the Mars-bound Mariners 3 and 4. But when Mariner 4 was successful, NASA decided to send a modified Mariner 5 to Venus with a launch in 1967. Mariner 5 flew

Figure 9.5. Mariner 2, the first spacecraft to successfully fly-by Venus. The spacecraft had two different antennae to communicate with Earth: an omnidirectional antenna, which is the small drum at the very top of the spacecraft, and a high-gain dish antenna at the bottom. The microwave radiometer, which measured the surface temperature of Venus, is the small dish just above the main spacecraft platform. (Courtesy NASA.)

past Venus on 19 October 1967, at an altitude of about 4,000 km, just one day after the Soviet Union's Venera 4. Mariner detected a surface temperature of at least 430°C and a surface atmospheric pressure of between 75 and 100 bar.

Mariner 10, the first spacecraft to visit Mercury, was also the first to use the gravitational attraction of one planet (Venus) to direct it to another (Mercury). In December 1969 NASA approved the Mariner 10 mission for launch during the 1973 launch window. Shortly after approval, it became clear that, if launch occurred around the middle of that window, the spacecraft could visit Mercury not once but three times as, after its first Mercury fly-by, its orbital period would be almost exactly twice that of the planet. Unfortunately, Mercury rotates three times for every two of its years, and so the same half of the planet would be illuminated at each intercept.

The main problem facing the Mariner 10 designers was the extra heat the spacecraft would receive when flying as close to the Sun as Mercury. As a result, the design included a large sunshade and most of the scientific instruments were placed on the shadow side. The most important of these instruments, a pair of television cameras, was mounted on a scan platform that enabled them to follow their target during each fly-by.

The 503 kg Mariner 10 was launched on 3 November 1973. On 5 February 1974 it flew within 5,800 km of Venus. The ultraviolet images, which had the most contrast, showed chevron-shaped cloud patterns that rotated around Venus once every four days. While these clouds had been observed fleetingly from Earth, Mariner 10's images provided far more detail, allowing Venus' atmospheric dynamics to be examined for the first time.

Mariner 10 then flew within 703 km of Mercury on 29 March 1974, followed by two further Mercury intercepts at about six-month intervals. The third encounter was the closest, at 327 km. Images showed a surface that looked, at first sight, like that of the Moon with a large number of impact craters, although extensive lunar-like maria were noticeably absent. In addition, the distance of the ejecta blankets from the crater rims on Mercury was less than on the Moon, because of the planet's higher surface gravity.

Mercury's atmosphere, which was found to be extremely tenuous, appeared to consist mainly of atomic hydrogen, helium and oxygen. The largest feature on the surface was the 1,500 km diameter Caloris basin, which was ringed by 2,000 to 3,000 m high mountains. Antipodal to Caloris was an area called the 'weird' or 'hilly and lineated' terrain, consisting of hills and mountains arranged in a strange pattern. They appeared to have been formed by the shock wave generated by the Caloris impact. Surprisingly, Mariner 10 found Mercury, despite its small size, to have a significant magnetic field and a magnetosphere, although it had no measurable radiation belts.

Bibliography

Dunne, James A., and Burgess, Eric, *The Voyage of Mariner 10; Mission to Venus and Mercury*, NASA SP−424, 1978.

Jet Propulsion Laboratory, *Mariner-Venus 1962, Final Project Report*, NASA SP−59, 1965.

Reiff, Glenn A., *Mariner-Venus 1967, Final Project Report*, NASA SP−190, 1971.

Wheelock, Harold J. (ed.), *Mariner Mission to Venus*, McGraw-Hill, 1963.

MARS EXPLORATION ROVERS

Mars Exploration Rovers were two identical spacecraft launched by NASA about a month apart in 2003. Each spacecraft included a 185 kg rover, called Spirit and Opportunity, which landed on the surface of Mars in January 2004 using airbags to cushion their landings. Spirit landed in Gusev crater, which was thought to have been flooded by liquid water early in its history, whilst Opportunity landed in the Meridiani Planum, where the Mars Global Surveyor had detected a large amount of grey hematite, which is normally formed in the presence of

water. The main mission was to characterise the surface in the vicinity of the landers, looking particularly for signs of past water activity.

Each lander included a rock abrasion tool to expose fresh rock for analysis. In addition, each one had a Mössbauer spectrometer to identify iron-bearing materials, a panoramic camera, a thermal emission spectrometer, an X-ray spectrometer, and a microscopic imager.

It took the science operations team some time to find any evidence of past water near Gusev crater with the Spirit rover. Most of the rocks nearby were found to contain olivine, indicating that they were of volcanic origin. However, when Spirit reached the Columbia hills, about 3 km away, it found hematite and sulphates, indicating the previous presence of water. Opportunity, on the other hand, soon found a great deal of evidence of previous water when it detected large concentrations of sulphates and abundant jarosite. Opportunity also discovered spherules of hematite that had probably been precipitated in a water-rich environment. Finally its images of layered rocks were very reminiscent of those layered rocks that had been formed in shallow water on Earth.

Bibliography

Bell, Jim, *In Search of Martian Seas*, Sky and Telescope, March 2005, pp. 40–47.

Chaikin, Andrew, *A Passion for Mars: Intrepid Explorers of the Red Planet*, Abrams, 2008.

Squyres, Steve, *Roving Mars: Spirit, Opportunity, and the Exploration of the Red Planet*, Hyperion, 2005.

Various authors, Special Section, *Journal of Geophysical Research*, 113 (2008), E6 and E12.

MARS EXPRESS

The 1,100 kg Mars Express spacecraft, launched on 2 June 2003, was the first European Space Agency (ESA) probe to visit Mars. It consisted of an ESA orbiter and a UK lander called Beagle 2. A number of the orbiter instruments were derived from those on the failed Russian Mars−96 mission. The orbiter was to provide global colour and stereo high-resolution imaging at 10 m resolution, with selected areas at 2 m, global mineralogical mapping in the visible and infrared, and radar sounding to measure the distribution of water and water ice in the top few kilometres of the Martian surface. The orbiter also measured the atmosphere, whilst Beagle 2, which was named after Charles Darwin's ship, was designed to search for evidence of life.

Mars Express was put into orbit around Mars on 25 December 2003, a few days after it had released the Beagle 2 lander, which failed in its attempted soft landing. The orbiter was put into its operational polar orbit of 260 × 11,600 km in late January 2004.

The Mars Express orbiter found that the mineralogical changes in Mars' early history were indicative of abundant water, whereas those of a later date suggested that the planet was, by then, colder and dryer. It found that the interior of the south polar cap was almost completely water ice. Buried water ice at the south polar cap extended down by as much as 3.5 km in places and that, together with buried water ice in the Dorsa Argentea nearby, would be enough to produce a 11 m deep global ocean. Mars Express also discovered methane, sparking a debate as to whether it was caused by some form of life or by volcanic activity.

Bibliography

Chicarro, A. F., et al, *MARS EXPRESS, The Scientific Investigations*, ESA SP−1291, June 2009.

Mars Express Project Team, *Mars Express – Closing in on the Red Planet*, and Chicarro, A., Martin, P., and Trautner, R., *Mars Express – Unravelling the Scientific Mysteries of the Red Planet*, both in ESA Bulletin No. 115, August 2003, pp 10–25.

Various articles, Science, **307** (2005), 1575–1597.

MARS GLOBAL SURVEYOR

Mars Global Surveyor (MGS) was a 1,060 kg NASA spacecraft launched to Mars on 7 November 1996. It was a partial replacement for the Mars Observer spacecraft that had failed just before it was to enter Mars' orbit in 1993. MGS was designed to take high resolution images of the surface of Mars, determine its geology, topography and gravitational field, establish the nature of its magnetic field, and measure the weather and thermal structure of its atmosphere. Its instruments included a camera, laser altimeter and thermal emission spectrometer. The camera produced images at about 1.4 m/pixel in the narrow-angle mode and 250 m/pixel in the wide-angle mode at nadir.

MGS was inserted into a 45 hour elliptical orbit on arrival at Mars in September 1997. It was then planned to use aerobraking in the Martian atmosphere to enable the spacecraft to reach its final Sun-synchronous orbit using a minimum of on-board propellant. This had significantly reduced the size of the spacecraft, and hence its launch costs. But a problem with MGS's solar array caused an extension of the aerobraking timescale from about 4 to 18 months. So the spacecraft did not reach its operational orbit, with an altitude from 368 to 438 km, until March 1999.

The first significant scientific results were produced during the aerobraking process when MGS confirmed that Mars did not have a dipole magnetic field. But it detected a strong remanent magnetic field over the older parts of the planet's surface. The spacecraft found that Mars' northern plains were remarkably flat, which led some astronomers to conclude that this area was the dried-up bed of an ancient ocean. This idea

was reinforced when it was found that much of the border of this area was at the same altitude, within ±280 m, around this hypothesised former ocean.

MGS detected large amounts of grey hematite, which normally forms in the presence of water. It also found a number of new impact craters in 2006, mostly in the range of from 10 to 30 m in diameter, that had not been present seven years earlier, and found bright new deposits of some material in two gullies. In both cases the bright material was thought to have been carried by a relatively slow-moving fluid that behaved like liquid water.

Bibliography

Albee, Arden L., et al., *Overview of the Mars Global Surveyor Mission*, Journal of Geophysical Research, **106** (2001), 23,291–23,316.

Chaikin, Andrew, *Global Surveyor's Last Hurrah*, Sky and Telescope, April 2007, pp. 38–41.

Chaikin, Andrew, *A Passion for Mars: Intrepid Explorers of the Red Planet*, Abrams, 2008.

Godwin, Robert (ed.), *Mars: The NASA Mission Reports*, Apogee Books, 2000.

Morton, Oliver, *Mapping Mars: Science, Imagination and the Birth of a World*, Fourth Estate, 2002.

Raeburn, Paul, and Golombeck, Matt, *Mars: Uncovering the Secrets of the Red Planet*, National Geographical Society, 1998.

MARS ODYSSEY

Mars Odyssey was an orbiting spacecraft primarily aimed at characterising the geology of Mars and measuring the radiation environment to assist planning of human exploration. It was the surviving half of the Mars Surveyor 2001 project, which originally consisted of an orbiter and lander. The lander was cancelled following the failure of the Mars Polar Lander in December 1999.

The 725 kg Mars Odyssey carried three groups of instruments. The Thermal Emission Imaging System (THEMIS), which mapped the surface minerology, consisted of a camera operating in the visible band with a resolution of 18 m/pixel, and a thermal infrared imaging spectrometer of 100 m/pixel. The Gamma Ray Spectrometer (GRS) suite, which included two neutron detectors, mapped the elemental composition of the surface, and determined the abundance of water ice in the upper metre of soil, via the proxy of measuring hydrogen. Finally the Mars Radiation Environment Experiment (MARIE) measured the variation in the radiation environment over time. (MARIE was to fail during a powerful solar storm in October 2003.)

The spacecraft was placed into an interim orbit around Mars in October 2001. It then used aerobraking to achieve its operational 400 km altitude polar orbit in January 2002. It found extensive surface areas high in hydrogen abundance polewards of about 60°N and 60°S. This was believed to show that the surface there was well over 50% water ice by volume. In fact, when the Mars Phoenix lander landed in one of these regions in 2008 it found ice just below the surface.

Bibliography

Russell, C. T. (ed.), *2001 Mars Odyssey*, Kluwer Academic Publishers, 2004.

Squyres, Steve, *Roving Mars: Spirit, Opportunity, and the Exploration of the Red Planet*, Hyperion, 2005.

Various authors, Space Science Reviews, **110** (2004), 1–130.

MARS PATHFINDER

Mars Pathfinder, which was launched on 4 December 1996, was the second spacecraft in NASA's Discovery Program, designed using Daniel Goldin's 'Faster, Better, Cheaper' philosophy. The 890 kg spacecraft, which was a technology demonstration mission, was the first successful mission to Mars in twenty years. Its landing site in the Ares Vallis had been chosen as it was thought to be the site of an ancient flood plain. After landing, Mars Pathfinder released a six-wheeled, 11 kg rover vehicle called Sojourner, which was the first vehicle to be placed on Mars. Earth-based operators controlled the movement of Sojourner, its stereo imaging system, and its X-ray spectrometer which was designed to analyse the Martian soil and any interesting surface rocks in the vicinity.

To keep the costs as low as possible, NASA decided to launch Pathfinder directly towards its landing site, without initially putting it into orbit around Mars. In addition, Pathfinder relied on airbags to cushion its landing, instead of more complex retro-rockets. These and other simplifications allowed the programme to be undertaken at considerably less cost than previous Mars spacecraft missions. On the other hand, the Pathfinder spacecraft, which operated on the surface for only 12 weeks, had very limited scientific goals.

Most of the rocks analysed by Sojourner had a high silica content, like andesites on Earth. Surface images showed that large rocks nearby were generally leaning in the same direction as each other, indicating that they may have been subjected to a strong flow of water in the past. But this was circumstantial, rather than really convincing evidence of previous water flows.

Bibliography

Bell, Jim, *Mars Pathfinder: Better Science?*, Sky and Telescope, July 1998, pp. 36–43.

Godwin, Robert (ed.), *Mars: The NASA Mission Reports*, Apogee Books, 2000.

McCurdy, Howard E., *Faster, Better, Cheaper: Low-Cost Innovation in the U.S. Space Program*, Johns Hopkins University Press, 2001.

Mishkin, Andrew, *Sojourner: An Insider's View of the Mars Pathfinder Mission*, Berkley Publishing, 2003.

Morton, Oliver, *Mapping Mars: Science, Imagination and the Birth of a World*, Fourth Estate, 2002.

Raeburn, Paul, and Golombeck, Matt, *Mars: Uncovering the Secrets of the Red Planet*, National Geographical Society, 1998.

Squyres, Steve, *Roving Mars: Spirit, Opportunity, and the Exploration of the Red Planet*, Hyperion, 2005.

MARS RECONNAISSANCE ORBITER

Mars Reconnaissance Orbiter (MRO) was a 2,200 kg NASA spacecraft launched on 12 August 2005 to study the surface, subsurface and atmosphere of Mars. It was placed into orbit around Mars in March 2006, using aerobraking to achieve its operational 255×320 km Sun-synchronous polar orbit six months later.

The spacecraft's payload included three cameras, two spectrometers and a radar. HiRISE, one of the cameras, had a resolution of 0.3 m and operated in three colour bands from blue-green to near infrared. It could produce stereo pairs allowing topography to be determined to about 0.25 m. The CRISM spectrometer produced a spectrum from 370 to 3,920 nm in 544 channels to enable various minerals to be identified. SHARAD, the subsurface radar, was designed to operate with the subsurface radar on the European Space Agency's Mars Express that had a lower resolution, but which could penetrate to a much greater depth.

Mars Reconnaissance Orbiter detected subsurface water ice between 35° and 60° latitude in both northern and southern hemispheres. This was the first time that subsurface water ice had been detected at such relatively low latitudes. Mars Reconnaissance Orbiter also detected repetitive layering in Mars' polar caps, and repeating patterns in sedimentary layers which seemed to be linked to Mars' climate cycles. It also found carbonates in the Nili Fossae region, which were probably formed 3 to 4 billion years ago from the alteration of olivine by water.

Bibliography

Various articles, Special Section, Journal of Geophysical Research, **112** (2007), E5.

MESSENGER

The MESSENGER (MErcury Surface Space ENvironment GEochemistry and Ranging) spacecraft, which was launched in 2004, was the first spacecraft to visit Mercury since Mariner 10 undertook three fly-bys in 1974 and 1975. MESSENGER's mission was to help determine Mercury's geology, its internal structure and the source of its magnetic field. The spacecraft was also expected to detect any polar deposits and determine the structure of Mercury's exosphere and magnetosphere.

When the MESSENGER space programme was approved in 2001 it was planned to launch the spacecraft in March 2004. It would then have had two fly-bys of Venus and Mercury, followed by injection into orbit around Mercury in April 2009. But various problems resulted in the launch being delayed to August 2004. This required the orbit schedule to be modified to include fly-bys of Earth (once), Venus (twice) and Mercury (three times), before the spacecraft was put into orbit around Mercury in March 2011.

The spacecraft had a large ceramic-fabric sunshade and radiators to keep it as cool as possible so close to the Sun. Its scientific payload provided multispectral imaging using a wide-field camera, and high-resolution monochromatic imaging using a narrow-field camera. In addition there were a laser altimeter, various spectrometers, and a magnetometer. Mariner 10 had imaged only half of Mercury, whereas MESSENGER was planned to image the whole planet at much higher resolution.

MESSENGER's fly-bys of Mercury took place all at about 200 km altitude in January and October 2008 and September 2009. The spacecraft was then inserted into a $200 \times 15,200$ km 12 hour near-polar orbit in March 2011. This orbit was chosen to reduce the amount of heat reflected onto the spacecraft from the planet's surface.

The spacecraft confirmed that Mercury's magnetic field was dipolar, with its magnetic axis closely aligned to its spin axis. But its magnetic equator was found to be offset relative to its geographic equator by a few hundred kilometres. MESSENGER found water group ions in Mercury's exosphere. The spacecraft's images also cleared up a longstanding problem following the Mariner 10 mission as to whether Mercury's planes were volcanic or not. MESSENGER found a large expanse of volcanic planes surrounding the north polar region and covering more than 6% of the total surface of Mercury. In places the lava was estimated to be about 2 km thick.

Bibliography

Beatty, J. Kelly, *Reunion with Mercury*, Sky and Telescope, May 2008, pp. 24–27.

Domingue, D. L., and Russell, C. T., (eds.), *The MESSENGER Mission to Mercury*, Springer, 2007.

Various articles, Science, **321** (2008), 59–94, and Science, **333** (2011), 1847–1868.

NEAR SHOEMAKER

NEAR (Near Earth Asteroid Rendezvous) Shoemaker was the first spacecraft to orbit an asteroid, the S-type asteroid Eros (433), and land on its surface. Costing just over $200 million,

it was the first of NASA's 'Faster, Better, Cheaper' Discovery spacecraft to be launched.

The 800 kg NEAR Shoemaker spacecraft included a multispectral camera, a laser altimeter, and an X-ray/gamma ray spectrometer, along with 320 kg of propellant to put it into orbit around Eros. It was launched on 17 February 1996, flew within 1,200 km of the C-type asteroid Mathilde (253) in June 1997 and was due to be placed into orbit around Eros in January 1999. But spacecraft problems caused the orbit insertion to be aborted and have the spacecraft fly-by Eros instead. NASA then put NEAR Shoemaker into a 13 month heliocentric orbit that closely matched that of Eros before it was put into orbit around Eros on 14 February 2000. It was later decided to try to land the spacecraft on Eros at the end of the mission in a completely unscheduled manoeuvre. This was successful on 12 February 2001. The spacecraft then continued to transmit from the surface whilst resting on two of its solar wings.

The NEAR Shoemaker spacecraft showed that Mathilde was a 59×47 km asteroid with four large craters on the 65% of the asteroid's surface that was visible during the encounter. Mathilde's density of about 1.3 g/cm^3 indicated that it was extremely porous, and crater counts indicated that it was at least 2 billion years old.

The spacecraft found that the $33 \times 13 \times 13$ km Eros had one large impact crater, and spacecraft trajectory changes indicated that Eros has a density of about 2.7 g/cm^3. Images from orbit showed a variety of geological features, ranging from odd-looking, squarish craters, to flat 'ponds' of very fine dust in low-lying areas, and boulders, some of which had small craters on their surface. There was a general lack of very small craters on the surface of Eros, and when NEAR Shoemaker landed on one of the flat dust 'ponds' it confirmed that the lack of small craters extended down to those of centimetre size. The spacecraft also showed parts of two small channels, which indicated some sort of slumping of surface material, possibly due to underground fractures.

Bibliography

Bell, Jim, and Mitton, Jacqueline (eds.), *Asteroids: NEAR Shoemaker's Adventures at Eros*, Cambridge University Press, 2002.

Russell, C. T. (ed.), *The Near Earth Rendezvous Mission*, Springer-Verlag, 1998.

Various articles, Science, **289** (2000), 2085–2105.

ORBITING SOLAR OBSERVATORY (OSO) SPACECRAFT

The first Orbiting Solar Observatory (OSO) spacecraft was proposed in 1959 by John Lindsay of the Goddard Space Flight Center as a low-budget Earth-orbiting, solar observatory spacecraft, which could observe the Sun in wavebands unobservable from Earth. Just three years later this spin-stabilised spacecraft, OSO-1, was launched to study the Sun in the ultraviolet, X-ray and gamma-ray wavebands. Other experiments measured neutrons, protons, electrons and micrometeorites. Unfortunately, the spacecraft worked properly for only 76 days before a problem with its spin control system resulted in intermittent operation. Nevertheless, it observed more than 75 flares and subflares, mapped the sky in gamma radiation, and monitored the inner Van Allen belt at an altitude of about 550 km.

By the time that OSO-1 was launched in March 1962, NASA had agreed to transform this single spacecraft mission into an ongoing programme of similar spacecraft. In the event, OSO-1 to -6 were successfully launched, weighing between 200 and 300 kg each, followed by OSO-7 and -8 that were appreciably heavier.

OSO-3 measured the spectrum of a solar flare in 1967 that implied that it had heated the local plasma to a temperature of about 30 million K. This spacecraft also detected 621 cosmic gamma rays over its 16 month lifetime. Although OSO-3 could not detect individual gamma-ray sources, it showed that the galactic equator was a clear source, with the galactic centre the most intense part.

On 14 December 1971, the white-light coronagraph on OSO-7 detected clouds of protons leaving the Sun in what are now called coronal mass ejections (CMEs). These clouds were hundreds of thousands of kilometres in diameter, with a temperature of about 1 million K, and were leaving the Sun at velocities of about 1,000 km/s.

OSO-7 discovered gamma-ray emission lines from a solar flare on 4 August 1972. The main line at 0.5 MeV was produced by electron/positron annihilation, showing that positrons had been produced in the flare. The other strong emission line was at 2.2 MeV, which was caused by the capture of neutrons by protons in the flare.

In 1973 OSO-7 found that Cen A is an X-ray source that varied in intensity over a few days. Subsequent observations with OSO-8 and Ariel 5 detected X-ray variations over less than one day, indicating that the source was less than one light day in diameter. This was the first indication that there was a supermassive black hole at or near the centre of the galaxy Cen A.

Bibliography

Hufbauer, Karl, *Exploring the Sun: Solar Science since Galileo*, Johns Hopkins University Press, 1991.

PHOENIX MARS LANDER

It had originally been intended to have a lander as part of the Mars Surveyor 2001 mission, but this lander had been cancelled following the failure of the Mars Polar Lander in December 1999. However, the Surveyor 2001 Lander was in an advanced state of development when it was cancelled, so it was mothballed for potential future use. In the event, it was used as the basis of the Phoenix Mars Lander.

The 350 kg Phoenix lander was the first spacecraft to land in the sub-polar regions of Mars when it touched down in Vastitas Borealis at 68° N in May 2008. This region was chosen as it was thought to contain the largest concentration of subsurface water ice outside of the poles. Phoenix, like the Vikings, was a fixed lander which used retrorockets for landing, unlike Sojourner and the Mars Exploration Rovers which were rovers delivered to the surface in airbags. Its mission was to examine the local environment to see if it had been suitable for life, to uncover the history of water and water ice there, and to monitor the polar weather. It included a robot arm to excavate dust, rock and ice that could be analysed by other instruments on the lander.

The Phoenix Mars Lander was the first to directly detect water ice when it heated a soil sample from only 5 cm below the surface. It also detected falling snow which vaporised before it reached the surface. Phoenix found that the soil was alkaline, and detected calcium carbonate and phyllosilicates, which are formed in water, and perchlorate salt.

Bibliography

Beatty, J. Kelly, *Polar Prospector*, Sky and Telescope, September 2008, pp. 22–24.

Various authors, Journal of Geophysical Research, **115** (2010).

PIONEER 10 AND 11

NASA's Pioneer 10 and 11 spacecraft were the first to visit the outer planets. They had originally been expected to monitor the solar wind outside the Earth's orbit, but in 1969 their missions and designs were changed to enable them to fly by Jupiter instead.

NASA decided to launch the two 260 kg Pioneer spacecraft in 1972 and 1973. If Pioneer 10 failed in its Jupiter fly-by mission, then NASA would send Pioneer 11 to repeat that mission. Otherwise NASA would consider redirecting Pioneer 11 to fly by Saturn after its Jupiter intercept. Because of the large distance of their target planets from the Sun, the Pioneers used radioisotope thermionic generators to provide electrical power, rather than the usual solar arrays. After the Voyager programme was approved in 1972, the Pioneers were not only perceived as important missions in their own right, but also seen as pathfinder missions to assist the later more complex Voyager missions.

To get to Jupiter, both spacecraft had to fly through the asteroid belt with an unknown risk of a collision. In addition, high-energy particles within Jupiter's very active magnetosphere may possibly destroy solid-state devices onboard the spacecraft. In the event, Pioneer 10 was launched toward Jupiter on 3 March 1972, followed 13 months later by the nearly identical Pioneer 11. Pioneer 10 survived and successfully flew past Jupiter on 4 December 1973, just 21 months after launch. After evaluating Pioneer 10's results, NASA announced in March 1974 that it would adjust Pioneer 11's intercept trajectory with Jupiter to allow it to fly past Saturn in September 1979. The new trajectory would send Pioneer 11 just 43,000 km (0.6 Jupiter radii) above Jupiter's clouds in December 1974.

Pioneer 10 first detected high-energy electrons from Jupiter some distance in front of its bow shock, indicating that they had crossed this shock wave. As the spacecraft continued to approach Jupiter, it crossed both Jupiter's bow shock and magnetopause. But three days later, an increase in the solar wind caused Jupiter's bow shock to overtake the spacecraft. Jupiter's radiation belts were found to be about 10,000 times as intense as Earth's Van Allen belts. Jupiter's magnetic field was partly dipolar, with the angle between Jupiter's magnetic and spin axes being about 10°, and with a dipole offset from the centre of about 0.10 radii.

Pioneer 10 detected helium in Jupiter's atmosphere for the first time and showed that the atmosphere consisted of about 99 percent hydrogen and helium. This, like the ratio of helium to hydrogen measured on Jupiter, is similar to that of the Sun. Pioneer also confirmed that Jupiter emitted about twice as much energy as it received from the Sun.

At Jupiter, Pioneer 11 provided similar information to Pioneer 10. Before the Pioneer intercepts, no magnetosphere had been detected for Saturn. Pioneer 11 showed that Saturn's magnetic field was dipolar with its magnetic axis being indistinguishable from its spin axis. Titan was found to be generally within Saturn's magnetosphere, but varying pressure of the solar wind caused the magnetopause to move back and forth across Titan, sometimes leaving the satellite exposed to the solar wind.

Pioneer's images of Saturn and its satellites showed virtually no more detail than visible from Earth. One new ring was found, the narrow F ring, just outside the A ring, and a new small satellite, Epimetheus.

Bibliography

Fimmel, R. O., Swindell, W., and Burgess, E., *Pioneer Odyssey*, NASA SP−396, 1977.

Fimmel, R. O., Van Allen, J., and Burgess, E., *Pioneer: First to Jupiter, Saturn and Beyond*, NASA SP−446, 1980.

Godwin, Robert and Whitfield, Steve (eds.), *Deep Space: The NASA Mission Reports*, Apogee Books, 2005.

Kraemer, Robert S., *Beyond the Moon: A Golden Age of Planetary Exploration 1971−1978*, Smithsonian Institution Press, 2000.

Morrison, David, *Voyages to Saturn*, NASA SP−451, 1982.

Morrison, David and Samz, Jane, *Voyage to Jupiter*, NASA SP−439, 1980.

PIONEER LUNAR SPACECRAFT

On 27 March 1958, only two months after the United States had successfully launched their first Earth-orbiting spacecraft, Neil McElroy, the Secretary of Defense, approved the nation's first lunar programme. This consisted of five Pioneer spacecraft. It was solely designed to beat the USSR to reach the Moon, in which they were ultimately unsuccessful. Three of the spacecraft were to be built by STL (Space Technology Laboratories) for the U.S. Air Force and launched using Thor-Able rockets, whilst the other two were to be built by JPL (Jet Propulsion Laboratory) for the Army and launched using a Juno 2 launcher.

The launcher failed for two of the Pioneers. Pioneer 1 reached a distance of about 110,000 km before falling back to Earth, Pioneer 3 discovered the outer Van Allen radiation belt about 15,000 km from Earth, and only one, Pioneer 4, reached the vicinity of the Moon. In March 1959 Pioneer 4 flew past the Moon at a distance of 60,000 km, about twice that planned. It showed that the particle radiation levels were low, and would not be a safety problem for future human exploration of the Moon.

Bibliography

Leverington, David, *New Cosmic Horizons: Space Astronomy from the V2 to the Hubble Space Telescope*, Cambridge University Press, 2000.

Reeves, Robert, *The Superpower Space Race: An Explosive Rivalry through the Solar System*, Plenum Press, 1994.

PIONEER-VENUS

In 1968 the Space Science Board of the National Academy of Sciences recommended that a multiple probe, or multiprobe, mission be sent to Venus in 1975. By 1972 NASA's Pioneer-Venus Science Steering Group had recommended that two such multiprobe spacecraft should be launched, each consisting of one large probe and three small probes, to be followed a year or two later by a Venus orbiter. In the event, one Pioneer-Venus Multiprobe and one Pioneer-Venus Orbiter (PVO) were launched in 1978. Although the PVO was launched about two and a half months before the Multiprobe, the two spacecraft arrived at Venus within days of each other. This slower trajectory of the PVO was to make it easier to place it into orbit around the planet.

The main instrument on the 550 kg PVO was a radar altimeter designed to penetrate Venus' all-enveloping clouds from a highly eccentric 24 hour near-polar orbit. The Earth-based

Arecibo radar had already provided the first relatively crude maps of parts of Venus. But the PVO was to provide the first detailed map of the planet with a maximum surface resolution of 20 km and altitude resolution of 100 m. Radar mapping took place around periapsis to give a maximum surface resolution.

Over the next two years the PVO radar mapped 93% of the surface. It showed a relatively flat surface, with 65% being within ±500 m of the mean level, although there were exceptions like the 11,500 m high Maxwell Montes and the 2,000 m deep Diana Chasma. Two large shield volcanoes, called Rhea Mons and Theia Mons, were also found in the Beta Regio highland region. Infrared images from the PVO showed a circumpolar collar of very cold air, about 2,500 km from the pole.

The Multiprobe included four atmospheric probes designed to descend to Venus's surface. None were expected to survive impact. The 316 kg main probe and the three identical 90 kg probes were targeted to enter the atmosphere at widely separated locations, to get some idea of Venus's atmospheric dynamics. They confirmed the theoretically based predictions made in 1973 by Godfrey Sill, Andrew Young and Louise Young that Venus' clouds were composed mainly of sulphuric acid.

Bibliography

Beatty, J. Kelly, *Pioneers' Venus: More than Fire and Brimstone*, Sky and Telescope, July 1979, pp. 13–15, and 27.

Fimmel, R. O., et al., *Pioneer-Venus*, NASA SP−461, 1983.

Kraemer, Robert S., *Beyond the Moon: A Golden Age of Planetary Exploration 1971–1978*, Smithsonian Institution Press, 2000.

Marov, Mikhail Ya., and Grinspoon, David H., *The Planet Venus*, Yale University Press, 1998.

POLAR

POLAR was the second of two NASA spacecraft to be launched in the American Global Geospace Science (GGS) programme. This was part of the International Solar Terrestrial Physics (ISTP) programme, which included a number of spacecraft making simultaneous measurements from different locations in near-Earth space. In particular, the WIND spacecraft measured the upstream interplanetary medium, Geotail measured the geomagnetic tail, and POLAR measured the energy received by the Earth's polar regions.

The POLAR spacecraft was launched on 24 February 1996 into an Earth orbit inclined at 86° to the equator. Its apogee was 9 R_E (Earth radii) and perigee 1.8 R_E. Initially the apogee was over the north polar region, but it moved towards the equator at about 16°/year.

Three of the twelve instruments on-board the spacecraft were designed to image the aurora in the UV, visible and X-ray bands when the satellite was near apogee. Other instruments measured magnetic and electric fields and the flux of charged particles and ions. X-rays are produced when electrons from the magnetosphere impinge on the upper atmosphere. The X-ray images and measured X-ray energies enabled the fluxes and characteristic energies of the parent electrons to be determined.

In 2000 POLAR detected Alfvén waves propagated along magnetic field lines from the geomagnetic tail towards Earth. These waves carried enough energy to create the aurora during magnetospheric substorms. In the following year the spacecraft simultaneously imaged both the aurora borealis and aurora australis. This is the first time that both aurorae had been imaged simultaneously by the same instrument on any spacecraft. The images showed that, although the aurorae are basically mirror images of each other, there were differences of detail. POLAR also provided a great deal of data on the high altitude polar cusp, where the magnetosheath plasma that had originated in the solar wind had direct access to the Earth's ionosphere. The spacecraft found that the latitude of the centre of the polar cusp varied from 86° to 70° depending on solar wind conditions, the angle of the solar wind to the Earth's axis and the time of day.

Bibliography

Acuña, M. H., et al., The Global Geospace Science Program and Its Investigations, Space Science Reviews, 71 (1995), 5–21.

RANGER

Ranger was a series of spacecraft launched to the Moon by NASA between 1961–65. Initially the programme consisted of five spacecraft, two Block Is and three Block IIs. The Block I's were designed to traverse a highly eccentric orbit with an apogee well beyond the Moon's orbit, whilst each of the Block IIs included a capsule to land on the Moon. However, before any of these could be launched, in May 1961 President John F. Kennedy committed the United States to a manned lunar programme. As a result, the Ranger spacecraft that had, until the president's decision, been designed mostly with science in mind, were redesignated as precursors to the manned programme.

The Ranger programme started badly when all five spacecraft failed between August 1961 and October 1962. In the meantime, a Block III Ranger programme had been approved, the spacecraft being designed to take high-resolution images of the Moon as they crashed onto its surface.

The first Block III spacecraft, designated Ranger 6, was launched in January 1964. Problems with its television cameras resulted in the spacecraft impacting the lunar surface with them switched off. However, Ranger 7 (see Figure 9.6), launched just six months after Ranger 6, was successful, impacting the Moon just 15 km from its target in the Mare Nubium. Rangers 8 and

Figure 9.6. Ranger 7, the first successful American spacecraft to impact the Moon, which took images of the Moon's surface as it crash-landed. The dark area near the top of the spacecraft is the aperture for the television camera. Communications with Earth were via an omni-directional antenna (the can at the top of the spacecraft), and a high-gain antenna, part of which can be seen behind the solar array. (Courtesy NASA.)

9 were also successful, landing in the Mare Tranquillitatis and crater Alphonsus, respectively.

There had been concern, prior to the Ranger programme, that the surface of the Moon may be covered by a thick layer of dust, thus making the landing of a manned spacecraft problematic. The Ranger spacecraft showed that this was not so, at least in the areas that they observed.

The Ranger images showed craters of all sizes down to the minimum resolved of about 1 m. The edges of most of the smaller craters were rounded, indicating that even on the Moon there was some form of weathering, probably due to the impact of meteorites. But some of the smaller craters were sharp and fresh looking, indicating that the surface was still changing, albeit very slowly. Some craters were clearly secondary craters, not caused by meteorite impact, but caused by the impact of material ejected when other, larger craters were formed.

Bibliography

Hall, R. Cargill, *Lunar Impact: A History of Project Ranger*, NASA SP–4210, 1977.
Ley, Willy, *Ranger to the Moon*, Signet, 1965.

Reeves, Robert, *The Superpower Space Race: An Explosive Rivalry through the Solar System*, Plenum Press, 1994.

RHESSI

The Reuven Ramaty High Energy Solar Spectroscopic Imager (RHESSI) spacecraft was the sixth NASA Small Explorer (SMEX) mission. Launched in 2002, and weighing just 290 kg, it was designed to image solar flares simultaneously, for the first time, in the hard X-ray and gamma-ray wavebands with high resolution spectroscopy.

RHESSI's main mission was to detect the rapid release of energy stored in unstable magnetic configurations that cause solar flares, measure the rapid conversion of this energy into heating plasma and accelerating particles, and show how these particles move from the Sun into space. The spacecraft's images covered 3 keV to 20 MeV with 2 to 30 arcsec resolution. High resolution spectroscopy was also undertaken up to 20 MeV. Other spacecraft and ground-based observatories provided complementary images in other wavebands.

The RHESSI spacecraft was launched on 5 February 2002 into a low Earth orbit, inclined at 38° to the equator, with a 3 year target lifetime. It was still operational in 2007, five years after launch. Over this period RHESSI had detected over 11,000 solar flares and 25,000 microflares. Thirty of the flares had detectable emission above 300 keV, and 18 showed gamma-ray line emission. The hard X-ray emission of the quiet Sun was found to be at least two orders of magnitude lower than that the weakest microflare.

RHESSI's data confirmed that solar flares were caused by magnetic reconnection. It also found that the vast majority of microflares occurred in active regions. It had been thought that large numbers of microflares occurring quasi-continuously over the Sun could provide the necessary energy for coronal heating, but RHESSI showed that their energy was insufficient. When the Sun was occulting solar flares, as seen from Earth, the spacecraft found that non-thermal hard X-ray sources were present very high in the corona accompanying all fast coronal mass ejections.

The spacecraft was able to measure the oblateness of the Sun to within about 0.1 milliarcsec, equivalent to about 100 m on the Sun, although it had not been designed to do this. This showed that the Sun's surface had bright ridges, which were magnetic in nature, arranged in a network pattern. It increased the Sun's apparent equatorial diameter above that expected from the Sun's axial rotation.

Bibliography

Dennis, B. R., Hudson, H. S., and Krucker, S., *Review of Selected RHESSI Solar Results*, and Brown, J. C., Kontar, E. P., and

Veronig, A. M., RHESSI Results – Time for a Rethink?, *Lecture Notes in Physics*, **725** (2007), 33–80.

Lin, R. P., Dennis, B. R., and Benz, A. O. (eds.), *The Reuven Ramaty High-Energy Solar Spectroscopic Imager (RHESSI) - Mission Description and Early Results*, Kluwer Academic Publishers, 2010.

Lin, R. P., et al., Heliophysics Senior Review 2008, http://hesperia.gsfc.nasa.gov/senior_review/2008/senior_review_proposal.2008.pdf

SAMPEX

The Solar, Anomalous and Magnetospheric Particle Explorer (SAMPEX) was the first of NASA's Small Explorer (SMEX) missions, weighing just 158 kg. It was designed to measure the elemental and isotopic composition of energetic particles from the Sun, of anomalous cosmic rays, and of galactic cosmic rays. SAMPEX was also to measure magnetospheric electrons precipitating in the Earth's upper atmosphere which affect its chemistry and dynamics. The spacecraft was launched in 1992 into a low Earth orbit inclined at 82° to the equator. Its instruments, which operated over the range from about 1 to 300 MeV, had sensitivities two orders of magnitude greater than those of previous low orbiting spacecraft.

Over its 12 year lifetime SAMPEX measured the elemental composition of trapped anomalous cosmic rays (ACRs) and discovered that they formed a belt within the inner Van Allen belt. It found that ACR nitrogen, oxygen and neon were generally singly charged. The ionisation results confirmed theories of ACR origin as being neutral interstellar material that is singly ionised near the Sun by ultraviolet light, or by charge exchange with the solar wind, and which is subsequently accelerated in the outer heliosphere. SAMPEX confirmed that the isotopic abundance of ACRs was different from that of galactic cosmic rays.

SAMPEX found that, whilst the inner Van Allen belt protons appeared to vary only with the solar cycle, the outer belt electrons varied on various timescales including that of the solar cycle.

Bibliography

Baker, D. N., et al. (eds.), *Solar Dynamics and its Effects on the Heliosphere and Earth*, Springer, 2007.

Bothmer, Volker, and Daglis, Ioannis A., *Space Weather: Physics and Effects*, Springer-Praxis, 2007.

SKYLAB

Skylab, which was launched on 14 May 1973, was the United States' first manned space station. Its main mission was to show that human beings could live and work in space over prolonged periods. Its astronomical mission was to study the Sun in the X-ray, ultraviolet and visible wavebands.

NASA had set up the Apollo Applications Program in 1965 to use Apollo-designed hardware for extended manned Earth-orbiting and lunar surface missions. Initially the programme had been very large, including six orbital workshops and four Apollo Telescope Mounts that would be attached to the workshops in orbit. But progressive budget cuts reduced the programme to just one orbital workshop, called Skylab, with an integral Apollo Telescope Mount (ATM).

Unfortunately, an explosion during Skylab's launch ripped off a heat shield and one of its two solar panel wings, and the other solar panel wing only partially deployed. But emergency repairs by astronauts that were launched eleven days later enabled Skylab to be used, albeit with a much reduced power supply. Eventually three teams of three astronauts operated Skylab for a total of 171 days over the next nine months.

The ATM contained most of the solar instruments which were much better in performance than those of earlier, unmanned spacecraft, as most of the Skylab instruments used film, rather than electronic imaging. In addition they were built with much less stringent mass and power constraints.

Skylab, which was operational just before solar minimum, detected six low-latitude coronal holes, most of which lasted a number of months. These coronal holes, which were imaged by the X-ray detectors, were found to have a strong association with large unipolar magnetic field regions in the Sun's photosphere. In addition, the coronal hole boundaries were marked by magnetic inversion lines where the magnetic field changed from one polarity to another. These results helped to prove that the high speed solar wind emanates from unipolar regions where the magnetic field lines run freely into space.

Ultraviolet images from Skylab showed large, bright, active regions in the chromosphere. These correlated with large, bright, arch-like features, called coronal loops, in the very high temperature corona imaged in X-rays. These coronal loops were found to link areas of opposite polarity in the photosphere.

X-ray bright points had first been detected by Giuseppe Vaiana using a sounding rocket experiment in 1969. Skylab showed approximately 100 X-ray bright points on each image spread evenly across the solar disc. Surprisingly there were even bright points in coronal holes, which are otherwise dark in X-rays. The lifetimes of bright points were typically about 8 hours, but a few of them showed exceptional brightening over a few minutes. Bright points were found to be associated with localised magnetic fields of opposite polarity in bipolar regions.

Skylab was also used to undertake a detailed examination of coronal transients, most of which were found to be coronal mass ejections. These in turn were found to be associated with eruptive prominences or flares.

Bibliography

Eddy, John A., *A New Sun: The Solar Results from Skylab*, NASA SP-402, 1979.

SOHO

The Solar and Heliospheric Observatory (SOHO) was a European Space Agency (ESA)/NASA project to study the Sun from its core outwards. On-board were twelve experiments that involved 39 scientific institutes from 15 countries. Their main objectives were to study the structure and dynamics of the solar interior using helioseismology, study the potential heating mechanisms of the solar corona, and detect where the solar wind is produced and how it is accelerated.

SOHO had its origins in two earlier ESA missions, GRIST and DISCO, which were cancelled in the early 1980s. GRIST was to have provided solar spectroscopy in the extreme ultraviolet. DISCO, on the other hand, was to have undertaken helioseismology, and to have measured variations in solar irradiance and in the in-ecliptic solar wind in support of what was to become the ESA Ulysses programme. ESA had planned to locate DISCO near to the L_1 Lagrangian point between the Sun and Earth, to provide uninterrupted views of the Sun from outside the Earth's magnetosphere. In the event SOHO undertook a composite of the GRIST and DISCO missions from a halo orbit around this L_1 Lagrangian point.

The 1,850 kg SOHO spacecraft was launched by an American Atlas launcher in December 1995, and inserted into its L1 halo orbit two months later. The launch and initial orbit operations were so successful that there was enough fuel on-board the spacecraft to maintain it in its halo orbit for at least ten years, or about twice as long as originally expected.

SOHO enabled detailed estimates of the Sun's internal temperature to be made, along with its density, atomic abundances, and interior mixing. It provided the first images of structures and flows below the surface, and confirmed that the decrease in angular velocity with latitude measured at the surface extended through the convective zone. At the base of this zone, about one-third of the way into the Sun, was an adjustment layer or tachocline leading to the more orderly radiative interior. It was thought that this tachocline was the source of the solar dynamo which created the Sun's magnetic field. SOHO also showed that strong converging downflows stabilised the structure of sunspots, and that sunspots were relatively shallow. SOHO imaged sunquakes or seismic waves produced by solar flares, and also enabled features on the far side of the Sun to be imaged using holographic reconstruction.

The spacecraft produced a great deal of new, detailed information on the temperature distribution of material in coronal holes and polar plumes. It discovered new dynamic phenomena such as coronal waves, and compressible waves in both polar plumes and coronal loops. It also discovered solar tornadoes about 10,000 km in diameter with steady wind speeds of 15 km/s and gusts some ten times faster. SOHO also provided evidence of the upward transfer of magnetic energy from the surface to the corona through a magnetic carpet, in which the observed magnetic flux was found to be highly mobile. This upward energy transfer was thought to be a major source of coronal heating.

SOHO found evidence that the solar wind streamed out from the Sun carried on waves produced by vibrating magnetic field lines. It found a clear connection between the flow speed of the solar wind and the chromosphere's magnetic network, with the largest outflow velocities coming from network boundaries.

The spacecraft observed coronal mass ejections (CMEs) leaving the Sun, which provided up to three days' warning of magnetic storms. CMEs were found to vary with the solar cycle, increasing from 0.5/day around solar minimum to over 6/day around solar maximum. SOHO enabled 3D images to be produced of CMEs for the first time.

By 2011 SOHO had imaged over 2,000 mostly Sun-grazing comets, many of which had been found by amateurs accessing SOHO images on the Internet. Further analysis yielded data on water vapour and dust production rates for comets.

Bibliography

Fleck, Bernhard, and Švestka, Zdeněk (eds.), *The First Results from SOHO*, Kluwer Academic Publishers, 1997.

Fleck, Bernhard, et al., *10 Years of SOHO*, ESA Bulletin, No. 126, May 2006, pp. 24–32.

Khol, John L., and Cranmer, Steven R. (eds.), *Coronal Holes and Solar Wind Acceleration*, Kluwer Academic Publishers, 1999.

Lang, Kenneth R., *The Sun from Space*, 2nd ed., Springer Verlag, 2008.

SOLAR DYNAMICS OBSERVATORY

The Solar Dynamics Observatory (SDO) was the first spacecraft in NASA's Living With a Star programme. SDO's mission was to understand the nature and source of solar variability that affects life on Earth. To do this it made measurements of solar parameters associated with those mechanisms that were thought to produce solar variability on timescales ranging from seconds to decades. It also monitored those aspects of the Sun's variable radiative, particulate and magnetic plasma emissions that have the greatest impact on the terrestrial environment, including the surrounding heliosphere.

The SDO programme, which began in the early 2000s, was designed to build on the successes of the SOHO and other similar solar missions, observing the Sun with greater spatial resolution and with a faster time cadence. The spacecraft, which was launched in February 2010 into an inclined geosynchronous orbit, had three primary instruments. They were the

Helioseismic and Magnetic Imager (HMI), the Extreme Ultraviolet Variability Experiment (EVE), and the Atmospheric Imaging Assembly (AIA). The HMI imaged the plasma flows inside the Sun that generate the Sun's magnetic field, and measured the strength and direction of the magnetic field emerging from the photosphere. The EVE measured variations in the Sun's extreme ultraviolet emissions from 0.1 to 105 nm. The AIA imaged the Sun at ten different wavelengths almost simultaneously, showing how the corona responded to magnetic fields that the HMI observed near the Sun's surface.

At the time of writing (early 2011) the SDO is producing exceptional images. But it is premature, as yet, to report on any fundamental improvements in our understanding of the Sun's variability.

Bibliography

Bobra, Monica, *New Scrutiny of the Sun's Secrets*, Sky and Telescope, February 2011, pp. 22–27.

Pesnell, W. Dean, *Opening a New Window on the Sun*, Astronomy, May 2011, pp. 24–29.

SDO; Solar Dynamics Observatory, *Our Eye on the Sun: A Guide to the Mission and Purpose of NASA's Solar Dynamics Observatory*, NASA, NP−2009-10-101-GSFC.

Solar Dynamics Observatory − Report of the Science Definition Team, July 2001.

SOLAR IRRADIANCE MISSIONS

It required the use of accurate spacecraft radiometers to detect the first unambiguous variations in solar irradiance (or the solar constant). This was in 1980 when Richard Willson showed using the active cavity radiometer, ACRIM (Active Cavity Radiometer Irradiance Monitor) I, on Solar Max that solar irradiance *reduced* slightly as large sunspot groups crossed the central solar meridian. This implied that sunspots blocked energy leaving the Sun. But over the next few years it became clear that solar irradiance *reduced* slightly as the Sun approached sunspot *minimum*. In 1988 Peter Foukal and Judith Lean explained this apparently inconsistent result as being due to faculae overcompensating the blocking effect of sunspots over the course of a solar cycle.

It was also important to detect any longer term variations in solar irradiance. The ACRIM I radiometer, which was operational for most of 1980, and from 1984 to 1989, provided a baseline. This instrument was followed by ACRIM II which was launched on the Upper Atmosphere Research Satellite (UARS) in 1991, and which produced similar data until 2001. In the meantime the ACRIMSAT spacecraft had been launched in 1999. This carried just one instrument, ACRIM III, whose design was an improved version of ACRIM II.

A comparison of the ACRIM I and ACRIM II measurements showed that these instruments could not measure solar irradiance accurately enough to determine long-term trends in absolute terms. The solar irradiance had varied by only about 0.1% during a solar cycle, and ideally absolute measurements should be made of an order of magnitude better than that, which ACRIM I and II could not do. Even the absolute accuracy of ACRIM III was only about 0.1%, although it had a stability of about 0.01%.

The 290-kg SORCE (Solar Radiation and Climate Experiment) spacecraft was launched in January 2003 into low Earth orbit some three years after ACRIMSAT. SORCE was designed not only to improve the measurement of total solar irradiance, but to obtain a better understanding of the solar irradiance variability in different wavebands. There were four instruments onboard this spacecraft including a Total Irradiation Monitor (TIM) instrument based on ACRIM III, but with considerable improvements. This gave an estimated absolute irradiance accuracy of 0.035%, and a relative accuracy of better than 0.001%/yr.

Attempts have recently been made to produce an accurate record of total solar irradiance using ACRIM I, II and III and TIM instrument data, and similar data produced by instruments on other, earlier spacecraft. These included the Hickey-Frieden (HF) radiometer on Nimbus 7, the Earth Radiation Budget Experiment (ERBE) on ERBS (Earth Radiation Budget Satellite), and the VIRGO (Variability of Solar Irradiance and Gravity Oscillations) instrument on SOHO (Solar and Heliospheric Observatory). So far, no clear long-term trend has been found.

Bibliography

Calisesi, Y., et al. (eds.), *Solar Variability and Planetary Climates*, Springer, 2007.

Rottman, G. J., Woods, T., and George, V. (eds.), *The Solar Radiation and Climate Experiment (SORCE): Mission Description and Early Results*, Springer, 2005.

See also: Solar Max.

SOLAR MAX

Solar Max (see Figure 9.7), otherwise called Solar Maximum Mission or SMM, was a spacecraft designed to observe the Sun, and solar flares in particular, during the 1980–81 solar maximum. This 2,400 kg NASA spacecraft included the first telescope designed to image the Sun in high-energy X-rays. A key feature of the spacecraft was the on-board coordination of instrumental observations in response to solar flares. Two of the instruments were also used to study gamma-ray bursts.

Solar Max was launched into a low Earth orbit on 14 February 1980. But less than a year later the spacecraft lost its fine pointing capability, reducing it to a non-imaging mission. It was

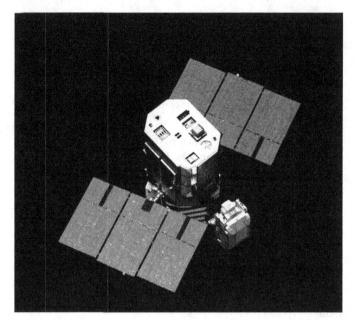

Figure 9.7. Solar Max, the first spacecraft to be repaired in orbit, consisted of a multimission modular spacecraft, contained all housekeeping functions, and an experiment module. There were seven experiments on board, some of which, like the Hard X-ray imaging spectrometer (HXIS), were designed to image active areas on the Sun, whereas others, like ACRIM (see text), simply measured the total energy emitted by the Sun in their wavebands of sensitivity. (Courtesy NASA/Marshall.)

eventually repaired in orbit in April 1984 by astronauts from the Space Shuttle Challenger, making it the first spacecraft to be retrieved, repaired, and redeployed in orbit. It collected data until November 1989, covering nearly an entire eleven-year solar cycle.

Shortly after launch, the ACRIM (Active Cavity Radiometer Irradiance Monitor) instrument detected a small but clear decrease in the solar constant when two large sunspot groups crossed the central solar meridian. Over the next few years the solar constant was found to go through a minimum at solar minimum in data produced by both the Solar Max and Nimbus 7 spacecraft. So, although the solar constant decreased when individual sunspot groups crossed the solar meridian, it also decreased as the Sun approached sunspot minimum.

Solar Max showed that electrons and protons were accelerated simultaneously in solar flares. It found that hard X-rays were emitted in short bursts during the impulsive stage of a flare, brightening simultaneously at both ends of the magnetic loop containing the flare. Soft X-rays were produced over a longer period of time. Intensity variations of the ionised 137.1 nm oxygen line, the ultraviolet continuum at 139 nm, and hard X-rays all occurred simultaneously, proving that they were all emitted in the Sun's transition region. On the other hand soft X-rays were emitted by active regions that extended well into the solar corona.

Solar Max detected new peaks in the Sun's gamma-ray spectrum at 1.3 and 1.7 MeV which were caused by the de-excitation of magnesium and neon nuclei following their collisions with protons. The spacecraft also allowed correlations to be established in the behaviour of coronal mass ejections, eruptive prominences and solar flares.

Bibliography

Hufbauer, Karl, *Exploring the Sun: Solar Science since Galileo*, Johns Hopkins University Press, 1991.

Lang, Kenneth R., *The Sun from Space*, 2nd ed., Springer-Verlag, 2008.

Strong, Keith, et al. (eds.), *The Many Faces of the Sun: A Summary of the Results from NASA's Solar Maximum Mission*, Springer-Verlag, 1999.

Various articles, Sky and Telescope, June 1984, pp. 494–503.

SOVIET MARS PROGRAMME

The Soviet Mars Programme began with the launch of two spacecraft to Mars in October 1960, but both spacecraft were lost when the third stage of their Molniya launcher failed. The Soviet Union then launched three spacecraft to Mars during the next launch window in 1962, but two of these were lost when their launch vehicles exploded. Contact with the third, called Mars 1, was lost when it was 106 million km from Earth, still some three months away from Mars. More Soviet failures to reach Mars followed, until in 1971 Mars 2 and 3 succeeded in reaching Mars, just as a major, planet-wide dust storm was covering the planet.

Mars 2 and 3 both consisted of an orbiter and lander. Unfortunately, their automated design made it impossible to wait for the dust storm to blow itself out before attempting the landings. The Mars 2 lander crashed on landing, while the Mars 3 lander successfully landed, but failed a few seconds later. The orbiters, on the other hand, were relatively successful, measuring a very weak magnetic field and detecting small amounts of water vapour in the atmosphere. They also measured the surface temperature, showing that it had cooled by about 25°C during the dust storm, while the atmospheric temperature had increased.

Unlike their successful Venus programme, the Soviet Mars programme was to suffer even more failures, with only Mars 5 in 1974 and Phobos 2 in 1989 being of limited success.

Bibliography

Perminov, V. G., *The Difficult Road to Mars: A Brief History of Mars Exploration in the Soviet Union*, NASA Monographs in Aerospace History, No. 15, July 1999.

Reeves, Robert, *The Superpower Space Race: An Explosive Rivalry through the Solar System*, Plenum Publishing, 1994.

SPUTNIKS 1–3

On 4 October 1954 the Special Committee for the International Geophysical Year (IGY) suggested that spacecraft should be launched during the eighteen-month IGY which was due to start in July 1957. These spacecraft were to help explore near-Earth space. Within a year both the Soviet Union and the United States had indicated that they would launch such spacecraft, although the Soviet programme was not approved by the USSR Council of Ministers until January 1956.

The payload of the 85 kg Sputnik 1, the first-ever artificial satellite, launched on 4 October 1957, consisted of just pressure and temperature sensors. But analysis of its orbit allowed the density of Earth's upper atmosphere to be determined. In addition, Sputnik 1's orbit did not drift westward as rapidly as expected, indicating that the Earth's gravity field was not as anticipated. A month later Sputnik 2 was launched. It carried a pair of Geiger-Muller tubes that detected the inner Van Allen radiation belt three months before it was detected by the American Explorer 1. But Soviet scientists had difficulty in analysing their results and so missed the evidence. Likewise the 1,300 kg Sputnik 3 could have mapped the inner Van Allen belt in May 1958, two months before the American Explorer 4, but its tape recorder failed and it was difficult to separate out local effects from global effects. It, like Explorer 4, also detected the outer Van Allen belt, but this was not recognised at the time for either spacecraft.

But the significance of these three spacecraft was not in their scientific results, which were incidental, but as a demonstration to the world, and the United States in particular, that the Soviet Union was a superpower with highly effective military capabilities.

Bibliography

Harford, James, *Korolev: How One Man Masterminded the Soviet Drive to Beat America to the Moon*, Wiley, 1997.

Van Allen, James A., *Origins of Magnetospheric Physics*, University of Iowa Press, 2004.

STARDUST

Stardust was the first spacecraft to capture cometary material and return it to Earth for analysis. It was launched in 1999 and used a gravity assist from Earth in 2001 to fly to within 240 km of Comet Wild 2 in 2004. En route it also flew within about 3,300 km of the $7 \times 5 \times 3$ km triangular prism-shaped asteroid Annefrank (5535), taking a number of relatively indistinct images. Comet Wild 2 had been chosen as it had only been recently diverted into the inner solar system. As a result, it was expected that its composition would have been little altered since it had originally been formed.

The spacecraft was designed to collect cometary dust particles using a collector tray that contained blocks of ultra-low-density aerogel foam. This aerogel was to slow down and capture dust particles, without damaging them, during Stardust's 6.1 km/s fly-by of the comet. The rear of the aerogel collector also collected dust grains, thought to be of interstellar origin, whilst en route to the comet. The collector tray was contained in a sample return capsule that was released from Stardust just before the capsule's re-entry to Earth, where it landed in the Utah desert in 2006. The main Stardust spacecraft was diverted from Earth impact and put into heliocentric orbit. Stardust also included a camera to produce high-resolution colour images of the comet.

Wild 2's nucleus was found to be about 4 km in diameter with an albedo of 0.03. Analysis of the samples returned to Earth was complicated by the high-speed collection which severely modified them. However it was clear that the comet consisted of a mixture of rocky material, which had been formed at very high temperatures, and ices that had been formed at very low temperatures. The existence of such high-temperature material in a comet, that had apparently been formed in the outer regions of the solar system, was a big surprise. The amino acid glycine was also found in the comet's samples, which was the first time that an amino acid had been found in a comet. Analysis of its carbon isotope abundance indicated that the glycine really had come from space and was not an earthly contaminant.

In 2007 NASA decided to redirect the main Stardust spacecraft to image the crater produced on Comet Tempel 1's nucleus by the impact of the Deep Impact probe in 2005. In the event Stardust flew past Tempel 1 at a distance of 180 km in February 2011. Stardust did image the crater, which was about 150 m in diameter, although it was difficult to identify owing to the amount of material that had settled back into it after impact.

Bibliography

Various articles, Science, 314 (2006), 1708–1739.

Various papers, *40th Lunar and Planetary Science Conference, 2009*, Lunar and Planetary Institute.

STEREO

Solar Terrestrial Relations Observatory (STEREO) was the third mission in NASA's Solar Terrestrial Probes (STP) programme. The main objective of the STP programme was to understand the fundamental physical processes in the space environment from the Sun, through the solar system, to the interstellar medium. STEREO's mission was to understand the causes and evolution of coronal mass ejections (CMEs), discover the mechanisms and sites of particle acceleration in the low corona and interplanetary medium, and enable three-dimensional models to be produced of the solar wind. To do this, two 630 kg STEREO spacecraft were placed into heliocentric orbits ahead of (STEREO A) and behind (STEREO B) the Earth, thus enabling stereoscopic images to be produced.

The two spacecraft were launched by a Delta II in October 2006 into highly elliptical geocentric orbits. On 15 December mission controllers used the Moon's gravity to inject STEREO A into a heliocentric orbit inside that of the Earth, which meant that it was ahead of the Earth in its orbit at a gradually increasing distance. A month later mission controllers used the Moon's gravity to inject STEREO B into a heliocentric orbit outside that of the Earth, so it was behind the Earth at a gradually increasing distance. The angle STEREO A/Sun/STEREO B increased at about 44° per year, with the Earth approximately halfway between the two spacecraft.

The instruments on both STEREO spacecraft included an extreme ultraviolet imager, two white light coronagraphs to image the solar corona, and two heliospheric imagers to image the space between the Sun and Earth. Other instruments observed the solar wind and radio disturbances travelling from the Sun.

Simultaneous observations with STEREO A, WIND and ACE (Advanced Composition Explorer) showed that STEREO A had detected ions streaming off the Earth's bow shock "upstream" from Earth up to 4,000 R_E (Earth radii) away. In April 2007, STEREO A imaged the collision between a CME and Comet Encke, which caused the comet's tail to be detached from the comet. A new tail was visible a few days later.

STEREO provided stereoscopic imaging of coronal loops, polar coronal jets and CMEs. The polar coronal jet images of 7 June 2007 showed a helical structure that placed a strong constraint on its possible initiation mechanism. STEREO also provided a wealth of data on ion acceleration in co-rotating interaction regions. For example it showed the continuous increase of the maximum energy of He^+ pick-up ions with increasing solar wind speed, and the temporal development of a suprathermal tail.

Bibliography

The Report of the Science Definition Team for the STEREO Mission, 1997, http://stereo.gsfc.nasa.gov/img/stdt.pdf.

STEREO: Beyond 90 Degrees: A Proposal to the Senior Review of Heliophysics Operating Missions, February 2008, accessed via the Internet http://stereo-ssc.nascom.nasa.gov/publications/STEREO_SRP08_Web.pdf.

Various articles, Space Science Reviews, **136**, Issue 4, 2008.

SURVEYOR

Surveyor was a series of spacecraft designed to soft-land on the Moon and find safe landing areas for the manned Apollo spacecraft. In total, NASA launched seven spacecraft between 1966 and 1968, of which just two were failures, Surveyors 2 and 4.

Surveyor, then called the Surveyor Lunar Orbiter, had been conceived in 1959 as a purely scientific programme. It consisted of a main spacecraft to orbit the Moon and a smaller detachable probe that would soft-land. In the following year, NASA decided that the soft-lander and lunar orbiting spacecraft should be autonomous spacecraft, based on a common bus design, and launched separately. As NASA added Apollo requirements, however, it became clear that this common bus concept was too demanding. As a result, the orbiter part of the mission was cancelled, and the Surveyor programme continued as a soft-landing mission only.

By the time that Surveyor 1 was launched on 30 May 1966, the programme was three years late and vastly over budget. The spacecraft included just one experiment, namely a television system designed to take panoramic images from the Moon's surface. In the event, Surveyor 1 performed almost flawlessly. Images from the surface showed that it had landed on a dark, level area of the Oceanus Procellarum, which was littered with craters ranging from a few centimetres to a few hundred meters in diameter.

Surveyor 3 had a soil scoop added to the payload, which showed that the lunar soil where it had landed was soft and clumpy like course damp sand. Surveyors 5 and 6 included an alpha-particle scattering experiment which found that the chemical composition of the lunar soil in their landing areas of the Mare Tranquillitatis and the Sinus Medii was similar to that of terrestrial basalt produced by volcanic activity. As on Earth, the most abundant element in the lunar samples was oxygen, followed by silicon.

Spacecraft images and on-board data indicated that the surface in the relatively flat areas of the Moon so far visited by Surveyor could support the weight of the Apollo lander. So NASA decided to make Surveyor 7 a scientific spacecraft, landing it near the rim of the crater Tycho. It had a soil scoop that weighed rocks to determine density and an alpha scattering experiment. The latter found that the content of the iron group of elements in the soil was about half of that previously measured in the maria regions.

Bibliography

Mellberg, William F., *Moon Missions: Mankind's First Voyages to Another World*, Plymouth Press, 1997.

Nicks, Oran W., *Far Travelers: The Exploring Machines*, NASA SP−480, 1985.

Reeves, Robert, *The Superpower Space Race: An Explosive Rivalry through the Solar System*, Plenum Press, 1994.

Various authors, *Surveyor Program Results*, NASA SP–184, 1969.

THEMIS

The Time History of Events and Macroscale Interactions during Substorms (THEMIS) spacecraft was the fifth of NASA's Medium Class Explorer (MIDEX) missions. It was designed to answer fundamental questions on the nature of substorm instabilities that explosively release energy stored in the Earth's magnetotail. To do this THEMIS consisted of five identical spacecraft placed in various orbital configurations in Earth orbit over time, plus a network of ground stations in auroral regions of the northern United States and Canada. Instruments at these ground stations included specially installed magnetometers and cameras.

The five small THEMIS spacecraft, each weighing 126 kg, were launched together on a Delta II in February 2007. Almost 40% of each spacecraft was fuel to allow for orbital manoeuvres. The instruments on-board included magnetometers, electric field instruments, electrostatic analyzers and particle detectors.

Initially the five THEMIS spacecraft followed one after another in a "string of pearls" configuration in the coast phase, with along-track separations varying from hundreds of kilometres to 2 R_E (Earth radii). In September 2007 they were moved to more distant orbits with apogees on the dawn side of the magnetosphere. Then in December 2007 the first tail science phase began with apogees in the magnetotail.

The THEMIS spacecraft detected their first substorm on 23 March 2007. They measured the rapid expansion of the plasma sheet at a speed consistent with the simultaneous expansion of the visible aurora. This was the first-ever simultaneous observation of the rapid westward expansion of such a disturbance in space and nearer the ground. During the two-hour event the disturbance moved westward much faster than expected at about 15° longitude/minute, or about 10 km/s at auroral altitudes.

On 30 May 2007 the THEMIS spacecraft detected a transient event in the magnetopause. At the time two of the spacecraft were just in the magnetosheath, one was in the magnetopause, and two were just in the magnetosphere. Their magnetic field measurements showed that they had detected a magnetic rope, which had a strong core field encircled by spiralling magnetic field lines.

Nine months later the THEMIS spacecraft constellation and associated ground stations were the first observatories to track the effects of a solar substorm in detail. They detected magnetic reconnection about 20 R_E from Earth in the magnetic tail, followed about 90 seconds later by the brightening of the aurora. About 90 seconds after that one of the spacecraft detected near-Earth current disruption.

Bibliography

Angelopoulos, Vassilis, et al., *Tail Reconnection Triggering Substorm Onset*, Science, **321** (2008), 931–935.

Burch, J. L., and Angelopoulos, V., *The THEMIS Mission*, Springer, 2009.

Various articles, Space Science Reviews, **141** (2008), 1–583.

TIMED

The Thermosphere Ionosphere Mesosphere Energetics and Dynamics (TIMED) mission was the first mission in NASA's Solar Terrestrial Probes (STP) programme. The spacecraft's primary objective was to examine the way that energy is transferred into and out of the mesosphere and lower thermosphere/ionosphere (MLTI) region of the Earth's upper atmosphere about 60 to 180 km above the surface. This region, which is sensitive to changes in the Sun above and the Earth's atmosphere below, had not been thoroughly investigated by spacecraft previously as its altitude was too low for continuous, in-situ spacecraft observations. TIMED observed this region from above using remote sensing techniques.

The 590 kg TIMED spacecraft was launched on 7 December 2001 by a Delta II into a 625 km circular orbit. On board its instruments included an ultraviolet imager/spectrograph to measure the composition and temperature profiles of the MLTI region and its auroral energy inputs, an extreme ultraviolet spectrometer/photometer system to measure solar radiation from the ultraviolet to soft X-rays, an infrared radiometer to measure temperature profiles, and a Doppler interferometer to measure wind and temperature profiles.

TIMED provided an unprecedented amount of continuous data on the MLTI structure and its spatial and temporal changes, and showed how they responded to external influences from both above and below. For example, TIMED observed the response of the MLTI environment to the superstorms of October and November 2003, and found clear evidence that there was a direct coupling between the Earth's ring current and its upper atmosphere. It appeared as though energetic oxygen atoms, produced by charge exchange of the O^+ in the ring current, precipitated directly from the ring current into the upper atmosphere.

In January 2005 there was a series of four long-duration X-class solar flares and fast coronal mass ejections. During this period TIMED measured the ionospheric response to long intervals of high-energy intense polar rain, which was correlated with extremely low solar wind densities in the high-speed streams. It found that the intense polar rain was confined to the northern hemisphere during the first high-speed stream period, and to the southern hemisphere for the second.

Bibliography

Abdu, M. A., Pancheva, D., and Bhattacharyya, A. (eds.), *Aeronomy of the Earth's Atmosphere and Ionosphere*, Springer, 2011.

Kusnierkiewicz, David Y., *A Description of the TIMED Spacecraft*, American Institute of Physics Conference Proceedings 387, Part One, 1997, pp. 115–121.

TIMED Program Senior Review 2005, http://www.timed.jhuapl. edu/WWW/science/meetings/2005_S3C_Senior_Review_11_ 15/TIMED_2005_Senior_Review_Presentation.pdf.

TRACE

The Transition Region and Coronal Explorer (TRACE) was the fourth of NASA's Small Explorer (SMEX) missions. Weighing just 210 kg, the spacecraft was designed to study the three-dimensional magnetic structure of the Sun from the photosphere through the transition region to the corona.

TRACE employed a 30 cm diameter telescope which imaged the Sun in various ultraviolet and extreme ultraviolet wavebands from about 250 nm to 17 nm. Images taken in the longest waveband showed the effect of magnetic fields in the 6,000 K photosphere, whilst those in the shortest waveband imaged the million-degree corona. A major objective was to explore the relation between changes in the surface magnetic fields and changes in the heating and structure of the transition region and corona.

The spacecraft was launched into a Sun-synchronous low Earth polar orbit in April 1998. This ensured that it would be in permanent sunlight, enabling it to observe almost continuously. The spacecraft was unique at the time in being able to produce both high spatial and high temporal resolution images of the Sun in the extreme ultraviolet, with a maximum spatial resolution of about 1 arcsec, and a maximum temporal resolution of about 1 second. It had been planned to operate TRACE for one year. In the event it was still operating in 2008, over ten years after it was launched.

TRACE discovered a solar feature in extreme ultraviolet images called solar "moss" that had a sponge-like mossy appearance at the base of some coronal loops. This moss occurred in large patches about 10,000 to 20,000 km in extent in the Sun's transition region ∼2,000 km above the photosphere. It consisted of very hot gas at a temperature of about one million K coming from the corona, interspersed with cooler (and therefore darker) gas from the chromosphere.

TRACE also found that the Sun's atmosphere was filled with ultrasound-like waves at a frequency of about 100 millihertz in the Sun's upper atmosphere. It was thought that these waves may have been produced by magnetic reconnections, or by lower-frequency sound waves coming from the solar surface.

Bibliography

Antia, H. M., Bhatnager, A., and Ulmschneider, P. (eds.), *Lectures on Solar Physics*, Springer, 2003.

Handy, B. N., et al., The Transition Region and Coronal Explorer, *Solar Physics*, **187** (1999), 229–260.

Phillips, K. J. H., Feldman, U., and Landi, E., *Ultraviolet and X-ray Spectroscopy of the Solar Atmosphere*, Cambridge University Press, 2008.

ULYSSES

Ulysses was a joint European Space Agency (ESA)/NASA mission to explore the heliosphere, or the region of space outwards from the solar corona, at almost all heliographic latitudes. ESA provided the spacecraft, and NASA the launcher and the spacecraft's radioisotope power generator. Launched in 1990, it was the first spacecraft to fly over both poles of the Sun.

The concept of a spacecraft to measure the solar wind, and the interplanetary environment away from the ecliptic plane, was first proposed by astrophysicists on both sides of the Atlantic in the first few years of the Space Age. NASA and ESRO (ESA's predecessor) had other priorities at the time, but in 1974 they agreed on a joint study into what was to become the Out-of-Ecliptic programme. Its title was later changed to the International Solar Polar Mission (ISPM), and finally to Ulysses.

Over the years this programme was the subject of a great deal of acrimony between ESRO (later ESA) and NASA, mainly caused by NASA's high-handed attitude. For example, NASA unilaterally rejected the electric propulsion system for getting the spacecraft out of the ecliptic. They then unilaterally changed the launch vehicle, then the launch date, and finally the name of what was supposed to be a joint programme. At that time there were to have been two spacecraft which would observe opposite poles of the Sun, using Jupiter's gravity to swing them out of the ecliptic. One spacecraft was to have been provided by NASA and one by ESA, with American and European experiments on both. Then in 1981 NASA unilaterally cancelled their spacecraft, which left a number of European experimenters with no spacecraft on which to put their experiments, and a great deal of money wasted. Eventually relations between ESA and NASA were patched up, but not before high-level protests at ambassadorial level had taken place.

The 370 kg Ulysses spacecraft was finally launched into low Earth orbit by the Space Shuttle Discovery on 6 October 1990. The spacecraft then used two propulsion modules to send it to intercept Jupiter on 8 February 1992, where it used Jupiter's gravity to swing it out of the ecliptic. The spacecraft carried five European and five American experiments. These were to investigate the solar wind, the solar and heliospheric magnetic field, solar emissions, gas and dust in interplanetary space, cosmic rays and cosmic gamma-ray bursts.

Ulysses' heliocentric orbit, following its Jupiter intercept, was inclined at about 80° to the solar equator, with an orbital period of 6.2 years. Its aphelion and perihelion were at about 5.4 and 1.3 AU from the Sun. Over its more than 17 years of operation the spacecraft passed over both poles of the Sun three times, first in 1994–95, then in 2000–01, and finally in 2006–08. Although Ulysses had been launched at solar maximum, the first and third pair of passes over the Sun's poles were at about solar minimum, and the middle pair at solar maximum.

Ulysses found that around solar minimum the slow solar wind (∼350 km/s) was constrained to equatorial regions. Whereas the fast (∼750 km/s) solar wind, which originated from cooler coronal regions near the Sun's poles, fanned out to fill about 70% of the heliosphere. At low heliographic latitudes, where the solar wind velocity was generally low, its density was relatively high, whereas at high latitudes, where the velocity was high, its density was low. The fast solar wind was found to have quite a different isotopic and atomic composition from the slow wind. At solar maximum the clear differences of solar wind velocity with heliographic latitudes broke down.

The spacecraft also found that the heliospheric magnetic field was more complicated than expected, as it did not follow an ordered spiral, but was more chaotic, undergoing large excursions in latitude. As a result, particles emitted during solar storms could reach higher and lower heliographic latitudes as they got further from the Sun.

Ulysses found that the Sun's magnetic dipole was aligned with the Sun's rotational axis at solar minimum. By solar maximum the dipole had moved to lie perpendicular to the rotational axis, and by the next solar minimum it was aligned with the rotational axis once more, but oriented in the opposite direction. Ulysses also provided valuable information on dust and gas in interplanetary space, the distribution and lifetimes of cosmic rays, and gamma-ray bursts.

Bibliography

Balogh, A., Marsden, R. G., and Smith, E. J., *The Heliosphere Near Solar Minimum: The Ulysses Perspective*, Springer-Praxis, 2001.

Bonnet, Roger M., and Manno, Vittorio, *International Cooperation in Space: The Example of the European Space Agency*, Harvard University Press, 1994.

Lang, Kenneth R., *The Sun from Space*, 2nd ed., Springer Verlag, 2008.

Marsden, R. G., *Ulysses Explores the South Pole of the Sun*, ESA Bulletin, No. 82, May 1995, pp. 48–55.

Marsden, Richard G., and Smith, Edward J., *News from the Sun's Poles Courtesy of Ulysses*, ESA Bulletin, No. 114, May 2003, pp. 60–67.

VEGA

Jacques Blamont, of the French space agency CNES, suggested to the Soviet Union in 1967 that they undertake a joint programme to send a balloon payload to float in Venus' atmosphere. It took about ten years for this joint Soviet/French programme to be approved for a 1984 launch. Then in 1980 Blamont pointed out that the main spacecraft bus could fly past Halley's comet after it had released an aeroshell containing the balloon into Venus' atmosphere. The Soviet authorities accepted this idea, but their modified spacecraft design resulted in the large French balloon system being cancelled, and being replaced by a smaller Soviet-made system.

In the event the Soviet Union launched two virtually identical spacecraft called Vega 1 and 2, both of which were to release an aeroshell, containing a balloon and lander, into Venus' atmosphere en route to Halley's comet. Vega 1, which was launched in December 1984, released its aeroshell two days before closest approach to Venus in June 1985. The balloon canister separated from the lander at an altitude of 61 km, just after the aeroshell had been jettisoned. The radio-transparent Balloon 1 then inflated and settled at an altitude of 54 km, in the middle of the planet's most active cloud layer, where the pressure and temperature were 0.54 bar and 32°C. It communicated directly with Earth every half hour. Twenty antennae were used around the world, including those of NASA's Deep Space Network, to receive the signals and track the balloon.

Balloon 1, which was released at a latitude of 7° N, continued operating for two days, drifting about a quarter of the way around the planet until its batteries went flat. Its average horizontal velocity was an unexpectedly high 240 km/h. Downward gusts of up to 12 km/h indicated good vertical mixing of the atmosphere.

Balloon 2 was released from the Vega 2 aeroshell at 7° S at about the same altitude as Balloon 1. It operated for about the same length of time as Balloon 1, and travelled about the same distance. Thirty-three hours after release, Balloon 2 came into a very turbulent region as it passed over a 5 km high mountain. Surprisingly, the turbulence continued for another 2,000 km.

As the landers descended, they detected sulphur, chlorine, and possibly phosphorus in the clouds. Neither of the two landers included cameras because they landed at night, but both carried a drilling system and an X-ray fluorescence spectrometer to analyze the surface material. The drilling system failed on Lander 1, but that on Lander 2, which landed on the edge of Aphrodite Terra, worked perfectly. It detected a type of rock similar to that found in Precambrian stratified massifs on Earth and in the lunar highlands. It was the oldest rock yet found on Venus.

The Halley's comet part of the Vega missions is included in the *Halley's Comet Intercepts* article.

Bibliography

Marov, Mikhail Ya., and Grinspoon, David H., *The Planet Venus*, Yale University Press, 1998.

Reeves, Robert, *The Superpower Space Race: An Explosive Rivalry through the Solar System*, Plenum Press, 1994.

Sagdeev, Roald Z., *The Making of a Soviet Scientist: My Adventures in Nuclear Fusion and Space from Stalin to Star Wars*, John Wiley and Sons, 1994.

Surkov, Yuri, *Exploration of Terrestrial Planets from Spacecraft: Instrumentation, Investigation, Interpretation*, 2nd ed., Wiley Praxis, 1997.

See also: Halley's comet intercepts.

VENERA PROGRAMME

All early Soviet attempts to reach Venus failed, including the first three Venera spacecraft, the first of which had been launched in 1961. But on 12 June 1967, the 1,090 kg Venera 4 was successfully launched to Venus. It had a 380 kg landing capsule with a parachute system to slow its descent through the planet's atmosphere. Four months after launch, Venera 4 arrived at Venus, but contact was lost with its landing capsule about 100 minutes after it hit the top of the atmosphere. The Soviet Union maintained that the capsule had reached the surface, but later analysis showed it had failed with about 25 km to go. Nevertheless, the capsule provided the first in situ measurements of Venus' atmosphere. When it stopped transmitting, Venera 4's capsule had measured an atmospheric pressure of 20 bar and a temperature of 270°C. In addition, it detected an atmosphere of about 96% carbon dioxide. Venera 5 and 6 confirmed the Venera 4 results eighteen months later.

The next pair of Soviet spacecraft was launched in August 1970 but only one, Venera 7, reached its target. Its landing capsule was significantly different from its predecessors, as it was built to withstand a much higher external pressure. In addition, its interior was cooled before the capsule was separated from the main spacecraft, and its parachute system was a new design. The latter was only partially successful, however, and the lander crashed onto the surface at about 70 km/h. Even so, spacecraft controllers detected a feeble signal for 23 minutes.

The Venera 7 capsule was the first artificial object to land and transmit data from the surface of another planet, measuring a ground-level atmospheric temperature of about 470°C. It did not transmit any atmospheric pressure data from the surface, but extrapolating data transmitted during the capsule's descent gave a ground-level pressure of about 90 bar.

The Venera 7 capsule had landed on the night side of Venus, but the Venera 8 capsule, launched in March 1972, landed on the daylight side. It measured surface atmospheric temperatures and pressures that were, within error, the same as those on the night side. The wind speed varied from 360 km/h at 48 km altitude to just 4 km/h below about 10 km. Because of Venus' cloud cover only 2 to 3% of the Sun's illumination reached the ground, where a gamma-ray spectrometer indicated that the surface rocks were probably volcanic in origin.

The Soviet Union then decided to design a completely new spacecraft, to be launched by their larger Proton rocket. Consequently, the mass of Venera 9, of about 5 tons, was more than four times that of its predecessor. The Venera 9 lander was mounted inside an aeroshell to protect it during its initial entry into Venus' atmosphere. After releasing the aeroshell, the main spacecraft was to orbit Venus, the first spacecraft to do so, acting as a relay for the lander. The lander included a camera to take the first photograph of the surface.

Both Venera 9 and its twin Venera 10 were launched in June 1975. On arrival, the Venera 9 lander touched down on the eastern slope of a shield volcano, Rhea Mons, while Venera 10 touched down at the foot of another shield volcano, Theia Mons. The Soviet Union received data from both landers for about an hour after landing, until their respective relay orbiters disappeared below their local horizons. The surface photographs taken by both landers showed numerous rocks with little or no sand. A simple radar system on the Venera 10 orbiter showed that the surface of Venus was relatively flat, varying by only a few kilometres in elevation from a perfectly smooth surface.

The orbital geometry was less favourable at the next launch window in September 1978. So engineers needed to significantly reduce the masses of the next two Soviet spacecraft, Venera 11 and 12. As a result, they replaced the orbiter with a fly-by spacecraft, although both Veneras still included a lander. During their descent, the landers showed that the droplets in the upper-level clouds were composed mostly of chlorine, while those in the other clouds appeared to largely consist of sulphuric acid. The camera covers did not eject on either lander, and the soil analyzers also failed.

The Venera 13 and 14 spacecraft that were launched in 1981 consisted of a fly-by spacecraft and a landing module. This time the landers each contained a drilling and analysis system, to analyze surface material. The surface material in the rolling plains region of Venera 13 was found to be like leucitic basalt, which is often found on the slopes of terrestrial volcanoes. The material in the lowland region of Venera 14 appeared to be like tholeiitic basalt found on the Earth's ocean floor.

The next two Soviet spacecraft, Venera 15 and 16, launched in 1983, were the final spacecraft of the Venera series. They did not carry landers, replacing their normal landing module with a synthetic aperture radar (SAR). In addition to the SAR, which had a surface resolution of about 2 km, both spacecraft included a radar altimeter that had a height resolution of about 50 m.

The images returned by Veneras 15 and 16 showed many ridges and narrow valleys. Several ten to forty metre diameter craters were found in the elevated Lakshmi Planum, but few craters were seen in the lowland regions suggesting that they may be young lava plains. Crater counts in some lowland regions gave an age as young as about 300 million years. Two unusual circular so-called 'corona' features were observed, which appear to have been formed by a mixture of both volcanic and tectonic processes. These two large corona features and a number of smaller ones appear to be unique to Venus.

Bibliography

Leverington, David, *New Cosmic Horizons: Space Astronomy from the V2 to the Hubble Space Telescope*, Cambridge University Press, 2000.

Marov, Mikhail Ya., and Grinspoon, David H., *The Planet Venus*, Yale University Press, 1998.

Reeves, Robert, *The Superpower Space Race: An Explosive Rivalry through the Solar System*, Plenum Press, 1994.

Surkov, Yuri, *Exploration of Terrestrial Planets from Spacecraft: Instrumentation, Investigation, Interpretation*, 2nd ed, John Wiley, 1997.

VENUS EXPRESS

Venus Express, the first European Space Agency (ESA) spacecraft to visit Venus, was a low-cost mission based on the reuse of ESA's Mars Express spacecraft bus. The designs of four of its seven scientific instruments were based on Mars Express instruments, with the remainder based on those from Rosetta, ESA's comet mission. The Venus Express programme was approved in November 2002, and the spacecraft launched just three years later by a Russian Soyuz-Fregat rocket from Baikonur, Kazakhstan.

Previous Soviet/Russian and American spacecraft had been designed to image the surface of Venus, while most of the atmospheric data was provided by in situ measurements from entry probes. Venus Express, on the other hand, was a Venus orbital spacecraft designed to measure the atmosphere below, within and above the clouds virtually continuously over two sidereal Venus days of 243 days each. As Venus had no significant magnetic field, the top of its atmosphere interacted directly with the solar wind. Venus Express was designed to observe this interaction which was expected to result in the solar wind depleting the top of Venus' atmosphere.

Venus Express arrived at Venus on 11 April 2006, about five months after launch. It was put into orbit and adjustments made over the next few days to achieve the desired 24 hour 250 × 66,600 km polar orbit, with apoapsis over the south pole.

Venus Express for the first time imaged both the horizontal and vertical structure of the double-eyed vortices over both poles. It found that the variation of wind velocity with height disappeared above 70° S, with all atmospheric layers co-rotating with the south polar vortex. The spacecraft also detected the loss of various ions from Venus' upper atmosphere caused by its interaction with emissions from the Sun. Venus Express provided high resolution images of the atmospheric 'ultraviolet absorbers' which absorbed large amounts of the incident solar ultraviolet. It generated images of Venus' sulphuric acid clouds in three dimensions, and produced temperature maps of the surface and of the atmosphere at different altitudes.

Bibliography

Svedhem, Håkan, et al., *ESA Bulletin*, August 2008, pp. 2–9.

Sivac, P., and Schirmann, T., *The Venus Express System Design*, ESA SP−1295, 2007.

Various authors, *ESA Bulletin*, November 2005, pp. 8–37.

Various authors, *Nature*, **450** (2007), 629–662.

VIKING

NASA's Viking programme originated with studies in 1960–67 for a large spacecraft called Voyager that could be sent to Mars or Venus, but this was cancelled because of cost. A scaled-down mission to Mars was then proposed in November 1967, which became the basis for Viking.

NASA's Viking programme included two spacecraft, each consisting of an orbiter and a lander (see Figure 9.8), to search for life on Mars. Both orbiters were to be put into orbit around Mars, with their landers still attached, searching for suitable places to land. Only when satisfactory places were found would the landers be released.

Each orbiter included a camera system, a water detector, and an infrared instrument to map the surface temperature. Initially NASA was to use the cameras and water detectors to try to locate a smooth, moist surface on which to land the landing modules, as a moist surface was most likely to sustain life.

Each lander was contained in a double-skinned cover attached to the underside of the orbiter. The cover's outer skin was designed to protect the lander from biological contamination while it was on the launcher, so as to avoid contaminating Mars. The inner skin was an aeroshell to protect the lander from the high temperatures caused by aerodynamic braking in the Martian atmosphere. Each lander included a camera system, soil scoop, mass spectrometer, seismometer, and detectors to measure atmospheric temperature, pressure and wind velocity. The most sophisticated instrument package on the lander was, however, a miniature biological laboratory to test the soil for signs of life.

Figure 9.8. The Viking spacecraft showing the lens-shaped bioshield, which contained the lander, at the bottom, and the retro-rocket at the top. The design of the orbiter, which is in the centre, was based on that of the successful Mariner 9 spacecraft. The tip-to-tip distance along the Viking solar arrays was about 9.8 m (32 ft). (Courtesy NASA/JPL/Caltech.)

The 3,500 kg Viking 1 and 2 spacecraft were both successfully launched to Mars in 1975. Viking 1 was put into orbit around Mars on 19 June 1976. Unfortunately, the images of its planned Chryse landing site showed that it was on the floor of what looked like a river channel that, although scientifically interesting, would be a risky place to land. Instead a new landing site was chosen at 22.5° N in western Chryse.

On 20 July 1976, the Viking 1 lander touched down just 28 km from its target. Twenty-five seconds after landing, it began transmitting the first-ever image from the surface of Mars. During the next few months, numerous images of the red, rock-strewn, sandy surface and pink sky were received and analyzed on Earth. It was early summer at its landing site when the Viking 1 lander touched down, and on the first day it detected an atmospheric temperature varying from −86°C at dawn to −33°C in the afternoon. The atmospheric pressure was 7.6 millibars, and the wind velocity gusted up to 52 km/h. The soil was found to be iron-rich clay, and the atmosphere was found to consist of 95.3% carbon dioxide, 2.7% nitrogen, 1.6% argon, and 0.13% oxygen. Interestingly, the mass spectrometer found no evidence of organic matter on the surface. Both the soil and the very fine dust were found to be magnetic.

Viking 2 arrived at Mars on 7 August 1976, but imaging of its proposed landing site showed it was too rough for a safe landing. Eventually, a new site was found at about 48° N in Utopia Planitia, some 4° farther north than originally planned. Scientists chose this more northerly position hoping its soil would have more moisture as it was nearer the north polar cap. After landing, the Viking 2 lander showed that its landing site had a similar red, rock-strewn surface to that of Viking 1.

It was initially thought that the miniature biological laboratories on both Viking landers had detected signs of life on Mars. But as time went by doubts began to multiply, so it is now generally accepted that the landers found no such signs of life.

Throughout their lifetimes, the Viking 1 and 2 orbiters together imaged virtually the whole Martian surface to at least 300 m resolution, with 2% being imaged at about 25 m resolution. NASA later reduced the periapsis of both Viking orbiters to 300 km, allowing imaging of selected areas at a resolution of 8 m.

The Viking orbiter images showed clear signs of flash floods, dendritic features resembling terrestrial river systems, and many landslides. Temperature and water vapour sensors showed that the north polar cap was made of water ice in the northern summer at a temperature of about −65°C. It was subsequently covered by carbon dioxide ice in the northern winter. In the case of the southern cap, however, the carbon dioxide ice did not completely melt in the southern summer. The water vapour in the Martian atmosphere was found to be highest in low-lying regions, and more water vapour was found during the summer than during the winter when it was presumably frozen out of the atmosphere.

Bibliography

Chaikin, Andrew, *A Passion for Mars: Intrepid Explorers of the Red Planet*, Abrams, 2008.

Corliss, William R., *The Viking Mission to Mars*, NASA SP−334, 1975.

Ezell, Edward C., and Ezell, Linda N., *On Mars: Exploration of the Red Planet 1958–1978*, NASA SP–4212, 1984.

Godwin, Robert (ed.), *Mars: The NASA Mission Reports*, Apogee Books, 2000.

Kraemer, Robert S., *Beyond the Moon: A Golden Age of Planetary Exploration 1971–1978*, Smithsonian Institution Press, 2000.

Raeburn, Paul, and Golombeck, Matt, *Mars: Uncovering the Secrets of the Red Planet*, National Geographical Society, 1998.

Spitzer, Cary R. (ed.), *Viking Orbiter Views of Mars*, NASA SP–441, 1980.

Viking Lander Imaging Team, *The Martian Landscape*, NASA SP–425, 1978.

VOYAGER

Voyager was conceived in the mid-1960s as a 'Grand Tour' mission in which one spacecraft would fly past Jupiter, Saturn, Uranus, and Neptune, using the gravity of one planet to accelerate and redirect it on to the next. Previously, travel to the outer solar system had been thought to require very powerful rockets. Then in 1965 Gary Flandro at JPL found that the four planets were perfectly aligned for an optimum gravity-assist trajectory if launch took place during the period 1976–80. Such an alignment, which occurred only once every 175 years, would enable the mission time from Earth to Neptune, for example, to be reduced from about 30 to 12 years.

A NASA working group suggested in 1969 that this Grand Tour mission could be split in two, with one spacecraft visiting Jupiter, Uranus and Neptune, and the other Jupiter, Saturn and Pluto. These could be supplemented by individually launched Jupiter and Saturn orbiters and entry probes. The American astronomical community was unsure of whether to support this programme, as it required the design of a new, sophisticated spacecraft to operate for more than ten years. This was virtually unheard of at the time when most missions were designed with an estimated lifetime of only a few years.

After further consideration NASA finally approved a simplified version of the original Grand Tour programme. Two 800 kg spacecraft (see Figure 9.9), to be called Voyager 1 and 2, were approved based on the Mariner design. Voyager 1 was to be launched in 1977 to fly past Jupiter in 1979 and Saturn in 1980, and Voyager 2 was to be launched in the same year to fly past Jupiter (in 1979), Saturn (1981), and possibly Uranus (1986) and Neptune (1989). Voyager 2 was designed to visit Uranus and Neptune, even though there was no funding for such visits at that time.

An important part of Voyager 1's mission was a close encounter with Saturn's largest satellite, Titan, which was thought to have an atmosphere like that of the early Earth. This intercept was considered so important that, if it failed, NASA would retarget Voyager 2 for a close fly-by of Titan, even though that would prevent it from visiting Uranus and Neptune.

The Voyager spacecraft had an impressive array of scientific instruments, including wide-angle and narrow-angle television cameras that had best surface resolutions at the Jupiter and Saturn systems of about 1 km. Many experiments, including the television cameras, were mounted on a scan platform at the end of a 2.3 m deployable boom, which allowed them to scan their targets. Other boom-mounted experiments included an infrared interferometer, spectrometer, and radiometer, a photopolarimeter, an ultraviolet spectrometer and a plasma detector. All spacecraft functions, except for trajectory changes, were to be controlled automatically on-board, as the round-trip time for signals from Earth to Neptune, for example, would be more than eight hours.

The 815 kg Voyager 2 was launched on 20 August 1977 followed by Voyager 1 two weeks later. Voyager 1 was launched on a slightly different trajectory so that it would overtake Voyager 2 en route to Jupiter.

Both Voyager spacecraft had their share of problems during the early part of their missions. For example, Voyager 2 automatically switched off a navigation gyroscope and switched on its backup during take-off, and there appeared to be a problem with the experiment boom that would not lock in place. The later launch of Voyager 1 allowed it to have an extra spring fitted to its boom to ensure full deployment. Five months later Voyager 1's scan platform stopped, but, fortunately, it later responded to commands from Earth. Voyager 1 crossed Jupiter's bow shock on 28 February 1979 and recrossed that and Jupiter's magnetopause several times throughout the next few days, as their positions responded to the varying intensity of the solar wind.

Pioneer 11 had detected a reduction in the number of high-energy particles near its closest approach to Jupiter, which had been attributed to an undiscovered ring or satellite. So, just before Voyager 1's closest approach, a single exposure was taken looking to the side of Jupiter where a very faint ring could be seen. Before the Voyager 1 encounter, scientists generally thought that Io would look like a reddish version of Earth's Moon, but covered with sulphur-coated impact craters. However, as Voyager 1 closed in, it resolved a fresh-looking surface of volcanic calderas and rivers of lava. Eight of its volcanoes were found to be active during the fly-by.

Systematic imaging of Saturn by Voyager 1 started on 25 August 1980, some 80 days before closest encounter. The images revealed cloud features, which showed that the equatorial jet stream was much broader than on Jupiter, with a maximum wind velocity of about 1,800 km/h, or about three times that on Jupiter.

As Voyager 1 approached Saturn, radial "spokes" were imaged on the B ring, and each of the main rings was found to consist of hundreds of thin concentric rings. Even the Cassini

Figure 9.9. Voyager in its orbital configuration. The longest boom was for a magnetometer, whilst power for the spacecraft was produced by radioisotope thermoelectric generators (RTGs) in the short stubby structure pointing directly downwards. The 3.7 m (12 ft) diameter high-gain antenna dominates this image. The scan platform, which supported the television cameras and spectrometers, is at the top left. (Courtesy NASA/JPL/Caltech.)

division contained rings. Two new satellites, Prometheus and Pandora, were found, one just inside and one just outside the narrow F ring, which appeared to be shepherding or stabilizing the ring. Another small satellite, Atlas, was found orbiting Saturn just 800 km outside the outer edge of the A ring, apparently restricting its outward expansion. The narrow ring F that had been discovered by Pioneer 11 was found to be three intertwined or braided rings.

Although Voyager 1's intercept with Titan was successful, its images were virtually featureless because Titan was completely covered in cloud. Nevertheless, Voyager 1 measured Titan's surface-level atmospheric pressure as about 1,500 millibars at a temperature of 94 K. Its main constituents appeared to be nitrogen and methane.

In the meantime, Voyager 2 had flown past Jupiter and was on its way to fly past Saturn in August 1981. Because of Voyager 1's success at Titan, NASA announced in January 1981 that Voyager 2 would fly by Uranus and Neptune after Saturn. The resulting gravity assist manoeuvre would cause the spacecraft to cross Saturn's ring plane just 1,200 km outside the newly discovered G ring.

Although Voyager 2 performed well during its encounters with Jupiter and Saturn, it was not in the best of health. For example, one axis of the scan platform had jammed during the

Saturn encounter and, although the fault had been corrected, NASA decided to restrict the scanning speed to avoid further problems at Uranus. Instead, engineers decided to pan the cameras by rotating the whole spacecraft.

Voyager 2 approached the Uranian system in a trajectory almost perpendicular to the orbital plane of its satellites, as these are almost perpendicular to the ecliptic. This meant that the whole close encounter sequence would last for little more than 5 hours, compared with 34 hours for the Jupiter system and even longer for Saturn. The trajectory of Voyager 2 would take it through Uranus' ring plane between Miranda, the nearest known satellite to the planet, and the outermost ring.

As Voyager approached Uranus, scientists realized that its satellites were not exactly where they were expected to be, because the planet was about 0.25 percent heavier than previously thought. So spacecraft operators slightly modified Voyager's trajectory with a 14-minute rocket burn; otherwise Voyager would also have missed Neptune by about 4 million km.

Five days before closest approach to Uranus in January 1986, Voyager detected a strong burst of polarised radio signals, confirming the presence of a magnetic field. Analysis of these radio emissions showed that the interior of Uranus rotates with a period of 17 hr 14 min. This was the first unambiguous measurement of Uranus' rotation period. Four days before

Voyager's closest approach to Uranus, two small shepherding satellites, Cordelia and Ophelia, were found on either side of the ε ring. The highest-resolution images of all the Uranian satellites were those of Miranda. Its complex structure was a big surprise with its chevron-shaped feature, two large ovoids and enormous cliffs up to 20 km high.

Voyager 2 was now on course for its close encounter with Neptune in August 1989. Evidence had started to accumulate in the early 1980s indicating that Neptune may possess one or more partial rings. Clearly the spacecraft had to avoid such rings from a safety point of view. In addition, in 1985 scientists concluded that Neptune was nearly 1,000 km larger than previously thought and its mass was 1.5 percent smaller. The estimate of Neptune's inclination was also increased and Triton's orbit modified. Voyager's new trajectory, taking all these factors into account, would take it only 4,800 km above Neptune's cloud layer and, just over five hours later, it would fly-by Neptune's largest satellite, Triton at a distance of about 40,000 km.

Voyager observations during the encounter phase showed a Great Dark Spot on Neptune, a little like the Great Red Spot on Jupiter. Eight days before closest approach, Voyager 2 detected radio signals from Neptune that were varying with a period of 16.11 hours, which was assumed to be the spin rate of Neptune's interior. Relative to this, Neptune's clouds showed that there was an equatorial jet of 1,800 km/h, which is surprisingly large considering that Neptune receives little solar radiation.

About three weeks before closest approach, Voyager 2 confirmed that Neptune apparently had ring arcs. But, as the spacecraft drew closer to Neptune, it became clear that these ring arcs were just the brightest segments of two complete rings, the Adams and Le Verrier rings. After its closest approach to Neptune, the spacecraft also imaged a third, diffuse ring and a broad band of relatively large particles and dust. Two new satellites, Galatea and Despina, were found, one just inside each of the two narrow rings. These satellites could not cause clumping of the rings on their own, however, and neither could stop either ring from expanding outward. To do this a further satellite was required just outside each of the rings, but no such satellites were found.

Voyager arrived at Triton during early summer in its southern hemisphere and revealed a bright, pinkish-white, southern polar cap of nitrogen ice extending three-quarters of the way from the pole to the equator. Dark streaks of up to 150 km long were seen all over the south polar cap, probably the result of geyser-like eruptions, two of which seemed to be in progress during the Voyager encounter.

As of 2008, Voyagers 1 and 2 were still working well beyond the orbit of Pluto. One is south and the other is north of the ecliptic, participating in the Voyager Interstellar Mission. The Voyager 1 and 2 spacecraft crossed the termination shock and entered the heliosheath in December 2004 and August 2007, respectively. Both spacecraft are being monitored to find out what happens when they leave the influence of the Sun's magnetic field, and of the solar wind, at the heliopause. After that they will be in the interstellar environment, giving the first direct measurements of that, if they survive that long.

See also: Jupiter, Saturn, Uranus and Neptune.

Bibliography

Dethloff, Henry C. and Schorn, Ronald A., *Voyager's Grand Tour: To the Outer Planets and Beyond*, Smithsonian Books, 2003.

Godwin, Robert and Whitfield, Steve (eds.), *Deep Space: The NASA Mission Reports*, Apogee Books, 2005.

Kraemer, Robert S., *Beyond the Moon: A Golden Age of Planetary Exploration 1971–1978*, Smithsonian Institution Press, 2000.

Littmann, Mark, *Planets Beyond: Discovering the Outer Solar System*, updated and rev., John Wiley, 1990.

Morrison, David, *Voyages to Saturn*, NASA SP–451, 1982.

Morrison, David and Samz, Jane, *Voyage to Jupiter*, NASA SP–439, 1980.

WIND

The main problem encountered with trying to understand the behaviour of fields and particles in the near-Earth environment was trying to separate time-dependent phenomena from spatial structures. So the International Solar Terrestrial Physics (ISTP) programme included a number of spacecraft making simultaneous measurements from widely separated locations in near-Earth space. In particular, the WIND spacecraft measured the upstream interplanetary medium, Geotail measured the geomagnetic tail, and POLAR measured the energy received by the Earth's polar regions.

The 1,200 kg WIND spacecraft included 300 kg of propellant which allowed it to make a number of substantial orbit changes over its lifetime. It was launched in November 1994 into a double lunar swing-by orbit with apogees, sunward of Earth, varying between 80 and 250 R_E (Earth radii), and perigees between 5 and 10 R_E. Two years later WIND was placed into a small halo orbit about the L_1 Lagrangian point about 250 R_E away from Earth on the Sun-Earth line. In this halo orbit WIND continuously measured the incoming solar wind, magnetic fields and particles, and provided about a one-hour warning to near-Earth ISTP and other spacecraft of changes in the solar wind.

In late 1998 WIND was placed into a 80 R_E apogee $17\frac{1}{2}$ day petal orbit. Its apogee gradually moved from the day to night side of Earth, so that by April 1999 it was in the geomagnetic tail. Later orbits included fly-bys of both the L_1 and L_2 Lagrangian points, before returning to the L_1 halo orbit in 2004.

On 1 April 1999 WIND was in the geotail at a distance of about 60 R_E from Earth when it became the first spacecraft to detect magnetic reconnection in the Earth's geomagnetic tail. This magnetic reconnection effect, first proposed some decades earlier, allowed energy to be transmitted along magnetic field lines towards Earth. In the event WIND detected the results of this magnetic reconnection in April 1999 while flying along the geotail away from Earth. At first it detected jets of plasma flowing towards the Earth. But as the spacecraft continued to fly away from Earth, the jets of plasma stopped and, further out, were observed to be flying in exactly the opposite direction.

Bibliography

Acuña, M. H., et al., *The Global Geospace Science Program and Its Investigations*, Space Science Reviews, **71** (1995), 5–21.

Lang, Kenneth R., *The Sun from Space*, 2nd ed., Springer Verlag, 2008.

Ogilvie, K. W., and Desch, M. D., *The WIND Spacecraft and Its Early Scientific Results*, Advances in Space Research, **20** (1997), 559–568.

Sibeck, D. G., and Kudela, K., *Interball in the ISTP Program: Studies of the Solar Wind – Magnetosphere – Ionosphere Interaction*, Kluwer Academic Publishers, 1999.

YOHKOH

Yohkoh, which was called Solar−A before launch, was a Japanese spacecraft designed to study high-energy emissions from the Sun, and solar flares in particular. It was launched on 30 August 1991 into a low Earth orbit during a period of solar maximum, when solar flares were more frequent. The 390 kg spacecraft included a soft X-ray telescope operating from 0.2 to 4 keV, a hard X-ray telescope operating from 20 to 100 keV, and two spectrometers covering the X-ray and gamma-ray wavebands up to about 30 MeV. Scientists from the United States and the UK contributed to the scientific payload.

Yohkoh observations covered almost a complete solar cycle. Over this period it imaged the constantly changing inner solar corona in X-rays. It found that hard X-rays were emitted not only from the feet of flaring loops, but also from their apex, most likely as a result of magnetic reconnection. The soft X-ray telescope imaged numerous microflares. In some active regions more than one hundred were observed, brightening and fading over timescales of a few tens of minutes, whereas in other active regions no microflares were seen. White light solar flares appeared to be caused by electrons beamed down magnetic loops into the chromosphere.

Yohkoh detected long X-ray jets moving outwards in the outer corona at velocities of up to 300 km/s, with the material apparently being ejected along open field lines. It was thought that these jets, which were also probably caused by magnetic reconnection, could contribute significantly to mass loss from the Sun.

Bibliography

Lang, Kenneth R., *The Sun from Space*, 2nd ed., Springer Verlag, 2008.

Martens, Petrus C. H., and Cauffman, David P. (eds.), *Multi-Wavelength Observations of Coronal Structure and Dynamics: COSPAR Colloquia Series Volume 13*, Elsevier Science, 2002.

Watanabe, T., Kosugi, T., and Sterling, A. C. (eds.), *Observational Plasma Astrophysics: Five Years of Yohkoh and Beyond*, Kluwer Academic Publishers, 1998.

PART 10
OBSERVATORY SPACECRAFT

1 · Overview – Spacecraft observatories

High-energy spacecraft

The 37 kg Explorer 11, launched in 1961, was the first dedicated high-energy observatory spacecraft. Its payload consisted of a simple gamma-ray telescope, with an anti-coincidence arrangement to discriminate against particles from the Van Allen radiation belts. Explorer 11 detected a total of just 22 cosmic gamma rays over its four-month operational lifetime. OSO (Orbiting Solar Observatory)-3, launched in 1967, whose main mission was to observe the Sun, also had a gamma-ray detector on-board. It detected about 600 cosmic gamma rays over its 16 month lifetime, or just over one gamma ray per day.

Riccardo Giacconi had suggested in 1963 that NASA should launch a small, dedicated, X-ray spacecraft that had conventional, non-focusing X-ray sensors. He further suggested that it should be followed by a spacecraft with a focusing, Wolter-type X-ray telescope. In 1965 NASA agreed to the simpler spacecraft, which was to become Explorer 42, otherwise known as SAS-1 (Small Astronomy Satellite 1), or Uhuru. It was launched by a Scout in December 1970 from the San Marco platform off the coast of Kenya, just 3° south of the equator. This enabled the launcher to put the spacecraft into a near-equatorial orbit using the minimum of propellant. Uhuru detected 339 discrete X-ray sources, about 100 of which had accurate enough locations determined to enable their counterparts to be identified in other wavebands. Two years later, NASA launched SAS-2, which produced the first detailed gamma-ray survey of the sky.

1974 saw NASA launch two foreign spacecraft involved in high-energy research, namely the Dutch ANS (Astronomische Nederlandse Satelliet) and British Ariel 5 spacecraft. ANS was operational for about 18 months, and made measurements in both the X-ray and ultraviolet wavebands. It is jointly credited, with the American military Vela 5 spacecraft, with the discovery of X-ray bursts. Ariel 5, which was the first British spacecraft dedicated to X-ray research, discovered numerous X-ray transients during its five-year lifetime.

These spacecraft basically completed the early exploratory phase of high-energy astronomy. In the meantime, NASA had decided to build four very large, 11,000 kg High Energy Astronomy Observatories (HEAOs), pushing their state-of-the-art designs to the absolute limit. This was a very bold move considering that previous high-energy observatory spacecraft had

each been of only about 100 to 200 kg in mass. Subsequent financial constraints resulted in the four HEAO spacecraft being reduced to three, with their target masses being reduced to 3,000 kg. The first of these, HEAO-1, was launched in 1977.

HEAO-1 was the last large American X-ray observatory to use proportional counters. This was followed by HEAO-2, otherwise called the Einstein Observatory, which was the first X-ray spacecraft to have an imaging capability. It used a Wolter-type focusing telescope of nested cylinders, as advocated by Giacconi in 1963. Einstein had a maximum resolution of about 2 arcsec, and a sensitivity several hundred times that of any previous spacecraft. HEAO-3, which was launched in 1979, was dedicated to gamma-ray and cosmic-ray research.

In the meantime, the European Space Agency's (ESA's) cosmic ray observatory, COS-B, had been launched in 1975. It detected over 200,000 cosmic gamma-rays in total, compared with the 8,000 that had been detected by SAS-2. ESA's X-ray observatory, Exosat, was launched eight years later. It was, in many ways, a smaller version of the Einstein Observatory, having both imaging Wolter X-ray telescopes and other instruments. But, because Exosat was in a highly eccentric orbit, it could, unlike Einstein, undertake uninterrupted observations of about 76 hours per orbit, which was invaluable for monitoring variable sources.

Japan launched three small to medium-sized X-ray spacecraft, Hakucho, Tenma and Ginga, over the period 1979 to 1987. These were less performant than the USA and European spacecraft of the same era, as they were relatively simple spacecraft with no imaging capability, although Hakucho provided important information on the variability of the Rapid Burster, and Ginga detected X-ray emissions from the supernova SN 1987A.

The fall of the Soviet Union in 1989 had a big effect on the operation of the Soviet Union/France's 4,400 kg X-ray and gamma-ray observatory, Granat, launched that year. The main control centre was in the Ukraine which had become independent of Russia, and there were also financial problems. The latter were eventually solved with French assistance, and the spacecraft operated satisfactorily for nine years, undertaking both imaging and spectroscopy of sources. Like Exosat it was initially placed into a highly eccentric orbit to enable long-term monitoring of sources. For the first five years it undertook pointed operations, but it was put into survey mode in 1994 when its attitude control gas was exhausted.

An American Delta II launcher put the 2,400 kg German Rosat X-ray observatory into orbit in 1990. Rosat's main strengths were its combination of high spatial resolution, low background noise, and soft X-ray and extreme ultraviolet imaging. It undertook the first X-ray and extreme ultraviolet all-sky survey using an imaging telescope system with a high X-ray sensitivity. Rosat increased the number of known X-ray sources from 6,000 to over 150,000, and significantly increased the number of known extreme ultraviolet sources.

The first major attempt to detect and understand gamma ray bursts, which had been detected by the Vela spacecraft in 1967, was made with the 16,000 kg Compton Gamma Ray Observatory (CGRO) launched in 1991. CGRO also had on-board an imaging telescope that operated over the range 1 to 30 MeV, and another instrument to map the sky from 20 MeV to 30 GeV. It produced the first all-sky survey above 100 MeV, discovering 271 sources, 170 of which were unidentified. CGRO, which was the second of NASA's Great Observatories to be launched (the Hubble Space Telescope being the first), was essentially a follow-on to HEAO-3. In 1999 NASA launched the Chandra X-ray Observatory, which was the follow-on to the Einstein Observatory. Chandra, the third Great Observatory to be launched, had started life in the 1970s as AXAF (Advanced X-ray Astrophysics Facility), providing both X-ray imaging and spectroscopy. But in the early 1990s the spectroscopy mission was simplified to save money. Chandra's X-ray images were the sharpest cosmic images yet produced in X-rays, having a resolution of about 0.5 arcsec.

ESA's XMM-Newton X-ray observatory, which was launched in the same year as Chandra, was a complementary facility, providing high-resolution X-ray spectroscopy, and the detection of fainter sources than Chandra. Three years later, ESA's INTEGRAL (International Gamma-Ray Astrophysics Laboratory) spacecraft was launched by a Russian Proton launcher from Baikonur, providing simultaneous observations in gamma ray, X-ray and visible wavebands.

The Italian BeppoSAX spacecraft, which had been launched in 1996, was, amongst other things, designed to detect gamma-ray bursts in both the gamma-ray and X-ray wavebands. The X-ray detection was to allow more accurate position fixes, of the order of 1 arcmin, of gamma-ray bursts to be given to ground-based observatories to provide follow-up observations. In the event, BeppoSAX was the first spacecraft to detect a gamma ray burst in other wavebands when it simultaneously detected a burst in February 1997 in both gamma-rays and X-rays. HETE-2 (High Energy Transient Explorer-2), launched four years later, was designed to provide position fixes of gamma-ray bursts to better than 10 arcsec and relay these to ground-based observatories within seconds or minutes. The Swift spacecraft

launched in 2004 undertook a similar programme to HETE-2, but providing position fixes to a higher accuracy.

By the twenty-first century the USA was no longer as dominant in the field of high-energy spacecraft observatories as they had been in the 1960s and 1970s. ESA and the European scientific community had had a capability in this area since the launch of ANS and Ariel 5 in 1974 as national spacecraft, and of COS-B as an ESA spacecraft in the following year. But since then the gap between Europe and the USA has been gradually reduced and Japan has also entered the field. In addition, more and more spacecraft of all nationalities now carry experiments from other countries as part of an international astronomical research effort.

A selection of high-energy observatory spacecraft is listed in Table 10.1A.

Low-energy spacecraft

OAO-2 (Orbiting Astronomical Observatory-2) was NASA's first successful dedicated ultraviolet observatory. Launched in 1968, this 2,100 kg spacecraft had a comprehensive array of imaging telescopes, spectrometers and photometers. It imaged about 10% of the sky, providing observations of over 5,000 stars. This was followed, four years later, by the European TD-1A spacecraft that had a combination of ultraviolet, gamma ray and X-ray detectors. Its ultraviolet detectors measured the ultraviolet intensities of over 31,000 stars.

NASA's OAO-3, or Copernicus, spacecraft was the premier ultraviolet observatory of this initial exploratory phase. Launched in 1972, its main instrument was an 80 cm diameter ultraviolet telescope that fed a scanning spectrometer covering the range from 70 to 330 nm, with a spectral resolution of 0.01 nm. The spacecraft's attitude was controlled to within 0.1 arcsec. Copernicus observed the hot outer coronae of stars, and strong stellar winds, as well as producing key results on the constituents and structure of the interstellar medium.

The Copernicus spacecraft remained operational for over 8 years, but the NASA/UK/ESA IUE (International Ultraviolet Explorer) spacecraft, launched in 1978, remained operational for an astonishing 18 years, 15 years longer than planned. IUE was placed into a much higher orbit than Copernicus, allowing long-duration, continuous observations. And because its orbit was geosynchronous, IUE was always visible from its American ground station, which made observations very convenient to schedule and down-link. IUE produced over 114,000 spectra during its lifetime.

Copernicus and IUE had operated at a minimum ultraviolet wavelength of 70 and 115 nm, respectively. But the EUV (extreme ultraviolet) telescope on the Apollo-Soyuz Test Project of the mid-1970s had operated between 5 and 60 nm. It showed that there were a number of interesting objects detectable in this EUV region. The Rosat X-ray observatory

Table 10.1. *Selected space astrophysical observatory missions*
(A) X-ray and gamma-ray[1]

Years operational	Mission	Main country or organisation	Main wavebands
1961	Explorer 11	USA	γ-ray
1969–79	Vela 5 and 6	USA	γ-ray, X-ray
1970–73	Uhuru (SAS-1)	USA	X-ray
1972–73	SAS-2	USA	γ-ray
1974–76	ANS	Netherlands	X-ray, UV
1974–80	Ariel 5	UK	X-ray
1975–79	SAS-3	USA	X-ray
1975–82	COS-B	ESA	γ-ray, X-ray
1977–79	HEAO-1	USA	X-ray, γ-ray
1978–81	Einstein (HEAO-2)	USA	X-ray
1979–85	Hakucho	Japan	X-ray
1979–81	HEAO-3	USA	γ-ray, X-ray
1983–88	Tenma	Japan	X-ray
1983–86	Exosat	ESA	X-ray
1987–91	Ginga	Japan	X-ray, γ-ray
1989–98	Granat	USSR, France	X-ray, γ-ray
1990–99	Rosat	Germany	X-ray, EUV
1991–2000	Compton GRO	USA	γ-ray
1993–2000	ASCA	Japan	X-ray
1995–	RossiXTE	USA	X-ray
1996–2002	BeppoSAX	Italy	X-ray, γ-ray
1999–	Chandra	USA	X-ray
1999–	XMM-Newton	ESA	X-ray
2000–	HETE-2	USA/Japan/France/Italy	γ-ray, X-ray
2002–	INTEGRAL	ESA	γ-ray, X-ray, optical
2004–	Swift	USA	X-ray, γ-ray, UV/optical
2005-	Suzaku	Japan/USA	X-ray
2008-	Fermi	USA	γ-ray, X-ray

[1] Excluding OSO series, as they were solar observatory spacecraft.

spacecraft, launched in 1990, was the next spacecraft to carry an EUV detector. Its Wide-Field Camera, which could operate down to 6 nm, detected about 480 sources, about half of which were hot coronae of cool low-mass stars. Two years later, NASA's EUVE (Extreme Ultraviolet Explorer) was launched, operating over the 7 to 75 mm band. It spent its first six months undertaking an all-sky survey, which produced a catalogue listing 410 point sources, of which 372 had plausible identifications. This all-sky survey was followed by pointed observations.

ESA's Hipparcos satellite, the first dedicated to astrometry, was launched in 1989. It was designed to measure the position of hundreds of thousands of stars to unprecedented accuracy. In the event, it enabled ESA to publish the Hipparcos Catalogue of about 120,000 stars with milliarcsecond astrometry in 1997, and the Tycho Catalogue with more than a million stars with 20 to 30 milliarcsecond astrometry. The Tycho-2 Catalogue

covering 2.5 million stars was later published in 2000. This work allowed the local distance scale of the universe to be determined to much higher accuracy than previously possible.

The long and tortuous history of the Hubble Space Telescope (HST) goes back to suggestions made by Lyman Spitzer in 1946 about the benefits of a large optical telescope in space. Eventually, in 1977 the Space Telescope, as the HST was known at the time, was approved. But technical and financial problems, followed by shuttle delays caused by the Challenger disaster, resulted in numerous launch delays. Then, shortly after its successful launch in 1990, it was discovered that the HST's 2.4 m primary mirror was flawed. This caused major problems with the imaging instruments, but less problems with the spectrometers, although they did require longer exposure times. Three years after launch, astronauts replaced the High-Speed Photometer with a corrective optics package

Table 10.1 *(cont.)*

(B) Ultraviolet

Years operational	Mission	Main country or organisation	Main wavebands (nm)
1968–73	OAO-2	USA	105–425
1972–73	TD-1A	ESRO	135–280, γ-rays
1972–81	Copernicus (OAO-3)	USA	70–330, X-rays
1978–96	IUE	USA/UK/ESA	115–320
1983–89	Astron	USSR/France	110–350, X-rays
1992–2001	EUVE	USA	7–75
1999–2007	FUSE	USA/Canada/France	90–120
2003–	GALEX	USA	135–280

(C) Visible

Years operational	Mission	Main country or organisation	Main wavebands (nm)
1989–93	Hipparcos	ESA	375–750
1990–	Hubble Space Telescope	USA/ESA	100–2500

(D) Infrared, microwave and radio

Years operational	Mission	Main country or organisation	Main wavebands
1983	IRAS	USA/Netherlands/UK	12–100 microns
1989–93	COBE	USA	1 μm–1 cm
1995–98	ISO	ESA	2–200 microns
1997–2005	HALCA	Japan	1.6–22 GHz
1998–2005	SWAS	USA	487–557 microns
2001–10	WMAP	USA	23–94 GHz
2003–	Spitzer	USA	3.6–160 microns
2006–	Akari	Japan	2–180 microns
2009–	Herschel	ESA	55–670 microns
2009–	Planck	ESA	30–857 GHz

at the HST's first servicing mission. This largely solved the imaging problem, at the expense of losing one of the HST's four primary focus experiments.

The HST made enormous contributions in many areas of astronomy from planetary science to galactic astronomy to estimating the age of the universe. Its impressive images have also been used for public relations purposes and as an educational tool to great effect.

Problems in producing infrared detectors had delayed attempts to build an infrared astronomical spacecraft. Also, because infrared detectors detect heat, either the sensors or the whole infrared telescope had to be cooled. For example, on IRAS (Infrared Astronomical Satellite), the first dedicated infrared spacecraft launched in 1983, the focal plane detectors were cooled to 2.5 K, whilst the telescope was kept below 10 K,

both by superfluid helium. But the IRAS images were relatively crude as there were only 62 infrared detector elements in the telescope's focal plane. Nevertheless, by the time that the helium coolant had been exhausted 300 days after launch, IRAS had observed 95% of the sky over the range 12 to 100 μm with a resolution of 4 arcmin.

The Infrared Space Observatory (ISO) was an ESA spacecraft designed as a follow-on to IRAS. The main IRAS mission had been to find infrared sources, whereas that of ISO was to observe designated infrared sources over relatively long periods of time. Its imaging camera operated from 2.5 to 17 μm with two 32 × 32 element detector arrays. It also had two spectrometers covering the range 2.4 to 200 μm. Launched in 1995, ISO found, surprisingly, that water molecules were commonplace in the Milky Way.

Spitzer, the fourth and last of NASA's Great Observatories, was initially expected to be a shuttle-based observatory, but in 1984 NASA changed it into the free-flying Space Infrared Telescope Facility, or SIRTF. Then budget problems in the early 1990s caused NASA to drastically reduce its target mass from 5,700 kg to just 750 kg. Modifying its orbit to an Earth-trailing heliocentric orbit helped by significantly reducing the mass of coolant required.

Launched in 2003 Spitzer was built around an 85 cm diameter telescope and three cryogenically cooled science instruments: an imager, spectrometer and photometer. The imager operated simultaneously in four narrow wavebands from 3.6 to 8 μm, each waveband using a dedicated 256×256 element detector. Spitzer's mission was to observe all types of cool objects, including dust, protoplanetary discs, brown dwarfs, and ultra-luminous infrared galaxies, to assist in our understanding of their formation and development.

The Cosmic Background Explorer, or COBE, had begun life in the 1970s when NASA suggested that competing experiment groups collaborate on designing a joint spacecraft to investigate the cosmic microwave background (CMB) radiation. The resulting COBE spacecraft was launched in 1989 into a polar orbit from Vandenberg Air Force Base in California. It included instruments to measure spatial anisotropies in the CMB, measure the CMB spectrum, and map the absolute sky brightness in the band 1 to 240 μm. The hoped-for anisotropies were detected at a level of about one part in 100,000, and the CMB spectrum was found to be that of a perfect black body at a temperature of 2.735 K, to a very high accuracy.

The WMAP (Wilkinson Microwave Anisotropy Probe) spacecraft undertook a follow-on mission to COBE, using differential microwave radiometers to measure temperature differences. It was launched in 2001 into an orbit around the L_2 Lagrangian point, some 1.5 million km from Earth, enabling it to have a view to space largely unobstructed by the Sun, Earth and Moon. WMAP's results extended those of COBE and provided significant constraints on the age and constitution of the universe.

A selection of low energy observatory spacecraft is listed in Table 10.1B, C and D.

Bibliography

Davies, John K., *Astronomy from Space: The Design and Operation of Orbiting Observatories*, Wiley-Praxis, 1997.

Giacconi, Riccardo, *Secrets of the Hoary Deep: A Personal History of Modern Astronomy*, Johns Hopkins University Press, 2008.

Kondo, Y. (ed.,), *Observatories in Earth Orbit and Beyond*, Kluwer Academic Publishers, 1990.

Leverington, David, *New Cosmic Horizons: Space Astronomy from the V2 to the Hubble Space Telescope*, Cambridge University Press, 2000.

Logsdon, John M. (ed.,), *Exploring the Unknown, Selected Documents in the History of the U.S. Civil Space Program; Volume V, Exploring the Cosmos*, NASA SP-2001–4407, 2001.

Seward, Frederick D., and Charles, Philip A., *Exploring the X-ray Universe*, 2nd ed., Cambridge University Press, 2010.

2 · Individual spacecraft observatories

AKARI

Akari (previously known as ASTRO-F) was the first Japanese spacecraft dedicated to infrared astronomy. This 950 kg spacecraft was launched in February 2006 into a 700 km Sun-synchronous polar orbit. Its main mission was to perform an all-sky survey in six infrared bands from 9 to 180 μm. It was also designed to undertake deep, pointed observations over the 2 to 180 μm band.

Akari's infrared telescope was a 69 cm diameter Ritchey-Chrétien cooled to 6 K by liquid helium. The detectors were also cooled to 2 K with liquid helium. There were two focal plane instruments: (i) the Far-Infrared Surveyor (FIS), which surveyed the sky simultaneously in four bands from 50 to 180 μm. Its sensitivity was an order of magnitude better than IRAS, the last spacecraft to undertake such an all-sky survey twenty years earlier; (ii) the Infrared Camera (IRC), which was used for pointed observations down to 2 μm, also extended the all-sky survey to 9 and 18 μm. The FIS and IRC were both used for imaging and spectroscopy.

Akari undertook more than 5,000 pointed observations of a wide variety of sources. It observed asteroids and comets, and detected radiation from proto-planetary discs. The spacecraft also detected numerous brown dwarfs and measured the mass loss from red giants. Akari found that the density of star formation in the universe was an order of magnitude higher about 10 billion years ago than it is now.

The helium coolant lasted until August 2007, by which time 94% of the sky had been surveyed. After then observations were carried out in the 1.8 to 5.5 μm band.

Bibliography

Onaka, T., et al., *AKARI: A Light to Illuminate the Misty Universe*, Astronomical Society of the Pacific Conference Series, **418**, 2009.

ANS

ANS (Astronomische Nederlandse Satelliet) was a joint Netherlands/USA space programme to observe the universe in both the ultraviolet and X-ray wavebands. The ANS spacecraft was three-axis stabilised with a pointing accuracy of 1 arcminute. Both the ultraviolet spectrophotometer (150–330 nm), which used a 22 cm Ritchey-Chrétien telescope, and the soft X-ray experiment (0.16–7 KeV) were provided by the Dutch, while the hard X-ray experiment (1.5–30 KeV) was an American contribution.

The 120 kg spacecraft was launched on 30 August 1974 by an American rocket into a Sun-synchronous, polar orbit. Unfortunately, a first-stage guidance failure injected the spacecraft into a 260 × 1,170 km orbit, instead of the planned circular 500 km orbit. This complicated observational scheduling and caused initial problems with background radiation.

ANS found that the X-ray source 4U 1820–30, which appeared to be in globular cluster NGC 6624, emitted two very brief X-ray bursts in the 1 to 30 keV range. At this time this was thought to be the first so-called X-ray burster to be discovered. But, unknown to the astronomical community, X-ray bursters had previously been detected by the American military Vela 5 spacecraft. Astronomers had thought that flare stars may produce measurable X-ray emission, so a survey of flare stars was also undertaken by ANS. This was successful when the flare stars UV Ceti and YZ Canis Minoris were found to be X-ray emitters.

Bibliography

Grindlay, J., et al., *Discovery of Intense X-ray Bursts from the Globular Cluster NGC 6624*, Astrophysical Journal, **205** (1976), L127–L130.

Leverington, David, *New Cosmic Horizons: Space Astronomy from the V2 to the Hubble Space Telescope*, Cambridge University Press, 2000.

Van Kasteren, Joost, *An Overview of Space Activities in the Netherlands*, ESA, 2002.

ARIEL 5

Ariel 5 was the fifth spacecraft to be launched for the UK by NASA. It followed an offer made by the Americans in 1959 to launch scientific equipment from other countries free of charge, subject to certain conditions. Ariel 5, which was the first UK spacecraft dedicated to X-ray research, was first proposed in 1968, some two years before the launch of the first such American spacecraft, Uhuru.

The 130 kg Ariel 5 was launched on 15 October 1974 by an American rocket from the Italian San Marco platform off the equatorial coast of Kenya. This enabled NASA to use the smallest possible launcher to put the spacecraft into the desired near-equatorial, circular orbit. The spacecraft contained six experiments, five British and one American. They measured the position, spectrum and polarisation of X-ray sources in the range 0.3–30 keV, and detected high-energy X-ray sources up to 1.2 MeV. In addition an all-sky monitor, operating in the 3–6 keV range, detected transient X-ray events. Source positions were measured to within about 1 to 10 arcmin over the 0.3 to 30 keV range, and about 2° up to 1.2 MeV.

For more than five years Ariel 5 undertook long-term monitoring of many X-ray sources. About forty X-ray transients were discovered, one or two of which were thought to contain black holes. The X-ray intensity of some Seyfert galaxies was found to vary in about a day, indicating that these X-ray sources were only about one light day across. The 6.7 keV resonance line of Fe XXV was found in the Tycho and Cas A supernova remnants, indicating the presence of shock-heated interstellar gas. Similarly the discovery of Fe XXV and XXVI resonance lines in the spectrum of some clusters of galaxies indicated extremely high plasma temperatures.

Bibliography

Davies, John K., *Astronomy from Space: The Design and Operation of Orbiting Observatories*, Wiley-Praxis, 1997.

Leverington, David, *New Cosmic Horizons: Space Astronomy from the V2 to the Hubble Space Telescope*, Cambridge University Press, 2000.

Massey, Harrie, and Robins, M.O., *History of British Space Science*, Cambridge University Press, 1986.

ASTRON

Astron was a Soviet astrophysical spacecraft whose main instrument was an 80 cm Ritchey-Chrétien ultraviolet telescope operating over the range 110–350 nm. At its focus was a three-channel spectrometer built in France. Astron also carried a small X-ray spectrometer covering the range 2–25 keV.

The 3 ton Astron spacecraft design was based on that of the Venera interplanetary spacecraft, with the astrophysical experiments in place of the Venus entry capsule. Astron was launched into a 2,000 × 200,000 km orbit in March 1983 with a period of 98 hours to enable long, uninterrupted observations to be made. But, in practice, the observing periods were just 3 to 4 hours per orbit.

Inevitably, even though Astron was launched five years after IUE, comparisons were made of their relative performance. Astron's light gathering power was three times that of IUE, both had similar spectral resolutions, but, although

Astron's lifetime was six times that specified, its achieved lifetime of six years was still significantly less than IUE's 18 years. Nevertheless, Astron provided many useful observations ranging from those of comets to stars and galaxies. It made particularly valuable observations of Halley's comet, supernova 1987A, the binary pulsar Her X-1, and the Rapid Burster, MXB 1730–335.

Bibliography

Boyarchuk, A. A., *UV Observations by ASTRON Satellite*, Irish Astronomical Journal, **17** (1986), 352.

Boyarchuk, A. A., *The Ultraviolet Telescope on the ASTRON Satellite*, Astrophysics and Space Physics Reviews, **5** (1987), 225–239.

Davies, John K., *Astronomy from Space: The Design and Operation of Orbiting Observatories*, Wiley-Praxis, 1997.

BEPPOSAX

BeppoSAX, or Satellite per Astronomia X, was named after Giuseppe (Beppo) Occhialini, a pioneer of cosmic-ray research and gamma-ray astronomy. It was an Italian spacecraft, with Netherlands participation, which was launched in April 1996 into a low Earth orbit. Its X-ray mission was to study the spectral behaviour of sources over a wide range of energies, to study the long-term variability of sources and to detect X-ray transient phenomena. BeppoSAX was also designed to detect gamma-ray bursts. It was assumed that some gamma-ray bursts would simultaneously trigger one of the X-ray detectors, which would then enable the source location to be determined sufficiently accurately for its optical counterpart to be detected.

The payload of BeppoSAX consisted of spectrometers covering the range 0.1 to 10 keV, a proportional counter sensitive to 4–120 keV, and a detection system with a very wide field of view covering the range 15–300 keV. BeppoSAX also had a gamma-ray burst monitor sensitive over 60–600 keV, with a time resolution of 1 ms. Finally there were two wide-field X-ray cameras that monitored the variability of sources down to 1 mCrab, and detected and localised X-ray transients.

BeppoSAX made the first major breakthrough in our understanding of gamma-ray bursts when it detected a gamma-ray burst, now known as GRB 970228, on 28 February 1997. One of the wide-field cameras also detected a simultaneous X-ray burst that enabled the position of the gamma-ray burst to be measured to within about 3 arcmin. Subsequent observations with the other X-ray instruments on-board allowed its position to be determined more accurately. The source was then identified optically with the William Herschel Telescope on La Palma. This was the first time that a gamma-ray burst source had been detected in another waveband. Over its lifetime of six years, BeppoSAX detected over 50 gamma-ray bursts, of which many were subsequently observed optically. This showed for the first time that gamma-ray bursts originated from well out-side the Milky Way.

Bibliography

Costa, E., Frontera, F., and Hjorth, J. (eds.), *Gamma-Ray Bursts in the Afterglow Era, Proceedings of the International Workshop Held in Rome, October 2000*, ESO Astrophysics Symposia, Springer, 2001.

Scarsi, L., et al., (eds.), *The Active X-Ray Sky: Results from BeppoSAX and RXTE*, Elsevier Science, 1998.

Schilling, Govert, *Flash: The Hunt for the Biggest Explosions in the Universe*, Cambridge University Press, 2002.

Vedrenne, Gilbert, and Atteia, Jean-Luc, *Gamma-Ray Bursts: The Brightest Explosions in the Universe*, Springer-Praxis, 2009.

CHANDRA X-RAY OBSERVATORY

The Chandra X-ray Observatory (see Figure 10.1) was a NASA X-ray spacecraft named after the Nobel laureate and theoreti-cal astrophysicist Subrahmanyan Chandrasekhar. Launched in 1999, as the third of NASA's Great Observatories, it was, at the time, the largest civilian spacecraft ever launched by the space shuttle. Chandra was complementary in performance to ESA's XMM-Newton X-ray spacecraft that was also launched in 1999. Chandra's main mission was high resolution imag-ing, whereas that of XMM-Newton was high sensitivity spec-troscopy.

Riccardo Giacconi and Harvey Tananbaum submitted a brief proposal to NASA in 1976 for an X-ray spacecraft to succeed the Einstein Observatory, which, at the time, was in the process of construction. Their proposal was favourably received, and they were awarded a contract to study the fea-sibility of what was called at the time the Advanced X-ray Astrophysics Facility or AXAF. But cost overruns with the Hubble Space Telescope, and arguments of priorities within NASA and the astronomical community, delayed its approval for twelve years. Even then, full approval by Congress was conditional on the production of a satisfactory set of X-ray mirrors.

The X-ray mirrors passed their test in 1991, but in the fol-lowing year there was another budget crisis at NASA, caused mainly by cost overruns on the International Space Station. This resulted in severe financial pressure on the AXAF budget, which eventually led to the deletion of the originally-foreseen in-orbit servicing by the space shuttle and splitting the mission in two. One spacecraft would primarily provide imaging and the other, with lower resolution mirrors, would provide spec-troscopy. In 1993 the latter was cancelled, although NASA's X-ray spectrometer was eventually flown on the Japanese Suzaku spacecraft.

Chandra's orbit was a 10,000 × 140,000 km, 64 hour orbit, which allowed long duration exposures of X-ray sources. It had a single Wolter Type I telescope about 10 m long operating over the 0.1–10 keV band. Its spatial resolution, which was the best of any X-ray telescope at the time, was 0.5 arcsec, its time resolution 16 μs, with a 30 arcmin field of view. The telescope could also be used to feed two spectrometers covering the energy range of 0.1–10 keV between them.

Chandra detected X-ray emission from Jupiter's auroral zones, and found that the X-ray emission from near Jupiter's north magnetic pole appeared to be due to a single hot spot that pulsed with a period of about 45 minutes. The spacecraft found a neutron star whose progenitor star appears to have had a mass of at least 40 solar masses, which was much heavier than theory predicted, and it resolved at least 90% of the enigmatic X-ray background radiation into individual X-ray sources. Chandra detected a group of massive stars less than a light year away from the Milky Way's central black hole Sgr A*, implying that they were formed in situ, rather than having migrated there from further out. The spacecraft also found plumes of high-energy particles coming from the vicinity of a hypothesised black hole in the central galaxy of the Perseus cluster. Finally, Chandra found direct evidence of the dark energy accelerat-ing the expansion of the universe by observing 26 clusters of galaxies at distances of from one to eight light years.

Bibliography

Fabian, A. C., Pounds, K. A., and Blandford, R. D. (eds.), *Frontiers of X-Ray Astronomy*, Cambridge University Press, 2004.

Schlegal, Eric M., *The Restless Universe: Understanding X-Ray Astron-omy in the Age of Chandra and Newton*, Oxford University Press, 2002.

Seward, Frederick D., and Charles, Philip A., *Exploring the X-Ray Universe*, 2nd ed., Cambridge University Press, 2010.

Tucker, Wallace, and Tucker, Karen, *Revealing the Universe: The Making of the Chandra X-Ray Observatory*, Harvard University Press, 2001.

COBE

The Cosmic Background Explorer (COBE) spacecraft (see Figure 10.2) was the first spacecraft to be dedicated to cosmol-ogy. It performed an all-sky survey at microwave and infrared wavelengths to measure the structure of the cosmic microwave background (CMB) radiation.

In 1974 NASA turned down three proposals to study the CMB during a competition for Explorer-class missions. Two years later, however, NASA suggested that the three CMB teams get together and propose a joint mission. This resulted in COBE, which was launched into a polar orbit from Vandenberg Air Force Base in California on 18 Novem-ber 1989. It included three instruments, the Differential Microwave Radiometer (DMR), the Far Infrared Absolute Spectrophotometer (FIRAS), and the Diffuse Infrared Back-ground Experiment (DIRBE).

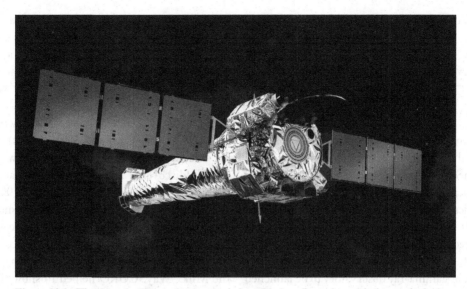

Figure 10.1. The Chandra X-ray observatory in its orbital configuration with the sunshade door open. The circles at the near end of the spacecraft were the entrance to the high-resolution Wolter-type X-ray telescope. At about 10 m in length this telescope determined the length of the spacecraft. (Courtesy NASA/CXC/NGST.)

It was argued at the time that, if the microwave or cosmic background radiation was a true indication of the early universe, it should have structure, as otherwise it was difficult to see how the galaxies could then have formed. The DMR instrument was designed to detect this anisotropy or spatial variation of the CMB, showing the overall structure of the early universe about 300,000 years after the Big Bang. FIRAS was to measure the spectrum of the CMB. If it was found to be that of a black body, this would give added credibility to the Big Bang theory.

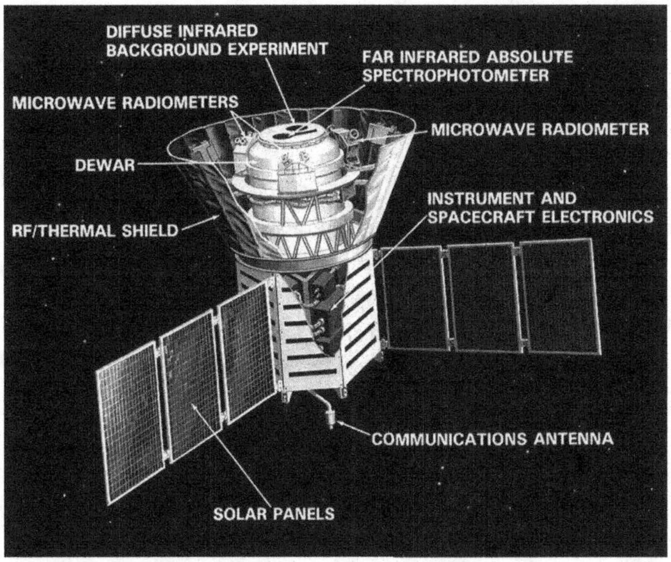

DIFFUSE INFRARED
BACKGROUND EXPERIMENT
FAR INFRARED ABSOLUTE
SPECTROPHOTOMETER
MICROWAVE RADIOMETERS
MICROWAVE RADIOMETER
DEWAR
INSTRUMENT AND
SPACECRAFT ELECTRONICS
RF/THERMAL SHIELD
COMMUNICATIONS ANTENNA
SOLAR PANELS

Figure 10.2. The COBE spacecraft was built around a Dewar that contained 650 litres of superfluid helium, with a design based on that of IRAS. The COBE Dewar was to keep the FIRAS detectors at about 4 K, and those in the DIRBE instrument as cool as possible. The conical shield at the top was to protect sensitive instruments from direct sunlight and Earth-based radiation. (Courtesy NASA.)

Finally, DIRBE was to map the absolute sky brightness in the band from 1 to 240 μm and detect early infrared galaxies.

As expected, the FIRAS instrument ran out of liquid helium, which had kept its sensors at about 4 K, ten months after launch. This helium depletion also degraded the performance of the DIRBE, but the DMR continued working normally until NASA switched it off in December 1993.

After analysing the FIRAS measurements, John Mather and colleagues announced in January 1990 that the CMB radiation was that of a black body at a temperature of 2.73 K, to a very high accuracy, so supporting the Big Bang theory. Then, two years later, George Smoot and colleagues announced their discovery of the anisotropy in the CMB, of just one part in 100,000 using the DMR. This showed, for the first time, that the universe was not perfectly smooth shortly after the Big Bang, thus providing the seeds for galaxy formation. The DIRBE instrument detected both the infrared radiation from the early universe and that from more recent galactic sources. In addition, it showed that the plane of the Milky Way was slightly warped, and even measured the distribution of dust in the solar system.

In 2006 John Mather and George Smoot received the Nobel Prize in Physics for their discoveries of the black body form and anisotropy of the CMB.

Bibliography

Calzetti, D., Livio, M., and Madau, P. (eds.), *Extragalactic Background Radiation*, Space Telescope Science Institute Symposium Series 7, Cambridge University Press, 1995.

Mather, John C., and Boslough, John, *The Very First Light: The True Inside Story of the Scientific Journey Back to the Dawn of the Universe*, Penguin Books, 1998.

Smoot, George, and Davidson, Keay, *Wrinkles in Time: The Imprint of Creation*, Abacus, 1995.

See also: Radiometry.

COMPTON GAMMA RAY OBSERVATORY

The Compton Gamma Ray Observatory (CGRO) (see Figure 10.3) was the second of NASA's Great Observatories to be launched. It was named after Arthur Holly Compton, the Nobel Prize–winning physicist who carried out cutting edge research in high-energy physics in general and cosmic rays in particular.

In the early 1970s the gamma-ray spacecraft HEAO-3 had been severely reduced in size to save money. So, in parallel with completing HEAO-3, NASA had begun to consider a large, more performant gamma-ray observatory to be launched in the following decade. The design work on this new Gamma Ray Observatory or GRO, as it was then called, was started in 1977, and in the following year NASA selected its complement of five instruments. GRO was to be launched by a space shuttle into a 300-km altitude orbit, from which it would climb to its operational altitude using an orbital manoeuvring engine. Like the Hubble Space Telescope, it was designed for on-orbit servicing, at which time the Shuttle would also replenish its propellant supply.

In 1980 Congress approved the program for a 1981 start, but funding constraints caused NASA to move the launch date from 1987 to the following year, and delete the on-orbit servicing requirement as well as one of its five instruments. In 1986 the launch date was delayed once more because of the Challenger disaster, and because no other launch vehicle was powerful enough to launch the 16,000 kg spacecraft. So NASA had to wait until a suitable shuttle launch became available.

The CGRO carried four instruments: the Burst and Transient Source Experiment (BATSE), the Oriented Scintillation Spectrometer Experiment (OSSE), the imaging Compton Telescope (COMPTEL), and the Energetic Gamma-Ray Experiment Telescope (EGRET). BATSE operated from about 30 keV to 1.9 MeV. It was designed to quickly locate the position of short-duration gamma-ray bursts and enable the other CGRO instruments, and other observatories, to locate them and hence determine their characteristics. When CGRO was launched, the nature of these gamma-ray burst sources was a complete mystery. OSSE was designed to observe discrete sources over the range 0.1 to 10 MeV. COMPTEL, which was provided by a European consortium, operated over the range 1 to 30 MeV. EGRET's main function was to complement COMPTEL and map the sky over the energy range from 20 MeV to 30 GeV.

The CGRO was deployed from the Space Shuttle Atlantis on 5 April 1991, orbiting the Earth at the relatively low altitude of 450 km to avoid the Van Allen radiation belts, which would have compromised the experiments. This low altitude would cause the orbit to decay relatively quickly, however, so on-board propellant was to be used to boost it back to its normal orbital altitude every year or so. The CGRO spent until November 1992 undertaking an all-sky survey, followed by pointed observations.

At the time of launch, there were only two gamma-ray pulsars known (the Crab and Vela pulsars), CGRO discovered 10; one gamma-ray quasar, or blazar, was known (3C 273), CGRO discovered about 70; and 500 gamma-ray bursts had been detected, CGRO detected over 2,700 evenly distributed across the sky. CGRO also detected gamma-ray emission from supernova remnants and from black hole candidates. It mapped the emission of aluminium 26 across the celestial sphere, showing relatively high concentrations in star formation regions of the Milky Way. CGRO helped to show that soft gamma-ray repeaters were neutron stars with very strong magnetic fields. It also detected gamma rays and neutrons emitted by solar flares.

One of CGRO's gyroscopes failed in December 1999, which made the spacecraft highly vulnerable to another gyroscope failure, which could have resulted in an uncontrolled re-entry. NASA ruled out the possibility of a space shuttle repair or retrieval mission and decided to end the project. So on 4 June 2000, they undertook a controlled re-entry of the spacecraft into the Pacific Ocean.

Bibliography

Kniffen, Donald A., *The Gamma Ray Observatory*, Sky and Telescope, May 1991, pp. 488–492.

Leonard, Peter J.T., and Wanjeck, Christopher, *Compton's Legacy: Highlights from the Gamma Ray Observatory*, Sky and Telescope, July 2000, pp. 48–54.

Leverington, David, *New Cosmic Horizons: Space Astronomy from the V2 to the Hubble Space Telescope*, Cambridge University Press, 2000.

Schilling, Govert, *Flash! The Hunt for the Biggest Explosions in the Universe*, Cambridge University Press, 2002.

Various articles, Astronomy and Astrophysics Supplement Series, **97** (1993), 5–29.

Vedrenne, Gilbert, and Atteia, Jean-Luc, *Gamma-Ray Bursts: The Brightest Explosions in the Universe*, Springer-Praxis, 2009.

COS-B

COS-B (Cosmic Ray Satellite-B) was a 280 kg European gamma-ray spacecraft launched by an American launcher in 1975. It was launched into a $440 \times 100,000$ km highly elliptical orbit to spend most of its time above the Van Allen radiation belts.

In 1969 the European Space Research Organisation (ESRO), a forerunner of the European Space Agency (ESA), had

Figure 10.3. Compton Gamma Ray Observatory in its orbital configuration. When it was launched in 1991 it was, at 17 tons, the heaviest unmanned civilian spacecraft so far launched by NASA. The OSSE instrument was located at the top left, COMPTEL was inside the circular drum in the middle of the top, and EGRET at the top right. The eight BATSE detectors were mounted at the corners of the spacecraft, four on the top surface and four on the bottom. (Courtesy NASA Goddard.)

considered eight alternatives for its second generation of scientific spacecraft missions. Eventually two of these spacecraft were built, COS-B and a magnetospheric spacecraft called GEOS. COS-B's payload consisted of its main gamma-ray telescope and a small X-ray telescope to allow simultaneous monitoring of pulsars in both X-rays and gamma rays. The spacecraft continued to operate for about $6\frac{1}{2}$ years, over three times its design lifetime.

COS-B, which covered the energy range from 30 MeV to 5 GeV, detected over 200,000 cosmic gamma-rays in total, compared with 8,000 for the previous American gamma-ray spacecraft, SAS-2. It observed about half of the celestial sphere and produced gamma-ray maps of the Milky Way in three energy bands. They showed the Cygnus, Carina and Perseus spiral arms, the Orion molecular cloud, and Gould's belt. In addition, COS-B located about 25 'point' sources, of which only a few could be clearly identified in other wavebands, because of its limited angular resolution of about 1° to 2°. These were the quasar 3C273, the Vela and Crab pulsars, the ρ Ophiuchi dark cloud and the enigmatic source Geminga.

Bibliography

Krige, J., and Russo, A., *A History of the European Space Agency 1958–1987: Volume I, The Story of ESRO and ELDO 1958–1973*, European Space Agency SP-1235, 2000.

Mayer-Hasselwander, H.A., et al., *Large-Scale Distribution of Galactic Gamma Radiation Observed by COS-B*, Astronomy and Astrophysics, **105** (1982), 164–175.

Murthy, Poolla V. Ramana, and Wolfendale, Arnold W., *Gamma-Ray Astronomy*, 2nd ed., Cambridge University Press, 1993.

Swanenburg, B.N. et al., *Second COS B Catalog of High-Energy Gamma-Ray Sources*, Astrophysical Journal, **243** (1981), L69–L73.

EUVE

The Extreme Ultraviolet Explorer (EUVE) was the first dedicated spacecraft to observe in the extreme ultraviolet (EUV) between about 10 and 100 nm. The programme, which had originally been proposed to NASA by Stuart Bowyer of the University of California in 1975, was approved in 1984 for a space shuttle launch. But the Challenger disaster of two years later caused NASA to change the launch vehicle from the shuttle to a Delta, which delayed the launch to 1992.

In the early 1960s astronomers had generally believed, based on radio telescope and sounding rocket observations, that absorption by hydrogen in the interstellar medium (ISM) would generally restrict observations in the EUV to objects within a few parsecs of the Sun. But in the early 1970s astronomers had found evidence, using Mariner 9, OAO-2 and Copernicus, for what is now called the Local Bubble, in which the interstellar medium was appreciably less dense than normal. Although the Local Bubble was observed to be a region ~100 parsec in diameter surrounding the Sun, it was not spherical. In fact there appeared to be a region virtually devoid of gas for over 300 parsec in one part of the sky. There was also a suggestion that in small areas of the sky the interstellar gas may be tenuous enough to allow us to detect some galaxies. Nevertheless it was thought, when EUVE was launched, that its observations would be largely limited to the local region of the Milky Way.

The 3,400 kg EUVE had four 40 cm diameter Wolter-Schwarzschild grazing incidence telescopes. Three of these were to provide a complete sky survey within the first six months between about 7 and 75 nm, whereas the fourth was to undertake a deep survey of a two degree-wide strip centred on the ecliptic. Surprisingly, in view of the logic of the above paragraph, EUVE detected its first extra-galactic object, the BL Lac object PKS 2155–304, whilst the spacecraft was still being checked out in orbit.

The first EUVE source catalogue that was published in 1994 listed 410 point sources, of which 372 had plausible identifications. Most of these were either the coronas of F- to M-type stars or the very hot photospheres of white dwarfs. The other objects were mainly very hot O- to A-type stars, cataclysmic variables, or active galactic nuclei (AGNs). After the six-month survey phase, EUVE undertook pointed observations, concentrating on EUV spectroscopy. By the end of its third year, EUVE had increased the number of extragalactic objects to 37, which were a mixture of AGNs, BL Lac objects, and Seyfert galaxies. Later observations even showed EUV-emitting regions of nearby clusters of galaxies.

Bibliography

Barstow, Martin A., and Holberg, Jay B., *Extreme Ultraviolet Astronomy*, Cambridge University Press, 2003.

Bowyer, S., et al., *The First Extreme Ultraviolet Explorer Source Catalog*, Astrophysical Journal Supplement Series, **93** (1994), 569–587.

Chien, Philip, *EUVE Probes the Local Bubble*, Sky and Telescope, February 1992, pp. 161–163.

EXOSAT

The European X-ray Observatory Satellite (Exosat) was a medium-sized European Space Agency (ESA) X-ray spacecraft launched by an American rocket in May 1983. Its mission, when originally conceived in the late 1960s, was to measure the position of X-ray sources by lunar occultation. But by the time that the program was approved in 1973, its original mission was obsolete. As a result, ESA decided to produce a more sophisticated spacecraft with a mass three times that originally envisaged.

Exosat's payload was designed to measure the precise location and time variability of X-ray sources, the mapping of extended sources, and broadband spectroscopy. Its instruments covered the range 0.05–50 keV, with spatial and time resolutions of up to 10 arcsec and about 20 microsec, respectively.

A key element of Exosat was its high-eccentricity (190,000 × 350 km) Earth orbit, which allowed it to undertake uninterrupted observations of up to 76 hours. This was crucial to the understanding of many sources. During its lifetime, Exosat observed active galactic nuclei, stellar coronae, white dwarfs, cataclysmic variables, X-ray binaries, galactic bulge X-ray sources, supernova remnants, and clusters of galaxies. It was the first spacecraft to observe quasi-periodic oscillations in the galactic bulge X-ray source, GX 5–1, and in other low-mass X-ray binaries, Scorpius X-1, Cygnus X-2, as well as in X-ray pulsars. In addition, Exosat clearly showed for the first time that not all flickering X-ray sources were black holes.

Bibliography

Altmann, G., et al., ESA Bulletin No. 31, August 1982, pp. 6–33.

Seward, Frederick D., and Charles, Philip A., *Exploring the X-ray Universe*, 2nd ed., Cambridge University Press, 2010.

Leverington, David, *New Cosmic Horizons, Space Astronomy from the V2 to the Hubble Space Telescope*, Cambridge University Press, 2000.

White, N. E., and Peacock, A., *The EXOSAT Observatory*, Societa Astronomica Italiana, Memorie, **59** (1988), 7–31.

EXPLORER 11

Explorer 11 was a 37 kg spacecraft launched by NASA in April 1961 to detect gamma rays between about 50 and 500 MeV.

The Explorer series of spacecraft were the earliest Earth-orbiting spacecraft in the American space programme, with Explorer 11 being the first one dedicated to gamma-ray observations. Explorer 11 spun round its longitudinal axis, but the fourth stage of the launch vehicle was deliberately left attached to cause the spacecraft/fourth stage combination to tumble about the transverse axis also, and so scan the celestial sphere. The orientation of the gamma-ray detector could only be determined to about 5°.

The payload of Explorer 11 consisted of a simple gamma-ray telescope, with an anti-coincidence arrangement to discriminate against particles from the Van Allen radiation belts. During its four-month operational lifetime, Explorer 11 detected only 22 cosmic gamma rays. With such a small number of events no clear source of cosmic gamma rays could be determined.

Bibliography

Kraushaar, W. L., and Clark, G. W., *Search for Primary Cosmic Gamma Rays with the Satellite Explorer XI*, Physical Review Letters, 8 (1962), 106–109.

Kraushaar, W., et al., *Explorer XI Experiment on Cosmic Gamma Rays*, Astrophysical Journal, 141 (1965), 845–863.

Melin, Marshall, *A Satellite for Gamma-Ray Astronomy*, Sky and Telescope, June 1961, pp. 330–332.

FERMI

Fermi was an international space observatory designed to study the universe in the X-ray and gamma-ray energy range from about 8 keV to 300 GeV. It was launched in June 2008 as a follow-on to NASA's Compton Gamma Ray Observatory (CGRO) which had been launched seventeen years earlier.

CGRO had carried four instruments, including EGRET (Energetic Gamma-Ray Experiment Telescope) which had mapped the sky up to 30 GeV. EGRET had been designed around particle detectors developed during the 1960s, using spark chambers as gamma-ray detectors. So after the CGRO had been safely launched, Peter Michelson of Stanford University, William Atwood of the Stanford Linear Accelerator Center and others began to consider the design of a follow-on spacecraft using more up-to-date technology.

Their proposed new spacecraft was named GLAST (for Gamma Large Area Silicon Telescope, but later for Gamma-ray Large Area Space Telescope). It used silicon strip detector technology. As a result it was more performant and could detect much fainter gamma-ray sources than EGRET. In addition, its design was more robust and stable than EGRET's, operating at much lower internal voltages and requiring no consumables.

The GLAST spacecraft design, operating from about 20 MeV to 300 GeV, was developed over the 1990s. Then an extra instrument was added to the spacecraft; this was the GLAST Burst Monitor, later called the Gamma-ray Burst Monitor (GBM), which operated from about 8 keV to 30 MeV. It was an important addition, as it complemented the initial GLAST instrument, now called the LAT (Large Area Telescope), allowing the investigation of gamma-ray bursts below 20 MeV into the X-ray region. The GLAST spacecraft was renamed the Fermi Gamma-ray Space Telescope shortly after launch in honour of Enrico Fermi, a pioneer in high-energy physics.

In its first year of operation the Fermi spacecraft detected about 1,500 gamma-ray sources which included active galaxies, supernova remnants and pulsars. In September 2008 it detected the gamma-ray burst GRB 080916C which follow-up ground-based observations determined was at a distance of about 12.2 billion light years. This burst had the fastest jet motions and the highest-energy initial emissions ever detected up to that date. In the following year GRB 090510 was detected which exhibited even faster material flows.

Fermi solved the problem of the supernova remnant CTA 1 which had been known to be a source of both X-rays and gamma-rays but which, surprisingly, had not been found to pulse at any wavelength. Fermi found that the remnant pulsed in gamma rays, so it was housing the first pulsar found to pulse in gamma rays only. Fermi also found two gamma-ray-emitting bubbles that extend 25,000 light years north and south of the centre of the Milky Way. The origin of these bubbles is currently unknown, but they have fairly sharp edges, indicating that whatever caused them occurred relatively suddenly.

Bibliography

Lichti, G. G., et al., *The GLAST Burst Monitor (GBM)*, American Institute of Physics Conference Proceedings, 662 (2003), 469–472.

Mattox, J. R., et al., *GLAST: The Gamma Ray Large Area Space Telescope*, Memorie della Società Astronomia Italiana, 67 (1996), 607–617.

Naeye, Robert, *NASA's New Gamma-Ray Trailblazer – GLAST*, Sky and Telescope, June 2008, pp. 18–22.

Reddy, Francis, *New Visions Reveal the Violent Universe*, Astronomy, March 2010, pp 24–29.

FUSE

The Far Ultraviolet Spectroscopic Explorer (FUSE) was a spacecraft operating over the far ultraviolet range from 90 to 120 nm. It was developed and operated for NASA by Johns Hopkins University, the development being in collaboration with the Canadian and French space agencies.

The Copernicus mission of the 1970s had shown the potential of a spacecraft operating in the far ultraviolet. But the sensitivity of Copernicus was such that it had a limited range of only about 1,000 parsec. Because of this, there was pressure on NASA to build a follow-on spacecraft with much higher sensitivity. This was to become FUSE, which had a sensitivity several orders of magnitude better than Copernicus.

The FUSE mission was selected by NASA in 1988 for a feasibility study. But the mission design was subjected to major changes over the next six years, with the planned orbit changing from near-Earth to geosynchronous, and back to near-Earth again. The design of the telescope optics was also completely changed from Wolter optics to a more conventional mirror system, to reduce costs. And NASA's costs were also reduced by including Canada and France as partners, who provided various hardware items. Eventually the 1,400 kg FUSE spacecraft was launched into a low Earth orbit in June 1999.

FUSE discovered more deuterium in the Milky Way than expected, indicating that less deuterium than expected had been consumed in stars since the Big Bang. Although the halo of hot gas surrounding the Milky Way had been known for some time, FUSE provided valuable new information on its constitution, and found that it extended far more into space than previously realised, possibly as far as the Magellanic Clouds. The spacecraft also found that there were large amounts of hot gas between the galaxies. FUSE was first to detect molecular nitrogen in interstellar space, large amounts of carbon in the disc surrounding the young star β Pictoris, and molecular hydrogen in Mars' upper atmosphere, providing evidence of early oceans.

Bibliography

Hoard, D. W., and Szkody, P., *Observations of Cataclysmic Variables with the Far Ultraviolet Spectroscopic Explorer*, in Cheng, K. S., Leung, K. C., and Li, T. P. (eds.), *Stellar Astrophysics – A Tribute to Helmut A. Abt*, Kluwer Academic Publishers, 2010.

Sonneborn, George (ed.), *Astrophysics in the Far Ultraviolet: Five Years of Discovery with FUSE; Proceedings of a Conference Held at the University of Victoria, Canada, 2004*, Astronomical Society of the Pacific Conference Series, 348, 2006.

GALEX

The Galaxy Evolution Explorer (GALEX) was a NASA satellite designed to undertake the first all-sky imaging and spectroscopic surveys in the ultraviolet.

It had become evident towards the end of the twentieth century that the types of galaxies in the universe had changed over time. It also appeared as though the star formation rate density peaked about $5-8$ Gyr ago (redshift, $z \sim 1-1.5$). The main goal of GALEX was to study star formation rates (SFRs) and star formation mechanisms in galaxies and their evolution over time.

Massive stars produce most of their energy in the ultraviolet (UV) and have relatively short lifetimes, so they are very useful in studying SFRs in galaxies. But the cosmological redshift limits the use of UV stars to $0 < z < 2$, or about the last 9 Gyr. By surveying large numbers of galaxies at varying distances from Earth, the history of star formation can be determined over this period of time.

NASA selected the GALEX mission in 1997. On 28 April 2003 an aircraft-launched Pegasus rocket placed the small 280 kg spacecraft into a low Earth orbit. On board was a 50 cm modified Ritchey-Chrétien telescope able to provide imaging and spectroscopy over the 135–280 nm band. The all-sky imaging survey went down to about 21 magnitude, and an ultra-deep survey of about 1 square degree went down to about 26 magnitude.

By combining the GALEX results with those of other space- and ground-based observatories, it appears as though the SFR has declined monotonically since $z = 3$, rather than having a previously-determined peak at $z \sim 1-1.5$. GALEX also discovered UV emission far outside the optical disc of a number of spiral galaxies and in the extended tidal tails of a number of interacting galaxies.

Bibliography

Dagostino, M. C., et al. (eds.), *New Quests in Stellar Astrophysics II: Ultraviolet Properties of Evolved Stellar Populations*, Astrophysics and Space Science Proceedings, Springer, 2009.

Martin, D. Christopher, et al., *The Galaxy Evolution Explorer: A Space Ultraviolet Survey Mission*, Astrophysical Journal Letters, 619, (2005), L1–6.

GRANAT

Granat was a 4,400 kg Russian/French X-ray and gamma-ray spacecraft launched in December 1989 into a highly eccentric ($2,000 \times 200,000$ km) orbit with a period of 96 hours. This was to allow long-term monitoring of sources. For the first five years it undertook pointed operations, but it was then put into survey mode in 1994 when its attitude-control gas became exhausted. By that time its orbit had been modified by solar-lunar perturbations into a $60,000 \times 140,000$ km orbit.

The fall of the Soviet Union in 1989 had a large effect on the operation of Granat. The main control centre was in the Ukraine, which became independent of Russia, and there were financial problems. These were eventually solved with French assistance.

The main instrument on Granat was the 900 kg SIGMA coded-mask, hard X-ray imaging telescope, provided by France, operating over the range $30-1,300$ keV. Russia supplied two instruments to undertake imaging, spectroscopy and timing in the $3-100$ keV band, and Denmark provided an all-sky monitor to alert the other instruments of new or interesting transient X-ray sources. There were also three X-ray/gamma-ray burst detectors, supplied by France and Russia, that, in total, covered the range 2 keV−100 MeV.

Granat undertook a very deep 100 day imaging of the galactic centre. It also detected electron-positron annihilation lines from the microquasar 1E1740−2942 which was believed to be a black hole near the centre of the Milky Way. The spacecraft discovered the ultra-soft X-ray transient called X-ray Nova Muscae, or GRS 1124−684, which was thought to be a binary containing a black hole. In all, Granat detected over 200 gamma-ray bursts. But probably its most significant observation was that the sky was highly variable in both hard X-rays and soft gamma rays.

Bibliography

Mandrou, P., et al., *Overview of Two-Year Observations with SIGMA on Board GRANAT*, Astronomy and Astrophysics Supplement Series **97** (1993), 1–4.

Revnivtsev, M. G., et al., *A Hard X-Ray Sky Survey with the SIGMA Telescope of the GRANAT Observatory*, Astronomy Letters, **30** (2004), 527–533.

HALCA

HALCA (Highly Advanced Laboratory for Communications and Astronomy), previously called Muses B, was a Japanese radio astronomy spacecraft launched on 12 February 1997. It was the world's first radio astronomy spacecraft dedicated to make observations with a ground-based very long baseline interferometer (VLBI).

Basic tests of such a space VLBI (SVLBI) system had been carried out in 1986 by linking an American TDRSS (Tracking and Data Relay Satellite System) communications spacecraft, which carried a 4.9 m diameter antenna, with 64 m ground-based radio antennas in Japan and Australia. Fringes were successfully detected using the quasars 1510−089, 1730−130, and 1741−038 at 2.3 GHz.

Three years later the Muses B spacecraft was approved by Japan as a technological experimental spacecraft operating as part of an extensive SVLBI system. This was to use ground-based radio telescopes, correlators and ground tracking stations provided by the USA, Japan, Europe, Australia, South Africa and Canada. Radio telescopes from other countries were also used from time to time. The observatory programme produced using HALCA as part of this configuration was called the VLBI Space Observatory Programme (VSOP).

The 830 kg HALCA was launched into a $560 \times 21{,}000$ km orbit to provide interferometric baselines up to 3 times longer than possible using just Earth-based radio telescopes. It had been planned to undertake observations at 1.6, 5 and 22 GHz, but onboard problems at 22 GHz limited observations to the first two frequencies. The effective diameter of the HALCA antenna was 8 m.

A few months after launch, HALCA observed the active galaxy PKS 1519−273 in parallel with the ground-based VLBA and VLA in America. The signal trains from this one space- and two ground-based observatories were subsequently combined and processed to produce the first-ever image made using a radio observatory in space.

About two-thirds of the scientific observation time in the VSOP was used by members of the international astronomical community after peer review of their proposals. Most of these general observing time (GOT) observations were of quasars, radio galaxies, BL Lac objects, pulsars and hydroxyl masers. The other third of observation time was devoted to the VSOP sub-milliarcsec survey of a few hundred bright, compact, extragalactic radio sources at 5 GHz. This was to determine their brightness temperature and basic structure, and to compare radio observations with those in other wavebands. The best resolution achieved in the VSOP programme was of the order of 0.2 milliarcsec.

Bibliography

Edwards, P. G., and Hirabayashi, H., *Highlights from Space VLBI Observations with the HALCA Satellite*, ASP Conference Series, **251** (2001), 348–349.

Hirabayashi, H., *The VSOP Mission – A New Era for VLBI*, IAU Colloquium 164, ASP Conference Series **144**, (1998), 11–15.

Hirabayashi, H., et al., *The VLBI Space Observatory Programme and the Radio-Astronomical Satellite HALCA*, Publications of the Astronomical Society of Japan, **52** (2000), 955–965.

HERSCHEL SPACE OBSERVATORY

Herschel was an infrared space observatory of the European Space Agency (ESA) designed as a follow-on to their Infrared Space Observatory (ISO). It also complemented NASA's Spitzer Space Telescope by extending infrared coverage into the submillimetre region.

Initial work on Herschel, then called FIRST (Far Infrared Space Telescope), began in the early 1980s. The idea was to build an infrared space telescope with the largest possible mirror to observe in the relatively poorly explored far infrared and submillimetre wavebands. The mission was eventually approved by ESA's Science Programme Committee in 1993 as part of ESA's Horizon 2000 scientific research programme.

The Herschel spacecraft had a 3.5 m diameter primary mirror in a Cassegrain configuration which fed three instruments: a very high resolution spectrometer, called HIFI, and two combined imaging photometers/spectrometers, called PACS and SPIRE. In total, the detectors of these instruments covered the far infrared and submillimetre range from 55 to 670 μm. (This compared with Spitzer, which had a 0.85 m diameter primary mirror and detectors covering the range from 3.6 to 160 μm.) The Herschel telescope, which was made almost completely of silicon carbide, was passively cooled to about 85 K, whilst the focal plane instruments were mounted inside a cryostat containing superfluid helium below a temperature of 1.7 K. In addition, the PACS and SPIRE bolometer arrays were cooled to 0.3 K using sorption coolers. The spacecraft's nominal lifetime of 3 years after the end of its commissioning phase was limited by the rate at which its helium coolant evaporated.

The 3.4 ton Herschel spacecraft was launched by an Ariane 5 in May 2009 towards the L_2 Lagrangian point of the Sun-Earth system. This was a double launch with ESA's Planck spacecraft. About two months later Herschel entered a Lissajous orbit around the L_2 point at a distance of 1.5 million km from Earth on the Earth's night side. This provided a stable thermal environment with good sky visibility. The HIFI instrument became inoperable shortly after launch, but it was successfully reactivated 5 months later.

Within the first year Herschel had provided valuable new data on planetary atmospheres, stars and star formation regions, the interstellar medium, molecular clouds, and extragalactic objects over a wide range of redshifts. Herschel discovered that there was about one thousand times as much dust in the Large Magellanic Cloud (LMC) as previously thought. Somewhat surprisingly the spacecraft was also able to detect supernova 1987A in the LMC in the far infrared.

Bibliography

Anon., *ESA's Report to the 34th COSPAR Meeting, Houston, USA, October 2002*, ESA SP−1259.

Clery, Daniel, *Herschel Will Open a New Vista on Infant Stars and Galaxies*, Science, **324** (2009), 584–586.

Harwit, Martin, *The Herschel Mission*, Advances in Space Research, **34** (2004), 568–572.

Various articles, Proceedings of SPIE, **4013** (2000).

Various articles, Astronomy and Astrophysics, **518** (2010).

HETE-2

HETE-2 (High Energy Transient Explorer−2) was a small American high-energy spacecraft built in collaboration with Japan, France and Italy. It was launched into a low Earth equatorial orbit on 9 October 2000. HETE-2's main mission was to detect and locate gamma-ray bursts (GRBs) to better than 10 arcseconds. The coordinates were then automatically notified to ground-based optical, infrared and radio observatories within seconds or minutes to allow them to find the GRB counterparts and their afterglows. A secondary mission was to undertake an X-ray sky survey, covering about 60% of the celestial sphere, and detect X-ray bursts and bursts from soft gamma-ray repeaters.

The launch of HETE-1, HETE-2's predecessor, had failed in 1996. Shortly afterwards, NASA agreed to the launch of a replacement spacecraft, HETE-2, using flight spare hardware from the first spacecraft. In the meantime, experience with BeppoSAX had caused the HETE team to replace the ultraviolet cameras of HETE-1's design with soft X-ray and optical cameras. The optical cameras were used as star trackers.

The scientific payload of HETE-2 consisted of a French Gamma-Ray Telescope (FREGAT), a Japanese Wide-Field X-ray Monitor (WXM) and an American Soft X-ray Camera (SXC). The FREGAT instrument, which operated over the range 6–500 keV, was designed to detect the GRB, then one or both of the X-ray instruments would be used to define its exact position in celestial coordinates using on-board processing.

In the first two and a half years of operation HETE-2 detected about 250 GRBs, of which it localised 43. Twenty-one of these led to the detection of X-ray, optical or radio afterglows, of which 11 had measurable redshifts. The mission confirmed the connection between long-duration GRBs and Type Ic core collapse supernovae. HETE-2 also observed numerous X-ray bursts, bursts from soft gamma-ray repeaters, and a limited number of X-ray flashes.

In 2005 HETE-2's data allowed the Chandra X-ray Observatory and the Hubble Space Telescope to identify the X-ray afterglow and, for the first time, the optical afterglow of a short-duration GRB (GRB 050709) that had been detected on 9 July. This showed the cosmological origin of this short GRB, and indicated that it was caused by the merging of two compact binaries.

Bibliography

Costa, E., Frontera, F., and Hjorth, J. (eds.), *Gamma-Ray Bursts in the Afterglow Era, Proceedings of the International Workshop Held in Rome, October 2000*, ESO Astrophysics Symposia, Springer, 2001.

Lamb, D.Q., et al., *Scientific Highlights of the HETE-2 Mission*, New Astronomy Reviews, **48**, April 2004, pp. 423–430.

Tamagawa, T., et al., *Prompt Gamma-Ray Burst Alert System of the HETE-2 Spacecraft*, Proceedings of the 28th International Cosmic Ray Conference at Tsukuba, Japan, 2003, pp. 2741–2744.

Vedrenne, Gilbert, and Atteia, Jean-Luc, *Gamma-Ray Bursts: The Brightest Explosions in the Universe*, Springer-Praxis, 2009.

HIGH ENERGY ASTRONOMY OBSERVATORIES (HEAOs)

NASA's High Energy Astronomy Observatory (HEAO) programme originally consisted of four 11,000 kg spacecraft. But in the early 1970s financial constraints caused the four spacecraft to be reduced to three, and their target masses reduced to about 3,000 kg. The first two spacecraft were dedicated to X-ray astronomy. HEAO-1 was a survey spacecraft, while HEAO-2 was the first X-ray imaging astronomical spacecraft. HEAO-3 was dedicated to gamma- and cosmic-ray astronomy. All three spacecraft were launched into almost-circular, low Earth orbits.

HEAO-1, which was launched in August 1977, was the last American X-ray observatory of the old design using proportional counters. The spacecraft rotated approximately once every 33 minutes about the Sun–Earth line, allowing its experiments to slowly scan the sky and cover the whole celestial sphere in six months. It surveyed the whole sky almost three times over the 0.2 keV to 10 MeV energy band, and produced a number of pointed observations of selected sources. HEAO-1 measured the X-ray background radiation from 2 to 60 keV, in addition to the variability of a variety of sources, including active galactic nuclei, X-ray binaries, and cataclysmic variables. It discovered the Cygnus Superbubble, the intermediate polar H2252−035, and a number of strong, soft X-ray sources, some of which were found to be RS CV_n−type binary stars.

HEAO-2, otherwise called the Einstein Observatory, was launched in November 1978. It was the first of a new kind of X-ray observatory using a Wolter-type focusing telescope of nested cylinders, as advocated some years earlier by Riccardo Giacconi. Its X-ray telescope was able to image X-rays in the 0.15–3 keV energy band with a maximum spatial resolution of about 2 arcsec, a temporal resolution of 8 msec, and a sensitivity several hundred times that of any previous spacecraft. Three spectrometers were also included in the payload. Einstein imaged supernova remnants for the first time, and detected X-ray jets from the radio galaxies Centaurus A and M87. It discovered three X-ray pulsars in supernova remnants, and found that ordinary red dwarfs, O-, B-, and A-type stars, and flare stars in their quiescent state were strong X-ray emitters. Additionally, it measured many X-ray transients and undertook medium and deep X-ray surveys of portions of the celestial sphere, to try to resolve the X-ray background radiation into discrete objects. As a result it detected about 1,000 new sources, including quasars, Seyfert galaxies, BL Lac objects, and clusters of galaxies. But a significant amount of the X-ray background radiation still remained unresolved.

HEAO-3, which was launched in September 1979, was a survey spacecraft like HEAO-1 but operating in the hard X-ray and gamma-ray bands from 50 keV to 10 MeV. Its results were much more modest than those of HEAO-1 and -2 because it was more difficult to detect and locate gamma-ray sources. HEAO-3 undertook a sky survey of narrow-line gamma-ray emission. It detected 511 keV gamma rays from the central region of the Milky Way produced by electron-positron annihilation. It observed the black hole candidate Cygnus X-1 for 170 days and detected gamma rays in the 400 keV to 1.5 MeV band and flickering 100 keV X-ray emission. The gamma-ray radiation gradually decreased in intensity as the hard X-rays increased.

Bibliography

Giacconi, Riccardo, et al., *The Einstein Observatory and Future X-Ray Telescopes*, in Burbidge, G., and Hewitt, A. (eds.), *Annual Reviews Monograph: Telescopes for the 1980s*, Annual Reviews, 1981.

Gursky, H., Ruffini, R., and Stella, L. (eds.), *Exploring the Universe: A Festschrift in Honor of Riccardo Giacconi*, Advanced Series in Astrophysics and Cosmology, Vol. 13, World Scientific Publishing, 2000.

Figure 10.4. ESA's Hipparcos astrometry spacecraft. It was built around a Schmidt telescope in which a beam combining mirror superimposed two fields of view 58° apart onto the common focal plane. That, and spacecraft scans at a fixed angle to the Sun direction, allowed the position of the stars to be located to high accuracy by successive approximations. (Courtesy ESA.)

Seward, Frederick D., and Charles, Philip A., *Exploring the X-Ray Universe*, 2nd ed., Cambridge University Press, 2010.

Tucker, Wallace, and Giacconi, Riccardo, *The X-Ray Universe*, Harvard University Press, 1985.

Tucker, Wallace, and Tucker, Karen, *The Cosmic Inquirers: Modern Telescopes and Their Makers*, Harvard University Press, 1986.

HIPPARCOS

Hipparcos (High Precision Parallax Collecting Satellite) (see Figure 10.4) was a European Space Agency (ESA) spacecraft designed to measure the position, parallax, proper motion, and intensity of stars to an unprecedented accuracy. This first astrometry spacecraft was named after Hipparchus of Nicaea, who published a catalogue of the positions of 850 stars in about 129 BC.

The initial concept of an astrometry satellite was proposed in 1966. This was developed and extended over the years until the Hipparcos programme was approved by ESA in 1980. At the core of the one-ton Hipparcos spacecraft was a folded Schmidt telescope with a 290 mm primary mirror accurate to λ/60. A focal plane grid, etched to an accuracy of 0.03 μm, was curved to match the field curvature of the telescope. The spacecraft was designed to allow a much better understanding of the movement of relatively nearby stars, the physics of stars and the age of the universe.

Hipparcos was successfully launched into a geosynchronous transfer orbit in August 1989. But the spacecraft's apogee motor failed, leaving it in its transfer orbit of 220 by 35,600 km, instead of the required geosynchronous orbit. This would have subjected Hipparcos to unacceptable air drag and to particle radiation in the Van Allen belts, which would have significantly degraded the spacecraft. But shortly after launch, engineers were able to use Hipparcos's on-board thrusters to increase its perigee to about 530 km, and so significantly reduced these effects. However, this new compromise orbit required the use of three ground stations, instead of the one originally envisaged, and significantly complicated the calculation of stellar positions to the required accuracy. The spacecraft operated until August 1993.

The results were published by ESA in the Hipparcos Catalogue in 1997, which contained about 120,000 stars with milliarcsecond astrometry. Also, using the Hipparcos star mappers, the Tycho Catalogue was published with more than one million stars with 20 to 30 milliarcsecond astrometry and two-colour photometry. The Tycho-2 Catalogue covering 2.5 million stars was published in 2000. Amongst other things, this astrometry work allowed the local distance scale of the universe to be determined to much higher accuracy than previously possible, causing the local scale to be increased overall by about 10%.

There were some problems with the distances of some of the stars in the 1997 Hipparcos catalogue which appeared to be less than those obtained by ground-based observatories. These differences were eventually found to be due to spacecraft thermal effects caused by eclipses, and to the impact of dust particles. Correcting for these reduced the formal errors of stellar positions from 0.5 milliarcsec to 0.09 milliarcsec. The new Hipparcos data was published in 2007.

Bibliography

Davies, John K., *Astronomy from Space: The Design and Operation of Orbiting Observatories*, Wiley-Praxis, 1997.

Perryman, M.A.C., *Ad Astra Hipparcos: The European Space Agency's Astrometry Mission*, ESA BR-24, 1985.

Perryman, Michael, *Astronomical Applications of Astrometry: Ten Years of Exploitation of the Hipparcos Satellite Data*, Cambridge University Press, 2008.

Sinnott, Roger W., *Hipparcos's Star Distances Get Better*, Sky and Telescope, January, 2008, pp. 31–32.

Turon, Catherine, *From Hipparchus to Hipparcos: Measuring the Universe, One Star at a Time*, Sky and Telescope, July 1997, pp. 28–34.

HUBBLE SPACE TELESCOPE

The Hubble Space Telescope (see Figure 10.5), which operated in the near ultraviolet, visible and infrared wavebands, was the first of NASA's Great Observatory missions to be launched. Its history can be traced back to Lyman Spitzer who wrote in 1946 about the potential of large optical telescopes in space above the distorting effects of the Earth's atmosphere.

In the early 1960s Boeing received a contract from NASA to undertake the outline design of a Large Space Telescope (LST) with a mirror diameter of 3.0 m, as that was the largest that could be launched by a Saturn launcher. But the astronomical community was generally sceptical about the case for spending vast sums of money on a 3.0 m space telescope, when a similar sum would have purchased many, much larger ground-based instruments. But what most astronomers failed to recognise was that the money could not be transferred to a ground-based programme even if the LST was not built.

Over the next few years there were many discussions about the size of the LST, including the possibility of an intermediate spacecraft with a smaller mirror. Various launch vehicles were also examined, including the proposed space shuttle. In addition, the question was raised about the role of astronauts in space in the LST programme. Should the LST be a manned facility linked to a space station, or should it be a stand-alone facility using astronauts just for periodic updating and repair?

The decision to build the space shuttle was announced in 1972, and two years later Congress agreed to fund the Phase B studies of an LST launched and serviced by the shuttle. But Congress required NASA to find international collaborators who would contribute substantially to the overall cost of the programme. Then in 1975 NASA decided to reduce the mirror size to 2.4 m and change the spacecraft's name from the Large Space Telescope to the Space Telescope, to try to deflect accusations that their plans were too grandiose. Congress approved the programme two years later.

ESA agreed to contribute one of the five scientific instruments, the Faint Object Camera, plus the solar arrays, and some staff for the Space Telescope Science Institute, which was to be responsible for managing the Space Telescope in orbit. In return European astronomers were guaranteed at least 15% of the observing time. In 1983 the Space Telescope was renamed the Hubble Space Telescope (HST). Its planned lifetime was 15 years.

Initially the five instruments on board the 12 ton HST consisted of two cameras (the Wide Field and Planetary Camera [WF/PC] and the Faint Object Camera), two spectrographs (the Faint Object Spectrograph and the High Resolution Spectrograph) and the High-Speed Photometer. All of these instruments were designed to be replaceable in orbit to enable the HST to be kept up to date. The spacecraft's fine guidance sensors, whose main role was to point the spacecraft, also served as a sixth scientific instrument.

The HST programme was beset by numerous technical, managerial and financial problems, followed by shuttle delays

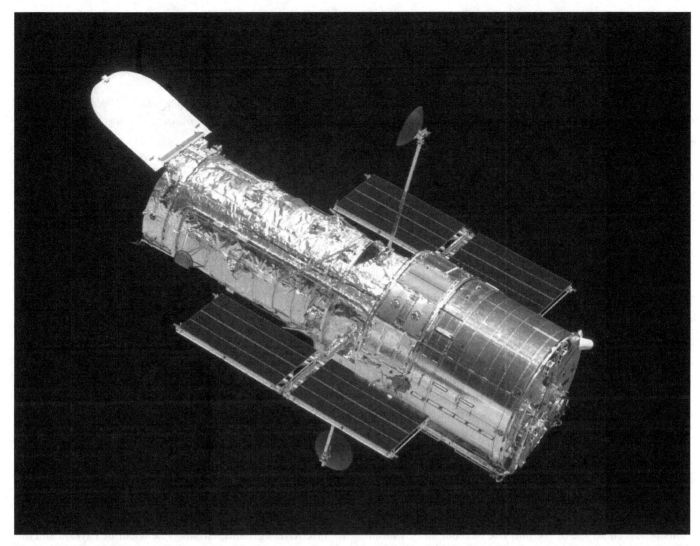

Figure 10.5..The Hubble Space Telescope in orbit with its aperture door open and solar arrays deployed. The five original instruments were located behind the 2.4 m primary mirror near the right-hand end of the telescope in this image. COSTAR, with its corrective optics for the faulty primary mirror, replaced one of the four axially located instruments, the High-Speed Photometer, during the first service mission in late 1993. The fifth original instrument, the Wide Field and Planetary Camera (WF/PC), was also replaced during this service mission. Its new version, WF/PC 2, had been modified to correct for the primary mirror's spherical aberration. (Courtesy NASA/ESA/STScI.)

caused by the Challenger disaster. As a result the projected launch date was constantly delayed. Then, shortly after its successful launch in April 1990, it was discovered that the HST's 2.4 m primary mirror was badly flawed. This was a serious setback to NASA's reputation, and caused major problems with the imaging instruments. There were fewer problems with the spectrometers, although they did require longer exposure times than planned.

In 1993 at the HST's first servicing mission, astronauts replaced the High-Speed Photometer with a corrective optics package, COSTAR (Corrective Optics Space Telescope Axial Replacement). This largely solved the imaging problem, but at the expense of losing one of the four primary focus instruments. In addition, the astronauts replaced the WF/PC by a WF/PC 2 of modified design. Three further servicing missions were

flown, but in 2004 NASA cancelled the fifth servicing mission because of safety concerns. However the follow-on James Webb Space Telescope was many years from completion at that time. So, following intense pressure, NASA announced in 2006 that it would fly one final shuttle servicing mission.

This final servicing mission, which took place in 2009, involved the installation of two new instruments; the Wide Field Camera 3 (WFC 3) and the Cosmic Origins Spectrograph (COS). The WFC 3, which replaced the WF/PC 2, had much better performance than its predecessor across all wavebands from the ultraviolet to the near infrared, whilst the COS operated exclusively in the ultraviolet. The COS replaced COSTAR which was no longer required as all the instruments then on-board had been designed to enable them to compensate individually for the spacecraft's defective mirror. In addition, the astronauts replaced all the HST's batteries, installed six new gyroscopes, and replaced a number of 'house-keeping' instruments to enable the spacecraft to continue operating for as long as possible, pending the launch of the James Webb Space Telescope.

The HST has made enormous contributions to most areas of astronomy, from planetary science to galactic astronomy. For example, it discovered two small satellites of Pluto, proplyds in the Orion nebula, cometary knots in the Helix nebula, and evaporating gaseous globules in the Eagle nebula. In addition, it has helped astronomers to understand gamma-ray bursts, black holes and other exotic objects in the universe.

In 1995 astronomers produced the most detailed optical view of the universe then achieved by pointing the HST to a blank area of sky about one arcminute in diameter, and imaging this area with the WF/PC 2 for ten days over 150 consecutive orbits. It took a total of 342 exposures in four wavebands centred from 300 to 814 nm. The resulting Hubble Deep Field showed about 2,000 galaxies down to magnitude 30 in the red region, some of them with redshifts of about 6. In 2004 an even deeper image, called the Hubble Ultra Deep Field, was released using images with exposures totalling about 16 days over 400 orbits with the Advanced Camera for Surveys and the Near Infrared Camera and Multi-object Spectrometer. It showed images of galaxies at a redshift of 10, or less 500 million years after the Big Bang. These images were a crucial source for the study of the evolution of galaxies.

In 2001 Wendy Freedman and colleagues published a value of the Hubble constant of 72 ± 8 km/s/Mpc, using the HST. This produced an age of the universe of about 13 ± 1 billion years for a flat universe. In the meantime, astronomers had also used the HST to observe distant supernovae. The results surprisingly implied that the expansion of the universe was accelerating.

The designs of ground-based telescopes did not stand still during the long HST programme of design, development and in-orbit operation. Adaptive optics systems were developed that enabled ground-based telescopes to compensate to a large degree for atmospheric turbulence. This resulted in large ground-based telescopes producing much sharper images than the HST for a small fraction of the cost. But adaptive optics systems operated best at relatively long wavelengths and over a limited field of view. So the HST's ability to produce high-resolution images, over relatively wide fields at optical wavelengths, still made it a vital facility.

Bibliography

Chaisson, Eric J., *The Hubble Wars*, Harvard University Press, 1998.

Fischer, Daniel, and Duerbeck, Hilmar, *Hubble Revisited: New Images from the Discovery Machine*, Springer-Verlag, 1998.

Livio, M., Noll, K., and Stiavelli, M. (eds.), *A Decade of Hubble Space Telescope Science*, Cambridge University Press, 2003.

O'Dell, C. R., *The Space Telescope*, in Burbidge, G., and Hewitt, A. (eds.), *Annual Reviews Monograph: Telescopes for the 1980s*, Annual Reviews, 1981.

Petersen, Carolyn Collins, and Brandt, John C., *Hubble Vision: Astronomy with the Hubble Space Telescope*, Cambridge University Press, 1995.

Smith, Robert W., *The Space Telescope: A Study of NASA, Science, Technology and Politics*, Cambridge University Press, 1993.

Tucker, Wallace, and Tucker, Karen, *The Cosmic Inquirers: Modern Telescopes and Their Makers*, Harvard University Press, 1986.

INTEGRAL

INTEGRAL (International Gamma-Ray Astrophysics Laboratory) (see Figure 10.6) was a 4,000 kg high-energy spacecraft of the European Space Agency. It was launched by a Russian Proton rocket on 17 October 2002. Russia provided the launcher free of charge in return for observing time on the spacecraft. The initial geosynchronous, 72 hour orbit was highly eccentric (9,000 × 154,000 km), to allow for long-duration observations outside of the Earth's Van Allen belts, which compromise gamma-ray observations.

The INTEGRAL mission, which had originally been proposed to ESA in 1989, had been approved in 1993 as part of ESA's Horizon 2000 scientific spacecraft programme. To reduce costs, ESA decided to use the XMM-Newton spacecraft service module design, the free Russian launcher and a Goldstone DSN ground station provided by NASA as one of INTEGRAL's ground stations. The scientific payload and the Science Data Centre in Geneva were provided by various other national and international organisations. The programme went through a particularly difficult phase in 1994 and 1995 when a number of ESA's payload partners reduced their involvement. This resulted in new payload teams being formed with ESA taking on a larger role than originally envisaged.

INTEGRAL's mission was fine-source imaging and high-resolution spectroscopy of high-energy sources up to 10 MeV, with spatial and spectral resolutions an order of magnitude better than with the Compton Gamma Ray Observatory. To do this its payload consisted of an Italian gamma-ray imager (IBIS) and a Franco/German gamma-ray spectrometer (SPI), both operating in the 15 keV–10 MeV band, two identical Danish/Finnish X-ray monitors (JEM-X), and a Belgian optical monitoring camera (OMC). All of these instruments were co-aligned allowing for simultaneous observations in the gamma-ray, X-ray and visible wavebands. IBIS, which had a resolution of 12 arcmin, was able to locate point sources to within ~1 arcmin. It had an energy resolution of about 9 keV at 100 keV, whereas the SPI had an energy resolution of 2.5 keV at 1 MeV. JEM-X had a spatial resolution of 3 arcmin, whereas the

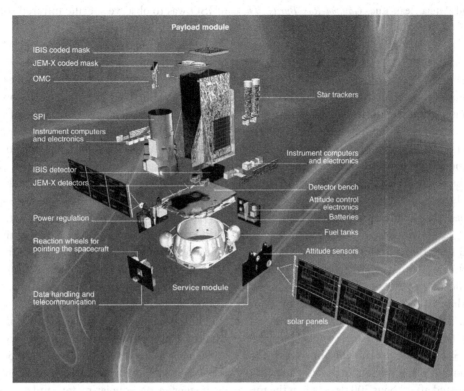

Figure 10.6. ESA's INTEGRAL gamma-ray observatory spacecraft. Its two main instruments, the IBIS gamma-ray imager and SPI gamma-ray spectrometer, were supported by the JEM-X X-ray monitor and the OMC optical monitoring camera. (Courtesy ESA.)

OMC's resolution was about 20 arcsec, with source location within 6 arcsec. Burst timing accuracy was about 120 μs.

The third INTEGRAL IBIS source catalogue, published in 2007, contained a total of 421 soft gamma-ray sources. Of the 74% that were identified, most were either low-mass or high-mass X-ray binaries, or active galaxies. INTEGRAL found two new classes of high-mass X-ray binaries (HMXBs), both of which probably contained a neutron star. The first were highly obscured HMXBs, of which IGR J16318−4848 was probably the most extreme, in which surrounding obscuring material was absorbing low- and medium-energy X-rays. This is why these highly obscured HMXBs had not been detected before. The second were supergiant fast X-ray transients (SFXTs), IGR J17544−2619 being a typical example, which were generally quiescent, but exhibited occasional X-ray flares lasting less than a day. INTEGRAL also detected hard X-ray emission from anomalous X-ray pulsars (AXPs) helping to show that they are magnetars, or pulsars with an extremely powerful magnetic field.

The IBIS instrument showed that nearly all the diffuse gamma-ray continuum emission from the centre of the Milky Way could be accounted for by about 90 point sources. INTEGRAL also provided the first map of the distribution of 511 keV gamma rays resulting from electron-positron annihilation. This showed a clear peak at the centre of the Milky Way and a very symmetrical distribution about the peak. The spacecraft found a hard X-ray source (IGR J17456−2901) just 0.9 arcmin from Sgr A* at the centre of the Milky Way. It was not clear whether this source, which showed no variability, was connected to Sgr A*, but it was the first source of significant hard X-ray emission to be found within 10 arcsec of Sgr A*.

Although it was not specifically designed to detect gamma-ray bursts (GRBs), INTEGRAL did detect a number, the position of one of which, GRB 031203, was determined and provided to the astronomical community in the then record time of just 18 seconds. This enabled its identification as a weak GRB in a galaxy less than 400 Mpc (1.3 billion light years) away. It was one of the closest GRBs ever observed.

Bibliography

Costa, E., Frontera, F., and Hjorth, J. (eds.), *Gamma-Ray Bursts in the Afterglow Era, Proceedings of the International Workshop Held in Rome, October 2000*, ESO Astrophysics Symposia, Springer, 2001.

Five Years of INTEGRAL in Space, Newsletter of the INTEGRAL Science Operations Centre, No. 18, ESA, October 2007.

Various Articles, ESA Bulletin Number 111, August 2002, pp. 37–78.

Various Articles, INTEGRAL Special Number, Astronomy and Astrophysics, 411 (2003).

Vedrenne, Gilbert, and Atteia, Jean-Luc, *Gamma-Ray Bursts: The Brightest Explosions in the Universe*, Springer-Praxis, 2009.

IRAS

The InfraRed Astronomical Satellite (IRAS) was the first dedicated infrared astronomical spacecraft. It was a Dutch-inspired spacecraft that, because of Dutch funding difficulties, became a NASA/Netherlands/UK programme. NASA provided the infrared telescope, cooling system and launcher, the Netherlands the spacecraft bus, and the UK the ground station and a 'quick-look' data analysis facility.

IRAS was designed to provide an all-sky survey in four wavebands from 12 to 100 μm with a resolution of 4 arcmin, and to make detailed spectroscopic and photometric observations of selected sources. The payload consisted of a 57 cm Ritchey-Chrétien reflecting telescope, which was cooled to a temperature of less than 10 K by superfluid helium to reduce its infrared emission. The focal plane detectors were cooled to 2.5 K by the same system.

The 1,070 kg IRAS spacecraft was launched in January 1983 into a Sun-synchronous, polar orbit. As expected the helium coolant became exhausted after about 300 days, when the mission was terminated. By then the spacecraft had surveyed 95% of the sky and made thousands of pointed observations. The main IRAS catalogue, which was published in 1984, contained about 245,000 sources.

IRAS discovered six new comets and additional components of the zodiacal light above and below the main dust band. It detected protostars in interstellar dust clouds, plus stars like Vega and β Pictoris still surrounded by remnants of their original dust clouds, and dust shells emitted by stars like Betelgeuse near the end of their lives. It found infrared patterns, now called infrared cirrus, radiating at 35 K all over the sky, and discovered a number of ultraluminous galaxies, some of which emit more than 90% of their energy in the infrared.

Bibliography

Davies, John K., *Astronomy from Space: The Design and Operation of Orbiting Observatories*, Wiley-Praxis, 1997.

Rowan-Robinson, Michael, *Ripples in the Cosmos; A View behind the Scenes of the New Cosmology*, W. H. Freeman, 1993.

Tucker, Wallace and Karen, *The Cosmic Inquirers: Modern Telescopes and Their Makers*, Harvard University Press, 1986.

ISO

The Infrared Space Observatory (ISO) (see Figure 10.7) was a European Space Agency (ESA) spacecraft designed as a follow-on to IRAS, which had detected some 245,000 infrared sources in 1983. IRAS's mission had been to find new sources, but that of ISO was to observe specific infrared sources over relatively long periods of time.

The study phase of ISO started in March 1983, just two months after the launch of IRAS. ISO was eventually launched by a European Ariane on 17 November 1995 into an orbit that was modified, by the spacecraft's on-board thrusters, to its planned operational 24-hour orbit of 1,000 × 70,600 km. This was chosen to minimise the spacecraft's exposure to the Earth's Van Allen radiation belts, and to allow scientific data to be downloaded to Earth in real time, eliminating the need for on-board data storage. Instrument observations were limited to about 17 hours per day when the spacecraft was above the radiation belts, as these degraded instrument observations. Two ground stations were used during the operational phase. One was ESA's ground station at Villafranca, Spain, and the other was NASA's at Goldstone, California, which was funded by the United States and Japan in return for guaranteed access to ISO.

The spacecraft payload consisted of a 60 cm diameter Ritchey-Chrétien reflecting telescope, which was cooled with 2,300 litres of superfluid helium at a temperature of 1.8 K. The telescope's field of view was shared among four instruments: an imager, a photopolarimeter, and two spectrometers. The imager operated from 2.5 to 17 μm with two 32 × 32 element detector arrays, while the photopolarimeter and spectrometers covered the range from about 2.4 to 200 μm. Routine operations, which began in February 1996, continued for about 26 months until ISO ran out of coolant. During this time the spacecraft carried out over 26,000 scientific observations, with average observing times of 24 minutes, and a maximum of 8 hours observing Titan.

The ISO spacecraft detected water in Titan's atmosphere and in the upper troposphere of Jupiter, Saturn, Uranus and Neptune. It also discovered crystalline silicates in both long and short period comets and in the environment surrounding young stars, as well as in red giants and planetary nebulae. The spacecraft found water molecules in the gaseous envelopes surrounding newly formed O- and B-type stars, as well as in stars near the end of their lives. In fact, ISO found that water molecules were common in the Milky Way.

ISO detected a number of star-forming regions in both normal and colliding galaxies, like the Antenna galaxies, where star formation had been triggered by the collision shock wave. The earliest phase of cloud collapse was found in a two-solar mass dark cloud, called Lynds 1689, that had a temperature of just 13 K. ISO also found that most ultraluminous infrared galaxies (ULIRGs) derive the majority of their energy from enormous bursts of star formation while only a fraction, mainly the most luminous ULIRGs, are powered by active galactic nuclei (AGNs).

The ISO spacecraft observed distant galaxies by viewing them through the Lockman hole. The furthest galaxy that it

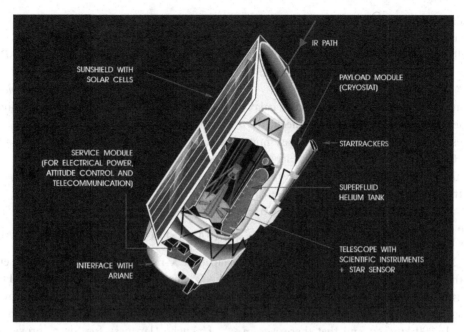

Figure 10.7. The Infrared Space Observatory (ISO) of the European Space Agency. This spacecraft was built around a 60 cm diameter telescope enclosed in a cryostat that was cooled with 2,300 litres of superfluid helium. (Courtesy ESA.)

observed, BR 1202−0725, was seen when the universe was only a tenth of its current age. This galaxy was found to be dusty, indicating that star birth and death had already occurred such a short time after the Big Bang.

Bibliography

Emery, R., *The Infrared Space Observatory (ISO) − A Study for a Cooled Telescope in Space for Infrared Astronomy*, ESA Bulletin 32, November 1982, pp. 19–21.

Kessler, M. F., et al., *ISO's Astronomical Harvest Continues*, ESA Bulletin 99, September 1999, pp. 6–15.

Kessler, M. F., et al., *Looking Back at ISO Operations*, ESA Bulletin 95, August 1998, pp. 87–97.

Leverington, David, *New Cosmic Horizons: Space Astronomy from the V2 to the Hubble Space Telescope*, Cambridge University Press, 2000.

IUE

The European Space Research Organisation's (ESRO's) Large Astronomical Satellite (LAS) had been cancelled in 1968 due to escalating costs. But the UK astronomer Robert Wilson suggested to NASA that a modified version of LAS, which was eventually to become the International Ultraviolet Explorer (IUE) (see Figure 10.8), could fill the gap between NASA's last Orbiting Astronomical Observatory and what was to become the Hubble Space Telescope. NASA agreed in principle, but increased the new spacecraft's planned orbit from low Earth to geostationary orbit, so that the Earth obscured less of the spacecraft's sky and allowed continuous communications with a given ground station. Eventually mass problems resulted in a 26,000 × 46,000 km altitude geosynchronous orbit being chosen instead.

In the event, the 670 kg IUE spacecraft was built around a 45 cm diameter Ritchey-Chrétien reflecting telescope which fed a spectroscope operating in the ultraviolet from 115 to 320 nm. A high-resolution spectrum was produced for objects down to 12th magnitude, or a low-resolution spectrum for dimmer ones. There was no imaging capability.

IUE was a joint programme of NASA, the UK and ESA, led by NASA. The UK provided the detectors, whilst ESA provided the solar arrays and one of the two ground control stations. As IUE was always in view of one of these ground stations, it allowed the spacecraft to be used like a ground-based observatory, which was a novel idea at the time, with astronomers directing their observations from the ground stations. In its 18-year lifetime, 15 years longer than planned, IUE produced more than 114,000 spectra.

The spacecraft, which was launched by an American Thor-Delta in January 1978, was used to study many different types of objects from comets to galaxies. For example, it showed that the terminal wind velocities from hot, massive, Wolf-Rayet stars were in the region of 1,500 to 3,500 km/s. IUE was the first to detect absorption lines in a ring nebula (NGC 6888) associated with a hot Wolf-Rayet star (HD 192163). Analysis of the nebula's spectrum showed that it was being heated up to 60,000 K by the shock wave caused by the interaction of the Wolf-Rayet's stellar wind and the interstellar gas. In addition, IUE provided valuable data on the chromosphere of cool stars and on their stellar winds. IUE also observed Supernova 1987A, providing key data that allowed the progenitor star to be identified.

IUE gave crucial insights into the structure of active galaxies like Seyfert galaxies. For example, astronomers decided to undertake a long series of coordinated observations to examine the variability of NGC 4151 in detail, as it is the nearest and brightest Seyfert 1 galaxy. The spacecraft produced a total of over 300 spectra of this Seyfert. They showed that the continuum emission from its nucleus doubled in intensity over periods varying from 5 to 30 days, implying a source size of the order of 5 to 30 light days. But in X-rays the rise time was often less than 12 hours, so clearly the X-ray source was much smaller than the UV source. Detailed analysis of the timings and variations in the widths of various emission lines from the surrounding gas showed that the gas was moving at velocities of up to 16,000 km/sec, and that there were a number of shells of gas, each with a different chemical composition. Similar but less complete observations were made by IUE of a number of other Seyfert galaxies which were found to have a broadly similar structure to that of NGC 4151.

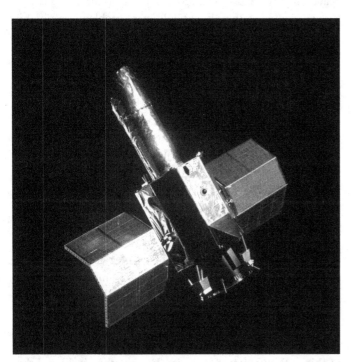

Figure 10.8. The International Ultraviolet Explorer (IUE) spacecraft which had a lifetime of 18 years, 15 years longer than planned. It had been built around an ultraviolet telescope which had a 45 cm diameter mirror made of beryllium to save weight. Either a high-resolution spectrum could be produced for bright objects, or a low-resolution one for dim ones. There was no imaging capability. (Courtesy ESA.)

Bibliography

Davies, John K., *Astronomy from Space: The Design and Operation of Orbiting Observatories*, Wiley-Praxis, 1997.

Freeman, Helen Ruth, *International Ultraviolet Explorer (IUE) Case Study in Spacecraft Design*, American Institute of Aeronautics and Astronautics, 2001.

Kondo, Yoji (ed.), *Exploring the Universe with the IUE Satellite*, Kluwer Academic Publishers, 1987.

Krige, J., et al., *A History of the European Space Agency 1958–1987, Vols. I and II*, ESA SP-1235, 2000.

JAPANESE X-RAY ASTROPHYSICAL SPACECRAFT

Japanese X-ray astrophysical spacecraft were, as of 2011, Hakucho, Tenma, Ginga, ASCA and Suzaku. The Japanese followed the United States, the USSR, and the Europeans in undertaking X-ray astrophysical research using spacecraft.

Hakucho (Japanese for Swan, previously called Corsa-B) was a small 100 kg spacecraft launched into a low Earth orbit in 1979. Its X-ray detectors were designed to survey the sky and monitor X-ray transients in the range 0.1–100 keV. Hakucho discovered eight new X-ray burst sources, and monitored the change in pulse rate of X-ray pulsars like Vela X-1. It also monitored the X-ray variability of the Rapid Burster and of the black hole candidate Cyg X-1. Hakucho was still operational when the second Japanese astrophysical spacecraft Tenma was launched in 1983.

Tenma (Japanese for Pegasus, previously called Astro-B) was designed to provide improved spectral and temporal resolution of X-ray sources in the range 2–60 keV. Further observations were undertaken down to 0.1 keV. Failure of the on-board battery had a significant effect on the operational efficiency of the spacecraft. Nevertheless, because of its good spectral resolution, Tenma was able to distinguish between iron line emission at 6.4 keV and 6.7 keV, which was apparently caused by different physical processes. The first line was found in pulsars and active galactic nuclei, and the second in low-mass X-ray binaries and the galactic ridge.

The 420 kg Ginga (Japanese for Galaxy, previously called Astro-C), launched in 1987, was designed to observe the variability and spectra of all types of X-ray sources, particularly active galactic nuclei. Its main instrument, the Large Area Proportional Counter, which operated in the 1.5–35 keV range, was built in collaboration with the UK. Ginga's Gamma Ray Burst Detector, which operated in the range 2–500 keV, was developed in collaboration with the USA.

Ginga discovered a number of X-ray transients, as well as an intense 6.7 keV iron line emission from the galactic centre. It detected cyclotron features in three X-ray pulsars, and X-ray emission from the supernova SN 1987A in the Large Magellanic Cloud. This SN 1987A X-ray emission was detected about six months after the explosion had been detected optically. By this time the surrounding gas cloud had apparently dispersed enough to allow X-rays to escape and be detected.

ASCA, the Advanced Satellite for Cosmology and Astrophysics (previously called Astro-D) was launched in 1993. The 420 kg spacecraft combined a broad-band imaging capability with good spectral resolution over the range 0.5–12 keV. The four grazing incidence X-ray telescopes were provided by NASA Goddard in collaboration with Nagoya University. The United States also provided two of the four X-ray detectors. ASCA observed X-rays emitted by SN 1993J located in M81, and found that the iron abundance in the coronae of active stars is less than in the Sun. It also detected non-thermal X-rays from SNR 1006, which showed how cosmic rays were accelerated in the supernova remnant.

The 1,600 kg Suzaku (previously called Astro-E2) was a Japanese/American spacecraft launched in 2005. It was the recovery mission for Astro-E that had failed to achieve orbit in 2000. Suzaku had five grazing incidence X-ray telescopes, with an X-ray Spectrometer (XRS) in the focal plane of one of these, and individual CCD X-ray imaging spectrometers, which were part of the XIS instrument, in the focal planes of each of the other four. The complete X-ray Imaging Spectrometer (XIS) instrument was sensitive over the range 0.2–12 keV, with an energy resolution of about 130 eV at 6 keV. Suzaku also contained a non-imaging Hard X-ray Detector (HXD) sensitive over the range 10–600 keV.

NASA had initially planned to fly the XRS on their Chandra X-ray observatory spacecraft, but financial problems in the early 1990s had resulted in it being taken off the Chandra manifest. It was eventually flown on the ill-fated Astro-E. The Suzaku XRS was considered to be the prime instrument to fly on Suzaku, as its microcalorimeters had an exceptional energy resolution of about 6 eV at 6 keV. Unfortunately the XRS failed shortly after launch because of the loss of its liquid helium coolant.

Suzaku detected and separated Fe band features in cataclysmic variables, X-ray binaries, active galactic nuclei, and the plane of the Milky Way. It measured C, N and O abundances in the interstellar medium and supernova remnants, and detected Cr and Mn lines in the Tycho supernova remnant. More locally, Suzaku investigated geocoronal and heliospheric soft X-ray charge exchange emission.

Bibliography

Davies, John K., *Astronomy from Space: The Design and Operation of Orbiting Observatories*, Wiley-Praxis, 1997.

Gursky, H., Ruffini, R., and Stella, L. (eds.), *Exploring the Universe: A Festschrift in Honor of Riccardo Giacconi, Advanced Series in Astrophysics and Cosmology*, Vol. 13, World Scientific Publishing, 2000.

Ho, C., Epstein, R. I., and Fenimore, E. E. (eds.), *Gamma-Ray Bursts: Observations, Analyses and Theories*, Cambridge University Press, 1992.

Seward, Frederick D., and Charles, Philip A., *Exploring the X-Ray Universe*, 2nd ed., Cambridge University Press, 2010.

ORBITING ASTRONOMICAL OBSERVATORIES (OAOs)

NASA's series of Explorer-class spacecraft, which had been available since the start of the space age, were rather small and unsuitable for missions, like those devoted to ultraviolet astronomy, that required relatively high pointing accuracy. So when it was possible a few years later to launch a large spacecraft, NASA thought that it would be more efficient to stabilise one large spacecraft, and attach as many instruments as possible, than produce a series of small, stable spacecraft. This was the philosophy behind NASA's series of four 2,000 kg (2 ton) Orbiting Astronomical Observatories (OAOs) which were designed to operate mainly in the ultraviolet.

The first Orbiting Astronomical Observatory, OAO-1, which was launched in 1966, failed the day after launch. However OAO-2, which was launched two and a half years later, was a success, Its payload consisted of four 32 cm diameter ultraviolet telescopes to image the sky over the range 105 to 320 nm, and ultraviolet photometers and spectrometers operating over the range 105 to 425 nm. Over its lifetime, OAO-2 produced images of about 10% of the sky, providing useful data on over 5,000 stars. The spacecraft also detected a one million kilometre diameter spherical cloud of neutral hydrogen around the comet Tago-Sato-Kosaka. OAO-2 had a significant effect on how astronomical research was conducted, with astronomers dependent for the first time on the effective transfer, analysis, storage, and handling of large quantities of high-speed data.

The launch of the next OAO spacecraft was a failure. But OAO-3, or Copernicus, which was launched in 1972, was a success. Its main instrument was an 80 cm diameter ultraviolet telescope with a pointing accuracy of about 0.1 arcsec. It fed a scanning spectrometer covering the range 70 to 330 nm, with a spectral resolution of about 0.01 nm. The instrument was designed to measure absorption lines of light from both bright stars and the interstellar medium. Copernicus also included an X-ray experiment from the UK to observe already-known X-ray sources at lower energies than previously possible.

Copernicus took high-resolution spectra of hundreds of stars. It detected deuterium in the interstellar medium by finding its Lyman-γ to Lyman-ε absorption lines superimposed on the spectra of luminous blue stars such as β Centauri. This was the first time that deuterium had been detected anywhere in the universe, except on Earth. Its observed concentration, of 15 parts per million by mass, is an important constraint on whether the universe is open or closed. The spacecraft also measured thin, highly ionized gas at a temperature of about 200,000 K in the regions between interstellar hydrogen clouds. In the Orion region, Copernicus detected clouds with a velocity of about 100 km/s, having a temperature of about 10,000 K. These were thought to have been produced either by old supernovae or by shock fronts created by high-velocity stellar winds, with velocities up to about 3,000 km/s, from the hottest stars. The X-ray experiment provided valuable data on a number of known X-ray sources, including Cygnus X-1, Cygnus X-3, and the radio galaxy Centaurus A.

Bibliography

Code, A. D., et al., *Ultraviolet Photometry from a Spacecraft*, Sky and Telescope, November 1969, pp. 290–293.

Leverington, David, *New Cosmic Horizons: Space Astronomy from the V2 to the Hubble Space Telescope*, Cambridge University Press, 2000.

Snow, Theodore P., Jr., *Ultraviolet Spectroscopy with Copernicus*, Sky and Telescope, November 1977, pp. 371–374.

Watts, Raymond N., Jr., *An Astronomy Satellite Named Copernicus*, Sky and Telescope, October 1972, pp. 231–232 and 235.

PLANCK

Planck was a European Space Agency (ESA) spacecraft designed to detect anisotropies in the cosmic microwave background (CMB). It was a successor to NASA's COBE (Cosmic Background Explorer) spacecraft which had been launched in 1989, and NASA's WMAP (Wilkinson Microwave Anisotropy Probe) launched in 2001.

What is now the Planck mission began in 1993 as two separate spacecraft missions. Both spacecraft included a 1 m class telescope, with detectors working in the range from 30 to 130 GHz for one spacecraft, and 140 to 800 GHz for the other. Shortly afterwards ESA decided to combine both missions into one spacecraft named Planck. This was after Max Planck who won the Nobel Prize for Physics in 1918 for his explanation of the black body radiation curve using his concept of energy quanta.

The Planck spacecraft, which was launched with ESA's Herschel spacecraft in May 2009, measured the whole celestial sphere in microwaves to an unprecedented sensitivity and angular resolution. The cold payload module included an off-axis Gregorian telescope with a 1.9×1.5 m carbon-fiber reinforced plastic primary mirror, coated with a reflective layer of aluminium. It fed microwave radiation to both a Low-Frequency Instrument (LFI) and a High-Frequency Instrument (HFI). The LFI had an array of tuned radio receivers to observe the sky in three frequency bands between 30 and 70 GHz (wavelengths 1 cm to 4 mm), whereas the HFI had an array of bolometric detectors covering six frequency bands

between 100 to 857 GHz (wavelengths 3 to 0.35 mm). This frequency coverage of Planck was much broader than that of WMAP that operated from 23 to 94 GHz. Planck's detectors were also up to ten times more sensitive than those on WMAP, and Planck's telescope had an angular resolution about three times better.

Planck, like WMAP, was launched to operate around the L_2 Lagrangian point some 1.5 million kilometres from Earth in a direction opposite to that of the Sun. This enabled it to view space largely unobstructed by the Sun, Earth and Moon.

At present (2011) Planck's CMB data has not yet been evaluated. But Planck's Early Release Compact Source Catalogue (ERCSC) has been released of foreground sources using the spacecraft's first all-sky survey. These sources were basically contaminating the CMB and their effects had to be removed before the true CMB could be properly identified. But the detection of these ERCSC sources was valuable in its own right.

The ERCSC contained details of over 15,000 sources both within and outside our galaxy. The galactic sources included features in the interstellar medium, cold molecular cloud cores, stars with dust shells, and H II regions. Extragalactic sources included radio galaxies, blazars, infrared-luminous galaxies, and galaxy clusters.

Bibliography

Anon., *First Scientific Results from Planck*, ESA Bulletin, No. 145, February 2011, pp. 58–59.

Bersanelli, M., et al., COBRAS/SAMBA: A Mission Dedicated to Imaging the Anisotropies of the Cosmic Microwave Background, Report on the Phase A Study, ESA D/SCI(96)3, February 1996.

Numerous authors, *Planck; The Scientific Programme*, ESA-SCI (2005) 1.

Tauber, Jan, et al., *Looking Back to the Dawn of Time: The Science behind ESA's Planck Observatory*, ESA Bulletin, No. 138, May 2009, pp. 20–26.

ROSAT

Rosat (Röntgen Satellit) was a German X-ray and extreme ultraviolet (EUV) spacecraft, with participation by the USA and UK. It was named after Wilhelm Röntgen, the German physicist who discovered X-rays.

Joachim Trümper of the Max Planck Institute had originally proposed Rosat as a German national X-ray spacecraft in 1975. But it was too expensive for Germany to undertake on its own. So the USA agreed to provide a shuttle launch free of charge, together with an improved version of the High-Resolution Imager (HRI), which had been flown on the Einstein Observatory spacecraft, as one of the X-ray telescope's detectors. In addition, the UK agreed to provide a Wide-Field Camera (WFC) operating in the EUV from 6 to 30 nm. The

Challenger disaster in 1986 resulted in the free shuttle launch being changed to a free Delta launch, as the former could not be provided until 1994, at the earliest.

The 2,400 kg Rosat spacecraft was launched in June 1990 into a 580 km altitude orbit. Its X-ray telescope was a Wolter-type with two position-sensitive proportional counters (PSPCs) and the American HRI detector all mounted on a carousel in its focal plane, enabling one of the three detectors to be used at a time. All three detectors operated over the 0.1 to 2.4 keV soft X-ray band, with the HRI having the best spatial resolution of 2 arcsec and time resolution of 60 microseconds. It was planned to operate the PSPCs and the WFC during the all-sky survey phase, with the HRI being used only for pointed observations during the subsequent observatory phase.

After a period of in-orbit calibration, Rosat's mission started with a six-month deep all-sky survey for soft X-ray and EUV sources. The spacecraft had almost finished the all-sky survey when its attitude control system failed. Although the problem was soon resolved, Rosat had, in the meantime, accidentally scanned too close to the Sun, destroying one of the PSPCs and severely degrading the WFC. Nevertheless, the remaining instruments were still operational, and the spacecraft completed the survey phase and undertook pointed observations until 1994 when its remaining PSPC ran out of gas. Rosat then continued observations using the HRI and ceased operating completely in 1999, some years after the end of its design lifetime.

Rosat undertook the first X-ray and EUV all-sky survey using an imaging telescope system with a high X-ray sensitivity. Rosat increased the number of known X-ray sources from 6,000 to over 150,000, about $\frac{1}{3}$ of which were stars, and $\frac{1}{3}$ active galaxies. It also significantly increased the number of known EUV sources of which 479 were included in its 2RE catalogue. Of these about half were the hot coronae of cool low-mass stars, with most of the remainder being hot white dwarfs.

Rosat produced the first high-resolution X-ray image of the Cygnus Superbubble and Monogem Ring, and enabled maps to be produced of the X-ray brightness of nearby clusters of galaxies. It also detected isolated neutron stars, and discovered X-ray emission from comets. The Vela and Cygnus Loop supernova remnants were well resolved in the EUV, and a few Seyfert galaxies and quasars were also identified.

Bibliography

Aschenbach, B., Hahn, H-M., and Trümper, J., *The Invisible Sky: ROSAT and the Age of X-Ray Astronomy*, Springer-Verlag, 1998.

Barstow, Martin A., and Holberg, Jay B., *Extreme Ultraviolet Astronomy*, Cambridge University Press, 2003.

Gursky, H., Ruffini, R., and Stella, L. (eds.), *Exploring the Universe: A Festschrift in Honor of Riccardo Giacconi, Advanced Series in Astrophysics and Cosmology*, Vol. 13, World Scientific Publishing, 2000.

Leverington, David, *New Cosmic Horizons: Space Astronomy from the V2 to the Hubble Space Telescope*, Cambridge University Press, 2000.

Seward, Frederick D., and Charles, Philip A., *Exploring the X-Ray Universe*, 2nd ed., Cambridge University Press, 2010.

RXTE

The Rossi X-ray Timing Explorer (RXTE) spacecraft was launched into low Earth orbit in December 1995. It was named after Bruno Rossi, a pioneer in X-ray astronomy.

Initially called the XTE, the RXTE was a 3,200 kg spacecraft designed to observe the time variability of X-ray sources on timescales from microseconds to years. It was manoeuvrable, being able to change orientation at 6°/minute, so that it could be rapidly moved to observe new events. In addition, it could detect sources as faint as 1 millicrab with an integration time of only a few seconds. RXTE was designed to study known sources, and to detect new transient events, including X-ray bursts.

The spacecraft included three experiments, two of which were pointed. The Proportional Counter Array, which operated over the range 2−60 keV, had a time resolution of 1 μs, whilst the High Energy X-ray Timing Experiment operated over the range 15−250 keV, with a time resolution of 8 μs. The All Sky Monitor, which was the only non-pointed instrument, operated in the range 2−10 keV. It was able to observe 80% of the sky over one 90 minute orbit. This enabled astronomers to detect new or interesting X-ray sources quickly, and so direct the RXTE or other spacecraft to study them further.

RXTE discovered a number of kilohertz quasi-periodic oscillators (QPOs), and observed a three-hour-long superburst from the compact low-mass binary 4U 1820−30. It showed that the galactic X-ray background came from innumerable cataclysmic variables, active stellar coronas and other individual sources. It detected the X-ray afterglow from gamma-ray bursts, and detected a black hole with a mass of about 3.8 suns, which was very near the minimum mass possible. RXTE also confirmed the frame-dragging effect predicted by Einstein's theory of relativity.

Bibliography

Davies, John K., *Astronomy from Space: The Design and Operation of Orbiting Observatories*, Wiley-Praxis, 1997.

Kaaret, P., Lamb, F. K., and Swank, J. H. (eds.), *X-Ray Timing 2003: Rossi and Beyond*, American Institute of Physics Publishing, 2004.

Scarsi, L., et al. (eds.), *The Active X-Ray Sky: Results from BeppoSAX and RXTE*, Elsevier Science, 1998.

Seward, Frederick D., and Charles, Philip A., *Exploring the X-Ray Universe*, 2nd ed., Cambridge University Press, 2010.

SAS-2 AND 3

Small Astronomy Satellites 2 and 3 (SAS-2 and 3) were the second and third in a series of small NASA astrophysical spacecraft.

SAS-2, which was designed to provide the first detailed gamma-ray map of the sky, was launched in November 1972 from the Italian San Marco platform off the equatorial coast of Kenya. This enabled NASA to use the smallest possible launcher to put the spacecraft into a near-equatorial, circular orbit. The SAS-2 detector covered the range from ∼25 MeV to 1 GeV. In the event, the spacecraft made observations of about half of the sky, including most of the galactic plane, and detected about 8,000 gamma rays before it failed after seven months. Even with a resolution of only about 1° or so, SAS-2 was able to show that the galactic centre, and the Crab and Vela pulsars were all strong gamma-ray emitters. It also provided evidence for a diffuse extragalactic gamma-ray back-ground.

SAS-3, which was launched in May 1975, also from the San Marco platform, operated for four years. Its mission was to detect and locate X-ray sources over the range 0.2 to 60 keV to a high degree of accuracy, and monitor the intensity and spectra of specific sources. The spinning spacecraft could have its spin stopped to allow the satellite to undertake longer-term observations of particular objects.

The SAS-3 spacecraft discovered the Rapid Burster, MXB 1730−335, with pulses varying in time from once every 8 seconds to once every 5 minutes. The energy in the strongest bursts was up to 1,000 times that in the weakest. SAS-3 also helped to untangle the nature of the source 3U 1809 + 50, which had originally been detected by Uhuru. This was the first so-called 'polar' to be identified, in which gas falls directly onto the magnetic poles of a white dwarf instead of via an accretion disc.

Bibliography

Davies, John K., *Astronomy from Space: The Design and Operation of Orbiting Observatories*, Wiley-Praxis, 1997.

Fichtel, Carl E., *Gamma Ray Astronomy*, Proceedings of the International Cosmic Ray Conference, Munich, West Germany, August 1975, published by the Max-Planck-Institut für extraterrestrische Physik, 1975, pp. 3678–3697.

Lewin, W. H. G., et al., *The Discovery of Rapidly Repetitive X-ray Bursts from a New Source in Scorpius*, Astrophysical Journal, **207** (1976), L95–99.

SPITZER SPACE TELESCOPE

The Spitzer Space Telescope (see Figure 10.9), named after the astrophysicist Lyman Spitzer, was a space-based infrared spacecraft observatory launched on 25 August 2003. It was the fourth and last of NASA's Great Observatories.

Figure 10.9. The infrared Spitzer Space Telescope was launched by a Delta II rocket in 2003. Its innovative Earth-trailing orbit substantially reduced the amount of coolant required to cool its detectors. The ideal position for the Sun was perpendicular to the telescope's axis to reduce heating to the absolute minimum. As a result body-mounted solar arrays could be used. (Courtesy NASA/JPL/Caltech.)

Spitzer, originally known as the Shuttle InfraRed Observatory (SIRO), had first been proposed in the late 1960s as a shuttle-launched spacecraft. At that time it was thought that maintaining infrared detectors at about 2 K would be a major problem, consuming too much coolant. So it was planned to retain SIRO connected to the shuttle, which would bring it down to Earth for more coolant and refurbishment after a few weeks.

This shuttle-based concept was developed over the next ten years or so, and the spacecraft's name changed to the Shuttle InfraRed Telescope Facility (SIRTF). But the successful operation of IRAS in 1983 showed that it was feasible to launch a free-flying infrared spacecraft that would have an acceptable lifetime. So in the following year NASA changed SIRTF to a free-flyer, with the "S" in its name now standing for 'Space' instead of 'Shuttle'. At that time the 3,500 kg SIRTF was expected to out-perform ESA's planned ISO spacecraft, as SIRTF was to have an 85 cm mirror, instead of ISO's 60 cm, and SIRTF was due to cover a broader infrared spectrum. In fact, NASA had tried to interest ESA to collaborate on SIRTF, but the relations between the two organisations were very poor at that time, because of the way NASA had treated ESA on another joint project, ISPM.

In 1991 the United States National Research Council's Bahcall report endorsed the SIRTF project, but shortly afterwards the cost estimates had escalated by about 70%. Subsequent redesigns substantially reduced its estimated mass from 5,700 kg in 1990, to 2,500 kg in 1993, to 750 kg in 1995, with consequent reductions in cost. A major contribution to the mass reduction was provided by the decision to only cool the detectors with superfluid helium, and rely on exposure to space and the evaporating helium to cool the telescope structure. This cooling was assisted by launching the spacecraft into an Earth-trailing heliocentric orbit, which reduced the heating from Earth.

Spitzer had three instruments on-board: an imager, spectrometer and photometer. The imager operated simultaneously in four narrow wavebands from 3.6 μm to 8 μm, each waveband using a dedicated 256 × 256 element detector. The spectrometer operated over 5.3 to 40 μm, whilst the photometer covered the range from 24 to 160 μm. Although these wavebands were not as broad as those on ISO, Spitzer had a better resolution and a longer lifetime, although it was launched 8 years later.

The Spitzer Space Telescope's main mission was to study brown dwarfs, protoplanetary and debris discs, the Milky Way, nearby galaxies, ultraluminous infrared galaxies (ULIRGs), active galactic nuclei (AGNs), and galaxy formation in the early universe.

Spitzer was the first observatory to detect light emitted by an exoplanet in orbit around a normal star. It also appears to have imaged, in the molecular cloud L1014, the youngest star ever observed up to that time. The spacecraft found evidence of large amounts of dust in the ejecta of the Cassiopeia A explosion. This was the first time that a supernova explosion had been shown unambiguously to produce large amounts of dust, which was key to the formation of new stars and planets. Spitzer imaged possible planet-forming gas clouds around stars, and observed all prominent star-forming regions within about 1,500 light years of the solar system as part of the Gould's Belt Survey.

Bibliography

Armus, L., and Reach, W. T. (eds.), *Spitzer Space Telescope: New Views of the Cosmos*, Astronomical Society of the Pacific Conference Series, 357, 2006.

Rieke, George H., *The Last of the Great Observatories: Spitzer and the Era of Faster, Better, Cheaper at NASA*, University of Arizona Press, 2006.

SWAS

The Submillimetre Wave Astronomy Satellite (SWAS) was a NASA Small Explorer (SMEX) mission designed to study the composition of interstellar clouds and determine how they cool as they collapse to form stars and planets. The 290 kg spacecraft was launched into a low Earth orbit in December 1998 using a Pegasus-XL rocket released at altitude by a Lockheed L1011 aircraft.

The spacecraft observed the spectral line of water at 557 GHz, molecular oxygen (487 GHz), neutral carbon (492 GHz), isotopic carbon monoxide (551 GHz) and isotopic water (548 and 557 GHz). It made detailed 1° × 1° maps of many giant molecular and dark cloud cores during the first five years of its mission. In addition, it observed planetary nebulae, circumstellar envelopes, supernova remnants, external galaxies, and various solar system objects. In June 2005 it was temporarily reactivated to observe the effects of the Deep Impact probe's collision with Comet Tempel 1.

SWAS found that there appeared to be much less water vapour than expected in clouds at low temperature. But in gas clouds where stars were being born, the temperatures were several thousand degrees K and the water concentration very high. This would help the clouds to eventually cool. The amount of oxygen in molecular clouds was found to be much less than expected.

Bibliography

Various articles, Astrophysical Journal, **539** (2000), L77–L154.

SWIFT

Swift was a spacecraft in NASA's Medium Explorer Program (MIDEX) designed to study gamma-ray bursts (GRBs) and their afterglows. It had been selected by NASA from a number of competing spacecraft in 1999. The Swift programme, which also included British and Italian on-board and ground station contributions, was a successor to HETE-2, NASA's previous GRB spacecraft programme.

The 1,470 kg Swift was launched into a low Earth orbit on 20 November 2004. Its mission was not only to detect and characterise GRBs, but also to examine the effect of the explosions on the objects' surroundings. Because GRBs are very distant, it was also expected that the results would aid our understanding of the early universe. In addition, Swift undertook an all-sky survey in hard X-rays when not observing GRBs.

There were three co-aligned instruments on Swift, the Burst Alert Telescope (BAT) that operated over the 15–150 keV range, the X-ray Telescope (XRT) operating over 0.2–10 keV, and the Ultraviolet/Optical Telescope (UVOT). When the wide-angle BAT detected a burst, it relayed its position to the ground with an accuracy of better than 4 arcmin within about 15 seconds. This allowed other ground- and space-based observatories to locate and study the source. At the same time, Swift automatically reoriented itself in about a minute to point its fine-resolution instruments at the burst location. If they detected the burst, its fine position was relayed to the ground, and thence to other ground- and space-based observatories. The XRT and UVOT, which also measured the afterglow's spectrum, were able to measure source positions within about 4 arcsec and 0.4 arcsec, respectively.

Swift detected about 100 GRBs/year, of which about a third had their redshifts determined. On 9 May 2005 it detected a short-duration GRB (GRB 050509B), and about 50 seconds later the XRT detected its fading afterglow, enabling its source position to be accurately located. This was the first time that an accurate position fix had been achieved for a short-duration GRB. Unfortunately its optical image could not be unambiguously detected. Then on 4 September 2005 Swift detected a long-duration GRB (GRB 050904) with a redshift of 6.29, indicating that it was about 13 billion light years away. So the explosion had occurred when the universe was only about 900 million years old.

Bibliography

Costa, E., Frontera, F., and Hjorth, J. (eds.), *Gamma-Ray Bursts in the Afterglow Era, Proceedings of the International Workshop Held in Rome, October 2000*, ESO Astrophysics Symposia, Springer, 2001.

Naeye, Robert, *Dissecting the Bursts of Doom*, Sky and Telescope, August 2006, pp. 30–37.

Seward, Frederick D., and Charles, Philip A., *Exploring the X-Ray Universe*, 2nd ed., Cambridge University Press, 2010.

Vedrenne, Gilbert, and Atteia, Jean-Luc, *Gamma-Ray Bursts: The Brightest Explosions in the Universe*, Springer-Praxis, 2009.

TD–1A

TD-1A was an ESRO (European Space Research Organisation) spacecraft launched by a Thor Delta rocket (hence the spacecraft's name) in 1972. It was the sole survivor of an ESRO programme that, in 1965, envisaged launching six three-axis stabilised Thor Delta spacecraft between 1969 and 1972. TD-1A's primary scientific mission was to undertake a sky survey in the ultraviolet, X-ray and gamma-ray wavebands.

TD-1A's main payload consisted of two ultraviolet telescopes. One fed a spectrometer operating between 210 and 280 nm which observed about 200 bright stars. The other undertook a spectrophotometric sky survey between 135 and 255 nm and a photometric survey at about 275 nm. It measured the ultraviolet fluxes of about 31,000 stars down to ninth visual magnitude, and enabled the absorption by interstellar dust to be estimated. These stars, which generally had surface temperatures in the range 15,000 to 50,000 K, were found to be concentrated in the direction of Cygnus and Orion, marking the position of the local spiral arm of the Milky Way.

The spacecraft's X-ray payload had to be switched off due to electrical interference, and, although the gamma-ray payload detected gamma rays, no gamma-ray source could be identified.

Bibliography

Krige, John, and Russo, Arturo, *A History of the European Space Agency 1958–1987: Volume I, The Story of ESRO and ELDO 1958–1973*, ESA SP-1235, 2000.

Krige, John, and Russo, Arturo, *Europe in Space 1960–1973*, ESA SP-1172, 1994.

Massey, Harrie, and Robins, M. O., *History of British Space Science*, Cambridge University Press, 1986.

UHURU

Uhuru (see Figure 10.10), also called SAS-1 (Small Astronomy Satellite 1), was the first satellite dedicated to X-ray astronomy. Its aim was to produce a catalogue of sources, including their positions, intensities, and spectra. Riccardo Giacconi proposed such a spacecraft to NASA in 1963, just after the discovery of the first non-solar X-ray sources.

The 140 kg Uhuru (Swahili for freedom) was launched on 12 December 1970 (Kenyan Independence Day) by an American rocket from the Italian San Marco platform off the equatorial coast of Kenya. This location, just 3° south of the equator, enabled NASA to use the smallest possible launcher to put the spacecraft into a near-equatorial, circular orbit.

Uhuru's payload consisted of two X-ray telescopes, covering the range 2 to 20 keV, that scanned the sky as the spacecraft spun about its axis. Strong X-ray sources were located to within about 1 arcmin and weak ones to about 15 arcmin. This accuracy, which was good for the time, was important as, although sounding rockets had detected a number of X-ray sources, their positions were not known with sufficient accuracy to enable their identification with known optical or radio sources.

During its lifetime of just more than two years, Uhuru detected 339 discrete sources. It identified the locations of about 100 of these sufficiently accurately to enable their counterparts to be identified at other wavelengths. Uhuru measured X-ray emission from supernova remnants, globular clusters, spiral and Seyfert galaxies, and clusters of galaxies. The spacecraft

Figure 10.10. Uhuru, the first dedicated X-ray spacecraft, during check-out. Its two X-ray collimators were mounted back-to-back. One of them is shown in this image as the large black rectangle near the top of the spacecraft. Uhuru was launched by a Scout rocket from Italy's San Marco platform, which was close to the equator. This was to maximise the mass that could be launched into an equatorial orbit. (Courtesy NASA.)

also found that X-ray sources tended to be highly variable in their X-ray intensity. Detailed observations of this variability helped to show that Centaurus X-3 and Hercules X-1 were both pulsars in a binary system, and Cygnus X-1 was a probably black hole, also in a binary system.

Uhuru heralded the start of a new era in X-ray astronomy as no longer did astronomers have to rely on a few minutes of sounding rocket data. They could now observe objects over longer periods of time, which not only improved the accuracy of position measurements, but crucially enabled them to observe the intensity variability of sources.

Bibliography

Culhane, J. Leonard, and Sanford, Peter W., *X-Ray Astronomy*, Faber and Faber, 1981.

Giacconi, R., et al., *An X-Ray Scan of the Galactic Plane from Uhuru*, Astrophysical Journal, **165** (1971), L27–L35.

Hirsh, Richard F., *Glimpsing an Invisible Universe: The Emergence of X-ray Astronomy*, Cambridge University Press, 1983.

Tucker, Wallace, and Giacconi, Riccardo, *The X-Ray Universe*, Harvard University Press, 1985.

VELA

Vela was a series of twelve classified United States spacecraft designed to detect nuclear explosions, and so monitor compliance with the Nuclear Test Ban Treaty. They were launched in pairs (designated A and B) between 1963 and 1970 into a ~110,000 km circular Earth orbit of period ~110 hours. The spacecraft of each pair were separated by 180° longitude.

The Vela spacecraft of the first three pairs were relatively small and unsophisticated, each spacecraft weighing just 150 kg. The capabilities of the next three pairs, each of which weighed about twice that of the first three pairs, were progressively improved with each launch. For example, unlike the earlier Vela spacecraft, Vela 5 A and B, launched in May 1969, and Vela 6 A and B, launched in April 1970, had sufficient timing accuracy that they could determine the approximate direction of any triggering source. The X-ray detectors of Vela 5 A and B and Vela 6 A and B had sensitivities of 3–12 keV, whilst the gamma-ray detectors of the Vela 5 pair were sensitive over the band 0.2–1.0 MeV and of the Vela 6 pair over 0.3–1.5 MeV.

In 1965 Ray Klebesadel of Los Alamos took action to ensure that data on any events that triggered any of the Vela detectors, which were clearly not related to nuclear explosions, were filed away for future study. As a result, four years later he and his colleague Roy Olson found what appeared to be a cosmic gamma-ray burst that had been recorded by both the Vela 3 and 4 pairs of satellites on July 2nd 1967. But they could not determine its direction well enough to be sure of its cosmic origins.

Then in 1972 Klebesadel, Olson and their colleague Ian Strong re-examined the data from the Vela 5 and 6 pairs of spacecraft. As a result, they were able to announce in the following year that they had detected 16 cosmic ray bursts of cosmic origin between July 1969 and July 1972. Eventually the four Vela 5 and 6 spacecraft detected a total of 73 gamma-ray bursts over the period from July 1969 to April 1979.

In addition to the above, the Vela 5 spacecraft were the co-discoverers with ANS of X-ray bursters, but the Vela discovery was not announced until after that of ANS. The Vela 5 B spacecraft was also the first to detect a soft X-ray transient, Cen X-4, in 1969.

Bibliography

Klebesadel, R. W., Strong, I. B., and Olson, R. A., *Observations of Gamma-Ray Bursts of Cosmic Origin*, Astrophysical Journal, **182** (1973), L85–L88.

Evans, W. D., Belian, R. D., and Conner, J. P., *Observations of Intense Cosmic X-Ray Bursts*, Astrophysical Journal, 207 (1976), L91–L94.

Schilling, Govert, *Flash! The Hunt for the Biggest Explosions in the Universe*, Cambridge University Press, 2002.

WMAP

The Wilkinson Microwave Anisotropy Probe (WMAP) was launched on 30 June 2001 to continue the work started by COBE just over ten years earlier in measuring the anisotropy of the cosmic microwave background (CMB) radiation over the celestial sphere. Originally called MAP, the spacecraft's name was changed to WMAP in 2002 in honour of David Wilkinson, a pioneer in the study of the CMB and a senior member of the team, who died that year. The mission had been proposed to NASA in 1995, and approved for development just two years later. The planned orbit for WMAP was around the L_2 Lagrangian point, some 1.5 million kilometres from Earth in the anti-Sun direction, to enable it to have a view to space largely unobstructed by the Sun, Earth and Moon.

The WMAP spacecraft used differential microwave radiometers operating over five separate frequency bands from 23 to 94 GHz to aid rejection of foreground signals from the Milky Way and other sources. Its aim was to produce a full-sky map of the CMB with a resolution of at least $0.3°$ and a sensitivity of 20 μK per $0.3°$ pixel, both at least an order of magnitude better than COBE. The spacecraft arrived on station at L_2 in October 2001 after a lunar fly-by. It measured the whole sky every six months, completing its first full-sky survey in April 2002.

WMAP not only produced a more accurate map of the variation of the CMB over the sky, but also allowed astronomers to deduce the value of various cosmological parameters. After analysing five years of in-orbit results, Gary Hinshaw and colleagues concluded in 2008 that its results were consistent with a flat universe dominated by vacuum or dark energy and dark matter. In particular, the universe was 13.69 ± 0.13 billion years old, and the Hubble constant H_0 was $71.9^{+2.6}_{-2.7}$ km/s/Mpc. The universe today consists of just $4.4 \pm 0.3\%$ normal matter, $21.4 \pm 2.7\%$ dark matter, and $74.2 \pm 3.0\%$ dark energy. WMAP also provided important constraints on cosmic inflation theories of the early universe, eliminating a number of them.

Bibliography

Hinshaw, Gary, and Naeye, Robert, *Decoding the Oldest Light in the Universe*, Sky and Telescope, May 2008, pp. 18–23.

Lemonick, Michael D., *Echo of the Big Bang*, Princeton University Press, 2003.

Longair, Malcolm, *The Cosmic Century: A History of Astrophysics and Cosmology*, Cambridge University Press, 2006.

Peiris, Hiranya, First Year WMAP Results: Implications for Cosmology and Inflation, in Thompson, J. M. T. (ed.), *Advances in Astronomy from the Big Bang to the Solar System*, Royal Society Series on Advances in Science – Vol. 1, Imperial College Press, 2005.

XMM-NEWTON

XMM-Newton (X-ray Multi Mirror–Newton) was a European Space Agency (ESA) X-ray spacecraft named after Sir Isaac Newton, arguably the most influential physicist of all time and the initiator of spectroscopy. It was launched in 1999 into a $7,000 \times 114,000$ km, 48 hour orbit, to allow long-duration observations. At the time, XMM-Newton was the largest scientific spacecraft ever built in Europe. Its performance was complementary to NASA's Chandra spacecraft, which was launched the same year. The ESA spacecraft was dedicated to high-sensitivity spectroscopy, whereas Chandra's main mission was high-resolution imaging.

The XMM spacecraft, the early name of XMM-Newton, had initially been proposed to ESA in 1982 as a possible X-ray spacecraft to follow Exosat, which was launched the following year. Over the years the XMM spacecraft's design gradually evolved, and by 1987 the number of X-ray telescopes on-board had been significantly reduced. But an optical monitor had been added, sensitive from about 160 to 600 nm, to allow simultaneous observations in the visible, ultraviolet and X-ray bands.

XMM-Newton's X-ray payload consisted of three mirror modules co-aligned with the optical monitor. Each mirror module, which had a focal length of 7.5 m, had 58 nested Wolter Type I shells. These fed three European Photon Imaging Cameras (EPICs) for high-throughput, non-dispersive, imaging/spectroscopy, and two reflection grating spectrometers (RGSs) for high-resolution dispersive spectroscopy. The EPIC instruments, which had a resolution of about 6 arcsec, were sensitive over the $0.1-15$ keV band, whereas the RGSs were sensitive over $0.35-2.5$ keV.

XMM-Newton detected quasi-periodic oscillations from the hypothesised supermassive black hole at the centre of the galaxy RE J1034+396. These were the first such oscillations detected from a supermassive black hole. The spacecraft found new supporting evidence that ultraluminous X-ray sources (ULXs) were intermediate mass black holes, and detected the gravitational redshift in light emitted by a neutron star. It detected individual hot spots on rotating neutron stars, found the first black hole candidate in a globular cluster, and detected periodic X-ray emission from a rotating radio transient (RRAT). It also found the most distant cluster of galaxies (XMMXCS 2215–1738) then known, which the Keck telescope showed had a redshift of 1.45, equivalent to a distance of about 3 Gpc (10 billion light years). The cluster contained hundreds of galaxies surrounded by gas at a temperature of about 10 million K.

Bibliography

Credland, J., et al., *XMM Special Issue*, ESA Bulletin, No. 100, Dec. 1999, pp. 7–86.

Fabian, A. C., Pounds, K. A., and Blandford, R. D. (eds.), *Frontiers of X-Ray Astronomy*, Cambridge University Press, 2004.

Schlegal, Eric M., *The Restless Universe: Understanding X-Ray Astronomy in the Age of Chandra and Newton*, Oxford University Press, 2002.

Seward, Frederick D., and Charles, Philip A., *Exploring the X-Ray Universe*, 2nd ed., Cambridge University Press, 2010.

Name index

Subject index

Bold type indicates the main reference or references when a number of pages are listed

Optical/infrared telescopes and observatories index

Bold type indicates the main reference or references when a number of pages are listed

Radio/submillimetre telescopes and observatories index

Bold type indicates the main reference or references when a number of pages are listed

Spacecraft index

Bold type indicates the main reference or references when a number of pages are listed

Printed in the United States
by Baker & Taylor Publisher Services